ENVIRONMENTAL SCIENCE

FOR A CHANGING WORLD

THIRD EDITION

Susan Karr
Carson-Newman University

Jeneen Interlandi
Science Writer

Anne Houtman
California State University, Bakersfield

Vice President, STEM: Ben Roberts
Program Director: Andrew Dunaway
Program Manager: Jennifer Edwards
Developmental Editor: Erica Champion
Marketing Manager: Shannon Howard
Marketing Assistant: Savannah DiMarco
Senior Media and Supplements Editor: Amy Thorne
Lead Content Developers: Heather Held and Emily Marino
Editorial Assistant: Alexandra Hudson
Manager of Publishing Services: Andrea Cava
Content Project Manager: Louis Bruno
Director of Content Management Enhancement: Tracey Kuehn
Managing Editor: Lisa Kinne
Director of Design, Content Management: Diana Blume
Cover and Text Design: Blake Logan
Senior Media Project Manager: Chris Efstratiou
Photo Editor: Sheena Goldstein
Photo Researcher: Lisa Passmore
Art Manager: Matthew McAdams
Senior Workflow Project Supervisor: Susan Wein
Production Supervisor: Lawrence Guerra
Project Management and Composition: Lumina Datamatics, Inc.
Text Illustrations: Nicole R. Fuller; Lumina Datamatics, Inc.
Cartography: Lumina Datamatics, Inc.
Printing and Binding: LSC Communications
Cover Photo: Claudiad/E+/Getty Images

Library of Congress Control Number: 2017933713
ISBN-13: 978-1-319-05962-0
ISBN-10: 1-319-05962-7

©2018, 2015, 2013 by W. H. Freeman and Company
All rights reserved

Printed in the United States of America

First printing

W. H. Freeman and Company
One New York Plaza
Suite 4500
New York, NY 10004-1562
www.macmillanlearning.com

ML Harris/Getty Images

BRIEF CONTENTS

DETAILED CONTENTS

Time Life Pictures/NASA/The LIFE Picture Collection/Getty Images

Monty Rakusen/Cultura/Getty Images

CHAPTER 3 EVOLUTION AND
BIODIVERSITY   126

© Henk Meijer/Alamy

CHAPTER 4 HUMAN POPULATIONS
AND ENVIRONMENTAL HEALTH   166

REUTERS/K. K. Arora/Newscom

Nick Brundle Photography/Getty Images

zlikovec/Getty Images

shotbydave/Getty Images

CHAPTER 10 AIR QUALITY AND CLIMATE CHANGE 430

Michael Bryant/Tribune News Service/
Philadelphia/PA/USA/Newscom

CHAPTER 11 ALTERNATIVES TO FOSSIL FUELS 480

ML Harris/Getty Images

ONLINE MODULE 11.3  SaplingPlus
SUSTAINABLE ENERGY:
MOBILE SOURCES
GAS FROM GRASS
Will an ordinary prairie grass become the next biofuel?

ABOUT THE AUTHORS

SUSAN KARR, MS, is an assistant professor in the biology department of Carson-Newman University in Jefferson City, Tennessee, and has been teaching for more than 25 years. She has served on campus and community environmental sustainability groups and has helped produce an annual "State of the Environment" report on the environmental health of her county. In addition to teaching non-majors courses in environmental science and human biology, she teaches an upper-level course in animal behavior in which she and her students train dogs from the local animal shelter in a program that improves the animals' chances of adoption. She received degrees in animal behavior and forestry from the University of Georgia.

Stephen Karr

Finbarr O'Reilly

JENEEN INTERLANDI, MA, MS, is a journalist and science writer who covers health and environmental issues for a wide range of publications, including *The New York Times Magazine*, *Pacific Standard*, and *Scientific American*. She is the lead investigative health reporter for *Consumer Reports* and was a 2013 Nieman Fellow at Harvard University and the recipient of a 2014 Pulitzer Traveling Grant. Before becoming a journalist, Jeneen worked as a researcher at both Harvard Medical School and the Lamont Doherty Earth Observatory at Columbia University. She holds master's degrees in environmental science and journalism, both from Columbia University.

ANNE HOUTMAN, DPHIL, is Dean of the School of Natural Sciences, Mathematics, and Engineering and Professor of Biological Sciences at California State University, Bakersfield. Her research interests are in the behavioral ecology of birds. She is strongly committed to evidence-based, experiential education and has been an active participant in the national dialogue on science education—how best to teach science to future scientists and future science "consumers"—for almost 20 years. Anne received her doctorate in zoology from the University of Oxford and conducted postdoctoral research at the University of Toronto.

Dear Reader,

For more than 25 years as an environmental science and biology instructor, I've found that "stories" capture the imagination of my students. Students are genuinely interested in environmental issues—using stories to teach these issues make the science more relevant and meaningful to them. Many leave the class with an understanding that what they do really matters, and they feel a willingness to act on that knowledge. This is why I am enthusiastic about our textbook, *Environmental Science for a Changing World*.

In each module of this text, you will be introduced to one or more major issues and concepts of environmental science through the context of a central case study—a story—threaded throughout the module. These stories are current, relevant, and captivating. Discrete sections and key concepts within each module help identify the environmental science content that students need to master, while the case study example helps put those concepts in context for better understanding.

In this third edition, we have broken the main topics down into 11 chapters, each containing two or more modules. These modules (identified as chapters in the second edition) focus on specific topics within each chapter. For example, Chapter 8, Food Resources, includes four modules, each related to a different aspect of feeding the world. This allows instructors to choose the modules to assign and cover that best meet their teaching goals. The most popular modules (those used by most instructors) are found in the print book, and additional modules are found online in SaplingPlus, the textbook's online resource.

A new addition to this edition is the inclusion of an Interactive Map at the end of each module. The map expands the text's tradition of teaching through case studies by presenting short summaries of additional case studies or advances in environmental science related to the module topic.

As with the previous editions, the text focuses on building core competencies for the non-major: environmental literacy, science literacy, and information literacy. End-of-module questions and activities, Infographic questions, and online exercises provide further opportunities to develop these competencies, as well as critical thinking skills.

Environmental Literacy: The scientific, social, political, and economic facets of contemporary environmental issues are examined with a focus on the scientific concepts and drivers of underlying issues.

Science Literacy: Each module includes experimental evidence and graphical data representation and describes the day-to-day work of scientists, giving students many opportunities to evaluate evidence and understand the process of science.

Information Literacy: Every individual needs the skills to find information and assess its quality. We explain how to effectively search for and find scientific information and how to critically analyze that information.

Every person involved in this book—the writers, illustrators, editors, and fellow instructors—has one sincere objective: to help students become informed citizens able to analyze issues, evaluate arguments, discuss solutions, and recognize trade-offs as they make up their own minds about our most pressing environmental challenges.

Sincerely,

Susan Karr

Susan Karr

CAPTIVATING STORIES

STUDENTS FOLLOW ONE RIVETING STORY THROUGH THE ENTIRE MODULE

FROM THE OPENING PAGE TO THE END-OF-MODULE QUESTIONS, AND EVERYWHERE IN BETWEEN, EACH MODULE FOLLOWS AN ENGAGING REAL STORY THAT ILLUSTRATES AND MOTIVATES CORE SCIENCE CONCEPTS. HERE ARE FOUR OF MANY PLACES IN MODULE 1.2 WHERE THE STORY UNFOLDS.

MODULE 1.2 SCIENCE LITERACY AND THE PROCESS OF SCIENCE

David Blehert and Melissa Behr gather data from North American bat caves looking for the cause of death for over six million bats.

White-nose syndrome (WNS) is a deadly, epidemic disease affecting some species of North American bats; it is caused by a non-native fungus, *Pseudogeomyces destructans*. Even species that were abundant only a few years ago have experienced dramatic population crashes and could potentially go extinct in the near future.

Blehert, Behr, and others had no trouble coming up with hypotheses. The three they thought most likely were:

1. The white-nose infection caused by the fungus was secondary and opportunistic; it was only able to grow on the bats because their defenses had been weakened by another pathogen.

2. The fungus had always been present, but had recently mutated to become deadlier.

3. The fungus had been transported from some other place and was new to the region.

To figure out which of these ideas was correct, they would have to test each one.

Blehert's team tested several hypotheses relating WNS and the bat deaths. They used comparative, observational, and experimental studies. They also had to determine whether the appearance of WNS was causing bat deaths or simply correlated to bat deaths.

Effective environmental policy depends on sound scientific data. After finding that WNS was introduced to North America by traveling humans, the U.S. Fish and Wildlife Service implemented a plan to protect bats from further spread of the disease.

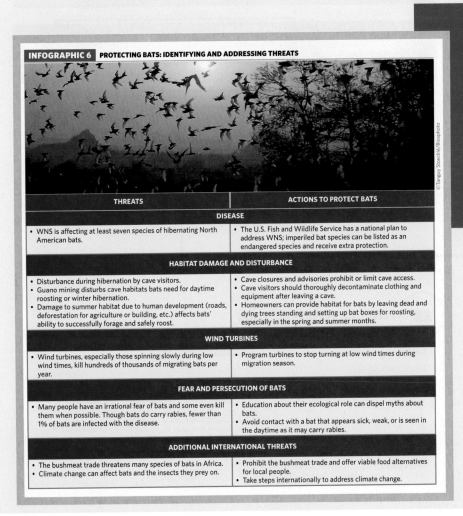

INFOGRAPHIC 6 PROTECTING BATS: IDENTIFYING AND ADDRESSING THREATS

©Tanguy Stoecklé/Biosphoto

THREATS	ACTIONS TO PROTECT BATS
DISEASE	
• WNS is affecting at least seven species of hibernating North American bats.	• The U.S. Fish and Wildlife Service has a national plan to address WNS; imperiled bat species can be listed as an endangered species and receive extra protection.
HABITAT DAMAGE AND DISTURBANCE	
• Disturbance during hibernation by cave visitors. • Guano mining disturbs cave habitats bats need for daytime roosting or winter hibernation. • Damage to summer habitat due to human development (roads, deforestation for agriculture or building, etc.) affects bats' ability to successfully forage and safely roost.	• Cave closures and advisories prohibit or limit cave access. • Cave visitors should thoroughly decontaminate clothing and equipment after leaving a cave. • Homeowners can provide habitat for bats by leaving dead and dying trees standing and setting up bat boxes for roosting, especially in the spring and summer months.
WIND TURBINES	
• Wind turbines, especially those spinning slowly during low wind times, kill hundreds of thousands of migrating bats per year.	• Program turbines to stop turning at low wind times during migration season.
FEAR AND PERSECUTION OF BATS	
• Many people have an irrational fear of bats and some even kill them when possible. Though bats do carry rabies, fewer than 1% of bats are infected with the disease.	• Education about their ecological role can dispel myths about bats. • Avoid contact with a bat that appears sick, weak, or is seen in the daytime as it may carry rabies.
ADDITIONAL INTERNATIONAL THREATS	
• The bushmeat trade threatens many species of bats in Africa. • Climate change can affect bats and the insects they prey on.	• Prohibit the bushmeat trade and offer viable food alternatives for local people. • Take steps internationally to address climate change.

EMPOWERING SCIENCE

ENVIRONMENTAL SCIENCE FOR A CHANGING WORLD OFFERS A CONSISTENT METHODOLOGY FOR TEACHING THE FIELD'S ESSENTIAL SCIENTIFIC CONCEPTS, WITH EACH MODULE CENTERED AROUND GUIDING QUESTIONS.

THESE QUESTIONS ESTABLISH A CLEAR, STEP-BY-STEP PATHWAY THROUGH THE MODULE FROM THE OPENING STORY; THROUGH THE NARRATIVE, KEY CONCEPT CALLOUTS, PHOTOS, INFOGRAPHICS, AND THE END-OF-MODULE ASSESSMENT; TO THE ASSESSMENTS AND MEDIA IN SAPLINGPLUS. THE GUIDING QUESTIONS AND KEY CONCEPTS BRING THE SCIENCE IN THE MODULE TO THE FOREFRONT, SO STUDENTS NEVER LOSE SIGHT OF FUNDAMENTAL CONCEPTS WHILE READING THE STORY.

MODULE 1.2 SCIENCE LITERACY AND THE PROCESS OF SCIENCE

FUNGAL ATTACKER THREATENS BATS

Unraveling the mystery behind bat deaths

Little brown bat (Myotis lucifugus) in flight. Edward Kinsman/Getty Images

After reading this chapter and studying the KEY CONCEPTS and INFOGRAPHICS, you should be able to answer these GUIDING QUESTIONS

CORE MESSAGE
The world faces many environmental challenges, some predictable and some unexpected, such as the sudden decline of bats in North America. Physical evidence, systematically collected and logically analyzed, helps scientists understand environmental issues, and this understanding can guide policy decisions that address environmental problems.

1. How do scientists study the natural world?
2. What influences the degree of certainty in a scientific explanation?
3. How do comparative studies contribute to the understanding of an environmental issue?
4. Why are observational and experimental studies needed to investigate the natural world?
5. What is the difference between a *correlation* and a *cause-and-effect relationship*?
6. How can science help policy makers address environmental problems?

---

1 Each module begins with a Core Message and a series of Guiding Questions that focus students on the chapter's central scientific content.

2 The module sections and Key Concepts correspond to the Guiding Questions, re-emphasizing the module's essential scientific ideas.

5 CORRELATION VERSUS CAUSATION

Key Concept 5: Two events that occur together are correlated, but this does not necessarily mean one caused the other or even that they are related in a meaningful way.

Observational and experimental studies differ not only in their approach, but also in the type of information they ultimately provide. Winifred Frick and her team at Boston University conducted an observational study that compared bat colony sizes over time in areas with and without WNS. They observed a sharp decline in colony size, but only in the colony experiencing WNS infections.

Observational studies like this (and the Reeder study) reveal **correlations**. That is, they can show that two things occur together: A colony was somehow exposed to WNS and most bats in that colony died shortly thereafter. But observational studies can't necessarily tell us if one thing

die-off). So although scientists suspected from the very start that the white fuzz was related to the bat die-offs, they needed more evidence to determine whether the two events were connected in a **cause-and-effect relationship**. For that, they needed experimental studies like Warnecke's that intentionally exposed bats to the fungus. If all other factors are the same between two groups except the one variable that is manipulated (in this case, exposure to the fungus) and we see differences in the dependent variable we are monitoring (in this case, bat

correlation Two things occurring together but not necessarily having a cause-and-effect relationship.

cause-and-effect relationship An association between two variables that identifies one (the effect)

3 Each module section includes an infographic, where students will find the information they need to think critically about the Guiding Questions. Additional questions within the infographics prompt students to reflect on the content and stretch their understanding.

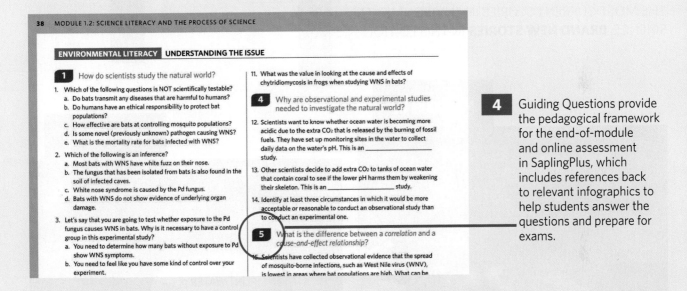

INFOGRAPHIC 5 CORRELATION VERSUS CAUSATION

Because they cannot control the extraneous variables in a research study, observational studies can only provide correlational evidence that two or more factors are related. Experimental studies that can manipulate and control variables can establish whether a cause-and-effect relationship exists.

A: POPULATION SIZE IN COLONIES WITH AND WITHOUT WNS

Y-axis: Winter count (# of bats), 0 to 700. X-axis: Year, 1985 to 2010.
- Colony A (WNS present after 2006)
- Colony B (uninfected)
Appearance of WNS

B: EXPOSURE TO WNS FUNGUS AND BAT SURVIVAL

Y-axis: Number of individuals surviving, 0 to 20. X-axis: Day of study, 0 to 140.
- Control (no fungal exposure)
- Test (exposed to fungus)

Study A: Observational study that used two colonies in years before and after WNS affected one of the colonies.

This study provides correlational evidence that suggests WNS caused the population crash. However, it does not provide cause-and-effect evidence since something else might have caused population size to fall in that colony.

Study B: Experimental study that monitored survival in bats exposed to the fungus that causes WNS (Pd) compared to bats with no Pd fungal exposure.

In this study, a cause-and-effect relationship can be established since the only difference between the groups was Pd exposure.

? Other than WNS, what else could have decreased the number of bats in Colony A after 2008?

---

38 MODULE 1.2: SCIENCE LITERACY AND THE PROCESS OF SCIENCE

ENVIRONMENTAL LITERACY UNDERSTANDING THE ISSUE

1 How do scientists study the natural world?

1. Which of the following questions is NOT scientifically testable?
   a. Do bats transmit any diseases that are harmful to humans?
   b. Do humans have an ethical responsibility to protect bat populations?
   c. How effective are bats at controlling mosquito populations?
   d. Is some novel (previously unknown) pathogen causing WNS?
   e. What is the mortality rate for bats infected with WNS?

2. Which of the following is an inference?
   a. Most bats with WNS have white fuzz on their nose.
   b. The fungus that has been isolated from bats is also found in the soil of infected caves.
   c. White nose syndrome is caused by the Pd fungus.
   d. Bats with WNS do not show evidence of underlying organ damage.

3. Let's say that you are going to test whether exposure to the Pd fungus causes WNS in bats. Why is it necessary to have a control group in this experimental study?
   a. You need to determine how many bats without exposure to Pd show WNS symptoms.
   b. You need to feel like you have some kind of control over your experiment.

11. What was the value in looking at the cause and effects of chytridiomycosis in frogs when studying WNS in bats?

4 Why are observational and experimental studies needed to investigate the natural world?

12. Scientists want to know whether ocean water is becoming more acidic due to the extra CO_2 that is released by the burning of fossil fuels. They have set up monitoring sites in the water to collect daily data on the water's pH. This is an _____ study.

13. Other scientists decide to add extra CO_2 to tanks of ocean water that contain coral to see if the lower pH harms them by weakening their skeleton. This is an _____ study.

14. Identify at least three circumstances in which it would be more acceptable or reasonable to conduct an observational study than to conduct an experimental one.

5 What is the difference between a *correlation* and a *cause-and-effect relationship*?

15. Scientists have collected observational evidence that the spread of mosquito-borne infections, such as West Nile virus (WNV), is lowest in areas where bat populations are high. What can be

4 Guiding Questions provide the pedagogical framework for the end-of-module and online assessment in SaplingPlus, which includes references back to relevant infographics to help students answer the questions and prepare for exams.

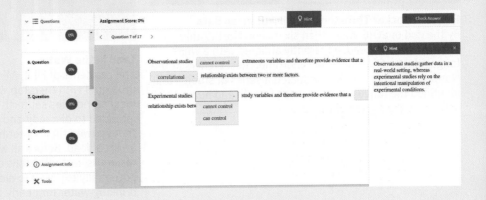

5 SaplingPlus provides students with an opportunity to check their understanding; all student responses are recorded in the instructor gradebook.

Questions — Assignment Score: 0%

Question 7 of 17

Observational studies [cannot control] extraneous variables and therefore provide evidence that a [correlational] relationship exists between two or more factors.

Experimental studies [ ] study variables and therefore provide evidence that a [ ] relationship exists between [ ]
- cannot control
- can control

Hint: Observational studies gather data in a real-world setting, whereas experimental studies rely on the intentional manipulation of experimental conditions.

WHAT'S NEW

ONE UNIFIED BOOK THAT COVERS THE MOST ESSENTIAL CONTENT IN THE COURSE

REVISED, LEARNING-FRIENDLY TABLE of CONTENTS

IN THE THIRD EDITION, we have broken the main topics down into 11 chapters, each containing two or more modules. These modules (identified as chapters in the second edition) focus on specific topics within each chapter. For example, Chapter 8, Food Resources, includes four modules, each related to a different aspect of feeding the world. This allows instructors to choose the modules that best meet their teaching goals. The most popular modules (those used by most instructors) are found in the print book, and additional modules are found online in SaplingPlus, the textbook's online resource.

THE MOST CURRENT STORIES IN ENVIRONMENTAL SCIENCE. **BRAND NEW STORIES** IN THIS EDITION:

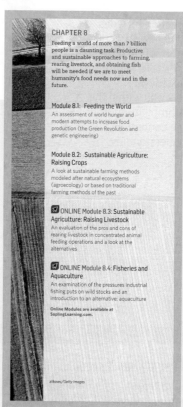

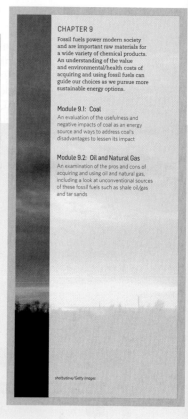

MODULE 1.2

STEPHEN ALVAREZ/National Geographic Creative

Fungal Attacker Threatens North American Bats
By the end of 2016, more than six million North American bats were found dead, apparently afflicted with a white fungus dubbed "white-nose syndrome." Researchers at the National Wildlife Center conducted a series of studies that identified the fungus as the cause of death and human activity as the source of the fungus.

MODULE 1.3

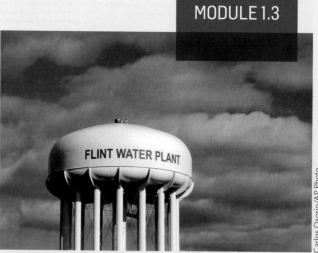

FLINT WATER PLANT

Carlos Osorio/AP Photo

Lead in the Water
The residents of Flint, Michigan, began reporting problems with their tap water in early 2015. State and local authorities set forth an anti-information campaign intended to discredit and distract from scientific studies that showed their water had been contaminated with toxic levels of lead. Before all was said and done, some 100,000 residents, including tens of thousands of children, were potentially exposed to elevated levels of lead in their drinking water for nearly two years.

Rafiq Maqbool/AP/REX/Shutterstock

The Kerala Model

A 1988 literacy campaign in Kerala, India, taught its citizens to read and pursued other actions that focused on improved health care—both of which helped empower women. The result, economists and demographers show, is decreased population growth and greater quality of life.

Climate Refugees

Climate change isn't just impacting polar bears anymore. People around the world are feeling the effects of global warming, drought, rising sea levels, and severe weather. At particular risk are the climate refugees, entire communities forced to leave their homes due to the changing climate.

Joe Raedle/Getty Images

RESOURCES TARGET THE MOST CHALLENGING CONCEPTS AND SKILLS IN THE COURSE

Classroom activities, animations, tutorials, and assessment materials are all built around the concepts and skills that are most difficult for students to master.

SaplingPlus

Created and supported by educators, SaplingPlus's instructional online homework drives student success and saves educators time. Every homework problem contains hints, answer-specific feedback, and solutions to ensure that students find the help they need. With SaplingPlus, every problem counts.

Fully loaded with our interactive e-Book and all student and instructor resources, SaplingPlus is organized around a series of pre-built units designed for either a traditional lecture or an active classroom—carefully curated, ready-to-use collections of material for each module of *Environmental Science for a Changing World*.

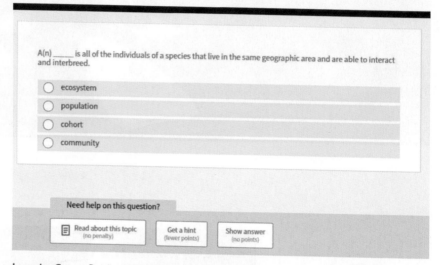

LearningCurve Put "testing to learn" into action. Based on educational research, LearningCurve really works: Game-like quizzing motivates students and adapts to their needs based on their performance. It is the perfect tool to get them to engage before class and review after class! Additional reporting tools and metrics help teachers get a handle on what their class knows and doesn't know. Available in SaplingPlus.

Interactive Maps Because local, regional, and global events and issues are all crucial to the course, each module in the third edition includes an interactive map highlighting related stories from regions in the United States and around the world. Each map is accompanied by assessments in SaplingPlus.

Scientific American **Quizzes** Paired with curated articles and podcasts, *Scientific American* Quizzes allow instructors to assign material and assess students' understanding. Available in SaplingPlus.

Tutorial Style Homework A set of ten assessment questions per module help students understand the more difficult concepts within each chapter. These assessment questions include hints and feedback to help guide students toward comprehension. Available in SaplingPlus.

Pre-Lecture Reading Quizzes Available in SaplingPlus for each module, these quizzes help ensure that students have read the material before attending class.

LECTURE TOOLS

The Hub for Active Learning For the third edition, we've developed a suite of active learning resources to help instructors make their classes more engaging. Each module has a series of short activities based on the content in the text. The resources available include handouts for students and instructor guides (including a PowerPoint slide show for each activity).

We've also created a series of critical-thinking activities based on current events for instructors to use. These critical-thinking activities are designed to develop students' critical-thinking and metacognition skills so they may be applied in all their coursework and in life. Resources available for these activities include an instructor guide, classroom activities, and assessment questions.

Optimized and Unlabeled Art
Infographics are optimized for projection in large lecture halls and provided in both labeled and unlabeled versions for effective presentation.

Layered or Active PowerPoint Slides
PowerPoint slides for select figures deconstruct key concepts, sequences, and processes in a step-by-step format, allowing instructors to present complex ideas in clear, manageable parts.

Lecture Outlines for PowerPoint Adjunct professors and instructors who are new to the discipline will appreciate these detailed companion lectures, perfect for walking students through the key ideas in each chapter. These rich, prebuilt lectures make it easy for instructors to transition to the book.

Clicker Questions Designed as interactive in-class exercises, these questions reinforce core concepts and uncover misconceptions.

ACKNOWLEDGMENTS

From Susan Karr...

I've discovered that an undertaking this big is truly a collaborative effort. It is amazing what you can accomplish when you work with talented and highly skilled people. I want to thank W. H. Freeman acquisitions editor Jennifer Edwards for her support and guidance and developmental editor Erica Champion and content developer Heather Held for their patience, insights, and outstanding editorial skills. I also want to thank Jeneen Interlandi, the accomplished and gifted writer who has made these chapters such a pleasure to read. I gratefully acknowledge the entire team at Lumina Datamatics for their skills, vision, and patience in the production of the book. Thanks also go to Shannon Howard, the marketing manager, and the sales force, who work tirelessly to see that the book is a success in the marketplace

I've had tremendous support from my biology department colleagues at Carson-Newman University and from focus group participants, who offered advice, answered questions, and helped track down elusive information—thanks for sharing your expertise. I also owe a debt of gratitude to my environmental science students over the years for their questions, interests, demands, and passion for learning that have always challenged and inspired me. Finally, I want to thank my husband, Steve, for supporting me in so many different ways it is impossible to count them all.

From Jeneen Interlandi...

Each of these chapters is a story—of scientists and everyday people, often doing extraordinary things. It has been my great pleasure to tell those stories here. For that, I thank each and every one of my sources. Their time and patience are what made this book possible.

I would also like to thank Susan, with whom it has been an honor to work, and the entire W. H. Freeman team for their tireless efforts.

REVIEWERS

We would like to extend our deep appreciation to the following instructors who reviewed, tested, and advised on the book manuscript at various stages of development.

Chapter Reviewers

Matthew Abbott, *Des Moines Area Community College*

David Aborn, *University of Tennessee at Chattanooga*

Michael Adams, *Pasco-Hernando Community College*

Shamim Ahsan, *Metropolitan State College of Denver*

John Anderson, *Georgia Perimeter College*

Walter Arenstein, *San Jose State University*

Deniz Ballero, *Georgia Perimeter College*

Marcin Baranowski, *Passaic County Community College*

Brad Basehore, *Harrisburg Area Community College*

Sean Beckmann, *Rockford College*

Ray Beiersdorfer, *Youngstown State University*

Tracy Benning, *University of San Francisco*

David Berg, *Miami University*

Bert Berquist, *University of Northern Iowa*

Joe Beuchel, *Triton College*

Aaron Binns, *Florida State University*

Karen Blair, *Pennsylvania State University*

Barbara Blonder, *Flagler College*

Steve Blumenshine, *California State University, Fresno*

Ralph Bonati, *Pima Community College*

Polly Bouker, *Georgia Perimeter College*

Richard Bowden, *Allegheny College*

Jennifer Boyd, *University of Tennessee at Chattanooga*

Scott Brame, *Clemson University*

Allison Breedveld, *Shasta College*

Mary Brown, *Lansing Community College*

Brett Burkett, *Collin College*

Alan Cady, *Miami University*

Elena Cainas, *Broward College*

Daniel Capuano, *Hudson Valley Community College*

Deborah Carr, *Texas Tech University*

Mary Kay Cassani, *Florida Gulf Coast University*

Michelle Cawthorn, *Georgia Southern University*

Niccole Cerveny, *Mesa Community College*

Lu Anne Clark, *Lansing Community College*

Jennifer Cole, *Northeastern University*

Eric Compas, *University of Wisconsin, Whitewater*

Jason Crean, *Saint Xavier University and Moraine Valley Community College*

Michael Dann, *Pennsylvania State University*

James Dauray, *College of Lake County*

Michael Denniston, *Georgia Perimeter College*

Robert Dill, *Bergen Community College*

Craig Dilley, *Des Moines Area Community College*

Michael Draney, *University of Wisconsin, Green Bay*

JodyLee Estrada Duek, *Pima Community College, Desert Vista*

Don Duke, *Florida Gulf Coast University*

James Dunn, *Grand Valley State University*

John Dunning, *Purdue University*

Kathy Evans, *Reading Area Community College*

Brad Fiero, *Pima County Community College*

Linda Fitzhugh, *Gulf Coast Community College*

Stephan Fitzpatrick, *Georgia Perimeter College*

April Ann Fong, *Portland Community College, Sylvania Campus*

Steven Forman, *University of Illinois at Chicago*

Nicholas Frankovits, *University of Akron*

Phil Gibson, *University of Oklahoma*

Michael Golden, *Grossmont College*

Sherri Graves, *Sacramento City College*

Michelle Groves, *Oakton Community College*

Myra Carmen Hall, *Georgia Perimeter College*

Sally Harms, *Wayne State College*

Stephanie Hart, *Lansing Community College*

Wendy Hartman, *Palm Beach State College*

Alan Harvey, *Georgia Southern University*

Keith Hench, *Kirkwood Community College*

Robert Hollister, *Grand Valley State University*

Tara Holmberg, *Northwestern Connecticut Community College*

Jodee Hunt, *Grand Valley State University*

Meshagae Hunte-Brown, *Drexel University*

Kristin Jacobson, *Illinois Central College*

Jason Janke, *Metropolitan State College of Denver*

David Jeffrey, *Georgia Perimeter College*

Thomas Jurik, *Iowa State University*

Charles Kaminski, *Middlesex Community College*

Michael Kaplan, *College of Lake County*

John Keller, *College of Southern Nevada*

Myung-Hoon Kim, *Georgia Perimeter College*

Elroy Klaviter, *Lansing Community College*

Paul Klerks, *University of Louisiana at Lafayette*

Janet Kotash, *Moraine Valley Community College*

Jean Kowal, *University of Wisconsin, Whitewater*

John Krolak, *Georgia Perimeter College*

James Kubicki, *Pennsylvania State University*

Diane LaCole, *Georgia Perimeter College*

Katherine LaCommare, *Lansing Community College*

Andrew Lapinski, *Reading Area Community College*

Kim D. B. Largen, *George Mason University*

Jennifer Latimer, *Indiana State University*

Kurt Leuschner, *College of the Desert*

Stephen Lewis, *California State University, Fresno*

Jedidiah Lobos, *Antelope Valley College*

Eric Lovely, *Arkansas Tech University*

Marvin Lowery, *Lone Star College System*

Steve Luzkow, *Lansing Community College*

Steve Mackie, *Pima Community College*

Nilo Marin, *Broward College*

Eric Maurer, *University of Cincinnati*

Costa Mazidji, *Collin College*

DeWayne McAllister, *Johnson County Community College*

Vicki Medland, *University of Wisconsin, Green Bay*

Alberto Mestas-Nunez, *Texas A&M University, Corpus Christi*

Chris Migliaccio, *Miami Dade College*

Jessica Miles, *Palm Beach State College*

Dale Miller, *University of Colorado, Boulder*

Kiran Misra, *Edinboro University of Pennsylvania*

Scott Mittman, *Essex County College*

Edward Mondor, *Georgia Southern University*

Zia Nisani, *Antelope Valley College*

Ken Nolte, *Shasta College*

Kathleen Nuckolls, *University of Kansas*

Segun Ogunjemiyo, *California State University*

Bruce Olszewski, *San José State University*

Jeff Onsted, *Florida International University*

Kathleen O'Reilly, *Houston Community College*

Nancy Ostiguy, *Pennsylvania State University*

Blair Page, *Le Moyne College*

Daniel Pavuk, *Bowling Green State University*

Dexter Perkins, *University of North Dakota*

Barry Perlmutter, *College of Southern Nevada*

Chris Petrie, *Brevard College*

Craig D. Phelps, *Rutgers, The State University of New Jersey*

Neal Phillip, *Bronx Community College*

Greg Pillar, *Queens University of Charlotte*

Thomas E. Pliske, *Florida International University*

Keith Putirka, *California State University, Fresno*

Bob Remedi, *College of Lake County*

Erin C. Rempala, *San Diego City College*

Eric Ribbens, *Western Illinois University*

Angel M. Rodriquez, *Broward College*

Mary Rosenthal, *Nashville State Community College*

Dennis Ruez, *University of Illinois at Springfield*

Robert R. Ruliffson, *Minneapolis Community and Technical College*

Melanie Sadeghpour, *Des Moines Area Community College*

Jay Sah, *Florida International University*

Seema Sah, *Florida International University*

Eric Sanden, *University of Wisconsin – River Falls*

Shamili Sandiford, *College of DuPage*

Judith Scherff, *Washburn University*

Waweise Schmidt, *Palm Beach State College*

Jeffery Schneider, *State University of New York at Oswego*

David Serrano, *Broward College*

Joe Shaw, *Indiana University*

William Shockner, *Community College of Baltimore County*

Patricia Smith, *Valencia College*

Dale Splinter, *University of Wisconsin, Whitewater*

Jacob Spuck, *Florida State University*

Theodore L. Steck, *University of Chicago*

Craig Steele, *Edinboro University*

Michelle Stevens, *California State University, Sacramento*

Rich Stevens, *Monroe Community College*

Michelle Pulich Stewart, *Mesa Community College*

John Sulik, *Florida State University*

Donald Thieme, *Valdosta State University*

Jamey Thompson, *Hudson Valley Community College*

Jonathan Titus, *State University of New York at Fredonia*

Gail Anderson Tompkins, *Wake Technical Community College*

Susanna T.Y. Tong, *University of Cincinnati*

Michelle Tremblay, *Georgia Southern University*

Karen Troncalli, *Georgia Perimeter College*

Thomas J. Vaughn, *Middlesex Community College*

Francis J. Veale, Jr., *Massachusetts Maritime Academy*

Carlos Villalobos, *Texas Tech University*

Caryl Waggett, *Allegheny College*

Daniel E. Wagner, *Eastern Florida State College*

Meredith Wagner, *Lansing Community College*

Alexander Wait, *Missouri State University*

Xianzhong Wang, *Indiana University – Purdue University Indianapolis*

Deena Wassenberg, *University of Minnesota*

Kelly Watson, *Eastern Kentucky University*

Edward Wells, *Wilson College*

Nancy Wheat, *Hartnell College*

Frank Williams, *Langara College*

Justin Kirk Williams, *Sam Houston State University*

Jennifer Willing, *College of Lake County*

William E. Winner, *North Carolina State University*

Danielle Wirth, *Des Moines Area Community College*

Lorne Wolfe, *Georgia Southern University*

Janet Wolkenstein, *Hudson Valley Community College*

Bethany L. Woodworth, *University of New England*

James Yount, *Eastern Florida State College*

Douglas Zook, *Boston University*

Focus Group Participants

John Anderson, *Georgia Perimeter College*

Tom L. Arsuffi, *Texas Tech University*

Teri Balser, *University of Wisconsin, Madison*

Tracy L. Benning, *University of San Francisco*

Kimberly A. Bjorgo-Thorne, *West Virginia Wesleyan College*

Elena Cainas, *Broward College*

Kelly Cartwright, *College of Lake County*

Mary Kay Cassani, *Florida Gulf Coast University*

Michelle Cawthorn, *Georgia Southern University*

Mark Coykendall, *College of Lake County*

JodyLee Estrada Duek, *Pima Community College, Desert Vista*

Rachel Goodman, *Hampden-Sydney College*

Jason Janke, *Metropolitan State College of Denver*

Catherine Kleier, *Regis University*

Charles Knight, *California Polytechnic State University, San Luis Obispo*

Janet Kotash, *Moraine Valley Community College*

Jean Kowal, *University of Wisconsin, Whitewater*

Nilo Marin, *Broward College*

Edward Mondor, *Georgia Southern University*

Brian Mooney, *Johnson & Wales University, North Carolina*

Barry Perlmutter, *College of Southern Nevada*

Matthew Rowe, *Sam Houston State University*

Shamili Sandiford, *College of DuPage*

Ryan Tainsh, *Johnson & Wales University*

Michelle Tremblay, *Georgia Southern University*

Kelly Watson, *Eastern Kentucky University*

Comparative Reviewers

Buffany DeBoer, *Wayne State College*

Dani DuCharme, *Waubonsee Community College*

James Eames, *Loyola University Chicago*

Bob East, *Washington & Jefferson College*

Matthew Eick, *Virginia Polytechnic Institute and State University*

Kevin Glaeske, *Wisconsin Lutheran College*

Rachel Goodman, *Hamdpen-Sydney College*

Melissa Hobbs, *Williams Baptist College*

David Hoferer, *Judson University*

Paul Klerks, *University of Louisiana at Lafayette*

Troy Ladine, *East Texas Baptist University*

Jennifer Latimer, *Indiana State University*

Kurt Leuschner, *College of the Desert*

Quent Lupton, *Craven Community College*

Jay Mager, *Ohio Northern University*

Steven Manis, *Mississippi Gulf Coast Community College*

Nancy Mann, *Cuesta College*

Heidi Marcum, *Baylor University*

John McCarty, *University of Nebraska at Omaha*

Chris Poulsen, *University of Michigan*

Mary Puglia, *Central Arizona College*

Michael Tarrant, *University of Georgia*

Melissa Terlecki, *Cabrini College*

Jody Terrell, *Texas Woman's University*

Class Test Participants

Mary Kay Cassani, *Florida Gulf Coast College*

Ron Cisar, *Iowa Western Community College*

Reggie Cobb, *Nash Community College*

Randi Darling, *Westfield State University*

JodyLee Estrada Duek, *Pima Community College, Desert Vista*

Catherine Hurlbut, *Florida State College at Jacksonville*

James Hutcherson, *Blue Ridge Community College*

Janet Kotash, *Moraine Valley Community College*

Offiong Mkpong, *Palm Beach State College*

Edward Mondor, *Georgia Southern University*

Anthony Overton, *East Carolina University*

Shamilli Sandiford, *College of DuPage*

Keith Summerville, *Drake University*

ENVIRONMENTAL SCIENCE

FOR A CHANGING WORLD

INTRODUCTION TO ENVIRONMENTAL, SCIENCE, AND INFORMATION LITERACY

CHAPTER 1

Informed citizens have an understanding of how the environment works, how science is used to acquire that understanding, and how to find and evaluate information. These tools can help us identify and pursue sustainable actions that benefit the environment and society.

Module 1.1: Environmental Literacy and Sustainability

An introduction to the scope and focus of environmental science and what it means to live sustainably

Module 1.2: Science Literacy and the Process of Science

A primer on the scientific method and how it is used to investigate environmental science questions

Module 1.3: Information Literacy and Toxicology

An introduction to critical thinking and evaluating information sources, using an examination of toxic substances in our environment as an example

LESSONS FROM A VANISHED SOCIETY

What can we learn about sustainability from a vanished Viking society?

The remains of Hvalsey, a Viking settlement church, in southern Greenland. Paul Souders/WorldFoto/Aurora Photos

After reading this chapter and studying the KEY CONCEPTS and INFOGRAPHICS, you should be able to answer these GUIDING QUESTIONS

CORE MESSAGE

Humans are a part of the natural world and are dependent on a healthy, functioning planet. We put pressure on the planet in a variety of ways, but our choices can help us move toward sustainability.

1. What is the purpose and scope of environmental science?
2. Why are both empirical and applied approaches useful in environmental science?
3. What characteristics make an environmental dilemma a "wicked problem"?
4. What does it mean to be sustainable?
5. Why do scientists think we are living in a new geologic epoch, the Anthropocene?
6. What are the characteristics of a sustainable ecosystem?
7. What can human societies and individuals do to pursue sustainability?
8. What challenges does humanity face in dealing with environmental issues?
9. Distinguish between anthropocentric, biocentric, and ecocentric worldviews.

Although not a tourist hotspot, Greenland offers some spectacular sights—colossal ice sheets, a lively seascape, rare and impressive wildlife (whales, seals, polar bears, eagles). But on his umpteenth trip to the island, Thomas McGovern was not interested in any of that. What he wanted to see was the garbage—specifically, the ancient, fossilized garbage that Viking settlers had left behind some seven centuries before.

McGovern, an archaeologist at the City University of New York, had been on countless expeditions to Greenland over the preceding 40 years. Digging through layers of peat and permafrost, he and his team had unearthed a museum's worth of artifacts that, when pieced together, were beginning to tell the story of the Greenland Vikings. But as thorough as their expeditions had been, that story was still maddeningly incomplete.

Here's what they knew so far: A thousand or so years ago, Norse settlers arrived in Greenland, possibly in search of new walrus populations to support the ivory trade. Ivory was a lucrative commodity and walrus populations had been overhunted from outposts in Iceland, another colony of Norway. An infamous Norse Viking by the name of Erik the Red, exiled from Iceland as a sentence for a murder conviction, supposedly led a small group of followers to a vast expanse of snow and ice that he had dubbed Greenland, a name, legend says, he felt would attract more settlers.

Most of Greenland was not, in fact, green. It was a forbidding place marked by harsh winds and sparse vegetation. But tucked between two fjords along the southwestern coast, protected from the elements by jagged, imposing cliffs, the Norse settlers found a string of verdant meadows, brimming with wildflowers. They quickly set up camp here and proceeded to build a society similar to the one they had left behind in Iceland and their ancestral home of Norway. Though their main enterprise might have been walrus hunts, they also farmed, hunted, and raised livestock. They built barns and churches as elaborate as the ones back home. They established an economy and a legal system, traded goods with mainland Europe, and, at their peak, reached a population of 3,000.

And then, after 450 years of prosperity, they disappeared, leaving little more than the beautiful, tragic ruins of a handful of barns and churches in their wake.

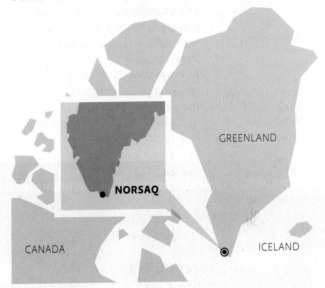

◉ **WHERE IS THE VIKING SETTLEMENT IN GREENLAND?**

GREENLAND

NORSAQ

CANADA

ICELAND

The how and why of this vanishing act remained a tantalizing mystery, one that has drawn scientists like McGovern to Greenland each summer. Researchers suspected that disturbances in the natural environment—a cooling climate, loss of soil, problems with the food supply—may have been the deciding factors.

While other researchers probed ice sheets and soil deposits in search of clues, McGovern stuck to the garbage heaps, or *middens*, as Vikings called them. Every farmstead had one, and every generation of the farmstead's owners threw their waste into it. The result was an archaeological treasure trove: fine-grain details about what people ate, how they dressed, and the kinds of objects with which they filled their homes. It gave McGovern and his team a clear picture of how they lived.

If they dug deep enough, McGovern thought, it might also explain how they died.

1 ENVIRONMENTAL SCIENCE

Key Concept 1: Environmental science draws on science and nonscience disciplines to understand the natural would and our relationship to it.

From a modern developed society like the United States, it can be difficult to imagine a time and place when the natural world held such sway over our fate. Our food comes from a grocery store, our water from a tap; even our air is artificially heated and cooled to our liking. These days, it seems more logical to consider societal conflict, or even collapse, through the lens of politics or economics. But, as we will see time and again throughout this book, the natural environment—and how we interact with it—plays a leading role in the sagas that shape human history; this is as true today as it was in the time of the Vikings.

environment The biological and physical surroundings in which any given living organism exists.

environmental science An interdisciplinary field of research that draws on the natural and social sciences and the humanities in order to understand the natural world and our relationship to it.

Environment is a broad term that describes the surroundings or conditions (including living and nonliving components) in which any given organism exists. **Environmental science**—a field of research that is used to understand the natural world and our relationship to it—is extremely interdisciplinary. It relies on a range of natural sciences (such as ecology, geology, chemistry, and engineering) to unlock the mystery of the natural world, and to look at the role and impact of humans in the world. It also draws on social sciences (such as anthropology, psychology, and economics) and the humanities (such as art, literature, and music) to understand the ways that humans interact with, and thus impact and are impacted by, the ecosystems around them. **INFOGRAPHIC 1**

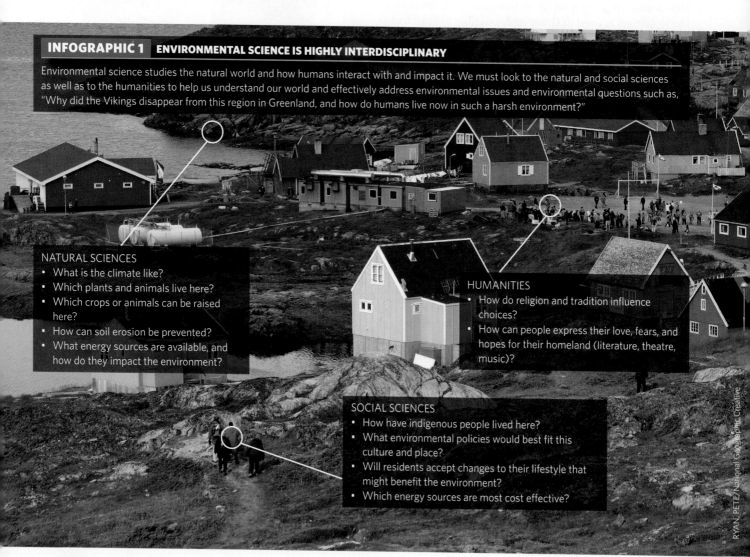

INFOGRAPHIC 1   ENVIRONMENTAL SCIENCE IS HIGHLY INTERDISCIPLINARY

Environmental science studies the natural world and how humans interact with and impact it. We must look to the natural and social sciences as well as to the humanities to help us understand our world and effectively address environmental issues and environmental questions such as, "Why did the Vikings disappear from this region in Greenland, and how do humans live now in such a harsh environment?"

NATURAL SCIENCES
- What is the climate like?
- Which plants and animals live here?
- Which crops or animals can be raised here?
- How can soil erosion be prevented?
- What energy sources are available, and how do they impact the environment?

HUMANITIES
- How do religion and tradition influence choices?
- How can people express their love, fears, and hopes for their homeland (literature, theatre, music)?

SOCIAL SCIENCES
- How have indigenous people lived here?
- What environmental policies would best fit this culture and place?
- Will residents accept changes to their lifestyle that might benefit the environment?
- Which energy sources are most cost effective?

RYAN, PETE/National Geographic Creative

? How would you use your particular college major to help address an environmental problem?

2 EMPIRICAL AND APPLIED SCIENCE

Key Concept 2: Empirical investigations provide information about the natural world; applied science focuses on the practical application of scientific knowledge.

Environmental science is an **empirical science**: It scientifically investigates the natural world through systematic observation and experimentation. The studies of Viking middens and ice cores are examples of empirical science seeking to understand Greenland and its history. Today, soil studies might be done to reveal the composition and water-retention abilities of Greenland's soil. Environmental science is also an **applied science**: We use its findings to inform our actions and, in the best cases, to bring about positive change. For example, an understanding of the suitability of soil for agriculture could help gardeners make choices about what to crops to plant. **INFOGRAPHIC 2**

Early on, researchers thought that the Greenland Colony's demise was largely due to the choices the settlers made—poor land management and rigid conservatism kept them from changing their ways. This view was popularized in *Collapse*, a best-selling book written by Jared Diamond, a biologist from the University of California at Los Angeles. But like any good scientist, McGovern followed the data; recently clues have begun to surface that paint a different picture.

Greenland's interior is covered by vast ice sheets that stretch toward the horizon—3,000 meters (10,000 feet) thick and more than 250,000 years old. To climate scientists these expanses of ice are a treasure trove. As snow falls, it absorbs various particles from the atmosphere and lands on the ice sheets. As time passes, the snow and particles compact into ice, freezing in time perfect samples of the atmosphere as it existed when that snow first fell. By analyzing those ice-trapped particles scientists can get a pretty good idea of what was happening to the climate at any given time. "It's like perfectly preserved slices of atmosphere from the past," says Lisa Barlow, a geologist and climate researcher at the University of Colorado at Boulder. "It gives us additional clues as to what was going on."

To uncover those clues, a team of scientists and engineers picked an accessible segment of ice sheet, not far from one of the Viking settlements, drilled from the surface all the way down to the bedrock below, and extracted a

empirical science A scientific approach that investigates the natural world through systematic observation and experimentation.

applied science Research whose findings are used to help solve practical problems.

INFOGRAPHIC 2 DIFFERENT APPROACHES TO SCIENCE HAVE DIFFERENT GOALS AND OUTCOMES

Environmental science is used to systematically collect and analyze data to draw conclusions and use these conclusions to propose reasonable courses of action.

EMPIRICAL SCIENCE IS USED TO INVESTIGATE THE NATURAL WORLD

Through observation, glaciologists study and record the rate of glacier melt in Greenland; it is increasing dramatically in some far-flung places.

IN APPLIED SCIENCE, KNOWLEDGE IS USED TO ADDRESS PROBLEMS OR NEEDS

Engineers use their understanding of flowing water's potential energy to harness its power; glacial meltwater can be diverted to produce hydroelectric power.

 If we don't have an application in mind for an empirical research topic, is it worth pursuing?

NSIDC courtesy Ted Scambos and Rob Bauer

Scientist from the National Snow and Ice Data Center at the University of Colorado at Boulder working with an ice core drill.

12-centimeter-wide (5 inches), 3,000-meter-long (~10,000 feet) cylinder of ice, which they then divided and dispersed among a handful of labs around the globe, including Jim White's light stable isotope lab—also at the University of Colorado at Boulder. Analysis of thin sequential segments of the ice core showed that when the Vikings first arrived in Greenland, the temperature was anomalously higher than the average over the preceding 1,000 years. By the time the Vikings had vanished, temperatures had lowered so much that scientists call the period the Little Ice Age—a time when all the seasons were cooler than normal and winters were exceptionally cold.

In addition to that single ice core, scientists have analyzed hundreds of mud cores taken from lake beds around the Viking settlements. These mud samples—which contain large amounts of soil that was blown into the lakes during Viking times—indicate that soil erosion had become a significant problem long before the region descended into a mini ice age. "This wasn't a climate problem," says Bent Fredskild, a Danish scientist who extracted and

studied many of the mud cores. "This was self-inflicted. It happened the same way that soil erosion happens today—they overgrazed the land, and once it was denuded, there was nothing to anchor the soil in place. So the wind carried it away."

Overgrazing wasn't their only mistake. The Greenland Vikings also used 2-meter-thick (about 6 feet) slabs of grassland to insulate their houses against the cold of winter; a typical home took about 4 hectares (10 acres) of grassland to insulate. On top of that, they chopped down the forests, harvesting enough timber to not only provide fuel and build houses but also to make the innumerable wooden objects to which they had become accustomed back in Norway.

Greenland's ecosystem was far too fragile to endure such pressure, especially as the settlement grew from a few hundred to a few thousand residents. The short, cool growing season meant that plants developed slowly, which in turn meant that the land could not recover quickly enough from the various assaults to protect the soil.

The Greenland settlers responded by changing some of their ways to better fit their new and, it turns out, changing environment. They depended less on livestock for food, whose numbers were dropping, and more on the sea mammals they hunted—walrus and seals. But this wasn't easy. Unlike the native Inuit who hunted for walrus in protected fjords, the Greenlanders pursued them out in the open ocean, a much more difficult and dangerous task. Hunting these animals in the open ocean required teamwork and led to a new social order. All members of the community worked together to support these seal and walrus hunts, harvesting enough food to get them through the hard winters. To manage this, they developed laws and court systems well before mainland Norway. But, it turns out, the path that they chose was only useful in the short term and may have prevented them from adapting further when things got even harder.

3 ENVIRONMENTAL ISSUES AS "WICKED PROBLEMS"

Key Concept 3: Environmental problems are difficult to solve because there are multiple causes and consequences, different stakeholders prefer different solutions, and potential solutions come with trade-offs.

environmental literacy A basic understanding of how ecosystems function and of the impact of our choices on the environment.

The ability to understand environmental problems is referred to as **environmental literacy**. Such literacy is crucial to helping us become better stewards of Earth.

This is especially important because many environmental problems are multifaceted and hard to solve. Scientists refer to them as "wicked problems." Wicked problems can be extremely complicated because they tend to have multiple causes, each one difficult to address. These multiple causes

INFOGRAPHIC 3 | **WICKED PROBLEMS**

Wicked problems are difficult to address because they have many causes, many consequences, and, in many cases, each stakeholder hopes for a different solution. Solutions that address wicked problems usually involve trade-offs, so there is no clear "winner." One example of a wicked problem is climate change. Many causes exist for the current climate change we are experiencing, some of which are shown below. There are multiple consequences of climate change; these effects will be varied for different species and people, depending on where they live and their ability to adapt to the changes. While there are many solutions that can help address climate change, each brings new problems that must also be addressed.

CLIMATE CHANGE

MULTIPLE CAUSES:

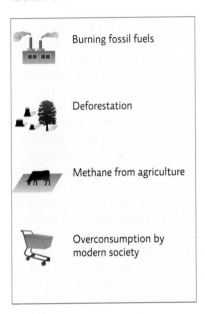

- Burning fossil fuels
- Deforestation
- Methane from agriculture
- Overconsumption by modern society

MULTIPLE CONSEQUENCES:

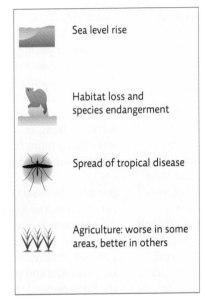

- Sea level rise
- Habitat loss and species endangerment
- Spread of tropical disease
- Agriculture: worse in some areas, better in others

SOLUTIONS COME WITH TRADE-OFFS:

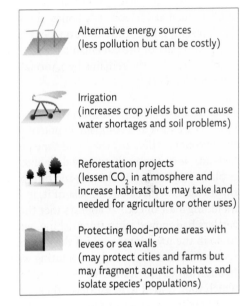

- Alternative energy sources (less pollution but can be costly)
- Irrigation (increases crop yields but can cause water shortages and soil problems)
- Reforestation projects (lessen CO_2 in atmosphere and increase habitats but may take land needed for agriculture or other uses)
- Protecting flood-prone areas with levees or sea walls (may protect cities and farms but may fragment aquatic habitats and isolate species' populations)

 What are some other environmental "wicked problems" we face?

lead to multiple consequences, requiring that we fight a battle against the wicked problem on several fronts.

Water and air pollution, the loss of biodiversity, climate change, how to feed an ever-growing world: These are all wicked problems that must be addressed. But, because of their complexity, any given response to an environmental problem involves significant **trade-offs**. No one response is likely to present the ultimate solution, and each potential solution may create new problems that must be solved. In addition, when confronting wicked problems, we must consider not only their environmental but also their economic and social causes and consequences. Scientists refer to this trifecta as the **triple bottom line**: Solutions must be good for the environment, good for society, and affordable.

Things are further complicated by the fact that different groups of people affected by the issue (stakeholders) may prefer a different solution. For example, a citizen of a landlocked state or nation may feel the sea level rise expected with climate change is a minor problem, whereas a coastal resident may feel threatened by the sea level rise already occurring. These two individuals might differ in what they feel is a reasonable response to climate change. **INFOGRAPHIC 3**

trade-offs The imperfect and sometimes problematic responses that we must at times choose between when addressing complex problems.

triple bottom line The combination of the environmental, social, and economic impacts of our choices.

4 SUSTAINABLE DEVELOPMENT

Key Concept 4: Living sustainably means living within the means of one's environment in a way that does not diminish the environment's ability to support life in the future.

Back in his Manhattan lab, McGovern sorts through hundreds of animal bones collected from various Greenland middens. By examining the bones from different layers, McGovern can tell what the people ate and how their diets changed over time. "This is a pretty typical set of remains for these people from this region and time period," he says, leaning over a shiny metal tray of neatly arranged bone fragments. Some are the bones of cattle, sheep, and goats. Others are the remains of local wildlife such as caribou. The bones of seal and walrus are also present, and as time went on, became an increasingly important part of the Greenlander's diet, making up as much as 80% of the remains by 1,300 (about 300 years after settlement).

Conspicuously absent, McGovern says, are fish of any kind. "If we look at a comparable pile of bones from [Norwegian settlers of] the same time period, from Iceland, we see something very different," McGovern explains. "We have fish bones, and bird bones, and little fragments of whale bones. Most of it, in fact, is fish—including a lot of cod." It appears that the Greenland settlers had moved away from fishing, which was difficult to do in the icy winter seas off the Greenland coast, and focused almost exclusively on hunting walrus and seal. In addition to providing ivory—the main "currency" with which they traded with Europe—these sea mammals could provide a larder of food that would last throughout the winter, and at the time, was the best choice. However, this choice led to problems later that set the Greenland colony on a path to decline and eventual extermination.

Some of the most telling clues to the mystery of the Greenland Vikings' demise come not from the Viking colonies but from another group of people who lived nearby: the Inuit. The Inuit arrived in the Arctic centuries before the Vikings. They were expert hunters of ringed seal—an exceedingly difficult-to-catch but very abundant food source. They knew how to heat and light their homes with seal blubber (instead of firewood).

The Vikings might have learned from their Inuit neighbors; by adapting some of their customs, they might have survived the Little Ice Age. But excavations show that virtually no Inuit artifacts made their way into Viking settlements. And according to written records, the Norse detested the Inuit,

sustainable development
Development that meets present needs without compromising the ability of future generations to do the same.

who, on more than one occasion, attacked the Greenland settlement and walrus hunting expeditions. They called the Inuit *skraelings*, which is Norse for "wretches," and never sought their friendship or their counsel.

Living in harsh northern climates has never been easy. The inhabitants of any region must learn how to "live in place"—learn to access and use available resources and to do so within the means of the environment. At some point in time, settlers in both Greenland and Iceland realized that their newly discovered land could not sustain their cow-farming, wood-dependent ways. Icelanders had cleared most of the region's forests and allowed their cows, sheep, and goats to chew the grasslands down to almost nothing before finally noticing how profound the differences between Iceland and Norway actually were: Growing seasons in Iceland were shorter, thin soils were much more fragile, and because the land could not rebound quickly, cow farming was unsustainable.

But once they saw that their old ways would not work in this new country, the Icelandic Vikings made changes. Not only did they switch from beef to fish, they also began conserving their wood and abandoned the highlands, where soil was especially fragile. As a result of these and other adaptations, they survived and prospered. The Icelanders responded to the limitations of their natural environment in a way that allowed them to meet present needs without compromising the ability of future generations to do the same—an approach known today as **sustainable development**.

The United Nations advocates sustainable development and offers guidance to achieve sustainability worldwide. In 2015, they published their *2030 Agenda for Sustainable Development*, presenting 17 goals that seek to protect the planet and ensure health prosperity for all people. Goals focus on human concerns (e.g., poverty, hunger, equity, and justice) and environmental issues (e.g., resource use, pollution, ecosystem health, and climate change). **INFOGRAPHIC 4**

Greenland settlers too made changes: They reduced their livestock herds, fertilized pastures with animal manure, and dug irrigation ditches. And they focused most of their energy and expertise on sea mammal hunts. But, it turns out, this might have been the decision that sealed their fate. The dangerous seal and walrus hunts required most of the men (and boats) in the settlement. Men would come from farms near and far during the spring

INFOGRAPHIC 4   **U.N. SUSTAINABLE DEVELOPMENT GOALS**

In 2015, the United Nations released its 2030 Sustainable Development Agenda and put forth 17 Sustainable Development Goals in the pursuit of their overarching goal to "end poverty, protect the planet, and ensure prosperity for all people." Specific 15-year targets are set for each goal.

Goals:
Eliminate poverty by 2030; implement social protection systems for everyone and create sound regional, national, and international policies that are "pro-poor and gender-sensitive."

Goals:
End hunger and all forms of malnutrition; double the agricultural productivity of small-scale farmers, the main source of food and income for many of the world's poor and hungry.

Goals:
Provide access to clean water and adequate sanitation for all; reduce water pollution; increase the efficient use of water to reduce waste; strengthen the ability of local communities to reach these goals.

Goals:
Produce national and international policies to reduce future climate change; take steps to address climate change that has occurred or will inevitably occur; establish a monetary fund to help developing countries respond.

Goals:
Prevent or significantly reduce ocean pollution; sustainably manage ocean fisheries and ecosystems; set aside at least 10% of marine areas for conservation; increase the ability of small scale fishers to sustainably use their ocean resources.

 Which of the sustainability goals above do you feel will be the hardest to reach? What kinds of things can be done to help achieve that goal?

and summer months when it was possible to head out in search of seal and walrus. One bad storm or attack by hostile Inuit could seriously deplete the number of able-bodied men in the settlement and time saw their numbers slowly erode. At a time when the Little Ice Age was making life exceedingly difficult, the Greenlanders were so specialized as walrus and seal hunters, they may have lacked the skills or equipment needed to pursue fishing on a large scale. Changing course may have proved impossible.

Adding to their plight, the Greenland Vikings soon found themselves more and more alone and on their own. As the productivity of the Viking colonies declined (and the price of walrus ivory dropped when elephant ivory entered the market), so did visits from European ships. As Europe itself faced the challenges of the Little Ice Age, the abundant trade of dried fish offered by Iceland was much more appealing to the merchants of Europe. As time wore on, it became apparent that the Greenland Vikings could expect very little in the way of trade; ships that had visited every year came less and less often. After a while, they did not come at all. For the Greenland Vikings, who depended on the Europeans for iron, timber, and other essential supplies, this loss proved devastating.

5   HUMAN IMPACT AND THE ANTHROPOCENE

Key Concept 5: Human impact on Earth may be so great that we may be ushering in a new geological epoch: the Anthropocene.

When it comes to the environment, modern societies are not as different from the Vikings as one might assume. The Vikings initially tried to use farming methods that were ill suited to Greenland's climate and natural environment. We, too, use farming practices that strip away topsoil and diminish the land's fertility. We have overharvested our forests and in so doing have triggered a cascade of environmental consequences: loss of vital habitat and biodiversity, soil erosion, and water pollution. We have overfished and overhunted, and have allowed invasive species to devastate ecosystems. And whereas the Greenlanders showed signs of trying to adjust to their environment amidst changes, we seem less inclined to do so, as these environmental assaults continue.

For a time, the Greenland Vikings did learn to live in their new home, adjusting their way of life and society itself to meet the challenges they faced. They learned how to survive "in place" by acquiring what anthropologists call *traditional ecological knowledge*. They learned the seasonal cycles and responded accordingly to exploit resources during seasons of plenty and prepare for lean times during the long winters. What about modern humans? Do we have the ecological knowledge needed to live sustainably? Many of our own problems stem from a disconnect in our understanding of the relationship between our actions and their environmental consequences. For example, many people in the United States don't realize that entire mountains are being leveled to produce their electricity; thousands of acres of habitat and miles of streams and rivers have been destroyed to access coal seams beneath the Appalachian Mountains. Likewise, we are often ignorant of the connection between the burning of that coal and mercury-contaminated fish or increased asthma rates.

We also face a suite of new problems that did not trouble the Vikings. A global population poised to top 11 billion come 2100 will strain Earth's resources like never before.

Anthropogenic climate change is another serious consequence of larger populations, increasing affluence, and more sophisticated technology. While the Vikings had to contend with climate change that was part of the natural climate cycle, the vast majority of scientists today conclude that modern humans are faced with rapidly warming temperatures caused largely by our own use of greenhouse gas— emitting fossil fuels.

---

anthropogenic Caused by or related to human action.

Anthropocene A proposed new geologic epoch that is marked by modern human impact.

> Many of our own problems stem from a disconnect in our understanding of the relationship between our actions and their environmental consequences.

In fact, the impact of humans on Earth is so great that many scientists suggest we have entered a new geologic time interval. They argue that distinctive geological evidence of our existence is accumulating and will be left behind long after we are gone—evidence such as the global distribution of radioactive elements from nuclear weapons, the profusion of plastic pollution, or soot from power plants. According to the current Geological Time Scale, we are presently in the Holocene Epoch (which began at the end of the last ice age, almost 12,000 years ago) but many believe that it is time for a new epoch designation—the **Anthropocene**. INFOGRAPHIC 5

The United Nations' Millennium Ecosystem Assessment is a scientific appraisal of current research that evaluates environmental problems and makes recommendations about addressing those problems. The report concludes that human actions are straining the ability of the planet's ecosystems to sustain future generations. But there is hope: If we act now, the report's authors write, we can still reverse much of the damage. Some of the best lessons about how we can do this come from the natural environment itself.

Current climate change is linked to human activities, most notably burning fossil fuels. The effects on ecosystems, species, and human societies are far reaching and will continue to escalate if we don't take steps to address it.

jrphoto6/Getty Images

INFOGRAPHIC 5 **THE ANTHROPOCENE: A NEW EPOCH?**

Human impact increased dramatically after the Industrial Revolution and again after World War II. This second surge in resource use and pollution generation has been dubbed the *Great Acceleration*. This effect is so great, many scientists believe it will leave a discernable mark on Earth's rock layers, distinct from the layers that separate previous geologic time periods. Research is underway to determine if we have entered a new geologic time period: the Anthropocene Epoch.

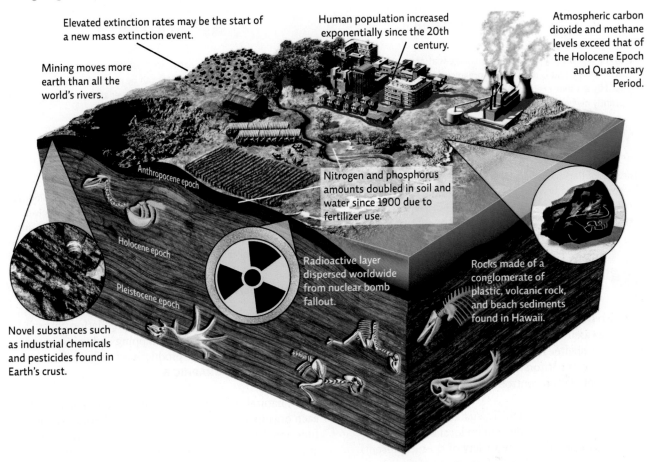

Elevated extinction rates may be the start of a new mass extinction event.

Mining moves more earth than all the world's rivers.

Human population increased exponentially since the 20th century.

Atmospheric carbon dioxide and methane levels exceed that of the Holocene Epoch and Quaternary Period.

Anthropocene epoch

Holocene epoch

Pleistocene epoch

Nitrogen and phosphorus amounts doubled in soil and water since 1900 due to fertilizer use.

Radioactive layer dispersed worldwide from nuclear bomb fallout.

Rocks made of a conglomerate of plastic, volcanic rock, and beach sediments found in Hawaii.

Novel substances such as industrial chemicals and pesticides found in Earth's crust.

 What factors have been instrumental in increasing human impact in this time we call the *Great Acceleration*?

6 THE CHARACTERISTICS OF A SUSTAINABLE ECOSYSTEM

Key Concept 6: Natural ecosystems are sustainable because of the way they acquire energy, use matter, control population sizes, and depend on local biodiversity to meet their needs.

Compared to their counterparts in Greenland, the Icelandic Vikings were more successful at responding to their environment as they adopted more **sustainable** practices. They didn't have to look far for a model of sustainability: Natural ecosystems are sustainable. This means they use resources—namely, energy and matter—in a way that ensures that those resources continue to be available.

To survive, organisms need a constant, dependable source of energy. But as energy passes from one part of an ecosystem to the next, the usable amount declines; therefore, new inputs of energy are always needed. A sustainable ecosystem is one that relies on **renewable energy**—energy that comes from an infinitely available

or easily replenished source. For almost all natural ecosystems, that energy source is the Sun. Photosynthetic organisms such as plants trap solar energy and convert it to a form that they can readily use or that can be passed up the food chain to other organisms.

Unlike energy, matter (anything that has mass and takes up space) can be recycled and reused indefinitely; the key is not using it faster than it is recycled. Naturally sustainable ecosystems recycle matter so that the waste from one organism ultimately becomes a resource for another.

sustainable Capable of being continued indefinitely.

renewable energy Energy that comes from an infinitely available or easily replenished source.

INFOGRAPHIC 6 **FOUR CHARACTERISTICS OF A SUSTAINABLE ECOSYSTEM**

Ecosystems found on Earth today have the capacity to be naturally sustainable—those that were not died out long ago. They all share characteristics that allow the capture of energy and use of matter in a way that allows them to persist over time, all without degrading the environment itself.

SUSTAINABLE ECOSYSTEMS

 RELY ON RENEWABLE ENERGY

Ecosystems must rely on sources of energy that are replenished daily because energy that is used by one organism is "used up" and cannot be used by another; energy is NOT recycled. This means new inputs are constantly needed.

 RECYCLE MATTER

No new matter arrives on Earth, so ecosystems must make do with what they have. Fortunately, matter can be recycled, and organisms in ecosystems use matter resources over and over again; the waste of one becomes resource for the next.

 HAVE POPULATION CONTROL

The sizes of the various populations in an ecosystem are kept in check by disease, predators, and competitors. This prevents a population from getting too large and damaging the ecosystem it, and others, depends on.

 DEPEND ON LOCAL BIODIVERSITY

Ecosystems access energy, recycle matter, and control population sizes largely through the actions of their resident species. Higher biodiversity (greater number of species and more variation of individuals within a species) generally means more energy can be captured, more matter can be recycled and at a faster pace, and population sizes can be better controlled.

 Identify an example of each of these sustainability characteristics from the ecosystem where you live.

It is also important to keep populations in check so that resources are not overused and there is enough food, water, and shelter for all. In sustainable ecosystems, predators, competitors for resources, and disease all provide population control.

Finally, sustainable ecosystems depend on local **biodiversity** (the variety of species present) to perform many of the jobs just mentioned; different species aid in population control; others have different ways of trapping and using energy and matter which boosts productivity and efficiency (see Module 2.1). **INFOGRAPHIC 6**

Thus, natural ecosystems live within their means, and each organism contributes to the ecosystem's overall function. This is not to imply that natural ecosystems are perfect places of total harmony, but those that are sustainable meet all four of these characteristics.

biodiversity The variety of life on Earth.

7 NATURE AS A MODEL FOR SUSTAINABLE ACTIONS

Key Concept 7: Human societies can become more sustainable by mimicking the way natural ecosystems operate.

Human ecosystems are another story. Humans tend to rely on **nonrenewable resources**—those whose supply is finite or is not replenished in a timely fashion. The most obvious example of this is our reliance on fossil fuels such as coal, natural gas, and petroleum, culled from deep within the earth, to power our society. Fossil fuels are replenished only over vast geologic time—far too slowly to keep pace with our rampant consumption of them. On top of that, we have a hard time keeping our population size under control, despite (or maybe because of) all the advances of modern technology. We also generate volumes of waste, much of it toxic, and have yet to fully master the art of recycling.

Increasingly, however, we humans are looking to nature to help us learn how to change our ways. In a growing and varied field known as *Biomimicry*, scientists are using nature as a *model*, *mentor*, and *measure* for our own systems. Emulating nature (nature as model) gives us an example of what to do; it can also teach us how to do it (nature as mentor) and the level of response that is appropriate (nature as measure). As we'll see in later chapters, scientists are using biomimicry to design more sustainable methods of growing crops and livestock for human consumption, of designing products, and of trapping and using energy. **INFOGRAPHIC 7**

Despite such efforts, some concerned scientists and environmentalists say that modern global societies are not acting nearly as quickly as they could or should. What prevents us from changing our ways, even in the face of brewing calamity?

nonrenewable resources A resource that is formed more slowly than it is used or that is present in a finite supply.

SUSTAINABLE ECOSYSTEMS:	WE CAN MIMIC THIS IN OUR OWN SOCIETIES BY:	
RELY ON RENEWABLE ENERGY	**USING SUSTAINABLE ENERGY SOURCES** We can move away from nonrenewable fuels such as fossil fuels by turning to sustainable energy sources such as solar, wind, geothermal, and biomass (harvested at sustainable rates).	
RECYCLE MATTER	**USING MATTER CONSERVATIVELY AND SUSTAINABLY** We can reduce our waste by reducing our use of resources and by recovering, reusing, and recycling matter that we do use; we would also benefit from minimizing the toxins we create or release into the environment that degrade our natural resources.	
HAVE POPULATION CONTROL	**GETTING HUMAN POPULATION GROWTH UNDER CONTROL** While predation controls many natural populations, there are many ways to reduce human birth rates without increasing the death rate through war or disease.	
DEPEND ON LOCAL BIODIVERSITY	**DEPENDING ON LOCAL HUMAN CONTRIBUTIONS AND BIODIVERSITY** Protecting biodiversity will help us achieve the three above goals; we can also regard the use of diversity as a metaphor and emulate nature by using a variety of local energy sources, building materials, and crops and by exploring the many ideas and innovations that come from a diverse human community.	

? What obstacles stand in society's way in pursuing the actions mentioned in this graphic?

8 CHALLENGES TO SOLVING ENVIRONMENTAL PROBLEMS

Key Concept 8: Impediments to solving environmental problems include short-term thinking and social traps.

Decisions by individuals or groups that seem good at the time and produce a short-term benefit but hurt society in the long run are called **social traps**. The **tragedy of the commons** is a social trap that often emerges when many people are using a commonly held resource, such as water or public land. Each person will act in a way to maximize his or her own benefit, but as everyone does this, the resource becomes overused or damaged. Herders might put more animals on a common pasture because they are driven by the idea that "if I don't use it, someone else will." We do the same thing today as we overharvest forests and oceans and as we release toxins into the air and water. Other social traps include the **time delay** and **sliding reinforcer** traps—actions that, like the tragedy of the commons, seem reasonable at first but have a negative effect later on. **INFOGRAPHIC 8**

Environmental literacy is our best hope for avoiding such traps—understanding how ecosystems operate and how our actions impact society and the environment. When people are aware of the consequences of their decisions, they are more likely to examine the trade-offs to determine whether long-term costs are worth the short-term gains and, hopefully, to make different choices when necessary. For example, the problem of lead in drinking water that Flint, Michigan is experiencing is related, in part, to the earlier decision not to invest in upgraded water infrastructure, a problem shared by many U.S. communities (see Module 1.2).

Scott Warren/Aurora/Getty Images

A replica of the first church built in Brattahlid, Greenland—a Viking settlement founded by Erik the Red more than 1,000 years ago.

social traps Decisions by individuals or groups that seem good at the time and produce a short-term benefit but that hurt society in the long run.

tragedy of the commons The tendency of an individual to abuse commonly held resources in order to maximize his or her own personal interest.

time delay Actions that produce a benefit today but set into motion events that cause problems later on.

sliding reinforcer Actions that are beneficial at first but that change conditions such that their benefit declines over time.

INFOGRAPHIC 8 | **SOCIAL TRAPS**

Social traps are decisions that seem good at the time and produce a short-term benefit but that hurt society (usually in the long run).

TRAGEDY OF THE COMMONS When resources aren't "owned" by anyone (they are commonly held), individuals who try to maximize their own benefit end up harming the resource itself.

This common resource (pasture) can support four animals sustainably. Each of the four farmers benefits equally.

If one farmer adds two more cows, the commons becomes degraded. All four farmers share in the degradation (less milk produced per cow), but the farmer with three cows gets all the benefit from adding the extra cows.

The other farmers also must add cows to return to the same level of production as before. **Over time, the commons degrades even more, and at some point, it will no longer support any animals.**

TIME DELAY Actions that produce a benefit today set into motion events that build over time and cause problems later on.

Modern fishing techniques that use giant nets can harvest large numbers of fish. Fishers try to get the most fish possible in the short term.

If more fish are taken over a number of years than are replaced naturally through fish reproduction, the population decreases, Other populations also decline as the fish they depend on are depleted.

After a decade or so, overfishing may so deplete the population that fishers cannot catch enough to meet needs. **The effects of decisions about how many fish to harvest are not felt until later.**

SLIDING REINFORCER Actions that are beneficial at first may change conditions such that their benefit declines over time.

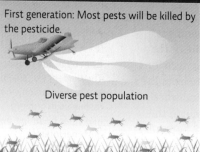 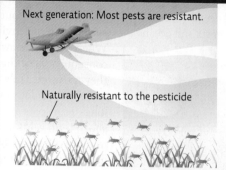 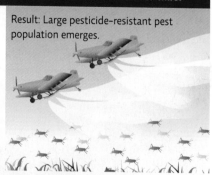

First generation: Most pests will be killed by the pesticide.

Diverse pest population

Next generation: Most pests are resistant.

Naturally resistant to the pesticide

Result: Large pesticide-resistant pest population emerges.

Pesticide application can reduce pest numbers on a crop, but a few pests might survive.

The surviving pests reproduce. The pesticide is only helpful if more is applied or a more toxic pesticide is used. This can be harmful to other organisms and to humans.

In addition to becoming resistant to the pesticide, the pest population may even become bigger if the pest's predators are also killed. **The helpful action changes how the system operates; the action loses its benefit or may even be harmful at a later time.**

 Why do you suppose humans are so prone to being caught in these social traps?

9 WORLDVIEWS AND ENVIRONMENTAL ETHICS

Key Concept 9: Our worldview reflects how we value the natural world and influences the ethical choices we make regarding the environment.

Conflicting worldviews is another challenge to sustainable living. Because our **worldviews**—the windows through which we view our world and existence—are influenced by cultural, religious, and personal experiences, they vary across countries and geographic regions, even within a society. People's worldviews determine their **environmental ethic**, or how they interact with their natural environment; worldviews also impact how people respond to environmental problems. When different people or groups, with different worldviews, approach environmental problems, they are bound to draw very different conclusions about how best to proceed.

The Vikings may have had an **anthropocentric worldview**—one where human lives and interests are most important. They may have viewed other species as having *instrumental value*—meaning the Vikings valued them only for what they could get out of them. Forests were nothing more than a source of timber; grasslands a source of home insulation and a feeding ground for cattle.

A **biocentric worldview** values all life. From a biocentric standpoint, every organism has an inherent right to exist, regardless of its benefit (or harm) to humans; each organism has *intrinsic value*. This worldview would lead us to be mindful of our choices and avoid actions that indiscriminately harm other organisms or put entire species in danger of extinction.

An **ecocentric worldview** values the ecosystem as an intact whole, including all of the ecosystem's organisms and the nonliving processes that occur within the ecosystem. Considering the same forests and grassland from an ecocentric worldview, the Vikings might have decided to protect both, not just for the resources they could harvest but to protect the complex processes that can produce those resources only when they remain intact.
INFOGRAPHIC 9

Back in Greenland, in the silt-covered ruins of a Viking farmhouse, archaeologists found the bones of a hunting dog and a newborn calf, dating back several centuries. The knife marks covering both indicate that the animals were butchered and eaten. "It shows how desperate they were," says McGovern. "They would not have eaten a baby calf, or a hunting dog, unless they were starving." By then, it was probably far too late for the Greenland Vikings to adapt in any meaningful way. Their tale had already been written—in ice cores, and mud cores, and in hundreds of years' worth of midden heaps.

worldview The window through which one views one's world and existence.

environmental ethic The personal philosophy that influences how a person interacts with his or her natural environment and thus affects how one responds to environmental problems.

anthropocentric worldview A human-centered view that assigns intrinsic value only to humans.

biocentric worldview A life-centered approach that views all life as having intrinsic value, regardless of its usefulness to humans.

ecocentric worldview A system-centered view that values intact ecosystems, not just the individual parts.

An ecocentric worldview values all living creatures and nonliving processes of an ecosystem, from animals and wildflowers to nutrient cycling in the soil and water flow from snowmelt. Sassalb mountain range, Switzerland.

Andreas Strauss/LOOK-foto/Getty Images

INFOGRAPHIC 9 **WORLDVIEWS AND ENVIRONMENTAL ETHICS**

People's environmental worldviews describe how they see themselves in relation to the world around them. Their worldview influences their environmental ethic, which in turn influences how they interact with the natural world. We present three common worldviews here (there are others).

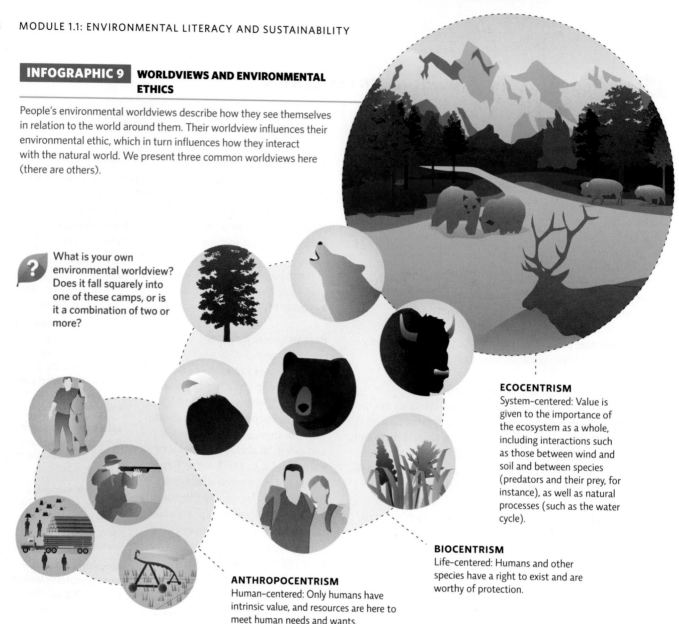

? What is your own environmental worldview? Does it fall squarely into one of these camps, or is it a combination of two or more?

ECOCENTRISM
System-centered: Value is given to the importance of the ecosystem as a whole, including interactions such as those between wind and soil and between species (predators and their prey, for instance), as well as natural processes (such as the water cycle).

BIOCENTRISM
Life-centered: Humans and other species have a right to exist and are worthy of protection.

ANTHROPOCENTRISM
Human-centered: Only humans have intrinsic value, and resources are here to meet human needs and wants.

The loss of the Norse settlement on Greenland had once been seen as a failure of the community to adapt to their surroundings—a collapse caused by unfortunate circumstances and bad choices. However, McGovern's research now suggests it was more of a *decline*, which came about due to an inability of the Norse community to "anticipate an unknowable future." Though they were able to adapt to their new surroundings and to the change that came their way in the first few centuries of their settlement of Greenland—an admirable feat considering the harsh conditions of their environment—in the end, they became too specialized to adapt further and too isolated to survive.

Throughout human history, societies have come and gone. For some, their environments had changed for the worse—sometimes at their own hands—contributing to their disappearance. The initial evidence that suggested the Greenland colony vanished due to poor choices—self-inflicted environmental damage and failure to adjust to a changing environment—made a tidy, cautionary tale about how we should avoid making the same mistakes and seek to live sustainably. But the reality—the settlers did try to make changes and live within the means of their

environment, but failed—is a more sobering reminder that even if we try to adapt and adjust to a changing environment, there are no guarantees. This suggests we should not take our environment for granted or degrade it in ways that diminishes its ability to support us. Neither should we assume there will always be time "later" to change our ways.

Our own story is being written now, in much the same way. But unlike our forebears, we still have enough time to shape our own narrative. Will it be dug up, 1,000 years hence, from the ruins of what we leave behind? Or will it be passed down by the voices of our successors, who continue to thrive long after we are gone? Ultimately, the answer is up to us.

Select References:

Benyus, J. (1997). *Biomimicry*. New York: William Morrow and Company.

Dugmore, A. J., et al. (2012). Cultural adaptation, compounding vulnerabilities and conjunctures in Norse Greenland. *Proceedings of the National Academy of Sciences*, 109(10), 3658–3663.

Kintisch, E. (2016). The lost Norse. *Science*, 354(6313), 696–701.

Lewis, S. L., & Maslin, M. A. (2015). Defining the Anthropocene. *Nature*, 519(7542), 171–180.

INTERACTIVE MAP SOCIETAL COLLAPSE   ANIMATED INFOGRAPHIC

Greenland Colony was not the only once-thriving society to collapse due, at least in part, to environmental problems. Read the case studies provided with this map for a look at some other notable examples from history.

MAYA

KHMER

MESOPOTAMIA

EASTER
ISLAND

  BRING IT HOME

PERSONAL CHOICES THAT HELP

The concept of sustainability unites three main goals: environmental health, economic profitability, and social and economic equity. All sorts of people, philosophies, policies, and practices contribute to these goals; concepts of sustainable living apply on every level, from the individual to the society as a whole. In other words, every one of us participates. Throughout this book, you will have the opportunity to learn about personal actions that can help address environmental issues, but a good starting place is to learn about your own environment and the place you call home.

Individual Steps
• Discover your local environment. What parks or natural areas are close by? Does your campus have any natural areas? Visit one and spend a little time observing nature and your own reactions. Write down your thoughts or share you experiences with a friend.
• Are there restaurants, grocery stores, or other retail venues accessible through public transportation or within walking distance of your campus or home? For a week or two, try walking or riding a bike or bus to these businesses instead of driving to others farther away. Is this a reasonable option for you? Why or why not?

Group Action
• Discover your own interests. There is a group for every interest—from outdoor recreation, wildlife viewing and preservation, and environmental education, to transportation and air quality issues. Get involved with organizations working to improve environmental issues or address social change and human rights. One person can make a difference, but a group of people can cause a sea of change.

Policy Change
• Discover what's happening in your community. Read the newspapers and monitor blogs covering environmental and quality-of-life issues. Alert your local, regional, and national representatives about the issues you care about and vote for government officials who support the causes you support.

ENVIRONMENTAL LITERACY **UNDERSTANDING THE ISSUE**

1 What is the purpose and scope of environmental science?

1. Environment is a broad term that includes one's _____ and _____ surroundings.

2. Why is environmental science considered an interdisciplinary field that includes the natural and social sciences and the humanities?

2 Why are both empirical and applied approaches useful in environmental science?

3. True or False: Research that seeks to understand the sources of water pollution for a given body of water is an example of applied science.

4. Some scientists study peatlands to look for clues about how ancient civilizations lived and died. Others work with peatlands to manage these ecosystems for their water purification services, or as sources of fuel or soil additives (peat moss). Which of the activities above describes empirical science, and which describes applied science? Explain.

3 What characteristics make an environmental dilemma a "wicked problem"?

5. Why is it that many problems are considered to be *wicked problems*?
   a. Their potential solutions almost always come with trade-offs.
   b. They arise out of the greed and malice of other people.
   c. These problems cannot be solved or even effectively addressed.
   d. They are hard to study using empirical methods, and so science can't help find solutions.

6. Consider the wicked problem of air pollution. Identify some of the causes and consequences of air pollution. What are some possible solutions, and what are the trade-offs that come with those solutions?

4 What does it mean to be sustainable?

7. Which of these is an example of acting sustainably?
   a. Using available resources to support our current lifestyle
   b. Using resources now as we need them but also researching new ways to meet future needs
   c. Dramatically reducing our use of resources, even if it means a lower quality of life for most people
   d. Meeting today's needs without compromising the ability of future generations to meet theirs

8. Identify three Sustainable Development Goals and explain how accomplishing each would bring the world population closer to sustainability.

5 Why do scientists think we are living in a new geologic epoch: the Anthropocene?

9. Which of these has been proposed as a possible unique marker that would distinguish the Anthropocene from the Holocene?
   a. Radionuclides from nuclear bombs
   b. Plastics
   c. Novel chemicals such as industrial chemicals and pesticides
   d. All of the above

10. What is the significance of the Anthropocene as a new geologic epoch?

6 What are the characteristics of a sustainable ecosystem?

11. Why is it important that sustainable ecosystems only rely on a renewable energy source such as the Sun?
   a. Solar energy can be captured and used by all organisms.
   b. It is clean and nonpolluting.
   c. Energy cannot be recycled, so new supplies are always needed.
   d. It is a natural form of energy.

12. Why must ecosystems recycle and reuse matter?

7 What can human societies and individuals do to pursue sustainability?

13. If we were to mimic the way nature uses matter resources, we would:
   a. reduce the waste we produce by reusing and recycling the matter resources we do use.
   b. use the most abundant resources first before turning to less abundant ones.
   c. focus on reducing what we "take" from the environment but not need to worry about what we put back into the environment.
   d. rely on one or two ways to use and recycle material globally rather than using a variety of methods that might differ in different places.

14. Identify at least one action we could take as a society to follow each of the four characteristics of a sustainable ecosystem.

8 What challenges does humanity face in dealing with environmental issues?

15. When you drive your car, it releases a small amount of air pollution. You know air pollution is bad, but if you stop driving, you'll be forced to find another way to get around, and the air will still be polluted because everyone else will still keep driving. So you keep driving, too. This is an example of the social trap known as _____.

16. The use of a pesticide helps control pests at first but their use sets into motion changes that make that pesticide less beneficial later on. This is the social trap known as the:
   a. tragedy of the commons.
   b. time delay social trap.
   c. sliding reinforcer social trap.

17. Why do we fall prey to social traps and how can we avoid them?

9 Distinguish between an anthropocentric, a biocentric, and an ecocentric worldview.

18. Jack is opposed to the selective killing of deer that have vastly overpopulated the local forest. Jill argues that the ecosystem will be destroyed if some are not removed. While Jack is _____ as he sees deer as having intrinsic value, Jill is _____ as she values not just the species but the ecosystem processes as well.

19. A classmate argues that she has an ecocentric worldview because human well-being is her top priority. Explain how she can be both ecocentric and anthropocentric.

SCIENCE LITERACY · WORKING WITH DATA

The following data show the number of different types of livestock grazed in the world and in Nigeria from 1961 to 2008. Use these graphs and the following table to answer the next five questions.

WORLD GRAZING LIVESTOCK BY TYPE, 1961–2008

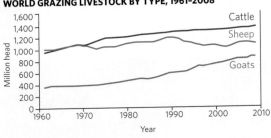

GRAZING LIVESTOCK IN NIGERIA, 1961–2008

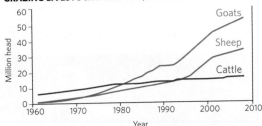

Table 1

Grazing Livestock Population (in millions)

Type of livestock	World in 1961	World in 2008	% change	Nigeria in 1961	Nigeria in 2008	% change
Cattle	942	1,372		6	16.3	
Goats	349	864		0.6	53.8	
Sheep	994	1,086		1	33.9	

Interpretation

1. How many different livestock types are included in the graphs, and what are the trends in the numbers?

2. Which animals constituted the bulk of the grazing herds around 2008 on a worldwide basis? In Nigeria? Was this true in 1961 (the first year for which data are reported)? Provide data to explain your responses.

3. Using the data from the table, calculate the percentage change from 1961 to 2008 for each type of animal. Which animal changed the most, worldwide and in Nigeria? (Show your calculations.)

Hint: percentage change $= \dfrac{(\text{2nd value} - \text{1st value})}{\text{1st value}}$

Advance Your Thinking

4. Unlike cattle and sheep, goats are much more flexible in what they eat, but their sharp hooves also pulverize soil more easily. The data show that the growth in goat populations is particularly dramatic in a developing country like Nigeria. What might explain this pattern? What might be some potential consequences?

5. How might the increase in the size of the goat herd be a potential social trap? What are some ways that we could avoid this trap?

INFORMATION LITERACY · EVALUATING INFORMATION

The Lorax, a children's book by Dr. Seuss, tells the story of the Lorax, a fictional character who speaks for the trees against the greedy Once-ler who represents industry. Written in 1971, *The Lorax* was banned in parts of the United States for being an allegorical political commentary. Today the book is used for educating children about environmental concerns (see www.seussville.com/loraxproject/). Even so, some people consider the book inappropriate for young children due to its "doom and gloom" environmentalism.

The book *The Truax*, by Terri Birkett, involves a forest industry representative offering a logging-friendly perspective to an anthropomorphic tree, known as the Guardbark. This story was criticized for containing skewed arguments, and in particular a nonchalant attitude toward endangered species. About 400,000 copies of the book have been distributed to elementary schools nationwide.

Read both books. You can find *The Lorax* at your local public library, and you can download *The Truax* as a PDF from http://woodfloors.org/truax.pdf.

Evaluate the stories and work with the information to answer the following questions:

1. What are the credentials of the author of each book? In each case, do the person's credentials make him or her reliable/unreliable as a storyteller? Explain.

2. Connect each story to the key concepts in the chapter:
   a. What are the underlying attitudes and worldviews of each story?
   b. Does each story reflect social traps and, if so, in what way?
   c. How might each story contribute to environmental literacy? Explain.
   d. What does each story have to say about sustainability? Explain.

3. What supporting evidence can you find for the main message in each story? In the story itself? From doing some research?

4. What is your response to each story? What do you agree and disagree with in each case? Explain.

 Additional study questions are available at SaplingLearning.com.

FUNGAL ATTACKER THREATENS BATS

Unraveling the mystery behind bat deaths

Little brown bat (Myotis lucifugus) *in flight. Edward Kinsman/Getty Images*

After reading this chapter and studying the KEY CONCEPTS and INFOGRAPHICS, you should be able to answer these GUIDING QUESTIONS

CORE MESSAGE

The world faces many environmental challenges, some predictable and some unexpected, such as the sudden decline of bats in North America. Physical evidence, systematically collected and logically analyzed, helps scientists understand environmental issues, and this understanding can guide policy decisions that address environmental problems.

1   How do scientists study the natural world?

2   What influences the degree of certainty in a scientific explanation?

3   How do comparative studies contribute to the understanding of an environmental issue?

4   Why are observational and experimental studies needed to investigate the natural world?

5   What is the difference between a *correlation* and a *cause-and-effect relationship*?

6   How can science help policy makers address environmental problems?

In the winter of 2007, David Blehert received a very troubling phone call from his colleague, Al Hicks. Blehert is chief of the Wildlife Disease Diagnostic Labs at the National Wildlife Health Center. Hicks is a biologist tasked with surveying, among other things, bat populations in New York State. Every other year, Hicks and his team visit caves across the state where bats are known to hibernate (called hibernacula) and conduct a census. They check the size of the colonies and look for any potential problems with either the bats themselves or with their preferred habitats. This year, they found a big problem: thousands of bats dead at several different sites and near total colony collapse at some of them.

But the bats hadn't merely died; they had suffered, badly. Many of them were huddled at the opening of the hibernacula, a sign that they had been roused from their deep slumber and had been looking for insects to replenish their fat stores. Of course, because it was the dead of winter, there were no insects to be found. The bats' skin was dry and flaking, and their bodies so emaciated that bones pushed easily through skin. "In thirty years of doing this work, he said he had never seen anything like it," Blehert recalls.

Before long, other surveyors were reporting similar troubles—including cases where entire colonies had died—at bat hibernacula across the region. Nobody knew what was killing the bats, but because most biologists noted a peculiar white fuzz on the victims' muzzles, the mystery affliction earned the moniker "white-nose syndrome" (WNS).

◉ WHERE IS WHITE-NOSE SYNDROME FOUND?

Howe's cave, first WNS cases

NY

Albany, NY

▢ WNS confirmed (2016)

STEPHEN ALVAREZ/National Geographic Creative

These dead little brown bats at Hellhole Cave in West Virginia show signs of white-nose syndrome.

In the next years, WNS spread, at the rate of about 320 km (200 miles) per year, to 29 states and 5 Canadian provinces. It has infected at least seven different species of bat and is driving at least two of them to the edge of extinction. By the end of 2016, some six million bats had died. For scientists like Blehert, those grim numbers would fuel a race against time to answer two urgent questions: What is causing the disease? And can it be stopped?

1 THE NATURE OF SCIENCE AND THE SCIENTIFIC METHOD

Key Concept 1: Scientists investigate the natural world using a transparent method of inquiry, and their findings are evaluated by other scientists.

Science is both a body of knowledge (including basic facts and complex explanations) and the process used to get that knowledge. The "body of knowledge" part of science is ever changing; facts, figures, and understandings are revised as scientists learn more about any given topic through their research. But the "process" part of science is fixed. This process is best thought of as a tool (and a powerful one at that!); it enables us to test our ideas by gathering evidence and then to evaluate the quality of that evidence.

Of course, not all questions fall under the purview of science. The scientific method is based on **observations**, known as **empirical evidence**, meaning information that can be detected with the five senses or with equipment used to extend those senses (like microscopes or sonar). For that reason, only phenomena that can be *objectively* observed (that is, the observation can be verified by anyone in the same place, using the same equipment, etc.) are fair game for science. We can even learn about past events by studying the physical evidence left behind; for example, the type and position of fossils in layers of rock tell us something about the organisms that lived in that location at the time the fossils were deposited—and, by extension, something about their ecosystem and the climate at that time. Phenomena that are not objectively observable (What is my dog thinking? Do ghosts exist?) and ethical or religious questions (Is the death penalty wrong? What is the meaning of life?) cannot be empirically studied and therefore are not within the purview of science.

Like all good scientists, Blehert and his colleagues applied the **scientific method** to solve the mystery of the dying bats. They worked logically from previous knowledge and observations of the natural world toward new questions and possible answers. They started with observations: thousands of dead bats and a mysterious white fuzz on most of their noses. They combined these observations with existing knowledge to make some

science A body of knowledge (facts and explanations) about the natural world and the process used to get that knowledge.

observations Information detected with the senses—or with equipment that extends our senses.

empirical evidence Information gathered via observation of physical phenomena.

scientific method The procedure scientists use to empirically test a hypothesis.

inferences Conclusions drawn based on observations.

hypothesis A possible explanation for what we have observed that is based on some previous knowledge.

inferences (explanations of what else might be true based on what they saw and what they know): Because bats hibernate in cold, dark, wet places—the kind of places where fungi also thrive—they inferred that the fuzz was a fungus. And because most of the bats that died had it on their noses and bodies, they inferred that it was somehow related to the deaths.

Next, they gathered more evidence to investigate these inferences. Melissa Behr, a pathologist on Blehert's team, collected samples of the white fuzz. Back in the lab, she confirmed it was a fungus and noted it looked quite different from most commonly known fungi. Scientists eventually came to identify it as *Pseudogymnoascus destructans* (Pd).

"It was an obscure little fungus, that hardly anyone even knew existed," Blehert says. "And it was part of a family that was not known to be pathogenic at all."

> **"**It was an obscure little fungus, that hardly anyone even knew existed.**"**
> —David Blehert

Was this fungus causing the bat deaths? Had it always been there, and if so, why was it suddenly more dangerous? If it's new to the caves, where did it come from? To begin answering those questions, researchers studying the problem would need to generate some hypotheses. A **hypothesis** is an inference that proposes a possible explanation for what we have observed and is based on some previous knowledge (previously published research, preliminary observations or experience).

Blehert, Behr, and others had no trouble coming up with hypotheses. The three they thought most likely were:

1. The white-nose infection caused by the fungus was secondary and opportunistic; it was only able to grow on the bats because their defenses had been weakened by another pathogen.

2. The fungus had always been present, but had recently mutated to become deadlier.

3. The fungus had been transported from some other place and was new to the region.

To figure out which of these ideas was correct, they would have to test each one.

To test a hypothesis, a researcher designs an observational or experimental study and makes an experimental prediction, a statement that identifies what is expected to happen if the hypothesis being tested is correct. A scientific study must be designed to be a "fair test" —this means the results of the test could support *or* falsify (prove wrong) the prediction. If evidence repeatedly falsifies an experimental prediction, the hypothesis is rejected and alternative hypotheses can be tested. If a hypothesis is supported, researchers repeat the study to validate the data and generate new predictions that test the same hypothesis from different angles. As supporting evidence mounts from replicate studies and from multiple predictions, we become more confident in our data and conclusions.

Before being published, scientific reports are subjected to **peer review**, meaning they are reviewed by a group of third-party experts. Studies that are not well designed or conducted are not accepted for publication. Therefore, peer-reviewed published research represents high-quality scholarship in the field. This process might slow down the dissemination of information but adds a needed check on the validity of that information. There are even formal processes for the retraction of a study that has come into question after publication. **INFOGRAPHIC 1**

peer review A process whereby researchers' work is evaluated by outside experts to determine whether it is of a high enough quality to publish.

INFOGRAPHIC 1   SCIENTIFIC METHOD

Scientists work from previous knowledge and observation to ask new questions and pose possible explanations (hypotheses) for what they observe. They then design a study to gather evidence to test predications made from their hypotheses.

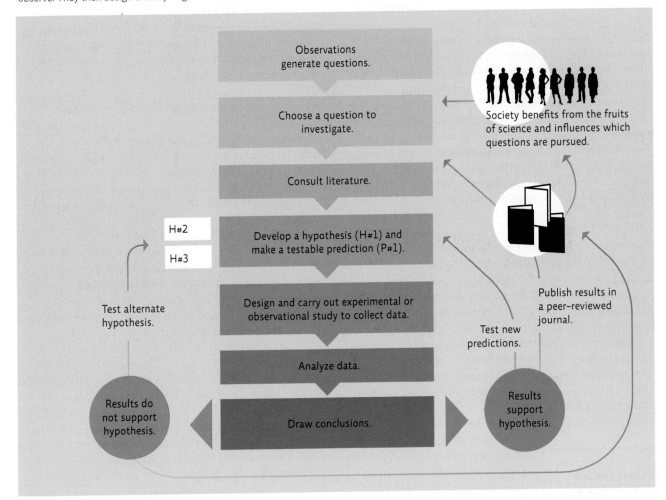

 If the data support a hypothesis, new predictions are still tested. Why is this useful?

2 CERTAINTY IN SCIENCE: FROM HYPOTHESIS TO THEORY

Key Concept 2: Scientific certainty increases as evidence mounts and can be quite strong, but science never claims or even expects it will reach a level of absolute, unquestionable proof because there are always new questions to ask and new experiments to perform.

Hypothesis testing is at the heart of science. For example, Blehert's team investigated the hypothesis that the fungal infection was opportunistic and the bats were already weakened from another infection. Close examination of internal organs (heart, liver, lungs, and kidneys) did not reveal any abnormalities—no sign of a problem or previously undetected infection that would make the bats more susceptible to the fungus. The skin infections caused by WNS were occurring in the absence of other infections. The *opportunistic hypothesis* was rejected, and scientists moved on.

The second hypothesis was that WNS was caused by a local strain of Pd that had recently mutated to become more deadly. "If it were a native species, you would find a lot of genetic diversity within the population," says Blehert. In other words, one would expect individual fungal samples from affected caves and bats to be slightly different from each other—and this variety would be due to inherited differences, accumulated over time as the native fungus occupied this habitat for generations. (See Module 3.2 for an introduction to genetic diversity.) It was expected that the fungus in infected caves would differ from samples taken from uninfected caves, but the *recent mutation hypothesis* led to the additional prediction that samples of the fungus from different WNS-infected caves should have genetic diversity indicative of a long-standing population.

The researchers did find that the fungus from caves with WNS outbreaks was different from the native species found in other caves whose bat occupants showed no signs of the disease. However, they also found that fungal samples from WNS caves and bats were all nearly identical genetic clones of each other. This lack of genetic diversity between samples—even ones taken from caves hundreds of miles apart—suggested to researchers that the fungus was not a native species but rather was one that had been introduced only once and had spread from that original introduction.

Meanwhile, as word of the colossal bat die-offs spread throughout the scientific community, researchers in Europe and China identified the same fungus in their bat caves, only the bats there were perfectly healthy. It appears that Pd had been native to those regions for a very long time. Because the bats in those areas had evolved with the fungus, they likely had adaptations to help them survive the infections (see Module 3.1 for an explanation of coevolution). Scientists are currently investigating the nature of these adaptations to determine why European and Asian bats do not succumb to the disease.

This new information lent some weight to the third hypothesis about the origin of the fungus—i.e., that it had been brought to North America from elsewhere (*novel pathogen hypothesis*). When evidence mounts like this to support a given hypothesis, scientists suspect they are on the right track. But even then, they don't stop. All conclusions in science are considered tentative and open to revision because our understanding of a concept or process may change as more observations and evidence are gathered.

That is not to say that all conclusions are equally valid. There are degrees of certainty in science; we know some things better than others. The more evidence we have in support of an idea, especially from different lines of investigation (the hypothesis tested in many different ways, not just tested the same way repeatedly), the more certain we are that we are on the right track.

If a hypothesis survives repeated testing by numerous research teams and in numerous ways, it may be incorporated into a **theory**: a widely accepted explanation that has been extensively and rigorously tested. Theories represent the highest level of certainty a scientific explanation can attain. But, in keeping with the tentative nature of science, even a well-substantiated theory is open to further study and can be revised, sometimes significantly, if new data strongly support a new conclusion. For this reason, scientists do not expect or require "absolute or unquestionable proof" for their conclusions and explanations. This is, in fact, not possible, as there are always new questions to ask and new ways to investigate those questions. But neither is this a weakness—it is a way to strengthen our understanding of a topic, uncover new questions, and continually add to and revise the body of knowledge that is science. **INFOGRAPHIC 2**

theory A widely accepted explanation of a natural phenomenon that has been extensively and rigorously tested scientifically.

INFOGRAPHIC 2   CERTAINTY IN SCIENCE

There are degrees of certainty in science; we know some ideas are better than others. The more evidence we have in support of an idea, especially when the evidence comes from different lines of inquiry, the more certain we are that we are on the right track. But since all scientific information is open to further evaluation, we do not expect or require "absolute" proof.

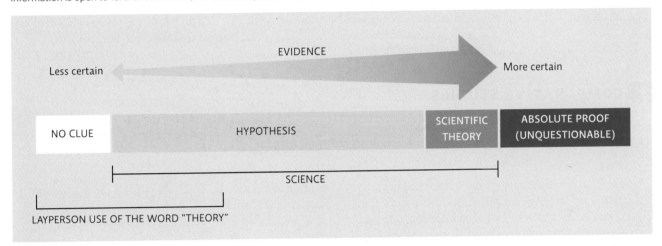

EVIDENCE

Less certain   More certain

| NO CLUE | HYPOTHESIS | SCIENTIFIC THEORY | ABSOLUTE PROOF (UNQUESTIONABLE) |

SCIENCE

LAYPERSON USE OF THE WORD "THEORY"

? Why do scientists say that a hypothesis can be disproved but never proven?

This definition differs significantly from the lay use of the word "theory," which typically refers to a speculative idea without much substance. Nonscientists, including some politicians, have been known to dismiss some scientific theories (like climate change) as "just a theory," by which they mean, "This is unproven and therefore we shouldn't have to consider it." But that line of thinking represents a serious misunderstanding of what a scientific theory

Many bat species live together by the hundreds or even thousands in caves or other protective sites. This communal living greatly increases the chance of spreading infection. Here, a colony of hibernating gray bats is being inspected for signs of WNS.

STEPHEN ALVAREZ/National Geographic Creative

really is. Theories in science represent the highest level of certainty assigned to an explanation. They should not be rejected without substantial scientific evidence to the contrary.

When it came to the bat die-offs, scientists were nowhere near a theory. But they did have enough information to hypothesize further: Since bats don't travel across oceans, the fungus was probably introduced to U.S. caves by humans—most likely cave explorers who carried fungal spores overseas on their clothing and gear—and then spread to multiple colonies by bat-to-bat contact. Bats live in large groups that move from summer to winter homes on a regular seasonal schedule; they could easily have carried the fungus hundreds of kilometers in a single season.

3 COMPARATIVE STUDIES

Key Concept 3: Scientists often conduct comparative studies—looking at an issue in different species or regions—to gain insights about the phenomenon they are investigating.

Because it was present on so many of the dead bats, scientists suspected the fungal infection was the culprit in the bats' woes. But could a fungal infection, one that barely bothered bats in other countries, really be wiping out entire colonies here? Sometimes scientists use comparative studies that examine similar phenomena in other species or regions for clues about what might be happening in the event that they are investigating. It turns out, there was already a historical precedent of how devastating a new fungal invader can be.

Back in the 1970's, scientists studying amphibians around the world first reported similar devastation: In Australia and the Americas, whole communities of frogs began dying en masse. Scientists noted that in places where frogs had been abundant just a year or so earlier, they were suddenly absent. It took researchers more than a decade to identify the offending pathogen; it was a fungus from the Chytrid (pronounced ki-trid) family. By then, it had spread to every continent on the planet except Antarctica, and the disease it caused, chytridiomycosis, had driven some 200 amphibian species to extinction. It has been called the worst infectious disease ever recorded in vertebrates. **INFOGRAPHIC 3A**

Like Pd, the fungus that causes WNS, chytrid is believed to have been spread around the world by human activities—namely the trade in amphibians. From the 1930's to the 1960's, physicians used the African clawed frog for pregnancy tests. In addition, American bullfrogs, a popular species for the food delicacy, frog legs, were widely exported. Both species carry the fungus but don't succumb to it. But unlike WNS, chytrid infections seem to move through a given region much faster than

---

INFOGRAPHIC 3A   CHYTRIDIOMYCOSIS

Currently, amphibians are the most threatened vertebrate group in the world. One-third are facing extinction due to a variety of threats including habitat destruction, overharvesting, climate change, environmental contaminants, and infectious diseases such as chytridiomycosis, a skin infection caused by a fungus from the Chytrid family.

Joel Sartore, National Geographic Photo Ark/Getty Images

Rabb's fringe-limbed tree frog of Panama; believed to be extinct in the wild due to chytrid.

THE DISEASE

Infected skin cells thicken, impairing gas and electrolyte exchange. (Amphibians breathe through their skin.) Skin ulcerations (lesions) develop and hemorrhage, leading to lethargy, which may affect behavior—infected frogs are less likely to show anti-predator behaviors, making them easier prey. The skin damage can quickly lead to death from cardiac arrest.

As with WNS, humans had a hand in the spread of chytrid. The fungus probably originated in the African clawed frog and was spread worldwide by the use of this species as a food source and for medical purposes.

THE EXTENT

In terms of the number of species threatened or lost, chytridiomycosis is considered to be the most significant infectious disease of vertebrates in history. It is now found on every continent except Antarctica and is contributing to the decline of more than 280 species of frogs and salamanders; it is believed to have driven at least 100 species to extinction.

 What are the similarities between white-nose syndrome and chytridiomycosis?

would be expected by amphibian movement (amphibians don't move as far or as fast as bats). Researchers are still working to understand the other forces that drive chytrid transmission and are looking for ways to reduce the disease's impact on amphibian populations.

The chytrid example taught scientists studying WNS in bats that a novel fungal invader could wreak havoc on host populations. (Module 3.1 presents additional examples of the problems caused by invasive species.) Could that explain what was happening to North American bats?

Scientists like Blehert continued to dig. They collected more observational data from bat hibernacula and bat carcasses. The fungus, which is adapted to live in cold conditions, seemed to thrive on hibernating bats, but it did not appear to affect active bats. Researchers suspected that the distinction had to do with basic bat physiology. Bats were easier prey for the fungus during hibernation because in that state, their body temperatures were lower than normal and their immune systems were suppressed. In active (awake) bats, the fungus's growth would be slower because those bats were warmer, and their immune systems were fully active and could fight off the infection.

Other observations revealed that the fungus appeared to attack specific areas of the bats' bodies, namely their wings, ears, and noses. Those areas are hairless; they are also especially vital to bats' ability to maintain body temperatures and avoid dehydration during their winter-long hibernation. During hibernation, bats are in a state of inactivity known as *torpor*, conserving energy by reducing their metabolic rate

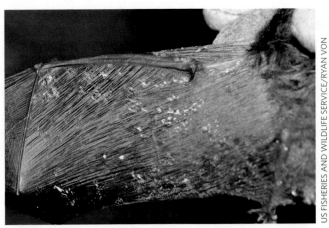

A close-up of the wing of a little brown bat infected with WNS. The fungus infiltrates hairless skin such as that found on the muzzle, ears, and wings of bats, damaging the tissue.

US FISHERIES AND WILDLIFE SERVICE/RYAN VON LINDEN, NEW YORK DEPARTMENT OF ENVIRONMENTAL CONSERVATION/SCIENCE PHOTO LIBRARY

and body temperature. Rousing from this state of torpor takes a lot of energy; they must raise their body temperatures from around 5°C up to 40°C (40°F to 104°F) each time. Normally, bats have enough energy (fat stores) saved up to rouse 8 to 12 times during hibernation; they drink water and excrete waste, then return to their torpid state.

Because bats with WNS appeared to die of starvation, Blehert and others hypothesized that the fungal infection was causing them to rouse more often, and thus burn up their fat reserves too early in the season. "They would scurry to the cave openings in search of insects to eat," Blehert explains. "But in the dead of winter, none would come." Without insect meals to replenish their stores, they would starve to death before spring arrived. **INFOGRAPHIC 3B**

INFOGRAPHIC 3B   WHITE-NOSE SYNDROME

White-nose syndrome (WNS) is a deadly, epidemic disease affecting some species of North American bats; it is caused by a novel fungus, *Pseudogeomyces destructans*. Even bat species that were abundant only a few years ago have experienced dramatic population crashes and could potentially go extinct in the near future.

US Fisheries and Wildlife Service/ Ryan Von Linden, New York Department Of Environmental Conservation/Science Photo Library

A little brown bat (*Myotis lucifugus*) from New York State infected with WNS.

THE DISEASE

Exposed skin areas such as wings and face are attacked; it also disrupts physiology and blood chemistry. It is only a threat to hibernating bats because it thrives in cold temperatures, and the bat's immune system is suppressed at this time.

NORMAL HIBERNATION: lower metabolic rate; live off fat stores; rouse occasionally to excrete waste; most calories are burned during these brief arousals.

HIBERNATION WITH WNS: increased frequency and duration of arousals as disease progresses; bats run out of fat stores before winter's end.

THE EXTENT

Bats with WNS have been found in 29 U.S. states and 5 Canadian provinces. By the end of 2016, it had killed more than six million bats.

 Though most hibernating bats infected with WNS die, not all succumb to the disease. Bats with WNS that survive to the end of hibernation and emerge in spring can recover. Why do these bats recover while others die?

4   OBSERVATIONAL AND EXPERIMENTAL STUDIES

Key Concept 4: Experimental studies manipulate conditions, whereas observational studies collect data without intentionally altering any conditions. Each supplies different lines of evidence for analysis.

As clear as it might have seemed at that point that an introduced fungus from overseas was indeed responsible for the bat die-offs, scientists still needed to test their hypothesis that death was caused by arousing too often and using up fat stores before the end of winter (and the return of insects). To do this, Blehert's team of researchers, led by DeeAnn Reeder, attached temperature-sensitive dataloggers to the backs of 504 free-ranging little brown bats (*Myotis lucifugu*) in hibernacula throughout the northeastern United States. By recording body temperature, these dataloggers would tell the researchers how often a bat aroused during hibernation and how long it "slept" in each of its torpor bouts. At the end of the hibernation season, bats were recaptured (or collected if dead)

observational study
Research that gathers data in a real-world setting without intentionally manipulating any variable.

and dataloggers retrieved. For each bat, researchers noted if it was infected with WNS and whether it was alive or dead. Here's what they found: Infected bats aroused more frequently (and so had shorter torpor bouts) than healthy ones. And among those that were infected, the severity of their condition (quantified by the degree of tissue damage caused by the fungus) corresponded to the length of their torpor bouts. In other words, the more tissue damage the bats had, the more often they woke. And the more often they woke, the more likely they were to die.

Reeder and Blehert's datalogger study is an example of an **observational study**—one where scientists collect data in the real world without intentionally manipulating the subject of the study. They did not intentionally inoculate bats with WNS; they simply followed bats, measured a parameter of interest, and

A biologist examines a bat for white-nose syndrome.

noted which bats were infected and which were not. In observational studies, researchers may simply gather data to learn about a system or phenomenon (like measuring the amount of CO_2 in the atmosphere over time to see if it is rising or falling), or they may be comparing different groups or conditions found in nature (like monitoring asthma cases in areas with high and low air pollution). These types of studies are very useful in environmental science because they enable us to look at whole systems that cannot be manipulated in a lab or field setting.

Experimental studies, by contrast, involve intentional manipulation of experimental conditions. Experimental studies have both an **independent variable** and a **dependent variable**. Scientists *manipulate* the independent variable and *measure* the dependent variable to see if it is affected. These types of studies rely on test groups and control groups. The **test group** subjects are exposed to a variable that scientists want to study, and their results are compared to a **control group** that was not exposed to the variable (or was exposed to the variable to a different degree). (Observational studies can also have independent and dependent variables as well as control and test groups—i.e., control bats in the study described above were collected from a site with no WNS and compared to test bats from sites where the disease was found.)

In an example of an experimental study, another team of researchers led by Lisa Warnecke deliberately inoculated bats with a solution containing Pd; control bats were given a sham inoculation containing the solution but no Pd (independent variable: type of exposure) in an effort to determine whether the fungus affected the bats' torpor (dependent variable: length of torpor bouts). Warnecke's results showed that bats that had been intentionally exposed to Pd experienced shorter torpor bouts due to awakening more often than control bats, providing evidence that Pd infections did interfere with hibernation. **INFOGRAPHIC 4**

Studies of this type must pass ethical review before beginning in order to ensure animals do not experience undue pain or stress; however, animal studies of any type raise ethical concerns. When scientists choose between experimental studies like this one, or observational studies that do not impose a potentially harmful manipulation, they weigh the ethics of intentionally making animals (or people) sick against the benefits that research might bring, such as finding a cure for a disease or its cause.

As mentioned earlier, in science there are degrees of certainty; we know some things better than others. These degrees of certainty are expressed in terms of probabilities with a branch of mathematics called **statistics**. Once data is collected from an observational or experimental study, statistical analysis is done to calculate the likelihood that the difference we observed between two groups is just a result of natural variation (i.e., not *statistically significant*) or if it represents genuine differences between the groups (i.e., the differences are *statistically significant*).

In statistics, certainty is expressed as a *P*-value, a calculation that represents the likelihood that the hypothesis would incorrectly be accepted. Scientists demand a high level of certainty to accept their hypothesis; typically they accept no more than a 5% chance that the wrong conclusion would be drawn—this represents a *P*-value no bigger than 0.05 (written as $P < 0.05$).

For example, in Reeder's study, the number of times a bat woke up early in hibernation differed between bats with or without WNS, but were these arousal frequencies different enough to attribute that difference to WNS? After all, there are often differences between individuals in a study—in this case, every bat didn't wake up the same number of times, even within the same group. A statistical analysis of the data answers that question. In this case, the answer was no—the *P* value calculated from this data was too high ($P = 0.601$). The average number of arousals might have been different, but it was not different enough to conclude it was a result of exposure to WNS rather than just natural variability in the two groups. However, the much higher arousal frequency late in hibernation in the WNS group was determined to be significantly greater than the other group ($P < 0.001$). In this case, we conclude that exposure to WNS fungus did increase arousal later in hibernation compared to uninfected bats. (For more on statistics and experimental design, see Appendix 3.)

experimental study Research that manipulates a variable in a test group and compares the response to that of a control group that was not exposed to the same variable.

independent variable The variable in an experiment that a researcher manipulates or changes to see if the change produces an effect.

dependent variable The variable in an experiment that is evaluated to see if it changes due to the conditions of the experiment.

test group The group in an experimental study that is manipulated such that it differs from the control group in only one way.

control group The group in an experimental study to which the test group's results are compared; ideally, the control group will differ from the test group in only one way.

statistics The mathematical evaluation of experimental data to determine how likely it is that any difference observed is due to the variable being tested.

Scientists collect evidence to test ideas. Experimental studies are used when the test subjects can be intentionally manipulated; observational studies allow scientists to look at entire ecosystems or other complex systems.

BACKGROUND KNOWLEDGE
Bats burn most of their winter calories during brief arousals from torpor bouts.

QUESTION
Do bats with WNS wake up more often during hibernation?

HYPOTHESIS
WNS causes bats to awaken more frequently, causing them to burn calories, depleting their fat stores.

SCIENTIFICALLY TEST THE HYPOTHESIS.

OBSERVATIONAL STUDY (Reeder, et al.)
Prediction: Bats with WNS will have abnormally shortened torpor bouts due to more frequent arousal.
Procedure: Researchers tracked bats in six different caves and determined how long each animal was awake or asleep by monitoring the bats' body temperature. Researchers compared the length of torpor bouts for uninfected bats to that of bats with WNS (survivors and nonsurvivors).

| #1 | #2 | #3 | #4 | #5 | #6 |
| WNS | No WNS | WNS | WNS | No WNS | No WNS |

RESULTS

TORPOR BOUT LENGTH AND WNS STATUS

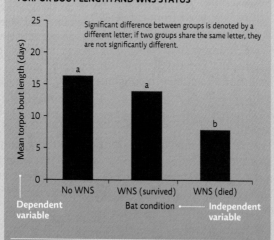

Significant difference between groups is denoted by a different letter; if two groups share the same letter, they are not significantly different.

Mean torpor bout length (days) — Dependent variable

Bat condition — Independent variable

No WNS (a), WNS (survived) (a), WNS (died) (b)

EXPERIMENTERS' CONCLUSION: Increased mortality/disease state is associated with shortened torpor bouts due to frequent arousal episodes.

EXPERIMENTAL STUDY (Warnecke, et al.)
Prediction: Bats inoculated with WNS will awaken from torpor more frequently than bats given a sham inoculation.
Procedure: Bats were either inoculated with WNS fungus (Pd) or given a sham inoculation; the length of torpor bouts was monitored at different stages of hibernation.

Randomly assign bats to groups.
Monitor arousal episodes using video footage.

CONTROL GROUP
Sham innoculation

TEST GROUP
Innoculated with WNS fungus

RESULTS

AROUSAL IN HIBERNATING BATS WITH OR WITHOUT WNS

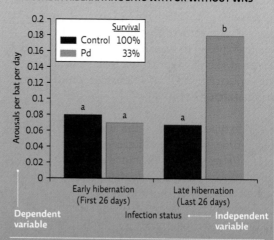

Survival	
Control	100%
Pd	33%

Arousals per bat per day — Dependent variable

Infection status — Independent variable

Early hibernation (First 26 days): Control (a), Pd (a)
Late hibernation (Last 26 days): Control (a), Pd (b)

EXPERIMENTERS' CONCLUSION: Bats with WNS have increased arousal frequencies later in hibernation, contributing to death by starvation.

 Why might waking up more frequently lead to death for a hibernating bat?

 Why do you think bats don't wake up more than usual early in hibernation but do experience more frequent arousals late in hibernation?

5 CORRELATION VERSUS CAUSATION

Key Concept 5: Two events that occur together are correlated, but this does not necessarily mean one caused the other or even that they are related in a meaningful way.

Observational and experimental studies differ not only in their approach, but also in the type of information they ultimately provide. Winifred Frick and her team at Boston University conducted an observational study that compared bat colony sizes over time in areas with and without WNS. They observed a sharp decline in colony size, but only in the colony experiencing WNS infections.

Observational studies like this (and the Reeder study) reveal **correlations**. That is, they can show that two things occur together: A colony was somehow exposed to WNS and most bats in that colony died shortly thereafter. But observational studies can't necessarily tell us if one thing caused the other, or even if the two factors are related in a meaningful way (because it could just be a coincidence—perhaps something else that wasn't evaluated caused the

die-off). So although scientists suspected from the very start that the white fuzz was related to the bat die-offs, they needed more evidence to determine whether the two events were connected in a **cause-and-effect relationship**. For that, they needed experimental studies like Warnecke's that intentionally exposed bats to the fungus. If all other factors are the same between two groups except the one variable that is manipulated (in this case, exposure to the fungus) and we see differences in the dependent variable we are monitoring (in this case, bat survival), then it is reasonable to conclude that the manipulation caused the effect. **INFOGRAPHIC 5**

correlation Two things occurring together but not necessarily having a cause-and-effect relationship.

cause-and-effect relationship An association between two variables that identifies one (the effect) occurring as a result of or in response to the other (the cause).

INFOGRAPHIC 5   **CORRELATION VERSUS CAUSATION**

Because they cannot control the extraneous variables in a research study, observational studies can only provide correlational evidence that two or more factors are related. Experimental studies that can manipulate and control variables can establish whether a cause-and-effect relationship exists.

A: POPULATION SIZE IN COLONIES WITH AND WITHOUT WNS

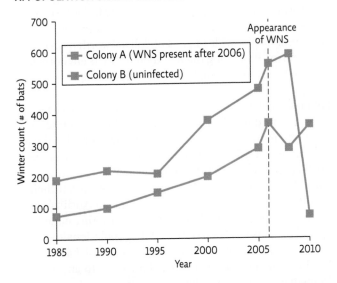

B: EXPOSURE TO WNS FUNGUS AND BAT SURVIVAL

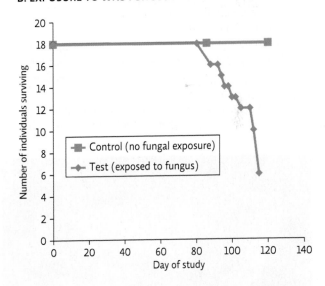

Study A: Observational study that used two colonies in years before and after WNS affected one of the colonies.

This study provides correlational evidence that suggests WNS caused the population crash. However, it does not provide cause-and-effect evidence since something else might have caused population size to fall in that colony.

Study B: Experimental study that monitored survival in bats exposed to the fungus that causes WNS (Pd) compared to bats with no Pd fungal exposure.

In this study, a cause-and-effect relationship can be established since the only difference between the groups was Pd exposure.

 Other than WNS, what else could have decreased the number of bats in Colony A after 2008?

It's important to understand that while experimental evidence is generally considered stronger, precisely because it can determine cause and effect, observational studies that amass a lot of evidence in support of a given correlation are also quite powerful. For example, we have only correlational evidence that smoking causes lung cancer in humans, but this relationship is not in question. In fact, observational studies that produce the same results as experimental studies offer particularly strong corroborating evidence. This is because they demonstrate the correlation between the variables, even when the populations are exposed to many other factors that might influence the outcome. Conversely, when the results from observational studies do not agree with those from controlled experimental studies, we can conclude that other factors are probably at play.

6 USING SCIENCE TO ADDRESS ENVIRONMENTAL PROBLEMS

Key Concept 6: Policy makers should base decisions for action on the best scientific evidence available; the more we know about a threat, the better equipped we are to fight it.

When environmental problems arise, we often establish **policies** to deal with them. These policies need to be informed by science. They also need to address the environmental problem, fit societal needs, and be economically viable—that is, they need to address the *triple bottom line* (see Module 1.1). Establishing such policies is never easy or straightforward. Even if a healthy environment is good for everyone in the long term, there are often conflicts between protecting the environment and serving the short-term interests of one societal group or another. Conflicts also arise between different stakeholders; the best answer for farmers might not be the best one for businessmen, and the needs of average consumers might conflict with both. Policymakers must balance all of these needs, and still factor in any economic considerations, all while keeping sight of their main goal: protecting human populations and the environment. (See Module 5.2 for more on environmental policy.)

policy A formalized plan that addresses a desired outcome or goal.

One dilemma faced by policy makers is the fact that gathering strong evidence about the cause of an environmental problem takes time, sometimes longer than we can wait if we are to take meaningful action. This means policies often are based on incomplete data. For example, we don't know yet what the loss of bats would mean for the areas threatened, but it could be quite severe—especially in the face of global warming that is expanding the range of disease-carrying mosquitoes. Like most predators, bats are vital to their ecosystem. They are important predators of nocturnal insects like mosquitoes. One estimate found that little brown bats eat 4 to 8 grams of insects a night when they aren't hibernating. Extrapolating these numbers to the more than 6 million bats that have died from WNS means that between 4,000 and 8,000 metric tons fewer insects are being consumed by bats annually. Put another way, every year bats provide billions of dollars' worth of free pest control. Protecting them is good for their environment and the many species that share it, including us.

Policies must also be flexible and able to adapt to new scientific findings. The U.S. Fish and Wildlife Service (USFWS) monitors bats and other species and can adjust the implementation of its policies as conditions change. Such adjustments include adding or removing species from the Threatened or Endangered Species List if their numbers increase or new threats emerge. In April 2016, the northern long-eared bat was listed as a threatened species because WNS has caused its populations to decline so precipitously. An emergency listing has been recommended for the little brown bat to give it immediate protection. The species was once abundant in the United States, but its population size has plummeted since WNS emerged.

Of course protection lists are not the only means by which officials are trying to protect bats from WNS. They have also closed hibernacula to the public in an effort to prevent humans from carrying fungal spores from one cave to the next. This approach seemed to be working

Joel Sartore/National Geographic/Getty Images

Bats are major insect eaters. A colony can consume thousands of mosquitoes and other small insects nightly, helping to keep these insect pests in check.

until the spring of 2016, when WNS was discovered in the state of Washington, a jump of more than 2,000 km (around 1,300 miles) from the closest WNS site. "That jump tells us that humans are still transmitting the fungus from cave to cave," says Blehert. "It means we need to redouble our infection control efforts."

Sadly, WNS is not the only threat bats face today. Human impact puts these animals at risk in a variety of ways. But as with WNS, we need to use science to investigate the threats to determine if they warrant attention and, if so, how best to proceed. The USFWS has compiled a document to help: *A National Plan for Assisting States, Federal Agencies, and Tribes in Managing White-Nose Syndrome in Bats.* The agency also makes recommendations, based on scientific research, for dealing with other threats to bats (and other wildlife.) **INFOGRAPHIC 6**

INFOGRAPHIC 6   PROTECTING BATS: IDENTIFYING AND ADDRESSING THREATS

©Tanguy Stoecklé/Biosphoto

THREATS	ACTIONS TO PROTECT BATS
DISEASE	
• WNS is affecting at least seven species of hibernating North American bats.	• The U.S. Fish and Wildlife Service has a national plan to address WNS; imperiled bat species can be listed as an endangered species and receive extra protection.
HABITAT DAMAGE AND DISTURBANCE	
• Disturbance during hibernation by cave visitors. • Guano mining disturbs cave habitats bats need for daytime roosting or winter hibernation. • Damage to summer habitat due to human development (roads, deforestation for agriculture or building, etc.) affects bats' ability to successfully forage and safely roost.	• Cave closures and advisories prohibit or limit cave access. • Cave visitors should thoroughly decontaminate clothing and equipment after leaving a cave. • Homeowners can provide habitat for bats by leaving dead and dying trees standing and setting up bat boxes for roosting, especially in the spring and summer months.
WIND TURBINES	
• Wind turbines, especially those spinning slowly during low wind times, kill hundreds of thousands of migrating bats per year.	• Program turbines to stop turning at low wind times during migration season.
FEAR AND PERSECUTION OF BATS	
• Many people have an irrational fear of bats and some even kill them when possible. Though bats do carry rabies, fewer than 1% of bats are infected with the disease.	• Education about their ecological role can dispel myths about bats. • Avoid contact with a bat that appears sick, weak, or is seen in the daytime as it may carry rabies.
ADDITIONAL INTERNATIONAL THREATS	
• The bushmeat trade threatens many species of bats in Africa. • Climate change can affect bats and the insects they prey on.	• Prohibit the bushmeat trade and offer viable food alternatives for local people. • Take steps internationally to address climate change.

 Why is it so important to stay out of caves with hibernating bats?

Humanity faces many environmental problems today, from rising levels of greenhouse gases to water and air pollution to biodiversity loss (and many others) that must be addressed if we are to reach the goal of living sustainably. (See Module 1.1 for an introduction to living sustainably.) It is vital that we base our policies on sound science rather than speculation or on what we'd like to be true.

Unfortunately, making logical decisions based on evidence, even when we have compelling evidence, may not be as easy as it sounds. Cognitive and social science research is revealing some interesting aspects of how the human brain operates. We would like to think we are rational creatures, weighing evidence to reach logical conclusions. But research shows that when presented with evidence that supports our current beliefs, this evidence tends to strengthen those beliefs. However, when presented with evidence that contradicts what we believe to be true, no matter how compelling, many of us simply reject it out of hand. To counter this "confirmation bias," we must be willing to consider other viewpoints and, in doing so, look at the evidence and evaluate its veracity. If we approach new information knowing we may be biased to our current position, we may be more open to the evidence and be willing to go where the evidence takes us—just like a scientist.

Despite the disappointing news of WNS's spread to Washington, scientists have other reasons to be hopeful.

Sybill Amelon, as part of a U.S. Forest Service Research team, is holding a bat treated for WNS that is about to be released into the wild after its recovery.

Katie Gillies/Bat Conservation International

In May 2016, nearly a decade after the bat die-offs had first been detected by one group of scientists in New York, another group of scientists in Georgia had found the very first inklings of a treatment—one that might help bats resist or even recover from a Pd infection. The root of this potential win is *Rhodococcus rhodochrous*, a bacterium that's both common and naturally occurring in North America. Previous research by Chris Cornelison, then of Georgia State University, had shown that organic compounds produced by this bacterium reduce the amount of fungus that grows on bananas. Cornelison had read about WNS and wondered if it might do the same for bats.

After some preliminary studies to make sure the bacterium did not harm bats, Cornelison worked with U.S. Forest Service scientists to test his idea in the field. They collected bats that were already infected with WNS from hibernacula, early in the hibernation period, and placed the bats in a cooler for 48 hours with Petri dishes containing the bacterium—enough time for the chemicals released by the bacterium to reach the bats. They then returned the bats to the hibernacula and left them there (in mesh bags) for the rest of winter. When they returned in spring, all 75 of their test bats were alive, though some still had wing damage from the fungus. After being given time to recover, the bats were released back into the wild.

Though it is too soon to know if the bacterium or the chemicals it produces might prove to be an effective treatment, teams of scientists are back in the lab and in the field, trying to find out, while, as Cornelison says, "there are still bats to treat."

Select References

Blehert, D. S., et al. (2009). Bat white-nose syndrome: an emerging fungal pathogen? *Science, 323*(5911), 227.

Cornelison, C. T., et al. (2014). A preliminary report on the contact-independent antagonism of *Pseudogymnoascus destructans* by *Rhodococcus rhodochrous* strain DAP96253. *BMC Microbiology, 14*(1), 246.

Cryan, P. M., et al. (2013). White-nose syndrome in bats: illuminating the darkness. *BMC Biology, 11*(1), 47.

Frick, W. F., et al. (2016). White-nose syndrome in bats. In *Bats in the Anthropocene: Conservation of Bats in a Changing World* (pp. 245–262). Cham (Zug), Switzerland: Springer International Publishing.

Reeder, D. M., et al. (2012). Frequent arousal from hibernation linked to severity of infection and mortality in bats with white-nose syndrome. *PLoS One, 7*(6), e38920.

Rogers, N. (2015). Bacteria may help bats to fight deadly fungus. *Nature, 522*(7557), 400–401.

Rosenblum, E. B., et al. (2010). The deadly chytrid fungus: a story of an emerging pathogen. *PLoS Pathogens, 6*(1), e1000550.

Warnecke, L., et al. (2012). Inoculation of bats with European *Geomyces destructans* supports the novel pathogen hypothesis for the origin of white-nose syndrome. *Proceedings of the National Academy of Sciences, 109*(18), 6999–7003.

INTERACTIVE MAP **FUNGAL INVADERS**   ANIMATED INFOGRAPHIC

Non-native pathogenic invaders can wreak havoc on species that don't possess adequate adaptations to fight off the invader or if the invader has no natural predators in its new environment. The way we plant crops (in large monocultures) also makes them vulnerable to attack. Here are some notable fungal invaders affecting plants and animals around the world.

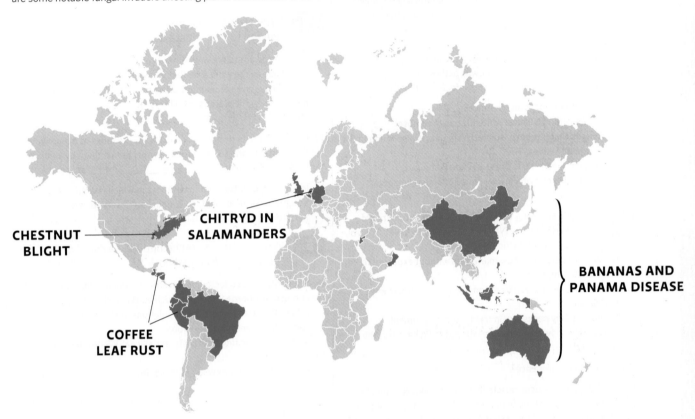

CHESTNUT
BLIGHT

CHITRYD IN
SALAMANDERS

BANANAS AND
PANAMA DISEASE

COFFEE
LEAF RUST

 BRING IT HOME

PERSONAL CHOICES THAT HELP
The story of how scientists unraveled the mystery of what was killing bats is a great example of how science documented a problem, uncovered its cause while discarding unsupported hypotheses, and informed public policy to address the problem. Anyone can be a scientist by logically and systematically collecting evidence to answer scientific questions. How scientific information is, or is not, put into action has far-reaching consequences, making science literacy a matter of importance of every individual.

Individual Steps
• Practice thinking like a scientist. Go outside for 10 minutes and observe the world around you. Make observations of what you see or hear. What predictions could you make

from your observations? How could you test them?
• Stay informed. Read or watch a science-related article or show once a month.

Group Action
• Demonstrate the importance of scientific literacy to your friends and family. Develop three additional questions from the material in the module and discuss them over dinner.
• Support science education: Find out about public lectures and programs in your area and attend one with your friends.
• Are there caves with bats in your area? Contact local wildlife officials to see if there are any projects that can be done to assist in the protection of the bats or caves.

Policy Change
• Attend a city council or county board meeting to see how policy issues are addressed in your area.
• Make knowledgeable voting decisions on ballot initiatives.
• Serve on local civic committees that address environmental issues in your community.

BartCo/iStock/Getty Images

ENVIRONMENTAL LITERACY   **UNDERSTANDING THE ISSUE**

1   How do scientists study the natural world?

1. Which of the following questions is NOT scientifically testable?
   a. Do bats transmit any diseases that are harmful to humans?
   b. Do humans have an ethical responsibility to protect bat populations?
   c. How effective are bats at controlling mosquito populations?
   d. Is some novel (previously unknown) pathogen causing WNS?
   e. What is the mortality rate for bats infected with WNS?

2. Which of the following is an inference?
   a. Most bats with WNS have white fuzz on their nose.
   b. The fungus that has been isolated from bats is also found in the soil of infected caves.
   c. White nose syndrome is caused by the Pd fungus.
   d. Bats with WNS do not show evidence of underlying organ damage.

3. Let's say that you are going to test whether exposure to the Pd fungus causes WNS in bats. Why is it necessary to have a control group in this experimental study?
   a. You need to determine how many bats without exposure to Pd show WNS symptoms.
   b. You need to feel like you have some kind of control over your experiment.
   c. It is not actually necessary in this type of experiment.
   d. You need to show that bats not exposed to Pd have healthy skin.

4. What is empirical evidence?

5. If an experiment produces results that confirm an experimental prediction and support the hypothesis, what are the next steps? What should be done if the results fail to support the hypothesis?

2   What influences the degree of certainty in a scientific explanation?

6. True or False: The realm of scientific certainty spans a spectrum of understanding from hypothesis to theory but excludes "no clue" and "absolute proof."

7. There is overwhelming evidence that the Pd fungus causes WNS—evidence gathered from many different studies that address many different aspects. Therefore, we can say that this explanation:
   a. has reached the level of scientific hypothesis.
   b. can be considered a scientific theory.
   c. is a confirmed scientific prediction.
   d. has been proven beyond doubt.

8. Why is it unreasonable to reject an accepted scientific explanation on the basis that it is "just a theory"?

3   How do comparative studies contribute to the understanding of an environmental issue?

9. True or False: Comparative studies are those that compare the response of a control group to the response of a test group.

10. What do chytridiomycosis and WNS have in common?
    a. Both occur in mammals, especially bats.
    b. Both are found only in North America.
    c. Both are caused by a fungus.
    d. Both occur in animals that have other illnesses.

11. What was the value in looking at the cause and effects of chytridiomycosis in frogs when studying WNS in bats?

4   Why are observational and experimental studies needed to investigate the natural world?

12. Scientists want to know whether ocean water is becoming more acidic due to the extra CO_2 that is released by the burning of fossil fuels. They have set up monitoring sites in the water to collect daily data on the water's pH. This is an _____ study.

13. Other scientists decide to add extra CO_2 to tanks of ocean water that contain coral to see if the lower pH harms them by weakening their skeleton. This is an _____ study.

14. Identify at least three circumstances in which it would be more acceptable or reasonable to conduct an observational study than to conduct an experimental one.

5   What is the difference between a *correlation* and a *cause-and-effect relationship*?

15. Scientists have collected observational evidence that the spread of mosquito-borne infections, such as West Nile virus (WNV), is lowest in areas where bat populations are high. What can be concluded from this observation?
    a. Mosquito-eating bats preferentially eat WNV-infected mosquitos.
    b. WNV does not develop in mosquitos when bat populations are high.
    c. The presence of bats is correlated with reduced transmission of WNV.
    d. Mosquitos that carry WNV avoid places where bats are present.

16. What is a *correlation* in science? How is it different from a *cause-and-effect* relationship?

17. What did it take to shift the conclusion that Pd and WNS were correlated to the conclusion that Pd causes WNS?

6   How can science help policy makers address environmental problems?

18. True or False: Policy makers need to wait until they have definitive scientific evidence regarding the best course of action before drafting policies to deal with a problem.

19. An environmental policy designed to help address WNS in bats should:
    a. consider the economic costs of implementing actions proposed by the policy.
    b. take any action needed to solve the problem in the short term.
    c. address concerns by wildlife officials rather than worries from the general public.
    d. have as its main goal the protection of the bat populations being affected.

20. The best policies are informed by science. What scientific evidence led to the policy of closing infected and uninfected caves to visitors?

SCIENCE LITERACY   WORKING WITH DATA

To determine how bats contract WNS in the wild, J. M. Lorch and colleagues used different methods to expose bats to the Pd fungus that causes WNS. Healthy bats were collected from the wild and the following groups set up:

- **Inoculation group:** 29 bats were directly treated with Pd from a laboratory culture of the fungus.

- **Contact exposure group:** 18 bats were housed in an enclosure with infected bats.

- **Airborne exposure group:** 36 bats were kept in cages close to, but not in direct contact with, WNS bats.

- Two control groups were used:

  - **Negative control group:** 34 bats with no exposure to Pd

  - **Positive control group:** 25 bats collected from the wild that already had WNS (compares the disease in the wild to that of experimental bats)

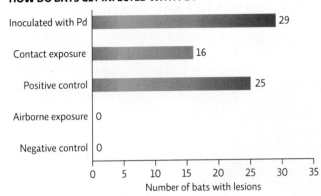

HOW DO BATS GET INFECTED WITH PD?

Number of bats with lesions

Interpretation

1. Look at the bars on the graph. Which exposure methods appear to transmit Pd and cause WNS lesions?

2. Compare the effectiveness of direct treatment with Pd (inoculation) and contact exposure with infected bats for causing WNS lesions. Does one route appear to be more effective at transmitting the disease than the other? How did you determine this?

3. Look at the sample size (number of bats) of each group. Now go back and look at your answer to question #2. Do you need to revise your answer? Explain.

4. Determine the percentage of each group that developed lesions and redraw the graph using "Percentage of bats with WNS lesions present" as your x-axis title. Do you think this is a better way to show these data? Explain.

Advance Your Thinking

5. The airborne exposure group had a slightly higher death rate than the contact exposure group ($P = 0.72$). Can you conclude that airborne exposure is more likely to lead to death than contact exposure? Explain.

6. What is the purpose of the positive and negative controls?

7. Why was this research done? Why might it be useful to know how the infection spreads?

INFORMATION LITERACY   EVALUATING INFORMATION

Use your understanding of the process of science to evaluate the threats and programs aimed at protecting bats. Go to the National Park Service (NPS) webpage, *Threats to Bats*, found at https://www.nps.gov/subjects/bats/threats-to-bats.htm.

Investigate the three categories of threats listed there: habitat loss, wind energy, and white-nose syndrome.

1. For each category, answer questions a and b:
   a. Summarize the threat posed by this category.
   b. What can be done to decrease the threat it poses?

2. Now, go back to the webpage for *Wind Energy* and select a link for one of the "papers" or "reports" posted there and answer the following questions:
   a. What is the title of the document you chose?
   b. What type of document is this (scientific report, general information, government report)?

   c. Look over the report. Read the abstract or introduction if one is provided. What is the focus of the report?
   d. Scan the document. Does the report offer scientific evidence in support of its claims or conclusions? Explain.
   e. Does this document appear to be a credible source of information? Explain.

3. Drawing from your reading of the report on the *Wind Energy* webpage and your understanding of WNS (from Module 1.2 and the NPS webpage on WNS), compare the threat to bats posed by wind turbines to that of white-nose syndrome in terms of severity of threat and our ability to respond. Which do you think we have a better chance of successfully addressing and why?

 Additional study questions are available at SaplingLearning.com.

LEAD IN THE WATER

A water crisis in Flint, Michigan

Lead contamination of Flint, Michigan's drinking water has caused an environmental health crisis. Carlos Osorio/AP Photo

After reading this chapter and studying the KEY CONCEPTS and INFOGRAPHICS, you should be able to answer these GUIDING QUESTIONS

CORE MESSAGE

We live in an environment full of chemicals, some of which are hazardous. The risk posed by those chemicals depends on a variety of factors, but knowing how to respond to the latest toxic scare is often difficult as we receive conflicting messages about the safety or risks of various chemicals. Developing information literacy skills enables us to evaluate information and then use the highest quality information we can find to make reasoned decisions about how to respond.

1. What classes of toxic substances are recognized, and what are common routes of exposure?

2. How do the chemical characteristics of a substance influence its toxicity?

3. How is toxicity affected by exposure route, chemical interactions, and the victim's characteristics?

4. How is toxicity determined?

5. How are toxic substances regulated in the United States?

6. What is information literacy, and why is it important in environmental science?

7. What is critical thinking, and how can it counter logical fallacies used in arguments?

In early 2015, the residents of Flint, Michigan—a city of about 99,000 people— began reporting a strange and grim litany of problems with their tap water. Some of them noted a blue tint to the water; others said it was brownish. Most everyone described strange odors—the water smelled like mothballs or like an overchlorinated swimming pool. Extra chlorination had been needed to address *Escherichia coli* (fecal bacteria) contamination in the water. And it tasted just awful: like metal. Before too long, some residents started experiencing symptoms: headaches, rashes, clumps of hair falling out. It didn't take long for folks to connect the water and the symptoms.

In April 2014, the city had switched its water supply from Lake Huron to the Flint River, which was long known to be polluted with industrial chemicals. The corrosive nature of this water could leach lead out of old pipes (new pipes are lead free). To combat this in other areas where it is a problem, water utilities treat the water supply with chemicals that lock the lead sediments in place on a pipe's inner surface. But the state told the Flint utility they did not have to do this—a baffling recommendation and violation of federal regulations. (The state later cited Flint's lack of funds to install the needed equipment.)

By October, the General Motors Flint Truck Assembly Plant was complaining that the new water was corroding car parts. The factory had requested to switch back to the original water supply, and now residents were asking for the same. At a community meeting with city officials, citizens held up bottles of discolored malodorous water and demanded explanations: What was in their drinking water? What health risks had they been exposed to by consuming it for the past several months? And what should they do to protect themselves?

Like the water problems faced by the residents of Flint, we often encounter environmental problems directly, in our own communities, or indirectly as larger-scale problems that affect regional or global populations. But we often receive conflicting information about these issues from news outlets, social media, television, and YouTube, to name a few. How can one distinguish between accurate information and the erroneous or downright fraudulent? In order to make informed choices about your life— down to the water you drink—you need the skills that allow you to logically analyze claims (critical thinking) and to evaluate the veracity of information (information literacy). To illustrate the importance and application of these skills, this module will provide an introduction to an environmental science issue closer to home than you might think: our exposure to toxic substances—the threat they pose and how we deal with them.

⊙ **WHERE IS FLINT. MICHIGAN?**

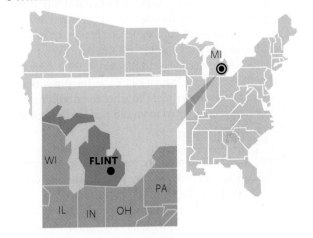

The National Guard delivers bottled water to Flint, Michigan residents.

1 TOXIC SUBSTANCES IN THE ENVIRONMENT

Key Concept 1: The environment is full of natural and synthetic toxic substances that have the potential to harm living things.

Toxic substances (also known as **toxics**) are chemicals that can harm living organisms. (A subset of toxics is *toxins*—toxic substances produced by an organism, such as snake venom.) A chemical's ability to cause harm is referred to as its *toxicity*. Toxics fall into two broad categories: natural and synthetic (chemicals introduced into the environment by humans).

Lead is a naturally occurring chemical (a chemical element) that was once used in a wide range of manufactured products, including gasoline, paint, and water pipes. It is considered toxic to humans at any level. When we ingest it, either through the air or water, it can cause serious permanent damage to our brains, livers, and kidneys. The lead in Flint's water didn't come from the river—it came from decades-old pipes that were made of lead. Many of these were "service lines"—the pipes that deliver water from a water main to a home or building. Even a new house connected to an old service line could have lead in its water.

The debate over how to regulate toxic chemicals—how to determine what quantity of any particular compound is safe for humans or the environment and how to ensure that exposure levels stay well below those quantities—began in 1962, with a book called *Silent Spring*. In this book, legendary environmental activist Rachel Carson asked her readers to imagine a world without the sounds of spring, a world in which the birds, frogs, and crickets had all been poisoned by toxic chemicals. Just 20 years had passed since the widespread introduction of pesticides (like DDT), she explained, but in that relatively short time, they had thoroughly permeated our society.

These chemicals were obviously great for killing weeds and pests; they had done an amazing job keeping mosquito-borne diseases like malaria at bay during World War II and combating hunger by boosting global food production. But no one had paid attention to the effects they might have on nontarget species or on their (or our) ecosystems. *Silent Spring* created an uproar, which led to much stricter regulations for chemical pesticides in general, and in the United States, a complete ban on DDT in particular. But half a century later, we are still struggling to effectively regulate the chemicals in our world—both synthetic and natural.

Toxic substances now pervade our environment thanks to the actions of humans that have created new toxics or released naturally occurring toxic substances much more readily than nature would. Toxics are found in building materials, fabrics and carpets, food containers, cleaning supplies, paints—even in cosmetics and personal care products. This issue is affecting organisms worldwide which, in turn, is affecting their ability to contribute to a healthy a functioning ecosystem. And of course, toxic substances are dangerous for humans too.

The residents of Flint didn't have the information they needed to fight for clean water. They didn't have scientific data to show them what toxic chemicals were in their water, and they didn't have the economic or political clout to force government regulators to step in. (Around 42% of Flint children live in poverty—the national average is just under 15%.) But they knew something was wrong and were raising their voices, asking for the basic human right of access to clean water—asking for **environmental justice**. As is often true with regard to environmental issues, it is minority and lower-income groups who most often bear the brunt of negative impacts of societal choices. (See Module 4.2 for an introduction to environmental justice.)

There are many types of toxic chemicals. *Carcinogens* are substances that cause cancer. Tobacco smoke contains more than 40 identified carcinogens; smoking causes lung, bladder, throat, and other types of cancer. The furniture in our homes and the cleaning products under our sinks may contain chemical carcinogens. (Radiation exposure, too, can cause cancer; that is considered a physical hazard, rather than a chemical one.) *Mutagens* damage DNA directly; they can cause cancer (in which case they would also be considered carcinogens) or disrupt normal body function. *Teratogens* are substances that cause birth defects by disrupting embryonic development; cautions to pregnant women to avoid alcohol, tobacco, and a wide variety of pharmaceuticals reflect the need to protect the developing child from teratogens.

---

toxic substance/toxic
A substance that causes damage when it contacts, or enters, the body.

environmental justice The concept that access to a clean, healthy environment is a basic human right.

Some chemicals, known as *poisons*, damage or kill cells or disrupt biochemical processes (the rat poison warfarin, for example, disrupts blood clotting, causing the animal to bleed internally). Lead is a neurotoxin that damages the brain and other nervous system tissue; pesticides are also poisons whose sole purpose is to kill the target organism. Unfortunately, humans and other nontarget species may also be affected by exposure. Other chemicals that cause local tissue damage, such as chlorine in pool water making your eyes red and itchy, are called *irritants*.

Chemicals don't have to be toxic or mutagenic to cause problems. *Sensitizers* are substances that can lead to the development of a skin or respiratory allergic response after repeated exposure. The urushiol oil in poison ivy is a natural example, but many chemical solvents in cleaners, degreasers, and glues contain sensitizing chemicals; the latex in surgical gloves is a strong sensitizer in some people, prompting an industry-wide shift away from latex gloves to less problematic materials such as nitrile. Many chemicals have multiple effects: Formaldehyde is an irritant, sensitizer, mutagen, carcinogen, teratogen, and poison.

Another class of hazardous substances are the *endocrine disruptors*—chemicals that interfere with the body's system of hormones. Hormones are chemical messengers in the body that coordinate all kinds of body functions from daily sleep/wake cycles to reproductive biology to development. Endocrine disruptors often mimic hormones and interfere with development or bodily functions. Hormones operate in the body at very low concentrations; this means very low concentrations of endocrine disruptors may have a large impact. Their effects (like that of mutagens and carcinogens) may not be felt for many years after exposure, making it difficult at times to link cause (exposure to the substance) with effect. Many are estrogenic—they

The first public test of an insecticidal fogging machine at Jones Beach State Park, New York, in July 1945. As part of the testing, a 6.5-kilometer (4 mile) area was blanketed with the DDT fog.

Bettmann/Getty Images

mimic the female hormone estrogen and thus have a feminizing effect on the body. Some fish and frog populations exposed to estrogenic chemicals in their water are predominately female due to a sex change that occurs when exposed to these chemicals. Petroleum-based chemicals such as those in many plastics (bisphenol A [BPA], phthalate) and pesticides and some pharmaceuticals are endocrine disruptors affecting humans and other species. Lead, too, has the potential to disrupt normal endocrine function by interfering with many of the glands that produce hormones—examples include decreased production of growth hormones and reproductive hormones such as testosterone. **INFOGRAPHIC 1**

acute effect Adverse reaction that occurs very rapidly after exposure to a toxic substance has occurred.

chronic effect Adverse reaction that happens only after repeated long-term exposure to low doses of a toxic substance.

Toxic exposure to high enough doses of poisons and irritants, and even a small dose of a chemical to which one has developed an allergy, leads to **acute effects**—those that occur very rapidly after exposure (the headache or breathing difficulty that comes from inhaling paint or solvent fumes are acute effects).

Chronic effects are those that result from repeated long-term exposure to low doses of a toxic. These types of effects usually take a long time to emerge. Examples of chronic toxicity abound: the lung cancer that develops after years of cigarette smoking, the health problems that emerge after a lifetime of breathing polluted air, the gradual decrease in sperm counts in men exposed to estrogen mimics.

Or the organ damage that occurs after a long stretch of consuming polluted water.

INFOGRAPHIC 1 TOXIC SUBSTANCES

We are exposed to a variety of toxic substances in our daily lives—at home, in our food, at our workplace, and the outdoor environment. Here, we show some of the different types of toxic substances, based on their harmful effects.

STOCKBACKGROUND/Alamy

TYPE OF TOXIC SUBSTANCE	EFFECT	EXAMPLES
Poison	Causes direct damage upon exposure at a high enough dose.	Pesticides, cleaning solutions, drain cleaner, pharmaceuticals, antifreeze
Irritant	Causes localized damage to tissue such as skin or eyes.	Household cleaners, chlorine, fabric softeners
Sensitizer	Can cause an allergic reaction to develop.	Formaldehyde, latex
Carcinogen	Causes cancer by causing mutations in DNA.	Components in tobacco smoke, paints, perchloroethylene (dry cleaning solvent)
Teratogen	Causes birth defects.	Alcohol, components in tobacco smoke, some pharmaceuticals, heavy metals (e.g., mercury, and lead)
Endocrine disruptor	Interferes with the hormones of the body.	Many chemicals in plastics such as bisphenol A (BPA) and phthalates; some pesticides; glycol ethers (in paints, cosmetics)

? Could a toxic substance be classified into more than one category?

2 FACTORS THAT AFFECT TOXICITY: CHEMICAL CHARACTERISTICS

Key Concept 2: Certain chemical traits, such as toxicity, persistence, and solubility, influence the danger of a given toxic substance.

In February 2015, Miguel Del Toral, a scientist at the U.S. **Environmental Protection Agency (EPA)** found the main cause of Flint's water woes when he detected lead levels of 104 ppb (parts per billion)—seven times higher than the agency's allowable limit—in the water of one resident's home. By June, similar lead levels had been found in several more homes. And by September, a team of scientists led by Marc Edwards from Virginia Tech had reported that 40% of homes in Flint had elevated water lead levels. That same team advised state officials that the water was unsafe for drinking or cooking.

In general, there are several key factors that determine how dangerous a given toxic will be; some of those factors have to do with the chemical's own characteristics. The first such characteristic is a chemical's **potency**—how much or little of it is needed to cause harm. In general, the more potent a chemical is, the less of it is needed to damage an organism. For example, sodium cyanide is very potent; a tiny amount will kill a person. Sodium chloride (table salt) is far less potent; it would take a great deal more of it to do the same amount of harm. But that doesn't mean sodium chloride is always harmless. As toxicologists are fond of saying, "the dose makes the poison." In other words, all things are toxic if the dose is large enough—even water. Conversely, most substances are safe, and some are even helpful, in low enough doses. Consider the rat poison mentioned earlier, warfarin. It is used as a blood-thinning drug in patients with high blood pressure (but in concentrations much, much smaller than found in the rat poison). However, there are some substances so toxic that even minuscule amounts will cause problems. Like lead.

Lead is so potent that the EPA has long held that there is "no safe level" of exposure, especially for children. Even a tiny amount of lead exposure can cause neurological deficits in both children and adults, but in children, it can permanently impair brain development, leading to lifelong learning disabilities and mental deficits. In 2012, the Centers for Disease Control and Prevention (CDC) strengthened the protections against lead, declaring that medical intervention is warranted when blood concentrations are at or above just 5 micrograms per deciliter (μg/dL). (The previous concentration was 10 μg/dL, or twice as high.) In Flint, most children with elevated blood levels had values between 5 and 9 μg/dL; a few had levels in the 30's. (The average of the U.S. population in 2014 was 0.8 μg/dL).

Another factor that affects a chemical's toxicity is its **persistence** or degradability—how easily it breaks down into its constituent parts. Because a persistent chemical is chemically stable, it lasts a long time, affecting ecosystems and the organisms that live there well after its initial release. Many synthetic chemicals are quite persistent, and sometimes that is by design (the chemicals in plastics are formulated to last a long time so the products made from them are durable). Lead is a metal, a very stable element, so it persists forever.

> Lead is so potent that the EPA has long held that there is "no safe level" of exposure, especially for children.

How long a chemical remains in our bodies is another factor that affects its toxicity, and that "residence time" is affected by the chemical's **solubility**—that is, whether it can dissolve in fat or water. Its solubility is important because it impacts whether or how readily we can excrete it. Water-soluble chemicals have the advantage of being a substance we can excrete—our kidneys (our main organs of excretion) can extract them from our bloodstream and expel them via urine. This means they don't linger in our bodies for very long. But water-soluble toxics are a particular problem for aquatic organisms that can easily take up these substances. This exposure can wreak slow havoc on aquatic environments and, by extension, on the ecosystems that surround them.

Fat-soluble chemicals present an extra level of complexity. Unlike water-soluble chemicals, fat-soluble substances pass easily through cell membranes, which means our cells can readily absorb these chemicals. And once they're inside, our bodies have a hard time expelling them—kidneys can only excrete water-soluble substances. Sometimes the liver can convert a fat-soluble molecule into a water-soluble one, so that it can be excreted in urine. But when our livers can't make this conversion, fat-soluble chemicals are stored in our fatty tissue where they

Environmental Protection Agency (EPA) The federal agency responsible for setting policy and enforcing U.S. environmental laws.

potency The dose size required for a chemical to cause harm.

persistence A measure of how resistant a chemical is to degradation.

solubility The ability of a substance to dissolve in a water or fat-based liquid or gas.

A caution sign warns people not to harvest or eat shellfish at a contaminated beach.

bioaccumulation The build-up of a substance in the tissues of an organism over the course of its lifetime.

biomagnification The increased concentration of substances in the tissue of animals at successively higher levels of the food chain.

can pile up in a process known as bioaccumulation.

Bioaccumulation refers to the buildup of substances in the tissue of an individual organism over the course of its lifetime. Most of the chemicals in plastics, for example, are fat soluble and

most if not all of us harbor these and other chemicals in our tissues. Bioaccumulation also leads to a phenomenon known as **biomagnification**, the fact that animals higher on the food chain bioaccumulate more chemicals than the organisms they eat. Here's why: When animals that are higher up on the food chain eat other animals that have bioaccumulated toxics, they consume their preys' entire lifetime dose of those toxics—and they do this every time they eat. Top predators, such as tuna, can have more than a million times the amount of a toxic as an organism at the bottom of the food chain. This impacts us as well: When we eat them, we consume all the toxics that they have picked up from preying on smaller animals. **INFOGRAPHIC 2**

Lead is water soluble so one might think we wouldn't have to worry about bioaccumulation; under natural circumstances, we absorb it in low enough amounts that it is excreted, slowly, in our urine. But it turns out that when we are exposed to more than natural background levels of lead, the metal bioaccumulates—mostly in our bones and teeth, but also in our blood and soft tissue. Lead does not appear to biomagnify up the food chain; scientists aren't sure why (perhaps the bones and teeth that harbor it are not eaten or readily digested). But when it comes to lead, there's plenty they do know, and the knowns were scary enough to turn Flint's water crisis into a national story.

INFOGRAPHIC 2 | **BIOACCUMULATION AND BIOMAGNIFICATION**

Animals can acquire fat-soluble toxic substances through air, water, or food sources. The substances build up in the tissue of the animal over its lifetime if it has continued exposure. These fat soluble substances are passed on to predators.

BIOACCUMULATION: (*occurs in the individual*)

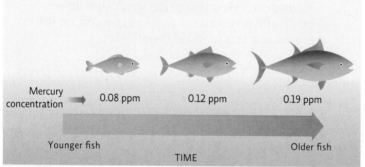

The fish accumulates some mercury every day; the longer it lives, the more it will accumulate and store in its tissues.

BIOMAGNIFICATION: (*a food chain phenomenon*)

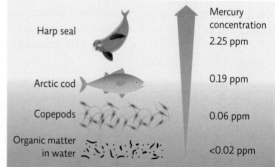

The harp seal ends up with a much higher dose of mercury than other organisms that are lower on the food chain because its prey have higher levels of mercury than prey eaten by consumers lower on the food chain.

 Why don't animals bioaccumulate or biomagnify water-soluble substances?

3 FACTORS THAT AFFECT TOXICITY: EXPOSURE, VICTIM TRAITS, AND CHEMICAL INTERACTIONS

Key Concept 3: It is difficult to determine safe exposure limits for a given chemical because the route of exposure, personal traits of individuals, and interactions with other chemicals all influence its potential for harm.

Throughout 2015, health officials in Flint continued to insist that the water was perfectly safe. But Dr. Mona Hannah-Attisha, a pediatrician at Hurley Medical Center, wasn't taking any chances with her patients. For months, she collected blood samples from the city's youngest residents and tested them for lead.

Her particular concerns had to do with the effects that lead has on children. Decades of research have shown that even low levels of lead exposure can have a devastating impact on children's physical, intellectual, and even emotional development. Children exposed to lead—in their drinking water, in paint chips, or in the air they breathe—can go on to develop learning disabilities, behavioral problems, and serious cognitive deficits.

It's not uncommon for certain toxins to have more serious effects on one person or group of people than another. In fact, no chemical operates in a vacuum. Other factors that affect toxicity pertain not to the chemical itself, but to the context of the exposure. For example, both the route of exposure (whether it was inhaled, consumed, injected, or just touched the surface of the skin) and the dose of that exposure (a little? a lot?) both play a role in determining how serious the effects of toxic exposure will be.

Personal characteristics of the individual being exposed to the toxic also influence its effect. Some people are genetically predisposed to be more or less sensitive to a chemical; illness may also make a person more vulnerable.

The type and amount of chemicals already in one's system can also influence how the newly introduced toxic behaves. Some chemicals in the body might combine to increase overall toxicity in a predictable manner; such effects are referred to as **additive effects**. For example, ampicillin and imipenem are two antibiotics that one could take for a bacterial infection. Both inhibit cell wall synthesis in bacteria (which inhibits bacterial reproduction), but each attacks a different binding site on bacteria cells. Taken together, they will boost the effectiveness of the drug treatment in a way that is consistent with the effect expected for each—an analogy would be two people attacking an invader: One hits high and another hits low.

Other chemicals may reduce toxicity due to interactions between the toxics that "cancel each other out," or at least lessen the damage; these are known as **antagonistic effects**. And others still may work together to produce **synergistic effects**, which are effects even bigger than either chemical would be expected to produce on its own. It is for this reason that we are warned not to mix drugs (or drugs and alcohol). For example, acetaminophen (Tylenol) is a safe, over-the-counter pain medicine. It is processed in the liver by two different enzyme pathways; the primary pathway breaks down acetaminophen to safe by-products that are then excreted, but the other pathway converts a small amount of the drug to a toxic by-product, even at the recommended dose. Normally, this is a very small amount and it doesn't harm the person. If a person drinks alcohol while taking acetaminophen, however, the primary pathway gets tied up processing the alcohol and the acetaminophen gets processed by the secondary pathway, potentially producing enough of the toxic by-product to cause death. (An overdose of acetaminophen will also do this, so always take the recommended amount and no more!)

INFOGRAPHIC 3

The reasons that lead impacts children more than adults are well understood. According to the World Health Organization, children absorb four to five times as much lead as adults do in the same environment because their gastrointestinal tracts are much more adept at absorbing it. The potential for neurological damage is also greater in children: Their brains and their blood-brain barriers are not yet fully developed, making them more vulnerable to permanent damage. Growing children also release stored lead from their bones into their bloodstream more readily as those bones grow and change in shape.

By September, Hanna-Attisha had enough data to publish her research. What she reported would finally tip Flint's managers into action: Among children younger than 5 years old, the incidence of elevated blood levels nearly doubled. In the worst-hit

additive effects Exposure to two or more chemicals that has an effect equivalent to the sum of their individual effects.

antagonistic effects Exposure to two or more chemicals that has a lesser effect than the sum of their individual effects would predict.

synergistic effects Exposure to two or more chemicals that has a greater effect than the sum of their individual effects would predict.

INFOGRAPHIC 3 FACTORS THAT AFFECT TOXICITY

Some chemicals are more toxic than others due to their mode of action. Other factors also affect how toxic a particular chemical will be for an individual.

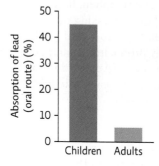

INDIVIDUAL FACTORS
Factors related to the individual may affect how toxic a chemical is. Some chemicals are more of a problem for the very young or very old, or for those who are ill. In some cases, genetic differences make a person more or less vulnerable to a given chemical.

EXPOSURE
Whether a toxic substance is inhaled, ingested, or contacts the skin may affect how much of a problem it causes. Dose is also important, since some substances have a threshold of toxicity. Frequency also matters: A single exposure may be tolerable at a given dose, but repeated exposure may cause problems.

AGE AND LEAD ABSORPTION

Absorption of lead (oral route) (%)
- Children: ~45
- Adults: ~5

manley099/E+/Getty Images

EXPOSURE ROUTE

Absorption of lead by children (%)
- Skin: ~1
- GI tract: ~40

CHEMICAL INTERACTIONS
We are never exposed to just one chemical. The fact that chemicals can interact in ways that increase or decrease their toxic effects complicates our efforts to determine a "safe dose." Exposure to heavy metals can have different effects depending on what other substances one is also exposed to, as shown by the results of these three experimental studies.

ADDITIVE EFFECTS (RATS)
Decrease in enzyme activity (%)

The effect of lead and cadmium is close to the sum of their individual effects.

- Lead: ~18
- Cadmium: ~14
- Lead plus cadmium: ~32

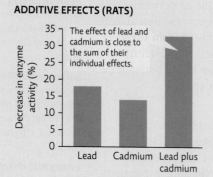

ANTAGONISTIC EFFECT (MICE)
Decrease in dopamine (%)

The combined effect is less than the sum of the individual effects.

- Lead: ~40
- Arsenic: ~19
- Lead plus arsenic: ~25

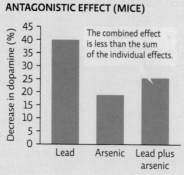

SYNERGISTIC EFFECTS (HUMAN CELLS)
Neuron death (%)

The combined effect of these two preservatives is much greater than the sum of their individual effects.

- Thimerosal (T) (mercury compound): ~10
- Aluminum hydroxide (AH): ~4
- T plus AH: ~59

 Based on your own individual factors (age, health, etc.) do you predict you are more or less vulnerable to toxics than the average person?

Children are more vulnerable to the effects of lead because their nervous system is still developing. In addition, pound for pound, they ingest more lead and absorb it more readily than adults.

neighborhoods, nearly 15% of children had elevated blood lead levels.

In a news article on Dr. Hanna-Attisha's work, *The New York Times* did not mince words:

Of all the concerns raised by the contamination of Flint's water supply, and the failure of the state and federal governments to promptly address the crisis... none are more chilling than the possibility that children in this tattered city may have suffered irreversible damage to their developing brains and nervous systems from exposure to lead.

"If you were going to put something in a population to keep them down for generations," Dr. Hanna-Attisha told the newspaper, "it would be lead."

4 STUDYING TOXIC SUBSTANCES

Key Concept 4: A variety of experimental and observational studies are needed to determine toxicity and safe doses for chemical exposure.

Determining the toxicity of any given chemical is tricky work that requires a combination of experimental and observational studies to get a full picture. (See Module 1.2 for more on experimental and observational studies.) Animal studies are particularly useful because they enable scientists to see the impact of a chemical in an intact organism with fully functioning systems.

One common experimental study exposes test subjects (living organisms or cells in a Petri dish) to various doses of the chemical to create a **dose-response curve**; this shows how the effect that is being tracked varies with dose.

dose-response curve A graph that shows the strength of an effect of a substance at different doses of that substance.

The data generated from experimental studies are often used to calculate the **LD$_{50}$ (lethal dose 50%)**—the dose that kills 50% of the population. This standard calculation helps compare toxics: the lower the LD$_{50}$, the more toxic the substance. Dose-response studies are also used to determine a "safe level of exposure."

Scientists can calculate the **NOAEL (no-observed-adverse-effect level)**, the highest dose where there is no adverse effect (one that impairs function), and the **LOAEL (lowest-observed-adverse-effect level)**, the lowest dose where an adverse effect is first seen. (At concentrations between the NOAEL and the LOAEL, an effect is seen but it does not unacceptably impair function.) The actual safety standard (i.e., how much of a chemical is acceptable for humans, or for environmental entities like waterways) is usually set at 100 to 1,000 times lower than the NOAEL, to account for uncertainty arising from the limitations of experimental studies (e.g., can't experiment on humans, can't account for all chemical interactions, etc.). **INFOGRAPHIC 4**

For ethical reasons, experimental studies on humans may be limited to cellular studies, but research on human populations can also be done with observational studies such as the ones conducted by Hanna-Attisha and Edwards that correlated lead levels in water with lead levels in children. These *epidemiological* studies ask questions like: Who gets sick and what do they have in common? Who doesn't get sick and what do they have in common? (See Module 4.3 for more on epidemiology.)

It was studies like these that led researchers to conclude that there was no acceptable level of lead in drinking water and to set the allowable limit at a vanishingly low level of 15 ppb. While it would be preferable to set it at zero, it is not economically feasible or probably physically possible to eliminate it entirely from the water supply since it is a naturally occurring chemical and still pervasive thanks to decades, if not centuries, of industrial use.

LD$_{50}$ (lethal dose 50%)
The dose of a substance that would kill 50% of the test population.

NOAEL (no-observed-adverse-effect level) The highest dose where no adverse effect is seen.

LOAEL (lowest-observed-adverse-effect level) The lowest dose where an adverse effect was first seen.

INFOGRAPHIC 4 **DOSE-RESPONSE STUDIES**   ANIMATED INFOGRAPHIC

Dose-response studies evaluate the effect a toxic substance has on test subjects at various doses. Charting the change in the response being measured (e.g., appearance of a rash, impaired kidney function, death) as the dose increases is one step in trying to determine a safe dose or level of exposure.

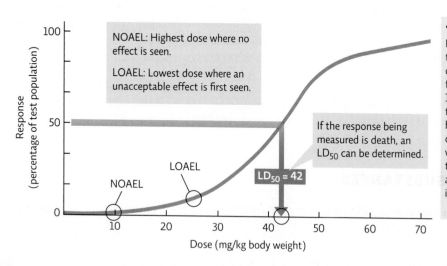

 Consider a second substance that has an LD$_{50}$ of 50 mg/kg. Is this substance more or less toxic than the one shown on the graph? Explain.

5 REGULATING TOXIC SUBSTANCES

Key Concept 5: A 2016 law changed United States regulatory policy so that the safety of a chemical must be demonstrated before it can enter the market.

In an ideal world, unbiased professional regulators would assess the safety of every new chemical before it entered our lives. They would discern all the potential consequences of excessive or continued long-term exposure and in so doing would protect us from any slow, unwitting poisoning. In reality, any policies that govern chemical safety must consider not only health but also a variety of other factors, including economics.

Financial decisions led to the Flint water crisis. The city was on the brink of financial collapse, and in an effort to save money, officials switched the city's water supply from the Detroit water system (Lake Huron water) to the Flint River. The Flint River water was cheaper and seen as a temporary water source that would tide them over until the city completed work on a project that would access water from Lake Huron directly (not through the Detroit system). But without the corrosion protection, lead soon infiltrated the water. Lead had been banned for use in water infrastructure (e.g., pipes and storage containers) in 1986 under the **Safe Drinking Water Act (SDWA)**, but in Flint, as in much of the country, many lead pipes were already in place in buildings or service lines before the ban was enacted.

The SDWA is one of many state and federal regulations meant to protect us from dangerous chemicals. In general, there are two basic approaches that public health officials and legislators use to regulate chemicals such as lead.

Until recently, regulators in the United States assumed that chemicals are safe until proven otherwise. Rather than thoroughly testing each individual compound, regulators made educated deductions about safety based on how other similar compounds have fared. As a result, toxic products were often discovered only after (sometimes long after) reaching the marketplace—usually after people suffer ill effects. Rather than preventing these products from reaching store shelves in the first place, we recalled them after the fact. This ad hoc regulation put the burden of exposing a chemical's risks on the public, not on the regulators.

In the United States, the **Toxic Substances Control Act (TSCA)** is the primary federal law governing chemical safety. It was first passed in 1976, but at that time, some 65,000 existing chemicals were "grandfathered in," meaning that they were not subject to the law's safety testing provisions. Of the roughly 20,000 chemicals that have been introduced to the marketplace since then, only about 200 have been rigorously tested; another 200 have been flagged for thorough testing under the TSCA.

The original TSCA had some major flaws that seriously restricted the EPA's ability to actually identify and regulate toxic substances. For example, the EPA was required to show that the benefit of regulation outweighed the costs of determining if it was hazardous and needed to be regulated. (This was a catch-22: How could they show benefits outweighed costs before they determined its safety?) Under the TSCA, the manufacturer did not have to show that a chemical was safe; the EPA had to prove it was unsafe. With 65,000 chemicals on the market, thousands of new ones entering the market each year, and protections that did not require companies to disclose "trade secrets" regarding chemical formulations, determining whether these chemicals were safe or hazardous was an impossible hurdle to overcome. The EPA was also required to choose the "least burdensome" method of regulation, which required such extensive evaluation of all options (to identify which was the least burdensome) that regulation of any type was difficult. The TSCA was in need of major reform.

That reform came in 2016 with the passage of the *Frank R. Lautenberg Chemical Safety for the 21st Century Act* (a law passed with substantial bipartisan support). The Lautenberg Act shifted the law into a more precautionary stance, an approach already taken by the European Union. Since so many factors influence the hazard imposed by any chemical (e.g., different vulnerabilities among people, interactions with

Safe Drinking Water Act (SDWA) Federal law that protects public drinking water supplies in the United States.

Toxic Substances Control Act (TSCA) The primary federal law governing chemical safety.

other chemicals, etc.), it makes sense to follow the **precautionary principle**, an approach that is appropriate when there is uncertainty about an outcome but the potential for serious problems. Once safety testing is done, regulatory agencies use a rule of thumb that calls for leaving a wide safety margin when setting the acceptable exposure limits.

Under the Lautenberg Act, a manufacturer must show a chemical is safe before it receives approval to go on the market. And that burden of proof focuses on health safety, not a cumbersome cost-benefit analysis to show that regulation is warranted. The EPA no longer has to prove regulation will be beneficial before the chemical is evaluated; they can simply require that manufacturers test any new or current chemical. On its end, the EPA must complete a risk assessment in a timely fashion to determine what type of testing is required. And what constitutes a trade secret is more strongly regulated, requiring much more transparency from chemical manufacturers. Implementing the new law, however, may be difficult

precautionary principle
Acting in a way that leaves a safety margin when the data is uncertain or severe consequences are possible.

with recent changes at the EPA. Supporters of the bill are concerned by the appointment of former chemical industry executives to key positions in the EPA, including to the Office of Chemical Safety and Pollution Prevention. **INFOGRAPHIC 5**

Of course, the TSCA is not the only law on the books that governs chemical safety. The *Clean Air Act* prompted U.S. regulators to begin phasing lead out of gasoline in the 1970's. And when it became clear that lead in paint posed a serious health hazard to children, the U.S. Congress passed a law devoted exclusively to addressing the problem—the *Lead-Based Paint Poisoning Prevention Act*. By 1978, lead-based paint was fully banned.

The SDWA should have protected the citizens of Flint from contaminated water. In addition to banning the use of lead in water infrastructure, the law sets standards for water quality that must be met. It requires public officials to take corrective action when lead levels in water reach 15 ppb. It also gives the EPA the authority to step in if it finds drinking water contaminated in a way that threatens people's health—especially if the state has not acted to resolve the issue.

INFOGRAPHIC 5 **REGULATING TOXIC SUBSTANCES IN THE UNITED STATES**

The U.S. Toxic Substances Control Act (TSCA) regulates all chemical substances manufactured, imported, distributed, or disposed of in the United States. It received a major update in 2016.

TOXIC SUBSTANCES CONTROL ACT (1976)

FRANK R. LAUTENBERG CHEMICAL SAFETY FOR THE 21ST CENTURY ACT (2016)

TOXIC SUBSTANCES CONTROL ACT (1976)		FRANK R. LAUTENBERG CHEMICAL SAFETY FOR THE 21ST CENTURY ACT (2016)
Must prove that the substance is NOT safe.	**APPROACH**	Must prove that the substance IS safe.
Used a *risk–benefit* safety standard for approval; new chemicals were only tested for health safety if shown to be harmful after they were on the market.	**STANDARD USED FOR ASSESSMENT**	Uses a *health risk* safety standard for approval before entering the market.
• ~65,000 chemicals already on the market were "grandfathered in" (never tested) when the TSCA was passed in 1976. • Manufacturers test chemicals for safety at EPA request (a slow and cumbersome process). • 200 chemicals of the 22,000 introduced since 1976 have been tested for safety.	**CHEMICAL TESTING**	• Requires testing by manufacturer for all new chemicals and for existing ones believed to pose a health risk. • A risk assessment will be done to identify high-priority chemicals for further testing. • Calls for extensive evaluation of 10 existing chemicals a year (out of a list of 90 designated as high priority).

CHEMICALS ON THE MARKET

 What plan has the 2016 reform of the Toxic Substances Control Act proposed for the testing of previously untested chemicals currently on the market? Using this approach, how long would it take to test the 65,000 chemicals grandfathered in?

The corrosive water flowing through Flint pipes not only caused lead to be released, it also caused the aging pipes to leach iron, often turning it a rusty-brown. If the water had been treated with the federally required phosphate solution that seals the lead on the pipe surface, it is unlikely that the problem would have developed.

But when Del Toral and the scientists from Virginia Tech issued their reports, and raised concerns about the way that city officials were testing and treating the water in Flint, no city, state, or federal government's regulators took action. Instead, the city and state embarked on a public relations campaign, insisting the water was safe, claiming that the studies showing otherwise were unreliable, and accusing the citizens who expressed concern of overreacting.

Was the city's and state's interpretation of the evidence valid or were they intentionally misleading Flint residents? What was a homeowner to think?

6 INFORMATION LITERACY: EVALUATING INFORMATION SOURCES

Key Concept 6: Published information about scientific topics abounds in our modern world. Information literacy skills can help individuals determine the reliability of that information.

The ability to distinguish between reliable and unreliable sources of information is referred to as **information literacy**. It's the key to drawing reasonable evidence-based conclusions about any given issue or topic; it is especially important in public health and science issues, because when it comes to these issues, misinformation abounds. One of the first steps in information literacy is identifying what type of information source one is consulting— is it a first-hand account from the original person or group that created or experienced it, or is it a retelling?

Our first choice for reliable information would be those that are **primary sources**, sources that present original data or information, including novel scientific experiments and first-hand accounts of any

information literacy The ability to find and evaluate the quality of information.

primary source Information source that presents original data or first-hand information.

given observation. Diary entries and interviews are primary information sources, as are original research articles published in scientific journals. These research reports are rigorously evaluated through **peer review**—a process whereby experts in the field assess the quality of the study's design, data, and statistical analysis, as well as the soundness of the paper's conclusions. Good studies are published; bad ones are rejected, and those that are later discredited are formally retracted. (See Module 1.2 for more on the design, execution, and publishing of scientific studies.)

The value of a primary information source is its authenticity; historians, for example, take great pains to find original documents when researching historical events because facts and interpretations can change with every retelling and vital information can be left out, sometimes intentionally to sway the reader to accept a point of view that might not be supported by the original evidence. It is no different for us as we wade through the sea of information that bombards us every day. Consulting a primary source reduces the chances that the original information will be misrepresented by retelling.

But most of us start a step or two lower than primary sources—perhaps it is a newspaper article about some recent research or event or a video we see on television or online. This information may be coming from a **secondary source**, one that presents and interprets information solely from primary sources. This could be a science journalist who reviews a recent research article or a reporter who interviews a scientist about his or her work. But most likely, our initial information comes from **tertiary sources**, those that use one or more secondary sources. Most books, including textbooks, and reports from the popular press are tertiary sources. Most blogs, websites, and news shows that provide additional commentary or foster debate also qualify as tertiary sources. But because tertiary references do not consult the primary information source, they may perpetuate errors made by a secondary source. For this reason, they are less dependable than secondary sources or primary sources.

peer review A process whereby researchers' work is evaluated by outside experts to determine whether it is of a high enough quality to publish.

secondary source Information source that presents and interprets information solely from primary sources.

tertiary source Information source that uses information from at least one secondary source.

That is not to say secondary and tertiary sources are not useful—they are a good starting place. But if you have unanswered questions or if there are inconsistencies in the reports you encounter, going back to the primary source is the best way to reconcile those differences or find answers to unanswered questions. For any claim you encounter, you should be asking, "what's the evidence for that claim?" Where do you find that evidence? Primary sources. **INFOGRAPHIC 6**

The research papers published by Hanna-Attisha, Del Toral, and Edwards are all primary sources; this means you can actually see the data (the evidence) on which the researchers based their conclusions and evaluate it yourself. Hanna-Attisha's and Edwards, were also peer reviewed, increasing the veracity of these information sources. And yet, officials managed to sow doubt and confusion in the wake of those reports. They accomplished this by acting through secondary and tertiary sources: the news media.

In July 2015, in response to Del Toral's EPA report, Brad Wurfel, a spokesperson for the Michigan Department of Environmental Quality (MDEQ), began planting seeds of doubt: Yes, he said in a public statement, the city of Flint needed to update its water infrastructure. But that did not mean the data of researchers like Del Toral and Edwards were accurate or reliable. The department cited its own tests that showed acceptable levels of lead. Wurfel was, in fact, "skeptical" of Del Toral's and Edwards' data, he said. Another official from the same agency went a step further. During an interview on Michigan Public Radio, he told listeners that the water in Flint was unequivocally safe. "Anyone who is concerned about lead in the drinking water in Flint can relax," he said. Flint Mayor Dayne Walling went further still: He drank a glass of Flint tap water on local television to demonstrate his confidence in the water's safety.

The attack on Hanna-Attisha's work would be even more severe. Hours after a news conference announcing her results, state officials said she was completely wrong. Their own data, which was much more substantial than hers, showed no such spike in blood levels. What's more, they suggested that Dr. Hanna-Attisha was deliberately creating hysteria. The force of their arguments—an entire department of epidemiologists arguing strongly against her findings—was enough to make some

INFOGRAPHIC 6 INFORMATION SOURCES

Information sources vary in their veracity. Consulting original information sources is the best way to ensure that information is accurate.

PRIMARY SOURCES

Primary sources are first-hand accounts of research and observations; they also include interviews or personal diaries.
In science, they include peer-reviewed articles in scholarly journals (these provide methods and data) and books based on primary research.

SECONDARY SOURCES

Secondary sources draw only from primary sources. Usually only scholarly review articles qualify as secondary sources.

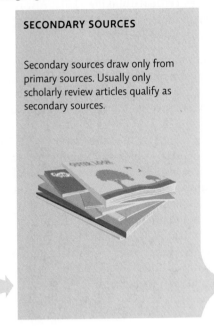

TERTIARY SOURCES

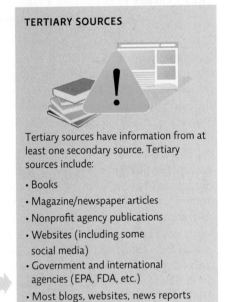

Tertiary sources have information from at least one secondary source. Tertiary sources include:

- Books
- Magazine/newspaper articles
- Nonprofit agency publications
- Websites (including some social media)
- Government and international agencies (EPA, FDA, etc.)
- Most blogs, websites, news reports

 The references cited at the end of this module include original research articles, an editorial article, a book, and newspaper article. Does this module qualify as a primary, secondary, or tertiary resource?

question Hanna-Attisha's work. Even Dr. Hanna-Attisha wondered, briefly, if she had drawn the wrong conclusion. But when her results, and those of Edwards, were validated by independent testing, the government agencies responsible for protecting the water supply came under fire.

This situation illustrates the importance of consulting the primary information source that contains a description of the actual study and its data. The data from the studies of Hanna-Attisha and Edwards were published and available for examination, as were the methods they used to collect their data. This allowed independent verification of that data. The same scrutiny revealed that the MDEQ test results were flawed because the method used to collect the water samples was inappropriate. (The water samples were collected only after letting the water flow through the tap for several minutes, a known way to decrease the amount of lead in the sample since the lead that would have collected in the water as it sat in the pipe would be flushed out with the initial flow. This meant less lead would be able to enter the water that was rapidly flowing through the tap—the water that was collected for the testing. (This was, in fact, a recommendation

to homeowners—always let the water run several minutes before using it to flush out the lead that had collected since the tap was last turned on.)

The lengths that officials at the local, state, and even federal levels took to try and convince residents the water was fine, despite mounting evidence to the contrary, was substantial. What began as a cost-cutting bad decision escalated to intentional cover-ups and attacks on those trying to uncover the truth. It's understandable that residents were confused. Was the water safe or was it dangerous? Could they wash with it but not drink it? And then there was an earlier *E. coli* (bacterial) outbreak that had caused many illnesses and had been traced to the water even before lead emerged as a problem, prompting the water department to add extra chlorine to the water. Now there was talk about the problem imposed by chlorine disinfectant by-products. Should they buy water filters or a new product called the Waterbug—a sponge purported to sop up disinfectant by-products—that was being endorsed by the nonprofit organization Water Defense? Would the Waterbug be helpful? Were disinfectant by-products even a problem?

7   INFORMATION LITERACY: CRITICAL THINKING AND LOGICAL FALLACIES

Key Concept 7: Evaluating a claim requires critical-thinking skills: Always ask for the evidence and evaluate it, look out for logical fallacies, and keep an open mind—be willing to go where the evidence leads you.

Sorting through all the rhetoric of this, or any complex environmental problem, requires diligence and the use of critical-thinking skills that allows one to logically evaluate the evidence—the same skills used by scientists to evaluate scientific studies (see Module 1.2).

In Flint, the situation was made worse by deliberate attempts to mislead the public. Literary devices used to confuse or sway the audience to accept a claim or position in the absence of evidence, or by twisting the evidence, are referred to as **logical fallacies**. "The water must be safe! Look! I'm drinking it myself!" is a logical fallacy known as a *hasty generalization*. Just because the mayor feels comfortable drinking one glass of water does not mean that the same water is safe for young children to drink and bathe in day after day.

Logical fallacies can persuade others to accept a conclusion—or to doubt one. Industry, for example, has been found to deliberately mislead consumers in order to make sales. Perhaps that comes as no surprise, but it is particularly disturbing when the products they sell are dangerous. For example, the tobacco industry argued for years that smoking cigarettes was safe, even though they had ample evidence from internal and external investigations that it was quite dangerous. To combat the growing mountain of evidence that smoking was linked to a number of serious health problems, the industry developed what science historian Naomi Oreskes calls a "playbook"—definitive steps that could be taken to confuse consumers. This included hiring scientists who would question studies that showed harm (*appeal to authority*) and claiming that diseases like cancer and heart disease were too complex to pin any particular individual's ailment on smoking (*appeal to ignorance*). In one of their own internal memos, a public relations firm working for the tobacco industry blatantly stated that "doubt is our product." If they could cast doubt on the science, they could delay regulation.

These same techniques were employed by those seeking to defuse fears in Flint. Residents were bombarded with assurances that were nothing more than *red herrings*—extraneous information meant to confuse the listener. For example, as mentioned earlier, extra chlorine had been added to deal with higher than normal bacterial levels, raising concerns about dangerous chlorination by-products in the water. After addressing the problem, letters from the water authority were sent out with this opening statement: "We are pleased to report that City of Flint water is safe and meets U.S. Environmental Protection Agency guidelines." This was true—for chlorination by-products—not for lead. This letter might have eased the fears of some residents—their water was now safe, right? But it told them nothing about other potential contaminants. Lead was not mentioned at all. Was it still a threat?

Concerned residents, scientists, and physicians were accused of being "anti-everything," implying that their concerns were unjustified and part of a fringe movement. *Ad hominem attacks* like this attack and try to discredit the person or group, thereby deflecting attention from the evidence. **INFOGRAPHIC 7**

Critical thinking is the antidote to logical fallacy. It enables individuals to logically assess and reflect on information and reach their own conclusions. Critical thinking is a skill that can be broken down into a handful of tenets of measures:

Be skeptical. Just like a good scientist, a nonscientist should not accept claims without evidence, even from an expert. This doesn't mean refusing to believe anything; it simply means requiring evidence before accepting a claim as reasonable, especially if it is counter to well-established scientific consensus.

Evaluate the evidence. Is the claim being made derived from anecdotal evidence (unscientific observations, usually relayed as secondhand stories) or from scientific studies? If it is based on studies, how relevant are those studies to the claim? Were they done on primates? Rodents? Cells in a Petri dish? Human populations?

Watch out for author biases. Is the author of the study or person making the claim trying to promote a position? Is that person financially tied to one conclusion or another? Is he or she trying to use evidence to support a predetermined conclusion?

logical fallacies Arguments that attempt to sway the reader without using reasonable evidence.

critical thinking Skills that enable individuals to logically assess information, reflect on that information, and reach their own conclusions.

INFOGRAPHIC 7   **LOGICAL FALLACIES**

Logical fallacies are arguments used to confuse or sway someone to accept a claim/position in the absence of evidence.

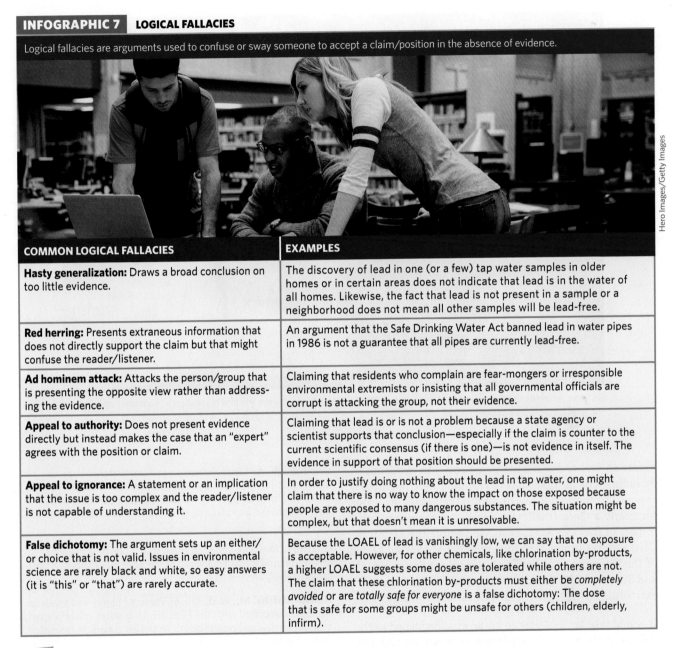

Hero Images/Getty Images

COMMON LOGICAL FALLACIES	EXAMPLES
Hasty generalization: Draws a broad conclusion on too little evidence.	The discovery of lead in one (or a few) tap water samples in older homes or in certain areas does not indicate that lead is in the water of all homes. Likewise, the fact that lead is not present in a sample or a neighborhood does not mean all other samples will be lead-free.
Red herring: Presents extraneous information that does not directly support the claim but that might confuse the reader/listener.	An argument that the Safe Drinking Water Act banned lead in water pipes in 1986 is not a guarantee that all pipes are currently lead-free.
Ad hominem attack: Attacks the person/group that is presenting the opposite view rather than addressing the evidence.	Claiming that residents who complain are fear-mongers or irresponsible environmental extremists or insisting that all governmental officials are corrupt is attacking the group, not their evidence.
Appeal to authority: Does not present evidence directly but instead makes the case that an "expert" agrees with the position or claim.	Claiming that lead is or is not a problem because a state agency or scientist supports that conclusion—especially if the claim is counter to the current scientific consensus (if there is one)—is not evidence in itself. The evidence in support of that position should be presented.
Appeal to ignorance: A statement or an implication that the issue is too complex and the reader/listener is not capable of understanding it.	In order to justify doing nothing about the lead in tap water, one might claim that there is no way to know the impact on those exposed because people are exposed to many dangerous substances. The situation might be complex, but that doesn't mean it is unresolvable.
False dichotomy: The argument sets up an either/or choice that is not valid. Issues in environmental science are rarely black and white, so easy answers (it is "this" or "that") are rarely accurate.	Because the LOAEL of lead is vanishingly low, we can say that no exposure is acceptable. However, for other chemicals, like chlorination by-products, a higher LOAEL suggests some doses are tolerated while others are not. The claim that these chlorination by-products must either be *completely avoided* or are *totally safe for everyone* is a false dichotomy: The dose that is safe for some groups might be unsafe for others (children, elderly, infirm).

 Which logical fallacy do you feel is the hardest to recognize and might slip by the average person? Why?

Be open minded. Try to identify your own biases or preconceived notions (the government can't be trusted, most people overreact to things like this, etc.) and be willing to follow the evidence where it takes you.

By late 2015, the scientific research showing lead in Flint's water and in the blood of its youngest citizens was well supported. In a complete course reversal, city and state officials acknowledged that the lead levels in Flint's water were too high, that the root of the problem was the switch in water supply, and that immediate corrective action was urgently needed. A state of emergency was declared—first by Michigan's governor and then by the President of the United States—and Michigan's national guard was activated to distribute water filters and bottled water to residents.

Dr. Marc Edwards conducted much of the research into Flint's water problems. He and his research team offer updates on the issue at www.flintwaterstudy.org.

Before all was said and done, some 100,000 residents, including tens of thousands of children, were potentially exposed to elevated levels of lead in their drinking water for nearly 2 years. It will take decades to assess the full impact of that exposure. Testifying before the Michigan Civil Rights Commission in September 2016, Paul Mohai of the University of Michigan's School for Environment and Sustainability called the lead crisis in Flint "an extraordinary example" of environmental injustice.

It would take far less time to know what happened to the folks responsible for the lapse and cover-up: By mid-2017, several city, state, and federal officials had resigned and some 13 criminal cases had been filed against state and local officials. Several of them would face jail time. On the upside, the city switched back to Lake Huron water and began a massive project to replace its lead water pipes, thanks to nearly $350 million in state and federal funds.

Select References:

Bellinger, D. C. (2016). Lead contamination in Flint—an abject failure to protect public health. *New England Journal of Medicine, 374*(12), 1101–1103.

Goodnough, A. (2016, January 29). Flint weighs scope of harm to children caused by lead in water. *The New York Times.* Retrieved from https://www.nytimes.com/2016/01/30/us/flint-weighs-scope-of-harm-to-children-caused-by-lead-in-water.html.

Gostin, L. O. (2016). Lead in the water: a tale of social and environmental injustice. *Journal of the American Medical Association, 315*(19), 2053–2054.

Hanna-Attisha, M., et al. (2016). Elevated blood lead levels in children associated with the Flint drinking water crisis: a spatial analysis of risk and public health response. *American Journal of Public Health, 106*(2), 283–290.

Oreskes, N., & Conway, E. M. (2012). *Merchants of doubt: how a handful of scientists obscured the truth on issues from tobacco smoke to global warming.* London: Bloomsbury.

Pieper, K. J., Tang, M., & Edwards, M. A. (2017). Flint water crisis caused by interrupted corrosion control: investigating "ground zero" home. *Environmental Science Technology, 51,* 2007–2014.

Zhang, N., et al. (2013). Early childhood lead exposure and academic achievement: evidence from Detroit public schools, 2008–2010. *American Journal of Public Health, 103*(3), e72–e77.

INTERACTIVE MAP   **LEAD POISONING IN THE UNITED STATES**   ANIMATED INFOGRAPHIC

This module briefly examined the issue of lead contamination of the water of Flint, Michigan. Lead unfortunately is not just a problem in Flint or with water supplies. Read the case studies presented by the online Interactive Map for more on the problem of lead contamination in the United States.

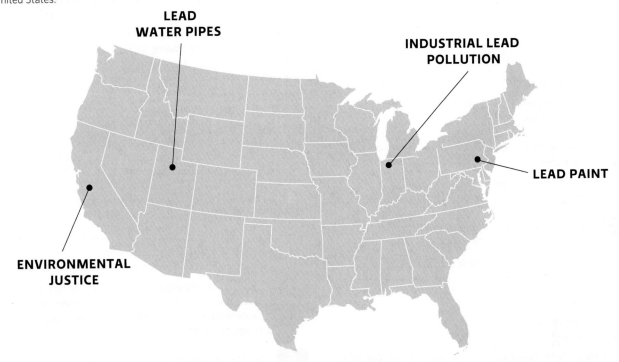

LEAD WATER PIPES

INDUSTRIAL LEAD POLLUTION

LEAD PAINT

ENVIRONMENTAL JUSTICE

 BRING IT HOME

PERSONAL CHOICES THAT HELP

While lead in water is making news, we are exposed to a bevy of other chemicals in our day-to-day lives—things like air fresheners, nail polishes, and plastic storage containers. Many of the products designed to improve our lives actually contain chemicals that may harm us in the long run. With just a few changes, you can dramatically reduce your long-term chemical exposure.

Individual Steps
• Have your water tested if you have reason to believe it does not meet safety standards.
• Buy products with nontoxic or less-toxic components such as organic produce, coffee, and chocolate (grown without the use of synthetic pesticides) or upholstered furniture that is not treated with toxic flame retardants. Visit the EPA website, www.epa.gov/saferchoice, to find safer product options.

• Check with your city's solid waste agency for guidelines about how to correctly dispose of household wastes, including paint, medication, cleaning products, and yard chemicals.

Group Action
• Talk to your roommates, family members, or employers to see if you can switch to products with fewer harmful chemicals.
• Arrange a viewing of the documentary *Merchants of Doubt* and lead a discussion about the techniques used to deliberately mislead the public and ways to combat those techniques.

Policy Change
• Learn about the laws controlling toxic substances for your home state and advocate for changes if you feel they are needed.

• Research the proposed Personal Care Products Safety Act (Senate Bill 1113) and contact your elected officials in support or opposition to the bill.

y-studio/iStock/Getty Images

ENVIRONMENTAL LITERACY   **UNDERSTANDING THE ISSUE**

1 What classes of toxic substances are recognized, and what are common routes of exposure?

1. A toxic substance that causes an allergic response after repeated exposure is a(n):
   a. irritant.
   b. teratogen.
   c. sensitizer.
   d. endocrine disruptor.

2. Which of the following is true of all toxic substances?
   a. They cause death.
   b. They harm living things.
   c. They are natural substances.
   d. They are poisons.

3. Distinguish between chronic and acute effects caused by exposure to a toxic substance.

2 How do the chemical characteristics of a substance influence its toxicity?

4. The dose of a toxic substance that is required to cause harm is known as its:
   a. potency.
   b. risk.
   c. persistence.
   d. solubility.

5. You are trying to choose a pesticide to use in your garden. Which of the following describes the best traits for that pesticide if you hope to reduce its negative environmental impacts?
   a. High potency and low solubility
   b. High persistence and ability to bioaccumulate
   c. Low persistence and water soluble
   d. Low potency and fat soluble

6. Distinguish between bioaccumulation and biomagnification.

3 How is toxicity affected by exposure route, chemical interactions, and the victim's characteristics?

7. True or False: The route of exposure to a toxic substance has little effect on its toxicity. It will be problematic whether it is inhaled, is ingested, or merely comes in contact with the skin.

8. Lead and cadmium are toxic metals that bind to the same receptor in the cell, so the effect of exposure to one or both is quite predictable: If one is exposed to 20 ppm lead and then also exposed to a comparable dose of cadmium, the effect is the same as it would be upon exposure to 40 ppm lead. This is an example of a(n):
   a. additive effect.
   b. antagonistic effect.
   c. synergistic effect.

9. Why are some people more vulnerable to toxic substances than others, even if exposed to the same dose?

4 How is toxicity determined?

10. True or False: The highest dose where no observed effect is seen is known as the NOAEL (no-observed-adverse-effect level).

11. The toxicity of a chemical is evaluated using cells in culture or live animal subjects by creating a(n) _____; a high LD_{50} indicates _____.
    a. observational study; a safe dose
    b. dose analysis; a threshold dose
    c. experimental study; high toxicity
    d. dose-response curve; low toxicity

12. Why is it so hard to determine a safe exposure for a given chemical? How do regulators setting safe exposure standards deal with the uncertainty associated with these factors?

5 How are toxic substances regulated in the United States?

13. True or False: In the past, the United States tested every chemical before it hit the market but budget cuts have caused a policy shift to test only products that have caused problems after being sold.

14. Distinguish between the regulatory approach used by the Toxic Substances Control Act of 1976 and the precautionary approach of the Laudenberg Act of 2016. What are the advantages and disadvantages of each?

6 What is information literacy, and why is it important in environmental science?

15. Which of the following is a secondary source of information?
    a. A blog in which the author enters personal observations on his day-to-day life
    b. A magazine or newspaper article that uses primary, secondary, and tertiary sources
    c. An article that references only peer-reviewed original research articles
    d. A government website that cites original research and scholarly review articles

16. Why are tertiary information sources considered less reliable than primary or secondary sources?

7 What is critical thinking, and how can it counter logical fallacies used in arguments?

17. Pesticides help increase agricultural yields, but some people oppose their use because of their inherent toxicity. An argument against pesticide use that attacks the pesticide maker on the grounds that he or she is simply profit driven is a(n):
    a. ad hominem attack.
    b. appeal to authority.
    c. appeal to ignorance.
    d. hasty generalization.

18. Identify and explain four tenets in critical thinking that can help you logically access and reflect on information in order to draw your own conclusions.

SCIENCE LITERACY **WORKING WITH DATA**

Lead causes a wide variety of health problems. Alessandra Stacchiotti and colleagues looked at the effect of lead on human kidney cells. (These cells will grow and divide in culture, increasing the population size as time goes on.) Petri dishes were initially plated with 0.45×10^6 cells, and after 24 hours, test cells were exposed to either 100 μM or 500 μM solutions of lead. Cell growth (the size of the population) was determined by counting the number of cells at 1, 2, and 3 days after treatment; this was compared to control cells that received no lead exposure.

CELL GROWTH AFTER LEAD EXPOSURE

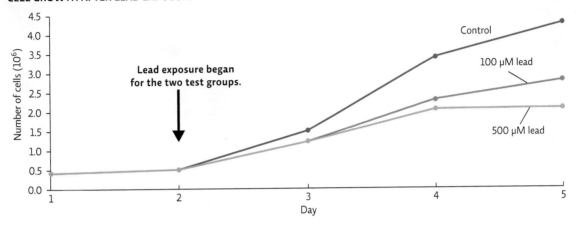

Interpretation

1. What relationship is seen between the growth rate of cells and the concentration of lead exposure?

2. How long did it take for lead exposure to begin to affect cell growth?

3. How much lower was the cell population size on Day 5 for each test group compared to the control group. (Express this as a percentage of the control group population size.)

Advance your Thinking

4. The researchers measured the number of cells in each dish each day. Other than cell death, what is another way that lead exposure could lead to fewer cells each day?

5. If these were kidney cells in your body, what effect do you predict lead exposure might have on kidney function? Identify the underlying assumption that influenced this prediction.

INFORMATION LITERACY **EVALUATING INFORMATION**

The Environmental Working Group (EWG) is a nonprofit group that makes recommendations for avoiding toxics in your environment.

Explore the EWG website (www.ewg.org) and work with the information to answer the following questions:

1. Determine if this is a reliable information source with a clear and transparent agenda:
   a. Explore the website and the topics covered. Briefly outline the scope and focus of the topics.
   b. Read about the board members on the *About Us* webpage. Do the credentials of the individuals align with the topics covered (i.e., Are there experts in toxicology, farming, water chemistry, medicine, etc.)? Does this make the information presented reliable or unreliable? Explain.
   c. Look on the website for information on water (under the tab *Key Issues*). Select and read one of the articles. Identify a claim made in this article.
      i. Is there sufficient evidence provided in support of this claim? Evaluate the credibility of that evidence: Is it reliable?
      ii. Are there any logical fallacies used? Explain.
   d. Overall, does this appear to be a reliable information source? Explain.

2. Open the *Tap Water Database* and enter your zip code for information on your community's tap water and evaluate the information found there.
   a. Identify any contaminants listed and record their potential source. How does this water sample rank for lead contamination? What is the source of this information? (If no contaminants are listed, choose another water utility that does contain contaminants.)
   b. Select the *Want to Filter These Contaminants Out* link.
      i. Explore the filters listed and choose a filter: Are data provided in support of its effectiveness? If so, is this data useful?
      ii. For some products, there are direct links to purchase the product on Amazon.com. Is this a reasonable way to donate to the organization or a questionable practice for a nonprofit organization? Explain.
      iii. Explore the *Water Filter Guide*. What kind of information is provided? Is this information useful? Does the guide appear to endorse any filters or filter types? Should it? Explain.
      iv. Overall, does this appear to be a reliable guide? Explain.

 Additional study questions are available at SaplingLearning.com.

ECOLOGY

CHAPTER 2

Ecology is the study of interactions organisms have with their environment and other organisms. Understanding these interactions and how species respond to changing conditions can help us as we try to prevent human-imposed ecosystem damage or attempt ecosystem restoration where needed.

Module 2.1: Ecosystems and Nutrient Cycling

A look at the living and nonliving components of ecosystems and the ways energy and matter are captured and moved through ecosystems

Module 2.2: Population Ecology

An examination of ecological populations and how ecologists study them and the biotic and abiotic factors that influence whether they thrive or decline

Module 2.3: Community Ecology

A look at the relationships between populations in a community, the impact of human actions that break community connections, and ways to protect or even restore damaged ecosystems

ENGINEERING EARTH

An ambitious attempt to replicate Earth's life support systems falls short

Biosphere 2 from afar. CAMERIQUE/ClassicStock/Alamy

After reading this chapter and studying the KEY CONCEPTS and INFOGRAPHICS, you should be able to answer these GUIDING QUESTIONS

CORE MESSAGE

Ecosystems are complex assemblages of many interacting living and nonliving components. Living organisms play irreplaceable roles in nature, supporting life and allowing ecosystems to function over the long term. It is important that we protect ecosystems and work to restore damaged areas to keep these connections intact so that we and other species can continue to live and thrive on this planet.

1. What is the hierarchy of organization recognized by ecologists, and why might it be useful to recognize such distinctions?

2. Why do ecosystems need a constant input of energy yet can handle the fact that they do not receive appreciable new inputs of matter

3. How do environmental factors affect the distribution and makeup of biomes?

4. What is a population's range of tolerance, and how does it affect the distribution of a population or its ability to adapt to changing conditions?

5. What role do biotic and abiotic factors play in matter cycles?

6. How does carbon cycle through ecosystems, how is this cycle disrupted, and what problems can this disruption cause?

7. How does nitrogen cycle through ecosystems, how is this cycle being disrupted, and what problems can this disruption cause?

8. How does phosphorus cycle through ecosystems, how is this cycle being disrupted, and what problems can this disruption cause?

On September 26, 1993, with their first mission complete, four men and four women emerged from Biosphere 2—a hulking dome of custom-made glass and steel—back into the Arizona desert, where throngs of spectators stood cheering. They had been sealed inside the facility, along with 3,000 other plant and animal species, for exactly 2 years and 20 minutes; it was the longest anyone had ever survived in an enclosed structure.

◉ WHERE IS BIOSPHERE 2?

AZ

TUCSON

● BIOSPHERE 2

TUCSON

The feat was part of a grand experiment, the goals of which were twofold. First, scientists wanted to prove that an entirely self-contained, humanmade system—the kind they might one day use to colonize the Moon or Mars—could sustain life. Second, they hoped that by studying this mini-Earth, which could be controlled and manipulated in ways the real Earth could not, they might better understand our own planet's delicate balance and how best to protect it.

Despite the fanfare surrounding the biospherians' emergence, it was tough to say whether the mission had been a success or a failure. More than one-third of the flora and fauna had become extinct, including most of the vertebrates and all of the pollinating insects. Morning glory vines had overrun other plants, including food crops. Cockroaches and "crazy ants" were thriving. Too little wind had prevented trees from developing stress wood—wood that grows in response to mechanical stress and helps trunks and branches shift into an optimal position; without stress wood, the trees were brittle and prone to collapse. And eating too many sweet potatoes had turned the biospherians themselves bright orange. (A string of plant diseases had decimated other crops.)

On top of that, nitrous oxide (laughing gas) had grown concentrated enough to "reduce vitamin B_{12} synthesis to a level that could impair or damage the brain," according to one interim report. And oxygen levels had plummeted from 21% (roughly the same as Earth's atmosphere) to 14% (just barely enough to sustain human life). To fix this, project engineers had been forced to pump in outside air, violating the facility's sanctity as a closed system.

Worst of all, missteps and course corrections had been mired in secrecy—each one leaked to the press only months after the fact. Rumors had begun to circulate that the eight people sealed inside—not to mention the ones from whom they took their orders—were more interested in creating a futuristic utopia than in conducting rigorous scientific research. As evidence for this rumor mounted, the scientific community grew suspicious. Was Biosphere 2 legitimate science, a publicity stunt, or some bizarre mix of the two?

To be sure, the eight biospherians had survived, and many experts agreed that, in principle at least, the facility still held enormous potential as a scientific tool. But before that potential could be realized, the

The eight biospherians emerge from Biosphere 2 after living there for 2 years.

scientific community and the public at large would need to know exactly what had happened inside the desert dome.

To answer that question, we need to answer a few others first: What exactly is a biosphere, and just how did Biosphere 2's creators set about building one?

1 THE ECOLOGICAL HIERARCHY: FROM BIOSPHERE TO INDIVIDUAL

Key Concept 1: Ecologists identify a nesting organization of life from individuals to the biosphere, often focusing on the levels of population, community, and ecosystem to examine how species respond to and affect the natural world.

species A group of plants or animals that have a high degree of similarity and can generally only interbreed among themselves.

biosphere The sum total of all of Earth's ecosystems.

biome One of many distinctive types of ecosystems determined by climate and identified by the predominant vegetation and organisms that have adapted to live there.

ecosystem All of the organisms in a given area plus the physical environment in which, and with which, they interact.

population All the individuals of a species that live in the same geographic area and are able to interact and interbreed.

habitat The physical environment in which individuals of a particular species can be found.

community All the populations (plants, animals, and other species) living and interacting in an area.

niche The role a species plays in its community, including how it gets its energy and nutrients, what habitat requirements it has, and with which other species and parts of the ecosystem it interacts.

biotic The living (organic) components of an ecosystem, such as the plants and animals and their waste (dead leaves, feces).

abiotic The nonliving components of an ecosystem, such as rainfall and mineral composition of the soil.

The field of ecology focuses on how **species** (organisms that have a high degree of similarity and can generally only successfully interbreed among themselves) interact with other components in their environment. In other words, it is about relationships. These relationships can be examined at different levels. The term **biosphere** refers to the total area on Earth where living things are found—the sum total of all its **biomes**. An **ecosystem** includes all the organisms in a given area (the **biotic** components) plus the nonliving (**abiotic**) components of the physical environment in which they interact. In the natural world,

ecosystems assume a range of shapes and sizes—a single, simple tide pool qualifies as an ecosystem; so does the entire Mojave Desert.

Having the entire biosphere as the focus of study is usually too expansive to manage, so some ecologists study how ecosystems function by focusing on interactions between individuals of the same species within a **population**. For example, the wolves of the Northern Rockies and Yellowstone National Park make up one population there; the elk make up a different population as do the bison, the beaver, and the willow trees. (See Module 2.1 for a closer look at this example and population ecology.)

It is important to note that these populations don't exist in isolation. For this reason, ecologists also study species' interactions with their physical surroundings (their **habitat**) and with other species in

Visitors can tour the facility. Here, they view the desert biome.

INFOGRAPHIC 1 ORGANIZATION OF LIFE: FROM BIOSPHERE TO INDIVIDUAL

Ecologists recognize a nesting hierarchy of organization from the biosphere down to the individual organism. Each category is made up of the smaller ones.

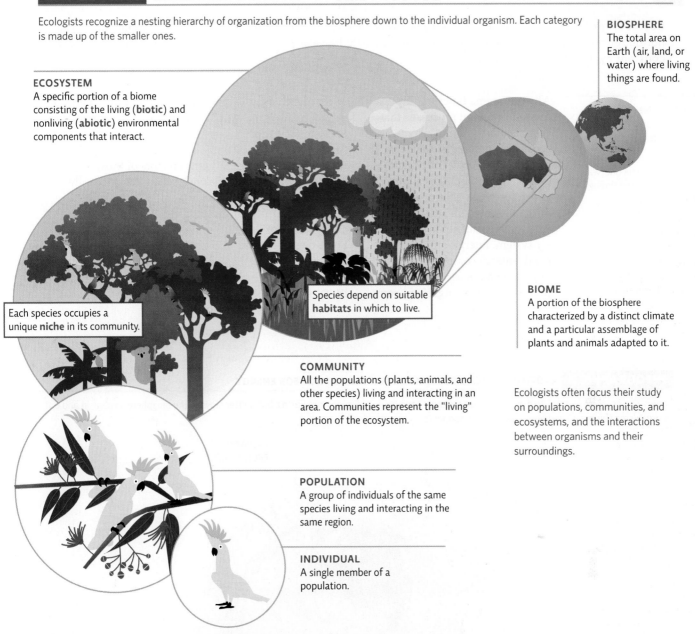

BIOSPHERE
The total area on Earth (air, land, or water) where living things are found.

ECOSYSTEM
A specific portion of a biome consisting of the living (**biotic**) and nonliving (**abiotic**) environmental components that interact.

Each species occupies a unique **niche** in its community.

Species depend on suitable **habitats** in which to live.

BIOME
A portion of the biosphere characterized by a distinct climate and a particular assemblage of plants and animals adapted to it.

COMMUNITY
All the populations (plants, animals, and other species) living and interacting in an area. Communities represent the "living" portion of the ecosystem.

Ecologists often focus their study on populations, communities, and ecosystems, and the interactions between organisms and their surroundings.

POPULATION
A group of individuals of the same species living and interacting in the same region.

INDIVIDUAL
A single member of a population.

 You are studying a pocket mouse that lives in an underground burrow where it stores seeds. Make a list of biotic (living) and abiotic (nonliving) things that might be important to its survival.

their ecological **community**. While many species in a community share the same habitat (e.g., forest floor or deep sea vent) and may even depend on some of the same resources, each individual species will utilize resources and interact with other components of its habitat in a way that is not completely like any other species—it has a unique ecological **niche**.
INFOGRAPHIC 1

Unfortunately, populations in an ecological community must deal with human actions that alter their ecosystem and disrupt vital connections. For example, human development of the Everglades has affected that entire community by altering water cycles, changing the availability of nutrients and resources for every population in the ecosystem, and changing the way they interact with each other—an example covered in depth in Module 2.3.

2 ENERGY AND MATTER IN ECOSYSTEMS

Key Concept 2: A constant input of energy is always needed because once it is used by one organism, it cannot be reused by another. New inputs of matter to Earth, however, are negligible, so life also depends on the constant cycling of matter resources.

matter cycles Movement of life's essential chemicals or nutrients through an ecosystem.

energy flow The one-way passage of energy through an ecosystem.

photosynthesis The chemical reaction performed by producers that uses the energy of the Sun to convert carbon dioxide and water into sugar and oxygen.

All ecosystems function through two fundamental processes that are collectively referred to as ecosystem processes: **matter cycles** and **energy flow**. Matter cycles are biogeochemical cycles that refer specifically to the movement of life's essential chemicals or nutrients through an ecosystem. Energy, on the other hand, enters ecosystems as solar radiation and is passed along from organism to organism, some

released as heat, until there is no more usable energy left. Therefore, we can say that matter *cycles* but energy *flows* in a one-way trip.

Earth—or "Biosphere 1," as the creators of Biosphere 2 liked to call it—is materially closed but energetically open. In other words, the plants and other organic material that make up an ecosystem, called *biomass*, cannot enter or leave the system, but energy can: Some energy leaves as heat or light, and new energy is absorbed from outside. In fact, plant biomass is produced with energy from the Sun through **photosynthesis**. **INFOGRAPHIC 2**

INFOGRAPHIC 2   **EARTH IS A CLOSED SYSTEM FOR MATTER BUT NOT FOR ENERGY**

Energy can enter and leave Earth as light (solar radiation) and heat (radiation from Earth), but matter stays in the biosphere, cycling in and out of organisms (biomass) and environmental components.

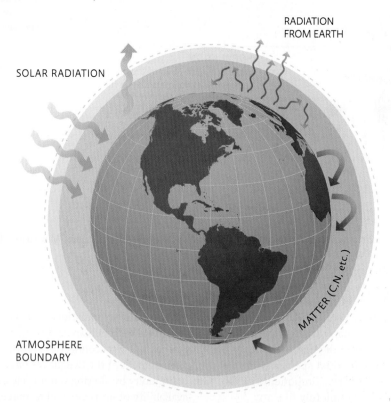

RADIATION FROM EARTH

SOLAR RADIATION

MATTER (C, N, etc.)

ATMOSPHERE BOUNDARY

 Why is it so important that species recycle matter and that they depend on a source of energy that is readily replenished (renewable)?

3 BIOMES

Key Concept 3: The biome that is present in a given area is influenced by the physical and climatic characteristics of its environment, particularly precipitation amount and temperature for terrestrial biomes.

Biomes are specific portions of the biosphere determined by climate and identified by the predominant vegetation and organisms adapted to live there. Biomes can be divided into three broad categories—terrestrial, marine, and freshwater. Within those three categories are several narrower groups, and within those are a variety of subgroups. An entire biome itself may be considered an ecosystem, as are the smaller groups and subgroups. For example, forests, deserts, and grasslands are the three main types of terrestrial biomes. Within the forest biome category are different types of forests, such as tropical, temperate, and boreal forests, and within each of those groups are subgroups (for example, dry tropical forest and tropical rainforest). **INFOGRAPHIC 3**

When ecologists study entire ecosystems, they are limited to making observations and trying to discern cause and effect from those observations. This is no small challenge; precise measurement of each and every relevant ecological factor is simply not possible, and even the simplest factor (for example, a change in rainfall, the loss of a single species, variations in solar radiation reaching the ground) can impact many of the other factors and affect the ecosystem as a whole. But this doesn't mean we can't gain insights from ecosystem-scale studies. Scientists measure important parameters of natural ecosystems to gather evidence to

try to determine cause and effect; they often compare those that are affected by natural disasters or human impact to less disturbed ecosystems in a kind of "natural experiment." From this, ecologists make their best estimations of how multiple factors affect one another (often via mathematical modeling). The more parameters that are properly measured and linked to one another, the better, but because all parameters and relationships cannot possibly be included, there may be a lot of room for interpretation when it comes to understanding what is happening at the ecosystem level.

Biosphere 2 offered ecologists an unprecedented research tool: a mini-planet where a variety of environmental variables—from temperature and water availability to the relative proportions of oxygen and carbon dioxide (CO_2) at any given moment—could be tightly controlled and precisely measured. "Manipulating these variables and tracking the outcomes could greatly advance our understanding of natural ecosystems and all the minute, complex interactions that make them work," says Kevin Griffin, a Columbia University plant ecologist who conducted research at the Biosphere 2 facility. "The plan was to use that knowledge to figure out how to repair degraded ecosystems in the real world, so that they continue to provide the services so essential to our survival."

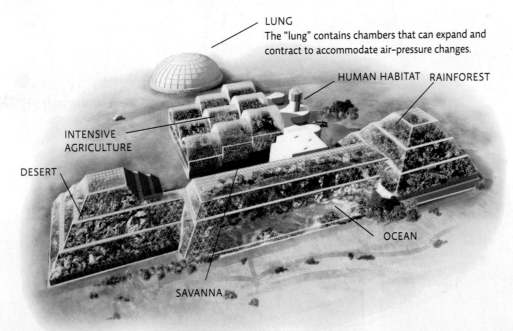

LUNG
The "lung" contains chambers that can expand and contract to accommodate air-pressure changes.

HUMAN HABITAT   RAINFOREST

INTENSIVE AGRICULTURE

DESERT

OCEAN

SAVANNA

Biosphere 2 houses several biomes under one roof, each contributing to overall function. One of the challenges faced by designers was how to include a variety of biomes in the close quarters of the 3-acre Biosphere 2 structure. For example, in nature, a tropical rainforest, whose temperature is fairly constant, would not be next to an arid desert, which drops in temperature at night. To deal with this, an ocean was placed between the desert and rainforest to serve as a temperature buffer.

INFOGRAPHIC 3   **GLOBAL TERRESTRIAL BIOMES**

What biome do you live in? Identify an area on another continent where you could travel and visit that same biome.

Terrestrial biomes are specific types of terrestrial ecosystems with characteristic temperature and precipitation conditions. Temperature varies with latitude (decreasing as one moves away from the equator) and altitude (decreasing as elevation increases); thus a cold climate can be found above 60° north and south latitudes, as well as on an equatorial mountaintop. Latitude also affects precipitation, with wet areas occurring at the equator and around 60° north and south and dry areas occurring around 30° north and south (due to global air circulation patterns).

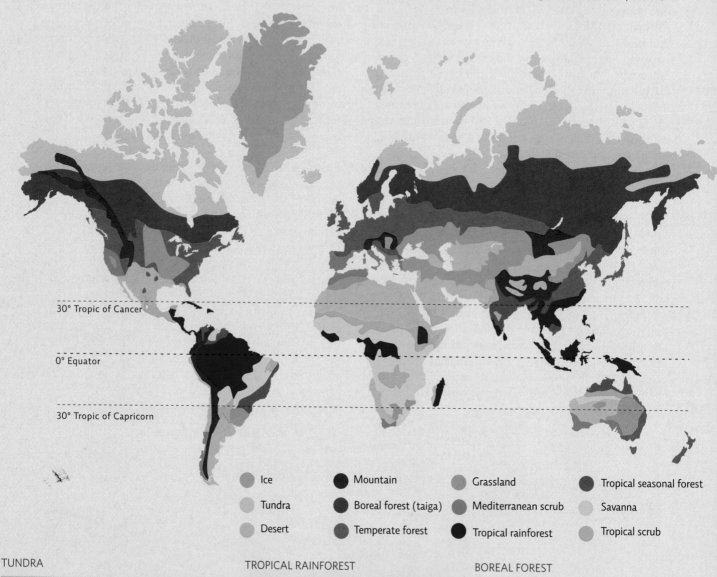

30° Tropic of Cancer

0° Equator

30° Tropic of Capricorn

- Ice
- Tundra
- Desert
- Mountain
- Boreal forest (taiga)
- Temperate forest
- Grassland
- Mediterranean scrub
- Tropical rainforest
- Tropical seasonal forest
- Savanna
- Tropical scrub

TUNDRA

George Burba/Shutterstock

TROPICAL RAINFOREST

Dr. Morley Read/Shutterstock

BOREAL FOREST

IDAK/Shutterstock.com

This biome climograph shows the approximate distribution of terrestrial biomes with regard to annual precipitation and temperature. As precipitation increases and more plant life is able to be supported, deserts, scrublands, or grasslands give way to forests. The temperature also influences what type of desert, grassland, or forest is present (for example, temperate forest versus boreal forest).

 Which biome exists across the widest range of temperatures? Precipitation?

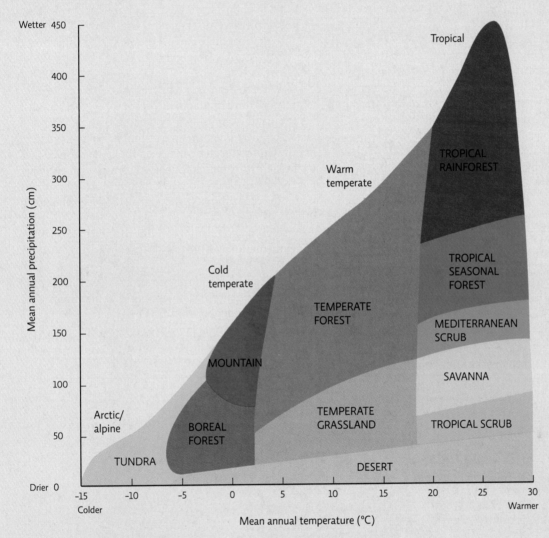

TROPICAL RAINFOREST

TROPICAL SEASONAL FOREST

MEDITERRANEAN SCRUB

SAVANNA

TROPICAL SCRUB

TEMPERATE FOREST

TEMPERATE GRASSLAND

MOUNTAIN

BOREAL FOREST

TUNDRA

DESERT

Tropical

Warm temperate

Cold temperate

Arctic/ alpine

Mean annual precipitation (cm)

Wetter 450
400
350
300
250
200
150
100
50
Drier 0

-15 -10 -5 0 5 10 15 20 25 30
Colder | Warmer

Mean annual temperature (°C)

MEDITERRANEAN SCRUB

DESERT

sborisov/Getty Images

SAVANNA

tonda/Getty Images

Kondrachov Vladimir/Shutterstock

4 RANGE OF TOLERANCE AND ITS IMPACT ON SPECIES DISTRIBUTION

Key Concept 4: Limiting factors determine the distribution and size of populations. Variability within a population increases its range of tolerance, expanding its distribution and increasing the chance it will be able to adapt to changing conditions.

Each biome required a mind-boggling array of considerations—not only how diverse plant and animal species would interact within and across biomes but also the nutrient requirements of each organism they planned to include. Termites, for example, would need enough dead wood at the beginning of closure to sustain them until some of the larger plants began dying off. Termites live in the soil, stirring it and allowing air to penetrate soil particles. If the termites ran out of dead wood and starved to death, organisms living in the soil would not get enough oxygen, and the entire desert would be jeopardized. Hummingbirds, on the other hand, would need nectar-filled flowers. "Try figuring out how many flowers a day a hummingbird needs," says Tony Burgess, a University of Arizona ecologist who helped design the biomes in Biosphere 2 and remained involved until 2004. "From there you need to know what the blooming season is, and then what the nectar load per flower is. And then you have to translate all of that into units of hummingbird support. Now imagine doing that sort of thing about 3,000 times."

In this context, dead wood, oxygen, and flowers are all examples of **limiting factors**—resources so critical that their availability controls the distribution of species and thus of biomes. The principle of limiting factors states that the critical resource in least supply determines the survival, growth, and reproduction of a given species in a given biome. In addition, living things can only survive and reproduce within a certain range (between the upper and lower limits that they can tolerate) for a given critical resource or environmental condition, referred to as their **range of tolerance**. INFOGRAPHIC 4

Ecologists routinely monitor a variety of limiting factors as a way of assessing ecosystem health, but anticipating what each individual organism would need to survive before the fact proved daunting.

limiting factor The critical resource whose supply determines the population size of a given species in a given ecosystem.

range of tolerance The range, within upper and lower limits, of a limiting factor that allows a species to survive and reproduce.

INFOGRAPHIC 4 · RANGE OF TOLERANCE FOR LIFE

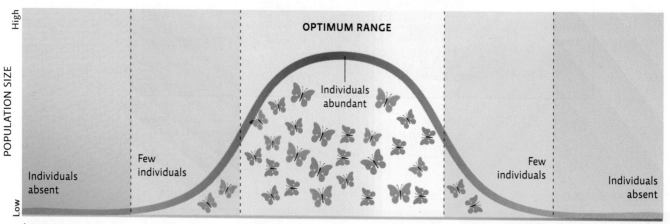

Populations have a range of tolerance for any given environmental factor (such as temperature). Every species has an upper and a lower limit beyond which it cannot survive (in this example, temperatures that are too cool or too hot). Most individuals, like the butterflies in this population, can be found around the optimum temperature, though what is "optimum" for each individual may differ slightly because of genetic variability. Genetic differences that allow some individuals to tolerate or even thrive at the edges of the population's tolerance offer the population a chance to adapt to changing conditions (such as a warmer climate) if needed. The more narrow the range of tolerance and the less genetically diverse the population, the less likely it will survive a change in conditions.

 Which population would have the greatest chance of adapting to an increase or decrease in the average temperature—one with a broad range of tolerance for temperature or one with a narrow range of tolerance?

A dedicated team of scientists—oceanographers, forest ecologists, and plant physiologists—spent 2 years sorting through these challenges. Drawing on their combined expertise, they set about choosing the combination of soils, plants, and animals that seemed most capable of working together to re-create the delicate balances that had made Biosphere 1 such a spectacular success. A summer-dormant desert, like the ones found in Baja, California, was chosen because it would reduce the desert's CO_2 demands when the savanna's productivity was at its highest. The ocean was situated between the desert and rainforest so that it could serve as a temperature buffer between the two. And each biome was created from a carefully selected array of species: Well water mixed with aquarium salt filled the ocean, to which were added coral reef sections culled from the Caribbean. The marsh biome was composed of intact chunks of swampland harvested from the Florida Everglades, and the savanna was composed of grasses from Australia, South America, and Africa. Each soil type contained its own array of **decomposers**—organisms that break down dead or decaying material, returning basic nutrients to the soil for plants to take up.

But it wasn't long before the rigor and pragmatism of good science began to clash with the idealism of Biosphere 2 financiers. And when that happened, critics say, science lost out.

Some scientists worried that the ocean wouldn't get enough sunlight to support plant and animal life. Others opposed the use of soil high in organic matter; soil microbes decompose the soil's organic carbon and release it into the air as CO_2. Although this organic soil might eliminate the need for chemical fertilizers, there were concerns that it would provide too much fuel for the soil microbes and would thus send atmospheric CO_2 concentrations through the roof. Despite these concerns, scientific advisers to the project were overruled.

The first few months of Mission 1 went smoothly enough, but eventually, plants and animals started dying. Humans grew hungry and mysteriously sleepy. And before long, tempers flared.

Like Earth, Biosphere 2 was designed as a materially closed and energetically open system: Plants would conduct photosynthesis fueled by sunlight that streamed through the glass, but no biomass would enter or leave. Temperature, wind, rain, and ocean waves would be controlled mechanically. "But the facility would have to be self-sustaining," says Burgess. "Everything would die, unless the biota met its most fundamental purpose—using energy flow for biomass production." Humans would have to survive exclusively on the plants and animals they could grow and harvest under the dome.

C. Kurt Holter/Shutterstock.com

Decomposers like these mushrooms (a fungus) growing on a decaying log are important consumers that help recycle nutrients in an ecosystem.

The system as a whole would have to continuously recycle every last bit of nutrient that was in the soil on day one.

At first, the carefully constructed agricultural biome seemed well suited to the challenge. Carrots, broccoli, peanuts, kale, lettuce, and sweet potatoes were grown on broad half-acre terraces that sat adjacent to the sprawling six-story human habitat. A bevy of domestic animals also provided sustenance—goats for milk, chickens for eggs, and pigs for pork. Indeed, eating only what they could grow made the biospherians healthier. Bad cholesterol and blood pressure went down; so did white blood cell counts. And slowly, the biospherians say, their relationship to food changed. "Inside, I knew exactly where my food came from, and I totally understood my place in the biosphere and how it impacted the food I ate," says Jayne Poynter, the biospherian responsible for tending the farm. "When I breathed out, my CO_2 fed the sweet potatoes that I was growing. When I first got out, I lost sense of that. I would stand for hours in the aisles of shops, reading all the names on all the food things and think 'where does this stuff come from?' People must have thought I was nuts."

But the glazed glass of the dome admitted less sunlight than had been anticipated. Less sunlight meant less biomass production. And that meant less food. Mites and disease also cut crop production. Biosphere 2's size precluded the use of pesticides and herbicides: In that small atmosphere, the toxins would build up rapidly and could have had a deleterious impact on air quality and human health.

Now, a few months in, food was starting to run out. And that wasn't the only problem.

decomposers Organisms such as bacteria and fungi that break organic matter all the way down to constituent atoms or molecules in a form that plants can take back up.

5 MATTER CYCLES AND SINKS

Key Concept 5: The matter cycles that move nutrients through ecosystems depend on living organisms and abiotic sinks of those resources.

After several months under the dome, the humans grew so tired they couldn't work. Nobody knew why, but scientists on the outside suspected that it had something to do with nutrient cycles.

On Earth, nutrients cycle through both biotic and abiotic components of an ecosystem—organisms, air, land, and water. They are stored in abiotic or biotic parts of the environment called **sinks** (or *reservoirs*). Organisms acquire nutrients from a sink, and those nutrients are then cycled through the food chain and eventually are

sinks Abiotic or biotic components of the environment that serve as storage places for cycling nutrients.

returned to a sink where they linger for various lengths of time, known as *residence times*. For example, carbon has an average residence time of about 5 years in the atmosphere, about 50 years in terrestrial soils, and more than 3,000 years in deep ocean sediments. This means the oceans are an important sink for carbon—a fact that is gaining significance because our actions are adding extra carbon to the atmosphere and this is contributing to climate change. But as more and more carbon is absorbed by the oceans, and as the oceans warm, their ability to absorb carbon declines, an observation that has climate scientists and policy makers worried. (See Module 10.2 for more on climate change.) **INFOGRAPHIC 5**

INFOGRAPHIC 5 | **MATTER CYCLES AND SINKS**

Matter cycles in and out of biotic and abiotic sinks. Different sinks have different residence times. Time spent in biotic components is generally shorter than in abiotic components (water, air, or underground) though fluxes in and out of the atmosphere, and surface waters tend to be quicker than those from soil, rock, or deep ocean areas.

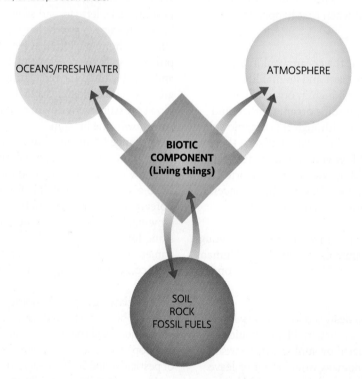

 Why is the residence time for matter such as carbon much longer in the deep ocean than in the atmosphere or soil?

6   THE CARBON CYCLE

Key Concept 6: Carbon cycles through the environment via photosynthesis and cellular respiration. Human actions are increasing the amount of atmospheric carbon, unbalancing this cycle.

For carbon, the atmosphere—where carbon is stored as CO_2—is the most important sink for terrestrial plants. (Oceans and soil are also abiotic sinks for carbon. Oceans absorb CO_2 directly from the atmosphere, and soils accumulate it during decomposition.) Plants and other photosynthesizers use carbon molecules from atmospheric CO_2 to build sugar, and they release oxygen in the process. Because they "produce" sugar (an organic molecule) from inorganic atmospheric CO_2, they are called **producers**.

This sugar molecule represents stored chemical energy that the producer can use. A **consumer**, the organism that eats the plant (or that eats the organism that eats the plant), also uses the chemical energy of sugar. This energy is released to the cell via the process of **cellular respiration**. All organisms—producers and consumers—perform cellular respiration. (For more information on producers, consumers, and the food chain, see Module 2.3.)

From its initial incorporation into living tissue via photosynthesis, to its ultimate return to the atmosphere through respiration, carbon cycles in and out of various molecular forms and in and out of living things as it moves through the **carbon cycle**.

As mentioned above, human activity is changing the amount of carbon in the atmosphere, unbalancing the carbon cycle. When we burn fossil fuels for energy, we add extra CO_2 to the atmosphere—about 25% is absorbed by the oceans and another 25% has been taken up by terrestrial plants (due to a boost in photosynthesis). We also add extra CO_2 to the atmosphere when we burn large swaths of forest, most often to clear land for agriculture. In addition, deforestation has the net effect of increasing atmospheric CO_2—not because we add more but because we remove the largest photosynthetic organisms on the planet that would be pulling CO_2 out of the atmosphere: trees. **INFOGRAPHIC 6**

In Biosphere 2, carbon cycled just like it does on Earth; carbon moved from living tissue to the atmosphere and back in the same predictable manner. But as the biospherians' energy waned, it became clear that something was going wrong.

It turned out that oxygen levels had fallen steadily—from 21% down to 14%. At such low concentrations, the biospherians were unable to convert the food they consumed into usable energy. "We were just dragging ourselves around the place," Poynter says. "And we had sleep apnea at night. So we'd wake up gasping for air because our blood chemistry had changed."

In just a few months, some 7 metric tons of oxygen—enough to keep six people breathing for 6 months—had gone missing. As scientists from Columbia University later discovered, soil microbes were gobbling up all that O_2 and converting it into CO_2 as they decomposed the organic matter in the soil.

The biospherians responded by filling all unused planting areas with morning glory vines, a pretty and fast-growing (but, as it turned out, invasive) species they hoped would maximize the amount of CO_2 converted back into O_2 by photosynthesis. But even with an abundance of plants and enough CO_2, photosynthesis was still limited by the availability of sunlight; even the morning glories couldn't keep up with the soil microbes in their warm, well-watered, highly organic soil.

Adding to the confusion, concrete used to build parts of Biosphere 2 was absorbing some of the CO_2 and converting it into calcium carbonate, trapping some of the carbon and oxygen in this unexpected sink.

producer An organism that converts solar energy to chemical energy via photosynthesis.

consumer An organism that obtains energy and nutrients by feeding on another organism.

cellular respiration The process in which all organisms break down sugar to release its energy, using oxygen and giving off CO_2 as a waste product.

carbon cycle Movement of carbon through biotic and abiotic parts of an ecosystem. Carbon cycles via photosynthesis and cellular respiration as well as in and out of other reservoirs, such as oceans, soil, rock, and atmosphere. It is also released by human actions such as the burning of fossil fuels.

INFOGRAPHIC 6 THE CARBON CYCLE

In photosynthesis, producers use solar energy to combine CO_2 and H_2O to make sugar, releasing O_2 in the process. When producers (or any consumer who eats another organism) need energy, they break apart the sugar via the reverse reaction, cellular respiration. Oxygen is required for this reaction (which is why it is called "respiration").

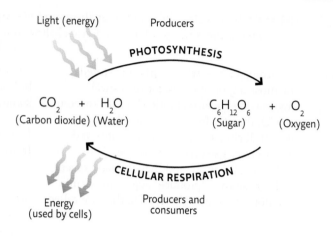

Light (energy)   Producers

PHOTOSYNTHESIS

CO_2 + H_2O $C_6H_{12}O_6$ + O_2
(Carbon dioxide) (Water)   (Sugar)   (Oxygen)

CELLULAR RESPIRATION

Energy   Producers and
(used by cells)   consumers

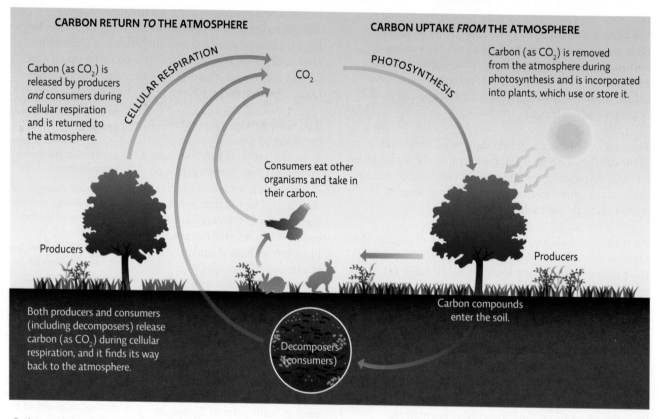

CARBON RETURN *TO* THE ATMOSPHERE

CARBON UPTAKE *FROM* THE ATMOSPHERE

Carbon (as CO_2) is released by producers *and* consumers during cellular respiration and is returned to the atmosphere.

CELLULAR RESPIRATION

CO_2

PHOTOSYNTHESIS

Carbon (as CO_2) is removed from the atmosphere during photosynthesis and is incorporated into plants, which use or store it.

Consumers eat other organisms and take in their carbon.

Producers

Producers

Carbon compounds enter the soil.

Both producers and consumers (including decomposers) release carbon (as CO_2) during cellular respiration, and it finds its way back to the atmosphere.

Decomposers (consumers)

Carbon cycles in and out of living things during photosynthesis and cellular respiration. As consumers (including decomposers) eat other organisms, carbon is transferred. Most of Earth's carbon is actually stored in rocks or dissolved in the planet's oceans, but some carbon is stored in the bodies of organisms and in soil. Without human interference, over the long term, the carbon cycle is balanced between photosynthesis and respiration.

Humans unbalance the carbon cycle via activities that increase the amount of CO_2 in the atmosphere.

Burning fossil fuels   Forest fire   Deforestation

 Explain how each of these human impacts (burning fossil fuels, forest fires, and deforestation) can result in a net increase in atmospheric carbon.

7 THE NITROGEN CYCLE

Key Concept 7: Nitrogen cycles through the environment in steps that depend on a wide variety of bacteria. Human impact is increasing the amount of usable nitrogen in the environment.

Besides carbon, other chemicals essential for life, such as nitrogen and phosphorus, cycle through ecosystems. Nitrogen, the most abundant element in Earth's atmosphere, is needed to make proteins and nucleic acids, but plants cannot utilize nitrogen in its atmospheric form (N_2). All plant life, and ultimately all animal life, too, depends on microbes (bacteria) to convert atmospheric nitrogen into usable forms as part of the **nitrogen cycle**.

In a process called **nitrogen fixation**, bacteria convert atmospheric nitrogen (N_2) into ammonia (NH_3), which plants subsequently take up through their roots; consumers take in nitrogen via their diet. In other steps of the nitrogen cycle, different species of bacteria are responsible for **nitrification**, the conversion of ammonia to nitrate (NO_3^-)—another form that can be directly used by the bacteria or taken up by plants. Still other bacteria convert nitrate back to molecular nitrogen (N_2), which finds its way back into the atmosphere in the process of **denitrification**.

Like the carbon cycle, human impact is also disrupting the nitrogen cycle. In fact, our use of fertilizers and vehicles emissions are adding as much fixed nitrogen to the environment as the natural nitrogen cycle does—a doubling of available nitrogen! This excess means nitrogen is no longer a limiting factor for plant growth, but different plants respond to this in different ways so the makeup of an ecological community can change with ripple effects felt throughout the ecosystem. **INFOGRAPHIC 7**

Thanks to an overabundance of soil microbes, levels of nitrous oxide, a normal by-product of denitrification, got high enough to interfere with vitamin B_{12} metabolism, affecting the Biospherian's nervous systems.

nitrogen cycle A continuous series of natural processes by which nitrogen passes from the air to the soil, to organisms, and then returns back to the air or soil.

nitrogen fixation Conversion of atmospheric nitrogen into a biologically usable form, carried out by bacteria found in soil or via lightning.

nitrification Conversion of ammonia to nitrate (NO_3^-).

denitrification Conversion of nitrate to molecular nitrogen (N_2).

INFOGRAPHIC 7   THE NITROGEN CYCLE

ANIMATED INFOGRAPHIC

Nitrogen, needed by all living things to make biological molecules like protein and DNA, continuously moves in and out of organisms and the atmosphere in a cycle absolutely dependent on a variety of bacteria.

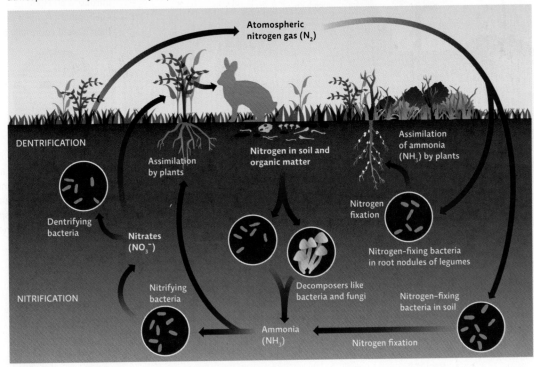

Nitrogen fertilizers promote plant growth, but this depletes other soil nutrients; they can also leach out of soils and pollute aquatic ecosystems.

Burning fossil fuels contributes to nitrogen pollution such as smog and acid rain.

 Look closely at the nitrogen cycle. How many different types of microbes are needed to complete the entire cycle?

8   THE PHOSPHORUS CYCLE

Key Concept 8: Phosphorus moves slowly through the environment, depending on physical and biological processes. Human impact has also unbalanced this cycle.

The phosphorous cycle was also disrupted in Biosphere 2. Unlike nitrogen and carbon, phosphorus—which is needed to make important biological molecules like DNA and RNA—is found only in solid or liquid form on Earth, so the **phosphorus cycle** does not move through the atmosphere but passes from inorganic to organic form through a series of interactions with water and organisms. Like the nitrogen cycle, the phosphorus cycle has been disrupted by the release of more phosphorus into the environment by human activities such as mining and the use of fertilizers. (See Module 8.2 for more on the environmental impacts of the agricultural use of nitrogen and phosphorus.) **INFOGRAPHIC 8**

phosphorus cycle A series of natural processes by which the nutrient phosphorus moves from rock to soil or water, to living organisms, and back to soil.

INFOGRAPHIC 8   **THE PHOSPHORUS CYCLE**

Phosphorus, needed by all organisms to make DNA, cycles very slowly. It has no atmospheric component but instead depends on the weathering of rock to release new supplies of phosphate (PO_4) into bodies of water or the soil, where it dissolves in water and can be taken up by organisms. Microbes also play a role when they break down organic material and release the phosphate to the soil.

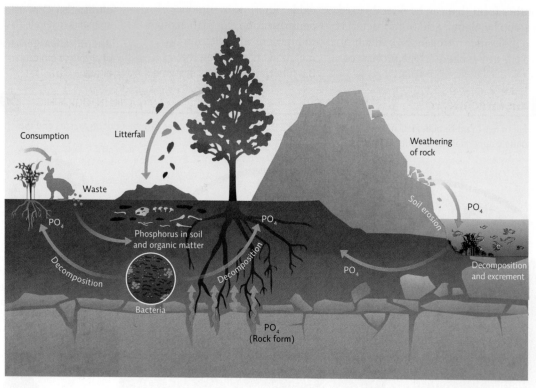

Dust released through mining or in eroded areas can introduce phosphorus into the environment much more quickly than it would normally enter.

Fertilizers and animal waste (including sewage) can alter plant growth and nutrient cycling, especially in aquatic ecosystems where phosphorus is usually a limiting nutrient.

 How might phosphorus from farms enter aquatic ecosystems (rivers, streams, lakes, and oceans)?

In Biosphere 2, phosphorus got trapped in the water system, polluting aquatic habitats, because the underwater and terrestrial plants there were dying off too quickly to complete this cycle. Biospherians removed excess nutrients from their water supply by passing the water over algal mats that would absorb the phosphorus and could then be harvested, dried, and stored.

As food reserves dwindled, the eight biospherians split into two factions. One group felt that scientific research was the top priority and wanted to import food so that they would have enough energy to continue with their experiments. The other group felt that maintaining a truly closed system—one where no biomass was allowed to enter or leave—was the project's most important goal; proving that humans could survive exclusively on what the dome provided would be essential to one day colonizing the Moon or Mars. To them, importing food would amount to a mission failure. "It was a heartbreaking split," Poynter says. "Just 6 months into the mission, and two people on the other side of the divide had been my closest friends going in." Eventually, Poynter snuck in food. That wasn't the only breach. To solve the various nutrient cycle conundrums, the project's engineers installed a CO_2 scrubber and pumped in 17,000 cubic meters (600,000 cubic feet or about 4 million gallons) of oxygen.

Three years after the first mission was completed, the editors of the respected journal *Science* deemed the entire project a failure. "Isolating small pieces of large biomes and juxtaposing them in an artificial enclosure changed their functioning and interactions, rather than creating a small working Earth as originally intended," they wrote. For the $200 million dome to survive as a scientific enterprise, they concluded, it would need dramatic retooling.

Biosphere 2 might not have met its original goal, but it was not a failure as some reports claimed. Negative results can be just as informative as positive ones—in some cases even more so because they uncover gaps in our knowledge and help us decide how to move forward. Biosphere 2 taught scientists that Earth may be far too complex, that ecosystem components intertwine in far too many complicated ways, for humans to re-create. Each is governed by a countless array of interacting factors, and a change in one can set off a whole chain of events that degrade the system's capacity to sustain life.

> Three years after the first mission was completed, the editors of the respected journal *Science* deemed the entire project a failure.

In fact, Biosphere 2's greatest liability—its skyrocketing CO_2 levels—is proving to be a valuable asset. "Now it's like a time machine," says Griffin, who points out that Biosphere 2 is allowing us a look at the consequences of elevated atmospheric CO_2 levels, the main contributor to climate change today. Recent research by Griffin has uncovered some of the complexities of carbon cycling. His group saw unexpected fluctuations in carbon release at various levels in the tree canopy, telling him there is much we still don't know about how carbon cycles—data that could only be gathered in an enclosed forest such as that found in Biosphere 2.

Today, school groups and tourists can visit Biosphere 2, and scientists from all over the world still use the facility to study a variety of research topics. Ultimately, though, the most valuable lesson Biosphere 2 has provided is how irreplaceable Biosphere 1 is.

Select References:

Cohen, J. E., & D. Tilman. (1996). Biosphere 2 and biodiversity: The lessons so far. *Science, 274*(5290), 1150–1151.

Griffin, K. L., et al. (2002). Canopy position affects the temperature response of leaf respiration in *Populus deltoides*. *New Phytologist, 154*(3), 609–619.

Marino, B., et al. (1999). The agricultural biome of Biosphere 2: Structure, composition and function. *Ecological Engineering, 13*(1), 199–234.

Poynter, J. (2009). TED Talk: *Life in Biosphere 2*, www.ted.com/talks/jane_poynter_life_in_biosphere_2?language=en

Justin Peterson, a Biosphere 2 undergraduate intern, assists PhD candidate Henry Adams with his Pinon Pine Tree Drought Experiment. The experiment's goal is to predict the effects of climate shifts on the trees.

INTERACTIVE MAP   BIOMES   ANIMATED INFOGRAPHIC

The different biomes of Earth reflect different climatic and physical conditions of the places where they are found. These case studies present more details about the physical and biological makeup of four very different biomes.

SIBERIAN TUNDRA

DEATH VALLEY
DESERT

AMAZONIAN
TROPICAL
RAINFOREST

SERENGETI
GRASSLAND

 BRING IT HOME

PERSONAL CHOICES THAT HELP

Nutrient cycling is critical for maintaining Earth's ecosystems, but we interfere with nutrient cycles through our daily activities. Driving a car interferes with the carbon cycle by releasing carbon from fossil fuel reservoirs. Applying synthetic chemical fertilizers to food crops interferes with the nitrogen cycle by adding soluble nitrogen compounds to aquatic ecosystems through runoff. The challenge is to figure out ways to work with nutrient cycles rather than against them—in other words, to return nutrients to the sinks from which they come. How might this be done in our daily lives and in our own communities?

Individual Steps

• Reduce your fossil fuel use to curtail carbon, nitrogen, and sulfur emissions. Take public transportation, walk, ride a bike, and drive a fuel-efficient vehicle.

• Compost food and yard waste. Then use this material to fertilize flowerbeds, trees, and garden plots. Composting will reduce or eliminate the need for inorganic fertilizers in your yard.

Group Action

• Participate in or organize an event to plant trees or native grasses. By doing so, you can help recapture the carbon put into the atmosphere by driving a car.

• Many urban areas are grateful for individuals being willing to plant native trees, shrubs, and wildflowers along roadways or in parks and other public spaces.

Policy Change

• Public policy currently prevents large-scale composting of municipal wastes in most areas. Working to change these policies will extend the life of our landfills and make use of valuable nutrient-rich materials.

• Support legislation to increase fuel efficiency of vehicles and subsidies for clean, renewable energy. More fuel-efficient cars and cleaner energy sources will reduce our carbon outputs from fossil fuel use.

Peter Essick/Aurora Photos

ENVIRONMENTAL LITERACY   **UNDERSTANDING THE ISSUE**

1 What is the hierarchy of organization recognized by ecologists, and why might it be useful to recognize such distinctions?

1. Which of the following is the correct hierarchy of life?
   a. Individual, community, biome, ecosystem, population, biosphere
   b. Individual, population, community, ecosystem, biome, biosphere
   c. Biosphere, community, biome, ecosystem, population, individual
   d. Biome, biosphere, population, ecosystem, community, individual

2. Why do ecologists focus mainly on the study of populations, communities, and ecosystems?

2 Why do ecosystems need a constant input of energy yet can handle the fact that they do not receive appreciable new inputs of matter?

3. True or False: Earth is energetically open but a closed system with regard to matter.

4. Why is it essential that sustainable ecosystems rely on an energy source that is readily replenished, like sunlight, rather than nonrenewable sources, such as fossil fuels?

3 How do environmental factors affect the distribution and makeup of biomes?

5. Which biome description is correct?
   a. Grasslands receive less rainfall than forests but more than deserts.
   b. Forests have freezing temperatures regularly.
   c. Grasslands are much warmer annually than forests.
   d. Deserts are hot year round—much hotter than forests.

6. Look at the location of temperate deciduous forests on Earth in the biome climograph in Infographic 3. What can you surmise about the climatic conditions in North America and Europe/Asia where these forests are found?

4 What is a population's range of tolerance, and how does it affect the distribution of a population within its ecosystem or its ability to adapt to changing conditions?

7. The population that would have the best chance of surviving an environmental change would be the one with the:
   a. narrowest range of tolerance for temperature.
   b. greatest genetic diversity.
   c. largest population size.
   d. least variation among individuals.

8. Using the example of spring wildflowers and the critical factor of rainfall, explain the term *environmental gradient* (for rainfall) and the term *range of tolerance* (in terms of the distribution of wildflowers within their range).

5 What role do biotic and abiotic factors play in matter cycles?

9. Which of these is an abiotic reservoir of matter?
   a. Trees
   b. Leaf litter
   c. The ocean
   d. Soil bacteria

10. Using the example of carbon, explain the concept of residence time for matter sinks.

6 How does carbon cycle through ecosystems, how is this cycle disrupted, and what problems can this disruption cause?

11. What is the purpose of photosynthesis?
    a. It produces a form of chemical energy that the plant can use as needed.
    b. It allows plants to remove excess CO_2 from the atmosphere.
    c. Plants do it to produce oxygen so animals (and humans) can breathe.
    d. It is a way for producers and consumers to release the energy stored in sugar molecules

12. Who performs cellular respiration?
    a. Producers
    b. Consumers
    c. Both producers and consumers.

13. Identify three ways that humans are unbalancing the carbon cycle.

7 How does nitrogen cycle through ecosystems, how is this cycle being disrupted, and what problems can this disruption cause?

14. True or False: The use of fertilizers in agriculture is unbalancing the nitrogen cycle.

15. The process of converting atmospheric nitrogen (N_2) to a form that organisms can use is known as:
    a. denitrification.
    b. decomposition.
    c. nitrogen assimilation.
    d. nitrogen fixation.

16. What effect would a wildfire that burned so hot that it sterilized the soil, killing all the microbes, have on the nitrogen cycle?

8 How does phosphorus cycle through ecosystems, how is this cycle being disrupted, and what problems can this disruption cause?

17. Two activities that are disrupting the phosphorus cycle are _____ and _____.

18. What is different about the phosphorus cycle as compared to the carbon or nitrogen cycle?
    a. It is the only one without a bacterial component.
    b. Unlike the other two, phosphorus does not cycle in and out of living organisms.
    c. It is the only one without an atmospheric component.
    d. It is much faster than the other two cycles.

19. What steps did the Biospherians take to address the imbalance in the phosphorus cycle in Biosphere 2? Why was this an effective solution?

SCIENCE LITERACY   **WORKING WITH DATA**

An experiment was done at Biosphere 2 to test the effect of drought and higher temperatures on pinyon pines. Estimates of photosynthetic activity and cellular respiration were made by measuring CO_2 uptake or output (measured in terms of μmoles of carbon exchange per meter square of leaf area per second).

Look at Graphs A and B and answer the following questions

A: MIDDAY NET PHOTOSYNTHESIS

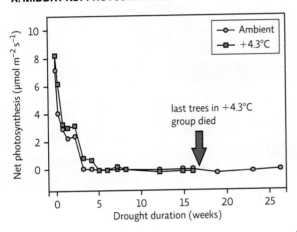

B: PREDAWN CELLULAR RESPIRATION

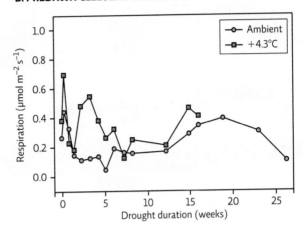

Interpretation

1. Look at Graph A. What is the general trend for photosynthetic activity over the course of the experiment for both the ambient temperature and the elevated groups?

2. Look at Graph B. What is the general trend for cellular respiration activity over the course of the experiment for both the ambient temperature and the elevated groups?

Advance Your Thinking

3. Why did cellular respiration also fall off toward the end of the experiment? (Hint: Link this back to the tree's photosynthetic activity.)

4. What can you conclude about the cause of death of the trees?

5. The researchers also measured the ability of the trees to take up and distribute water to their tissues in all the test trees and saw no difference—no breakdown in the ability to do this in the drought-stricken trees at either test temperature. Why did they collect this data?

INFORMATION LITERACY   **EVALUATING INFORMATION**

Earth is vastly larger than Biosphere 2 but its resources are just as finite. People must decide, for each biome, whether to leave some places untouched or use all of an area and its resources for human purposes. One example would be plowing under an entire prairie and growing wheat, displacing all of the native plants and animals. The rainforests are another such biome. Left alone, they produce huge amounts of oxygen for the whole planet and also support millions of species. Many humans want to use the trees for lumber and the land to grow crops or raise cattle.

Can we use rainforests in sustainable ways? Go to the Rainforest Alliance's website (www.rainforest-alliance.org) and read about sustainable agriculture and Rainforest Alliance–certified products such as coffee, tea, and cocoa.

Evaluate the website and work with the information to answer the following questions:

1. Evaluate the reliability of this information source:
   a. Does the organization give supporting evidence for its claims?
   b. Does it give sources for its evidence?
   c. What is the mission of this organization?

2. Note the range of publications available on the Rainforest Alliance website (www.rainforest-alliance.org/publications) and view the white paper entitled Cocoa Certification (search for "cocoa certification" using the "search publications" option).
   a. Do you agree that certification is an adequate tool to ensure sustainability in cocoa farming?
   b. Identify a claim the organization makes and the evidence it gives in support of this claim. Is it convincing?

3. The Global Canopy Programme (www.globalcanopy.org) is a group based in the United Kingdom that has joined with the Carbon Disclosure Project to ask companies to disclose their "forest footprint" (www.cdp.net/en). Should companies be required to participate?
   a. Should some intergovernmental group investigate companies and estimate impacts?
   b. Is it important to allow a "free market" for goods and services, assuming that businesses will take care of the lands they use but do not own?

 Additional study questions are available at SaplingLearning.com.

GRAY WOLVES RETURN TO YELLOWSTONE

Endangered gray wolves return to the American West

Biologists track wolves from a helicopter. Once they spot a wolf, they dart it with a tranquilizer so that they can fit it with a radio collar. The wolf is unharmed. BARRETT HEDGES/National Geographic Creative

After reading this chapter and studying the KEY CONCEPTS and INFOGRAPHICS, you should be able to answer these GUIDING QUESTIONS

CORE MESSAGE

Ecologists study populations to better understand what makes them survive and thrive. The size, distribution, and growth rate of populations is influenced by a variety of factors, such as the availability of resources and the presence of other species, like predators. Determining the factors that affect a given population is an important part of managing it, especially for populations that are endangered.

1 What is a population and why do ecologists study them?

2 What population distributions are seen in nature?

3 What is the importance of population size and density?

4 What is exponential growth and when does it occur in a population?

5 What is logistic growth and when does it occur in a population?

6 How do density-dependent and density-independent factors affect population growth?

7 What are the life-history strategies of *r*- and *K*-selected species, and how do they relate to population growth patterns and their ability to respond to environmental changes?

8 What are top-down and bottom-up regulation, and which is most important in determining the size of a population?

At least a half a dozen times each winter, Doug Smith climbs into a helicopter, gun in tow, and hunts wolves (*Canis lupus*) in Yellowstone National Park. He's not looking to kill them—just to put them to sleep for a little while so that he can outfit them with radio collars to track the sizes of their packs, what they eat, and where they go over the course of the following year. Smith, a biologist, spends about 200 hours per year on these "hunting" expeditions, which are part of the Yellowstone Gray Wolf Restoration Project that he leads. The project has been responsible for reintroducing a total of 41 wolves to Yellowstone since 1995, after their disappearance as a result of predator control programs implemented by the U.S. government in the early 20ᵗʰ century.

Sometimes, though, things go awry on Smith's radio collar missions. For instance, the tranquilizer dart doesn't fully sedate big wolves—some of which can reach 80 kilograms (175 pounds). Smith is forced to approach the animals while they're still awake. "I have to grab them on the scruff of the neck and manhandle them until I get the collar on," he explains. "They're typically not dangerous then—they've had enough drug to be kind of out of it—but they're still able to walk around. It's a wild experience." Sometimes the wolves—who are typically frightened of the helicopter—try to attack it while it's hovering with the doors open, just out of their reach. "I've had two females turn and run at the helicopter, teeth gnashing, jumping up trying to get me," Smith recalls. "I'm hovering above it, going back and forth, and I can't get a shot because all I'm seeing is face and teeth." Despite these adventures, Smith says a wolf has never actually bitten him—and he has tranquilized and collared more than 300 of them.

⊙ WHERE IS YELLOWSTONE?

MONTANA

YELLOWSTONE NATIONAL PARK

WY

IDAHO

WYOMING

---

1 POPULATION DYNAMICS

Key Concept 1: Ecologists study populations to better understand what makes them thrive, decline, or become overpopulated, in an effort to manage them and the other populations they impact.

Ecologists monitor **populations** of organisms in ecosystems around the world for a variety of reasons, whether to protect endangered species, to manage economically valuable species such as commercial fisheries or timber, or to control pests. (See Module 2.1 for an introduction to the concept of an ecological population.) In Yellowstone, Smith and others monitor wolf populations. Elk (*Cervus elaphus*) populations, a popular game species, are also closely monitored, as are the populations of aspen, willow, and cottonwood trees, the most important winter food for elk. In some cases, elk numbers and aspen growth are closely tied to the size of the wolf population—but not in all areas. It turns out that many factors play a role in the sizes of these populations.

population All the individuals of a species that live in the same geographic area and are able to interact and interbreed.

Doug Smith and Yellowstone Delta wolf 487M, the largest wolf collared in the winter of 2005.

The geographic area where a given population is found is its **range**. The historic range of the wolf encompassed most of North America, but human impact has reduced that range tremendously. Smith and his colleagues monitor the range of the Yellowstone wolves to understand how well the population is doing. As the wolf population grew and new packs formed, the wolves expanded their range into other areas of the Northern Rockies outside of the park boundaries. If their range begins to shrink, this might be a warning that habitats are declining in quality or that other pressures are restricting the movement of wolves. **INFOGRAPHIC 1**

range The geographic area where a species or one of its populations can be found.

2 POPULATION DISTRIBUTIONS

Key Concept 2: A population's distribution within its range is influenced by behavioral and ecological factors and is a reflection of how individuals interact with each other and their environment.

population distribution The location and spacing of individuals within their range.

clumped distribution A distribution in which individuals are found in groups or patches within the habitat.

In addition to the geographic range occupied by a population, another important feature is **population distribution**, or the location and spacing of individuals within their range. A number of factors affect distribution, including species characteristics, topography, and habitat makeup. Ecologists typically speak of three types of distribution. In a **clumped distribution**, individuals are found in groups or patches within the habitat. Yellowstone examples include social species like the prairie dog or beaver that are clustered around a necessary resource, like water. Wolves travel in packs and therefore have a clumped distribution. Elk, one of their prey species, congregate as well; living in herds offers some protection against the wolves.

INFOGRAPHIC 1   **GEOGRAPHIC RANGE OF THE GRAY WOLF**

The range of a species represents its geographic distribution. Distinct populations may exist within the range, especially if the species is broadly distributed. A small population of the Mexican wolf, a subspecies of the gray wolf that once ranged widely from central Mexico throughout the Southwest, is being restored to Arizona and New Mexico in a reintroduction program similar to the one that brought wolves back to the Northern Rockies.

The wolves in the Northern Rockies, the Great Lakes region, and in the Southwest represent distinct populations that rarely, if ever, mix

CANADA

U.S.A.

Wolf distribution

■ Current range   ■ Historic range

 What are some factors that might expand the range of the Mexican wolf? What might cause it to shrink?

In a **random distribution**, individuals are spread out in the environment irregularly, with no discernible pattern. Random distributions are sometimes seen in homogeneous environments, in part because no particular spot is considered better than another. Species that rely on wind and water to disperse their offspring—like wind-blown seeds or the free-floating larvae of coral—also often have a random distribution. **Uniform distributions**, rare in nature, include individuals that are spaced evenly, perhaps due to territorial behavior or mechanisms for suppressing growth of nearby individuals (seen in some plant species). **INFOGRAPHIC 2**

If a population ecologist understands the normal distribution pattern for the species he or she is studying, he or she can better track the species and determine if a problem is emerging. For example, one would not expect to find wolves everywhere in Yellowstone—they will be in areas with elk or other prey species. However, if a wildflower that normally has a random distribution over

random distribution A distribution in which individuals are spread out over the environment irregularly, with no discernible pattern.

uniform distribution A distribution in which individuals are spaced evenly, perhaps due to territorial behavior or mechanisms for suppressing the growth of nearby individuals.

INFOGRAPHIC 2 **POPULATION DISTRIBUTION PATTERNS**

The distribution of individuals in populations varies from species to species and is influenced by biotic and/or abiotic factors.

CLUMPED

Elk stay in herds, which offers some protection against predators.

RANDOM

The seeds of many Yellowstone flower species are distributed randomly and germinate where they fall.

UNIFORM

The creosote bush of the desert Southwest produces toxins that prevent other bushes from growing close by.

 In which population distribution pattern would individuals within the population experience the most competition with other individuals in their population? Why, then, is this distribution pattern ever seen?

a sunny hillside shows a clumped distribution one season, this might be an indication that something is causing that species to be lost in areas where it should be found—perhaps pollution has contaminated the area or a new predator or competitor has moved in. Knowing what to expect allows a scientist to look for the cause if a population's distribution changes. In addition, an understanding of a species' distribution within its range provides information about its habitat needs, which can be used when designing wildlife preserves or habitat corridors that allow animals to move from one region to another.

3 POPULATION SIZE AND DENSITY

Key Concept 3a: Populations require minimal sizes and densities to reproduce successfully and maintain social ties, but high population density can lead to problems such as disease and overuse of resources.

Key Concept 3b: By tracking population size and density, scientists can be alerted to problems that are causing a decline or be prepared to address or avert overpopulation issues.

Understanding how any population interacts with biotic and abiotic forces in its environment, through programs like the Wolf Restoration Project, is key to preventing species from disappearing from their ecosystem forever. That's because many factors influence the livelihood of a species and its ability to survive and reproduce. For wolves the main threat was humans.

In the early 20th century, humans set out to exterminate wolves from the American West. (They were already gone from most areas east of the Mississippi River.) As part of a government-sponsored program, Congress allocated $125,000 for the Predator Control Program, which employed poison—and later hunting—to eradicate predators and rodents that might harm crops or livestock. Wolves were one of the targeted species because wolves preyed upon livestock. Wolf eradication was also supported because it boosted deer and elk populations for hunting. "It was park policy to kill all predators, and wolves were their biggest objective," Smith explains. Between 1914 and 1926, at least 136 wolves were killed in Yellowstone. In 1944, the last known wolf in the Yellowstone area was killed.

Humans not only threatened wolves by hunting them but also destroyed the animals' natural habitat (and that of their prey) when they cut down forests to build farms and ranches. They starved the animals by hunting elk, deer, and bison, wolves' usual sources of food. At the time, people didn't think that this combination of changes would very nearly cause wolves to go extinct or result in exploding elk and deer populations. But now that scientists like Smith have spent years watching how the animals live, they have a much better understanding of what needs to be done to keep them alive.

Before humans started killing wolves in the early 1900's, the exact numbers of gray wolves living in Yellowstone were unknown, but estimates range from 300 to 400. Elk populations at that time hovered around 10,000 animals. The size of a population in a given geographic area is determined by the interplay of factors that simultaneously increase the number of individuals in a population (births and immigration) and those that decrease numbers (deaths and emigration). An understanding of these factors helps us predict population size at any given time. Ecologists who study changes in population size and makeup (e.g., the average age of the individuals in the population or the proportion of males to females) are studying its **population dynamics**. They find that the population size of some

species increases and decreases rather predictably (barring a catastrophic event), while others tend to fluctuate more randomly, affected by a variety of factors.

Every population has a **minimum viable population**, or the smallest number of individuals that would still allow a population to be able to persist or grow, ensuring long-term survival. This is an important concept when considering how to conserve endangered or threatened species. A population that is too small may fail to recover for a variety of reasons. For example, some species' courtship rituals require a minimum number of individuals for success. Other activities that depend on numbers—like flocking, schooling, and foraging—fail below certain population sizes. Genetic diversity (inherited variety between individuals in a population) is also important: A population with little genetic diversity is less able to adapt to changes and is therefore more vulnerable to environmental change. A small population is also subject to inbreeding, which allows harmful genetic traits to spread and weaken the population. Conversely, a population that has exceeded **carrying capacity**—the population size that a particular environment can support indefinitely without long-term damage to the environment—also causes problems.

Another important metric is **population density**—the number of individuals per unit area. Population density is an important feature that varies enormously among species, or even among populations of the same species in different ecosystems. Similar to problems encountered in a small population, if a population's density is too low, individuals may have difficulty finding mates, or the only potential mates may be closely related individuals, which can lead to inbreeding, loss of genetic variability, and, potentially, extinction. Density that is too high can also cause problems, such as increased competition, fighting, and spread of disease. Deer, elk, and moose populations in the United States, whose density has increased in recent years because of exploding numbers combined with shrinking habitats, now frequently suffer from an infectious disease known as chronic wasting disease.

INFOGRAPHIC 3

population dynamics
Changes over time in population size and composition.

minimum viable population
The smallest number of individuals that would still allow a population to be able to persist or grow, ensuring long-term survival.

carrying capacity (K) The maximum population size that a particular environment can support indefinitely.

population density The number of individuals per unit area.

INFOGRAPHIC 3 **POPULATION SIZE AND DENSITY INFLUENCE LONG-TERM POPULATION SUCCESS**

Population density influences how well a population thrives. When population density is too low, a population may suffer social or reproductive problems; if the size falls below the minimum viable population size for that species, the long-term viability of the population is in doubt. But when populations grow too large for an area, problems associated with high densities may also negatively impact the population.

PROBLEMS IF POPULATION SIZE AND DENSITY IS:

TOO LOW (below minimum population size)

Normal social behaviors are deficient (e.g., group foraging or defense).

Unable to find mates.

Normal courtship and mating behaviors don't occur.

Genetic diversity falls (inbreeding).

Important community connections may be lost, affecting other species.

TOO HIGH (above carrying capacity)

Social behaviors break down with overcrowding.

Spread of disease increases.

Food supplies are insufficient.

Increased chance of conflict with humans.

Damage to environment from overuse of resources.

 A major problem many species today face is a shrinking habitat due to habitat destruction and fragmentation by human actions. How would a shrinking habitat affect the density of a population and what problems might this lead to for that population?

4 EXPONENTIAL POPULATION GROWTH

Key Concept 4: Growth and resistance factors influence population growth. Exponential growth occurs when population growth is unrestricted; however, it will not continue indefinitely.

Critics began raising concerns about the Predator Control Program in the 1920's, recognizing that wolves played a key role in their ecosystem as a top predator. Without the wolf, elk and deer populations could grow unchecked, leading to a damaged and overgrazed habitat.

Scientists can monitor elk populations to determine whether they are growing, and how fast, using some simple mathematical models that describe population growth over time. The annual **population growth rate** is determined by *birth rate* (for example, the number of births per 1,000 individuals per year) minus the population *death rate* (for example, number of deaths per 1,000 individuals per year).

population growth rate The change in population size over time that takes into account the number of births and deaths as well as immigration and emigration numbers.

Immigration and emigration rates might also be factored into the equation, if appropriate.

Population growth rates can also be determined by taking a simple census of population size at the two time points in question. For example, in 1920, the elk population was 10,000; the next year, it was 10,500, an increase of 500 animals. Using the simple equation that divides the change in population size by the original population size, we find that this population increased by 5% (500/10,000 = 0.05).

Assuming growth remains constant, we can predict future population size using the growth rate calculated. For example, in 1932, the elk population was 16,000. If it still had an annual growth rate of 5%, we would predict that the population would be 16,800 in 1933 (16,000 × 0.05 = 800).

Lena Lir/Shutterstock

Rodents have a high biotic potential. A population that moves into a habitat with a good source of food and few predators, like a barn without a resident cat or rat snake, can quickly multiply.

Growth rates can also be negative, reflecting a shrinking population. Between 2009 and 2010, the elk population decreased from 6,070 to 4,635, giving this population a growth rate of -24% (4,635 − 6,070 = −1,435; −1,435/6,070 = −0.24).

Population growth is dependent on the presence of **growth factors** (resources individuals need to survive or reproduce). Conversely, **resistance factors** (things that reduce population size by directly or indirectly killing individuals or prompting emigration), such as predators, competitors, diseases, or pollution, will decrease population size. When there are no environmental limits to survival or reproduction, a population will reach its maximum per capita rate of increase (r), called its **biotic potential**. This occurs, theoretically, when every female reproduces to her maximal potential and every offspring survives. A population increasing in this manner will quickly grow to fill its environment. This period of growth, which can't go on indefinitely, is referred to as **exponential growth**, named for the mathematical function it represents.

Populations that have a high biotic potential have high *fecundity* (females typically produce lots of offspring, reach reproductive maturity quickly, and produce many "clutches" per year). The higher the biotic potential, the faster the population of a given species will grow under ideal conditions. Yellowstone species such as deer mice and the problematic non-native weed known as spotted knapweed have higher biotic potential than species such as grizzly bears and spruce trees.

In nature, exponential growth is typically seen when a species first enters a new environment or when there is an influx of new resources. The population must have a high birthrate—most individuals must have access to enough food, water, and habitat in which to reproduce—and a low death rate. The loss of predators can also lead to exponential growth among their prey species. For instance, without wolves to thin their ranks, elk numbers in Yellowstone doubled between

growth factors Resources individuals need to survive and reproduce that allow a population to grow in number.

resistance factors Things that directly (predators, disease) or indirectly (competitors) reduce population size.

biotic potential (r) The maximum rate at which the population can grow due to births if each member of the population survives and reproduces.

exponential growth The kind of growth in which a population becomes progressively larger each breeding cycle; produces a J curve when plotted over time.

1914 and 1932, after the Predator Control Program had been implemented. This led to the need to cull the herd (in spite of increased hunting pressure), a standard practice for many years that reduced the herd to around 6,000 in 1968. When this practice ended, elk population sizes climbed to a high of around 19,000 in the 1980's before wolves were reintroduced.

Population growth is often evaluated in terms of its *doubling time*: the time it takes for the population to double in size if the growth rate is constant. For a population growing exponentially, a simple mathematical equation can estimate doubling time: 70/r = doubling time in years, where r is the growth rate, entered as a whole number (e.g., if the growth rate is 2%, r = 2).

A population that is growing exponentially will have a *J-shaped curve* if plotted on a graph with time on the x-axis and population size on the y-axis. The J curve shows a slight lag at first and then a rapid increase. This is due to the fact that the larger the population, the faster it grows, even at the same growth rate. Think of it this way: Doubling a small number yields a number that is still small. Doubling a large number, on the other hand, produces a very large number. **INFOGRAPHIC 4**

As an example of how profound exponential growth can be, imagine if someone offered to give you a penny one day and then, each subsequent day for a month, doubled the amount given to you the previous day. On day 1, you would have 1 cent; on day 2, you would be given 2 cents; on day 3, you'd be given 4 cents, and on day 4, you would be given 8 cents. On day 31, you would receive almost $11 million, giving you a 31-day total of more than $21 million. Exponential growth can create large populations quickly.

INFOGRAPHIC 4   **EXPONENTIAL GROWTH OCCURS WHEN THERE ARE NO LIMITS TO GROWTH**

Because deer mice have a high biotic potential, even a single pair could produce thousands of descendants in their lifetime.

BIOTIC POTENTIAL OF DEER MICE

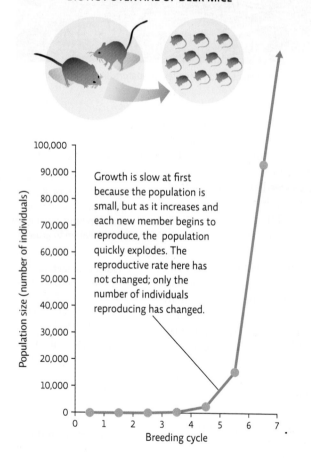

Growth is slow at first because the population is small, but as it increases and each new member begins to reproduce, the population quickly explodes. The reproductive rate here has not changed; only the number of individuals reproducing has changed.

REPRODUCTIVE RATE

Assume each pair produces 10 pups/litter, none of the pups die, and each adult female continues to reproduce. The per capita rate of increase would be:

$$r = \frac{\text{births} - \text{deaths}}{\text{original population size}}$$

For this example:

$$r = \frac{10 - 0}{2} = 5$$

This represents a 500% growth rate!

The generation time of a deer mouse is about 10 weeks (3 weeks gestation, 7 weeks to reach reproductive age). That means this single pair could produce more than 9,000 descendants in just 60 weeks (15 months)!

DOUBLING TIME

The doubling time of a population experiencing exponential growth can be calculated using the "Rule of 70": Divide 70 by the per capita growth rate (r) over a period of time such as per year or per breeding cycle. This equation is most useful when the population growth rate is small (but constant). A population increasing at 3.5% per generation would double in 20 generations.

$$\frac{70}{3.5} = 20$$

 Cockroaches thrive in some homes and apartments. What could trigger exponential growth in a cockroach population? What could prevent it?

5 LOGISTIC POPULATION GROWTH

Key Concept 5: As a population's size approaches carrying capacity, exponential growth may transition to logistic growth, slowing population growth rates.

Exponential growth can't last forever. As a population begins to fill its environment, resistance factors begin to slow its growth rate. Resources become scarce as more individuals use the available resources. This can lead to starvation for some; crowding may bring about an increase in disease and aggression for others. Predation pressure may also increase as the more numerous prey are easier to track and capture or simply because the predator population itself has increased. All of these stressors can increase death rates.

Birth rates may also decline. Underfed individuals may not be able to successfully reproduce; others may be unable to find a suitable area for rearing young in an overcrowded habitat. This kind of growth—in which as population size increases, growth rate decreases—is called **logistic growth**. A population that grows logistically will produce an *S-shaped curve* if plotted on a graph with time on the *x*-axis and population size

on the *y*-axis. The S is created by initial exponential growth that produces the J-shaped curve, followed by decelerating growth as the species approaches its carrying capacity (signified as *K* in population mathematical models)—its maximum sustainable population size, where it levels off.

Carrying capacity is determined by the presence of growth factors and varies between species; the same environment can support many more elk than wolves, for example. Over time, a population's carrying capacity can change. If resources are diminished at a faster rate than they are replenished, the carrying capacity will drop. If, on the other hand, new resources are added or become available, perhaps due to the loss of a competitor, the carrying capacity for a given species will rise. **INFOGRAPHIC 5**

> **logistic growth** The kind of growth in which population size increases rapidly at first but then slows down as the population becomes larger; produces an S-shaped curve when plotted over time.

INFOGRAPHIC 5   LOGISTIC POPULATION GROWTH   ANIMATED INFOGRAPHIC

Exponential growth turns into logistic growth (S curve) as population size approaches carrying capacity (*K*) and resistance factors begin to limit survival.

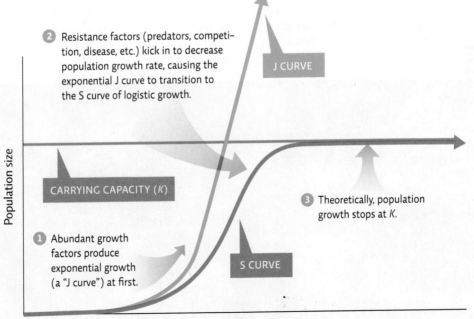

① Abundant growth factors produce exponential growth (a "J curve") at first.

② Resistance factors (predators, competition, disease, etc.) kick in to decrease population growth rate, causing the exponential J curve to transition to the S curve of logistic growth.

③ Theoretically, population growth stops at *K*.

CARRYING CAPACITY (*K*)

J CURVE

S CURVE

Population size (y-axis) / Time (x-axis)

 What could increase or decrease the carrying capacity for a particular population in an environment?

6 DENSITY-DEPENDENT AND DENSITY-INDEPENDENT GROWTH FACTORS

Key Concept 6a: The larger and more dense a population is, the greater the impact of density-dependent factors such as predators, competitors, or disease.

Key Concept 6b: Density-independent factors such as fire or flood will have the same effect on large and small populations alike.

Populations will grow as long as growth factors are available, but as the population gets larger, resources start to decline. The *limiting factor*, the resource that is most scarce, tends to determine carrying capacity. The effects that predators (a resistance factor) have on populations can vary widely, in part because predators are **density-dependent factors**—their effects on a prey's population go up as the size of that prey population goes up. Density-dependent factors are usually biotic—imposed by other species. In the case of wolves preying on elk, the denser an elk population is, the easier it is for wolves to successfully find and take a member of that population. In the same way, elk are density-dependent factors on the ability of young aspens to grow after sprouting (aspen is an important winter food for elk; they eat the young shoots). Competition for scarce resources among members of a population, or between members of two different populations vying for the same resource, also increases as the density of one or both populations increases. For the elk, this means less food for each member of the population when densities are high or when competing mule deer are present. Smith also found evidence that competition impacts the wolf population in a density-dependent fashion. As the size of wolf packs in Yellowstone increased, so did the incidence of wolf-on-wolf aggression, contributing to a subsequent decline in the wolf population. Disease, too, spreads more easily in larger populations with higher densities.

On the other hand, some factors (usually abiotic rather than biotic factors) affect a population no matter how large or small it is, such as droughts, storms, and fire. Human impact such as toxic pollution could also affect members of a population regardless of its size. These **density-independent factors** don't necessarily regulate population size in a predictable fashion, but they can increase or decrease it. For example, a population of elk trapped in a flood or a mudslide after an intense rain event will die regardless of how many individuals there are in the affected population.

Determining how these factors contribute to a population's increase or decrease gives scientists tools they can use to monitor and manage populations. For example, an elk herd that is decreasing in number would lead a manager to investigate potential causes such as greater predation or spread of a disease. An elk herd that is increasing in size might signal a loss of predators and lead to decisions to help control that population to avoid problems with overpopulation of the elk. **INFOGRAPHIC 6**

density-dependent factors Factors, such as predation or disease, whose impact on a population increases as population size goes up.

density-independent factors Factors, such as a storm or an avalanche, whose impact on a population is not related to population size.

Wild Horizon/UIG via Getty Images

The bare trunks of these aspen trees show the winter browse line from herbivores such as elk and deer, which eat leaves as high as they can reach during winter. As forage becomes more and more scarce, these animals will even strip off the bark, damaging the trees.

INFOGRAPHIC 6 · DENSITY-DEPENDENT AND DENSITY-INDEPENDENT FACTORS AFFECT POPULATION SIZE

Density-dependent factors exert more of an effect as population size increases. On the other hand, density-independent factors have the same effect regardless of population size.

DENSITY DEPENDENT

COMPETITION

Mule deer compete with elk for food; the larger the deer population, the greater the competition.

DISEASE

Infectious diseases, such as chronic wasting disease, which weakens and eventually kills the animal, spread more easily in large populations of elk, deer, or moose.

PREDATORS

Predators are more successful at capturing prey from larger populations than from smaller ones.

DENSITY INDEPENDENT

FIRE/FLOOD

Forest fire.

STORMS

Row of trees flooded by a river.

AVALANCHE AND OTHER NATURAL DISASTERS

Snow avalanche.

 Identify the following as either density-dependent or density-independent factors for an elk population: a tick infestation, building a dam that floods a valley in elk habitat, drought, bison, and a blizzard.

7 LIFE-HISTORY STRATEGIES: *r*– AND *K*-SPECIES

Key Concept 7: The population size of *r*-selected species can increase or decrease quickly if the environment changes. *K*-selected species' populations don't fluctuate as widely but they are less able to respond to environment changes.

life-history strategies
Biological characteristics of a species (for example, life span, fecundity, maturity rate) that influence how quickly a population can potentially increase in number.

r-selected species Species that have a high biotic potential and that share other characteristics, such as short life span, early maturity, and high fecundity.

K-selected species Species that have a low biotic potential and that share characteristics such as long life span, late maturity, and low fecundity; generally show logistic population growth.

The biology of a species (which reflects its adaptations for growth, reproduction, and survival) also affects its populations' growth potential. For instance, ecologists recognize a continuum of **life-history strategies** among species. Species whose members mature early, have high fecundity, and have relatively short life spans are known as **r-selected species**—so named because of their high rate of population increase (*r*). Yellowstone *r*-selected species, such as deer mice and spotted knapweed, are well adapted to exploit unpredictable environments and are able to increase quickly if resources suddenly become available.

On the other hand, **K-selected species**, which in Yellowstone include bears, wolves, and slow-growing trees like spruce, are found at the other end of the continuum. Individuals in these species have longer life spans, are slow to mature, and have lower fecundity. Because of this, their reproductive rates are lower, but this means their population growth rates are responsive to minor changes in environmental conditions; they decrease or increase slowly if resources show a gradual decrease or increase in availability. For example, females may have fewer offspring or reproduce less often when food resources are in short supply. This responsiveness tends to keep population sizes close to carrying capacity (*K*). **INFOGRAPHIC 7**

Some species, like elk and deer, have characteristics of both *r* and *K* species; they fall somewhere in the middle of the continuum. They are large organisms that have one or two offspring per year and provide parental care

A gray wolf on the prowl at Yellowstone National Park.

INFOGRAPHIC 7 LIFE-HISTORY STRATEGIES

Different species have different potentials for population growth, known as life-history strategies. A species' biology may place it anywhere along a continuum between two extremes—the *r*- and *K*-selected species.

CHARACTERISTICS OF *r*-SELECTED SPECIES

1. Short life
2. Rapid growth of individual
3. Early maturity
4. Many, small offspring
5. Little parental care
6. Adapted to unstable environment
7. Prey
8. Uses many habitats and resources

CHARACTERISTICS OF *K*-SELECTED SPECIES

1. Long life
2. Slower growth of individual
3. Late maturity
4. Few, large offspring
5. High parental care
6. Adapted to stable environment
7. Predators
8. Needs specific habitat and resources

r-SELECTED SPECIES   **K-SELECTED SPECIES**

DANDELIONS

NORBERT ROSING/National Geographic Creative

DEER MICE

Wayne Lynch/AGE Fotostock

ELK

Juniors Bildarchiv/F314/Alamy Stock Photo

BEARS

Tom Reichner/Shutterstock

SPOTTED KNAPWEED

John W. Bova/Science Source

SPRUCE TREES

TAYLOR KENNEDY -SITKA PRODUCTIONS/ National Geographic Creative

 Consider an aquatic ecosystem. Where would you place the following organisms on a life-history continuum: tuna, sperm whale, plankton, and jellyfish? (Hint: Identify the most extreme *r*- and *K*- species and then place the others in between those on the continuum.)

(*K* characteristics), but their population sizes can increase rapidly if conditions are favorable for growth and survival (*r* characteristics).

K-species and *r*-species often experience different types of population change. For instance, population sizes tend to be stable, especially for *K*-species, in undisturbed, mature areas. However, it may take a long time for a *K*-species to bounce back if sudden or major changes in their environment decrease their population size. For this reason, many of our endangered species are *K*-species.

On the other hand, *r*-species, with their high reproductive potential, sometimes have sudden, rapid population growth, characterized by occasional surges to very high population numbers, which may overshoot carrying capacity, followed by sudden crashes, especially in response to seasonal availability of food or temperature changes; their high rate of reproduction does not allow the population the time to adjust and produce fewer offspring as resources become scarce or conditions change. When this occurs, the population that exceeds carrying capacity will drop below carrying capacity and then increase again; some populations will eventually level off close to carrying capacity, while others continue to overshoot and crash. But this tendency to overshoot and crash is not a weakness—it is a strength that allows *r*-species to inhabit unpredictable or seasonally changing environments.

8   TOP-DOWN AND BOTTOM-UP REGULATION

Key Concept 8: Population size is influenced by factors that decrease it (top-down) and factors that increase it (bottom-up), but which one has the greatest impact varies from population to population.

Thanks in part to Smith's determination, the Yellowstone Gray Wolf Restoration project is going strong. In 2016, there were approximately 100 wolves in Yellowstone National Park (and more than 5,000 living in the lower 48 states). But one chief lesson hammered home by observing wolves in Yellowstone is that populations do not exist in isolation—they live in communities with other populations and influence each other greatly.

Elk populations dropped after the wolf introduction. The ability of a predator (or other resistance factor) to control the size of its prey populations is known as **top-down regulation**. However, ecologists have debated for years the relative importance of top-down regulation versus the regulation by growth factors that affect numbers such as the availability of water and food or sunlight (**bottom-up regulation**). The recent drought that has hindered plant growth likely contributed to the decline in elk abundance in Yellowstone from the "bottom up" as their food supply was diminished. Even physical disturbances like fire are considered bottom-up regulators when they free up nutrients, boosting plant growth. In Yellowstone, the picture that seems to be emerging is that both top-down and bottom-up regulation are important in controlling elk populations and the plant species on which they feed. Which is most important in a given place and time, or for a given or species, varies. Determining whether the return of the wolf has provided meaningful top-down regulation is turning out to not be as easy as researchers once expected it to be. **INFOGRAPHIC 8**

For example, after the *extirpation* (local extinction) of the wolf and the increase of the elk population that followed, aspen trees were overgrazed. Elk like to eat the tender young shoots, which made it difficult for the new trees to grow and become part of the mature stand of trees. Aspen are regrowing in some areas but not all. Elk are less likely to browse in aspen groves if wolves are in the area, but wolf presence does not always lead to aspen recovery. Oregon State University researcher Cristina Eisenberg discovered that in areas where there are many wolves, elk

> One chief lesson hammered home by observing wolves in Yellowstone is that populations do not exist in isolation—they live in communities with other populations and influence each other greatly.

returned to feeding on aspen. She surmised that when predation pressure is predictable and high everywhere (field, forest, and aspen groves), elk returned to feeding on their preferred food. Eisenberg found that aspen recovery was only seen in areas of high wolf population density where wildfire had swept through. Wildfires clear out competing plants (including adult aspen trees), spurring a burst of growth of young aspens. Fire that leaves behind fallen trees creates a habitat from which elk generally stay away (escaping the wolves would be difficult in these deadfall littered areas). Without elk present, young aspen shoots in these areas grow tall enough to escape elk browsing when the elk eventually return, allowing the aspen stand to recover.

The overall effects of returning wolves to Yellowstone are just beginning to unfold and be uncovered by researchers. The fact that the wolves were removed for so long may have changed the ecosystem for good. This is reason to pause and consider human actions today that threaten other species. Many populations of animals other than wolves are declining worldwide due to human impact, especially from habitat loss, the introduction of non-native species, and predator removals. In fact, the number one reason that species become endangered today is habitat destruction (see Online Module 3.3). People damage habitats, remove resources, and break needed connections within ecosystems, and populations respond. Some species may benefit from the change and their population could increase (spotted knapweed, a non-native plant in Yellowstone, spreads quickly in disturbed areas), while other species may decline in number because they are displaced by others or because needed conditions for growth are no longer present. (Songbirds may lose habitat if the woodland patches in which

top-down regulation The control of population size by factors that reduce population size (resistance factors) such as predation, competition, or disease.

bottom-up regulation The control of population size by factors that enhance growth and survival (growth factors) such as nutrients, water, sunlight, and habitat.

INFOGRAPHIC 8   TOP-DOWN AND BOTTOM-UP REGULATION

Population size is affected by both the presence of resistance factors (things that reduce population size) and growth factors (the availability of resources that allow the population to grow), but ecologists have long debated which one has the greatest influence on population size. In most cases, both impact population size, though which plays the greater role may vary from species to species or even within a species, depending on a wide variety of factors, such as relative population sizes of a population versus its predators, competitors or prey, or seasonal climate changes.

TOP-DOWN REGULATION

CONTROL IS FROM PREDATORS HIGH ON THE FOOD CHAIN

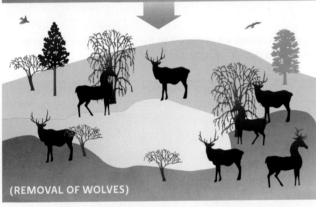

(REMOVAL OF WOLVES)

When wolves were removed from Yellowstone, the elk population increased, and the abundance of aspen, cottonwood, and willow decreased due to overgrazing by the elk.

According to this model, population size is primarily determined by resistance factors such as the control exerted by top predators that eat herbivores; this then limits herbivore consumption of plants, increasing the population size of plants in the ecosystem.

BOTTOM-UP REGULATION

CONTROL IS FROM THE BOTTOM OF THE FOOD CHAIN

(POOR GROWTH CONDITIONS FOR PLANTS)

In Yellowstone, willow shoots do not grow well in dry areas; this could limit the elk population, which in turn could limit the wolf population.

Conversely, growth factors could be the most important determinant of population size. The availability of needed resources—nutrients, sunlight, and water—influences the growth of plants, which determines the population size of animals that eat plants (herbivores), which in turn determines the population size of animals that eat the herbivores.

 In most unprotected areas of the United States, wolf population densities are much lower than they are inside the protected Yellowstone National Park. Do you think wolves exert top-down control on elk populations in these unprotected areas? Explain.

they nest are cut down.) As we will see in Module 2.3, understanding community interactions and population dynamics helps managers monitor, protect, and even restore populations and their communities.

The success of the reintroduction program led to the wolf being "delisted" in Montana and Idaho in 2011 (a delisted species is no longer protected by the federal Endangered Species Act, and management authority is returned to the state), though not without opposition from some conservation groups. (They were also delisted in Wyoming at this time but were relisted in 2014 in the face of litigation.) This delisting has allowed wolf hunts with quotas set by the U.S. Fish and Wildlife Service. To Smith, the management of wolves includes the protection of some packs as well as policies that allow hunting of others. He reasons that hunting wolves only in areas where they conflict with humans may be one of the best ways to protect wolves in wild places like Yellowstone. If people know they can protect their animals and livelihoods, they may be more amenable to allowing the wolves to remain in wilderness areas.

To ensure that the wolves reintroduced to Yellowstone are given a chance to really flourish, Smith and his colleagues diligently stay on the wolves' trails, studying their population dynamics. "We want to know their population size, what they are eating, where they are denning, how many pups they have and how many survive, and how the wolves interact with each other," he explains. Why is it so important to ensure that the wolves do well? Simply put: "They were here first," he says. "We want to restore the original inhabitants to the Park."

Select References:

Beschta, R. L., & W. J. Ripple. (2014). Divergent patterns of riparian cottonwood recovery after the return of wolves in Yellowstone, USA. *Ecohydrology.* doi: 10.1002/eco.1487.

Cubaynes, S., et al. (2014). Density-dependent intraspecific aggression regulates survival in northern Yellowstone wolves (*Canis lupus*). *Journal of Animal Ecology, 83*(6), 1344–1356.

Eisenberg, C., et al. (2013). Wolf, elk, and aspen food web relationships: Context and complexity. *Forest Ecology and Management, 299,* 70–80.

Smith, D., et al. (2013). *Yellowstone Wolf Project: Annual Report, 2012.* Yellowstone National Park, WY: National Park Service, Yellowstone Center for Resources.

Lynx and snowshoe hares go through boom-and-bust population cycles. Lynx populations are controlled by the size of the hare populations (bottom-up regulation), whereas hare populations are controlled by lynx predation (top-down regulation).

Tom & Pat Leeson/AGE Fotostock

INTERACTIVE MAP POPULATION GROWTH PATTERNS

ANIMATED INFOGRAPHIC

Different species exhibit different population growth patterns, a reflection of their biology and the species with which they interact. Read about the species presented here for a closer look at four examples of population growth.

REINDEER

THE SNOWSHOE HARE AND CANADA LYNX

HARBOR SEAL

CYANOBACTERIA

 BRING IT HOME

PERSONAL CHOICES THAT HELP

Understanding the factors that influence how populations change can help us manage species that are facing extinction or help us control (or even eliminate) non-native species that are causing problems. How people view species and their connection to our world has a large impact on how management plays out.

Individual Steps

• Learn more about wolves at the International Wolf Center (www.wolf.org).
• Use the Internet and books on wildlife to research what your area might have been like prior to human settlement. Which species have been extirpated, and which ones have been introduced? How have wildlife populations changed as a result of human action?
• See if you can recognize distribution patterns in the wild. Do some flowers or trees grow in clumps or patches? Can you find

species that appear to have a random or uniform distribution?

Group Action

• Explore organizations that support predator preservation, such as Defenders of Wildlife and Keystone Conservation, for suggestions on how you can help educate others about the importance of predators.
• Join a local, regional, or national group that works to monitor, protect, and restore wildlife habitats, such as the Defenders of Wildlife Volunteer Corps.
• Investigate predator compensation funds such as the Defenders of Wildlife Wolf Compensation Trust and the Maasailand Preservation Trust (for livestock losses due to lion predation). Do you feel this is a worthwhile approach?

Policy Change

• Check the U.S. Fish and Wildlife Service website, www.fws.gov, for updates on wolf management and protection status.
• Write a letter to the editor of your newspaper in support of or in opposition to the decision to delist the wolf in parts of the American West.

John E. Marriott/Getty Images

ENVIRONMENTAL LITERACY UNDERSTANDING THE ISSUE

1 What is a population and why do ecologists study them?

1. Which of the following describes a population?
   a. All the catfish in a pond
   b. All the gray wolves in North America and Europe
   c. All the species of deer in the United States
   d. The different types of fish, sponges, and starfish that live in a tidal pool

2. What is the field of population dynamics and why is it useful when studying populations?

2 What population distributions are seen in nature?

3. What might explain why a population is randomly distributed in its habitat?
   a. Individuals tend to stay away from others in a predictable pattern.
   b. Resources are scarce and hard to find.
   c. Individuals are dispersed through wind or flowing water.
   d. None of these would produce a random distribution.

4. Birds that tend to congregate near patches of suitable food sources have a _____ distribution.
   a. clumped
   b. predictable
   c. random
   d. uniform

5. Why is it useful for an ecologist to understand how a species she is studying is distributed within its ecosystem?

3 What is the importance of population size and density?

6. The number of individuals in a given area, such as an acre or a square mile, is a measurement known as _____ _____.

7. The concept of minimum viable population:
   a. predicts how many individuals can fit into a habitat.
   b. describes the potential number of individuals if there are no predators.
   c. describes the potential number of individuals if resources are limited.
   d. describes the smallest number of individuals needed to ensure the long-term continuation of a particular population.

8. Explain what problems can emerge if a population's density is too low or if it is too high.

4 What is exponential growth and when does it occur in a population?

9. Exponential growth of a population:
   a. is often seen if a population reaches a new environment that is favorable.
   b. is a J-shaped curve on a population graph.
   c. occurs when there are no environmental limits to survive or reproduce.
   d. All of the above are true.

10. What is doubling time? What would be the doubling time for a population whose annual growth rate is 5%?

5 What is logistic growth and when does it occur in a population?

11. Consider the roach. Why does its population size never reach its biotic potential?
    a. Resistance factors limit population size.
    b. Females produce few offspring at a time.
    c. Its tolerance limits are too broad.
    d. Too many growth factors are present.

12. Kangaroo rats eat seeds and are eaten by coyotes. Under what conditions might the kangaroo rat population increase logistically rather than exponentially?

6 How do density-dependent and density-independent factors affect population growth?

13. True or False: Density-independent factors tend to be biotic factors.

14. Lesser goldfinches are small, seed-eating birds. In cities, both wild hawks and domestic cats eat these birds. Discuss several density-dependent and density-independent factors, including both growth and resistance factors, that could affect their carrying capacity.

7 What are the life-history strategies of *r*- and *K*-selected species, and how do they relate to population growth patterns and their ability to respond to environmental changes?

15. True or False: The population of a *K*-selected species is less able to respond to environmental changes than that of an r-selected species.

16. What type of species is more likely to experience "overshoot and crash" population growth cycles?
    a. Species with low biotic potential
    b. Species with high parental care
    c. *r*-selected species
    d. *K*-selected species

17. Compare the life-history strategy of a deer mouse with that of a bear, and identify each as either an *r*- or *K*-selected species.

8 What are top-down and bottom-up regulation, and which is most important in determining the size of a population?

18. True or False: The return of wolves to Yellowstone affected elk foraging behaviors, but only in certain areas.

19. In a bottom-up regulation scenario, the size of an elk herd is determined by the:
    a. number of wolves in the area.
    b. number of wolf predators in the area.
    c. amount of grass, aspen, and other food sources.
    d. All of these are examples of bottom-up regulation.

20. How can it be that a single population might sometimes be controlled from the top down and other times be controlled from the bottom up? Give an example of an instance in which an elk population would be controlled by top-down regulation and another example of its control by bottom-up regulation.

SCIENCE LITERACY   WORKING WITH DATA

In order to understand the effect of the increased wolf population on the abundance of elk in the Yellowstone area, researchers have been careful to track a variety of factors that might also be influencing the size of elk population. The following graphs show the size of the elk population since the introduction of the wolf in 1995 relative to the wolf population (Graph A) and relative to the area's "dryness" for that area over than same time period (Graph B) as measured by the Palmer Drought Severity Index. [Values range from +10 (very wet) to –10 (very dry) with zero representing the long-term average.]

GRAPH A

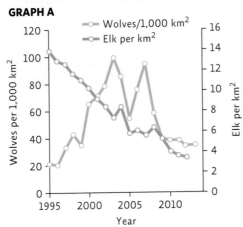

GRAPH B

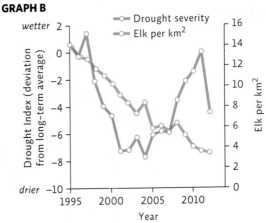

Interpretation

1. Look at Graph A. What is the relationship between the two population sizes over the timeframe shown?

2. Using only the data from Graph A, what conclusion could you draw regarding the effect of the wolf reintroduction on the elk population?

3. Now look at Graph B to determine if drought severity is correlated with elk population size. Using only these data, what conclusion could you draw regarding the effect of drought on the elk population?

Advance Your Thinking

4. Records show that elk harvests by hunters increased in the 1990's and 2000's. Additionally, the population size of grizzly bears, a major predator of elk calves, tripled in that timeframe. How might this have impacted overall elk abundance?

5. An evaluation of the *predation rate* (the proportion of the elk population killed by wolves) reveals wolves killed less than 4% of the elk population in the 7 years after introduction but they now kill around 12%–16% of the elk population. Does this offer evidence that the wolves are exerting predation pressure on the elk?

6. Considering all the data provided here, what is your conclusion regarding the effect of the wolf introduction on the abundance of elk?

INFORMATION LITERACY   EVALUATING INFORMATION

For over 100 years, up until the 1960's, wolves were extirpated in most of the United States. Over the past few decades, studies have indicated that wolves are quite important to the functioning of an ecosystem. Many groups have worked to reintroduce wolves in various U.S. states but they often encounter fierce resistance from other groups concerned with the safety of humans and livestock.

Read the article "Reintroducing the Gray Wolf in the U.S.," at www.actionbioscience.org/biodiversity/johnson.html. Then go to the main website for the group, at www.actionbioscience.org, and investigate the organization.

Evaluate the websites and work with the information to answer the following questions:

1. Determine if this is a reliable information source with a clear and transparent agenda:
   a. Does the organization give supporting evidence for its claims?

   b. Does it give sources for its evidence?
   c. What is the mission of this organization?

2. Visit two other websites: the Defenders of Wildlife website, www.defenders.org, and the Rocky Mountain Elk Foundation's website, www.rmef.org. In each site's search box, enter the word "wolves" and read at least two articles at each website. Answer all the items under Question 1 for each of these websites.

3. Do you believe that wolves should be reintroduced in a few isolated areas; in many areas, including those where human contact is frequent; or not at all? Justify your decision.

4. Do you believe that the hunting of wolves outside protected areas like Yellowstone National Park should be allowed? Justify your answer.

 Additional study questions are available at SaplingLearning.com.

THE FLORIDA EVERGLADES: A COMMUNITY IN CRISIS

A bird species in the Everglades reveals the intricacies of a threatened ecosystem

Wood storks in Florida. Stephen Vincent/Alamy

After reading this chapter and studying the KEY CONCEPTS and INFOGRAPHICS, you should be able to answer these GUIDING QUESTIONS

CORE MESSAGE

Ecological communities are complex assemblages of all the different species that can potentially interact in an area. All the pieces of the ecological community are connected; change one thing, and many others are affected. This means ecosystems are often negatively affected by human impact. Understanding the interconnections within the communities may allow us to better protect and even help restore damaged ecosystems.

1. What is an ecological community?

2. What can we learn about a community by identifying its food web?

3. What are trophic levels, how are they classified, and why are these illustrated as a pyramid?

4. How is species diversity measured and why is it important to a community?

5. What are core and edge habitats and how can habitat fragmentation affect the species that inhabit them?

6. What is a keystone species?

7. What species interactions are seen in a community and what problems can emerge if these interactions are disrupted?

8. What is the goal of restoration ecology and what obstacles do restoration ecologists face?

9. What is ecological succession and how can we use this knowledge to assist in ecosystem restoration?

James Rodgers steered his canoe toward a large cypress tree as sunlight trickled through the dizzy pattern of leaves overhead. The tree had several wood stork nests in it, and Rodgers and his assistant wanted to get a closer look at all of them. They were in the thick of a dense swamp near the northwestern edge of the Florida Everglades, and it was the height of breeding season for the storks—eggs had hatched, and nestlings everywhere were crying, loudly, for food. Rodgers was silent. He knew from experience that alligators patrolled the waters surrounding stork nests and that too much human disturbance could "flush" the wary adult storks—forcing them to flee in a hurry, which would leave their babies vulnerable to aerial predators, such as hawks or crows, that would feed on the chicks in the nest.

The wood stork is an unassuming sort of bird: more than 1 meter (3 feet) tall, yes, but also covered with a mottled black-and-white coat of feathers—bland compared to some of its tropical neighbors. Despite the lack of majesty of the wood storks, however, Rodgers and others at the Florida Fish and Wildlife Service keep close tabs on their ranks.

Here's why: In the late 1970's, the number of nesting pairs of the bird plummeted to an all-time low of 4,500 or so. By the early 1980's, the bird had earned a spot on the Endangered Species List. It was then that Rodgers and his colleagues were first tasked with determining which of several factors (Reduced nesting habitat? Health of females? Damaged feeding grounds?) was most responsible for the decline of this particular bird. And it was through those research efforts—focused intently on the wood stork—that they found an entire ecosystem on the brink.

◉ **WHERE ARE THE FLORIDA EVERGLADES?**

EVERGLADES
NATIONAL PARK

GULF
OF
MEXICO

FL
◉
EVERGLADES

IMBODEN, OTIS/National Geographic Creative

The inlets of Everglades National Park contain a unique mix of tropical and temperate plants and animals, including more than 700 plant and 300 bird species.

1   COMMUNITY ECOLOGY

Key Concept 1: Community ecologists study the many populations that live and interact within a given area.

Community ecology is the study of how all the populations in a given ecosystem function—how space is structured, why certain species thrive in certain areas, and how individual species in the same community interact with one another and with their **habitat**. Each species there occupies a unique **niche**—that is, its role and set of interactions in the community, how it gets its energy and nutrients, and its preferred habitat. Some species will be **niche specialists**—they have very specific habitat or resource requirements, and this restricts where they can live. However, specialists are often highly adapted to acquire these resources and can outcompete others who might be trying to utilize the same resources.

On the other end of a spectrum are the **niche generalists**. These species can use a wide variety of resources. The advantage to this adaptation is flexibility—if one food resource dwindles, the generalist can simply switch to another. In the Everglades, the alligator is a generalist with regard to food choices; it will eat just about anything. The wood stork is more of a specialist feeder and prefers fish of a certain size that are found in shallow, murky water. As long as these fish are present, the stork's exceptional fishing skills give it an advantage over other fish-predators. But if these fish are few in number, the stork has few other options, turning to the less-abundant salamanders and frogs or trying to catch larger or smaller fish.

Ecological communities are complex places defined by their many interconnected species. The Everglades is made up of a wide variety of ecosystems— marshes, prairies, swamps, and forests—each with an array of species that are adapted to the unique conditions of their habitat and that interact closely with each other. The wood stork is one of many different Everglades wading birds that feast on fish, crustaceans, frogs, and insects. Sawgrass, a mix of algae and

community ecology The study of all the populations (plants, animals, and other species) living and interacting in an area.

habitat The physical environment in which individuals of a particular species can be found.

niche The role a species plays in its community, including how it gets its energy and nutrients, what habitat requirements it has, and with which other species and parts of the ecosystem it interacts.

niche specialist A species with very specific habitat or resource requirements that restrict where it can live.

niche generalist A species who occupies a broad niche because it can utilize a wide variety of resources.

bacteria known as periphyton, and an assortment of small plants and water-tolerant trees support a wide variety of birds, reptiles, mammals, fish, and countless insects. The alligator serves as an important apex predator. All these populations together make up the communities of the Everglades. **INFOGRAPHIC 1**

Community ecologists investigate how various species contribute to ecosystem services like pollination, water purification, and nutrient cycling (see Module 2.1). For example, wetlands like the Florida Everglades help capture contaminants and excess nutrients, preventing them from reaching downstream fresh- and saltwater ecosystems. They also help refill groundwater supplies and provide flood control, vital ecosystem services for the people of South Florida. But human impact has drained much of the wetland and reordered the landscape, and pollution has disrupted nutrient cycles—actions that are altering the ability of communities to continue to provide these services.

Before giving way to a hodgepodge of resorts, sugar plantations, and dense urban centers, the region was defined by an uninterrupted web of natural ecosystems, collectively known as the Everglades, that stretched from Lake Okeechobee to the Florida Bay. It was here that wood storks flourished. In the 1930's, an estimated 15,000 to

Map of South Florida.

INFOGRAPHIC 1   THE EVERGLADES COMMUNITY

An ecological community consists of all the populations of species that live and interact in one area. In the Florida Everglades, different communities can be found in the lagoons, out in the sawgrass marshes, in the cypress swamps, and the many other unique habitats in this large wetland ecosystem—unique communities that are all interdependent and connected. The species that make up a given community fill specific niches and each contributes to the overall health of their ecosystem.

 Think of an ecological community in an ecosystem near you. Identify some of the populations that make up that community.

20,000 pairs nested throughout the southeastern United States—largely in South Florida, where foraging grounds were ideal.

But it was not long after they first discovered the Everglades that American explorers began hatching plans to drain and then develop them. Swamps and muddy rivers choked with grass were seen as having no inherent value. "From the middle of the 19th century to the middle of the 20th, the United States went through a period in which wetland removal was not questioned," says University of Florida geographer and historian Christopher Meindl. "In fact, it was considered the proper thing to do." The Central and Southern Florida Project,

authorized by the U.S. Congress in 1948, set out to systematically drain the Everglades.

As the human population swelled in the region, water that once fed swamps and marshes was rerouted to the faucets of burgeoning developments. And as water levels changed, becoming deeper in some areas and completely disappearing in others, the total wading bird population plummeted—by 90% between the 1930's and 1990's.

As ecologists would soon discover, these changes disrupted the entire ecosystem—from the health of giant wading birds right down to the movement of matter and energy.

2 THE FOOD WEB

Key Concept 2: The flow of matter and energy through a community is represented by a food chain that shows who eats whom and always begins with producers. The combination of all the food chains in one area make up a community's food web.

Energy is the foundation of every ecosystem; it is captured by photosynthetic organisms and then passed from organism to organism via the **food chain**—a simple, linear path that shows who eats whom. Any given ecosystem might have dozens of individual food chains. Linked together, they create a **food web**, which shows all the many connections in the community. Both food chains and webs help ecologists track energy and matter through a given community. They can vary greatly in length and complexity between different types of ecosystems. But most share a few common features, and all are made up of the same basic building blocks—namely, producers and consumers.

food chain A simple, linear path starting with a plant (or other photosynthetic organism) that identifies what each organism in the path eats.

food web A linkage of all the food chains together that shows the many connections in the community.

Florida wood storks sit near the top of a food chain that begins with sawgrass, other plants like cypress and mangrove trees, and periphyton. These photosynthetic organisms are all known as **producers**. Producers capture energy directly from the Sun and convert it to food (sugar) via photosynthesis. (See Infographic 6 in Module 2.1.) They are then eaten by a wide range of **consumers**—organisms that gain energy and nutrients by eating other organisms. Animals, fungi, and most bacteria and protozoa are consumers. **INFOGRAPHIC 2**

INFOGRAPHIC 2   **EVERGLADES FOOD WEB**

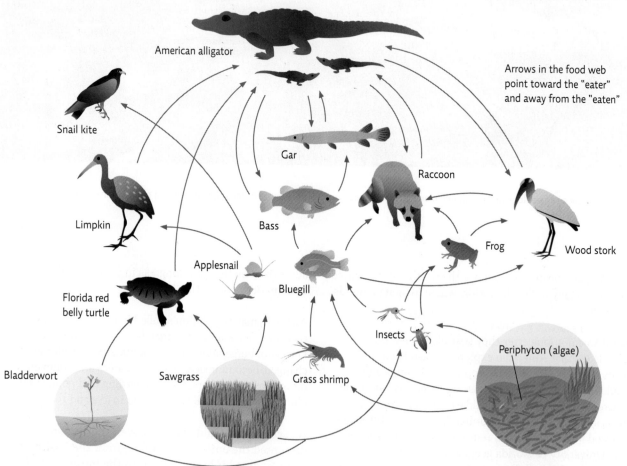

Arrows in the food web point toward the "eater" and away from the "eaten"

The food web of the Everglades is very complex and varies among the different ecosystems found there. Periphyton algae mats form the base of the food web and may be the most important producers in the ecosystem. The American alligator is the main apex predator, though when young is also prey to various birds, fish, mammals, and even other alligators.

 Identify two or three food chains within this web that end with the alligator.

The wood stork is a specialist feeder that hunts almost exclusively in shallow, muddy, plant-filled water that is so cloudy that fish cannot be seen. They inch their way through these waters at a steady, two-steps-per-second clip, sweeping their long, narrow bills—which are kept precisely 8 centimeters (3 inches) agape, and submerged all the way up to the breathing passage—side to side in a relentless hunt for food. When a bill's methodical searching meets the sensation of a wriggling fish, it snaps shut in just 25 milliseconds. It's the fastest reflex known in all vertebrates, and it enables the wood stork to capture prey that no other wading birds can access.

But for this tactile (or nonvisual) feeding method to work, the prey must be densely concentrated. This means wood storks need seasonally drying wetlands to forage—and lots of them. Even a small drop in feeding success can impact the ability of these colossal birds to successfully rear their young, a factor that can have serious repercussions for this *K*-selected species. Storks mature slowly and may not reproduce until they are 4 years old. They lay a single clutch of eggs per year, and it is only in years of abundant food that all the young (perhaps three or four) survive. In lean years, only one chick may survive, or none. (See Module 2.2 for more on *K*- and *r*-selected species.)

It's this sensitivity that makes the wood stork such a good **indicator species** for the Everglades. An indicator species is one that's particularly vulnerable

Wading birds like storks, heron, and ibis prefer to nest in trees at the water's edge, especially on islands surrounded by waterways patrolled by alligators, a deterrent to nest predators like racoons.

to ecosystem perturbations. Because even minor environmental changes can affect them dramatically, they can warn ecologists of a problem before it grows. "It is much easier to follow one or two species than to try and monitor an entire ecosystem," says Rodgers, who is a wood stork specialist. "So if an indicator species can be identified, this makes it much easier to keep tabs on the health of the ecosystem."

producer An organism that converts solar energy to chemical energy via photosynthesis.

consumer An organism that obtains energy and nutrients by feeding on another organism.

indicator species A species that is particularly vulnerable to ecosystem perturbations, and that, when we monitor it, can give us advance warning of a problem.

3 THE TROPHIC PYRAMID

Key Concept 3: The trophic levels of the food chain are shown as a pyramid; lower levels are larger than upper ones. Because energy only enters at the producer level and the organisms at each level use most of the energy they have taken in, only a small percentage is passed on to support the next level.

The different levels at which organisms capture energy or feed are known as **trophic levels**. Consumers are organized into trophic levels based on what they eat. *Primary consumers* eat producers; *secondary consumers* eat primary consumers; *tertiary consumers* eat secondary consumers, and so on, ending with the apex predators in the last trophic level. Of course, many consumer species often feed at more than one trophic level: Wood storks eating crayfish are feeding at trophic level 3, but when they eat small fish like bluegill, they are feeding at trophic level 4. Alligators eat a wide variety of animals—turtles, fish, birds, even mammals like the raccoon—making them the apex predator in many Everglades food chains.

When any of these organisms die, they are eaten by an army of consumers known as **detritivores**—animals like worms, insects, and crabs that feed on dead plants and animals—and **decomposers**—organisms like bacteria and fungi that break decomposing organic matter all the way down into its constituent atoms and molecules. This returns matter back to the water or soil where

trophic levels Feeding levels in a food chain.

detritivores Consumers (including worms, insects, crabs, etc.) that eat dead organic material.

decomposers Organisms such as bacteria and fungi that break organic matter all the way down to constituent atoms or molecules in a form that plants can take back up.

it can be taken up once again by producers, simply one step of many that make up the complex matter cycles that move matter through all the levels of the trophic pyramid.

As one moves up the food chain, energy and *biomass* (matter comprising all the organisms at that level) decrease, creating a trophic pyramid. The reason that any consumer trophic level is smaller than the one below it is simple: Nearly every organism uses up the majority of the energy and matter it has acquired in the complicated act of living. So when an organism is killed and consumed by a predator, it only passes on a small percentage of all the energy and matter it consumed during its lifetime. **INFOGRAPHIC 3**

As researchers discovered in the 1980's, it was a kink in the food chain that hurt the wood storks. While they feed on many things, they prefer fish—and not just any fish, but those between 2 and 15 centimeters (1 and 6 inches) long. Most fish need more than a single season to grow

this big; in fact, they need wetlands that are flooded for longer than a year and only very rarely go completely dry. It turns out that as humans altered water cycles in South Florida, there were fewer and fewer such areas, and thus fewer fish for the storks to feed their young.

Much of Florida is made up of low, flat land that floods during the rainy season (June through September), which delivers about 75% of its annual rainfall. Many wetland fish grow and reproduce in this expanding habitat. Then, as rains taper off, water begins to recede and the fish become concentrated in small ponds and *sloughs* (free-flowing channels of water that develop between sawgrass prairies). Foraging storks follow these receding waters, from upland ponds to lowland coastal areas, feeding on fish. They are so dependent on this water cycle, that their breeding cycle is regulated by water levels. Such profound connectedness—between landscape and life—is common in the Everglades.

INFOGRAPHIC 3 | **TROPHIC PYRAMID**   ANIMATED INFOGRAPHIC

Energy enters at the base of the food chain in the first trophic level (TL) via photosynthesis and is passed on to higher levels as consumers feed on other organisms. This is shown as a pyramid (smaller on top) because only a small percentage of the energy is passed on to each higher level, with the majority being "lost" to the environment (usually as heat from the energy that the organism burns in day-to-day life before it is eaten). Most food chains have only four or five levels due to this progressive loss of energy. Though the actual amount varies from ecosystem to ecosystem, for illustration purposes, we show 10% passing on to the next higher level.

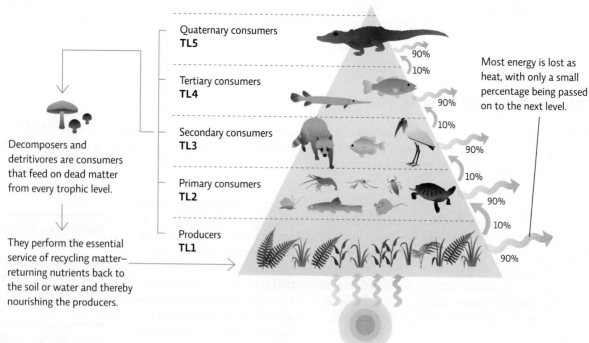

TROPHIC LEVELS (TL)

Quaternary consumers
TL5

Tertiary consumers
TL4

Secondary consumers
TL3

Primary consumers
TL2

Producers
TL1

Most energy is lost as heat, with only a small percentage being passed on to the next level.

Decomposers and detritivores are consumers that feed on dead matter from every trophic level.

They perform the essential service of recycling matter–returning nutrients back to the soil or water and thereby nourishing the producers.

 Why are there seldom more than five trophic levels?

4 SPECIES DIVERSITY

Key Concept 4: Species diversity is measured in terms of the number of different species present (richness) and the relative abundance of each species (evenness). High richness and evenness tend to make a community more resilient to environmental changes.

Such connections between species and their environment, and among species, give rise to ecological diversity, a measure of the number of species at each trophic level, as well as the total number of trophic levels and available niches. (See Module 3.2 for more on ecological diversity.) Greater ecological diversity means more niches and thus more ways for matter and energy to be accessed and exchanged. This generally increases a community's **resilience**—its ability to adjust to changes in the environment and return to its original state rather quickly.

Species diversity, which refers to the variety of species in an area, is measured in two different ways: species richness and species evenness. **Species richness** refers to the total number of different species in a community. **Species evenness** refers to the relative abundance of each individual species. In general, organisms at a higher trophic level will have fewer members than those at a lower trophic level, but populations within the same trophic level should have relatively similar numbers. If they do, the community is said to have high species evenness. If, on the other hand, one or two species dominate any given trophic level, and there are few members of other species, then the community is said to

have low species evenness. In such uneven communities, the less abundant species is at greater risk of dying out.

Both richness and evenness have an impact on the community. In general, higher species richness and evenness makes for a more diverse community and a more intricate food web. Greater intricacy enables more matter and energy to be brought into the system and also makes the community less likely to collapse in the face of calamity. Of course it isn't always as simple as "more species are better." When non-native species are introduced to an area, species richness may initially increase, but these non-natives may disrupt the community in ways that ultimately diminish diversity. The Everglades is facing a formidable invasive species that is even challenging alligators for the role of apex predator and changing the food web in the area—the Burmese python. (See Module 3.1 and its Interactive Map for more on invasive species and the Burmese python.)
INFOGRAPHIC 4

resilience The ability of an ecosystem to recover when it is damaged or perturbed.

species diversity The variety of species in an area; includes measures of species richness and evenness.

species richness The total number of different species in a community.

species evenness The relative abundance of each species in a community.

INFOGRAPHIC 4 **SPECIES DIVERSITY INCLUDES RICHNESS AND EVENNESS**

The species diversity in an area is a measure of species richness (the total number of species) and species evenness (a comparison of the population size of each species). The Everglades contains forested areas known as hardwood hammocks. Each forest plot shown here contains 15 trees, but they differ in terms of species richness and evenness.

● Hackberry ● Red Maple ● Gumbo Limbo
● Mahogany ● Live Oak ● Cocopalm

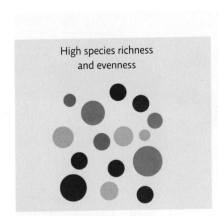

High species richness and evenness

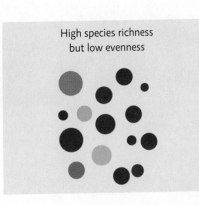

High species richness but low evenness

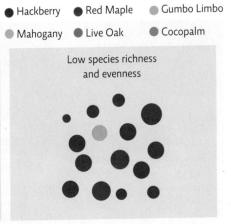

Low species richness and evenness

Which forest plot shown here would likely have the highest species diversity (richness and evenness) of birds? Explain.

5 HABITAT STRUCTURE: EDGE AND CORE REGIONS

Key Concept 5a: Community composition is affected by the physical structure of the habitat, with some species preferring to inhabit ecotone regions where one habitat meets another (the edge) and others staying deep within one habitat (the core).

Key Concept 5b: Human impact that fragments habitats may increase edge and decrease core areas, harming core species and disrupting community connections.

A community's composition and diversity is also heavily influenced by its physical features. As abiotic physical features like temperature and moisture change, so does community composition. This often happens in **ecotones**, places where two different ecosystems meet—like the edge between a forest and field or river and shore. The different physical makeup of these edges creates different conditions, known as **edge effects**, which either attract or repel certain species. For example, it is drier, warmer, and more open at the edge between a forest and field than it is further into the forest. This difference produces conditions favorable to some species but not others.

Ecotones may also attract some species that use different aspects of the two adjacent communities; fish such as young snapper or grunts, for example, prefer to live in areas where seagrass beds are fairly close to a shoreline populated by mangrove trees. The mangrove "prop" roots, which anchor the trees into the wet, sandy ground below, offer the fish safety from predators during the day but are close enough to the seagrass beds where the snapper and grunts feed at night for easy "commuting." These fish are not found in coastal areas without the combination of protective coastal mangrove trees and close-by, offshore seagrass beds.

Species that thrive in edge habitats like this are called *edge species*. Other species, those that can only be found deep within the core of a given habitat, are called *core species*. Some of the many species that find food and refuge in the seagrass, such as crustaceans, sea urchins, and worms, prefer to stay in core areas, where they are better hidden and protected from wave action or can make use of deeper sediment buildup in these inner areas.

Unfortunately, human actions often fragment large expanses of intact core habitat, increasing edge habitat. This fragmentation scenario plays out in ecosystems around the world, especially terrestrial ones such as forests and grasslands where urban/suburban, agricultural, and industrial development creates patchworks of formerly expansive habitats. Edge species may thrive in these patchy habitats, but because core species may not venture out across the edge in search of new habitat, they are easily trapped by habitat fragmentation; we may eliminate these core species altogether if we don't leave enough core area behind. In other instances, edge habitats are totally destroyed and replaced with other habitats, such as the removal of mangrove forests in areas undergoing coastal development, eliminating the edge species who resided there. (See Module 3.2 for more on habitat fragmentation and core species.) **INFOGRAPHIC 5**

ecotones Regions of distinctly different physical areas that serve as boundaries between different communities.

edge effect The different physical makeup of an ecotone that creates different conditions that either attract or repel certain species (e.g., it is drier, warmer, and more open at the edge of a forest and field than it is further in the forest).

Nicholas Reuss/Getty Images

Deer are an edge species—they feed in fields but need the cover of a forest when they bed down to rest.

INFOGRAPHIC 5 **EDGE EFFECTS**

MANGROVE EDGES

The mangrove-seagrass ecotone provides an example of an edge effect. Fish such as immature gray snapper and bluestriped grunt "commute" between the mangrove trees and the seagrass beds. The proximity of these two areas is vital to provide both the protection during the day and feeding opportunities at night that these young fish need.

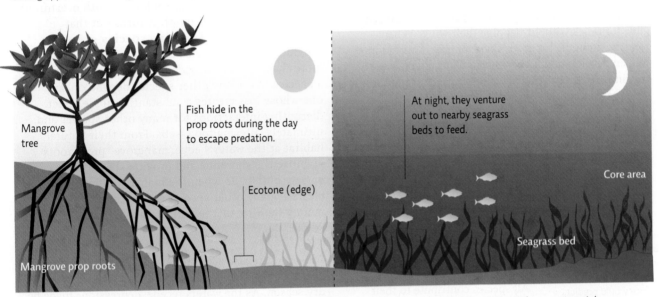

Mangrove tree

Fish hide in the prop roots during the day to escape predation.

At night, they venture out to nearby seagrass beds to feed.

Core area

Ecotone (edge)

Seagrass bed

Mangrove prop roots

 Many coastal mangrove areas are being fragmented as stretches of mangrove are removed for residential or commercial development. What might happen to the gray snapper and blue striped grunt populations if the mangrove trees are removed from part of the shoreline? What impact would this have on the seagrass beds?

FOREST EDGES

Habitat structure influences where species live. Edge species, like deer, prefer habitats with forest and field edges, whereas core species, like some warblers (small birds), prefer the inner areas of forest and do not readily venture into edge regions.

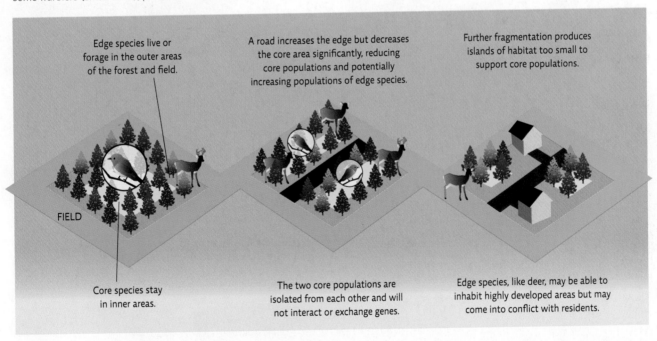

Edge species live or forage in the outer areas of the forest and field.

A road increases the edge but decreases the core area significantly, reducing core populations and potentially increasing populations of edge species.

Further fragmentation produces islands of habitat too small to support core populations.

FIELD

Core species stay in inner areas.

The two core populations are isolated from each other and will not interact or exchange genes.

Edge species, like deer, may be able to inhabit highly developed areas but may come into conflict with residents.

 Why might a wooded corridor that connects two habitat fragments need to be wider for a warbler than for a deer if we expect the bird to use it to travel between fragments?

6 KEYSTONE SPECIES

Key Concept 6: Keystone species are particularly important to other members of their community, and if their numbers decline, many other species may be negatively affected.

Wood storks are spectacular fliers. They can reach altitudes as high as 1,500 meters (5,000 feet) and can glide for miles without flapping their wings When foraging grounds dry up, or flood, or are converted into human developments, these aerial skills are pushed to the limit—flying as much as 120 kilometers (75 miles) in some cases in search of food.

But they weren't the only ones to struggle in the newly developed region. As cities replaced swamps and roads replaced rivers in the Florida Everglades, the flow of water was disrupted like never before. As sloughs ran dry, key detritivores and decomposers like worms, grass shrimp, and microbial communities that had thrived there were decimated. With each species lost, connections were broken, leading to the decline of the snakes, fish, alligators, turtles, and wading birds that fed on them. All species contribute to their ecosystem, but it turns out, some are more important than others.

keystone species A species that impacts its community more than its mere abundance would predict, often altering ecosystem structure.

The replacement of mangrove forests with oceanfront resorts illustrates this point. It turns out that mangrove trees are a **keystone species**—one that provides a unique service that impacts its community more than its mere abundance would predict. It's a species that many other species depend on, and one whose loss creates a substantial ripple effect, disrupting interactions for many other species and, ultimately, altering food webs. From their natural habitat at the water's edge, mangrove "prop" roots stabilize the shoreline and provide shelter for a wide variety of fish. So when the mangrove forests are cleared, many other species suffer: the fish that hide among their roots, the fish that feed on those fish, and so on.

Alligators are also a keystone species in the Everglades, one that a great many species depend on during the dry season. As the waters recede, depressions made by alligators (gator holes) are some of the few places that still hold standing water. These holes become refuges for

The gopher tortoise is a keystone species in the longleaf pine forests of Georgia and Florida. Its burrows provide refuge from weather and fire to more than 400 different species.

Shellphoto/Getty Images

fish, invertebrates, and aquatic plants; they also become very attractive to the animals who feed on these aquatic creatures. Without gator holes, many species would not survive the dry season. As an apex predator, alligators also help control the population of a wide variety of species. This generalist predator tends to prey on the most abundant populations at any given time. As one population declines in abundance, the alligator turns to other, more abundant prey. This allows the declining population to recover and keeps the abundant population in check. (For more examples of keystone species, see the Interactive Map at the end of this module.) **INFOGRAPHIC 6**

Wood storks also depend on the presence of alligators, but not just for the dry season gator holes. In the

1980's, Rodgers and his colleagues embarked on a comprehensive study of stork nests in an effort to see in which types of trees the storks preferred to nest and whether the availability of those trees was impacting their ability to breed. "We went to 20 stork colonies," Rodgers remembers. "We measured every tree, recorded its species, size, cored it for age, noted its branching structure." The conclusion, reached after 5 years of painstaking work, can be summed up in a single sentence, Rodgers says: Wood storks will nest in just about anything, as long as it's surrounded by water that is patrolled by alligators. "Without the alligators, raccoons swim across, and climb up and destroy everything," Rodgers says. "Without the alligators, when predators get in, we've seen [the storks] abandon entire colonies."

INFOGRAPHIC 6 | KEYSTONE SPECIES SUPPORT ENTIRE ECOSYSTEMS

Some species are especially important to their ecosystem. If a keystone species is lost or declines in number, the ecosystem could change drastically, and other species that depend on it may suffer or be lost.

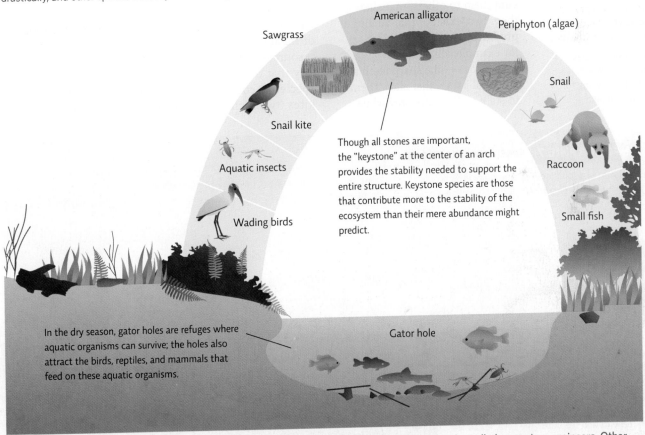

American alligator
Sawgrass
Periphyton (algae)
Snail kite
Snail
Aquatic insects
Raccoon
Wading birds
Small fish

Though all stones are important, the "keystone" at the center of an arch provides the stability needed to support the entire structure. Keystone species are those that contribute more to the stability of the ecosystem than their mere abundance might predict.

In the dry season, gator holes are refuges where aquatic organisms can survive; the holes also attract the birds, reptiles, and mammals that feed on these aquatic organisms.

Gator hole

 Keystone species whose actions alter the habitat in a way that benefits other species are also called ecosystem engineers. Other than the alligator, identify a species that acts as an ecosystem engineer in its ecosystem and explain how its actions benefit other species.

7   SPECIES INTERACTIONS

Key Concept 7a: Interactions within and between species may be beneficial, neutral, or harmful to participants but all are critical to energy capture and flow, and to matter cycling.

Key Concept 7b: Changes that interfere with these interactions can imperil many others and decrease the overall functioning of the ecosystem and the services it provides.

Communities are all about relationships. Successful communities are those where a balance has evolved between all the organisms living there. Species interactions serve many purposes; for example, they control populations and affect carrying capacity. Biodiversity (lots of species, lots of variety within a species) is important because more diversity means more ways to capture, store, and exchange energy and matter (see Module 3.2). But it is not sheer numbers that matter most; it is all the connections among species—how they help or hurt one another—that determine how and how well an ecosystem works. Each species is unique and thus interacts in its own unique ways with all the species around it.

> **predation** Species interaction in which one individual (the predator) feeds on another (the prey).
>
> **competition** Species interaction in which individuals are vying for limited resources.

Many species have adaptations that bind them to others, that allow them to coexist, or that facilitate **predation**. In the Everglades, for example, alligators have adaptations—like sharp teeth and powerful jaws—that allow them to stalk and capture prey while most of the fish they prey on have adaptations—like camouflage and a wary nature—that help them avoid capture.

Competition—the vying between organisms for limited resources—is another way that species interact. In general, it is subtle rather than outright fighting. *Intraspecific competition* (that which occurs between members of the same species) is generally stronger than *interspecific competition* (that which occurs between members of different species). This is because members of the same species share exactly the same niche and thus compete for all resources in that niche, whereas members of different species may compete for only a single resource, like water.

Other neighbors—those that prey on the same food or inhabit similar niches—find a way to partition resources. That is, they divvy up the goods in a way that reduces competition and allows several species to coexist. For example, limpkins and snail kites (two niche specialist birds that feed almost

The heart of a functioning community is its species interactions. Some interactions are beneficial and others cause conflict, but all are important in keeping matter and energy flowing through an ecosystem.

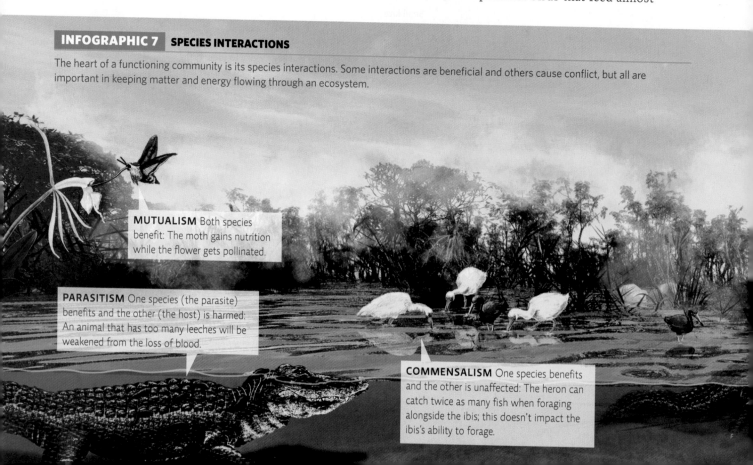

MUTUALISM Both species benefit: The moth gains nutrition while the flower gets pollinated.

PARASITISM One species (the parasite) benefits and the other (the host) is harmed: An animal that has too many leeches will be weakened from the loss of blood.

COMMENSALISM One species benefits and the other is unaffected: The heron can catch twice as many fish when foraging alongside the ibis; this doesn't impact the ibis's ability to forage.

exclusively on apple snails) hunt in different regions of the Everglades. This strategy—known as **resource partitioning**—increases the ecosystem's overall capture of matter and energy and thus benefits the entire community.

There are other strategies, too, that keep an ecosystem functioning and strong. Some of these interactions show a tremendous interdependency on the part of the participants. Known as **symbiosis**, these relationships can take one of three forms. The most commonly recognized form of symbiosis is **mutualism**, where both species benefit from the relationship. Recent research by Lucas Nell, now at the University of Georgia, has revealed that a surprising mutualistic relationship exists between storks and alligators. As mentioned earlier, nesting near patrolling alligators increases the nesting success of storks. But the alligators also benefit by feeding on nestlings that drop out of stork nests. Nell's research showed that alligators that patrolled the sloughs that surround breeding bird colonies were better fed and had larger fat reserves than those feeding in other areas.

In the symbiotic relationship known as **commensalism**, one species benefits from the relationship, and the other is unaffected. For example, in fields and pastures throughout the Southeast, cattle egrets follow grazing cattle, snatching up the bugs disturbed by the hooves of the passing cows. This has no effect on the cows—they don't eat the bugs and are not disturbed by the bugs—but it greatly enhances the foraging efficiency of the egrets.

Parasitism, where one species benefits from the relationship and the other is negatively affected, is also a form of symbiosis due to the close biological tie between the parasite and its host. Parasites may be external, such as ticks, fleas, and leeches, or internal, such as intestinal worms. Guinea worm disease, highlighted in Module 4.3, is just one example of a wide variety of parasitic diseases that afflict human populations.

INFOGRAPHIC 7

By ensuring that all populations persist, even as individuals die, these delicate checks and balances allow more energy to be captured and exchanged and thus increase the amount of biomass the ecosystem is able to produce. When individual species are lost, or when a landscape is physically altered, the balance is tipped. And when that happens, things can fall apart. Fast.

resource partitioning A strategy in which different species use different parts or aspects of a resource rather than compete directly for exactly the same resource.

symbiosis A close biological or ecological relationship between two species.

mutualism A symbiotic relationship between individuals of two species in which both parties benefit.

commensalism A symbiotic relationship between individuals of two species in which one benefits from the presence of the other, but the other is unaffected.

parasitism A symbiotic relationship between individuals of two species in which one benefits and the other is negatively affected.

? Choose an ecosystem other than the Everglades and give examples of mutualism, commensalism, parasitism, predation, competition, and resource partitioning.

RESOURCE PARTITIONING Both species benefit by partitioning a resource rather than competing for it: Though they both eat apple snails, snail kites and limpkins don't directly compete for them since each predator feeds in a different region of the Everglades.

PREDATION One species benefits (predator) and the other is harmed (prey): Alligators prey on a variety of animals and are prey themselves when young.

COMPETITION All participants are negatively affected: Apple snails, a variety of fish, and crustaceans all compete for periphyton algae as a food source.

8 RESTORATION ECOLOGY

Key Concept 8: Human impact often reduces species diversity. It may be difficult to restore all the species and their connections when we try to repair ecosystem damage; therefore, our best course of action is to avoid the damage in the first place.

Indicator species like the wood stork gave ecologists an early warning that the Everglades ecosystem was suffering from the drastic changes imposed by human impact and prompted action to investigate. The federal 1992 Water Resources Development Act enlisted the U.S. Army Corps of Engineers to investigate the damage to the Everglades that resulted from nearly 50 years of unchecked expansion. They found that the size of the Everglades had been reduced by 50% since the late 1800's. Constructed canals and levees had dramatically altered water levels, leaving some areas parched and others flooded. And poorly timed water releases were further starving ecosystems that had already been affected by hypersalinity, excessive nutrients (from agricultural runoff), and an ever-growing list of non-native species.

Ignoring these problems any longer could greatly imperil the 10 million people that had made their home in the region. "What folks finally realized when we reexamined the area was that the wetlands were this essential filter—they cleaned the water of pollutants," says Kim Taplin, a restoration ecologist who works for the U.S. Army Corps of Engineers, restoring the Florida wetlands. "So as the ecosystems have suffered, water quality has declined considerably. We're going to have millions of people with no clean water, unless we fix it."

> **"We're going to have millions of people with no clean water, unless we fix it."** —Kim Taplin.

Avoiding the ecosystem damage in the first place is almost always our best option, but when damage is done, fixing it is the work of restoration ecologists. **Restoration ecology** is the science that deals with the repair of damaged or disturbed ecosystems. Indicator species like the stork not only alert ecologists to a problem but also are monitored as a gauge of success in the restoration efforts: Is the population recovering as expected? It requires a special blend of skills—not only biology and chemistry but also engineering and a heavy dose of politics.

restoration ecology The science that deals with the repair of damaged or disturbed ecosystems.

In 2000, the U.S. Congress enacted the most comprehensive—and expensive—ecological repair project in history. The Comprehensive Everglades Restoration Plan, or CERP, included more than 60 construction projects to be completed over a 30-year period. The idea was to restore some of the natural flow of water through the Everglades, starting with the Tamiami Trail.

The Tamiami Trail, a 240-kilometer (150-mile) stretch of U.S. highway that connects the South Florida cities of Tampa and Miami, is a serious barrier to freshwater flow in the region. It's also heavily traveled. That means the U.S. Army Corps must not only tear down the road but also build something in its place. "Most of our restoration projects involve building even more structures," says Tim Brown, project manager for the U.S. Army Corps of Engineers' Tamiami Trail project. "It's a delicate balance. We of course want to restore as much of the natural system as possible. But we are also charged with protecting lives and property, and in this case, that means building bridges."

Each facet of CERP has brought its own fresh round of debate over how best to balance the needs of a swelling human population against the importance of restoring and protecting a heavily degraded ecosystem. Of course, no one knows for certain what will work and what won't. The Everglades landscape has changed dramatically, in ways that not even the best scientists can reverse; decades of development will do that. **INFOGRAPHIC 8**

Joe Cavaretta/Sun Sentinel/MCT via Getty Images

Canals that drained (and still drain) the Everglades allowed communities like Sunrise, Florida, to be built where Everglades once existed.

INFOGRAPHIC 8 | **THE COMPREHENSIVE EVERGLADES RESTORATION PLAN**

Restoration efforts attempt to return an ecosystem back to its original condition, or as close as possible. Restoring needed habitat is often the first step. Species can then return, or be brought in to repopulate the community but this is not an easy task—we may not know which species are missing or which are critical to re-establish lost mutualistic, commensal, or predator-prey connections. The Everglades Restoration Plan focuses on restoring some of the lost wetland area. If successful, this not only will help to restore some native populations but also will restore the important ecosystem services such as protection from hurricane storm surges and the capture and storage of rainwater, which provides flood control and freshwater to the ecological and human populations of South Florida.

On the maps below, darker green areas represent wetland areas or river floodplains; white arrows show overland water flow.

HISTORIC FLOW

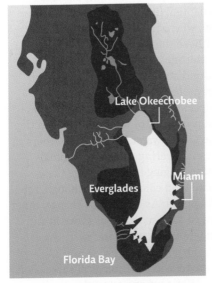

Historically, the Everglades covered most of South Florida—more than 10,000 square kilometers (4,000 square miles).

CURRENT FLOW

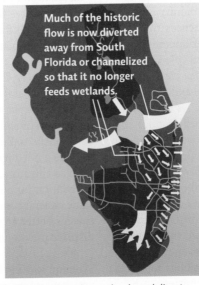

Much of the historic flow is now diverted away from South Florida or channelized so that it no longer feeds wetlands.

Projects to drain the wetlands and divert water to agricultural lands disrupted normal flow, drained about half of the wetlands, and resulted in water shortages for the downstream ecosystems and for people as well.

PROPOSED FLOW UNDER CERP

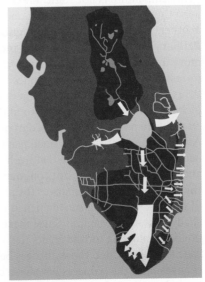

The goal of the Comprehensive Everglades Restoration Plan (CERP) is to restore the flow of water back to some historic wetland areas through the removal of some canals and levees, as well as to capture some of the freshwater that drains into the ocean, benefiting both the ecosystems and the residents of South Florida.

 Explain how rerouting some of South Florida's water flow back through the center of the state, through restored wetlands, will help provide residents with more drinking water.

For the wood stork, restoration success remains low with population sizes far below the targets set for this species, calling for continued action to address habitat loss and reproductive failures. The number of nests in 2014 was the lowest seen since 2004; nesting success in 2013 was only a little better. Though stork numbers are not increasing, neither are they declining, suggesting this smaller population is stable, for now.

"The bottom line, though, is that there's only so much we can do," says Rodgers. "It's a lot just to figure out what the baseline was or should be. Some plant species have probably gone extinct, and some non-natives are virtually impossible to remove. What we can do is figure out what some of the big obstacles to recovery

are, remove them, and after that, let nature take its course." For his part, Rodgers says he can't imagine that the great 1,000-breeding-pair wood stork colonies that early settlers described will ever return to South Florida. The landscape has been too dramatically altered, he says.

A new problem is making the timely restoration of the Everglades even more vital—climate change. Rising sea level and salt water intrusion into freshwater habitats are already occurring and are serving as an added incentive to restore the wetland so that it can capture more freshwater for the ecosystems, farms, and cities of South Florida. (See Module 10.2 for more on climate change and its effects on Florida.)

9 ECOLOGICAL SUCCESSION

Key Concept 9: Over time, ecosystems naturally transition from one community to another in response to changing environmental conditions. An understanding of this ecological succession process can guide our restoration efforts.

Though the changes the Everglades have experienced are extreme, changes to ecological communities are really the norm; nature is not static. Predictable transitions can sometimes be observed in which one community replaces another, a process known as **ecological succession**. **Primary succession** begins when **pioneer species** move into new areas that have not yet been colonized. In terrestrial ecosystems, these pioneer species are usually lichens—a symbiotic combination of algae and fungus. Lichens can tolerate the barren conditions. As time goes by and lichens live, die, and decompose, they produce soil. As soil accumulates, other small plants move in—typically sun-tolerant annual plants that live 1 year, produce seed, and then die—and the plant community grows. Gradually, the plant growth itself changes the physical conditions of the area—covering sun-drenched regions with broad, shady leaves, for example. Since these conditions are no longer suitable for the plants that created them, new species move in, and those changes beget even more changes until the pioneers have been completely replaced by a succession of new species and communities.

ecological succession
Progressive replacement of plant (and then animal) species in a community over time due to the changing conditions that the plants themselves create (more soil, shade, etc.).

primary succession
Ecological succession that occurs in an area where no ecosystem existed before (e.g., on bare rock with no soil).

pioneer species Plant species that move into an area during early stages of succession; these are often *r*-species and may be annuals—species that live 1 year, leave behind seeds, and then die.

secondary succession
Ecological succession that occurs in an ecosystem that has been disturbed; occurs more quickly than primary succession because soil is present.

Secondary succession describes a similar process that occurs in an area that once held life but has been damaged somehow; the level of damage the ecosystem has suffered determines which stage of plant community moves in. For example, a forest completely obliterated by fire may start close to the beginning with small herbs and grasses, whereas one that has suffered only moderate losses may start midway through the process with shrubs or sun-tolerant trees moving in. The stages are roughly the same for any terrestrial area that can support a forest: first annual species, then shrubs, then sun-tolerant trees, then shade-tolerant trees. Grasslands follow a similar pattern, with different species of grasses and forbs (small leafy plants) moving in over time.

Wetland areas also go through succession, responding to the presence of water and sediment depth. In the Everglades, each ecosystem is guided along this path by its own constellation of forces. Some, like the iconic sawgrass ecosystems, are fire adapted; fire returns them to early stages again and again, where the underwater roots of the emergent plants (those that are rooted underwater but grow above the waterline) survive and quickly regrow. Others, if left undisturbed, would pass through successional stages of pioneers (grasses) to shrubs or small trees to larger species of trees, depending on the deposition of soil and proximity of the water table to the surface (the top of the groundwater in the area). Others still are guided by the engineering changes of animals such as alligators, whose digging habits provide the foundation for an entire food chain. In each, though, the same general concept applies: As conditions change, other species better adapted to those conditions move in and displace previous residents. **INFOGRAPHIC 9**

We tend to view the progression of ecological succession as a "repair" sequence. While we can certainly step in to assist in this natural progression to help a damaged ecosystem recover to a former state, the ecosystem is simply doing what comes naturally—responding to changing conditions.

Intact ecosystems have a better chance at recovering from, and thus surviving, perturbations. More complex communities are also more resilient because it is less likely that the loss of one or two species will be felt by the community at large—even if some links in the food web are lost, other species are there to fill the void.

Of course, if keystone species are lost, the community will feel the effect. The loss of the alligator from the complex Everglades community would impact many species and change the face of the ecosystem.

However precarious their recovery might be, wood storks have rebounded in recent years in some areas. Some say this rebound is the result of careful conservation efforts. Others insist that it is merely the result of above-average rainfall in recent years. For his part, Rodgers sees another trend at work. Once again, he says, the storks are trying to tell us something. "They have shifted their center of distribution from South Florida to Central and North Florida," he says. "They're now spilling into Georgia and North Carolina—something we've never seen before."

INFOGRAPHIC 9 **ECOLOGICAL SUCCESSION**

ANIMATED INFOGRAPHIC

FOREST ECOLOGICAL SUCCESSION DEPENDS ON SOIL AND LIGHT AVAILABILITY

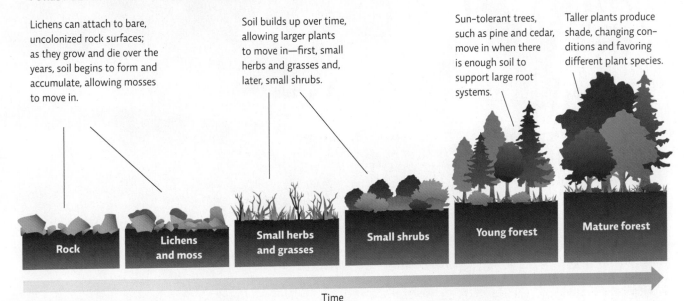

Lichens can attach to bare, uncolonized rock surfaces; as they grow and die over the years, soil begins to form and accumulate, allowing mosses to move in.

Soil builds up over time, allowing larger plants to move in—first, small herbs and grasses and, later, small shrubs.

Sun-tolerant trees, such as pine and cedar, move in when there is enough soil to support large root systems.

Taller plants produce shade, changing conditions and favoring different plant species.

Rock

Lichens and moss

Small herbs and grasses

Small shrubs

Young forest

Mature forest

Time

In terrestrial ecosystems, we see natural stages of succession occur whenever a new area is colonized or an established area is damaged. Sun-tolerant species give way to shade-tolerant ones as more soil is built up, supporting larger plant species.

EVERGLADES ECOLOGICAL SUCCESSION DEPENDS ON THE WATER LEVEL

When water is more than 50 centimeters (20 inches) deep, floating and submerged vegetation grow. As sediments collect, emergent grasses that are rooted underwater—but grow tall enough to emerge above the water surface—move in and establish the sawgrass marsh.

Water-tolerant cypress and willow move in and replace sawgrass as sediment becomes deeper and more stabilized.

Other species of trees, such as the pond apple, move in as sediment gets closer to the water surface.

If dry land emerges from the water, upland species such as oak move in and replace the water-tolerant species of the mixed swamp forest.

Water table (upper level of groundwater)

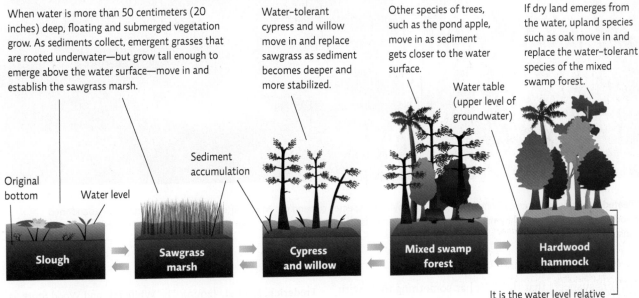

Original bottom

Water level

Sediment accumulation

Slough

Sawgrass marsh

Cypress and willow

Mixed swamp forest

Hardwood hammock

It is the water level relative to the land that determines which species move in.

Ecological succession in the Everglades doesn't necessary follow the tidy predictable sequence seen in terrestrial ecosystems; in fact, periodic fires and the cyclic rainfall patterns may not support a predictable progression at all. Water levels are also important and influence which species move in. Succession in this area can actually go both ways: As ground level changes relative to the water level, an area might flood anew, or sediment buildup might continue to raise the land relative to the water level. In the absence of disturbance, succession will progress to the hardwood hammock forest when sediment builds up enough to expose dry land.

 Look at the forest successional stages shown in the top part of this diagram. Why can't small shrubs or young trees grow on the land shown supporting small herbs and grasses?

Ron Erwin/Getty Images

Pileated woodpeckers are keystone species that many other species depend on. Many different species commandeer abandoned woodpeckers holes in a commensal relationship that benefits the new hole occupant but does not affect the woodpecker than originally constructed the nest hole.

Rodgers suspects that the shift has something to do with the way climate is changing in the region, though he says much more research is needed before anyone can say for certain. "We're still trying to figure out what that means," he says. "But we know it's a clue to something."

Select References:

Brandt, L.A., et al. (2014). *System-wide indicators for Everglades restoration.* Unpublished Technical Report. http://issuu.com/evergladesrestoration/docs/2014_indicator_report?e=8031892/12097978.

Frederick, P., et al. (2009). The White Ibis and Wood Stork as indicators for restoration of the everglades ecosystem. *Ecological Indicators*, 9(6), S83–S95.

Nell, L. A., et al. (2016). Presence of breeding birds improves body condition for a crocodilian nest protector. *PloS One*, 11(3), e0149572.

Rodgers, J. A., et al. (1996). Nesting habitat of wood storks in North and Central Florida, USA. *Colonial Waterbirds*, 19(1), 1–21.

INTERACTIVE MAP   **KEYSTONE SPECIES**

Just like the alligator in the Everglades, ecological communities everywhere have keystone species, each contributing to its ecosystem in unique ways. Here are some additional examples.

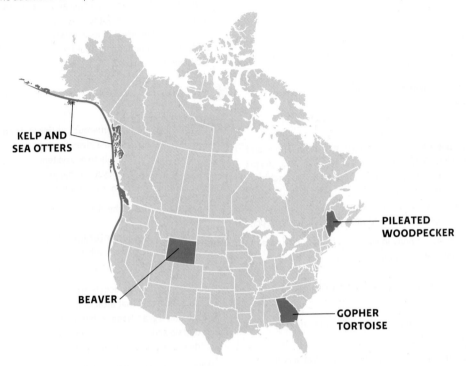

KELP AND
SEA OTTERS

PILEATED
WOODPECKER

BEAVER

GOPHER
TORTOISE

 BRING IT HOME

PERSONAL CHOICES THAT HELP

The world is full of weird and wonderful species. Every year we discover new information about how intricate our biological communities are. By restoring habitats and increasing our understanding of the relationships between species, we can better ensure their long-term survival.

Individual Steps

• Visit a park or nature preserve and watch for signs of species interactions. Do you hear animals or birds? Can you see signs of predation or herbivory?

• Buy a Duck Stamp. Usually purchased by waterfowl hunters for license purposes, nonhunters can purchase a stamp, which supports wetland conservation in the National Wildlife Refuge System.

Group Action

• The Everglades case study is an example of a very extensive restoration project. Call your local park district or nature preserve to see what restoration work is happening in your area and how you can become involved.

Policy Change

• Follow the U.S. Fish and Wildlife Service Open Space blog to learn more about wildlife and issues facing conservation (www.fws.gov/news/blog).

Chris Matula/ZUMA Press/Boyton Beach/FL/USA/Newscom

1 What is an ecological community?

1. Community ecologists study the:
   a. relationships between species in a given area.
   b. way species interact with their environment.
   c. things that increase or decrease the number of species in an area.
   d. All of the above

2. Which of the following might describe the ecological community of a freshwater pond?
   a. All the different populations of animals such as fish and invertebrates that live in the pond
   b. All the different populations of animals that live in the pond and the important abiotic factors such as water temperature and pH
   c. All the different plant, animal, fungal, protozoan, and bacterial populations that live in the pond
   d. All of the important physical features of the pond such as the water temperature and oxygen content, and variety of habitats in the pond

3. Identify an ecological community in your area by naming as many populations found there as you can.

2 What can we learn about a community by identifying its food web?

4. True or False: A food chain shows the flow of matter, but not energy, through a community.

5. Relate the concept of a food chain to a food web. Which concept might be more useful to a community ecologist?

3 What are trophic levels, how are they classified, and why are these illustrated as a pyramid?

6. A given ecosystem has five trophic levels. The number of organisms in each level is indicated in the choices below. Which level represents that of the secondary consumer?
   a. 1,000,000   c. 18,000   e. 60
   b. 40,000   d. 2000

7. The trophic levels on a trophic pyramid are shown as a pyramid because each level is smaller than the one below it. Why is this relationship seen?

4 How is species diversity measured and why is it important to a community?

8. True or False: Ecosystem complexity increases as the variety of habitat and the number of species increases.

9. The total number of species in an area is known as species _____, whereas the relative abundance of each individual species in that area is known as species _____.

10. Explain why both species richness and species evenness are important for a healthy ecosystem.

5 What are core and edge habitats and how can habitat fragmentation affect the species that inhabit them?

11. True or False: Edge species are more negatively affected by habitat fragmentation than core species.

12. Cowbirds lay their eggs in the nests of smaller forest birds such as bluebirds. The bluebirds then spend the next several weeks caring for a baby cowbird, which quickly kills the bluebirds' own young. Cowbirds prefer open, disturbed areas near a forest and seldom venture far into the forest for any reason, even to lay eggs. Use the concept of edge effect to explain what happens to the populations of bluebirds when humans build roads, recreation areas, homes, and businesses in a large forest.

6 What is a keystone species and why is it important?

13. A keystone species is:
   a. typically the most numerous species in its community.
   b. the species lowest on a given food chain.
   c. particularly vulnerable to ecosystem damage and will suffer with only minor ecosystem changes.
   d. very important to many other members of the community.

14. How do alligators fit the definition of a *keystone species*?

7 What species interactions are seen in a community and what problems can emerge if these interactions are disrupted?

15. An example of mutualism is a(n):
   a. dog and a flea that attacks it.
   b. ant and a grasshopper that feed on the same foods.
   c. buffalo and the oxpecker bird that eats ticks found on the buffalo.
   d. deer and a wolf that preys on it.

16. What is resource partitioning and how does it reduce competition?

8 What is the goal of restoration ecology and what obstacles do restoration ecologists face?

17. It might be hard to restore damaged ecosystems like the Everglades because:
   a. there are no economic benefits to restoring the places like the Everglades.
   b. there is no early warning system in place to alert us that the area is being damaged.
   c. some important species might no longer be present.
   d. All of these are correct.

18. What is an indicator species and why are they useful when monitoring the health of an ecosystem?

9 What is ecological succession and how can we use this knowledge to assist in ecosystem restoration?

19. Ecological succession is important because it:
   a. allows ecosystems to respond to environmental changes.
   b. promotes competition and fair use of resources.
   c. eliminates *K*-species.
   d. replaces invasive species with native species.

20. An example of when secondary succession would occur in a particular area would be after:
   a. lichens have started to grow on a bare rock surface.
   b. a flood has removed much of the vegetation.
   c. hot ash from a volcano has completely burned and buried the area.
   d. a disease has reduced the top predator's population.

21. How might a restoration ecologist use an understanding of ecological succession to help repair a damaged area?

SCIENCE LITERACY WORKING WITH DATA

The Mississippi River lies on three sides of the city of New Orleans, curving around it. To the fourth side of the city lies Lake Pontchartrain, 65 kilometers (40 miles) long and over 50 kilometers (30 miles) wide; the lake connects directly with the Gulf of Mexico. Until 140 years ago, much of southern Louisiana was swampland, lower than sea level and much lower than the level of the Mississippi River. As the swamps were drained, levees were built, beginning in the 1700's, speeding up after 1880, and continuing today.

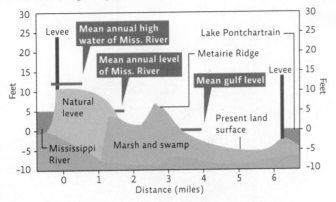

Interpretation

1. What is the difference in mean annual water level, in feet, between Lake Pontchartrain and the Mississippi River at New Orleans? Why is this?

2. New Orleans lies between the levees that hold back the Mississippi River and Lake Pontchartrain. What are the highest and lowest elevations of the city itself, in feet?

3. What type of former ecosystem lies underneath most of the city?

Advance Your Thinking

4. Without the levees throughout the southern half of the state, what would happen?

5. Why do southern Louisiana cemeteries feature above-ground tombs, crypts, and vaults rather than graves?

6. In 2005, Hurricane Katrina devastated New Orleans and much of southern Louisiana. In its aftermath, researchers studied the area, trying to understand what happened. What steps could Louisiana take to avoid a repeat of this flooding and damage in the future?

INFORMATION LITERACY EVALUATING INFORMATION

Woodlands are some of the most important communities in the United States. They are rich with wildlife, and we see them as places worth preserving. The United States began fighting wildfires in a systematic way around the turn of the 20th century. In the past 30 years, however, there have been increased discussions about the automatic response of immediately quelling all wildfires; some ecologists argue that some wildfires are helpful and should be allowed to burn. After all, change from storms, fires, and floods is part of the natural cycle of an ecosystem.

Learn about wildfires and decide for yourself what the best response should be.

1. Search the Internet for information about fire ecology to help develop an informed opinion about how we should respond to wildfires on wildlands. Topics to research include fire-adapted ecosystems, wildfire suppression, the cost of fighting wildfires, and prescribed fires. You may visit as many websites as you need, but you must visit at least four different websites for your research. Answer the following questions to evaluate whether each of the websites you visit is a reliable information source:
   a. Who are the authors of the information given in this article or on this webpage?

   b. Do the authors give supporting evidence for their claims?
   c. Do they give sources for their evidence? Are these reliable sources?
   d. Do you detect any strong biases for or against fighting wildfires? Explain.
   e. Summarize the position of this source regarding how we should respond to wildfires or what could be done to prevent wildfires.
   f. Based on your answers to the above questions, is this a reliable source of information? Explain.

2. Based on the information you obtained from the websites that you deemed to be reliable information sources, write an essay that addresses the following two questions. Provide evidence from your sources in support of your position.
   a. Should all wildfires on wildlands be fought immediately?
   b. What, if anything, should be done in an area to prevent or lessen a possible wildland wildfire?

 Additional study questions are available at SaplingLearning.com.

EVOLUTION AND BIODIVERSITY

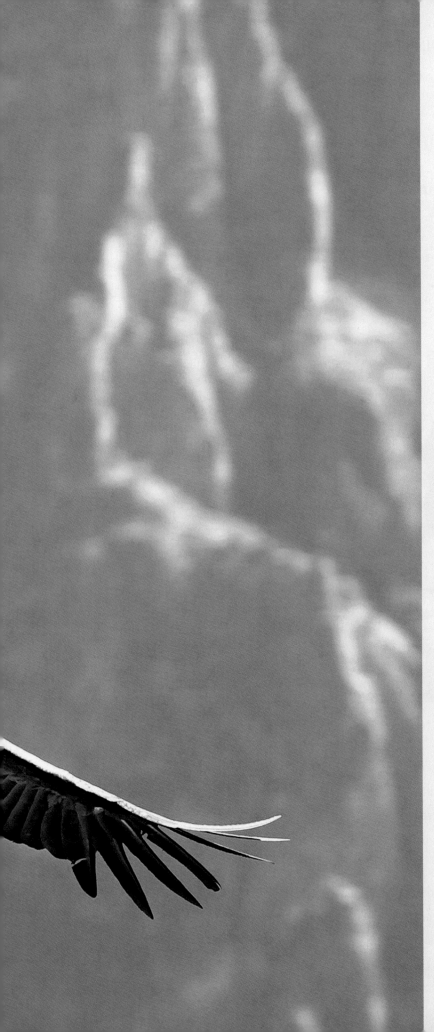

CHAPTER 3

The incredible variety of life on Earth is the result of evolution of species over time; however, human impact is causing a disturbing loss of biodiversity. We can take steps to protect and preserve species but this requires good science, adequate funding, and political will for success.

Module 3.1: Evolution and Extinction

A look at evolution by natural selection and the problems that result when drastic changes are imposed on populations that exceed their ability to adapt

Module 3.2: Biodiversity

An evaluation of the extent and importance of biodiversity and the impact of human actions on current biodiversity

☑ ONLINE Module 3.3: Preserving Biodiversity

A survey of the main threats to biodiversity and steps that can be taken to protect species

Online Modules are available at SaplingLearning.com.

© Henk Meijer/Alamy

A TROPICAL MURDER MYSTERY

Finding the missing birds of Guam

The brown tree snake (Boiga irregularis). James Balog/The Image Bank/Getty Images

After reading this chapter and studying the KEY CONCEPTS and INFOGRAPHICS, you should be able to answer these GUIDING QUESTIONS

CORE MESSAGE

The variety of life on Earth is a result of natural selection favoring the individuals within populations that are best able to survive in their particular environment. Given enough time, populations may be able to adapt to environmental changes, but extinction is also a natural part of this process. Today, human activities are introducing changes so quickly that some species cannot adapt fast enough to survive and are going extinct; many others are endangered and may soon follow.

1 What is biological evolution, and how do populations adapt to changes via natural selection?

2 Why is genetic diversity important to natural selection?

3 What is coevolution, and how does its presence or absence make some native species vulnerable to invasive predators or competitors?

4 How do random events influence the evolution of a population?

5 What factors impact the pace of evolution and extinction, and why are extinctions that occur quickly more of a concern than those that take a long time to unfold?

6 How do the mass extinction events of the past compare to extinctions during intervening times and today?

7 How do humans, intentionally or accidentally, affect the evolution of a population?

On a crisp December morning in 2013, representatives of several federal agencies met on Anderson Air Force Base in Guam—a South Pacific island and U.S. territory—to watch an experiment that sounded more like science fiction than science. As they looked on with binoculars, military personnel in a small fleet of helicopters dropped dead baby mice, one-by-one, into the surrounding jungle. The mice had been laced with acetaminophen and fitted with mini-parachutes.

The mice could be thought of as paratroopers in a war that the island has been fighting for half a century, against a most elusive, yet devastating, enemy: brown tree snakes. Introduced accidentally from other Pacific islands to Guam through ships sometime back in the 1940's, they have driven almost all of the island's native bird species to extinction.

The mouse airdrop was part of the effort to decrease the snake population. Acetaminophen is lethal to the snakes. The parachutes would ensure that the mice would catch in the trees, where the snakes live and eat. If it works, Operation Mouse Drop will help address a problem that has plagued wildlife biologists for half a century and that, for years, was shrouded in mystery.

⊙ WHERE IS GUAM?

GUAM
HAGÅTÑA
CHINA

PHILIPPINES

⊙ **GUAM**

North Pacific Ocean

INDONESIA   NEW GUINEA

AUSTRALIA

Dr. Julie Savidge

Dr. Julie Savidge holding a Mariana Fruit-Dove. This species only occurs on certain islands within the Mariana Islands and the last sighting on Guam was in 1985. This bird was caught as part of an early blood sampling effort to see if exotic diseases might be causing the bird decline on Guam.

It was the late 1960's when the birds of Guam began dying off with disturbing speed. By the early 1980's, four species had gone extinct, ten others were in danger of joining them, and wildlife experts had no clue why. Biologist Julie Savidge, a PhD student at the University of Illinois, took the project on and headed to Guam. Early hypotheses (diseases or pesticide) didn't pan out but the locals were certain that brown tree snakes (*Boiga irregularis*), non-native snakes up to 2 meters long (6 feet) were responsible for the birds' demise. Savidge began to investigate whether these reptiles might be causing the **extinctions**. Surely it was something else; could a few snakes really obliterate a whole island's worth of birds?

extinction The complete loss of a species from an area; may be local (gone from an area) or global (gone for good).

1 NATURAL SELECTION AS A MECHANISM FOR EVOLUTION

Key Concept 1a: Populations can adapt to a changing environment when individuals whose inherited traits make them better suited to survive or reproduce leave more offspring with those traits on average than other, less-suited individuals—a process known as natural selection.

Key Concept 1b: Individuals do not evolve; populations do. A population is said to have evolved if the frequency of genes in a descendant population is different from those in its ancestral population.

Before they started disappearing, Guam was home to 18 native species of birds, each specially suited to life on the island.

Populations usually contain individuals that are genetically different from one another. According to the evolutionary theory first put forth by Charles Darwin and Alfred Russel Wallace, a **selective pressure** on a population—a nonrandom influence that affects who survives or reproduces—favors individuals with certain inherited traits over others (such as better camouflage, tolerance for drought, or enhanced sense of smell). These individuals have *differential reproductive success* compared to other individuals: They leave more offspring than those who are less suited for their environment.

The process by which organisms best adapted to the environment survive to pass on their traits is **natural selection**. Evolutionary biology helps us understand the diversity of life on Earth and how populations change over time. It is one of the pillars of biological science, and the vast amount of evidence from a wide variety of scientific disciplines in support of both the occurrence of evolution and the mechanisms by which it happens has elevated this explanation to the level of scientific theory (see Module 1.2).

For most populations, more offspring are born than can survive, since resources are limited and many species produce large numbers of young. Since only some individuals will survive, over time, the population will contain more and more of these better-adapted individuals and their offspring. Ultimately, this changes how common certain variants of **genes** (these variants are called **alleles**) are in the population: The frequency (percentage in the population) of some alleles increases and that of others decreases. When this occurs, the population has experienced

evolution, or changes in the **gene frequencies** within a population from one generation to the next. Natural selection may be *stabilizing, directional,* or *disruptive,* depending on which genetic traits are favored or selected against. **INFOGRAPHIC 1**

It is important to note that *individuals* are selected, but *populations* evolve; individuals do not change their own genetic makeup to produce new necessary adaptations, such as bigger size or pesticide resistance. If they get the opportunity to reproduce, they pass on their traits to the next generation. If they cannot tolerate environmental changes, as was the case with the first bird species to disappear from Guam (the bridled white-eye), they die or fail to reproduce and do not pass on their genes. Individuals may be able to adjust their behavior to accommodate environmental changes, but if a trait is not genetically controlled, and therefore is not heritable, it cannot be passed on to the next generation.

---

selective pressure A nonrandom influence that affects who survives or reproduces.

natural selection The process by which organisms best adapted to the environment (the fittest) survive to reproduce, leaving more offspring than less well-adapted individuals.

genes Stretches of DNA, the hereditary material of cells, that each direct the production of a particular protein and influence an individual's traits.

alleles Variants of genes that account for the diversity of traits seen in a population.

evolution Differences in the gene frequencies within a population from one generation to the next.

gene frequencies The assortment and abundance of particular variants of genes relative to each other within a population.

U.S. Department of Agriculture wildlife specialist Tony Salas holds a brown tree snake at Anderson Air Force Base in Guam.

INFOGRAPHIC 1   **NATURAL SELECTION AT WORK**   ANNATED INFOGRAPHIC

When the environment presents a selective force (e.g., a new predator, changing temperatures, change in food supply), natural selection is the primary force by which populations adapt. The survivors are those who were lucky enough to have genetic traits that allowed them to survive in their changing environment. (Others who did not possess the trait were not as likely to survive to reproduce.) Because survivors pass on those adaptations to their offspring, the gene frequencies of the population change in the next generation, which means some traits are more common and others are less common than they used to be. When this happens, the population is said to have evolved.

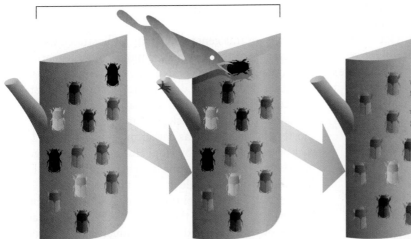

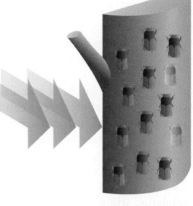

Original population · Next generation · Later generation

Beetles resting on this tree vary in color.

Genetic variation exists in the population: Individuals possess inherited differences.

Individuals with the less favorable trait (coloration that makes them stand out on a tree trunk) are more likely to be eaten.

Differential reproductive success: Not everyone will survive to reproduce.

Fewer dark individuals are born (though recombination might produce some from light or tan parents).

Gene frequencies have changed: The population is evolving.

Over time, the population may be mostly or solely made up of tan individuals.

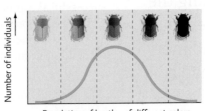

Number of individuals

Population of beetles of different colors

— Original population
— Evolved population

Natural selection can have different outcomes, depending on what varieties in the population the environment favors or selects against.

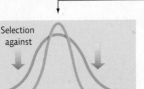
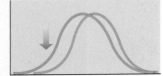
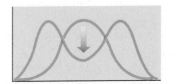

Selection against

Stabilizing selection favors the norm and selects against extremes.

Directional selection continually favors a particular extreme of the trait (bigger, darker, etc.).

Disruptive selection favors the extremes but selects against the intermediate forms.

 All trees are tan; tan beetles are favored.

 In areas with trees darkened from pollution, darker beetles are favored.

 In a forest with light and dark trees but no tan trees, tan beetles (the intermediate color) are not favored.

 Identify the gene frequencies of the "original population" for each color morph (dark gray, dark brown, light brown, dark tan, and light tan) by counting the number of each and expressing it as a percentage of the whole. Now do the same for the "later generation." Has evolution occurred? Explain.

2 GENETIC DIVERSITY AND NATURAL SELECTION

Key Concept 2: Genetic diversity in a population is the raw material on which natural selection operates. The more diverse a population, the more likely there will be individuals present who can withstand or even thrive if environmental conditions change.

The ability of a population to adapt is a reflection of its tolerance limits, which largely depend on **genetic diversity**—different individuals having different alleles. A population that is highly diverse (has individuals with many different traits) is likely to have wider tolerance limits (see Infographic 4 in Module 2.1), which increases the population's potential to adapt to changes. This means it is more likely that some individuals will exist that can withstand (or even thrive in) the changes and that the population as a whole will survive. If a change occurs that produces a condition outside the range where individuals can survive and reproduce (for instance, the climate becomes warmer than anyone can survive), the population will die out. Similarly, if a new challenge is presented, such as the introduction of a new predator or competitor, the survival of the population will depend on whether there are any individuals in the population who can effectively deal with the new species. If the snakes on Guam were indeed responsible for killing the birds, any birds that happened to have effective snake-avoidance behaviors would have had a greater chance of survival.

> If the snakes on Guam were indeed responsible for killing the birds, any birds that happened to have effective snake-avoidance behaviors would have had a greater chance of survival.

Two main sources of variation can increase genetic diversity in a population. The ultimate source of new variability is genetic *mutation*, a change in the DNA sequence in the sex cells that alters a gene, sometimes to the extent that it produces a new trait. Mutations are rare, but because DNA replication and repair occur all the time, these rare events add up. When a mutation produces traits that are beneficial, they can quickly be passed on to the next generation, allowing the population to evolve to be better adapted to its environment. A second source of genetic variety occurs as eggs and sperm are made: *Genetic recombination* shuffles alleles around and

genetic diversity The heritable variation among individuals of a single population or within a species as a whole.

sometimes produces individuals with new traits when a sperm fertilizes an egg.

The value of this genetic diversity is illustrated today in the example of the rock pocket mouse of the American Southwest desert. Animals of this species have coats that are either light tan or a darker color. It turns out that coat color corresponds to a population's environment: Areas of light-colored rock contain populations with mostly tan mice, whereas darker mice inhabit black lava rock regions. Research by Hopi Hoekstra and colleagues at the University of Arizona has shown that coat color is determined by a single gene that comes in two different alleles. The dominant allele is designated by the uppercase letter D; the recessive allele is designated by the lowercase letter d. All individuals have two copies of the gene, and the color of their coat is determined by which two alleles they possess. Darker mice have at least one dominant allele (DD or Dd). Tan mice possess two recessive alleles (dd).

It is likely that coat color provides camouflage and protection from visual hunters, but only if the mouse is on a background of the same color. A study on deer mice (a similar species) showed that predatory owls are more successful at capturing mice on a contrasting background. This gives support to the conclusion that coat color is adaptive as camouflage and therefore is responsive to natural selection. **INFOGRAPHIC 2**

These potato beetles show genetic diversity—the raw material on which natural selection works.

Roger Tidman/FLPA/Science Source

INFOGRAPHIC 2 · EVOLUTION IN ACTION

Natural selection produces populations with different gene frequencies (more or less of a particular gene variant or allele). For this to occur, there must be genetic variation (more than one allele for a given trait) and a selective pressure (a reason one variant is better than another in a given situation).

Different color morphs of the rock pocket mouse (*Chaetodipus intermedius*) are found on different-colored rocky outcroppings in the desert Southwest. An evaluation of the mice living on or near the Pinacate lava flow in southern Arizona represents the first documentation of the genetic basis (in this case, a single gene) for a naturally favored trait. The well-known peppered moth is another example in which different color variants are favored in different habitats but the genes responsible for that trait have not yet been identified.

Even though there is gene flow between dark and light populations that are close to one another, populations on tan rock have mostly tan individuals, and populations on dark rock have mostly dark individuals, suggesting a strong selective pressure that favors one color over the other.

Predatory owls are likely the selective pressure that favors different coat colors in different habitats.

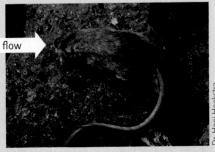

Gene flow

Tan mice (the recessive trait, dd) predominate in light-colored rocky outcroppings.

Darker mice (the dominant trait, DD or Dd) predominate on darker lava rocks.

A study done with dark and light varieties of deer mice revealed that owls caught twice as many opposite-colored mice (dark mice on a light background or tan mice on a dark background) as mice whose coloration matched their background, even in almost total darkness. Owl predation is therefore likely to be a strong selective pressure on coat color, driving directional selection that produces either light or dark populations of mice, depending on the background.

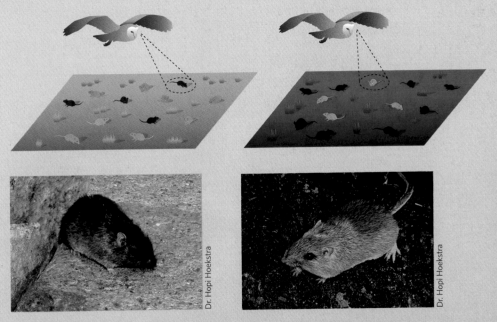

 If this mouse population migrated to an area with a red rock habitat with visual predators like hawks or owls, what could prevent the population from evolving into one with red coats?

3 COEVOLUTION

> **Key Concept 3a:** Two species can become highly adapted to one another when each becomes the selective pressure that favors certain traits in the other, a process known as coevolution.
>
> **Key Concept 3b:** Species that never coevolved with a particular predator or competitor may not have the traits needed to survive if that species invades their habitat.

A special type of natural selection is known as **coevolution**. In coevolution, two species each provide the selective pressure that determines which of the other's traits is favored by natural selection. Predator and prey species usually evolve together, each exerting selective pressures that shape the other. As predators get better at catching prey, the only prey to survive are those a little better at escaping, and it is those individuals that reproduce and populate the next generation. This game of one-upmanship continues generation after generation, with each species affecting the differential survival and reproductive success of the other. The result can be a predator extremely well equipped to capture prey and prey extremely well equipped to escape. **INFOGRAPHIC 3**

coevolution A special type of natural selection in which two species each provide the selective pressure that determines which traits are favored by natural selection in the other.

invasive species A non-native species (a species outside its range) whose introduction causes or is likely to cause economic or environmental harm or harm to human health.

endemic Describes a species that is native to a particular area and is not naturally found elsewhere.

Non-native species that cause ecological, economic, or human health problems and are hard to eradicate are considered **invasive species**, and they can cause significant damage in areas they invade. In fact, invasive species are the second leading cause of species endangerment, worldwide (see Online Module 3.3).

If the birds on Guam were indeed eradicated by the invasive snake species, it was because the speed at which the eradication happened prevented the bird populations from potentially coevolving survival strategies to deal with the new snake population. The brown tree snake was already well adapted to preying on birds. But Guam's bird populations had never faced such a predator and had no natural defenses. It was an unfair fight. And, like populations worldwide that are isolated on islands or mountaintops or by fragmented habitats, Guam's populations were further handicapped by their isolation; they rarely, if ever, received new individuals from populations elsewhere. Indeed, some of Guam's species were **endemic**—found nowhere else—so no other

Many ground-nesting birds and turtles in Hawaii have no defenses against this invasive mongoose, a skilled predator that eats their eggs and hatchlings.

Malcolm Schuyl/Alamy

INFOGRAPHIC 3   **COEVOLUTION ALLOWS POPULATIONS TO ADAPT TO EACH OTHER**

As selection favored beetles closest to the tree color, only birds with the keenest eyesight feed well enough to survive and reproduce.

Any beetle with an even better camouflage would escape predation and pass on its genes.

This then favors birds with even keener eyesight that would feed well and pass on the sharp eyesight trait to their offspring.

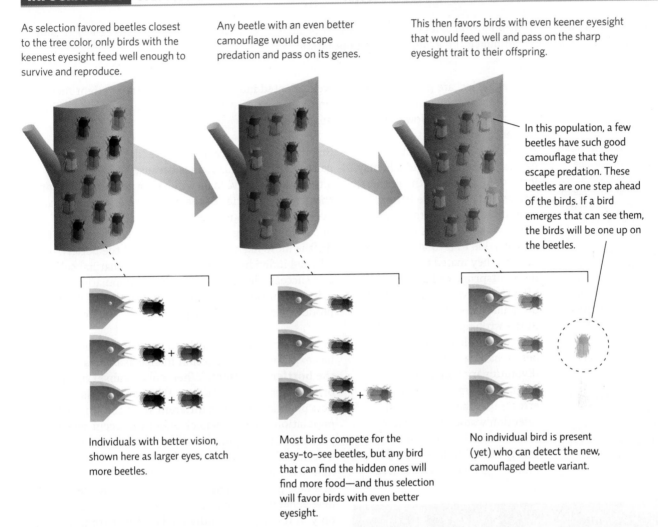

In this population, a few beetles have such good camouflage that they escape predation. These beetles are one step ahead of the birds. If a bird emerges that can see them, the birds will be one up on the beetles.

Individuals with better vision, shown here as larger eyes, catch more beetles.

Most birds compete for the easy-to-see beetles, but any bird that can find the hidden ones will find more food—and thus selection will favor birds with even better eyesight.

No individual bird is present (yet) who can detect the new, camouflaged beetle variant.

 What do you predict would happen to the original beetle population if a species of bird with extremely keen eyesight (like those shown on the right side of the diagram) were accidently introduced into the beetle's habitat?

populations even existed to contribute new members who might be better adapted to the snake.

Some of Guam's bird species went extinct sooner than others. For instance, the endemic bridled white-eye, the gregarious bird species that was extinguished first, happens to be very small, raising the possibility that the small size of these birds might have put them at a disadvantage. Larger species like flycatchers survived longer, though they, too, eventually disappeared. Other bird species experienced **extirpation**; the Guam rail, for instance, is gone from Guam, but other populations still live on the nearby island of Rota. If some individuals of the bridled white-eye or other extinct species had been able to avoid the snake (perhaps due to a heritable trait that made them more wary of the predator), they might have given rise to new populations that could cohabitate with the snake.

When populations diverge because of isolation, food availability, new predators, or habitat fragmentation that prevents the ability of population members to freely interbreed, new species may arise (*speciation*). This increases the number of species in a community and sometimes produces specialists that can exploit open niches. This separation may be physical (for example, geographic boundaries the individuals won't cross) or may arise when something prevents some individuals from choosing others as mates, as may happen when individuals spend their time in different parts of their habitat. However, not all evolution is driven in this manner. Random events play a role, too, typically by decreasing genetic diversity rather than increasing it.

extirpation Locally extinct in one geographic area, but still found elsewhere.

4 RANDOM EVENTS AND EVOLUTION

Key Concept 4: Along with natural selection, random events such as genetic drift, the bottleneck effect, and the founder effect also influence the evolution of a population.

In **genetic drift**, some traits (alleles) are passed on or lost by random chance, not because they were selected for (or against), as with natural selection. How could this happen? Even with natural selection at work, in each generation some individuals may leave more offspring than others, not because they were better adapted to their environment but because they "got lucky"—perhaps due to favorable external factors, they mated more, had more offspring, or had more offspring survive. Others might be "unlucky"—they might be in the wrong place at the wrong time (e.g., killed by a mudslide) before having a chance to mate. (Evolutionary biologists refer to this as a type of "sampling error"—just because natural selection would favor certain traits over others, doesn't mean every "better" individual will leave more progeny than all the less

genetic drift The change in gene frequencies of a population over time due to random mating that results in the loss of some gene variants.

bottleneck effect The situation that occurs when population size is drastically reduced, leading to the loss of some genetic variants, and resulting in a less diverse population.

founder effect The situation that occurs when a small group with only a subset of the larger population's genetic diversity becomes isolated and evolves into a different population, missing some of the traits of the original.

well-adapted individuals; the best traits might increase one's chances of survival or reproduction, but no trait guarantees it.)

Small populations are much more likely to experience genetic drift than large ones because the offspring of a few "lucky" individuals will have a greater impact on the gene frequencies of the next generation in a small population than a few "lucky" individuals in a large population. In large populations, the effect of genetic drift is more likely to be masked by all the better-adapted individuals who are favored by natural selection and successfully reproduce. In small populations, genetic drift can quickly lead to losses in genetic variability that produce major evolutionary changes in a population.

Genetic drift is more likely to happen if a population has experienced a reduction in size as occurs with the **bottleneck effect**. When only a subset of the original variants reproduces, they can give rise to a new population that is different from the original population. The bottleneck effect can occur when a portion of the population dies, perhaps because of a natural disaster like a flood or because of a strong new selective pressure, such as the introduction of a new predator. The survivors then produce a new generation, and any alleles that were present only in the deceased individuals are lost from the population forever.

The **founder effect** also reduces population size and therefore available alleles. Consider the situation in which a small subset of a population colonizes a new area. If this subset (the founding population) happens to be less genetically diverse than the original, and if the subset becomes completely isolated from the original group such that there is no mixing of the two populations (and no chance to reintroduce those missing alleles), the founding population will produce a population that has different genetic frequencies than the original population. (See Module 3.2 for more on the effect of isolation.)

Today, human impact increases instances of both the founder effect and the bottleneck effect. Much of what we do isolates populations into smaller groups, forcing them into these situations. **INFOGRAPHIC 4**

Susan Karr

The ancestors of the giant tortoises of the Galapagos islands, which might have arrived at these isolated islands on floating vegetation, have evolved into distinct populations. The larger domed-shelled tortoises, like this one living on Santa Cruz, are found in areas with abundant plant life. The saddle-backed tortoises are adapted to arid regions with less food; their shell is angled upward in the front, allowing the animal to stretch its neck higher in search of food.

INFOGRAPHIC 4   **RANDOM EVENTS CAN ALTER POPULATIONS THROUGH GENETIC DRIFT**

GENETIC DRIFT

Genetic drift occurs when random events eliminate some gene variants (alleles) from a population. This happens because, in addition to natural selection, chance also influences who survives or reproduces. Genetic drift is more likely to accumulate and have major effects in small populations.

A population contains a variety of individuals, but some gene variants are more common than others.

Random mating occurs, but some unlucky individuals don't find mates.

Subsequent generations may have different gene frequencies.

BOTTLENECK

If something causes a large part of the population to die, leaving the survivors with only a portion of the original genetic diversity, the population may recover in size but will not be as genetically diverse as the original population.

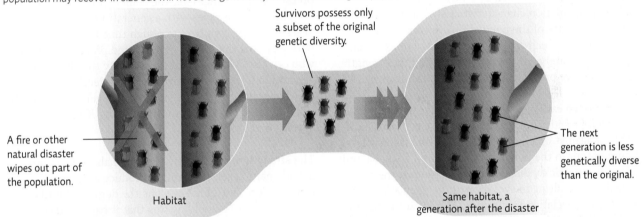

Survivors possess only a subset of the original genetic diversity.

A fire or other natural disaster wipes out part of the population.

Habitat

Same habitat, a generation after the disaster

The next generation is less genetically diverse than the original.

FOUNDER EFFECT

If a small subset of a population that possesses only a fraction of the genetic variability of the original population colonizes a new area, and the subset becomes completely isolated from the original group, the new population will likely produce descendant populations that have different gene frequencies than the original population.

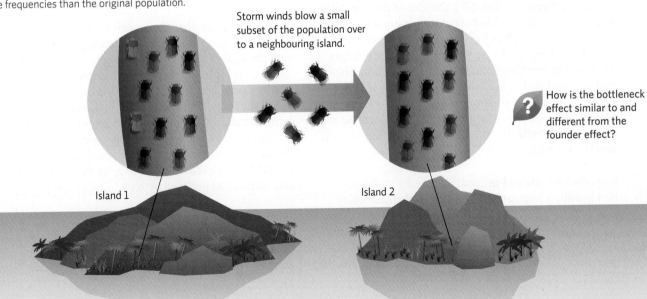

Storm winds blow a small subset of the population over to a neighbouring island.

Island 1

Island 2

? How is the bottleneck effect similar to and different from the founder effect?

5 THE PACE OF EVOLUTION AND EXTINCTION

Key Concept 5a: The pace of evolution and extinction is generally slow and is affected by population size and genetic diversity, reproductive rate, generation time, and the strength of the selective pressures at play.

Key Concept 5b: When extinctions unfold over long periods of time, community connections are not broken because better-adapted species replace their predecessors and the niche remains filled; rapid extinction events may eliminate well-adapted species and break important community connections.

The pace of evolution is not constant or the same for all species but in general it is slow—changes accumulate over generations and speciation events can take thousands or millions of years. (There is evidence that species may go through long periods of stasis punctuated by rapid episodes of speciation—"rapid" here means less than 100,000 years or so—in response to environmental change.)

Evolution's pace is affected by a variety of factors including population characteristics such as the genetic diversity and size of the population, aspects of the species' biology such as its biotic potential (maximum reproductive rate) and generation time, and the strength of the selective pressure a population is facing.

As mentioned earlier, genetic diversity is important because it provides options for natural selection to favor or select against. If a new selective pressure favors a less common trait in the population, or if a new favorable trait arises, it can quickly displace other variants in the next generation, resulting in evolution of the population. If this results in a speciation event, it can eliminate an ancestral species. When this happens—when a new species replaces an ancestral one—the outcome is unlikely to negatively impact the ecosystem because the niche is still filled.

In addition to genetic diversity, the size of the population also makes a difference in how quickly natural selection can produce a change in a population: Beneficial traits can spread more quickly in smaller populations simply because it is more likely that the individuals with the trait will find each other and mate (as long as it is not a population that is widely dispersed). Of course, as mentioned earlier, smaller populations could also be at a higher risk of extinction because they likely contain less genetic diversity.

Reproductive rate and generation time also influence how quickly a population can adapt to changes. Many problem species, like insect pests, are r-selected species and have high reproductive rates and fast generation times, which means they

endangered species Species at high risk of becoming extinct.

can often stay one step ahead of our efforts to control them. Populations of r-selected species decimated by a disturbance or a depleted resource can quickly bounce back and repopulate the area because the remaining individuals can produce so many offspring in a single breeding cycle. Many **endangered species**, on the other hand, are K-selected species, with slower reproductive rates and longer generation times; therefore, they take longer to recover if population numbers fall. Selective pressures that change over the course of just a few years can eliminate a species with a generation time of many years—there is simply not enough time for those able to withstand the stressor to grow up and produce progeny who can also withstand the stressor. By the time this next generation is of reproductive age, the selective pressure may have changed again, rendering this generation poorly adapted for survival. (For more on r- and K-selected species, see Module 2.2.)

The strength of the selective pressure also affects how quickly natural selection might produce a change in a population. One of the reasons the demise of birds in Guam was so stupefying was that it happened so quickly—particularly for the small birds, which were easiest for the snakes to eat. Larger birds disappeared later, when the snakes started eating their nestlings and eggs. **INFOGRAPHIC 5**

To determine whether the snakes were a strong selective pressure on the birds of Guam, Savidge first had to be sure the snakes actually *liked* eating birds. To test this, she set out bird-baited traps around the island and on nearby Cocos Island (which has no brown tree snakes). What she found shocked her: "In one area where the birds were extinct, 75% of my traps got hit within 4 nights," she recalls. On Cocos, all the birds used as bait survived. Surveys of Guam's abundance of small mammals also revealed heavy predation pressure by the snakes—mice and shrews had declined by 94%.

The hunting skill of the snake and the lack of antipredator behaviors in the birds made this new predator a strong selective pressure; a less-proficient predator or one whose hunting style was familiar to the birds would have been a

INFOGRAPHIC 5 **THE PACE OF EVOLUTION**

The speed at which evolution can occur is influenced by a variety of factors. While the relationships below are not guaranteed to affect the rate of evolution as shown (for example, even a population with high genetic diversity would not be able to evolve (or survive) if *no one* in the population could withstand the environmental disturbance experienced), in general, the following correlations are seen.

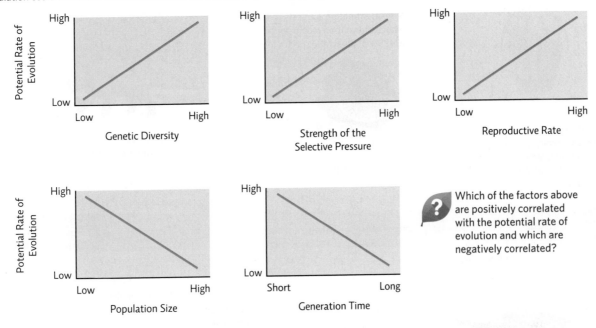

Which of the factors above are positively correlated with the potential rate of evolution and which are negatively correlated?

weaker selective pressure. Indeed, one lesson of the brown tree snake invasion is that while speciation typically occurs at a slow pace, extinction can occur much more quickly if the rate of change exceeds the ability of the population to adapt. These rapid extinction events are a concern because they can break community connections and leave unfilled niches, negatively impacting other species.

All in all, Savidge's three pieces of evidence—the fact that the geographic location of the snakes correlated strongly with the birds' disappearance, that brown tree snakes liked to eat birds, and that other small mammals also went missing after the snakes' arrival—convinced Savidge that she had finally solved the mystery of Guam's disappearing birds. Brown tree snakes, she concluded, were definitely the culprit.

6 MASS EXTINCTIONS: PAST AND PRESENT

Key Concept 6: Extinction rates were much higher in mass extinction events than at other times. Past mass extinctions are linked to natural causes; today, human impact appears to be causing another mass extinction.

Extinction is nothing new on Earth. By most estimates, more than 99% of all species that ever lived on the planet have gone extinct. Based on a critical analysis of the fossil record, scientists agree that there have been five *mass extinction events*—when species have gone extinct at much greater rates than during intervening times, each event leading to the loss of 75% or more of the species present on Earth. The most infamous of these was the *K-T boundary mass extinction*, which occurred at the transition from the Cretaceous to the Tertiary period, 65 million years ago. Most scientists agree the K-T extinction event

was set off by an asteroid impact in the Gulf of Mexico; 76% of all living species, including the dinosaurs, were wiped out. **INFOGRAPHIC 6A**

Earth's past mass extinctions were due to catastrophic events or physical changes to the atmosphere or oceans, which altered the environment faster than species' ability to adapt. Though most species loss wasn't overnight, taking from hundreds to millions of years to unfold, these kinds of events eventually led to the emergence of new species, as surviving populations adapted to the

INFOGRAPHIC 6A EARTH'S MASS EXTINCTIONS

There have been five mass extinctions in Earth's history, each believed to be caused by major environmental changes such as the meteor that struck Earth at the end of the Cretaceous period.

Permian extinction: Largest extinction event on record, with 90–95% of all marine species lost

Ordovician extinction: >20% of families (a taxonomic grouping that includes similar species) lost, including 85% of all marine species

Cretaceous extinction: 70% of species lost

Cambrian	Ordovician	Silurian	Devonian	Carboniferous	Permian	Triassic	Jurassic	Cretaceous	Tertiary	Quarternary
520	510	439	409	363	290	248	210	146	65	1.64

Millions of years ago

Triassic extinction: 20% of families lost, including most species in a prominent reptile group known as crurotarsans—ancestors to modern crocodiles—whose loss allowed the evolution of dinosaurs

Devonian extinction: >20% of families lost, including 80% of all marine species

 Why do we see mass extinction events occurring at transitions from one geologic time period to the next?

available niches (a process that took millions of years). Cycles of extinction and evolution ultimately gave rise to the diversity of life we see on Earth today—estimates range from 3 to 100 million species. Throughout most of Earth's history, the **background rate of extinction**—the average rate of extinction that occurs between mass extinction events—has been slow. The fossil record [the total collection of fossils (remains, impressions, traces of ancient organisms) found on Earth] tells us that, on average, 1 or 2 species out of every 1 million species goes extinct each year. In a world with 3 million species, this would be 3 to 6 species per year; if Earth is home to 100 million species, that would be 100 to 200 species per year.

A recent 2015 collaborative study by Mexican and U.S. researchers looked closely at vertebrate extinctions. They compared the number of species extinctions

---

background rate of extinction The average rate of extinction that occurred before the appearance of humans or that occurs between mass extinction events.

that have been documented since 1500 to a background rate that was determined by an extensive examination of the mammalian fossil record. When they compared the number of extinctions observed to the number of extinctions that would have been expected given the background rate of extinction, they found a sharp increase beginning about 200 years ago. They further calculated that if the number of species lost in the last 100 years had been lost at the background rate of extinction, it would have taken between 800 and 10,000 years. **INFOGRAPHIC 6B**

Most scientists agree that the accelerated extinction we are witnessing can be considered a sixth major extinction event, and that it is largely driven by human actions. As our use of resources increases, driven by population growth and affluence, our impact is becoming much more devastating for other species. We remove the resources they need to survive, minimize their habitat ranges, introduce new predators or competitors,

INFOGRAPHIC 6B   THE RATE OF EXTINCTION IS ACCELERATING

Recent rates of species extinctions are well above that expected by the background rate, a trend attributed to human impact such as habitat destruction, the introduction of invasive species, and climate change. This graph shows an estimate of extinctions since 1500 for vertebrates in comparison to the background rate of extinction. Accelerated extinction rates are seen for vertebrates, especially since the 19th century. This estimate may under-represent extinctions for fish, amphibians, and reptiles because less data is available for these groups.

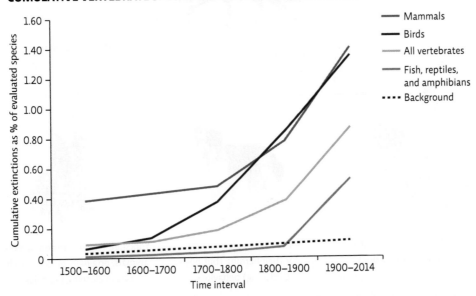

CUMULATIVE VERTEBRATE SPECIES REPORTED EXTINCT SINCE 1500

Why might we have more data on bird and mammal extinctions than those for fish, amphibians, and reptiles? Do you think more data would reveal higher rates of extinction or lower ones for these groups?

and strip them of their genetic diversity, all of which slowly eliminate them. British researchers Ian Owens and Peter Bennett analyzed the extinction risk for 1,012 *threatened* bird species (those at risk for becoming endangered) and found that habitat destruction was cited as a risk factor in 70% of the cases. Other human interventions, such as the introduction of non-native species or overharvesting were implicated in 35% of the cases. (See Module 3.2 and Online Module 3.3 for more on species endangerment.)

In Guam, the near-total disappearance of birds between the 1960's and 1980's was a biological murder mystery. The brown tree snake was the cause of their demise but the inadvertent introduction of this species by humans puts the blame squarely on us.

7 ARTIFICIAL SELECTION

Key Concept 7: In artificial selection, humans choose which traits to keep and which to eliminate from a population through selective breeding. Our actions have also inadvertently led to the evolution of antibiotic- or pesticide-resistant populations.

The introduction of the brown tree snake to Guam was an accident: A snake hitchhiker crossed the ocean on a human tanker—unbeknownst to the crew—and landed in a veritable bird buffet. But humans also directly affect the evolution of a population through **artificial selection**. Artificial selection works the same way as natural selection but the difference is that the selective pressure is us (humans). For many animal species, from pets to farm animals and plant species, humans choose who breeds with whom in an attempt to produce new individuals with the traits they desire. By doing this over many generations, people have accentuated certain plant and animal traits, sometimes to extremes. For instance, artificial selection created domestic dogs from their wolf ancestors. **INFOGRAPHIC 7**

artificial selection A process in which humans decide which individuals breed and which do not in an attempt to produce a population of plants or animals with desired traits.

INFOGRAPHIC 7 **HUMANS USE ARTIFICIAL SELECTION TO PRODUCE PLANTS OR ANIMALS WITH DESIRED TRAITS**

All dogs (*Canis lupus familiaris*) are descendants of the wolf (*Canis lupus*). By only breeding those males and females with the traits desired (size, herding ability, protective instinct, etc.), humans have created more than 170 dog breeds.

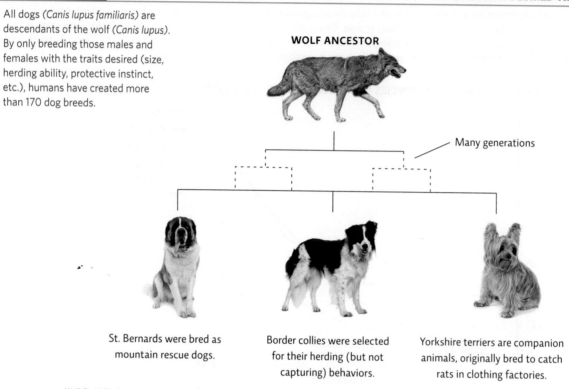

WOLF ANCESTOR

Many generations

St. Bernards were bred as mountain rescue dogs.

Border collies were selected for their herding (but not capturing) behaviors.

Yorkshire terriers are companion animals, originally bred to catch rats in clothing factories.

Wolf: DaddyBit/Getty Images; St. Bernards: GlobalP/Getty Images; Border collies: © GlobalP/Getty Images; Yorkshire terriers: jimmyjamesbond/Getty Images

 Why can we say that artificial selection is goal directed but natural selection is not?

But evolution is ever at work. Pesticide- and antibiotic-resistant populations can emerge as an inadvertent human-influenced selection. When we apply a chemical that kills a pest or pathogen, some individuals survive because of their natural genetic resistance; that is, the individuals were already resistant even though they had never encountered the chemical. These survivors are then the only individuals who reproduce, producing the next generation that is also pesticide resistant, ultimately changing the frequency of resistant genes in the population (see Infographic 2B in Module 8.2).

An understanding of how populations evolve, and how we can affect that process, can help us avoid actions that create problems such as untimely species extinctions or the emergence of antibiotic-resistant bacteria. By changing the environment, humans apply a number of new selective pressures on populations. Our changes have the capacity to be so rapid or so great that natural selection simply cannot keep up—perhaps because a new needed trait is not present in the population, or it cannot spread quickly enough to prevent a population collapse. The accidental introduction of the predatory brown tree snake is one such example; it introduced a major change, basically overnight,

that was able to eat its way through the vertebrate populations before those populations could adapt.

New snake control methods continue to be developed and tested in Guam. The goal is not to eradicate the snakes (which probably isn't possible) but to control and contain them so that they don't find their way to other islands, explains Diane Vice, a wildlife biologist with the Guam Department of Agriculture. "The hope is to create safe habitat," Vice says, "so that these beautiful native species can once again thrive."

Select References:

Ceballos, G., et al. (2015). Accelerated modern human–induced species losses: Entering the sixth mass extinction. *Science Advances,* 1(5), e1400253.

Clark, L., et al. (2012). Efficacy, effort, and cost comparisons of trapping and acetaminophen-baiting for control of brown treesnakes on Guam. *Human–Wildlife Interactions,* 6(2), 222–236. www.aphis.usda.gov/wildlife_damage/nwrc/publications/12pubs/clark122.pdf.

Hoekstra, H. E., et al. (2005). Local adaptation in the rock pocket mouse (*Chaetopidus intermedius*): Natural selection and phylogenetic history of populations. *Heredity,* 94(2), 217–228.

Savidge, J. A. (1987). Extinction of an island forest avifauna by an introduced snake. *Ecology,* 68(3), 660–668.

INTERACTIVE MAP   **INVASIVE SPECIES**

Non-native species that find their way to other ecosystems can wreak havoc on native species unprepared to deal with them. Here are a few notable invasive species that are taking their toll on U.S. ecosystems.

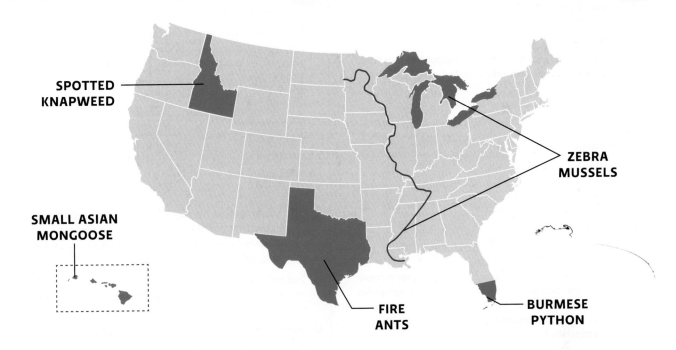

SPOTTED KNAPWEED

ZEBRA MUSSELS

SMALL ASIAN MONGOOSE

FIRE ANTS

BURMESE PYTHON

   BRING IT HOME

PERSONAL CHOICES THAT HELP

The astonishing variety of life found on Earth is a result of natural selection favoring those individuals within populations that are best able to survive in their particular environment. Given enough time, some populations may be able to adapt to environmental changes. However, human activities may disrupt natural ecosystems so that organisms cannot adapt fast enough to survive, and those organisms may go extinct. Conservation activities can help protect vulnerable organisms and ecosystems.

Individual Steps

• Live in an older, established area of your community. Suburban sprawl reduces habitat for wildlife, and reliance on cars causes greenhouse gas emissions that could result in species-threatening climate change.

• Save your pocket change and, at the end of every year, donate the money to a land, marine, or wildlife protection agency.
• Create a personal blog that includes photographs of wildlife, facts about current threats to plants and animals, and articles about conservation.

Group Action

• Throw a party in support of wildlife conservation. Take a collection at the door and donate the money to an organization that supports conservation.

Policy Change

• "Adopt an Organism." The U.S. Fish and Wildlife Service maintains a database of endangered plants and animals in every state. Research this database to find wildlife that interests you. Determine what agency, conservation group, or legislator

you could contact and then start your own protection campaign. See what meetings, petitions, and legislation could impact your organism and get involved.

ENVIRONMENTAL LITERACY **UNDERSTANDING THE ISSUE**

1 What is biological evolution, and how do populations adapt to changes via natural selection?

1. Evolution is defined as:
   a. a change in the gene frequencies in a population over time.
   b. the changes observed in individuals that are adapting to a new environment.
   c. any shift in the alleles that an individual possesses or passes on to its offspring.
   d. the acquisition of mutations in response to environmental pressures.

2. A population of butterflies used to have small, medium, and large individuals, but several years ago a non-native bird was introduced to the butterfly's habitat. It eats butterflies but only those that are medium size. Eventually, medium-sized butterflies became rare. This is an example of:
   a. disruptive selection.
   b. stabilizing selection.
   c. artificial selection.
   d. directional selection.

3. "When exposed to pesticides, some Japanese beetles become more pesticide resistant and harder to kill." There is something wrong with this statement. Reword it to more accurately represent what happens when pesticide application leads to a pesticide-resistant population.

2 Why is genetic diversity important to natural selection?

4. True or False: Populations with high genetic diversity will always be able to adapt to changing environments.

5. What are the main sources of genetic variation in a population?
   a. Mutation
   b. Genetic recombination
   c. A selective pressure that creates a need for a new trait
   d. A and B
   e. A, B, and C

6. Using the example of the rock pocket mouse, explain the importance of genetic diversity to a population.

3 What is coevolution and how does its presence or absence make some native species vulnerable to invasive predators or competitors?

7. Which of the following is an example of coevolution?
   a. Polar bears and Arctic foxes both are white for camouflage on snow.
   b. Dolphins and whales have flippers that are similar in shape and function to the fins of a fish, allowing them both to swim efficiently.
   c. Moths that are preyed upon by bats can hear the ultrasonic sounds the bats use in hunting.
   d. Humans and chimpanzees share 98% of their DNA.

8. Suppose a native Guam bird species began to show evasive behaviors that allowed it to avoid the brown tree snake. Describe a potential coevolution scenario that might have allowed this adaptation to emerge in the population.

4 How do random events influence the evolution of a population?

9. Ten thousand years ago, most members of Species A were killed by a series of volcanic eruptions. However, some members escaped; all modern members of Species A are descended from those 100 individuals. This is an example of:
   a. stabilizing selection.
   b. artificial selection.
   c. mass extinction.
   d. the bottleneck effect.

10. Why might a population, isolated by the founder effect, be more vulnerable to extinction than the original population from which they came?

5 What factors impact the pace of evolution and extinction and why are extinctions that occur quickly more of a concern than those that take a long time to unfold?

11. True or False: Extinctions that unfold slowly usually occur because one species replaces another, leaving the community intact and having little effect on the community at large.

12. Which factor below tends to decrease the potential pace of evolution?
    a. Longer generation times
    b. Stronger selective pressures
    c. Higher genetic diversity
    d. All of the above decrease the pace of evolution as they increase in magnitude.

13. Why are K-selected species more vulnerable to extinction than r-selected species?

6 How do the mass extinction events of the past compare to extinctions during intervening times and today?

14. Why do most scientists think that we are in the midst of a sixth mass extinction?
    a. The background extinction rate is approximately 20%.
    b. Most extinctions are occurring in the taiga and tundra.
    c. No new species are being discovered.
    d. Current extinction rates are greater than the background rate.

15. What human actions are endangering species today?

7 How do humans, intentionally or accidentally, affect the evolution of a population?

16. Artificial selection differs from natural selection in that:
    a. it cannot produce large changes over time, only minor ones.
    b. humans are the selective pressure, choosing which traits to favor.
    c. it does not depend on genetic variability in the population.
    d. it causes individuals to change their traits rather than causing a change in the population's gene frequencies.

17. Suppose you wanted to use artificial selection to produce a breed of hairless dogs for people who are allergic to dog hair. How would you go about doing this?

SCIENCE LITERACY   WORKING WITH DATA

The following graph depicts the relationship between numbers of extinctions and human population size since the 19th century.

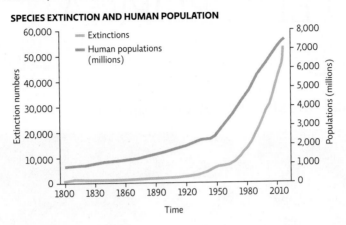

SPECIES EXTINCTION AND HUMAN POPULATION

Interpretation

1. Describe what is happening to:
   a. the extinction rate over time.
   b. human population growth over time.

2. The two curves have been graphed together. What is the implication of presenting these data in this manner?

Advance Your Thinking

3. The y-axis is labeled "Extinction numbers." What taxonomic units are being measured? What if the taxonomic unit being evaluated here had been genus (a taxonomic group that can contain more than one species) or family (a taxonomic group that can contain more than one genus)? Would you be more or less concerned about the trend of extinctions? Explain.

4. What type of relationship is suggested by the figure: correlational or causal? What additional data would you like to see to support the graph's main point?

INFORMATION LITERACY   EVALUATING INFORMATION

Life on Earth as we know it is a result of millions of years of evolutionary processes. These processes are not immune to changes in the environment; changes in evolutionary processes will ultimately result in changes in biodiversity, which will necessarily affect life on Earth, including humans.

Go to the website www.actionbioscience.org/evolution/myers_knoll.html and read the article "How Will the Sixth Extinction Affect Evolution of Species?" which discusses the effects of the sixth mass extinction on evolution. (If the URL provided does not take you to the article, search for it by title at www.actionbioscience.org.)

Evaluate the website and work with the information to answer the following questions:

1. Is this a reliable information source? Does it have a clear and transparent agenda?
   a. Who runs this website? Does this person's/group's credentials make this source reliable/unreliable? Explain.
   b. Who are the authors? What are their credentials? Do they have the scientific background and expertise that lends credibility to the article?

2. In your own words, explain why the authors think that the current mass extinction will change evolutionary processes.
   a. What changes in particular do they think will impact future evolution?
   b. What types of evidence do they provide to support their arguments? Give specific examples.

3. The sixth mass extinction is attributed to human impact. Go to the International Union for Conservation of Nature (IUCN) "Species of the Day" website (www.iucnredlist.org/species-of-the-day/archives).
   a. Choose five species at random that are identified as vulnerable, endangered, or critically endangered. Read about each species and list the reasons for their endangerment.
   b. Create one master list of all the threats you encountered for your five species and categorize them as either *human caused* or *caused by natural events*. Which list is longer?
   c. Does human impact play a role in the endangerment of these species? What actions would be most useful to address these threats?

 Additional study questions are available at SaplingLearning.com.

PALM OIL PLANTATIONS THREATEN TROPICAL FORESTS

Can we have tropical forests and our palm oil too?

In the past two decades, more than 20 million acres of rainforest have been cleared and planted as oil palm plantations. HOTLI SIMANJUNTAK/EPA/Redux

After reading this chapter and studying the KEY CONCEPTS and INFOGRAPHICS, you should be able to answer these GUIDING QUESTIONS

CORE MESSAGE

The variety of life on Earth is tremendous. This biodiversity provides important ecological services to ecosystems and we depend on these services for things like food, medicine, economic development, and recreational opportunities. The decline of biodiversity has serious ramifications for other species as well as human well-being; evaluating actions that threaten biodiversity and taking steps to reduce that impact can help protect the biodiversity that remains.

1. What is biodiversity?

2. Why is biodiversity important?

3. How do genetic, species, and ecosystem diversity each contribute to ecosystem function and services?

4. What are biodiversity hotspots, and why are they important?

5. What role does isolation play in a species' vulnerability to extinction?

6. How do habitat destruction and fragmentation threaten species?

7. How can we acquire the resources we need without harming the ecosystems and species that provide those resources?

Even before his plane landed in Sumatra, Laurel Sutherlin could tell the scene awaiting him would be worse than expected. They were still 30,000 feet over the Java Sea, and already a sickly haze of yellow smoke seemed to be enveloping the aircraft. The 30-something naturalist and environmental crusader had traveled all the way from California to see firsthand the impact the burgeoning palm oil industry was having on this South Pacific island.

Palm oil comes from the waxy orange fruits that sprout from oil palm trees. Oil can be extracted from both the flesh and the seed of the fruits, and it makes up nearly half of the planet's edible oil production. In recent years, it has made its way into a mind-boggling list of consumer products—foods like chocolate, peanut butter, cereal, and biscuits and also cosmetics, shampoos, and other household items such as mouthwash, diaper cream, and toilet cleaner. By some accounts, nearly half of all packaged food and household products in the United States contain palm oil. In the United States alone, palm oil consumption has increased sixfold since 2000, reaching 1.18 million metric tons in 2015. That's about as much as India uses and about half of what is used by the European Union, or the biggest user, China. Almost 95% of all that palm oil comes from Southeast Asia—in particular, from Malaysia and the Indonesian island of Sumatra, where Sutherlin's plane was now landing. This area now leads the world in rates of deforestation.

His eyes and nose were assaulted with smoke the instant he stepped off the plane. "I actually had to suppress an

⊙ **WHERE IS SUMATRA?**

MALAYSIA

SUMATRA

INDONESIA

initial panic that I would suffocate from the smoke," he would later recall. In fact, he had been lucky to land at all; air traffic would be canceled later that day due to poor visibility. Sutherlin had heard stories about the devastation wrought by deforestation. But he was not prepared for what he was about to see.

A young worker collects palm fruit clusters at an oil palm plantation in Sumatra. Oil can be extracted from both the flesh and the seed of the fruit.

Dimas Ardian/Bloomberg via Getty Images

1 BIODIVERSITY: THE VARIETY OF LIFE

Key Concept 1: We have identified only a fraction of the species that make up the tremendous variety of life (biodiversity) on Earth. We know much more about smaller groups such as plants and vertebrates than more diverse groups like invertebrates.

Rainforests, including those being cleared in Indonesia and Malaysia, contain the greatest concentration and variety of plant and animal terrestrial life-forms on Earth. This variety is called **biodiversity**, and it is the most unique and extraordinary feature of our planet. So great is the diversity of life on Earth that it is virtually impossible to know just how many species exist; in fact, many believe that the vast majority of all living species have yet to be discovered or identified by humans. So far, science has identified about

biodiversity The variety of life on Earth; it includes species, genetic, and ecological diversity.

1.9 million species, and common estimates for the total number of species on Earth range from 3 to 11 million (with one estimate as high as 100 million). Most scientists agree that we have identified only a small fraction of the species that live on Earth. "The numbers are mind boggling," says Jim Miller, a scientist at the Missouri Botanical Garden in St. Louis. "It's almost unfathomable." What we do know is that some life-forms are far more diverse than others; there are far more insects than there are vertebrates, for example, and there are relatively few mammals (fewer than 5,500 species) overall. **INFOGRAPHIC 1**

INFOGRAPHIC 1   BIODIVERSITY ON EARTH

We have identified about 1.8 million species so far, but our knowledge of Earth's total biodiversity is scant. Some researchers, especially those who work with tropical species and insects, believe the estimates of 3 to 11 million species are far too low; there may well be 5 million insect species alone. Of the species we have identified, insects far outnumber any other group of organisms. In fact, vertebrates (the group to which humans belong) likely make up only 1% of all creatures on Earth.

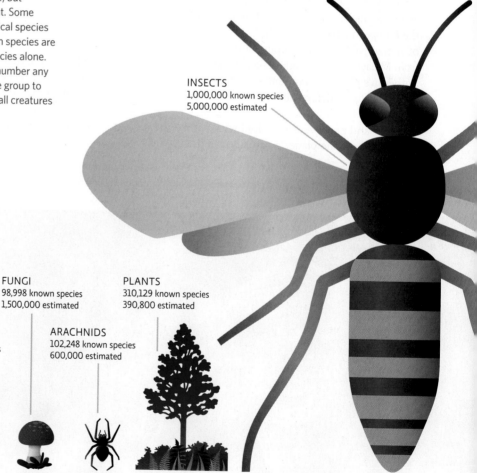

INSECTS
1,000,000 known species
5,000,000 estimated

VERTEBRATES
61,995 known species
80,500 estimated

FUNGI
98,998 known species
1,500,000 estimated

PLANTS
310,129 known species
390,800 estimated

ROUNDWORMS
25,000 known species
500,000 estimated

MOLLUSKS
85,000 known species
200,000 estimated

ARACHNIDS
102,248 known species
600,000 estimated

BACTERIA
7,643 known species
1,000,000 estimated

Size roughly equals the proportion of all known species.

 How important do you think it is it to get an accurate accounting of the number of species on Earth?

We also know that some regions of the world contain far greater concentrations of biodiversity—megadiversity—than other regions. Indonesia is one such region; it has more species of mammals, parrots, palms, swallowtail butterflies, and coral than any other country on the planet. Roughly 10% of all flowering plant species are found there. A recent census of a 3.4-square-kilometer (1.3-square-mile) plot on the Indonesian island of Siberut, for example, revealed 139 different species of trees. Indeed, tropical forests (and tropical coral reef ecosystems) are the most biodiverse in the world. John Terbough of the Duke University Center for Tropical Conservation reports that a 1-hectare (2.5-acre) research plot in an Amazonian rainforest contained close to 300 tree species. Compare this to the tree species diversity of the boreal forest that spans all of northern Canada—about 20 species. But habitat destruction and other pressures are threatening this biodiversity on a global scale.

Since the early 1990's, more than 8 million hectares (20 million acres) of rainforest (about the size of Maine) have been cleared in Southeast Asia to make way for oil palm plantations. Tropical forests in Africa and South America have also been lost to oil palms. Rubber plantations, too, account for significant deforestation in these areas—globally the land area converted to rubber plantations is about 57% of that used for oil palm plantations (it's around 70% in Southeast Asia) and the extent of rubber plantations is expected to increase fourfold by 2024.

The environmental costs of the clearing of tropical forests are expected to be far reaching. For one thing, such rapid and colossal loss of rainforest contributes to regional and global climate change, and is almost certain to accelerate global warming. (For much of 2015, Indonesia's greenhouse gas emissions exceeded that of the United States due to the massive forest clearing and fires.) For another, forest loss contributes to soil erosion, water pollution, and flooding—not to mention a scarcity of resources on which intricate forest communities depend. Rampant air pollution from forest clearing poses a significant threat to human health in the region.

But the most immediate consequences, by far, will be visited upon the region's wildlife. Fewer than 50% of the species normally found in the natural forest will be found in oil palm plantations. Mammals, in particular, tend to avoid the plantations; only about 10% of the original mammalian fauna are typically present. Scientists estimate that converting a forest into an industrial oil palm plantation results in the death or displacement of more than 95% of the orangutans living there. The habitat loss also impacts countless bird, insect, and plant species, many of which fail to thrive or even return at all because other species on which they depend are gone.

"When you look at a candy bar or package of crackers in a grocery store, or a jar of peanut butter in your kitchen, it's difficult to imagine that it has anything to do with orangutans going extinct," Sutherlin says. "But the link is actually quite direct."

Rescue workers from the animal charity Four Paws found this orangutan mother holding her daughter tightly as the pair was being surrounded by a group of young men paid to hunt and kill orangutans in the area. The orangutans were rescued by the charity workers and relocated to a remote area in the rainforest.

Vier Pfoten/Four Paws/RHOI/Rex/AP Images

2 THE VALUE OF BIODIVERSITY

Key Concept 2: Biodiversity contributes to the health and well-being of ecosystems, which in turn benefits human populations by providing ecosystem goods and services as well as cultural and health benefits.

Biodiversity is responsible for much more than the majesty of nature. For starters, it provides the key connections between individual species and between species and their environment. These connections help regulate the ecosystem as a whole: Insects pollinate flowers, for example. Photosynthetic organisms (plants on land, algae and phytoplankton in the sea) bring in energy, produce oxygen, and sequester carbon. Other organisms capture and pass along important nutrients like nitrogen and phosphorus. Others still help purify the air and water, and all species eventually become food for other creatures. Water also cycles through living things; in a forest, thousands of gallons a year are captured and passed along by each tree, releasing enough water vapor into the atmosphere to affect local rainfall. Life-giving soil forms as organisms decompose: Soil supports plant life whose roots, in turn, hold it in place, keeping it from being washed away in rains or floods. Meanwhile, predators and competitors keep each other in check, so that no single species grows too populous or gobbles up too many needed resources.

Biodiversity can also provide direct protection against disease in various ways. The incidence of Lyme disease in the United States, for example, is often lower in areas with higher biodiversity (especially areas with squirrels and possums, two species that effectively remove disease-carrying ticks and prevent the spread of the disease to other organisms). Likewise, in recent research, John Swaddle showed that fewer human cases of West Nile virus occur in eastern U.S. counties with higher bird biodiversity, possibly because some species are less effective at transmitting the virus than others—and the more species present, the less likely the virus will be transmitted to humans.

Biodiversity supplies cultural benefits as well—whether it is the enjoyment of a natural area for recreation or aesthetic appreciation, or a societal tradition rooted in nature.

ecosystem services
Essential ecological processes that make life on Earth possible.

instrumental value An object's or species' worth, based on its usefulness to humans.

intrinsic value An object's or species' worth, based on its mere existence; it has an inherent right to exist.

Biodiverse ecosystems have economic value, too. Forests provide not only food, fuel, and building materials but also pharmaceuticals. People use the chemicals they extract from plants and animals not only to attend to their individual human health but also as a source of income. Roughly half of all modern medicines—including medicine cabinet staples like aspirin and codeine, as well as most hypertension drugs and some cancer-fighting superstars—were originally derived from plants, many of them traditional remedies used for centuries. Over generations, rural forest-based communities in Sumatra have perfected a sophisticated management of forest resources that produces hillsides of coffee, elegant groves of cinnamon trees, acres of terraced rice paddies, and a healthy mixture of fruit trees, vegetables, tobacco, and other useful edible, commercial, and medicinal plants.

Ever since Robert Constanza's 1997 study that estimated the annual value of **ecosystem services** to be worth almost twice the annual gross domestic product of the entire world, environmental economists have set out to quantify the monetary worth of ecosystem services to bring attention to this often underappreciated value. A more recent estimate puts that value at $125 trillion. Even nature-based recreation is a multibillion-dollar business worldwide. Whale watching, a pastime in more than 80 countries, brings in $1 billion annually. Visitors spent more than $16 billion at U.S. national parks in

One ecosystem service of biodiversity is the provision of food, both through the food we eat and the other species that help that food grow, such as pollinators and soil decomposers.

Ron Boardman/Life Science Image/FLPA/Science Source

2015, visits that also pumped a similar amount into local economies. The U.S. Department of Fish and Wildlife Service estimated the economic value of hunting, fishing, and wildlife viewing in 2006 to be more than $122 billion. (For more on valuing ecosystem services, see Module 5.1.) **INFOGRAPHIC 2**

Of course, many people feel that the value of any given species goes beyond these **instrumental values** (i.e., the ecological, cultural, and economic benefits it brings). Many say that a species like the orangutan has **intrinsic value** (an inherent right to exist) and is therefore worth preserving, regardless of what benefits it might provide.

INFOGRAPHIC 2 ECOSYSTEM SERVICES

We depend on genetically diverse, species-rich communities to provide the goods and services we use every day. Impoverished ecosystems that have lost genetic, species, or ecological diversity cannot perform these tasks as well as highly diverse ecosystems can.

1. CULTURAL BENEFITS
Aesthetic
Spiritual
Educational
Recreational

2. HUMAN PROVISIONS
Food
Fiber products, such as cotton and wool
Fuel
Pharmaceuticals

3. ECOSYSTEM REGULATION AND SUPPORT
Nutrient cycling
Pollination and seed dispersal
Air and water purification
Flood control
Soil formation
Erosion control
Climate regulation
Population control

2. Many forest products can be harvested.

1. Many people enjoy spending time in nature.

3. Shoreline vegetation protects the bank from erosion.

Danita Delimont/Gallo Images/Getty Images

 In terms of reasons to value biodiversity, which of the beneficial ecosystem services listed here are the most important to you personally? Which ones do you think might be the most influential to society?

3 TYPES OF BIODIVERSITY: GENETIC, SPECIES, AND ECOLOGICAL

Key Concept 3: Genetic, species, and ecological diversity all enhance the ability of an ecosystem to function and for its members to adapt to changing conditions.

Biodiversity can be broken down into three types of diversity that really represent three different levels of diversity, from the population up to the ecosystem. **Genetic diversity** is the heritable variation among individuals of a single population or within the species as a whole. This is diversity at the population level. Genetic diversity provides the raw material that allows populations to adapt to their environment (potentially leading to the emergence of new species). Without genetic diversity, even the most well-adapted species could be in danger of facing extinction if conditions change. (See Infographic 4 in Module 2.1.)

The importance of genetic diversity was tragically illustrated in 19th-century Ireland. At the time, potatoes were the main food source for one-third or more of the population. Potatoes grew well in the Irish climate, despite the plant not being native to Ireland—it hails from South America. On the slopes of the Andes mountains, thousands of different genetic varieties of potato plants grew, each adapted to the slightly different niche found at different latitudes and elevations,

genetic diversity The heritable variation among individuals of a single population or within the species as a whole.

species diversity The variety of species, including how many are present (richness) and their abundance relative to each other (evenness).

ecological diversity The variety within an ecosystem's structure, including many communities, habitats, niches, and trophic levels.

traits that were selected for by nature and later enhanced with artificial selection by hundreds of generations of Andean farmers. But only a few varieties of potato were imported to Ireland, and by the 1840's, only two types were planted, predominately one variety called the "lumper." These varieties grew well and fed millions until a fungus arrived in the mid-1800's that attacked potatoes. In the Andes, this fungus would damage some crops, but because all the potato varieties were so different, even if some plants died, others

would survive. But in Ireland, without genetic diversity in the potato "population," most of the potato plants were identical—what killed one, killed them all. The vulnerable lumper succumbed to the fungus, and the potato crop was wiped out for several years. Between 1845 and 1851, more than 1 million people died of starvation and 1 million more emigrated (many to the United States).

We can also find biodiversity at the community level—that is, **species diversity**. This is an accounting of the number of species living in an area (richness) and their distribution (evenness) (see Module 2.3).

Some ecosystems naturally have higher species diversity than others. Areas with high species diversity often owe that variety to their **ecological diversity**—the variety of habitats, niches, and ecological communities in an ecosystem. Rainforests contain a wide variety of habitats and considerable physical complexity, creating lots of niches. Each of the many species living in a tropical rainforest occupies an individual niche, and they all contribute to the functioning of the ecosystem (including vital services like energy capture, nutrient cycling, and decomposition). For example, a wide variety of plant species occupy every level of the rainforest, from the forest floor to the canopy, and they capture varying amounts of sunlight. This large producer base supports many trophic levels and many species within those trophic levels, increasing the efficient use and transfer of nutrients in the ecosystem. Few matter resources will go unused. The loss of species (extinction) in an ecosystem removes contributing members from that ecosystem, potentially impacting other species and the ability of that ecosystem to function and provide ecosystem services.
INFOGRAPHIC 3

4 BIODIVERSITY HOTSPOTS

Key Concept 4: Protecting biodiversity hotspots, areas with high numbers of endangered endemic species, can be a cost-effective way to protect many endangered species.

Tropical forests tend to be particularly flush with both ecological and species diversity, thanks largely to the abundant sunlight and climatic conditions conducive to growth. The forests that are currently being laid low in Sumatra are no exception. They have been left unhampered for so many millennia that these steamy amphibious ecosystems swarm with a cornucopia of life: elephants, orangutans, tapirs, tigers, and every manner of

bird and beetle the human imagination can fathom. "The truth is, no one has any idea how many species used to live here," Sutherlin says. "Half the species in these forests have yet to be described to science."

Since areas with high ecological diversity offer so many unique habitats and niches, they often have a large number of **endemic species** that are specially adapted

INFOGRAPHIC 3   **BIODIVERSITY INCLUDES GENETIC, SPECIES, AND ECOSYSTEM DIVERSITY**

There is individual variation between these *Delia* butterflies.

BIODIVERSITY

GENETIC DIVERSITY
Variations in the genes among individuals of the same species

SPECIES DIVERSITY
The variety of species present in an area including the number of different species that are present as well as their relative abundance

Tropical forests have some of the highest plant and animal species diversity in the world.

Indonesia's rainforest ecosystems contain varied habitats and multiple niches that support a complex community.

ECOLOGICAL DIVERSITY
The variety of habitats, niches, trophic levels, and comuunity interactions

 Give an example of genetic diversity and species diversity from the grocery store produce section.

to that locale and naturally found nowhere else on Earth. They are most commonly found in small isolated ecosystems (e.g., islands, mountaintops) because the species' population members cannot easily disperse and share genes with other populations. **Biodiversity hotspots** are areas that have high endemism (greater than 1,500 endemic plant species) and have lost at least 70% of their original habitat. These areas contain a large number of **endangered species** (those at risk of becoming extinct).

Thirty-six land and aquatic regions have been designated as hotspots and many are found in the tropics. The latest biodiversity hotspot to be added to the list is the North American Coastal Plain—the coastal region from Massachusetts down to the Gulf coast of Texas and northern Mexico and extending up the Mississippi Basin into southern Missouri and Kentucky. With the number of endemic species topping 1,800 and having lost 85% of its original natural habitat, it more than qualifies as a hotspot. This region is especially rich in turtle species (39% of

its reptile species are endemic) and amphibians (47% are endemic). **INFOGRAPHIC 4**

Because they are pockets of high biodiversity and contain many unique species found nowhere else, biodiversity hotspots are conservation priorities. Investing time and money in protecting these areas can potentially provide a lot of return on that investment—many endangered species, or species vulnerable for endangerment, will be protected.

The island of Sumatra is one such hotspot; it is home, for example, to the endemic Sumatran Tiger. The pygmy elephant is likewise endemic to the nearby island of Borneo. And the orangutan, one of humans' closest relatives on the evolutionary tree, is endemic to the region, with different types of orangutan found on the islands of Sumatra and Borneo. All three of these species are being driven to the edge of extinction by our pursuit of palm oil.

endemic species A species that is native to a particular area and is not naturally found elsewhere.

biodiversity hotspot An area that contains a large number of endangered endemic species.

endangered species Species at high risk of becoming extinct.

INFOGRAPHIC 4 | BIODIVERSITY HOTSPOTS

Biodiversity hotspots, areas with high numbers of endemic but endangered species, cover a small percentage of land and water areas but hold more than 40%–50% of all plant and vertebrate endemic animal species. Most hotspots are located in tropical biomes or in isolated terrestrial ecosystems, such as mountains or islands. Even small disturbances, such as a small farm plot or road that cuts through the area, can threaten endemic species that populate specialized niches in these hotspots.

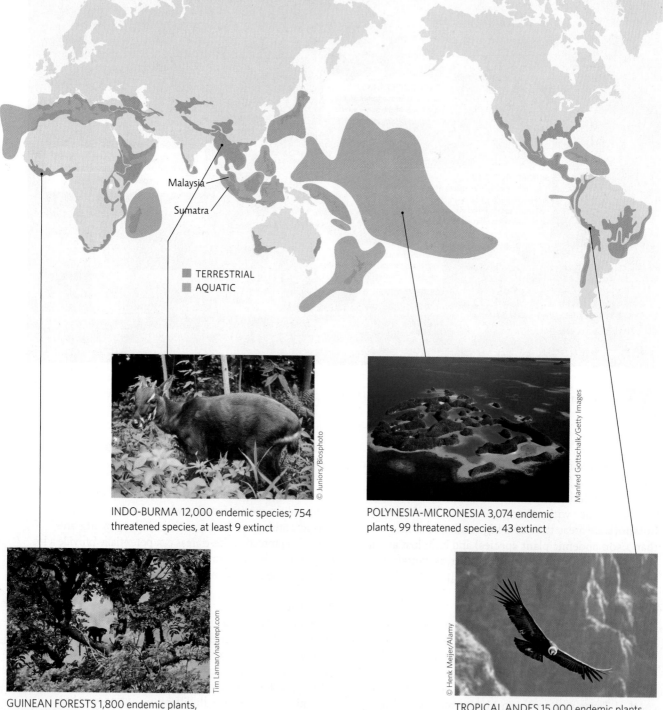

Malaysia

Sumatra

■ TERRESTRIAL
■ AQUATIC

INDO-BURMA 12,000 endemic species; 754 threatened species, at least 9 extinct

POLYNESIA-MICRONESIA 3,074 endemic plants, 99 threatened species, 43 extinct

GUINEAN FORESTS 1,800 endemic plants, 115 threatened species, 0 extinct

TROPICAL ANDES 15,000 endemic plants, 487 threatened species, 2 extinct

 Many hotspots are in tropical areas, but some are also found in areas further north and south of the equator. Why do you suppose so many nontropical coastal areas are biodiversity hotspots?

5 ISOLATION AND EXTINCTION RISK

Key Concept 5: Isolation increases the risk a population will be lost if conditions change because it is unlikely individuals from other populations might emigrate, bringing new, potentially helpful genetic diversity to the group.

Species come to islands such as Sumatra in various ways: They may be blown in on storms, they may arrive as lost migrators, or the island may have broken off from a larger landmass at some point. Because these are rare events, once a species arrives, it is unlikely to be joined by other members of its species. The founding population is therefore isolated, and as it adapts over time to its new island home, it may eventually evolve into a new species (see Module 3.1 for more on the *founder effect*). For this reason, the number of unique species (the degree of endemism) generally increases with isolation. On the Hawaiian Islands, almost 3,800 km (2,400 miles) from the nearest mainland, 90% of native species are endemic.

Their isolation and high endemism make remote islands particularly vulnerable to species loss. The smaller the population (often a reflection of the habitat size), the more likely it is that random events such as fires or floods might exterminate the entire group. And isolation reduces the chances of recolonization that might replace lost members or increase genetic diversity.

Today, human impact contributes to isolation in the form of **habitat fragmentation**—producing habitat "islands" where before there were larger expanses of uninterrupted habitat. Small, completely isolated ecological communities are at the highest risk for species extinction. Research by Luke Gibson and colleagues showed that after dam construction flooded a Thailand forest, leaving behind some small islands covered in forest, almost all small mammal species were lost in as little as 5 years on the smallest islands. Isolation can be a recipe for disaster in the face of rapid environmental change. (See the *Science Literacy* section at the end of this module for more on Gibson's research.)

Stuart Pimm of Duke University and his colleagues analyzed the fossil record and the living species on Pacific islands colonized by humans about 4,000 years ago and concluded that as many as half of the species might have been driven to extinction after humans arrived. In fact, even though they make up a tiny percentage of the land mass on Earth, islands have accounted for about half of all recorded extinctions in the past 400 years. In Hawaii alone, fully one-half of the indigenous flora faces immediate extinction. Such a high extinction rate threatens the fragile tapestry of life on these islands, from the soil and freshwater supplies to the health and economic future of the islands' residents.

INFOGRAPHIC 5

> **habitat fragmentation** The destruction of part of an area that creates a patchwork of suitable and unsuitable habitat areas that may exclude some species altogether.

More than 90% of the native species on the Hawaiian Islands are found nowhere else, making these populations more vulnerable to extinction; if they are lost there, they are lost forever. The smaller the island, the smaller the population of a given species, another vulnerability for extinction. Here, botanists work through the night setting up a protective fence around the last specimen of *Delissea undulata* found in the wild. This endangered plant was growing on the side of a collapsed lava tube, but had been knocked over by wind or animals and was dangling from its roots.

CHRIS JOHNS/National Geographic Creative

Isolation can increase the number of endemic species in an area because local populations do not "share" genes with other populations. Over time, an isolated population may diverge from its ancestral population as it becomes adapted to its immediate environment.

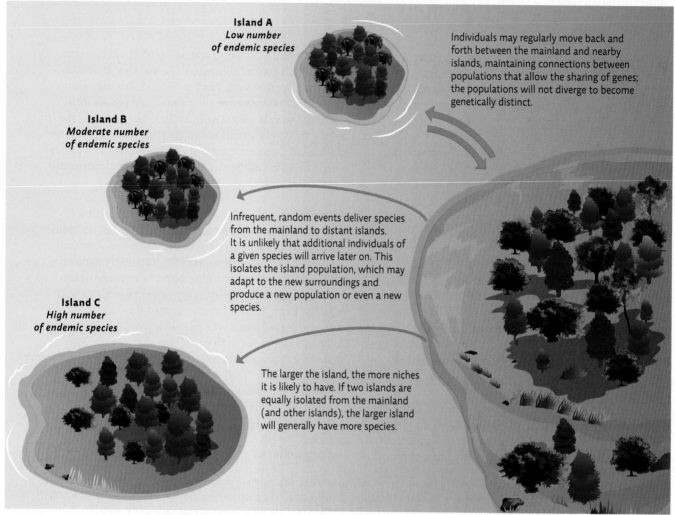

Island A
Low number of endemic species

Individuals may regularly move back and forth between the mainland and nearby islands, maintaining connections between populations that allow the sharing of genes; the populations will not diverge to become genetically distinct.

Island B
Moderate number of endemic species

Infrequent, random events deliver species from the mainland to distant islands. It is unlikely that additional individuals of a given species will arrive later on. This isolates the island population, which may adapt to the new surroundings and produce a new population or even a new species.

Island C
High number of endemic species

The larger the island, the more niches it is likely to have. If two islands are equally isolated from the mainland (and other islands), the larger island will generally have more species.

 Why might populations of small animals, like salamanders that live only at high elevations on adjacent mountains, remain isolated from one another, even though there are no barriers between the two mountains?

6 THREATS TO BIODIVERSITY

Key Concept 6: Human impact, especially habitat destruction, threatens biodiversity, endangering ecosystems and the species (including humans) that depend on the ecosystem services these areas provide.

The world's biodiversity is threatened from many angles. The scientific consensus is that human actions are behind the high extinction rates of the recent past, especially since 1900. These threats include habitat destruction, the introduction of invasive species (see Module 3.1 for a detailed example), pollution, overharvesting (for consumption, the pet trade, and legal and illegal body parts, such as hides, tusks, and bones), and anthropogenic

climate change. Many of these threats are related to overpopulation and affluence that leads to greater resource use. (See Infographic 1 in Online Module 3.3 for more on threats to biodiversity.)

At least part of the surge in palm oil use—and the habitat destruction it engenders—can be traced back to 2007, when then New York City Mayor Michael Bloomberg famously banned the use of hydrogenated oils, which contain high quantities of trans fats, in all New York City restaurants. The ban was well intentioned: Obesity rates were high in the city and in the United States as a whole, and trans fats are, well, fattening. They're also artery clogging, heart attack inducing, and all-around unhealthy to consume, and in 2007, they were pervasive. Seeing the writing on the wall, popular food brands, including Oreo and Dunkin' Donuts, followed Bloomberg's lead by voluntarily replacing hydrogenated oil with the next best alternative: palm oil.

Despite the sweet sugary tastes those brand names evoke, palm oil itself has no flavor. What it does have is a soft, thick texture that remains fantastically stable across many temperatures. This thickness and stability make palm oil useful. It keeps doughnuts firm and Oreo cookie filling from melting into drizzle. It keeps cosmetics and detergents and peanut butter and Girl Scout cookies consistent and solid at room temperature.

Palm oil is also a healthier alternative to the trans fats it replaced. It contains vitamins (notably A and E) and is a better option than trans fats regarding cardiovascular health. For example, volunteers who replaced trans fats in their diet with palm oil showed a decrease in cholesterol levels and other markers for cardiovascular disease. However, compared to unsaturated fats (e.g., olive oil), palm oil consumption raised total cholesterol and LDL cholesterol (the "bad" cholesterol). So, while palm oil is a better dietary fat than trans fats, research suggests it is not the healthiest fat for the human diet so should still be consumed in moderation.

Our palm oil obsession is contributing to the number-one cause of species endangerment: **habitat destruction**. Clearing land for agricultural purposes has historically been, and continues to be, the leading cause of habitat destruction. Currently, clearing land to raise cattle or their feed accounts for more habitat destruction than other agricultural endeavors. Brazil alone lost almost 20 million hectares (50 million acres) from 2000 to 2006 to deforestation, primarily for livestock rearing (pasture) and soybean production (also largely to produce animal feed). (Compare this to the 8 million hectares lost over 20 years in Southeast Asia for oil palm plantations.)

Other drivers of habitat destruction include urban and suburban development, resource extraction (not just timber harvesting but also mineral mining and fossil fuel extraction), and large water projects such as dams that destroy terrestrial and aquatic habitats. By obliterating certain parts of an ecosystem, habitat destruction makes that ecosystem physically less diverse, which then makes it less biodiverse. This loss of biodiversity, in turn, leads to a reduction of ecosystem services. **INFOGRAPHIC 6**

habitat destruction The alteration of a natural area in a way that makes it unsuitable for the species living there.

Wildlife corridors like this one over the Trans-Canada Highway in Banff National Park in Alberta, Canada allow animals safe passage over roads.

Robert McGouey/Wildlife/Alamy

INFOGRAPHIC 6   OIL PALM PLANTATIONS ARE NOTHING LIKE NATURAL TROPICAL FORESTS

The tropical forests of Indonesia and Malaysia are among the most biodiverse in the world, home to thousands of endemic species. The lowland tropical forests of Sumatra alone have an estimated 30,000 different plant species and as many as 1,000 different tree species.

INDONESIA'S FORESTS

HIGH BIODIVERSITY
- Complex forest structure produces many niches.
- Home to more than 15% of all the bird, reptile, and amphibian species on Earth.

NATURAL RESOURCES
- Provide food, wood, medicinal, and other resources for millions of residents.

ECOSYSTEM SERVICES
- Water production, collection, and purification.
- Erosion control.
- Important carbon sinks that help mitigate climate change.

Auscape/UIG/Getty Images

As mentioned earlier, the detrimental impacts of habitat destruction can also be influenced by habitat fragmentation, the loss of even part of a larger habitat. Deforestation that leaves patches of forest may not be suitable for species that need large expanses of forest. Some species will simply not cross over the deforested patch to disperse from one side of the patch to the other, so the patches effectively isolate them from other parts of their population. Even a narrow or little-used road may isolate individuals on one side or the other. Wildlife corridors that connect fragmented habitats can be effective *conservation* tools. Specially constructed road overpasses or underpasses, or ribbons of natural habitat that connect two habitat fragments, can allow individuals to travel more freely and are vitally important in maintaining genetic diversity in the populations—especially for species that range widely and live at low population densities, such as large carnivores like grizzly bears and wolves.

To reach a patch of rainforest that had yet to be touched by destruction, Sutherlin and his colleagues traveled 10 hours through the night from Riau to Jambi Province, then another 4 hours by car over horrendous dirt roads to South Sumatra. From there, they took motorcycles, over winding-ribbon trails, through a desolate oil palm plantation to the edge of the peat lands, and then continued by foot, "on a rough trail along a canal dug by loggers to remove logs from the forest." Sutherlin was thrilled when they arrived at the forest edge, sweaty and exhausted, to finally see some tall trees still standing. Monkeys howled in the distance. An electric blue butterfly swirled around his head. And spider-hunters, dollarbirds, and bulbuls circled overhead, flitting in and out of view. "Then, as if on cue," he says, "a chainsaw began to roar just out of sight, followed quickly by the terrible sound of trees crashing through trees to the ground."

Oil palm plantations have very low biodiversity, which means they are unable to provide the ecosystem services normally provided by the large, unfragmented forests they replace.

INDONESIA'S OIL PALM PLANTATIONS

LOW BIODIVERSITY
- One species of tree and a limited number of understory or canopy plants.
- Have only half as many bird species as the neighboring intact forest.
- House much less insect biodiversity.

NATURAL RESOURCES
- Leading cause of deforestation.
- Loss of food, fuel, and other resources needed by locals.
- Contaminated water supplies.

ECOSYSTEM SERVICES
- Prone to flooding and contribute to sediment pollution, damaging aquatic habitats.
- Release of carbon (fires) and reduced capacity for carbon storage.

blickwinkel/Sailer/Alamy

 Describe how tropical forests differ from oil palm plantations in terms of genetic, species, and ecosystem diversity.

Virtually all of the forest's inhabitants are facing annihilation. "The remaining populations of endemic Sumatran rhinos are widely considered to be the living dead," says Sutherlin. "Their habitat is too sparse, too fragmented, and too disturbed, their numbers too few."

Extinction doesn't have to be worldwide to have an impact. Species that become extinct locally but not everywhere are said to be **extirpated**. So, while a species may not be globally extinct (and perhaps not even identified as an endangered species), its extirpation can have serious impacts on its own ecosystem, potentially disrupting the connections on which the ecological community depends, triggering a cascade that threatens other species in the region. "Simple" ecological communities (ones with relatively few species) are much more affected by a single loss. But even "complex" ecosystems (those with many species and many filled niches) can be gravely imperiled if too many members are lost, especially if keystone species are lost. One recent evaluation found that terrestrial ecosystems have globally lost around 15% of their native biodiversity, a degree of loss the authors conclude is probably below the safe limit of no more than 10% loss, seriously impairing an ecosystem's ability to provide ecosystem services in a meaningful way.

extirpated Locally extinct in one geographic area, but still found elsewhere.

7 PROTECTING BIODIVERSITY

Key Concept 7: Effective biodiversity protection programs must address the needs of humans as well as the ecosystems and species that are in danger.

Technologies that improve the productivity of oil palm trees are seen as one part of the solution—more palm oil per tree means fewer trees need to the planted. It turns out, the three most common cultivars (strains) of oil palm trees that are routinely planted differ greatly in their productivity. Of the three—dura, pisifera, and a hybrid between the two called tenera—tenera is considered the best because it yields the most oil—30% more than either of its siblings, by most estimates. "They're known as farmers' gold," says Raviga Sambanthamurthi, a geneticist with the Malaysian Palm Oil Board.

Scientists knew for many years that these "golden" fruits were the result of variations of a single gene, Sambanthamurthi says. But they didn't know which gene or how to screen for it. That left farmers with some dismal options. The tenera variety is produced by manually cross-breeding the other two strains, which is easy enough. But farmers usually have to wait 5 or 6 years, until the oil palm plants bear fruiting bunches, before they can tell how successful they've been and how many tenera plants they've got.

So Sambanthamurthi was thrilled when she and a team of researchers discovered the gene that is directly responsible for the tenera fruit's thin shell and high oil content. The discovery equips farmers with the ability to identify the tenera seeds before planting. "That puts years back on the clock," she says. "It enables farmers to produce more oil per hectare. And that means it reduces the pressure to clear virgin forest." It was the dream of making such a significant contribution as this that drove Sambanthamurthi to return to her native Malaysia and join the research division of the country's Palm Oil Board, after completing her PhD in genetics at the University College of London. "I knew if I was going to make a difference, it would be in palm oil," she said. "Sustainable palm is the key to protecting our biodiversity, and our heritage."

> "Sustainable palm is the key to protecting our biodiversity, and our heritage."
> —Raviga Sambanthamurthi.

In Sumatra and elsewhere in Southeast Asia, the solution to protecting biodiversity will have to involve palm oil production. As Sambanthamurthi and others point out,

it's the most productive oil-bearing crop at our disposal, producing almost 4 metric tons of oil per hectare. It claims just 5% of the total vegetable oil acreage globally but accounts for the production of roughly one-third of all vegetable oil, making it the biggest single source of edible oil worldwide. Soybeans, by comparison, produce only 0.4 metric ton per hectare. "To get the same amount of soybean oil, you would have to use ten times as much land. So really, it's the best option we have environmentally as well."

There are ways to produce palm oil sustainably. To be certified sustainable by the nonprofit group Roundtable on Sustainable Palm Oil (RSPO), palm oil producers must adhere to certain practices that minimize the use of fire and pesticides, and they must take steps to prevent soil erosion and water pollution. The rights of workers (fair wages and good working conditions) are a high priority, and the local community is given an opportunity to weigh in on proposed oil palm plantations. In addition, growers cannot clear old-growth forests or those with high biodiversity or cultural value for conversion to plantations. Sutherlin and others say the reasons for adopting those methods are clear: If we factor in the value of services provided by intact ecosystems, sustainably produced palm oil is significantly cheaper than nonsustainably produced product (an example of *true cost accounting*; see Module 5.1). In fact, production methods that destroy ecosystems and drive species to extinction incur costs that we may never be able to repay. Only a careful and honest analysis of the trade-offs of using palm oil will help us decide how best to proceed. **INFOGRAPHIC 7**

But for oil producers to adopt better ways, demand will have to come from consumers. "Companies aren't hearing from their customers," Sutherlin says. "They tell us flat out 'We know it's bad. We know it's doing all these horrible things to the environment. But it's our cheapest option, and no one's threatening to boycott us over it.'" In 2013, Sutherlin's organization, the nonprofit Rainforest Action Network, launched the "Snack Food 20" campaign—a call on the top 20 global snack food companies, including Krispy Kreme, Kraft, Kellogg, Heinz, General Mills, and Nestlé, to use only sustainably produced palm oil in its products (certified by the RSPO). Many of these companies are responding but more transparency is needed, communicated with better labels, to allow consumers to choose products with sustainably sourced palm oil.

INFOGRAPHIC 7 **PROTECTING BIODIVERSITY REQUIRES A CONSIDERATION OF ECONOMIC GOALS AND ENVIRONMENTAL NEEDS**

Deciding how to proceed with regard to palm oil requires that we evaluate the trade-offs of our choices and that we strike a balance between meeting our desire for palm oil and the survival of the ecosystems and local communities affected by its cultivation.

JAMES P. BLAIR/National Geographic Creative

TRADE-OFF ANALYSIS OF PALM OIL

BENEFITS OF PALM OIL	DISADVANTAGES OF PALM OIL	ADDRESSING THE DISADVANTAGES
Palm oil is the preferred dietary replacement for trans fats.	Though better than trans fats for one's health, it is still a fat whose consumption should be limited.	Consumers should be educated so that they understand that consumption of palm oil (in foods), like consumption of any other fat, must be moderate.
Palm oil is already widely used in many products.	The high demand for palm oil in products increases the need for more oil palm plantations.	Manufacturers should only source palm oil that is certified as sustainably grown, and this information should be readily available to consumers.
Palm oil is inexpensive.	If palm oil is priced in a way that reflects its true costs, the prices of goods that contain it will increase.	Consumers should purchase products that contain certified sustainable or organic produced palm oil, even if more expensive, to support companies that use it.
Oil palm plantations can produce much more oil per acre than can other oil crops.	Oil palm plantations reduce biodiversity and decrease the ability of local ecosystems to provide ecosystem services.	Cultivation of oil palms should be done in a way to minimize damage to local ecosystems and biodiversity, leaving some areas uncultivated as refuges for local biodiversity. The most productive palm varieties should be used to increase production per acre.

 What role does the consumer play in the future course of the palm oil industry and what can manufacturers of products that contain palm oil do to help consumers make informed decisions? How can producing palm oil that is sustainably certified help address some of the disadvantages of using palm oil?

Others have taken up the cause as well. In 2010, New York Comptroller Thomas Di Napoli, who manages investments for the state's $153 billion pension fund, won pledges from companies like Dunkin' Donuts, Sara Lee, and Smucker's to only use sustainably produced palm oil. Unilever, maker of food and cleaning products, advertises that it is now sourcing only sustainable palm oil. And since at least 2007, two Michigan Girl Scouts have been lobbying the Girl Scouts U.S.A. to extract a similar pledge from the makers of Girl Scout cookies. Like so many other processed food manufacturers, the Girl Scouts switched to palm oil so the cookies would be free of trans fats. The organization says it would like to

switch to sustainably produced palm oil, but so far there simply isn't enough of it out there. In any given year, Girl Scouts sell hundreds of millions of dollars' worth of cookies. That's a lot of palm oil. But sustainable supplies are slowly increasing. In 2015, about 17% of the global supply was sustainably grown—up from only 6% in 2013.

One thing is certain: There is still much value in saving these forests. Orangutans still swing freely through the canopies, and new species of lizards and birds continue to be discovered. Despite the destruction plaguing Indonesia's forests, all hope is not yet lost.

Select References:

Fitzherbert, E. B., et al. (2008). How will oil palm expansion affect biodiversity? *Trends in Ecology and Evolution, 23*(10), 538–545.

Gibson, L., et al. (2013). Near-complete extinction of native small mammal fauna 25 years after forest fragmentation. *Science, 341*(6153), 1508–1510.

Newbold, T., et al. (2016). Has land use pushed terrestrial biodiversity beyond the planetary boundary? A global assessment. *Science, 353*(6296), 288–291.

Pimm, S., et al. (2006). Human impacts on the rates of recent, present, and future bird extinctions. *Proceedings of the National Academy of Sciences, 103*(29), 10941–10946.

Singh, R., et al. (2013). The oil palm SHELL gene controls oil yield and encodes a homologue of SEEDSTICK. *Nature, 500*(7462), 340–344.

Swaddle, J. P., & Calos, S. E. (2008). Increased avian diversity is associated with lower incidence of human West Nile infection: observation of the dilution effect. *PloS One, 3*(6), e2488.

INDO-MALAYAN archipelago 15,000 endemic plants, 162 threatened species, 4 extinct

Jiri Loun/Science Source

INTERACTIVE MAP **BIODIVERSITY HOTSPOTS**   ANIMATED INFOGRAPHIC

Species in each of the world's biodiversity hotspots are threatened by a unique constellation of forces, but most of these threats are human in origin. Read about four of these hotspots in more detail to see what threats they face and what is being done to protect them.

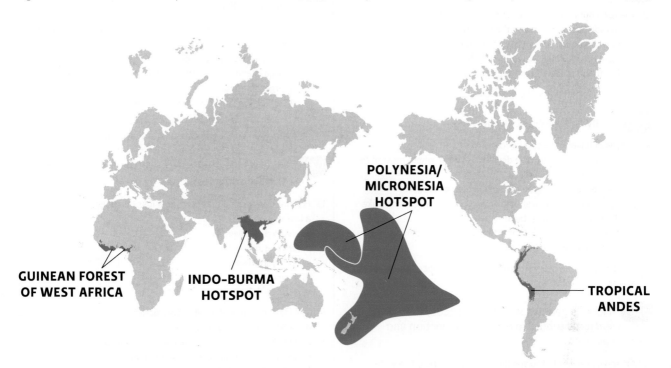

POLYNESIA/
MICRONESIA
HOTSPOT

GUINEAN FOREST
OF WEST AFRICA

INDO-BURMA
HOTSPOT

TROPICAL
ANDES

 BRING IT HOME

PERSONAL CHOICES THAT HELP

Species and habitats provide numerous benefits to people, including water and air purification, food sources, recreation, and medicine. Unfortunately, many species are facing threats at ever-increasing levels. The good news is that we as a society have a direct impact on these threats and can make changes to ensure the survival of many of our at-risk species.

Individual Steps
• Don't buy products made from wild animal parts such as horns, fur, shells, or bones. Only buy captive-bred tropical aquarium fish, not wild-caught fish.
• Visit the Rainforest Action Network website at www.ran.org/conflict-palm-oil and read about their Snack Food 20 campaign and visit www.ran.org/sj20scorecard for an evaluation of how the top 20 snack food producers are doing in their quest to source sustainable palm oil. Look for labels on products that identify

RSPO

the palm oil it contains as "Certified Organic" or "Certified Sustainable Palm Oil RSPO" with its trademarked logo, a globe-shaped palm.
• Make your backyard friendly to wildlife, using suggestions from the National Wildlife Federation (www.nwf.org).
• Install an Audubon Guide app on your smartphone, or buy a field guide to learn the plant and animal species in your area.

Group Action
• Work with faculty and other students to organize a bioblitz for a protected area in your region. A bioblitz, which is an intensive survey of all the biodiversity in the area, can generate a large amount of data to be used for habitat management and species protection.
• Join a citizen science program monitoring wildlife. Many regional conservation groups have monitoring opportunities and provide training. For national programs, see the Cornell Lab of Ornithology website at www.birds.cornell.edu and the Izaak Walton League of America website at www.iwla.org.

Policy Change
• The Endangered Species Act was the first U.S. legislation established to protect species diversity. To learn more about current challenges and updates to the program, visit the U.S. Environmental Protection Agency (www.epa.gov/espp).

ENVIRONMENTAL LITERACY | **UNDERSTANDING THE ISSUE**

1 What is biodiversity?

1. The total number of different species on Earth is:
   a. unknown, but insects are the most numerous species.
   b. a few million, mostly bacteria and fungi.
   c. more than 20 million, with half of them being plants.
   d. less than 1 million, mostly vertebrates.

2. Give a basic definition of biodiversity.

2 Why is biodiversity important?

3. Why is biodiversity loss a concern?
   a. It primarily occurs in the developed world, where most of the world's population lives.
   b. It increases the degree of endemism in an area.
   c. It primarily affects well-known and charismatic species like elephants and orangutans.
   d. It disrupts ecological connections, potentially diminishing ecosystem services.

4. Identify the three categories of ecosystem services and give examples of each.

3 How do genetic, species, and ecosystem diversity each contribute to ecosystem function and services?

5. All apples belong to the species *Malus domestica*. The wide variety of apples available in the produce department of your local grocery store is an example of _____ diversity.

6. An example of species diversity might be the:
   a. wide variety of coloration and tail size in guppies.
   b. diverse habitat types and organisms inhabiting a deep lake, its edges, and the surrounding meadow and forest areas.
   c. many different species inhabiting a swamp.
   d. None of the above

7. Define genetic diversity and use the example of the potato blight in Ireland to explain the importance of genetic diversity to a population.

4 What are biodiversity hotspots, and why are they important?

8. True or False: Biodiversity hotspots are areas with many endemic species that are well protected and not threatened with endangerment.

9. A species that naturally occurs in only one place is called a(n):
   a. endangered species.
   b. hotspot species.
   c. endemic species.
   d. threatened species.

10. What role do biodiversity hotspots play in efforts to protect biodiversity?

5 What role does isolation play in a species' vulnerability to extinction?

11. Which island is most likely to have the highest number of endemic species?
    a. A small island close to the mainland
    b. A large island close to the mainland
    c. A large island far from the mainland or other islands
    d. A small island far from the mainland but close to a large island

12. Why are isolated populations more vulnerable to extinction than populations that are not isolated from each other?

6 How do habitat destruction and fragmentation threaten species?

13. Habitat destruction that leads to a loss of niches can contribute to a loss of:
    a. ecological diversity.
    b. species diversity.
    c. ecosystem services.
    d. A and B
    e. A, B, and C

14. The leading cause of habitat destruction is:
    a. land clearing for agriculture.
    b. water projects such as the building of dams.
    c. fossil fuel extraction and other mining.
    d. suburban development.

15. Explain why habitat fragmentation may be just as serious a threat to a species' survival as total habitat destruction.

7 How can we acquire the resources we need without harming the ecosystems and species that provide those resources?

16. True or False: Sustainable solutions to growing oil palms must consider the needs of consumers and nearby communities.

17. In terms of its impact on species and their environment, producing palm oil may be a better option than producing soybean oil because:
    a. palm oil is less expensive to grow.
    b. oil palm trees are not grown in areas with very many endangered species.
    c. palm oil is easier to extract from the seed than soybean oil.
    d. growers can produce more oil per hectare with palm plantations than soybean fields.

18. Evaluate the trade-offs of using palm oil as a replacement for trans fats and make a recommendation regarding our future path that considers economic, environmental, and societal needs.

19. How can you, as an individual, help maintain biodiversity worldwide? Justify your choices.

SCIENCE LITERACY | WORKING WITH DATA

Habitat loss is currently the main driver of species endangerment and extinction, but habitat loss need not be complete to cause a problem; habitat fragmentation may also be an insurmountable problem for some species. Islands that are created when a river is dammed to form a reservoir provide instant habitat fragments. Luke Gibson and his team evaluated the number of small mammal species in large [10–56 hectares (25–140 acres)] and small [<10 hectares (<25 acres)] forested islands in Chiew Larn Reservoir of Thailand. Island sampling was done shortly after the reservoir was formed (about 6 years after isolation); the islands were sampled again about 26 years after isolation. Their results are below. [For comparison, on average, nine species were found on mainland (pre-reservoir) plots; the richness did not change in this mainland forest over the study period.]

Number of Small Mammal Species (Richness) Found on Islands 6 and 26 Years after Isolation

	Island (ID #)	Area of Island (hectares)	Species Richness (6 years)	Species Richness (26 years)
Large Islands	6	56.3	12	5
	5	12.1	9	3
	9	10.4	7	1
Small Islands	28	4.7	2	2
	7	1.9	3	2
	33	1.7	1	1
	3	1.4	2	1
	41	1.1	3	1
	39	1.0	3	1
	40	0.8	2	1
	2	0.4	2	1
	16	0.3	2	1

Interpretation

1. Before evaluating the data, draw a graph that compares species richness of large islands 6 years after isolation versus 26 years after isolation and that also shows the same for small islands. (Hint: Calculate the average species richness values for each group and draw a bar graph that allows you to directly compare the richness of large islands after both sampling periods to the richness of small islands after both sampling periods.)

2. Consider that species richness before isolation was 9. How does the species richness compare in large islands before isolation, 6 years after isolation, and 26 years after isolation? How does it compare for the small islands over those two sampling periods?

Advance Your Thinking

3. What might lead to the difference in species richness losses in large islands compared to small islands?

4. On all of the islands, the most common (and sometimes only) small mammal 26 years after isolation was a non-native rat. Could this have had an influence on the loss of the other, native species? Explain.

5. What conclusion can be drawn regarding the value of leaving behind fragmented forest landscapes for protecting species that live in the habitat fragments?

INFORMATION LITERACY | EVALUATING INFORMATION

Visit the World Wildlife Fund (WWF) website (www.worldwildlife .org). Explore the website to learn about the organization and their work to protect endangered species.

1. What is the mission of the WWF? How do you know this?

2. Read about several of the species that WWF is protecting: Do they use a single species conservation approach, an ecosystem conservation approach, or both? Does their approach support their stated mission?

3. What scientific evidence does the organization provide to indicate that their work is effective at protecting species?

4. Read about the WWF's work in Borneo and Sumatra. What threats do they identify for this region and how are they helping species and the local communities there? Do you think their work is helping the region?

5. What is your overall opinion of the WWF: Would you do anything differently if you were in charge?

6. Is this an organization you would support financially? Explain.

Additional study questions are available at SaplingLearning.com.

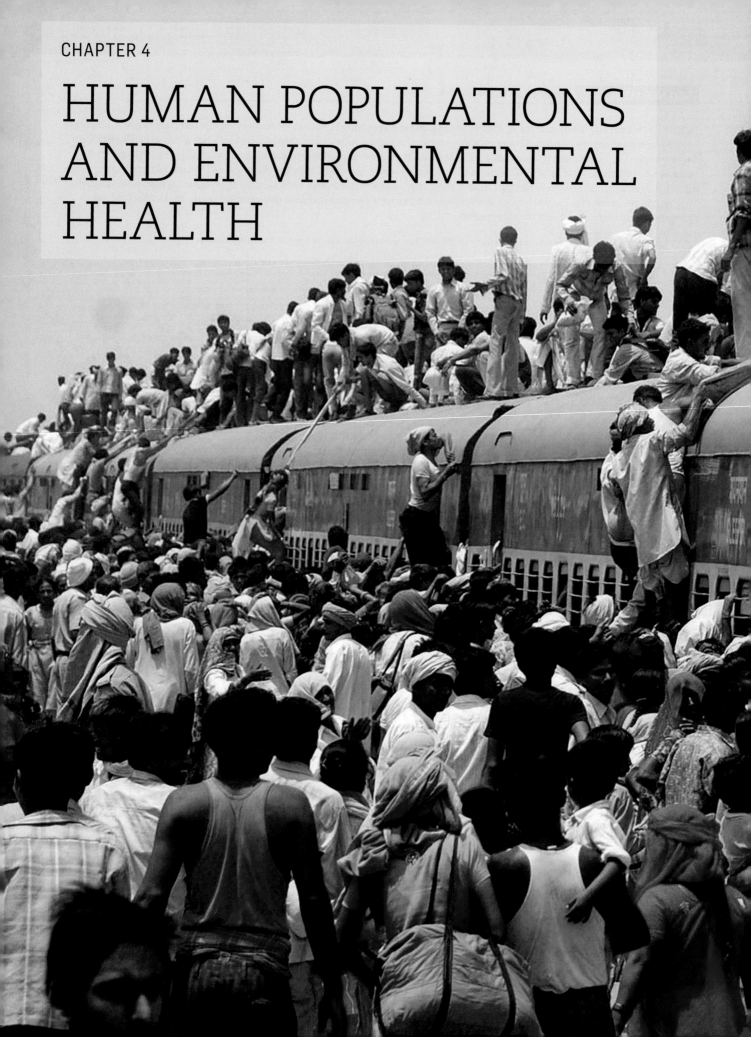

HUMAN POPULATIONS AND ENVIRONMENTAL HEALTH

CHAPTER 4

Many environmental problems arise from the sheer number of people, as well as high individual impacts of the human population. Some solutions lie in reducing population growth rates, in making our cities more sustainable, and in taking steps to create healthy environments.

Module 4.1: Human Populations

An examination of past and present human population growth and ways to reduce its current rate of increase

Module 4.2: Urbanization and Sustainable Communities

An introduction to smart urban development and ways to reduce the per capita impact of city residents while improving their quality of life

Module 4.3: Environmental Health

An assessment of the impact of the environment on human health and ways to address environmental health issues

THE KERALA MODEL

India's path to population control

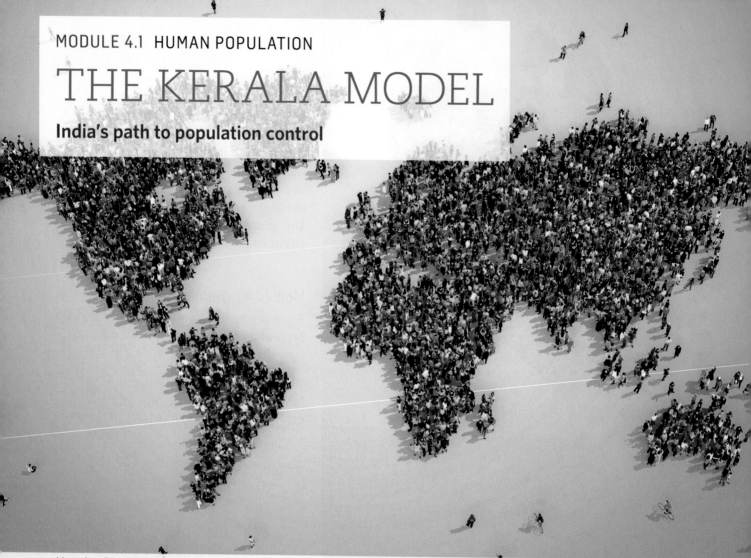

More than 7.5 billion people live on Earth and the human population is still growing, increasingly straining the resources needed to support us as well as other species.

Arthimedes/Shutterstock

After reading this chapter and studying the KEY CONCEPTS and INFOGRAPHICS, you should be able to answer these GUIDING QUESTIONS

CORE MESSAGE

The human population is increasing rapidly and is now more than 7.5 billion. Human impact on the environment is due both to our sheer numbers and to an increasing impact per person. To address population growth, we need to pursue a variety of approaches that address the factors that encourage high birth rates; many of these approaches focus on issues of social justice.

1 How and why have human population size and growth rate changed over time?

2 How big is the human population today, and where do most of these people live?

3 What factors affect population growth?

4 How does the age structure of a population influence its growth?

5 What is the demographic transition, and why is it important?

6 How can addressing social justice issues help achieve zero population growth?

7 What determines Earth's carrying capacity for humans, and can Earth support the current or future (projected) human population?

Back in 1988, a swarm of volunteers descended on Erakulam, a city of about three million people in the state of Kerala, India. They were on a mission to teach as many people as possible to read and write, and they were not leaving any stones unturned. "Classes were held in cowsheds, in the open air, in courtyards," an organizer told *The New York Times*. "For fisherman, we went to the seashore. In the hills, tribal groups sat on rocks. Leprosy patients were taught to hold a pencil in stumps of hands with rubber bands."

India is a developing country on track to surpass China as the world's most populous nation in the near future. Its huge and still rapidly growing population must find ways to stem population growth in a world and a country quickly running low on resources. Like much of India, Kerala, a tiny, bean-shaped state tucked into the southwest corner of India, is both very poor and very crowded. The average Keralan earns just 1/70th of what the average U.S. citizen earns and lives in just half the space of the average resident of New Jersey (the most densely populated U.S. state with about 1,200 people per square mile). But Kerala has some unique features that set it apart from other, equally poor and crowded places. Women tend to be more educated and have more power, for example. And decent health care is the norm for all citizens.

The Erakulam campaign made Kerala the first fully literate state in all of India. But it also brought about another success—one that stretched far beyond the volunteers' makeshift classrooms: As literacy rates went up in Kerala, the state's population stopped growing so quickly. The achievement was so unprecedented that economists came up with a new term to describe it: The Kerala Model. To understand the relationship between literacy and population growth, we must first take a closer look at the nature of human populations and the forces that cause them to grow and shrink.

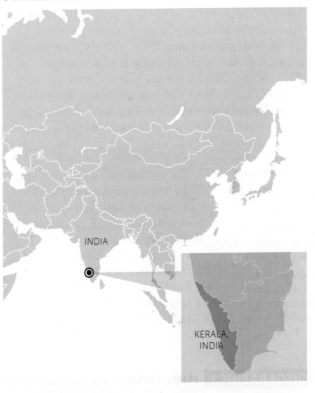

⊙ **WHERE IS KERALA?**

INDIA

KERALA, INDIA

Educational opportunities, especially for women and young girls, are highly correlated with decreased population growth rates.

Robert van der Hilst/Corbis Documentary/Getty Images

1 HUMAN POPULATION: PAST, PRESENT, AND FUTURE

Key Concept 1: The human population grew very slowly for most of human history. Only recently have accelerated growth rates sent our population soaring past 7 billion. World population is projected to stabilize around 11 billion by 2100.

Population growth rate is a measure of how quickly a population will increase in size. Right now, global population size is the largest it has ever been, and the growth rate is very high. For most of human history, our population size was orders of magnitude smaller than it is now. The Agricultural Revolution—the prehistoric transition from hunter-gatherer societies that were small and nomadic to farming societies that were stationary and larger—led to the first dramatic increase in the number of people, for two reasons: First, it enabled us to produce more food and feed more people. Second, it created a greater need for human labor. Because they take a lot of work to run, farms created an incentive for people to have more children. With farms, people also ate better and lived longer. As a result, death rates dropped, birth rates stayed high, and the number of people increased.

---

population growth rate The change in population size over time that takes into account the number of births and deaths as well as immigration and emigration numbers.

zero-population growth The absence of population growth; occurs when birth rates equal death rates.

But not indefinitely. Before long, the human population stabilized in size, eventually plateauing at around 100 million people, worldwide. That number held for centuries, until the next revolution—the Industrial Revolution—led to dramatic improvements in sanitation and health care. Those changes drove

the death rates down even further and spurred the next, even more extreme wave of population growth.

More recently, Earth has witnessed yet another surge in human population growth as the improvements in health care and sanitation brought on by the Industrial Revolution have spread from developed countries like the United States to developing ones like India and China. These advances have enabled people in all countries to live longer than ever before—life expectancy at birth has more than doubled since 1800, when it was less than 30 years of age. (According to the World Health Organization, in 2015, life expectancy was age 71.4 worldwide; age 69 in less developed countries). Many countries have life expectancies in the low 80s—the highest 2015 value was Japan at age 83.7; the United States came in at age 79.3. Even the country with the lowest life expectancy today (Sierra Leone at age 50.1) exceeds that of countries with the highest life expectancies in 1800!

Though still increasing—and very quickly in some areas—overall, human population is growing at a slower rate than it was in the mid-20th century, thanks to decreases in birth rates. Annual global growth rates peaked around 1963 at 2.2% and slowly decreased to the current level of just over 1% (1.13%

INFOGRAPHIC 1 | **HUMAN POPULATION THROUGH HISTORY**

Population size in 2050 and beyond will largely depend on future growth rates and may stabilize around 11 billion; if population growth rates do not decline as quickly as expected, a high-end estimate puts our population at more than 16 billion by 2100.

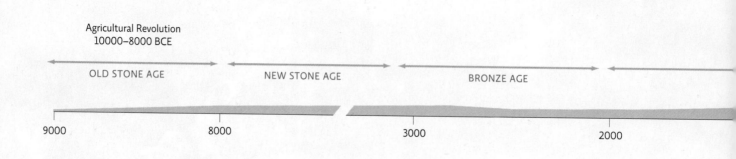

in 2016). Though a growth rate of barely more than 1% doesn't sound like much, if our growth rate stayed the same, by 2100, there would be more than 20 billion people on Earth.

Future projections that assume lesser growth rates predict human population size will stabilize somewhere between 7.3 billion (this would be a decline; population in mid-2017 was just over 7.5 billion) and 16.6 billion by 2100—a mid-range projection is 11 billion. (See the *Science Literacy* activity at the end of this module for a look at these projections.) Where we eventually stabilize will depend largely on how quickly we reach a stasis, where death and birth rates even out (a growth rate of 0%). That stasis is referred to as **zero-population growth. INFOGRAPHIC 1**

Because of concerns about how many people the planet can support and what will happen if population growth goes unchecked, zero population growth has long been the goal of many individual countries around the world, including India.

? What kinds of factors might result in a population of 16.6 billion by 2100 (the high estimate), and what might prevent our population from growing this large? For the medium and low estimates, what circumstances might lead to populations of those sizes?

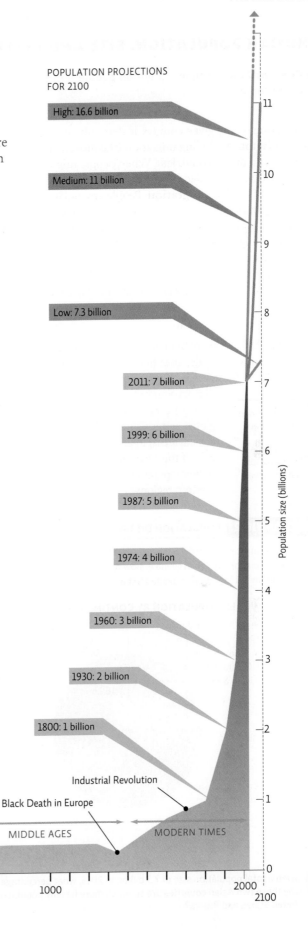

2 HUMAN POPULATION: SIZE AND DISTRIBUTION

Key Concept 2: More than 7.5 billion people inhabit Earth; 60% live in just 10 countries—most of those in China and India.

The population of any given country is determined by a myriad of factors, including migration (the movement of people in and out of countries). When people migrate into a population, it is called **immigration**; when they migrate out, it is called **emigration**. People are motivated to move away from situations (e.g., war, discrimination) or toward new opportunities (e.g., jobs, better farmland)—often both. Immigration is expected to drive most future population growth in the United States. Globally, most migration is immigration to cities—as of 2008, more people lived in urban areas than in rural ones (see Module 4.2).

In general, the distribution of humans around the world is wildly uneven. Worldwide, most human populations are located close to the ocean or major rivers. Population distribution in the United States is also quite variable—following the global pattern, around 50% live near coastal areas; in addition, regions east of the Mississippi River tend to be more population dense than western regions.

immigration The movement of individuals into a given population.

emigration The movement of individuals out of a given population.

In 2005, world population was 6.5 billion; by 2011, it reached 7 billion, and in 2017, human population topped 7.5 billion people; we are adding about a half billion people every 6 years! The vast majority live in just 10 countries; more than half live in Asia. The two most populous countries—China and India—stand out. Each of them has more than four times the population of the United States, the next largest country. **INFOGRAPHIC 2**

In recent history, leaders of China and India have employed some desperate attempts to curb population growth within their borders, in large part because they are (or were) growing beyond their ability to feed themselves. Many of those efforts were coercive: Beginning in 1979, the Chinese government prohibited families from having more than one child. The government vowed to deny state-funded education and health care to all but the first-born child; parents who didn't comply risked losing their jobs and faced severe fines and penalties, often several times their annual salary. And earlier that same decade, the Indian government established forced "vasectomy camps" across the country. They withheld food ration cards, business licenses, and even water for crop irrigation if men didn't

---

INFOGRAPHIC 2 | POPULATION DISTRIBUTION

The human population surpassed 7.5 billion in 2017. Around 60% of the world's population live in the 10 most populous countries, most of them in Asia. Population growth in China is slowing, and the population size of India should surpass it around 2022. Future population size increases in the United States are expected to be due to immigration.

WORLD POPULATION BY CONTINENT (2015)

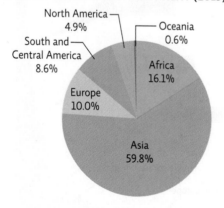

- North America 4.9%
- Oceania 0.6%
- South and Central America 8.6%
- Africa 16.1%
- Europe 10.0%
- Asia 59.8%

THE TEN MOST POPULOUS COUNTRIES

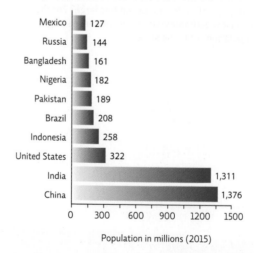

Country	Population in millions (2015)
Mexico	127
Russia	144
Bangladesh	161
Nigeria	182
Pakistan	189
Brazil	208
Indonesia	258
United States	322
India	1,311
China	1,376

Population in millions (2015)

? Given a total population size of 7.3 billion in 2015, what percentage of people lived in China and India that year? What might explain why these two Asian countries are so much more heavily populated than other countries with similar geographic sizes such as the United States and Russia?

Steve Raymer/National Geographic Magazines/Getty Images

A crowded city street in Kolkata, India. Global human population growth began to surge in the late 20th century and continues to grow at a fast pace, adding a billion people every 12 years or so.

"volunteer" for the sterilization surgery. Women too were coerced or paid (a form of coercion to those in poverty) to have sterilization surgery, and this practice continues today, sometimes with tragic consequences when hurried procedures done on women in poor health result in death.

Regardless of how well or poorly these efforts succeed in reducing population growth, most experts agree that they represent severe human rights violations and that it would be better to find a noncompulsory way to address human reproduction and population growth.

It turned out that in Kerala, a densely-packed society with a unique culture was doing just that.

3 FACTORS THAT AFFECT POPULATION GROWTH

Key Concept 3: A variety of pronatalist pressures, such as high infant mortality, lead to higher desired family size, which correlates well with the high population growth rates seen in less developed countries. It is in these areas where most future growth will occur.

Demography is the statistical analysis of populations. Demographers evaluate metrics known as **demographic factors** in an effort to understand how human populations change over time. In simple terms, populations grow when births outnumber deaths. As we saw in Module 2.2, factors that favor births and survival are considered population *growth factors* whereas factors that lead to death or reduced reproduction are termed *resistance factors*. And the factors that affect births and deaths aren't all that different between human populations and those of other species: disease, famine, and predators (or war in our case) are resistance factors; access to clean water, nutritious food, and protection from the elements are examples of growth factors. For humans, a decrease in resistance factors and an increase in growth factors have contributed to our phenomenal population growth over the past century or so.

demography The statistical analysis of the characteristics of a population.

demographic factors Population characteristics such as birth rate that influence changes in population size and composition.

Demographic factors tend to differ substantially between more developed countries (those with a moderate to high standard of living on average) and less developed countries (those with a weak economy and a lower standard of living than a developed country). For example, the *crude birth rate* (the number of offspring per 1,000 individuals) tends to be almost twice as high in less developed countries (21 births per 1,000 people) compared to developed ones, but in some countries, it is much higher than this average: Niger has a crude birth rate of 50—four times higher than that of the United States (12.4). Meanwhile, the *crude death rate* (the number of deaths per 1,000 individuals per year) is less variable worldwide. In 2015, it was 8 per 1,000 for the planet as a whole with an average crude death rate of 7 in developing countries and 10 in developed countries. A higher death rate is found in those developed countries that have an aging population (more elderly people in their population). Because of their high birth rates, by far, most current and future population growth is occurring, and will continue to occur, in developing nations.

But as much as demographic factors can differ between locations, they do tend to be influenced by the same forces everywhere. Health (especially of children), education (especially of females), economic conditions, culture, and religion all work together to determine a country's **desired fertility** (the number of children the average couple says they want to have). These same factors strongly affect the **total fertility rate (TFR)**—the number of children a woman actually has in her lifetime. As of 2016, the global TFR was 2.4, though like birth rates, these TFRs are highly variable from place to place.

Desired fertility and TFR tend to closely track one another, meaning that women tend to have the number of children they say they want to have. And both numbers are linked to population growth: As desired fertility increases, TFR increases; and as TFR increases, so does population size.

Forces that increase desired fertility (meaning they drive people to want more children) are called **pronatalist pressures**—things like cultural or religious views that favor large families and the need for children to work the farm or care for elderly parents. In many regions where birth rates are high, women may not have access to contraception (called "unmet need") due to the inability to afford birth control, cultural taboos against its use, or pressure from patriarchal societies that view children as a sign of male fertility. One of the most significant pressures in less developed countries like India is the **childhood mortality rate**—the number of children who die in their first 5 years of life per every 1,000 live births in that year. (Most of these deaths occur before age 1, so demographers often speak in terms of the *infant mortality rate*.) Statistics on childhood mortality vary greatly between more and less developed nations from a high of 157 deaths per 1,000 births in Angola to a low of 1.9 deaths per 1,000 births in Luxembourg. (The world average is 42.5.)

Higher childhood mortality rates tend to lead to higher desired fertility and thus higher TFRs, because couples will have more children to account for the likelihood that some of those children will die young. By the same token, decreasing childhood mortality will decrease birth rates because people tend to have fewer children when those children can be expected to survive. The bottom line: When childhood mortality decreases, so does desired fertility. And as desired fertility falls, so does TFR.

INFOGRAPHIC 3

> When childhood mortality decreases, so does desired fertility. And as desired fertility falls, so does total fertility rate.

In the 1970's, the consensus among development experts was that the best way to reduce a given country's TFR was to grow the economy: Raising a country (or state's) gross domestic product (GDP), the thinking went, would foster a drive among people toward material wealth, which would in turn improve living conditions, and lead to lower childhood mortality rates and lower TFRs. A rising tide, the reasoning went, would raise all boats.

But in the beginning of that decade, Indian economist Amartya Sen and his colleagues began arguing for a different approach: Build development policies around quality of life, not economic growth. Sen (who would go on to win a Nobel Prize for his work) used Kerala as a prime example. The state had a low GDP but high literacy rates, and even its poorest citizens enjoyed wide access to child and maternal health clinics (which offered birth control and nutritional assistance). Even as the economy stagnated, this unprecedented access to health care and education (especially for women) was driving population growth down and quality of life up.

In the 1950's, the population growth rate in Kerala had been one of the highest in India. But in the 1970's, it was beginning to fall. The rate at which children must be born to replace those dying in the population, **replacement fertility**

---

desired fertility The ideal number of children an individual indicates he or she would like to have.

total fertility rate (TFR) The number of children the average woman has in her lifetime.

pronatalist pressure Factor that increases the desire to have children.

childhood mortality rate The number of children under 5 years of age that die per 1,000 live births in that year.

replacement fertility The rate at which children must be born to replace those dying in the population.

INFOGRAPHIC 3   PRONATALIST PRESSURES: FACTORS THAT INFLUENCE POPULATION GROWTH

Most population growth in the recent past has occurred in developing countries, and it is there that most future growth will occur. The pronatalist pressures that increase population growth rates can be economic, social, cultural, or related to family health. Because actual fertility closely matches desired family size, steps that decrease desired family size are key to reducing high birth rates in many developing countries.

DESIRED FERTILITY CORRELATES WITH TOTAL FERTILITY

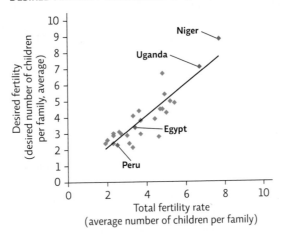

PRONATALIST PRESSURES

High infant mortality rates, which are closely tied to poverty

Lack of education or job opportunities for women

Valuing children (cultural and religious views); practical help (e.g., farms, care of elderly)

Unmet need for contraception

 Why does high infant mortality increase population growth rather than decrease it?

---

was reached around 1987, and within two decades the state had achieved sub-replacement fertility. (Global replacement fertility is 2.1 children per woman, rather than 2, because not all children survive and not all couples have children.) Kerala's achievements are impressive: the lowest birth and death rates, lowest childhood mortality rate, and highest life expectancy at birth (75) in all of India.

Sen might be on to something. More and more social scientists are arguing that monetary wealth or economic growth is the wrong gauge of societal progress and well-being. (The GDP of a nation goes up, for example, when air pollution increases health costs for the population—this may be good for the economy, but it is not good for society.) The goal of striving for a high quality of life for all citizens was formalized by Bhutan

in the 1970s —they seek to increase "Gross National Happiness" with a balanced approach that integrates economic development and cultural and environmental protections. The United Nations recognized the importance of this by signing a resolution that calls for the pursuit of a more equitable approach to economic growth that considers the "happiness and wellbeing of all people"; the resolution established March 20 as International Happiness Day.

Of course, reducing TFR did not immediately solve Kerala's overpopulation problem. Two decades after reaching replacement fertility, the state is still waiting to reach zero population growth; the reasons have to do with the breakdown of the population by age and its sex ratio (the proportion of males to females).

4 AGE STRUCTURE AND POPULATION MOMENTUM

Key Concept 4: Youthful populations have a great deal of population momentum and will continue to grow even if these young people reproduce at replacement rates.

To predict a population's future growth, demographers use **age structure** diagrams. These diagrams illustrate how a given population breaks down by age and gender, showing how many males and females are found in specific age groups (e.g., ages 0–4, ages 5–9, and so on). These data help demographers make predictions about where the population is headed and make inferences about current conditions. For example, a population with mostly young people, like that of Nigeria, has a greater potential for growth, because many people are of childbearing age or younger—this population cohort is still, or will soon be, having children. A population with similar numbers in all age classes is fairly stable as those who are born replace those who die. An aging population, like that of Japan, may even be shrinking if more of the population is found in older, post-reproductive age classes than in younger ones.

When a population is poised to continue growing for some time, even as TFR falls—usually because it has a high proportion of younger people compared to older people—that society is said to have **population momentum**. If there are a lot of young women in the population who still have not given birth to any children, they have the potential to contribute to a population size increase for another generation, even if these individuals reproduce at replacement rates. In other words, there will be far more births than deaths. Many countries in Africa have very young populations, and population momentum will contribute greatly to the expected high population growth in these regions.

Kerala has population momentum for two reasons: lots of younger people and lots of females. In fact, Kerala is one of the only states in all of India where women outnumber men. This feature is also a result of its unique approach to health and education: In many patriarchal societies where male children are preferred, girls are aborted or given up for adoption at much higher rates than boys. For example, a study of one Bombay (now Mumbai) clinic found that of 8,000 abortions performed in the early 1990's, all but one were of female fetuses. (In one report, India's Department of Women and Child Development wrote that "to be born female comes perilously close to being born less than human.") And it's not just

age structure The percentage of the population that is distributed into various age groups.

population momentum The tendency of a young population to continue to grow even after birth rates drop to replacement fertility (two children per couple).

India. In China, where until recently, one child per family was the strictly enforced rule, there were 113 boys born for every 100 girls. In Kerala, however, because women can receive education, earn income, and hold some power, families keep their girls.

Populations that skew older may grow slowly, not at all, or even shrink, since many or most of the people are past childbearing age. Once they have been industrialized for a while, more developed countries tend to transition to older age distributions. Currently, many industrialized nations—including the United States and many European countries—have top-heavy age structure diagrams with many older people. In many cases, the struggle to provide for these rapidly aging populations has turned prickly. In 2010, in order to relieve retirement age problems (and thus government pension payouts), France raised the retirement age from 60 to 62, sparking riots throughout the nation. In the United States, the prospect of baby boomers' retirement bankrupting Social Security has spurred intense and vitriolic debate. **INFOGRAPHIC 4**

> Most population growth is occurring and is expected to continue to occur in less developed countries, with African countries leading the way.

The workforce, too, is being affected in nations with an aging population. In the United States, for example, worries about retirement funding are keeping many older citizens in the workforce, making it harder for younger workers to get started in their careers. At the opposite end of the spectrum, in some areas, there are not enough young people to take over jobs left behind by retirees. China's one-child policy has led economists to predict that the annual size of the labor force aged 20–24 will shrink substantially by 2020—perhaps by as much as 50%. This shortfall prompted the Chinese government to end the one-child policy in 2015; families are now allowed to have two children.

Most population growth is occurring and is expected to continue to occur in less developed countries, with African countries leading the way. China will likely level

INFOGRAPHIC 4 | **AGE STRUCTURE AFFECTS FUTURE POPULATION GROWTH**

The fastest-growing regions are those with a youthful or very young population.

Median age (years)
- <22 Very young
- 22–29 Youthful
- 29.1–36 Transitional
- >36 Mature
- No data

Age structure diagrams show the distribution of males and females of a population in various age classes. The width of a bar shows the percentage of the total population that is in each gender and age class. The more young people in a population, the more population momentum it has; it will continue to grow for some time. The more people of reproductive age, the higher the growth rate, but this measure is also influenced by income; growth rates decline as income increases.

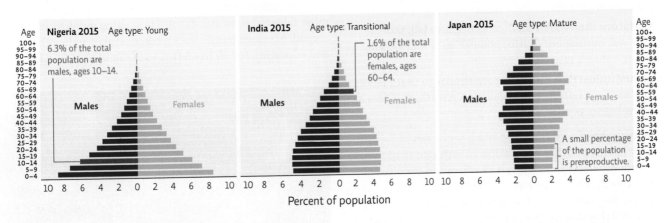

Percent of population

In Nigeria's young population, 71% of the people are under 30, and there is a very high capacity for growth (high population momentum).

India's transitional population is growing more slowly than Nigeria's because India's pre-reproductive age cohorts make up a smaller part of the population.

Japan's mature population has a fairly even distribution among age classes, with two slight bulges seen around 40 and 65. This population is fairly stable or may even be decreasing slowly as deaths start to outnumber births.

 What percentage of the total population (males and females) is between 0 and 4 years of age in each of the three countries shown here (Nigeria, India, and Japan)?

off around 1.4 billion (it was 1.376 billion in 2015) and perhaps even shrink slightly after 2030, due to an aging population. Demographers predict India will level off around 2060 at 1.75 billion.

Despite reaching replacement fertility in the 1990's, Kerala's population is still growing at a natural rate of 8.6 per 1,000 people, and experts say it will take several years more to reach zero population growth.

5 ADDRESSING POPULATION GROWTH: THE DEMOGRAPHIC TRANSITION

Key Concept 5: Helping low-income countries develop economically may spur a demographic transition that could decrease population growth rates.

Demographers and economists weren't entirely wrong for believing that economic stability would reduce population growth. In fact, there is so much evidence of this relationship that there's a name for it: the **demographic transition**. The demographic transition is a theoretical model that describes the expected drop in once-high population growth rates as economic conditions improve the quality of life in a population. It tends to happen as a country's economy changes from preindustrial to postindustrial, when low birth and death rates replace high birth and death rates.

The demographic transition helps demographers understand a population's growth patterns and predict future ones. It has four stages:

1. Preindustrial: Birth and death rates are high but similar (replacement fertility), so population growth is slow or stationary; population size is low.

2. Industrializing: Better conditions lead to lower death rates; however, birth rates remain high, resulting in rapid population growth.

3. Mature industrial: Birth rates begin to fall, though they still outnumber deaths; population is still growing but at a slower rate.

4. Postindustrial: Birth rates are similar to death rates (replacement fertility), so population growth stabilizes at a new higher population size.

Industrialized countries that reach replacement fertility after having a higher population growth rate tend to have strong economies, high levels of education for citizens, and good healthcare systems. The United States, Japan, and most of the European Union are examples of countries that are postindustrial. The reasons that economic stability begets population stability are believed to be tied to better health: Economic stability improves access to health care and prenatal care so that deaths rates (including childhood mortality) decline. As it becomes clear that more children will make it to adulthood, desired fertility declines, and TFR along with it. The lag between the drop in death rates and the subsequent drop in birth rates results in a higher overall population, but once birth rates decline, such that birth rates equal death rates, the population stabilizes at its new size. **INFOGRAPHIC 5**

demographic transition A theoretical model that describes the expected drop in once-high population growth rates as economic conditions improve the quality of life in a population.

To be sure, the transition is not uniform or smooth and steady. Within transitioning societies, birth rates tend to drop faster for those with higher incomes. And populations tend to increase for a while after economic conditions improve not only because of population momentum but also because it takes a while for cultural norms to shift toward lower desired fertility and smaller families. Recent evidence suggests that India is now passing through the demographic transition—moving into the later phase of stage 3 (mature industrial) as a result of major economic development and growth.

Kerala made it through the demographic transition in the early 1990's—well ahead of the rest of India and without first ushering in an era of economic growth. This was an amazing feat, to be sure, but it came with some serious side effects. Because the economy remains stagnant, university graduates have a hard time finding work that befits their educational status. As a result, unemployment remains high, and the state itself is suffering from brain drain, with many of its most talented sons and daughters seeking employment abroad. Meanwhile, the state's commitment to health and education for all has left it with significant budget

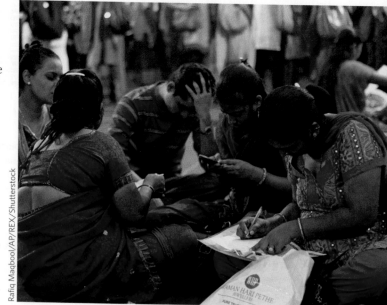

Education in India is producing many new graduates, but the job market in some areas lags behind, leaving many qualified individuals competing for scarce jobs like these hopeful candidates at a job fair in Hyderabad, India.

Rafiq Maqbool/AP/REX/Shutterstock

INFOGRAPHIC 5 **DEMOGRAPHIC TRANSITION**

ANIMATED INFOGRAPHIC

Some industrialized nations have gone through the demographic transition from high birth and death rates to low birth and death rates, giving them a stable population. Demographers are considering adding a fifth stage, *Declining*, for those populations experiencing more deaths than births. It is unclear whether industrialization will produce the same pattern in all developing countries. Still, steps to improve the quality of life and decrease death rates are important worldwide, even if other measures are needed to slow birth rates.

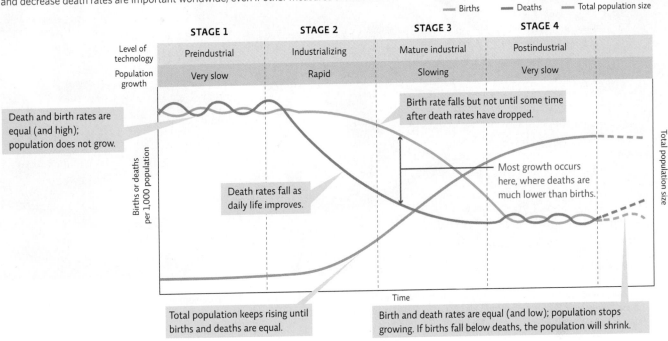

 Why might industrialization *not* lead to a demographic transition in all countries?

deficits. Indeed, even as development experts around the world look to Kerala as a model of how to do more with less, some wonder if the state's victory over population growth will prove sustainable. "Can the Kerala Model survive even in Kerala," asks noted environmentalist Bill McKibben. "Or will it be remembered chiefly as an isolated and short-term outbreak from a prison of poverty?"

6 ADDRESSING POPULATION GROWTH: SOCIAL JUSTICE

Key Concept 6: In many cases, addressing social justice issues, especially improving the well-being of women and children, is the key to reducing total fertility and reaching zero population growth.

Demographers have put forth many explanations for Kerala's unprecedented decline in population growth rate in the absence of substantial economic growth, but one factor stands out among the others: Kerala also had India's highest overall literacy rate (91%) and the highest literacy rate for women, by far. In fact, of all the factors that Sen and his colleagues cited for Kerala's high quality of life and decreasing population rates (land reform, educational access, child welfare, health care), education ranked as the most significant. And not just education: education of women in particular.

It turns out that education of girls and economic opportunities for women correlate with lower population growth in many regions, and it's not difficult to see why: Education empowers women to take more control over their lives and their fertility—by demanding and using birth control, for example, or by marrying later or delaying childbirth while pursuing a career. The typical Keralan woman marries somewhat later than her Indian counterpart (age 22 compared to 18).

And because women who earn more can better support their families, childhood mortality tends to decline as well. In Kerala, women's income has not necessarily increased with education, but that income is more dependable (making it easier for women to predictably care for their children) and women there make regular use of the free birth control that is offered to them. They tend to choose

Programs that provide microloans, such as this one offered by Bharat Financial Inclusion Ltd. (BFIL), allow women to start their own businesses and rise out of poverty. BFIL works with groups of five women who support each other and hold joint liability for the loans. Training and support from BFIL is strong, and loan repayment is greater than 99%.

smaller families now that they are less worried about some of their progeny dying young and because they want all of their own children to have private schooling, a university education, and so on (something that's much easier with two children than with seven).

Education for women is just one of many social justice issues that strongly influence desired fertility and TFR. (Others include access to health care and birth control and land and legal reform that enable women to own property and run businesses.) All of these factors affect the less privileged members of society the most. In the United States, the high infant mortality rate (6.1 deaths per 1,000 births) relative to other developed countries (between 2 and 4 in most developed countries) manifests itself mainly in those living in poverty—babies born into poverty are much more likely to die than those born to U.S. families that can afford decent health care. (Most of the countries with lower infant mortality rates have some form of national healthcare system).

The greatest value of Sen's work may have been to show how addressing these social issues can significantly reduce population growth. It was on this understanding that the *United Nations Development Programme* (UNDP) developed a new metric with which to shape sustainable development policies: *human development indicators*. These indicators have been instrumental in the drafting of the UN's Sustainable

Development Goals—a quality education and gender equality are goals 4 and 5. (See Infographic 4 in Module 1.1.)

But that correlation (between social justice and curbing population growth) depends heavily on culture. For example, access to contraceptives can help drive TFR down, but only in regions where there is unmet need for birth control. In some places, religious beliefs prevent increased access to birth control options from having an impact. In other places, other cultural forces that lead people to want bigger families (it's not always about high childhood mortality or the need for farm labor) weigh heavily. **INFOGRAPHIC 6**

Kerala's culture gave it three distinct advantages in spreading education and health care. First, the state was always matrilineal. That is, power, land, and money passed down through daughters rather than sons, and women (especially those of higher castes) were routinely educated. So when progressive leaders began extending that right to poorer communities, the culture at large was primed for it.

Second, there is no real divide between urban and rural areas in Kerala. So unlike other states in India (and across the developing world), schools and health clinics didn't get concentrated in cities, but instead were accessible to both groups. And third, the state boasted an above average concentration of missionary schools that catered to people of all faiths.

INFOGRAPHIC 6 **ACHIEVING ZERO POPULATION GROWTH IS LINKED TO SOCIAL JUSTICE** ANIMATED INFOGRAPHIC

Addressing pronatalist pressures can help reduce population growth rates. This often involves bringing social justice to women and families so they can better care for themselves and their children. The graphs below show the correlation between total fertility rate (the average number of children per female) and three important factors: childhood mortality, education for women, and access to birth control. Each point on a graph represents the data for one country.

CHILDHOOD MORTALITY AND TFR

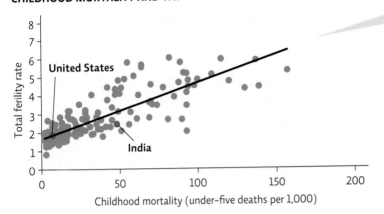

Infant mortality is closely correlated to poverty; therefore, actions that decrease poverty in a population should also reduce infant mortality and that population's growth rate.

EDUCATION OF WOMEN AND TFR

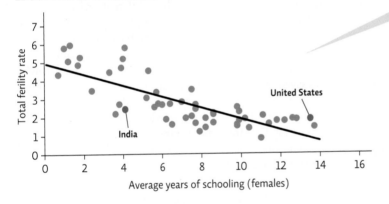

Fertility declines as educational opportunities for girls and women increase. This means that funding for education and job opportunities for women can be very effective at lowering TFR.

CONTRACEPTIVE USE AND TFR

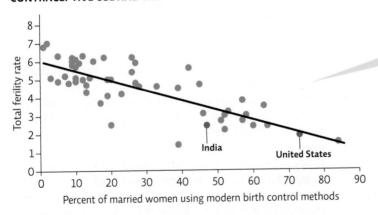

Access to contraceptives and family planning have been effective in many areas of the world. However, in countries with high desired fertility, such as Niger, providing contraceptives may have little impact on TFR.

 Which of the graphs shown here depict positive correlations? Which are negative correlations? Are positive correlations "good" and negative correlations "bad"? Explain.

7 HOW MANY PEOPLE CAN EARTH SUPPORT?

Key Concept 7: The number of people Earth can support depends on how many resources each uses. We may have already exceeded our planet's carrying capacity by using more resources than it can replace in the long term.

In the decades since Kerala achieved 100% literacy and sub-replacement fertility, the state has been heralded far and wide as a model of sustainable development. Demographers and development workers the world over have been studying and discussing the Kerala Model— there is even a yearly conference by that name—for what it might teach them about how to achieve similar successes in other countries with bulging populations and small economies.

But some experts say that Kerala's most valuable lessons may actually be for wealthy countries, not poor ones. Coercive, government-imposed ways to limit population growth are not necessary, even in rapidly growing regions. "Kerala suggests a way out of two problems simultaneously," McKibben writes. "Not only the classic development goal of more food in bellies and more shoes on feet, but also the emerging, equally essential task of living lightly on the Earth, using fewer resources, creating less waste."

In other words, not only has Kerala succeeded at reducing the number of people who consume resources, it did so without increasing the consumption of resources by each member of its society. In so doing, it maintained its carrying capacity.

Carrying capacity is the maximum population size that a given environment can support. The size of that capacity is determined by a range of forces, including the amount of crucial resources available and the rate at which those resources can be replenished (if they are renewable), as well as the rate of resource consumption by the population. The amount of waste generated plays a role, too; if waste is generated faster than the ecosystem can assimilate it, the ecosystem will suffer and the rate of resource replenishment will decrease. In other words, a degraded ecosystem can sustain far fewer people than a well-maintained healthy one.

When a population's size exceeds the carrying capacity of its environment, that environment is said to be **overpopulated**. Some parts of the world are certainly overpopulated today; they have exceeded the carrying capacity for their region, and as a result they face food and water scarcity issues as well as problems managing their waste.

But whether a given population qualifies as overpopulated is not a simple question of numbers. Take the planet as a whole: 7.5 billion is such a large number it can be difficult to comprehend. Is this too many? Are we overpopulated? The answer depends on several factors, including how many resources each of those people use and how much waste they produce.

One measure of our impact on the environment is the **ecological footprint**, expressed as the area needed to provide resources and assimilate waste (usually expressed in global hectares—a metric that normalizes different types of land and aquatic areas in terms of average global productivity). The more resources each person takes and the more waste each person generates, the greater their ecological footprint. And the greater the ecological footprint for an average person living in a given environment, the lower that environment's carrying capacity. (See Module 5.1 for a more in-depth look at the concept of ecological footprint and resource use by the human population.) In other words, more people can be supported if they consume less compared to a population where per capita consumption is high.

By comparing the ecological footprint to the amount of productive area available on Earth, we can determine if we are living sustainably—within the means of the planet to replace what we took. Researchers at the Global Footprint Network have done just that. Though the footprint varies widely from country to country, they have estimated that the overall ecological footprint of humans currently exceeds what Earth can sustainably support in the long run. We manage to live beyond our means by using resources that should remain in place to replenish what can be taken sustainably. By taking more now than Earth can replace at its own ecological pace, we are reducing Earth's carrying capacity for the future—in other words, if you cut down all the trees in your forest faster than they regrow, before long you no longer have a forest. Can you wait for it to recover? Have you left it in a condition that allows its recovery? **INFOGRAPHIC 7**

carrying capacity The maximum population size that a particular environment can support for the long term; for human populations, it depends on resource availability and the rate of per capita resource use by the population.

overpopulated The number of individuals in an area exceeds the carrying capacity of that area.

ecological footprint The land area needed to provide the resources for, and assimilate the waste of, a person or population.

INFOGRAPHIC 7   **HUMAN POPULATION AND EARTH'S CARRYING CAPACITY**

A comparison of humanity's ecological footprint (what we take) to Earth's productive capacity (what is available) can reveal whether our population has exceeded the long-term carrying capacity of Earth. Current assessments suggest we currently use resources faster than Earth's systems can replace, and our footprint is increasing. This overshoot is caused by both our sheer numbers and the rise in per capita resource use, especially in developed countries.

GLOBAL ECOLOGICAL FOOTPRINT

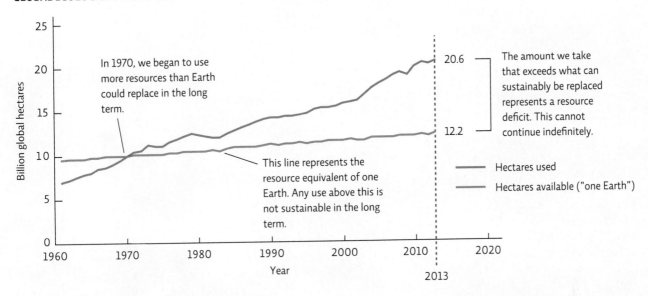

In 1970, we began to use more resources than Earth could replace in the long term.

20.6
12.2

The amount we take that exceeds what can sustainably be replaced represents a resource deficit. This cannot continue indefinitely.

This line represents the resource equivalent of one Earth. Any use above this is not sustainable in the long term.

—— Hectares used
—— Hectares available ("one Earth")

PER CAPITA FOOTPRINT COMPARISON

The graph above suggests the human population as a whole has overshot carrying capacity, but there is great disparity in the resources used and waste produced among the countries of the world. The per capita footprint of developing countries is much lower, in general, than that of developed ones.

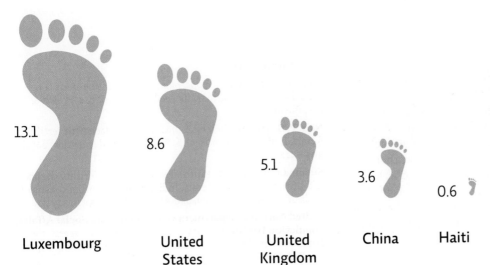

13.1 Luxembourg

8.6 United States

5.1 United Kingdom

3.6 China

0.6 Haiti

Luxembourg has the largest per capita footprint, but as a small country, it contributes little to the overall impact. The large U.S. footprint has a greater impact due to its large population size, but China's total footprint (around 5 billion global hectares [gha]) exceeds that of the United States, which is estimated to be around 2.7 billion gha. (The world footprint totals around 20.6 billion gha.)

Values shown are global hectares per person (2013).

 Using the data values shown for 2013 on the graph, calculate how much we exceeded Earth's capacity to support us and use this value to calculate how may "Earths" we were using in that year.

When humans overharvest resources or damage ecosystems, we reduce the carrying capacity of the area. Here, tracts of Amazonian rainforest are systematically destroyed for agriculture, reducing the area's ability to support native species and the land's long-term productivity.

Carrying capacity is itself a moving target. It tends to increase with technological innovations and decrease with environmental damage. In general, less developed regions tend to have a greater number of people but a smaller per capita ecological footprint, and more developed regions tend to have fewer people, each with a large ecological footprint. Right now, there are more than 7 billion people on Earth, and at least some experts say that that number already exceeds the planet's carrying capacity. In any case, whether we stabilize at 11 or 16 billion, or somewhere in between, will depend on how quickly we achieve replacement fertility worldwide.

Select References:

McKibben, B. (1995). *Hope, human and wild: true stories of living lightly on the earth.* Milkweed Editions. Little, Brown & Co., Boston.

Nair, P. S. (2010). Understanding below-replacement fertility in Kerala, India. *Journal of Health, Population, and Nutrition, 28*(4), 405–412.

Susuman, A. S., et al. (2016). Female literacy, fertility decline and life expectancy in Kerala, India: an analysis from census of India 2011. *Journal of Asian and African Studies, 51*(1), 32–42.

United Nations, Department of Economic and Social Affairs, Population Division. (2015). *World Population Prospects: the 2015 Revision, Key Findings and Advance Tables.* Working Paper No. ESA/P/WP.241.

INTERACTIVE MAP POPULATION GROWTH OR CONTROL IN OTHER REGIONS   ANIMATED INFOGRAPHIC

There are many reasons why populations grow and many ways to curb that growth. See the online interactive map for five examples of population growth or control in different regions of the world.

 BRING IT HOME

PERSONAL CHOICES THAT HELP

The impact of humans on the planet is created by a combination of population size and per capita resource use. The issue of population and carrying capacity is complex. We cannot have a truly sustainable society until key components such as poverty, lack of education, and basic human rights are addressed.

Individual Steps

• Buy Fair Trade Certified labelled products. These products provide a livable wage to workers and are often linked to education and community development.

• Research the products you buy to make sure that you are not supporting child slave labor, sweatshop facilities, or environmentally destructive actions.

Group Action

• Raise money and invest it in a socially responsible project. The Foundation for International Community Assistance (www.finca.org) and Kiva (www.kiva.org) are nonprofit organizations that provide microloans to help people start small businesses in less developed countries.

• Join an organization such as Habitat for Humanity, which builds housing for low-income families.

Policy Change

• We all know that our tendency as humans is to build up and out; however, revitalizing our current older downtown areas is important to prevent new habitat destruction as well as to make use of current infrastructure. Urge your local government to keep shopping and dining establishments in historic downtowns. If needed, work to develop a community partnership that starts clean-up programs and community gardens in abandoned lots.

ENVIRONMENTAL LITERACY **UNDERSTANDING THE ISSUE**

1 How and why have human population size and growth rate changed over time?

1. A graph that shows human population size over the course of human history:
   a. is a line that has steadily increased at a constant rate.
   b. produces an undulating curve that rises and falls repeatedly, with our current population similar in size to past peaks.
   c. resembles a J-shaped curve—flat at first, with a rapid upward rise since about 1800.
   d. is an S-shaped curve—flat at first, then increasing rapidly for a few hundred years and then leveling off in the past 50 years.

2. What factors led to the increase in population growth rate during and after the Industrial Revolution?

2 How big is the human population today, and where do most of the people live?

3. The migration of people into a population is called _____ whereas the migration of people out of a population is _____.

4. About how many people live on Earth today?
   a. 5 billion
   b. 70 million
   c. 7.5 billion
   d. 11 billion

5. Where does the United States rank globally in terms of population size? What will contribute to future growth in this county?

3 What factors affect population growth?

6. True or False: The vast majority of all future population growth will occur in developing countries.

7. In the long term, reducing childhood mortality generally:
   a. increases the rate at which a population grows.
   b. decreases the rate at which a population grows.
   c. has little effect on the rate at which a population grows.

8. Describe the link between desired fertility and the total fertility rate of a population and explain how they affect population growth.

9. Identify several factors that contribute to the high growth rate in developing nations.

4 How does the age structure of a population influence its growth?

10. True or False: A population with more old people than young people has a lot of population momentum.

11. An age structure diagram can be used to:
    a. distinguish between births and immigration.
    b. see the historical growth of a population.
    c. predict the future growth of a population.
    d. measure the quality of life of a population.

12. Why will a youthful population still be growing in size even after it has reached replacement fertility?

5 What is the demographic transition, and why is it important?

13. The term *demographic transition* refers to:
    a. slower growth as the population size approaches carrying capacity.
    b. the decline in death rates and then birth rates as a country becomes industrialized.
    c. the requirement for a population to reach a specific size before it becomes stable.
    d. migration from the overpopulated countryside to urban centers.

14. Why might a modern-day developing nation in an intermediate stage of the demographic transition (experiencing lower death rates) not complete the transition to lower birth rates as expected?

6 How can addressing social justice issues help achieve zero population growth?

15. True or False: Population growth rates tend to go down when job opportunities for women go up.

16. Why does education for girls and women usually reduce their total fertility?
    a. They are more likely to secure birth control so they can control their own fertility.
    b. They often marry later or delay childbirth to pursue a career.
    c. They can better care for their children, so childhood mortality decreases.
    d. All of the above

17. In terms of reducing population growth rates, how important is reducing desired family size?

18. Why must culture and religious traditions be taken into consideration when designing programs to reduce population growth?

7 What determines Earth's carrying capacity for humans, and can Earth support the current or future (projected) human population?

19. How is carrying capacity expressed?
    a. It is the amount of land (in hectares) needed to support a population.
    b. It is the amount of resources (in kilograms) needed to support a population.
    c. It is the population size (number of individuals) that an area can support.
    d. It is amount of time (in years) that a given area can support a given population.

20. How could it be possible that we are already living beyond the long-term carrying capacity of Earth?

SCIENCE LITERACY   WORKING WITH DATA

The United Nations (UN) Population Division predicts future population growth based on scenarios of different rates of fertility (fertility variants). The constant fertility projection shows population growth at current growth rates; the other variants assume lesser levels of the global TFR of 2.6 (high), 2.1 (medium), and 1.6 (low).

A: ESTIMATED AND PROJECTED WORLD POPULATION ACCORDING TO DIFFERENT SCENARIOS

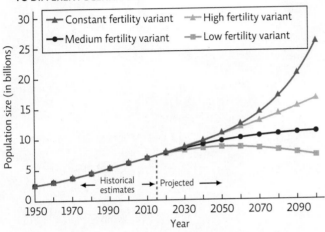

B: PROJECTED POPULATION CHANGE BETWEEN 2015 AND 2100 (MEDIUM VARIANT)

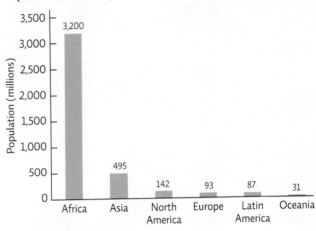

Interpretation

1. What are the population projections for each fertility variant in 2050? In 2100?

2. Describe the trend in population growth for each line seen in Graph A.

3. How does the information in Graph A relate to that in Graph B?

Advance Your Thinking

4. Why does the UN produce population projections based on different fertility rates?

5. Why is Africa projected to have such large growth if the medium variant population projections assume that global fertility rates will be at replacement? Use information from the module to support your answer.

INFORMATION LITERACY   EVALUATING INFORMATION

Many people consider providing education and job opportunities for women to be important not only for their effects on fertility but also as a matter of social justice. Two organizations that work on this are the Foundation for International Community Assistance (FINCA) and Kiva, which provide microfinance services to low-income people, mostly women. Explore the FINCA website (www.finca.org) and the Kiva website (www.kiva.org).

Evaluate the websites and work with the information to answer the following questions for each organization:

1. Evaluate each organization to determine if it is a trustworthy organization with a clear and transparent agenda:
   a. Who runs this organization? Do their credentials make the work of the organization and the information presented reliable or unreliable? Explain.

b. What are the mission and vision of this organization? What are its underlying values? How do you know this?

2. Explore the "About" links.
   a. How well does each explain the principle of *microfinance*? Give your own definition of microfinance.
   b. Does the website provide supporting evidence that its programs can help the poor? Is the evidence reliable? Explain.
   c. Do you think the business model of FINCA or Kiva is valid and effective? Explain.

3. How might the FINCA or Kiva model influence cultural, economic, and demographic factors that influence population growth?

4. Explore each website to see how interested people can participate.
   a. Who can be a part of FINCA's or Kiva's solution? How can they get involved?
   b. If you decided to contribute to a microloan program, which of these organizations would you choose? Why?

 Additional study questions are available at SaplingLearning.com.

CREATING GREEN CITIES

Building a better backyard in the Bronx

Clay Garden, built on the site of a burned-down home by a resident across the street, provides urban farming opportunities for local residents. In the background are the Webster Morrisania public housing projects, which provide homes for some of the poorest people in the Bronx.

nina berman/NOOR/Redux

After reading this chapter and studying the KEY CONCEPTS and INFOGRAPHICS, you should be able to answer these GUIDING QUESTIONS

CORE MESSAGE

Cities can be both an environmental blessing and a curse. Using green strategies to plan or retrofit cities can benefit citizens, businesses, and the environment—not to mention reduce environment-related health problems and degradation of natural resources. Since more than half of the world's population now lives in cities, and that proportion is rising, greening our cities is crucial if we are to lower humanity's environmental impact.

1. What is the pattern of global urbanization and megacity growth in recent decades?
2. What are the trade-offs associated with cities or urban areas?
3. What is environmental justice and what common problems does it address?
4. What environmental problems does suburban sprawl generate?
5. What is the value of urban green space?
6. What are "green cities"?
7. How does smart growth help make a city "green"?
8. What are some of the design features of a green building?

As she walked her dog Xena, Majora Carter considered her options. She had moved back home to save money while she attended graduate school. Initially, she had wanted as little to do with the decaying neighborhood as possible. But then she'd taken work at a community development center and now a colleague at the city parks department was offering her a $10,000 grant to come up with a waterfront development project for her neighborhood. Carter was balking.

At the moment, she and her neighbors were fighting the city's attempt to move a new waste facility to the East River waterfront. Their tiny parcel of New York already had some of the poorest air quality in the country thanks to current waste facilities, four power plants, and the largest food distribution center in the world. Carter wasn't sure she had the time or energy to take on another project. Besides, she knew the waterfront had long been claimed by industry. And then Xena began pulling her toward an abandoned lot, through a garbage-strewn path surrounded by towering weeds. There at the end, sparkling in the early morning light, was the East River. Carter stood in awe. Maybe the river wasn't so inaccessible after all.

◉ **WHERE IS THE BRONX, NEW YORK?**

1 GLOBAL URBANIZATION PATTERNS

Key Concept 1: Today, more people live in cities than in rural areas. The number of megacities has greatly increased since 1950 and will continue to do so in the future.

For the first time in human history, more than half the world's population (54% in 2014) lives in **urban areas**—densely populated regions that include both cities and the suburbs that invariably surround them. In the United States the proportion is even higher: More than 60% of Americans are urban dwellers. **Urbanization**, the migration of people to large cities, is happening around the world at an unprecedented rate. As global population swells, rural lands are morphing into urban and suburban ones, and ordinary cities are growing into *megacities*—those with at least 10 million residents. With more than 21 million inhabitants in 2016, the New York City metropolitan area qualifies as the largest city in the United States and one of the world's 34 megacities.

> For the first time in human history, more than half the world's population lives in urban areas.

In the coming decades, the world's share of urban dwellers is poised to grow. In 2016, about 55% of the world population, or 4 billion people, lived in a city of at least 1 million. The United Nations predicts that the next billion people added to the planet (by 2030) will all live in these big cities—there will actually be a slight decrease in the rural population. Cities everywhere will continue to grow, including megacities. By 2030, it is predicted that there will be 41 megacities—compare to only five in 1980 and 34 in 2015—and more of them will be amazingly large: 13 will have over 20 million residents. By 2030, Mexico City, Mexico, and Sao Paulo, Brazil, will be the only 20 million+ cities that are not in Asia. (New York is projected to be under 20 million in 2030.) **INFOGRAPHIC 1**

urban areas Densely populated regions that include cities and the suburbs that surround them.

urbanization The migration of people to large cities; sometimes also defined as the growth of urban areas.

INFOGRAPHIC 1 **URBANIZATION AND THE GROWTH OF MEGACITIES**

The world's population is becoming more urban. The growth of megacities, those with at least 10 million people, has increased dramatically over the last half century. In 1950, there were only two megacities: New York City with 12.3 million and Tokyo, Japan, with 11.3 million. In 2015, there were 34 megacities, with Tokyo's population of more than 38 million making it the largest city in the world. This map shows the growth, since 1980, of what are expected to be the 30 largest megacities in 2025.

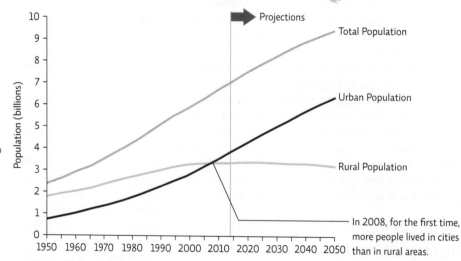

■ POPULATION 1980

■ POPULATION 2025

■ MORE DEVELOPED REGIONS

■ LESS DEVELOPED REGIONS

■ LEAST DEVELOPED REGIONS

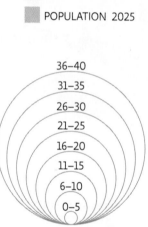

36–40
31–35
26–30
21–25
16–20
11–15
6–10
0–5

POPULATION (MILLIONS)

WORLD POPULATION GROWTH BY LOCATION

World population is continuing to increase, though at a slower rate than in the 20th century. Most future growth will be in urban areas.

Projections

Total Population

Urban Population

Rural Population

Population (billions)

10
9
8
7
6
5
4
3
2
1
0

1950 1960 1970 1980 1990 2000 2010 2020 2030 2040 2050

In 2008, for the first time, more people lived in cities than in rural areas.

 What are the advantages and disadvantages of living in a megacity?

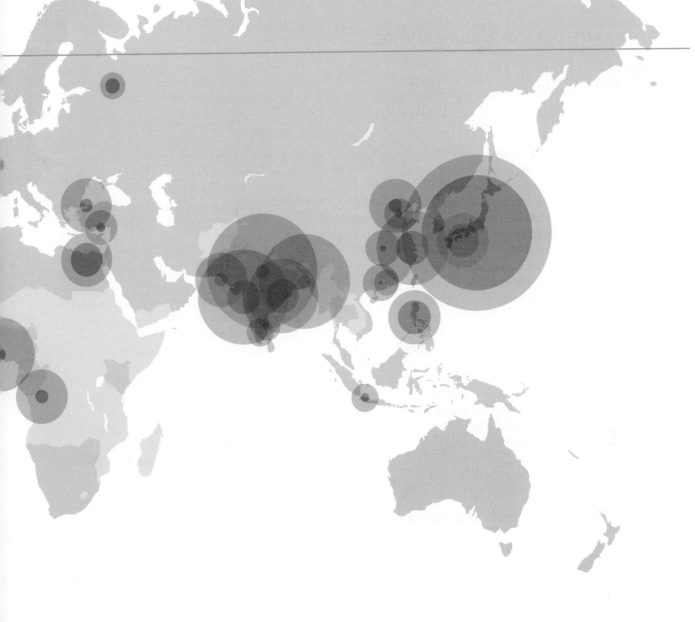

The United Nations predicts that by 2030, there will be 41 megacities. Most of those cities will be in Asia and Latin America.

FUTURE GLOBAL URBAN GROWTH WILL OCCUR IN CITIES OF ALL SIZES

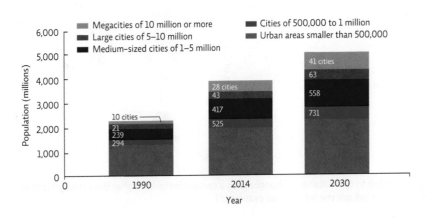

2 THE TRADE-OFFS OF URBAN LIVING

Key Concept 2: Cities offer many services and opportunities, and have a lower per capita carbon footprint, but have higher rates of crime and disease and problems dealing with waste and stormwater.

To be sure, cities bring some obvious advantages to their inhabitants: more job opportunities, better access to education and health care, and more cultural amenities, to name a few. But as far as the environment is concerned, urbanization is both a blessing and a curse. On the plus side, concentrating people in smaller areas (building *up* rather than *out*) can reduce the development of outlying agricultural land and wild spaces and thus protect existing farms and ecosystems. Higher population densities also make some environmentally friendly practices more cost effective. For example, it's easier to implement recycling and mass transit programs in cities because there are more people to share the costs of these services. Living in smaller homes that are closer to needed amenities and having access to mass transit also decreases the energy use—and the **carbon footprint**—of urban dwellers compared to those who live in suburban areas. **INFOGRAPHIC 2A**

carbon footprint The amount of carbon released to the atmosphere by a person, company, nation, or activity.

urban heat island effect The phenomenon in which urban areas are warmer than the surrounding countryside due to pavement, dark surfaces, closed-in spaces, and high energy use.

On the minus side, cities are *locally* unsustainable: They require the import of resources like food and energy and the export of waste. Because they are densely populated, most cities are also hotbeds of traffic congestion (which pollutes the air) and sewage overflow (which pollutes nearby lakes, rivers, and coastal waters).

Another problem stems from the way cities are designed and built—namely, the replacement of vegetation with pavement. Plants absorb water, filter air, and regulate area temperatures; asphalt pavement and concrete do not. In fact, the blacktop that covers most cities prevents rainwater from being absorbed into the ground, which in turn diminishes groundwater supplies and can lead to flooding (see Module 6.2). Cities also require an abundance of energy. This trifecta—too few plants, too much pavement, and high energy use—conspires to trap solar heat absorbed, and put off, by buildings, making most cities warmer than their surrounding countrysides. This phenomenon is known as the **urban heat island effect**. **INFOGRAPHIC 2B**

INFOGRAPHIC 2A **URBAN LIVING CAN REDUCE ONE'S ENVIRONMENTAL IMPACT**

Due to higher population densities, less personal vehicle travel, smaller homes, and efficiencies of scale, people living in large urban areas typically have a lower environmental impact than those in suburban areas. A 2009 study by geographer David Dodman compared the carbon footprints of various countries and large cities within those countries. Almost all of the cities evaluated had lower carbon footprints than their national average.

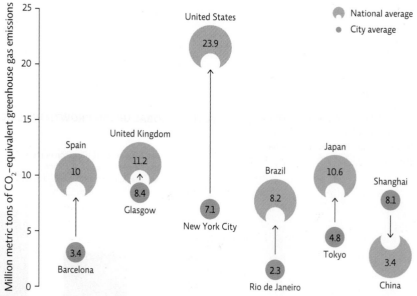

ANNUAL PER CAPITA CARBON FOOTPRINT

Million metric tons of CO_2-equivalent greenhouse gas emissions

- National average
- City average

United States 23.9 / New York City 7.1
Spain 10 / Barcelona 3.4
United Kingdom 11.2 / Glasgow 8.4
Brazil 8.2 / Rio de Janeiro 2.3
Japan 10.6 / Tokyo 4.8
Shanghai 8.1 / China 3.4

According to this data, only Shanghai has a carbon footprint higher than the national average. Why do you think this is true for Shanghai but not the other cities evaluated?

INFOGRAPHIC 2B   TRADE-OFFS OF URBANIZATION

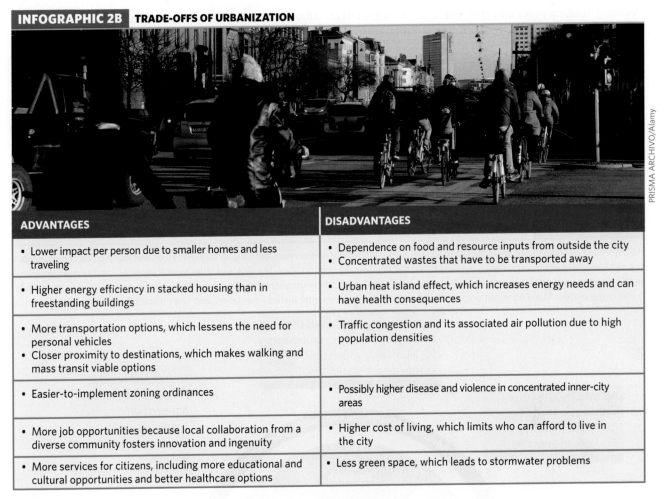

PRISMA ARCHIVO/Alamy

ADVANTAGES	DISADVANTAGES
• Lower impact per person due to smaller homes and less traveling	• Dependence on food and resource inputs from outside the city • Concentrated wastes that have to be transported away
• Higher energy efficiency in stacked housing than in freestanding buildings	• Urban heat island effect, which increases energy needs and can have health consequences
• More transportation options, which lessens the need for personal vehicles • Closer proximity to destinations, which makes walking and mass transit viable options	• Traffic congestion and its associated air pollution due to high population densities
• Easier-to-implement zoning ordinances	• Possibly higher disease and violence in concentrated inner-city areas
• More job opportunities because local collaboration from a diverse community fosters innovation and ingenuity	• Higher cost of living, which limits who can afford to live in the city
• More services for citizens, including more educational and cultural opportunities and better healthcare options	• Less green space, which leads to stormwater problems

 How might the disadvantages of urbanization be addressed to lessen their impact?

3   ENVIRONMENTAL JUSTICE ISSUES

Key Concept 3a: Environmental justice recognizes the right of everyone to a safe and healthy environment and a voice in policies that impact their environment.

Key Concept 3b: Minorities and low-income communities are more likely to suffer environmental injustice than other communities.

All urban dwellers are vulnerable to the health effects associated with pollutants. But in most cities, the pros and cons of city living are unevenly realized. For example, New York City is one of the wealthiest, most populous cities in the world, but most of the cultural amenities, top-notch healthcare facilities, and job opportunities are concentrated in Manhattan, while most of the garbage, sewage, and power plants are located in the Bronx. This imbalance has spawned a whole new area of activism known as **environmental justice**, based on the idea that no community should be saddled with more environmental burdens and fewer environmental benefits than any other.

Violations of environmental justice occur everywhere—in the Bronx, in low-income communities of the deep south, in minority communities of Midwest cities—inequities can be found in every U.S state and every country of the world. Minority and low-income communities are more likely to bear the burden of an unhealthy environment (e.g., close proximity to polluting or undesirable industry, poor quality housing, little access to green space, and limited transportation options). They are also less likely to have a voice in

environmental justice The concept that access to a clean, healthy environment is a basic human right.

decision-making processes that affect their community. In the United States, the Environmental Protection Agency (EPA) and other federal agencies seek to address issues of environmental injustice and offer programs and avenues for individuals or communities to address problems they face. **INFOGRAPHIC 3**

Environmental justice issues are particularly relevant in the most impoverished cities in the world. In Mumbai, the largest city in India, with 20.5 million people, more than 8.5 million are slum dwellers who live in horrid, overcrowded conditions—as many as 18,000 people per acre—without adequate sanitation or running water. Globally, more than 1 billion of the world's population lives in slums, mostly in large cities in developing countries.

With most future population growth predicted to occur in large cities, urban planners are desperately searching for ways to create cities where the basic needs of residents are met and where the environmental benefits outweigh the environmental costs. The story of how the South Bronx waterfront was lost and then reclaimed provides important lessons about how to do this.

INFOGRAPHIC 3 | ENVIRONMENTAL JUSTICE

All communities should have equal access to a clean, safe, and healthy environment but traditionally minorities and low-income communities are saddled with more environmental burdens than their neighboring communities. The EPA and other federal agencies identify goals (shown in this diagram) and provide programs that work to bring about equity in environmental quality and participation in decision-making processes for all communities. Many nonprofit organizations, such as the Environmental Justice Foundation, also work toward those same goals.

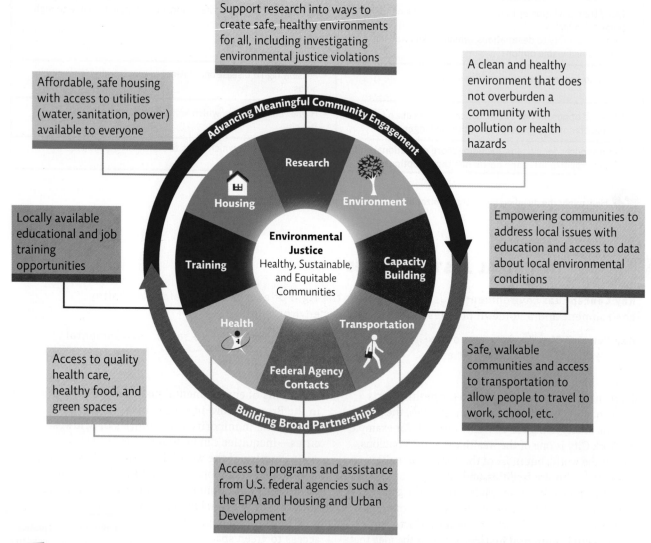

 The federal government has an ambitious plan to erase environmental injustice. Why then, is it still prevalent in our society?

4 SUBURBAN SPRAWL

Key Concept 4: Suburban sprawl displaces farmland and natural areas; residents often have a higher per capita environmental impact than urban dwellers.

In the late 1940's, the South Bronx was a mostly European-descended, white, working-class suburb of Manhattan. But as more Hispanic and black Americans moved to the area, whites moved to nearby commuter towns. Today, **urban flight**—the process of people leaving a city center for surrounding areas—is triggered by a variety of forces, including overcrowding, noise and air pollution, the high cost of city living, and, in some cases, racial tensions.

No matter what the cause, urban flight results in **suburban sprawl**—a slow conversion of rural areas outside of a city into suburban and exurban ones. **Exurbs** are more sparsely populated towns beyond the immediate suburbs whose residents also commute into the city for work.

As its name suggests, suburban sprawl tends to spread out over long corridors in an unplanned and often inefficient manner. By covering ever-greater swaths of terrain with concrete and pavement, sprawl reduces the amount of

land available for farming, wildlife, and *ecosystem services* (see Module 5.1). Because of the haphazard way in which they are developed, the resulting communities are heavily dependent on driving; unlike cities, which are densely populated and can accommodate mass transit systems, suburbs and exurbs require residents to drive almost everywhere they need to go. And because suburban homes are typically larger than urban ones (and exurban homes are often even larger than suburban ones), they tend to have a greater environmental impact. **INFOGRAPHIC 4**

Urban planners today are well aware of these perils and often work to mitigate them. But in the 1960's, urban planning tended to accommodate urban flight and

urban flight The process of people leaving an inner-city area to live in surrounding areas.

suburban sprawl Low-population-density developments that are built outside of a city.

exurbs Towns beyond the immediate suburbs whose residents commute into the city for work.

INFOGRAPHIC 4 | SUBURBAN SPRAWL

Urban flight often leads to suburban sprawl—low population density in developments outside a city. Homes typically get larger the farther they are from the city, and residents have a larger environmental impact (larger homes and more time spent driving). The suburbs now have their own suburbs—the exurbs, which are commuter towns that are beyond the traditional suburbs but whose residents still commute into the city, often an hour or more each way. Both suburbs and exurbs often displace farmland and wildlands.

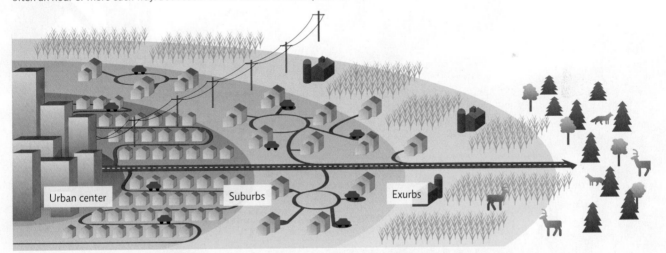

Urban center · Suburbs · Exurbs

Homes get larger (and less energy efficient) with distance from urban center. Single-family homes are more common and take up more space (on bigger plots of land). Commuting time increases.

Loss of arable land as developers purchase and subdivide fertile fields.

Loss of species habitat.

 If you worked in an urban center, would you rather live there, in the suburbs, or in the exurbs? Explain.

the sprawl that came with it. The Cross Bronx Expressway, completed in 1972, enabled suburban commuters to completely bypass the Bronx as they traveled in and out of Manhattan each day. But the expressway segregated the ailing borough from the rest of New York City. "The South Bronx was utterly cut off," says Marta Rodriguez, a lifelong Bronx resident and a colleague of Carter's. "We didn't stand a chance."

5   THE VALUE OF GREEN SPACE

Key Concept 5: Green spaces improve the local environment, mental and physical health, and have been linked to decreased crime rates.

Carter knew that replacing the abandoned lot with a park would be a big first step toward righting some of the wrongs that her community had endured. The trees and plants would trap pollutants from the air, preventing them from infiltrating people's lungs. The grass and soil would absorb rainwater so that it could no longer carry trash and detritus from the streets into the river. And claiming even a small patch of waterfront for themselves would give Carter and her neighbors a sense of ownership, not to mention a connection to nature and a place to stretch their legs.

green space A natural area such as a park or undeveloped landscape containing grass, trees, or other vegetation in an urban area, usually set aside for recreational use.

> Carter knew that replacing the abandoned lot with a park would be a big first step toward righting some of the wrongs that her community had endured.

Indeed, studies have shown that **green spaces** in a city improve both the physical and psychological health of people who live near them by providing more opportunities for physical activity, relaxation, and social interaction. Community gardens offer the chance for fresh produce—a healthy food option often lacking in low-income urban areas.

A volunteer gardener at Finca Del Sur, a garden in the South Bronx, tends the corn stalks while a passenger train goes by in the background. The garden was created on an empty plot of land bordered by a highway exit ramp and a commuter train line.

nina berman/NOOR/Redux

There is even evidence that the presence of green space makes an urban area safer. In Bogotá, Colombia, in the late 1990's, for example, a particularly environmentally conscious mayor noticed that while his city was designed to accommodate heavy automobile traffic, the vast majority of his electorate did not drive. So he narrowed municipal thoroughfares from five lanes to three, expanded bike lanes and pedestrian walkways, and established a string of parks and public plazas throughout the city. The result? People stopped littering. Crimes rates dropped. And slowly but surely, city residents reclaimed their streets.

Green spaces also provide other environmental and societal benefits. They provide habitat for wildlife, help mitigate against the urban heat island effect, reduce stormwater runoff after a rainfall, absorb air pollution, and help offset greenhouse gas emissions. Property values also tend to go up in urban areas close to parks or other green spaces, and community members are more likely to take pride in their surroundings and participate in community events in these areas. **INFOGRAPHIC 5**

Because of the growing list of green space benefits, many cities are actively putting in more parks, greenbelts, and other open spaces. For example, Vancouver, Canada has a goal of ensuring that all residents are within a 5-minute walk of a green space and has committed to planting 150,000 trees by 2020.

INFOGRAPHIC 5 | **THE VALUE OF GREEN SPACES**

Urban green spaces provide a wide variety of benefits to the residents and their community.

SOCIETAL BENEFITS
- Community pride and participation
- Higher property values
- Lower crime rate

HEALTH BENEFITS
- Physical activity opportunities
- Relaxation and rejuvenation
- Social interactions

ENVIRONMENTAL BENEFITS
- Trees absorb CO_2 and offset greenhouse gas emissions
- Unpaved areas reduce stormwater runoff and flooding
- Wildlife habitat
- Reduced urban heat island effect

 Explain how more green space might lead to a lower crime rate.

6 GREEN CITIES

Key Concept 6: Green cities actively pursue sustainability through design and city services that improve quality of life while decreasing per capita impact.

Carter and those who joined her project were on track to reclaim their slice of the Bronx. Starting with the $10,000 seed grant from the city parks department, they leveraged a small fortune in additional grants, donations, and private investment, until they had finalized plans to build a $3 million park, complete with gardens, grassy knolls, and East River kayaking. Hunts Point Riverside Park—the spot that Carter stumbled upon—would be the borough's first waterfront park in more than 60 years. But that was just the beginning.

Energized by their successful riverside park project, Carter and her neighbors formed a nonprofit called the Sustainable South Bronx (SSBx). The group immediately set its sights on an even grander vision: They would create a 2.5-kilometer (1.5-mile) greenbelt around the entire community, including a waterfront greenway and other open spaces, all connected by an interlinking system of bike and pedestrian pathways. They would also disassemble the Sheridan Expressway, a stretch of abandoned highway originally meant to cut across the entire northeast Bronx, and turn it into 11 hectares (28 acres) of additional parkland, some of which they would designate as *conservation easements*— tracts of land that the city would agree not to develop.

green city A city designed to improve environmental quality and social equity while reducing its overall environmental impact.

New Urbanism A movement that promotes the creation of compact, mixed-use communities with all the amenities of day-to-day living close by and accessible.

But a green city is more than just a city with green spaces. A **green city** is one that pursues sustainable options that make them more environmentally friendly and socially equitable. **New Urbanism**, as the movement is called, maintains that cities (both now and in the future) have the capacity to reduce our per capita environmental impact, even as they improve the quality of life for people, provided they are designed properly. The City University of New York Institute for Research on the City Environment estimates that if a city is designed and built with an eye toward sustainability, the impact of any given urban dweller could be trimmed to about half that of the average American.

Greening our cities becomes ever more vital as urban growth in many areas, especially the developing world, is increasing at unprecedented rates. And though we have a long way to go to make cities an environmental solution

Majora Carter received a MacArthur "genius" grant for her work with Sustainable South Bronx. She stresses that the environmental movement is not just one of the middle-class majority who can afford to buy organic food, drive hybrid cars, and live in areas with little pollution. Low-income families also deserve a clean and healthy environment.

nina berman/NOOR/Redux

rather than one of the environment's biggest problems, the consensus emerging from environmentalists, sociologists, and economists is that the future lies in cities—where most people will live and perhaps where most people should live. Cities promote interaction

among a diverse group of people. This in turn promotes the exchange of ideas and lessens cultural and economic barriers. Many cities around the world are pursuing sustainable development and paving the way for others to do the same. **INFOGRAPHIC 6**

INFOGRAPHIC 6 GREEN CITIES

The Green City Index is a research project conducted by the Economist Intelligence Unit and sponsored by the Siemens Corporation. A total of 31 indicators are evaluated from 9 categories to assess how far along on a path to sustainability a city is.

Sylvain Sonnet/Photodisc/Getty Images

CATEGORY	METRICS ASSESSED	POLICIES EVALUATED
Carbon dioxide	• CO_2 emissions (per capita) • CO_2 intensity (emissions per GDP unit)	• CO_2 emissions strategy (presence, ambitiousness)
Energy	• Energy consumption (per capita) • Energy intensity (consumption per GDP unit) • Renewable energy consumption (% of total)	• Clean energy and energy efficiency
Buildings	• Energy consumption of residential sector (per square foot)	• Energy-efficient buildings standards • Programs to promote energy efficiency
Transportation	• Non-car transport (% of population using public transport, bikes, or walking) • Extent of public transport and bike lanes (length per km^2 of city area)	• Programs to incentivize cleaner transportation • Programs to reduce vehicle traffic in the city
Water	• Water consumption (per capita) • Water leakage (% of water lost) • Wastewater treatment (% of homes with access)	• Programs to improve water use efficiency and wastewater treatment
Air	• Emissions of specific pollutants (nitrogen dioxide, ground level ozone, particulate matter, sulfur dioxide)	• Policies to improve air quality
Land use	• Amount of green space (per km^2 of city)	• Programs to increase green space • Policies to curtail urban sprawl
Solid waste	• Per capita solid waste production • Percentage of waste that is recycled or composted	• Polices to incentivize waste reduction • Polices to support recycling or composting
Environmental governance	• Management and financial support of environmental programs • Citizen participation in decision making	• Environmental strategic plan (baseline reviews, targets for the future, and a plan to achieve those targets)

 What steps would you recommend that your own city or community take in an effort to reduce its environmental impact? Justify your answer.

7 SMART GROWTH AS A WAY FORWARD

Key Concept 7: Smart growth allows cities to develop in a way that minimizes environmental impact while enhancing community living.

Sustainable cities are cities where the environmental pros outweigh the cons—where sprawl is minimized, walkability is maximized, and the needs of inhabitants are met locally. Strategies that help create walkable communities with lower environmental impacts are known as **smart growth**. To achieve self-sufficiency, for example, a sustainable city might maintain a mixture of open and agricultural land along its outskirts to provide a large part of the local food, fiber, and fuel crops, along with recreational opportunities and ecological services. Waste and recycling facilities could also be located nearby, along with other enterprises aimed at producing resources needed by area residents.

To stave off sprawl, the same city might establish urban growth boundaries—outer city limits beyond which major development would be prohibited. Keeping any outward growth that does occur as close to mass transit as possible minimizes the impacts of transportation. **Infill development**—developing empty lots within a city— and building "up" (a parking garage) rather than "out" (an expansive parking lot) also minimizes the amount of land used.

smart growth Strategies that help create walkable communities with lower environmental impacts.

infill development The development of empty lots within a city.

Smart growth also provides a range of options for reducing traffic congestion and the air pollution that comes with it: reliable public transportation, car-sharing programs that allow residents to use cars when needed for a monthly fee, and sidewalks and overhead passageways that allow pedestrians to safely cross busy roads. To encourage more walking and less driving, zoning laws might allow for mixed land uses, where residential areas are located reasonably close to commercial and light industrial ones. **INFOGRAPHIC 7**

Of course, building an ideal city from scratch is easy compared with the task of overhauling an existing city. Upgrading decaying infrastructure like roads, public places, and sewage and water lines can be more expensive than new construction, and the process is disruptive to residents. Even so, there are plenty of ways that American cities can push themselves into the environmental plus column.

Persuading people to support smart growth, as Carter and her colleagues soon discovered, is a matter of showing them that the benefits could be economic as well as environmental. "You need to show them what we call the triple bottom line," says James Chase, vice president of SSBx (and Carter's husband), referring to the economic, social, and environmental impacts of any decision. "Developers, government, and residents all need some tangible, positive return." A major park project would surely be a boon for all three. Developers would be guaranteed millions in waterfront development contracts. Residents could look forward to cleaner air and water, a prettier neighborhood, and better health. The state and city governments would save a bundle in healthcare costs. The greenbelt would also spur the local economy: Such a vast stretch of public space would attract street vendors, food stands, bicycle shops, and sporting goods stores.

8 GREEN BUILDING

Key Concept 8: Green building design focuses on efficient use of energy and water and on building materials with low environmental and health impacts.

Buildings in cities are major energy users—they consume an estimated 40% of global energy and generate 30% of the world's greenhouse gases—making them key targets in the quest to reduce our overall environmental impact. Addressing this opportunity is a movement known as **green building**—that is, the construction of buildings that are better for the environment and the health of those who use them.

In one program implemented by the U.S. Green Building Council, buildings that meet a minimum standard are awarded a

green building Construction and operational designs that promote resource and energy efficiency and provide a better environment for occupants.

Leadership in Energy and Environmental Design (LEED) A certification program that awards a rating (standard, silver, gold, or platinum) to buildings that include environmentally sound design features.

Leadership in Energy and Environmental Design (LEED) certification. Points are awarded based on how well the building meets certain criteria, such as energy and water efficiency. Of course, buildings can be built or retrofitted without seeking LEED certification (a costly procedure) but for those organizations that choose to participate, it offers publicity and an accountability that may encourage the pursuit of loftier goals.

The Bronx Library Center is a silver-certified LEED building. It earned the silver certification by recycling 90% of the waste materials created during the construction of the building, using architectural design and efficient heating and cooling systems to save 20% of energy costs, and using sustainably grown wood in 80% of the construction lumber.

INFOGRAPHIC 7 SUSTAINABLE CITIES AND SMART GROWTH

Smart growth can be applied to large cities or to smaller communities. It employs strategies that make efficient use of land to create pleasant livable communities with less environmental impact than current suburban areas.

Take advantage of compact building design and incorporate environmentally friendly technologies.

Create a range of housing opportunities and choices.

Renovate and develop existing communities (rather than build outside the city).

Foster distinctive, attractive communities with a strong sense of place.

Encourage community and stakeholder collaboration; make development decisions that are fair and cost-effective.

Mix land uses to place residential and commercial areas together.

Provide urban green space; preserve farmland and critical environmental areas.

Create walkable and bike-friendly neighborhoods.

Provide a variety of "clean" transportation choices into and around the city.

NATURAL GAS BUS

Which of these smart growth principles would be most appealing to you if you were looking for a place to live in a city?

In its own green-building initiative, Carter's team launched Smart Roofs, LLC, a green-roof installation company. Green roofs are one type of rain garden—an area with plants suited to local temperature and rainfall conditions (see Module 6.2). Research shows that consumers could save more than $5 million in annual cooling costs if green roofs were installed on just 5% of the city's buildings. That amount of green roofing could achieve an annual reduction of 350,000 metric tons of greenhouse gases. And Riverkeep, an environmental nonprofit, found that green roofs can retain 3,000 liters of storm water for every $1,000 of investment—easing pressure on the city's overburdened sewer systems and mitigating water pollution from stormwater runoff. **INFOGRAPHIC 8**

By the time the Hunts Point Riverside Park opened, dozens of cities across the country—from Madison, Wisconsin, to Miami, Florida—had taken up the mantle of sustainability and smart growth. In 2013, the United States led the world in green building, with more than 44,000 LEED-certified buildings or projects under construction. That same year, the Bronx received an award for Via Verde, a LEED gold mixed-income housing complex, which incorporates a wide variety of green building features such as solar panels, natural lighting, and green roofs. It also has easy access to transportation and a neighborhood medical clinic on site. The project was so successful it prompted New York City to change its green zoning rules to make it easier to implement similar projects throughout the city.

Select References:

Beckett, K., & Godoy, A. (2010). A tale of two cities: A comparative analysis of quality of life initiatives in New York and Bogotá. *Urban Studies, 47*(2), 277–301.

Carter, M. (2006 February). *Greening the Ghetto.* www.ted.com/talks/majora_carter_s_tale_of_urban_renewal.

Dodman, D. (2009). Blaming cities for climate change? An analysis of urban greenhouse gas emissions inventories. *Environment and Urbanization, 21*(1), 185–201.

Lee, A. C. K., et al. (2015). Value of urban green spaces in promoting healthy living and wellbeing: Prospects for planning. *Risk Management and Healthcare Policy, 8*, 131–137.

INFOGRAPHIC 8 **GREEN BUILDING**

Many steps can be taken to build or retrofit a building so that it has less environmental impact and is a healthier environment for those who live, work, or go to school there. The nonprofit group Green Building Council certifies buildings through its LEED (Leadership in Energy and Environmental Design) program. A building receives a standard, silver, gold, or platinum rating based on a variety of criteria that include energy efficiency, sustainable building material use, and innovative design.

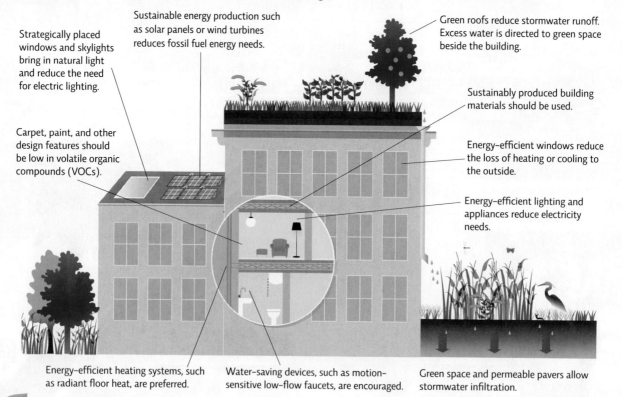

Strategically placed windows and skylights bring in natural light and reduce the need for electric lighting.

Sustainable energy production such as solar panels or wind turbines reduces fossil fuel energy needs.

Green roofs reduce stormwater runoff. Excess water is directed to green space beside the building.

Sustainably produced building materials should be used.

Carpet, paint, and other design features should be low in volatile organic compounds (VOCs).

Energy-efficient windows reduce the loss of heating or cooling to the outside.

Energy-efficient lighting and appliances reduce electricity needs.

Energy-efficient heating systems, such as radiant floor heat, are preferred.

Water-saving devices, such as motion-sensitive low-flow faucets, are encouraged.

Green space and permeable pavers allow stormwater infiltration.

? How could a LEED-certified building help address urban problems such as the urban heat island effect and stormwater issues?

INTERACTIVE MAP **GREEN CITIES**

Many cities of the world are taking steps to develop more sustainably in an effort to improve their local environments and their standards of living. Green cities have many things in common, such as recognizing the importance of civic involvement, having a government commitment to sustainable development, and pursuing a holistic approach that looks at all the ways the city can reduce its ecological footprint while still developing and growing. While there are many others, here are some notable examples.

SAN FRANCISCO, CALIFORNIA

COPENHAGEN, DENMARK

SINGAPORE

CURITIBA, BRAZIL

 BRING IT HOME

PERSONAL CHOICES THAT HELP
A sustainable community is one that promotes economic and environmental health and social equity. It is one in which the health and well-being of all citizens are considered, while those citizens help implement and maintain the community.

Individual Steps
• Investigate and support sustainable businesses in your area (see www .sustainablebusiness.com). Research products before you purchase them to understand the impact of your consumption choices (see www.goodguide.com).

• If you have a balcony or yard, plant flowers, vegetables, or trees.
• Support local businesses by shopping and dining close to home.

Group Action
• Join neighborhood clean-up days. If you can't find one, organize one.
• Reduce reliance on cars. Start a petition to get more bike lanes in your city. Ride public transit more often.
• Find out how colleges and universities are working toward sustainable practices at www. AASHE.org.

Policy Change
• Attend a meeting of your city council or county commission and ask members to look into smart growth opportunities.
• See how well you can plan for a sustainable community. Play the PC strategy game Fate of the World (www.gamesforchange.org/play/fate-of-the-world/) and see how policies you put in place impact global climate change, rainforest preservation, and resource use.

ENVIRONMENTAL LITERACY **UNDERSTANDING THE ISSUE**

1 What is the pattern of global urbanization and megacity growth in recent decades?

1. True or False: A megacity is defined as a city with more people than its local environment can support.

2. In 2025, where will most of the world's megacities be found?
   a. North America
   b. Europe
   c. Africa
   d. Asia

3. How has the distribution of human population between urban and rural dwellers changed over time?

2 What are the trade-offs associated with cities or urban areas?

4. True or False: Urban dwellers typically have lower carbon footprints than people who live in rural or suburban areas.

5. The urban heat island effect is caused by:
   a. minimal green space.
   b. lots of pavement and buildings.
   c. high energy use in the city as a whole.
   d. All of these contribute to the urban island effect.

6. Point out several reasons why the typical city dweller has a lower environmental impact than a suburban dweller. What are some problems that might be greater for a large city than a smaller suburban one?

3 What is environmental justice and what common problems does it address?

7. Which of the following is an example of environmental injustice?
   a. Locating industries away from where people live
   b. Building garbage dumps in high-poverty, low-income areas
   c. People chaining themselves to trees to prevent the trees from being cut down
   d. Preventing the construction of a dam to save an endangered species of fish

8. If urban centers have better health care and job opportunities than outlying areas, why are some city residents exposed to worse, not better living conditions?

4 What environmental problems does suburban sprawl generate?

9. True or False: As one moves from the city center to the suburbs and exurbs, home size tends to increase.

10. The exurbs are:
    a. discrete communities within the city center.
    b. regions too distant from the city for city commuters to live.
    c. areas outside of suburbs with even bigger houses and longer commutes.
    d. suburban areas that are losing residents to the city.

11. Compare the environmental impact of residents in urban centers, suburban areas, and exurban areas.

5 What is the value of urban green space?

12. True or False: The presence of urban green spaces is correlated with lower crime rates.

13. Which of the following health benefits are correlated with the presence of urban green space?
    a. Improved mental health
    b. Improved physical health
    c. Opportunities for social interactions
    d. A and B
    e. A, B, and C

14. Identify some of the environmental benefits of urban green spaces.

6 What are "green cities"?

15. Which of these is typically NOT a characteristic of a green city?
    a. Grass-roots citizen groups lead development because city government involvement is low.
    b. Effective transportation options are available.
    c. High-efficiency buildings are preferred, especially for new builds.
    d. Green space is included for city residents.

16. Explain the triple bottom line, using the Bronx waterfront restoration as an example.

7 How does smart growth help make a city "green"?

17. A city that promotes smart growth:
    a. encourages development at the city edges.
    b. provides tax incentives for people who own more than one car.
    c. allows vacant lots to accumulate in the city for a more open look.
    d. mixes land uses to place residential and commercial areas together.

18. What is infill development and how can it help revitalize an urban area and help a city pursue sustainable development?

8 What are some of the design features of a green building?

19. True or False: For a building to be considered sustainable, it must be a new build—retrofitting an existing building to make it sustainable is not a viable option.

20. "Green" buildings are those that:
    a. are LEED-certified at the level of silver or above.
    b. contain state-of-the-art energy efficiency equipment.
    c. are built in a way to be resource and energy efficient and provide a safe living environment.
    d. have a small footprint (are under 1,000 square feet) and contain multiple family units, such as apartment buildings.

21. Describe a LEED-certified building. What criteria are used to evaluate a building for LEED certification?

SCIENCE LITERACY | WORKING WITH DATA

One common measurement of how well a building performs with regard to energy usage is its energy use intensity (EUI). EUI is expressed as energy used (in gigajoules) relative to a building's size so that structures of different sizes can be compared. This graph plots data for 36 LEED-certified office buildings, comparing their actual performances to expected performances, based on their designs. (Each data point represents a different building's EUI.) The line shown is not a trend line for the actual data; it is a line that bisects the graph at a 45-degree angle. This line allows us to see how closely the expected values match the observed values.

MEASURED VERSUS EXPECTED EUI IN LEED-CERTIFIED BUILDINGS

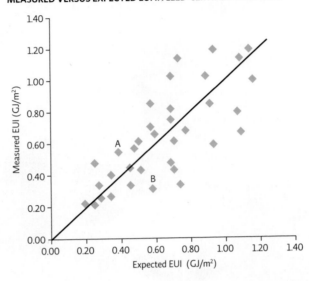

Interpretation

1. Look at the points labeled A and B. For each, what was the expected EUI predicted by the design plan? What was the actual measured EUI for this building?

2. Which of these two buildings, A or B, is performing better than expected?

3. Give the coordinate points for the one building that performed exactly as expected.

4. What does the line in the scatter plot tell you about those buildings above or below it?

Advance Your Thinking

5. About how well do LEED-certified buildings meet or exceed their predicted EUI?

6. The average EUI for non-LEED-certified office buildings is 2.19 GJ/m². How do LEED-certified buildings, even those that do not perform as well as expected, compare to this average?

INFORMATION LITERACY | EVALUATING INFORMATION

More and more people around the world live in cities, and the number of megacities is increasing. There are costs and benefits to living in cities, and in an effort to increase the benefits, there is a growing movement toward "greening" cities. Environmental and social scientists have published a number of studies documenting the effects of greener cities on environmental and human health.

Go to the website Green Cities: Good Health (http://depts .washington.edu/hhwb/). Read the introduction to the site.

Evaluate the website and work with the information to answer the following questions:

1. Determine if this is a reliable information source with a clear and transparent agenda:
    a. Who runs this website? Do the organization's credentials make it reliable or unreliable? Explain.
    b. Who are the authors? What are their credentials? Do they have the scientific background and expertise to lend credibility to the website?

Look under Research Themes to see what research the program pursues. Choose one of the links under Fast Facts, such as Crime & Public Safety, Active Living, or Mental Health & Function.

2. What type of information is provided on the page? What is the source of the information?

3. List a couple of the fast facts provided. Then scroll down the page and read each section. For the topic you chose, what is the primary claim? What data is provided to support the claim?

4. Do you find the data convincing? Why or why not? How does what you read relate to your own life? Give specific examples.

 Additional study questions are available at SaplingLearning.com.

ERADICATING A PARASITIC NIGHTMARE

Human health is intricately linked to the environment

Women gather water from Ogi, a sacred pond in Nigeria. Previously contaminated with water fleas that carried Guinea worm larvae, the pond was successfully decontaminated by local health officials. Guinea worm disease is on track to become the first disease since smallpox to be totally eradicated.

Vanessa Vick/The New York Times/Redux

After reading this chapter and studying the KEY CONCEPTS and INFOGRAPHICS, you should be able to answer these GUIDING QUESTIONS

CORE MESSAGE

Human health is impacted by the environment. Human actions that cause environmental changes can either facilitate disease or reduce its transmission. By understanding the various risk factors for environmentally mediated health problems, steps can be taken to reduce health risks, such as improved sanitation, access to clean air and water, and public health programs to reach people in affected areas.

1. What environmental hazards do humans face?

2. What are the most common environmental routes of transmission for infectious diseases and how has human impact increased the risk of transmission?

3. What is the focus of the field of public health?

4. What can be done to reduce the incidence and spread of infectious diseases?

5. How do factors that affect human health differ between more and less developed nations?

6. What role do environmental factors play in the global burden of disease?

7. What are zoonotic diseases, why is their incidence increasing, and what can be done to reduce their incidence?

8. What can be done to address environmental hazards and how will this contribute to our overall goal of sustainability?

Ernesto Ruiz-Tiben shook the tube of water and held it up to the sunlight so that the women who were gathered around him could see the tiny black flecks that had settled out. There was a soft, collective gasp at the spectacle. The black flecks—tiny "water fleas" known as copepods—offered the Nigerian women the first visible proof of what Ruiz-Tiben, director of the Carter Center's Guinea Worm Eradication Program, had been trying to explain to them: The water in which they drank and bathed was contaminated with tiny crustaceans, and these creatures were solely responsible for the searing worm infections that seemed to sweep through the village every year or so, usually right around harvest time.

Guinea worm infection begins when a person ingests water contaminated with copepods infected with Guinea worm larvae. The larvae burrow into the victim's abdominal tissue, mature, and mate. Male worms die, but the females remain and grow—up to 100 centimeters (3 feet) long—while migrating through the victim's tissue. About 1 year later, the females release acid just beneath the skin, creating a painful blister that drives victims to plunge the blister into water; the female worm then squirts out a dense cloud of milky white larvae, starting the cycle over again. The disease is not fatal, but recovery is both very slow and very debilitating. There are no medications or vaccines for *dracunculiasis*, or Guinea worm disease (GWD) as it is more commonly known, and the infection itself does not confer immunity; that means the same people can fall prey to the worms over and over again.

◉ **WHERE IS NIGERIA?**

NIGERIA

South Atlantic Ocean

1 ENVIRONMENTAL HAZARDS

Key Concept 1: Humans face a variety of environmental factors than negatively affect health. Air and water pollution are the leading environmental health threats.

Environmental factors contribute significantly to disease, injury, and death; worldwide, they account for about a quarter of deaths. Many of these environmental factors can be modified to reduce the incidence of environmentally mediated illness and death.

Some environmental hazards are *physical hazards*—things that are harmful when contacted (they cut, crush, cause choking or drowning, etc.) or that cause damage without direct contact (extreme heat and cold, loud noises or strong vibrations). Even the human-built environment—speeding

cars and overcrowded urban areas—are seen as physical hazards. *Chemical hazards* cause damage by virtue of their chemical makeup; air and water pollution, and occupational or household exposure to hazardous chemicals are examples. *Biological hazards* such as infectious agents also threaten human health and have a strong environmental link. And climate change is exacerbating many of these hazards as it affects water supplies, increases the incidence of extreme weather such as heat waves, and expands the range of pathogens and the vectors that spread disease, such as mosquitoes. **INFOGRAPHIC 1**

INFOGRAPHIC 1 TYPES OF ENVIRONMENTAL HAZARDS

A variety of environmental hazards impact human health. Though some of these threats are natural in origin, anthropogenic causes account for most of the hazards we currently face today.

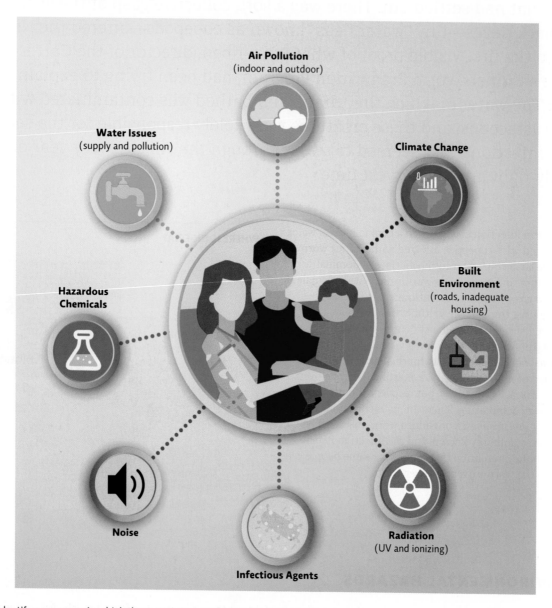

 Identify some ways in which these environmental hazards might overlap.

The leading environmental threat overall is air pollution. The small particles (soot) and chemicals in air pollution cause respiratory disease by damaging lungs; they contribute to cardiovascular disease by damaging blood vessels. In addition, impaired lungs that cannot bring in enough oxygen will stress the heart (which is tasked with delivering oxygen to the body's tissues), further contributing to cardiovascular disease.

Outdoor air pollution from industry, power plants, and vehicles is a problem throughout the world whereas indoor air pollution produced from burning fuel inside a poorly ventilated home is especially problematic in

low-income areas. But in low-income nations, biological hazards remain the biggest threat, and it is here that death rates due to infections and parasitic diseases, like GWD, are the highest.

Of course, we are never exposed to any hazard—be it physical, chemical, or biological—in isolation. Rather, different types of hazards interact with one another and with other elements of the human environment in ways that can make it tricky for healthcare workers to map cause-and-effect relationships. For example, someone negatively affected by a chemical hazard such as air pollution may be more susceptible to a biological hazard such as a lung infection.

2 ENVIRONMENTALLY MEDIATED INFECTIOUS DISEASES

Key Concept 2: Infectious diseases threaten many human populations; human actions that increase habitat for pathogens or their vectors can facilitate the spread of these diseases.

Efforts to reduce the incidence of GWD began as part of an international program to provide safe drinking water to all people. According to the WHO, well over 1 billion people per year fall victim to **waterborne diseases**—those acquired by consuming contaminated water. In fact, waterborne and **vector-borne diseases** are the main **infectious disease** threats to human health. (Vectors are organisms that transmit a **pathogen** from one host to another; in the case of GWD, copepods serve as the vector.) GWD is just one of countless such diseases, but by providing safe drinking water, Ruiz-Tiben and his colleagues

hoped that they might also eliminate others. **INFOGRAPHIC 2**

One of the biggest influences on the seriousness of any given hazard is the health of the natural environment, something we've had unprecedented influence over in recent centuries. "We humans have been remarkably effective at

waterborne disease An infectious disease acquired through contact with contaminated water.

vector-borne disease An infectious disease acquired from organisms that transmit a pathogen from one host to another.

infectious disease An illness caused by an invading pathogen such as a bacterium or virus.

pathogen An infectious agent that causes illness or disease.

INFOGRAPHIC 2 | INFECTIOUS DISEASE TRANSMISSION

Infectious diseases are illnesses caused by a pathogen (bacterium, virus, protozoan, fungi, or multicellular organism such as a worm). Many of these are communicable diseases—they can be passed on to others through direct contact or contaminated sources of food or water, or via vectors. Waterborne and vector-borne diseases affect populations in all countries but are especially problematic in less developed areas and for those in poverty.

WATERBORNE DISEASES

The pathogen, most commonly from human or animal waste, is acquired by consuming contaminated water. Environmental populations of nonfecal pathogens also transmit diseases such as Guinea worm disease.

780 million people lack access to clean water

2.5 billion people with inadequate sanitation

850,000 deaths due to infectious waterborne diseases

43% of those were in children, ages 0–4

ENVIRONMENTAL LINKS

Fecal contamination of water from sewage and animal waste; contaminated water used on crops; floods or rain event that deliver fecal waste to bodies of water or overwhelm sewage treatment plants.

EXAMPLES: *cholera, E. coli infections, typhoid fever, giardiasis, Guinea worm disease*

VECTOR-BORNE DISEASES

The pathogen is transmitted from one host to another by another organism (the vector) such as a mosquito, fly, tick, or snail.

17% of infectious disease are vector-borne

1 billion people infected yearly

40% of world population at risk for dengue

1 million deaths per year due to vector-borne diseases

>600,000 were from malaria

ENVIRONMENTAL LINKS

Events or actions that increase vector habitat: standing water in vessels, tires, etc. (mosquitoes); dams (snails and copepods); trash (flies and rodents)

EXAMPLES: *malaria and Zika (mosquito), Lyme disease (tick), schistosomiasis (snail)*

 Identify at least three reasons why waterborne and vector-borne diseases are especially problematic in less developed areas and for those in poverty.

rearranging the natural world to meet our own needs," says Sam Myers, a scientist at the Harvard University School of Public Health.

> **"We humans have been remarkably effective at rearranging the natural world to meet our own needs."**
> —Sam Myers

Myers estimates that between one-third to one-half of global resources produced by ecosystem functions are now diverted to human uses. In the past 300 years alone, we have completely deforested between 7 and 11 million square kilometers (2.7–4.2 million square miles) of land—an area the size of the continental United States—and converted some 40% of the planet's ice-free land surface to cropland or pasture. We have converted an additional 2 million square kilometers (0.8 million square miles) of forest into highly managed plantations with significantly less biodiversity. We are already using roughly half of the planet's accessible surface freshwater and fishing 90% of monitored fisheries at or beyond their sustainable limits. And with population rising, these numbers are only likely to increase.

Though this type of manipulation has made life better for many, Myers and others worry that by so dramatically altering the ecosystems around us, we not only have imperiled our access to some of the most

basic components of human health—namely adequate nutrition, safe water, and clean air—but also have increased our exposure and vulnerability to natural disasters (see Module 7.1) and disease.

In the northeastern United States, for example, habitat fragmentation has contributed to an increase in the incidence of Lyme disease. Habitat fragmentation alters the community structure by increasing the populations of field mice (which carry but don't effectively remove disease-carrying ticks) and decreasing the number of squirrels and opossums (which are much more successful at removing ticks before becoming infected). Meanwhile, in the Amazon basin, deforestation has increased the breeding habitat for mosquitoes that transmit malaria. In Cameroon, deforestation has altered aquatic habitats in ways that favor a schistosomiasis-carrying snail. This problem was made worse by an ineffective government response that failed to get medicine to affected areas or take steps to reduce the snail populations, two steps that have been shown to be very effective at controlling this debilitating disease. In Asia and South America, monsoon rains and agricultural runoff have conspired to alter the salt and nutrient levels of coastal waters in ways that favor cholera population explosions. In the United States, heavy rains that caused untreated sewage to wash into the rivers around Milwaukee, Wisconsin in 1993 resulted in a *Cryptosporidium* outbreak that sickened more than 400,000 residents and killed 69—the largest waterborne disease outbreak in U.S. history. Climate change, which is increasing sea surface temperatures, is linked to a rise in food poisoning cases from eating raw oysters. (See the Interactive Map at the end of the module for details on other examples of environmentally mediated diseases.)

Throughout the developing world, both dam building and urbanization have increased the incidence of a wide range of waterborne pathogens, including but not limited to GWD. In cities, not only do people live in much closer quarters, but a wide variety of human-made objects, such as old tires and discarded plastic food containers, find second life as vessels for rainwater that collects during wet seasons. This water provides an excellent habitat for a whole suite of vectors that transmit a host of diseases: mosquitoes that carry dengue and Zika viruses and the *Plasmodium* protozoan that causes malaria, black flies and snails that carry worms that cause debilitating diseases like river blindness (onchocerciasis) and schistosomiasis. Dams—especially those in tropical regions—do something similar: They create large bodies of standing water that have been associated with an uptick in the same cadre of diseases.

Floodwaters are likely to be contaminated, conduits of disease.

peeterv/Getty Images

3 PUBLIC HEALTH PROGRAMS

Key Concept 3: Public health officials work to improve the health of a population as a whole.

Solving community health problems like GWD is the job of **public health**, a field that deals with the health of human populations as a whole. Public health **epidemiologists** like Ruiz-Tiben work to gauge the overall health status of a population or even of a nation. They use statistical analysis (e.g., rates of infant mortality, incidence of various diseases) to identify specific health threats to groups of people; they then recommend ways to mitigate those threats. Devising a plan of action is often the trickiest part; it requires the study of a whole host of interacting variables—from cultural and social forces (which influence things like diet and smoking habits), to economic stability or instability (which determines a given population's access to resources),

to environmental factors like water cleanliness or changes in habitat that affect disease transmission.

INFOGRAPHIC 3

Environmental health is a branch of public health that focuses on potential health hazards in the natural world and the human-built environment; such hazards include not only things like contaminated water, air, and soil but also human behaviors—hand washing and water drinking, for example—that help determine whether those factors become hazards.

public health The science that deals with the health of human populations.

epidemiologist A scientist who studies the causes and patterns of disease in human populations.

environmental health The branch of public health that focuses on factors in the natural world and the human-built environment that impact the health of populations.

INFOGRAPHIC 3   PUBLIC HEALTH PROGRAMS

The goal of public health programs is to improve the health of human populations through prevention and treatment of disease at the community level.

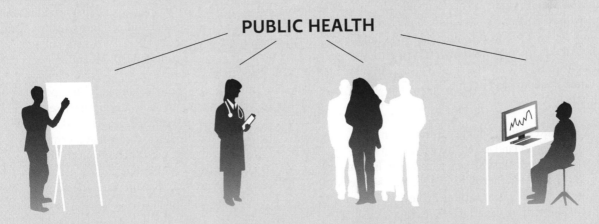

PUBLIC HEALTH

EDUCATES
Public health professionals provide information and healthcare advice to communities. Changing behaviors can be a critical part of improving public health.

PROVIDES HEALTH CARE
Public healthcare workers provide needed preventive medical care and treatment.

PROPOSES ACTIONS
Once risks are identified, public health professionals make recommendations to improve health in specific groups and in the population as a whole.

CONDUCTS RISK ASSESSMENTS
Epidemiologists analyze statistics related to a population's health to determine risk for various groups in the population (e.g., children, the elderly, the chronically ill) to a variety of environmental factors (e.g., sanitation, pollution, climate change).

 How does the U.S. childhood vaccination program protect the health of children too young or ill to receive the vaccinations?

4 REDUCING THE SPREAD OF INFECTIOUS DISEASE

Key Concept 4: An understanding of how an infectious disease is contacted and spread can help public health officials design programs to decrease disease incidence.

GWD has been around for centuries but persists today only in remote populations whose only drinking water comes from ponds or other standing water sources—ideal habitat for the copepod vectors. But Ruiz-Tiben and his staff at the Carter Center were aided in their quest by a simple fact: Unlike some of those other infectious organisms, Guinea worms seem to be utterly dependent on humans to complete their life cycle. When the Carter Center began its quest to eradicate GWD, humans were the only known reservoir for adult worms (larvae can only survive in copepods, and only for a few weeks). That meant Ruiz-Tiben and his colleagues could break the Guinea worm's life cycle—and thus obliterate the disease—simply by changing human behavior.

Convincing villagers to filter their water before drinking it would stop new infections. Getting the villagers to apply a mild pesticide would decontaminate area ponds. And teaching infected individuals to avoid communal swimming or bathing while worms were emerging from the skin, and treating new infections as soon as blisters emerge, would break the worm's life cycle once and for all. Their goal was total eradication of GWD; if they succeeded, it would be only the second disease in human history (after smallpox) to be completely wiped off the face of Earth.

As straightforward as it all sounded, Ruiz-Tiben had not had much luck so far. Despite his careful detailing of the science, most of the villagers with whom he had spoken still believed—rather fiercely—that the sickness was delivered by angry gods, as punishment for various misdeeds. And if that explanation seemed ludicrous to Ruiz-Tiben, well, his counterexplanation—that the worms actually came from water the villagers had consumed a year prior—seemed equally ludicrous to them. He thought he might make some headway by showing them the dead copepods in the water.

In the Nigerian village where Ruiz-Tiben was working, the main problem was not urbanization or habitat fragmentation or even deforestation. Rather, the main problem was a normal human trait: resistance to change. Even after the villagers saw the copepods in the water (which he had killed with a dash of the pesticide Abate), they continued to resist using both the water filters and pesticides. When Ruiz-Tiben discovered a hidden pond infested with copepods, the women of the village formed a human shield around it so that his team could not treat it with Abate. The pond was a sacred ancestral pool, they

insisted. To douse it with chemicals would invite the wrath of their gods.

The belief was anchored in an ancient history. In fact, GWD is itself ancient; it has been traced all the way back through the Old Testament and to the Egyptian mummies. Some say a symbol of modern medicine—a serpent coiled around a staff (known as a Rod of Asclepius)—derives from our treatment of GWD: To prevent the worm from breaking off at the blister, it is wound slowly around a stick as it emerges from the body—a practice that has not changed for thousands of years.

Both the worms and the copepods that carry them are native to Africa and Asia; both evolved across human history to exploit human hosts and human water sources. The copepods thrive in open, stagnant water sources—like ponds and pools formed by dams—whose availability varies by season and region. In the Sahelian zone, transmission generally occurs in the rainy season (from May to August), when shallow ponds grow deep enough to bathe in. In the humid savanna and forest zones, infections peak during the dry season (from September to January), when water holes shrivel and grow still, inviting copepods to multiply.

The only humans who come in contact with copepods are those who drink unpurified water from these sources. The villagers with whom Ruiz-Tiben was working were poor and lived in a naturally dry area; that meant that for generations, they had had no choice but to drink and bathe in whatever water they could find. Consistent water sources—those that do not dry up when the rainy season is over, or evaporate during particularly hot years—were sacred. And the women were understandably leery of polluting their most treasured supply with a foreign chemical that had been made in a distant land.

It was not until a revered general from the region intervened on Ruiz-Tiben's behalf that the women finally relented. The general, a former president of Nigeria, assured the women that the pesticide would not harm their fish but would instead keep their families from getting sick. Their ancestors, he said, would not want them to be sick. Reluctantly, the human shield broke up, and Ruiz-Tiben and his team were able to treat the pond with Abate, ridding it of copepods. **INFOGRAPHIC 4**

But as the Carter Center's army of public health workers would soon discover, Nigeria's challenges paled in comparison to the obstacles faced by other Guinea worm—infested countries.

INFOGRAPHIC 4 **DEVELOPING ERADICATION PROGRAMS FOR ENVIRONMENTALLY MEDIATED DISEASES**   ANIMATED INFOGRAPHIC

An examination of the life cycle of the Guinea worm demonstrates steps environmental public health officials can use to address environmentally mediated infectious disease.

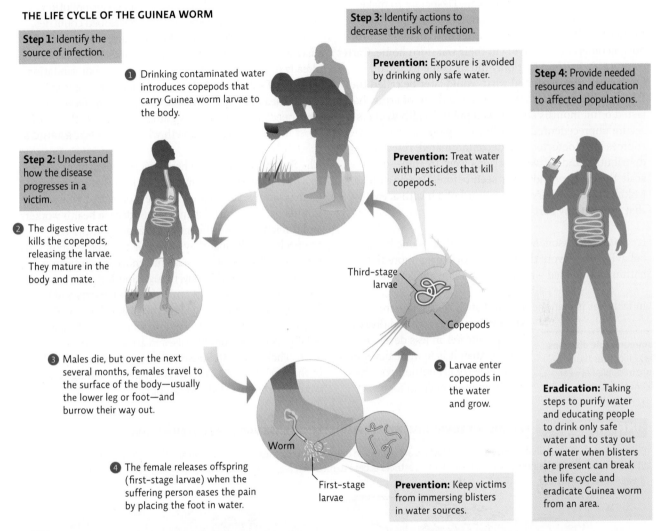

THE LIFE CYCLE OF THE GUINEA WORM

Step 1: Identify the source of infection.

Step 2: Understand how the disease progresses in a victim.

Step 3: Identify actions to decrease the risk of infection.

Step 4: Provide needed resources and education to affected populations.

1 Drinking contaminated water introduces copepods that carry Guinea worm larvae to the body.

2 The digestive tract kills the copepods, releasing the larvae. They mature in the body and mate.

3 Males die, but over the next several months, females travel to the surface of the body—usually the lower leg or foot—and burrow their way out.

4 The female releases offspring (first-stage larvae) when the suffering person eases the pain by placing the foot in water.

5 Larvae enter copepods in the water and grow.

Prevention: Exposure is avoided by drinking only safe water.

Prevention: Treat water with pesticides that kill copepods.

Prevention: Keep victims from immersing blisters in water sources.

Third-stage larvae

Copepods

First-stage larvae

Worm

Eradication: Taking steps to purify water and educating people to drink only safe water and to stay out of water when blisters are present can break the life cycle and eradicate Guinea worm from an area.

 Would it be technically possible to eradicate GWD if only step 4 of the Guinea worm's life cycle (delivery of eggs into water) were eliminated?

5 HEALTH ISSUES IN DEVELOPED VERSUS LESS DEVELOPED NATIONS

Key Concept 5: People in wealthier nations are more likely to die from lifestyle diseases, whereas in low-income countries, the leading causes of death are environmentally mediated infectious diseases.

Back in 1995, when Sudan was still one nation with two warring factions—north versus south—Nabil Aziz Mikhail bore witness to a quiet sort of miracle.

The country's Ministry of Health had long reported Guinea worm cases in the low thousands. But Mikhail, who had just assumed the role of Guinea Worm

Eradication Coordinator, had quickly discovered that the number was much, much higher than that: A better estimate was 100,000+ cases. Mikhail knew that simple measures, such as those the Carter Center was employing elsewhere, could stop the disease in its tracks. But he also knew that no such measures could be employed in Sudan, marred as it was by poverty and extreme violence. The most afflicted areas were simply too dangerous to venture into. Even if they could be reached, eradication programs—community education, latrine building, even pesticide application—would be impossible to implement under the circumstances.

Here's where the miracle comes in: Desperate to make a dent in the problem, Mikhail and his colleagues called former U.S. President Jimmy Carter and invited him to host a conference on GWD in their war-torn home. Carter went a step further: Not only did he come to Sudan, but he quickly negotiated what would later be called the "Guinea Worm Cease Fire"—a laying down of arms that lasted 6 full months and allowed public health workers to secure unprecedented gains in the region. Infected water sources were detected and decontaminated, filters were distributed, and active infections were treated—on a scale the country had never seen before. "It was a dream," Mikhail says now. "I've never heard of a health activity that brought any sort of cease fire."

It was also a lesson, Mikhail says: If there's one human behavior that favors the Guinea worm even more than bathing in infested water, it's war. Plain and simple.

In fact, war is just one reason that the death rate from environmentally mediated diseases is much greater in less developed nations than it is in more developed ones. Poverty is another; poor basic nutrition is another still.

---

noncommunicable diseases (NCDs) Illnesses that are not transmissible between people; not infectious.

As we discuss elsewhere in this text, the differences between more developed and less developed countries are vast. In more developed countries, **noncommunicable diseases (NCDs)**, such as cardiovascular disease and cancer, represent the bulk of the disease burden. (Also known as lifestyle diseases, NCDs are largely determined by choices about things like diet and exercise though air pollution is also a contributing factor.) In less developed countries, while NCDs like these are ticking upward, infectious diseases, caused by all the environmental factors we've already discussed (e.g., lack of clean water, poor sanitation, burning of solid fuels indoors for heat and energy), are still the leading cause of death. The problems faced by less developed areas has added significance because that is where most future population growth will occur. **INFOGRAPHIC 5**

The Sudan cease fire gave health workers a fighting chance to address the environmental conditions that favored Guinea worms. "Once the violence stopped, progress was imminent," says Ruiz-Tiben. "The Sudanese health workers and foreign nongovernment organizations surprised themselves with what they were able to accomplish in those 6 months. But when the cease fire ended, the infection rates crept up again." Today, though South Sudan became an independent nation in 2011, violence persists and has left the country one of the few still plagued by GWD. Encouragingly, the number of infections in South Sudan is falling rapidly, from 521 cases in 2012 down to 6 cases in 2016. Those 6 South Sudan cases represented only 24% of the world's GWD cases in 2016; the remaining 19 cases were reported in Chad and Ethiopia.

INFOGRAPHIC 5 **THE TOLL OF ENVIRONMENTALLY MEDIATED DISEASE VARIES BETWEEN NATIONS**

Death rates due to modifiable environmental factors are highest in developing countries; young children and the elderly are the most vulnerable to environmentally mediated diseases. Poverty that restricts access to medical care, clean water, and an adequate diet is perhaps the leading "health risk" worldwide.

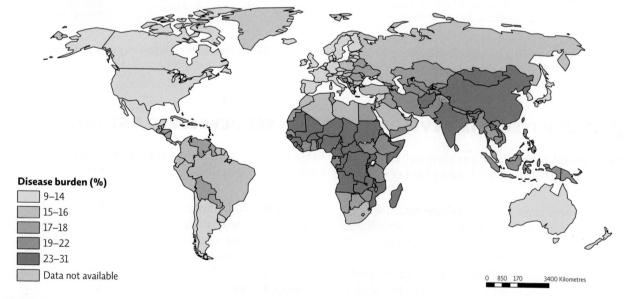

Disease burden (%)
- 9–14
- 15–16
- 17–18
- 19–22
- 23–31
- Data not available

0   850   170   3400 Kilometres

? Propose actions that would reduce the environmental disease burden in developing and developed countries.

6 GLOBAL ENVIRONMENTAL HEALTH

Key Concept 6: Environmental factors contribute to 23% of deaths annually. Most of these are due to non-communicable disease; about 20% are due to infectious and parasitic diseases.

Though GWD appears to be one its way out, environmental hazards are responsible for about 23% of disease and deaths worldwide, according to the WHO. Fortunately, many environmental hazards are *modifiable*—that is, we can take action to change them. Like GWD, they can be mitigated (indeed, at least 8 million deaths could be prevented each year) through reasonable measures. It is on these modifiable hazards—like contaminated drinking water—that environmental public health workers like Ruiz-Tiben focus their efforts.

To be sure, infectious and parasitic diseases account for less of the global burden of disease than NCDs like cardiovascular illness and diabetes. In fact, NCDs cause the most deaths globally. But infectious and parasitic diseases are still a major problem, especially in developing countries, and account for about 20% of deaths due to environmental factors worldwide each year. The main environmentally mediated infectious diseases are diarrheal diseases (due to environmental factors like unsafe drinking water, poor hygiene, and inadequate sanitation), lung infections (linked to bad air, especially indoor air pollution), and mosquito-transmitted diseases like malaria (a parasitic disease) and dengue fever (made worse in areas without access to preventative methods such as mosquito nets or pesticides to kill mosquitoes). Contaminated water also transmits parasitic diseases such as GWD and schistosomiasis. **INFOGRAPHIC 6**

INFOGRAPHIC 6 **THE ENVIRONMENTAL BURDEN OF DISEASE**

Environmental factors account for around 23% of all deaths. Air pollution is the leading environmental factor affecting health, followed by waterborne and vector-borne diseases.

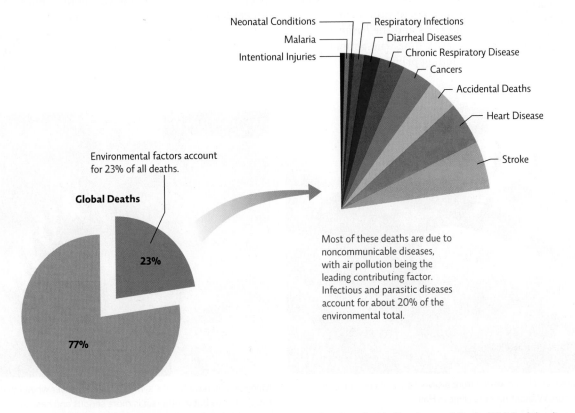

Environmental factors account for 23% of all deaths.

Global Deaths

23%

77%

Most of these deaths are due to noncommunicable diseases, with air pollution being the leading contributing factor. Infectious and parasitic diseases account for about 20% of the environmental total.

Neonatal Conditions — Respiratory Infections
Malaria — Diarrheal Diseases
Intentional Injuries — Chronic Respiratory Disease
— Cancers
— Accidental Deaths
— Heart Disease
— Stroke

 The United Nations has identified ending poverty and hunger as two of its Sustainable Development Goals. Which of the diseases listed above with environmental influences could become less common if these goals are met? Which might become more common? Explain.

7 ZOONOTIC AND EMERGING INFECTIOUS DISEASES

Key Concept 7: Diseases that pass from animals to humans (zoonotic) are increasing due to hygiene issues, the bushmeat trade, and climate change that expands the range of some vectors.

While some diseases, such as smallpox and HIV, are only found in humans and only transmitted between humans, many infectious diseases are **zoonotic**— that is, they can spread between infected animals and humans. In fact, 75% of all **emerging infectious diseases**—those that are new to humans or have rapidly increased their range or incidence in recent years—are zoonotic. This includes recent epidemics of Zika, chikungunya, and dengue—three infectious viral diseases spread by the *Aedes* mosquito—and Ebola, a viral disease spread by direct contact with infected animals or humans. Zoonotic pathogens can also be acquired by consuming contaminated food such as undercooked beef that contains the pathogenic strain of *Escherichia coli*, $O_{157}:H_7$ (see Online Module 8.3).

zoonotic disease An infectious disease of animals that that can be transmitted to humans.

emerging infectious diseases Infectious diseases that are new to humans or that have recently increased significantly in incidence, in some cases by spreading to new ranges.

About 72% of zoonotic diseases, including Ebola and rabies, comes from wildlife, a number that is increasing due to human encroachment into formerly wild areas and the *bushmeat trade*—killing wild animals, including primates (monkeys, chimpanzees, gorillas) for food. Many pathogens that infect wild primates can also infect humans, and the hunting and consumption of those animals increases the chances that individuals will be exposed to these pathogens.

Environmental changes are believed to play a role in many of these emerging infectious diseases. In the United States, recent increases in cases of West Nile virus may be linked to climate change that increases habitat for the mosquitoes that spread the disease. **INFOGRAPHIC 7**

While outbreaks of these diseases are certainly a problem, the zoonotic disease that has had the greatest impact on human health in the 20th and 21st century is influenza, with millions of cases annually and as many as 500,000 deaths per year. The Spanish Flu pandemic of 1918 infected a third of the world's entire human population and killed an estimated 50 million people; deaths attributed to the more recent H_1N_1 flu epidemic of 2009 may have exceeded 500,000. Influenza is particularly difficult to control because the influenza virus changes or mutates every year in its animal hosts, which then spread the virus to people.

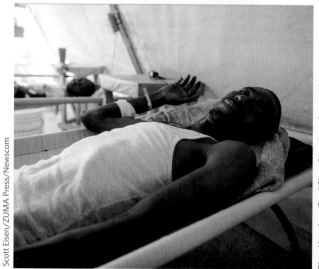

Scott Eisen/ZUMA Press/Newscom

A man who is recovering from a severe case of cholera at the Doctors Without Borders clinic in Haiti.

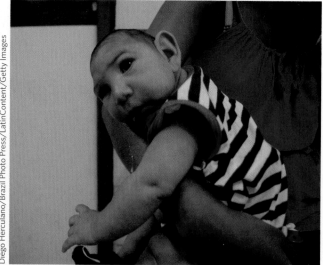

Diego Herculano/Brazil Photo Press/LatinContent/Getty Images

Many people have flu-like symptoms and recover when infected with the Zika virus but in others it is more serious and has been linked to severe birth defects if a woman is infected during pregnancy. This baby, infected with Zika virus in utero, shows signs of microcephaly.

Public health officials work every year to predict which strains of the influenza virus are most likely to become pandemic (spread around the world), and they produce a vaccine that protects against those strains. Some years the vaccine misses one or more strains, but even in those years the vaccine probably offers some level of protection.

Unfortunately, at this time there is no vaccine for Zika, chikungunya, or dengue but research is underway to develop vaccines; a Zika vaccine was approved for clinical trials in 2016. Currently, the best way to prevent the spread of these diseases is to control the mosquitoes that carry the viruses, especially in urban areas where outbreaks impact large numbers of people. Other environmental modifications aimed at decreasing the spread of zoonotic disease focus on improving hygiene when handling animals or their meat and milk products and providing better sanitation with regard to animal waste and livestock rearing conditions. Taking steps to protect biodiversity and combat climate change will also help; improving access to safe, high-quality protein food sources will reduce the need for the bushmeat trade.

INFOGRAPHIC 7 | **ZOONOTIC DISEASES: ROUTES OF TRANSMISSION AND RISK FACTORS**

Zoonotic diseases are infectious diseases that occur in animals and can be transmitted to humans. (Some can be transmitted from humans back to animals.) Transmission may be through direct contact with an infected animal or its waste, through eating contaminated animal food products, from a vector that transmits the pathogen, or through contact with contaminated surfaces. Person-to-person transmission is also possible with some zoonotic diseases.

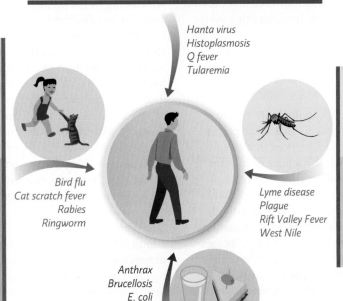

Airborne
Droplets or dust that contain pathogens from feces, urine, or other body fluids can be inhaled.

Risk factors
- Dry environments (dusty barnyards, desert areas, etc.)
- Inadequate containment or handling of animal waste

Hanta virus
Histoplasmosis
Q fever
Tularemia

Direct Contact
Infection can occur via direct contact with infected pets, farm or wild animals, or surfaces contaminated with their waste.

Risk factors
- Animals in overcrowded or unsanitary conditions
- Poor personal hygiene
- Failure to clean surfaces that may be exposed to animal waste
- Bite or scratch from infected animal

Bird flu
Cat scratch fever
Rabies
Ringworm

Vector-borne
Mosquitoes, ticks, and fleas can transmit diseases from animals to humans (and back).

Risk factors
- Presence of vector habitat
- Presence of problematic animal hosts
- Climate change that extends the range of vectors

Lyme disease
Plague
Rift Valley Fever
West Nile

Anthrax
Brucellosis
E. coli
Ebola

Consumption
Ingestion of contaminated water or food products can transmit a wide variety of pathogens from *E. coli* to the Ebola virus.

Risk factors
- Bushmeat trade
- Improper handling of meat or dairy products
- Drinking water contaminated with a zoonotic pathogen

 What could be done to address each of the risk factors noted in this Infographic?

8 ADDRESSING ENVIRONMENTALLY MEDIATED HEALTH PROBLEMS

Key Concept 8: Policies and actions that improve air and water quality, provide access to clean fuels, and empower communities to meet their own needs will reduce environmental health hazards.

In Ezza Nkwubor, in southeastern Nigeria, a team of elderly men—all local villagers—stand guard over a large, silent pond. The water has been treated with pesticides, and the men's job is to ensure that nobody with emerging worms comes into contact with it. Elsewhere, the sick have been quarantined—their wounds carefully tended and water and free food brought to them for the entire month that it takes the worm to emerge. Young boys—travelers and hunters—carry whistle-shaped cylinders tied to strings around their necks. The cylinders serve as portable filters so that the boys can drink directly from an environmental water source and still protect themselves from infection when they're out hunting. And season after season, women teach their protégés the importance of filtering water before giving it to their families.

It's the picture of success that Ruiz-Tiben and his colleagues have envisioned for decades. "It just shows what you can accomplish when the support is there," Ruiz-Tiben says, underscoring what has been a key lesson of the GWD eradication campaign: Implementing even the simplest technologies requires financial and political support—not only from the developed world or the international community but also from the countries themselves.

Yes, it is relatively cheap and technologically simple to build pit latrines and septic systems that make proper waste disposal possible or to build fences that can keep animals out of human water supplies or to plant vegetative buffers that can soak up runoff before it pollutes area streams. And yes, when combined, such a roster of straightforward measures might dramatically reduce the incidence of any number of life-threatening diseases (including, for example, the diarrheal diseases that kill so many children each year). But without the money to buy wood (for fences) or plants (for buffer zones) and without know-how and "local buy-in," such projects would never get off the ground.

Even education campaigns that help people understand how to avoid exposure to certain infectious agents—a measure that requires almost no material support—still take a concerted effort, and thus a well-trained workforce, knowledgeable about disease prevention and respectful of local cultures and traditions. "It's a lot of work to overcome preconceived notions," says Ruiz-Tiben. "You have to present the information in a way they can relate and respond to."

Ruiz-Tiben's work ties in closely to the UN's Sustainable Development Goals (SDGs): SDG 3 is Good Health and Well-Being. A close look at the 17 goals shows a clear link between improving environmental health and reaching these goals. Not only will improving the environment aid in reaching these goals, working to reach these goals in other ways will also help improve the environment. For example, access to clean water and sanitation will improve health, helping families move out of poverty. (Healthy individuals are more likely to be able to work and support their families.) Likewise, providing education and job opportunities will provide the funds a family needs to acquire clean water and pay for sanitation programs. (See Module 1.1 for more on the UN Sustainable Development Goals.) **INFOGRAPHIC 8**

To be sure, effective government policies and funding are needed at all levels to improve the health and well-being of people, but the example of Cameroon's inability to effectively address schistosomiasis shows that it is not always there. Nonprofit organizations like the Carter Center and the Gates Foundation pour funds and manpower into humanitarian efforts and help bridge these gaps. With a loftier goal, in 2016, the Chan Zuckerberg Initiative pledged $3 billion in support of scientific research to "cure, prevent, or manage all diseases within our children's lifetime." To generate further international support and local support, environmental health workers from the developed world need, urgently, to consider the perspectives of their developing-world brethren.

Consider the case of DDT: The pesticide can go a long way toward keeping insect vectors like mosquitoes from human hosts. But when scientists in the developed world linked it to a roster of poor health and poor environmental outcomes, the chemical was banned in many regions, including places where malaria is common. Leaders in those countries were, by many accounts, responding to pressure from the developed world. Today, a thorough risk assessment of the problem is making world leaders reconsider: In some malaria hotspots—certainly in places where mosquito-borne diseases kill tens of thousands of people every year—it turns out that judiciously applied DDT still provides the best mitigation strategy and may yet prove to be worth those risks.

INFOGRAPHIC 8   **REDUCING ENVIRONMENTAL HEALTH HAZARDS MOVES US CLOSER TO SUSTAINABILITY**

The 17 UN Sustainability Development Goals (SDGs) are closely tied to environmental health, especially Goal 3: Good Health and Well-Being. Steps that improve air and water quality and reduce exposure to disease-carrying vectors can reduce environmental health hazards but require education and effective public policy for success. Here are just some of the SDG environmental health links.

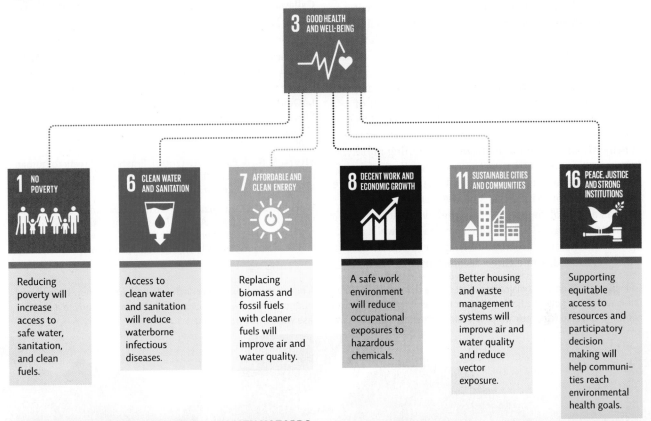

3 GOOD HEALTH AND WELL-BEING

1 NO POVERTY

Reducing poverty will increase access to safe water, sanitation, and clean fuels.

6 CLEAN WATER AND SANITATION

Access to clean water and sanitation will reduce waterborne infectious diseases.

7 AFFORDABLE AND CLEAN ENERGY

Replacing biomass and fossil fuels with cleaner fuels will improve air and water quality.

8 DECENT WORK AND ECONOMIC GROWTH

A safe work environment will reduce occupational exposures to hazardous chemicals.

11 SUSTAINABLE CITIES AND COMMUNITIES

Better housing and waste management systems will improve air and water quality and reduce vector exposure.

16 PEACE, JUSTICE AND STRONG INSTITUTIONS

Supporting equitable access to resources and participatory decision making will help communities reach environmental health goals.

ACTIONS THAT REDUCE ENVIRONMENTAL HEALTH HAZARDS

STEP	EXAMPLES
Provide access to clean water	Dig wells and filter surface water, including using personal filters such as the LifeStraw; provide financial assistance to low-income areas for these technologies.
Improve sanitation and hygiene	Keep sewage out of surface waters by building latrines, planting streamside vegetation to reduce runoff, and keeping animals out of water sources.
Reduce vector exposure	Remove vector habitat (like standing water for mosquitoes), provide barriers like mosquito netting, and apply pesticides to kill vectors; vaccinate pets.
Reduce air pollution	Use cleaner-burning fuels, better-ventilated indoor stoves, and solar ovens to reduce indoor air pollution; adopt and enforce air quality standards to reduce outdoor air pollution.
Provide education	Teach individuals how to avoid exposure to pathogens and how to protect themselves from infection and hazardous chemicals.
Establish effective public policy and/or funding	Pass laws and regulations aimed at reducing environmental hazards and improving health care. In areas without government backing or funding, seek funding from nonprofits.

 What do you think will be the biggest impediment to the eradication of GWD in Africa—the implementation of technical solutions to provide safe water or the education of people in local communities on how to reduce their exposure to contaminated water?

As with most other environmental problems, there are no easy solutions. To conquer GWD, environmental health workers like Ruiz-Tiben have had to battle indifference, poverty, human stubbornness, and now a new complication: It appears that dogs can act as a host for the species of Guinea worm that infects humans, and cases of dog GWD have increased dramatically in the past few years, most notably in Chad. Researchers have always known that dogs and other mammals could be infected with Guinea worm but it was believed to be a different species of worm or a rare occurrence. Genetic testing in 2014 confirmed that the dogs are infected with the same species that infects humans. So efforts now include steps to eradicate the infection from dogs as well as humans, including things like tying up infected dogs to prevent their access to water sources and experimental trials to test the effectiveness of heartworm drugs to kill the worms in infected dogs.

In the end, those battles are paying off: In 2009, Nigeria became the fifteenth African country to rid itself of the ancient worm. At that time, it was estimated that just 3,500 or so cases remained throughout the entire continent, and those numbers were dwindling rapidly. By 2012, only 542 cases were reported in all of Africa; in 2014, that number had dropped to 126. In 2016, there were only 25 confirmed cases of GWD. Concern about dogs as a host of the disease has fueled the urgent quest to eradicate GWD in those animals as well, and health officials are hopeful that these efforts are working. Indeed, some three decades after beginning

its quest, the Carter Center is finally closing in on its ultimate goal: eradicating Guinea worm disease, everywhere, once and for all.

Select References:

Callaway, E. (2016) Dogs thwart end to Guinea worm. *Nature,* 529(7584),10−11.

The Carter Center. *Health Programs,* www.cartercenter.org.

Guimarãe, R. M., et al. (2007). DDT reintroduction for malaria control: The cost−benefit debate for public health. *Cadernos de Saúde Pública,* 23(12), 2835−2844.

LoGiudice, K., et al. (2003). The ecology of infectious disease: Effects of host diversity and community composition on Lyme disease risk. *Proceedings of the National Academy of Sciences,* 100(2), 567−571.

Myers, S. S., & Patz, J. A. (2009). Emerging threats to human health from global environmental change. *Annual Review of Environment and Resources,* 34, 223−252.

World Health Organization. (2016). *Preventing Disease through Healthy Environments: A Global Assessment of the Burden of Disease from Environmental Risks.* http://apps.who.int/iris/bitstream/10665/204585/1/9789241565196_eng.pdf.

Guinea worm disease is a major impediment to a farmer's ability to work. A health volunteer in Ghana educates children on how to use pipe filters when they go to the fields with their families. Pipe filters, individual filtration devices worn around the neck, work similarly to a straw, allowing people to filter their water to avoid contracting Guinea worm disease while away from home.

Courtesy Louise Gubb/The Carter Center

INTERACTIVE MAP | **ENVIRONMENTALLY MEDIATED DISEASES**   ANIMATED INFOGRAPHIC

A wide variety of pathogenic diseases have a strong environmental component. Some of these are on the rise due to a confluence of conditions that favor their occurrence or spread. Though many of these diseases are predominately found in less developed countries, others are also found in the more developed countries. Several are highlighted on the map below.

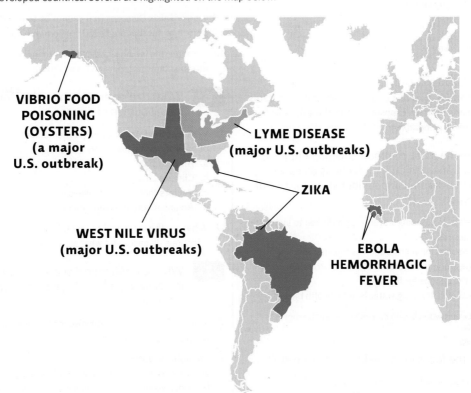

VIBRIO FOOD POISONING (OYSTERS) (a major U.S. outbreak)

LYME DISEASE (major U.S. outbreaks)

ZIKA

WEST NILE VIRUS (major U.S. outbreaks)

EBOLA HEMORRHAGIC FEVER

 BRING IT HOME

PERSONAL CHOICES THAT HELP
People in developed countries typically do not experience the same prevalence of infectious disease as do those in the developing world. However, outbreaks of illnesses like whooping cough, West Nile virus, bacterial food poisoning, and antibiotic-resistant bacterial infections do occur in the developed world and are largely preventable.

Individual Steps
• Many diseases are spread by contaminated hands. The most effective way to remove infectious bacteria and viruses is with 20 seconds or more of hand washing with soap and water. This is even more effective than using hand sanitizer.
• Reduce the likelihood that antibiotic-resistant bacteria will emerge by taking the entire prescription of any antibiotic you are prescribed.

Group Action
• Mosquitoes are responsible for spreading many diseases. Organize your neighbors to take preventive steps to reduce mosquito breeding, including removing containers that might trap rainwater, draining areas of standing water, and cleaning out rain gutters.

• Organize a fund-raising campaign to help purchase pipe filters such as the LifeStraw or finance well digging in areas that need access to clean water.

Policy Change
• While the Safe Drinking Water Act requires the Environmental Protection Agency to test public water supplies for contaminants and bacteria, the U.S. Food and Drug Administration is not empowered to require the same level of testing of bottled water. Ask your legislators what steps could be taken to hold bottled water to similar standards.

ENVIRONMENTAL LITERACY **UNDERSTANDING THE ISSUE**

1 What environmental hazards do humans face?

1. Which of the following can be pathogens?
   a. Bacteria and viruses
   b. Fungi
   c. Protozoa and worms
   d. All of the above

2. Environmental hazards can be divided into physical, chemical, and biological hazards. Give an example of each of these hazards. Describe a scenario in which exposure to one of the hazards could make a person more vulnerable to another type of hazard.

2 What are the most common environmental routes of transmission for infectious diseases and how has human impact increased the risk of transmission?

3. How is the increase in standing water in cities linked to increased human health hazards?
   a. It increases habitat for vectors such as mosquitoes.
   b. It reduces the amount of water available for irrigation.
   c. It increases the chance for flooding.
   d. None of the above; standing water is not a health hazard.

4. Distinguish between waterborne diseases and vector-borne diseases.

3 What is the focus of the field of public health?

5. Which of the following is NOT a tool used by a public health official?
   a. Statistical analysis of disease-related factors
   b. Imposing regulations to ensure compliance with public health directives
   c. Evaluation of the economic stability of a community
   d. Educational programs to teach communities about health problems

6. How do the focus and goal of public health programs differ from those of individual healthcare providers?

4 What can be done to reduce the incidence and spread of infectious diseases?

7. Which of the following steps is NOT necessarily needed to address an environmentally mediated infectious disease?
   a. An understanding of the way the disease is acquired
   b. High-tech equipment and medicines to diagnose or treat a disease
   c. Identification of ways to decrease the transmission of the disease
   d. Public health programs and education that take local culture and traditions into consideration

8. Guinea worm disease is still found in Africa. What can be done to eradicate it, and why has this not already been done?

5 How do factors that affect human health differ between more and less developed nations?

9. True or False: Diseases that cannot be transmitted from one person to another are called noncommunicable diseases.

10. Compare the leading causes of death in low- and high-income countries. Why do these differences exist?

6 What role do environmental factors play in the global burden of disease?

11. Which of the following environmental factors is the leading cause of environmental health problems?
    a. Air pollution
    b. Climate change
    c. Vectors that transmit diseases
    d. An unsafe urban environment

12. What are the main environmental factors contributing to infectious diseases and what areas are most at risk?

7 What are zoonotic diseases, why is their incidence increasing, and what can be done to reduce their incidence?

13. True or False: The zoonotic disease that probably affects the most people annually is malaria.

14. Zoonotic diseases:
    a. are increasing in frequency worldwide.
    b. come mainly from domesticated animal species.
    c. are diseases that spread between humans and plant species.
    d. All of the above

15. Which of the following would reduce the transmission of a zoonotic disease?
    a. Better hygiene such as hand washing
    b. Moving livestock farther away from communities to reduce *E. coli* infections
    c. Clearing habitat in areas with lots of squirrels and opossums to reduce Lyme disease
    d. Encouraging rural African villagers to hunt wild game instead of raising domesticated livestock

16. Identify the risk factors that increase *your* chance of acquiring a zoonotic disease.

8 What can be done to address environmental hazards and how will this contribute to our overall goal of sustainability?

17. Which of the following methods have been useful in reducing indoor air pollution in rural or low-income areas?
    a. Cleaner burning fuels
    b. Well-ventilated stoves
    c. Solar ovens
    d. All of the above

18. How will reducing poverty help to improve environments and reduce environmental health risks?

SCIENCE LITERACY   WORKING WITH DATA

Accurate estimates of deaths due to malaria are important for many reasons, including decisions regarding vector control and health interventions, and also to direct charitable donations effectively. A recent analysis estimated much higher annual death rates from malaria than did previous studies, with the majority of the increase coming from victims over age 5. The table below shows the annual number of deaths from malaria in different world regions, estimated by the Institute for Health Metrics and Evaluation (IHME) and the World Health Organization (WHO).

Comparison of IHME and WHO Estimates of Malaria Deaths by WHO Region

WHO Region	Number of Malaria Deaths		
	IHME	WHO	Difference
Africa	1,098,818	596,000	502,818
Americas	986	1,000	–14
Eastern Mediterranean	47,499	15,000	32,499
Europe	3	—	3
Southeast Asia	84,573	38,000	46,573
Western Pacific	5,596	5,000	596
Total	1,237,475	655,000	582,475
% malaria deaths under age 5	58%	86%	

Interpretation

1. Describe in one sentence what the table shows about the total number of deaths due to malaria.

2. What region of the world has the highest mortality due to malaria? What proportion of the world total comes from this region?

3. What percentage and what number of malaria deaths do IHME and WHO estimate are of children under age 5?

Advance Your Thinking

4. In which region are the new mortality estimates most enlarged? Why do you think this might be the case?

5. It is often extremely difficult to estimate the cause of death in a developing country; health workers rely on a "verbal autopsy," in which they ask surviving family members a series of questions about the deceased's symptoms. If it were possible to confirm that a person who died was or was not infected by malaria, rather than relying on verbal autopsies, would you predict that the estimates of mortality from malaria would increase or decrease? Explain your reasoning.

INFORMATION LITERACY   EVALUATING INFORMATION

Humanitarian organizations are increasingly interested in ensuring that their time and energy are invested in the most efficient and effective manner possible. The Bill & Melinda Gates Foundation has led the way with this approach and has donated a large amount of money to the Institute for Health Metrics and Evaluation (IHME) with this goal in mind.

Go to the IHME website (www.healthdata.org).

Evaluate the website and work with the information to answer the following questions:

1. What is the IHME's mission?
   a. Review the topics listed under the "About" tab. What is the overall mission of IHME? Do you believe this mission is reasonable? Explain.
   b. How is IHME funded? Could the funding source(s) influence the IHME's mission? Explain.

2. Select the "News" link under the "News & Events" tab and look over the titles shown. Choose two of the articles and read each. Identify each article by name and complete the following evaluation for each:
   a. Is this a primary, secondary, or tertiary information source? Justify your answer. (See Module 1.3.)
   b. Identify the authors of the article. Do their credentials qualify them as suitable information sources?

   c. Identify a claim made in the article. Does the article give supporting evidence for this claim? If so, identify the evidence.
   d. Based on your evaluation of the article you examined, do you feel that the news coverage is consistent with the IHME's stated goals? Explain.

3. Select the "Research Articles" link under the "Results" tab. Look over the titles and choose two or more articles to evaluate. For each article, answer a–g:
   a. What type of article is this—primary research, a review article, or an opinion piece (editorial or blog)? How do you know?
   b. How does this differ from the news article you read?
   c. Is this a primary, secondary, or tertiary information source? Justify your answer.
   d. Identify the authors of the article. Do their credentials qualify them as suitable information sources?
   e. Try to access the actual article: Click on the "Read the article" link above the list of authors. (You may then have to click on another link to access the "Full Text" or a PDF of the article.) Is the full article available for you to read?
   f. Explain what steps you would take to access the article if a full copy is not available from this website. (Do this even if the full article is available for the article you chose.)
   g. How useful are these research articles? Do they help fulfill the mission of the IHME? Explain.

 Additional study questions are available at SaplingLearning.com.

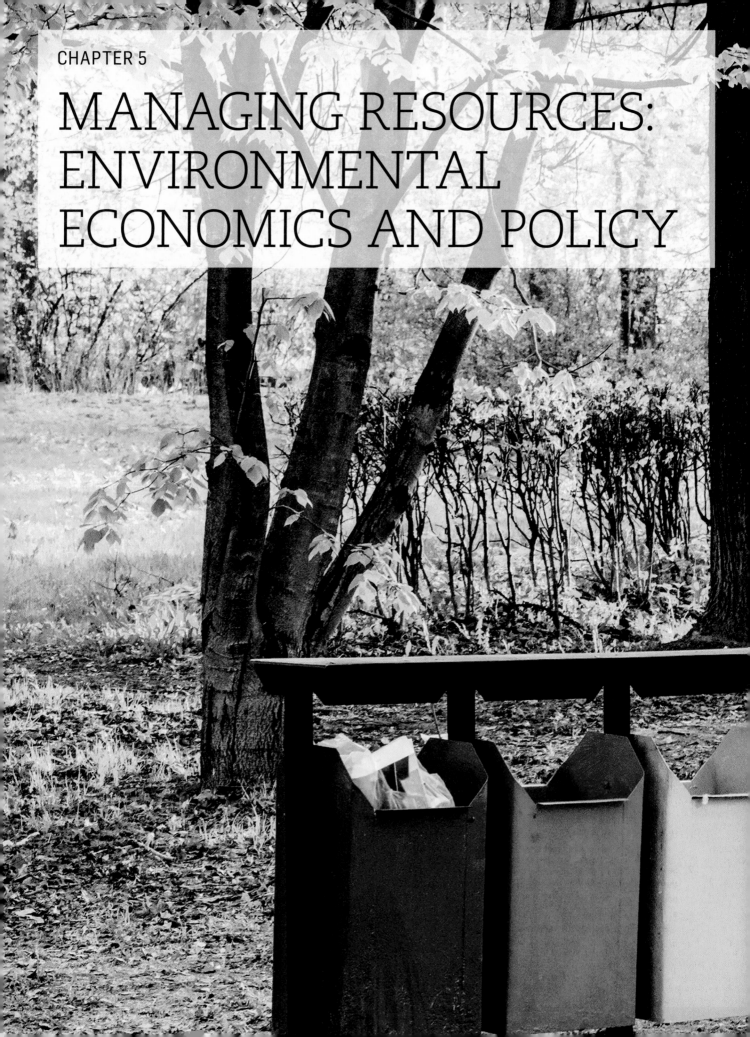

MANAGING RESOURCES: ENVIRONMENTAL ECONOMICS AND POLICY

CHAPTER 5

How we choose to manage resources greatly affects our environmental impact. These choices are influenced by economics and bound by policies. As an example of how to turn an environmental problem into a resource, Module 5.3 will look at our options for managing solid waste.

Module 5.1: Ecological Economics and Consumption

An examination of how economic choices impact human environmental impact and new economic models that may be able to reduce that impact

Module 5.2: Environmental Policy

A look at the importance of environmental policies and how they are made in the United States and globally

Module 5.3: Managing Solid Waste

An evaluation of the problems created by mismanagement of solid waste and ways to reduce the production of waste and its impact

WALL TO WALL, CRADLE TO CRADLE

A leading carpet company takes a chance on going green

Carpet tiles are paving the way for a more sustainable way to produce and market carpet. *Sophie James/Alamy*

After reading this chapter and studying the KEY CONCEPTS and INFOGRAPHICS, you should be able to answer these GUIDING QUESTIONS

CORE MESSAGE

Human impact on Earth is closely tied to the way we use resources. Our economic choices tend to focus on short-term gain rather than long-term sustainability, but we can make better and more informed decisions by taking all the costs—economic, social, and environmental—of a given action into account. Using nature as a model can help us make more sustainable choices while still supporting a viable economy.

1. What are ecosystem services, and why are they important to ecosystems and human populations?

2. What is an ecological footprint?

3. What are natural capital and natural interest, and how can they be sustainably used?

4. What does the IPAT equation tell us about our impact and the potential to reduce that impact?

5. What is true cost accounting, and why should we employ it?

6a. What concerns do environmental economists have with mainstream economics?

6b. What similarities and differences exist between linear and circular models of economics?

7. How can businesses make better choices that reduce their overall impact?

8. How can consumers make better choices that reduce their overall impact?

It was the summer of 1994, and Ray Anderson was feeling pretty good about things. His Atlanta-based company, Interface Carpet, was the world's leading seller of carpet tiles—small, square pieces of carpet that are easier to install and replace than rolled carpet—and it was raking in more than $1 billion per year. One day, though, an associate from Anderson's research division approached him with a question. Some customers apparently wanted to know what Interface was doing for the environment. One potential customer had told Interface's West Coast sales manager that, environmentally speaking, Interface "just didn't get it."

Anderson was dumbfounded. The carpet industry was not generally an ecoconscious industry; after all, synthetic carpet is made from petroleum in a toxic process that releases significant amounts of air and water pollution, along with solid waste. Indeed, Interface used more than 500,000 kg (1 billion pounds) of oil-derived raw materials each year, and its plant in LaGrange, Georgia, released 5.4 metric tons of carpet trimming waste to landfills each day. "I could not think of what to say, other than 'we obey the law, we comply,'" he recalled—in other words, his company did things by the book, in terms of the environment. Wasn't that enough? His research associate suggested that the company launch a task force to create a companywide environmental vision. Anderson agreed, albeit reluctantly.

Desperate for inspiration, Anderson began leafing through *The Ecology of Commerce*, a book by environmental activist, entrepreneur, and writer Paul Hawken, which one of his sales managers had lent him. The book told the story of a small island in Alaska, on which the U.S. Fish and Wildlife Service had introduced a population of reindeer during World War II. Although the reindeer thrived for a time on the available plants, eventually the population exploded beyond what the environment could support. The reindeer ultimately died out because, as Anderson explained, "you can't go on consuming more than your environment is able to renew." Yet that, he suddenly realized, was precisely what Interface was doing—using more resources than it could possibly renew.

◉ **WHERE IS LAGRANGE, GEORGIA?**

ATLANTA
◉

● LAGRANGE

GA

GEORGIA

Jorgen Caris/Hollandse Hoogte/Redux

Ray Anderson, founder of Interface. The background displays sample pieces of his carpet tiles.

1 ECONOMICS AND THE ENVIRONMENT

Key Concept 1: Life on Earth depends on ecosystem services provided by nature. Recognizing the value of these services may motivate us to protect them.

Anderson took his quest for sustainability to heart. Hawken's book was a turning point. "As I read the book, it became clear that, God almighty, we're on the wrong side of history, and we've got to do something."

The choices businesses (and, by extension, consumers) make have tremendous impacts on the environment. The amount and type of energy and water they use, the way they handle the waste they produce, the raw materials they use—these decisions affect not only business operations themselves but also Earth as a whole, especially considering the magnitude of the resources and waste that some large businesses use and produce.

Anderson realized that he had to make changes to Interface; he needed to build it into a **sustainable**, environmentally sound business. "I didn't know what it would cost, and I didn't know what our customers would pay, so it was a leap of faith," Anderson recalled.

sustainable Capable of being continued indefinitely.

economics The social science that deals with the production, distribution, and consumption of goods and services.

ecosystem services Essential ecological processes that make life on Earth possible.

Businesses that are environmentally mindful aren't limited to simply trying to minimize their impact on nature; they can actually look to nature as an economic model from which to learn and model their choices. After all, **economics**—the social science that deals with how we allocate scarce resources—is not just about money.

> "God almighty, we're on the wrong side of history, and we've got to do something."
> —Ray Anderson

Most of the resources we depend on actually come from the environment. Environmental resources like timber and water and ecological processes like water purification and pollination are essential and economically valuable **ecosystem services**. We take many of these services and resources for granted, but some are priceless because there are no substitutes—such as the oxygen produced by green plants, which we need to survive. **INFOGRAPHIC 1**

INFOGRAPHIC 1 VALUE OF ECOSYSTEM SERVICES

Robert Costanza and his colleagues evaluated Earth's ecosystem services and quantified their 2011 values to be around $125 trillion (in 2007 U.S. dollars). Though the figures are considered to be gross underestimates, especially for entities hard to quantify like habitat and genetic resources, these values show that ecosystems provide us with valuable, sometimes irreplaceable, services Costanza points out that ecosystems can continue to do this only to the degree that human impact will allow. When we degrade ecosystems, we reduce their ability to provide these services. He estimates that we have lost $20 trillion in ecosystem services since 1997 due to ecosystem degradation.

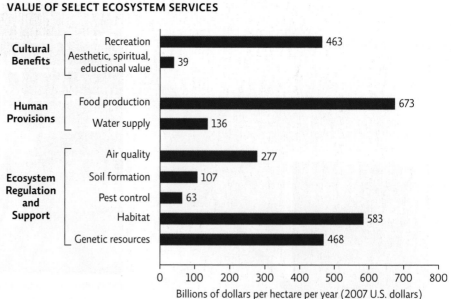

VALUE OF SELECT ECOSYSTEM SERVICES

Category	Service	Billions of dollars per hectare per year (2007 U.S. dollars)
Cultural Benefits	Recreation	463
	Aesthetic, spiritual, eductional value	39
Human Provisions	Food production	673
	Water supply	136
Ecosystem Regulation and Support	Air quality	277
	Soil formation	107
	Pest control	63
	Habitat	583
	Genetic resources	468

 How would the price of food be affected if we incorporated the ecosystem services of pest control, soil formation, and water supply into that price?

2 MEASURING OUR IMPACT: THE ECOLOGICAL FOOTPRINT

Key Concept 2: Human impact can be measured in terms of our ecological footprint—the amount of land needed to support our lifestyle.

When ecosystems are intact, they are naturally sustainable: They rely on renewable resources and also provide services that help replenish and recycle these resources. But ecosystems will only be able to provide us with their valuable goods and services as long as we let them. As Anderson came to realize, when we degrade ecosystems by using more from them than can be replenished, we threaten our planet's ability to provide the services we need, and this ultimately threatens our own future. By using nature as a model, businesses can lessen their impact on the environment and still make choices that support a viable industrial economy.

Like many other businesses, Interface Carpet has a large **ecological footprint**—that is, the land needed to provide its resources and assimilate its waste (typically expressed as hectares [ha] or acres [ac] per person or population). The ecological footprint is a value that businesses, individuals, and populations use to quantify their impact on the environment. **INFOGRAPHIC 2**

The United States, for instance, has a particularly high per capita (per person) footprint, in that it requires much more land area to support each person than it actually possesses. The country is forced to import resources from other countries and even to export some waste. In fact, if the more than 7 billion people who populate the planet all lived like the average person in the United States, we would need the landmass of almost five Earths to sustain everyone.

According to the World Wildlife Federation's 2016 *Living Planet Report*, humanity's combined footprint exceeds what is ultimately sustainable. This means that unless we stop using them so quickly, we are going to run out of some resources. But it's not just about using up some needed resources—it's also about not damaging the very ecological processes and species that provide them.

ecological footprint The land area needed to provide the resources for, and to assimilate the waste of, a person or population.

INFOGRAPHIC 2 ECOLOGICAL FOOTPRINT

The ecological footprint is the land area needed to provide the resources for, and assimilate the waste of, a person or population and may extend far beyond the actual land occupied by the person or population; it is usually expressed as a per capita value (hectares or acres/person). The current world footprint would require about 1.5 Earths to maintain, but obviously we just have one to work with.

Raw materials used in the city are imported from elsewhere.

Some waste is assimilated by areas outside the city.

Physical footprint of the city

Ecological footprint

 Identify some personal choices you could make to reduce your own ecological footprint.

3 NATURAL RESOURCES AS CAPITAL AND INTEREST

Key Concept 3: If we only harvest resources at or below the rate at which they are produced—that is, take only the natural interest—we will leave behind enough natural capital to replace what we took.

What kinds of essential resources does Earth provide us? Considered in financial terms, our **natural capital** includes the natural resources we consume, like oxygen, trees, and fish, as well as the natural systems—forests, wetlands, and oceans—that produce these resources. Our **natural interest** is what is produced from this capital, over time—more trees and oxygen, for example—much like the interest you earn with a bank account. Natural interest represents the amount of readily produced resources that we *could* use and still leave enough natural capital behind to, in time, replace what we took. Natural interest might be represented by an increase in a fish population, for instance, or new growth in a forest— basically, the extra that is added in a given time frame. **INFOGRAPHIC 3**

natural capital The wealth of resources on Earth.

natural interest Readily produced resources that we could use and still leave enough natural capital behind to replace what we took.

If we only withdraw resources equivalent to (or less than) the natural interest, we will leave behind enough natural capital to replace what we took. When Anderson spoke to his employees in the summer of 1994 about his new plan for sustainability, he stressed that his goal was to begin putting back more than the company took from the planet; in other words, he wanted Interface to be what he called a "restorative enterprise." Up to that point, the company was using up far more natural capital in the form of resources like petroleum and water than was ultimately sustainable. Anderson realized that if we take more than is replaced, capital will shrink and therefore produce less the next year. Essentially, by taking 50% more resources than is sustainable, we are taking resources away from the future, in what eco-architect Bill McDonough calls *intergenerational tyranny*. When we liquidate our natural capital more quickly than it can be replaced and call that "income," the question becomes this: Where will future income come from?

This can be an especially big problem with commonly held resources like water: Once we remove it from wells or rivers, we have to wait for the next rainfall to replenish it. When many users are accessing the resource, it can quickly become degraded if they do not work together to manage it—a tragedy of the commons (see Module 1.1).

Interface produces carpet made from yarns that contain up to 100% recycled content. The company has procedures in place to retrieve thread and carpet trimming from the production floor and recycle those into new product. They also reclaim used fishing nets and used carpet to recycle the components into new carpet.

Michiel Wijnbergh/Hollandse Hoogte/Redux

INFOGRAPHIC 3 CAPITAL AND INTEREST

Natural resources can be compared to the financial concepts of capital and interest. Natural capital is the wealth of resources on Earth and includes all the natural resources we use, as well as the natural systems that produce some of those resources (forests, wetlands, oceans, etc.). Natural interest is the amount produced regularly that we could use and still leave enough natural capital behind to replace what we took.

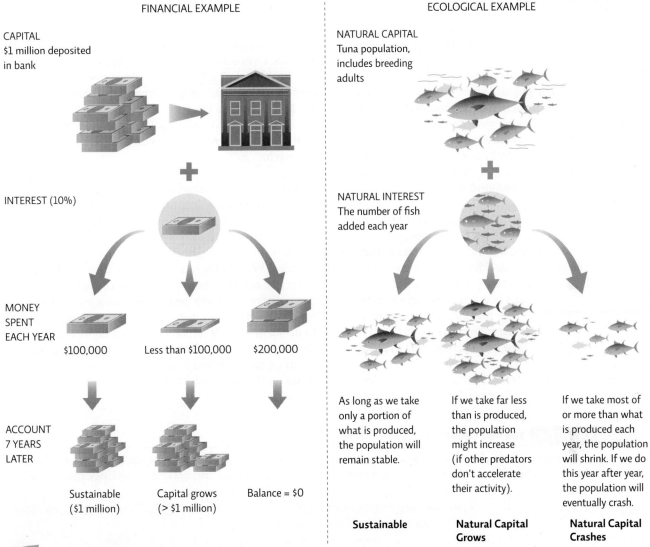

 Why might it be difficult to harvest a natural resource like a tuna population sustainably, even if we set that as our goal?

Starting in 1994, Interface made major changes in the pursuit of its new goals. By 2015, the company had cut the amount of energy it derived from fossil fuels by 84% and reduced its total energy use by 45%. It did this in part by maximizing energy efficiency in its facilities; installing skylights and solar tubes to replace artificial, electricity-dependent lighting; and installing more energy-efficient heating, ventilation, and air

conditioning systems. In one of its factories, Interface also installed a real-time energy tracker that displays energy use prominently for its employees to see, inspiring them to think of new ways to conserve energy. Although Interface declined to reveal how much money it invested in such improvements and technology, the company has ultimately recouped its costs in energy savings, according to a company spokesperson.

4 FACTORS THAT AFFECT OUR ECOLOGICAL FOOTPRINT: THE IPAT EQUATION

Key Concept 4: The impact of a population generally increases as its size, affluence, and use of technology increase. However, the right technology can reduce resource use and pollution generation, thus helping to decrease impact.

Researchers can use the **IPAT model** to estimate the size of a population's ecological footprint, or impact (I), based on three factors: population (P), affluence (A), and technology (T). The premise is that as population size increases, so does impact. More affluent and technology-dependent populations use more resources and generate more waste than do less affluent and technology-dependent populations; technology allows us to build more things, dig deeper, and fly higher, all of which drain the environment. **INFOGRAPHIC 4**

One caveat with regard to this model is that technology can have the opposite effect: Some technologies can decrease, rather than increase, environmental impact. In 2006, for instance, after deciding to become sustainable, Interface invented a new technology called TacTiles: 2.5 × 2.5-inch squares of adhesive tape that join carpet tiles together. The adhesive is made from the same plastic used to make soda bottles. In contrast with traditional "spread on the floor" adhesives, Interface's new tape

---

IPAT model An equation (I = P × A × T) that measures human impact (I), based on three factors: population (P), affluence (A), and technology (T).

---

does not contain any volatile organic compounds, which the U.S. Environmental Protection Agency recognizes as a health risk. TacTiles also make it possible for customers to replace single carpet tiles easily, when, for instance, there has been a spill. Although resources are still required to make TacTiles, they present fewer health risks and produce less waste than the traditional approach.

In 2007, to further reduce its impact, Interface launched a major carpet-recycling initiative called ReEntry 2.0. More than 2 million metric tons (2.5 million U.S. tons) of carpet are pulled up and discarded globally each year, and less than 5% of that has historically been reused or recycled. With ReEntry 2.0, Interface developed a way to recycle carpets—both its own and those made by its competitors—to make new carpet, using only a small amount of virgin raw materials (in this case, petroleum) to do so. Interface's ReEntry 2.0 program has diverted about 100,000 metric tons of material from landfills. Interface has promised to eliminate *any* negative impact it has on the environment by 2020, in a plan it calls "Mission Zero."

INFOGRAPHIC 4 **THE IPAT EQUATION**

The IPAT model suggests that the environmental impact of a society is based on the size of its *population* (P), its *affluence* (A), and its use of *technology* (T). As any or each of these factors increase, so does the population's overall impact. The right kind of technology, however, can lower overall impact.

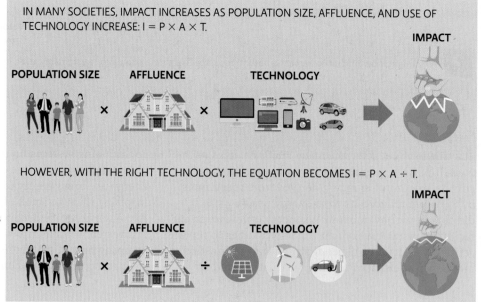

IN MANY SOCIETIES, IMPACT INCREASES AS POPULATION SIZE, AFFLUENCE, AND USE OF TECHNOLOGY INCREASE: I = P × A × T.

POPULATION SIZE × AFFLUENCE × TECHNOLOGY → IMPACT

HOWEVER, WITH THE RIGHT TECHNOLOGY, THE EQUATION BECOMES I = P × A ÷ T.

POPULATION SIZE × AFFLUENCE ÷ TECHNOLOGY → IMPACT

 Identify some technologies that you use that increase your impact. Are there alternative technologies you could use (or propose be developed) that would decrease that impact?

5 TRUE COST ACCOUNTING

Key Concept 5: When the price of a product does not reflect the social and environmental external costs, those costs are paid by others, rather than being passed on to the consumer. Internalizing these external costs better reflects the true cost of a product.

Current-day, or "mainstream," economics allows managers to evaluate possible resource-use choices and make the most profitable decisions, often by seeking to *maximize value*—that is, achieving the greatest benefit at the lowest cost. However, one of the complaints against mainstream economics is that it doesn't take into account *all* potential costs when trying to maximize value. For instance, a carpet tile might require a certain amount of material that has a particular monetary cost; but what about the environmental costs associated with drilling enough oil to make that material in the first place, or the costs associated with cleaning up the pollution it creates?

To determine how much to charge for its goods or services (in its quest to earn a profit), a business, such as a company that produces paper, must account for all its expenses. Wages, fees, insurance, building maintenance, and other expenses are all part of the **internal costs** of doing business. But nature also provides some ecosystem services that might be important to the business, such as nutrient cycles that support tree growth or the water cycle that provides water for the trees. Because these services are supplied by nature "free of charge" and are not part of the cost of doing business, they are classified as **external costs**. However, the price of a good or service that is only based on internal costs is often incomplete

because there can also be *negative* external costs—problems that result from doing business that are not accounted for in an internal cost assessment, such as the health costs associated with the waste produced by making a product or the environmental damage caused by pollution generated by the company.

Historically, economists have regarded these, too, as external to the business (the business doesn't pay for them), and they aren't reflected in the price the consumer pays for the good or service. But if the business doesn't pay for the costs or pass those costs on to the consumer, who does pay? Other people, present and future, and other species do. They pay in the form of degraded health, ecosystems, and opportunities.

An assessment of the cost of a good or service (or any of our choices) should include more than just the economic costs; it should also include the social and environmental costs—the **triple bottom line**.

internal cost A cost—such as for raw materials, manufacturing costs, labor, taxes, utilities, insurance, or rent—that is accounted for when a product or service is evaluated for pricing.

external cost A cost associated with a product or service that is not taken into account when a price is assigned to that product or service but rather is passed on to a third party who does not benefit from the transaction.

triple bottom line The combination of the environmental, social, and economic impacts of our choices.

East Valley Tribune, Tim Hacker/AP Photo

Brandon Sargent co-founded EcoScraps with two college friends in 2010. The company collects food waste from local restaurants and grocery stores which it composts into potting soil.

(See Module 1.1 for an introduction to the concept of the triple bottom line.) By ignoring the external costs, economies create a false idea of the true and complete costs of particular choices. A customer may pay $8 for every 50-square-centimeter (about 8 square inches) of carpet tile, but the **true cost** for that piece of carpet would be much higher if it included costs such as greenhouse gas emissions and the cost of treating people for asthma if they have fallen ill as a result of the particulate matter released during the carpet's production. The inadequate valuation of a product could eventually lead to the exploitation or overuse of resources needed to produce it—an example of market failure. When external costs are internalized, on the other hand, people (or species) who don't benefit from the transaction do not pay for it. In this case, the product

true cost The sum of both external and internal costs of a good or service.

or service can be more appropriately priced or valued; this new price more accurately reflects the true cost of the product or service.

Because we are unaccustomed to paying true costs, we would most likely be appalled at how much some goods and services would really cost if all externalities were internalized. Although it sounds discouraging, any time we purchase products that were made in a more environmentally or socially sound manner, we come a little closer to bearing the responsibility for our choices. We also create a demand for these products in the marketplace. And, if businesses are forced to internalize external costs, it then becomes profitable for them to take steps to lower those costs—for example, by installing pollution prevention technologies—a benefit that could lower the environmental and societal costs overall. **INFOGRAPHIC 5**

INFOGRAPHIC 5 | TRUE COST ACCOUNTING

Many environmental and health costs of our goods and services are externalized (not included in the price the consumer pays). But if consumers don't pay all the costs to produce a product, such as paper, who does?

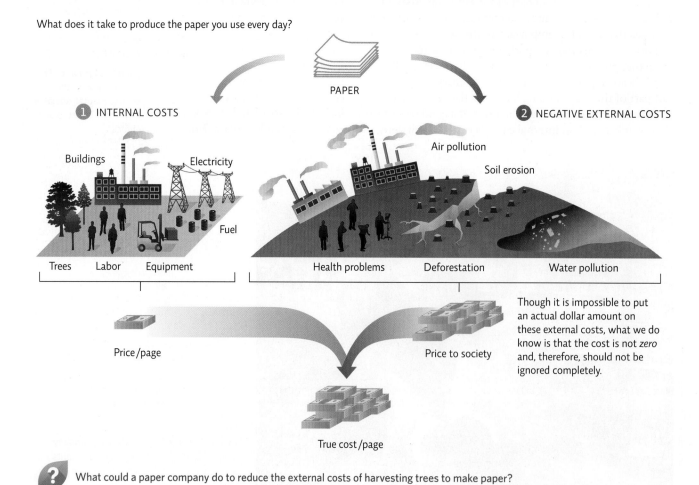

What does it take to produce the paper you use every day?

PAPER

1 INTERNAL COSTS

Buildings · Electricity · Fuel · Trees · Labor · Equipment

2 NEGATIVE EXTERNAL COSTS

Air pollution · Soil erosion · Health problems · Deforestation · Water pollution

Price /page

Price to society

Though it is impossible to put an actual dollar amount on these external costs, what we do know is that the cost is not *zero* and, therefore, should not be ignored completely.

True cost /page

? What could a paper company do to reduce the external costs of harvesting trees to make paper?

6 ENVIRONMENTAL ECONOMICS VERSUS MAINSTREAM ECONOMICS

Key Concept 6a: Environmental economists argue that mainstream economics will fail in the long run because it makes some assumptions that are inconsistent with the way nature operates.

Key Concept 6b: Linear economic production models use inputs and produce waste without regard to sustainability; circular systems depend on renewable resources and see waste as a useful input.

Ecologically minded economists advocate incorporating environmental considerations into economic decisions. **Environmental economics** is a discipline that considers the long-term impact of our choices on human society and the environment. (There are actually two camps—environmental economics and ecological economics—but their distinctions are not important for this discussion. For simplicity, in this text, we will combine them and use the term environmental economics.) Environmental economists argue that without including environmental considerations, mainstream economic theory will fail because it is based on several erroneous assumptions. **INFOGRAPHIC 6A**

One inaccurate assumption of mainstream economics is that natural and human resources are either infinite or that substitutes can be found if needed. This is true for some but not all resources. For instance, fossil fuels are finite and will run out, even with technological advances that allow us to access more of the fuel that is left. Crop productivity has limits—soil can be amended with fertilizer additives, but it cannot be replaced for most crops—and there is no substitute for water. In addition, our actions can degrade the quality of air and water resources faster than nature can restore them.

Mainstream economics also assumes that economic growth will go on forever. Since there are inherent limits to what Earth can provide, unlimited economic growth (at least that which depends on finite resources and the rate at which those resources are replaced) is not, in fact, possible. We

environmental economics
New theory of economics that considers the long-term impact of our choices on people and the environment.

INFOGRAPHIC 6A   ASSUMPTIONS OF MAINSTREAM AND ECOLOGICAL ECONOMICS: A COMPARISON

Environmental economists criticize several assumptions of mainstream economics that they say will fail in the long run.

MediaWorldImages/Mar Photographics/Alamy

FACTOR	MAINSTREAM ECONOMICS	ECOLOGICAL ECONOMICS
Resource availability	Natural resources are either infinite or substitutable.	Some natural resources are finite, and there are not suitable substitutes for all. Therefore, when using finite resources, we should use those that are renewable resources and use them at a rate that is equal to or less than the rate of replacement.
Economic growth	Economic growth will go on forever.	Because it is based on finite resources or the rate of their replacement, economic growth based on consumption of resources has limits.
Resource value	The future value of a resource (its value if we save it) is worth less than its value if used today.	Recognizing that short-term exploitation of a resource may diminish its potential to be replaced, more weight is given to long-term benefits and costs to than to short-term ones.
Production models	Models of production typically follow a linear sequence: Raw materials come in (inputs), products are made, and waste is discarded (outputs).	Production models should be closed loops like nature, focusing on renewable resources used at or below their rate of replacement and treating waste so that it becomes an input that creates new resources.

Not all economic growth requires the consumption of finite resources. Give an example of an economic pursuit that does have the capacity to keep growing every year without depending on higher use of natural resources.

have to work within the limits of available resources in ways that allow essential ecosystem services to continue.

Another problem with mainstream economics is that it **discounts future value**: It tends to give more weight to short-term benefits and costs than it does long-term ones. In other words, mainstream economics considers something that benefits or harms us today more important than something that might do so tomorrow. For instance, we value the tuna we can harvest today more highly than tuna we might harvest 10 years from now, so the value of taking a large harvest of tuna today outweighs the benefits of taking less now to ensure that there is still some later. If the money we could earn by using the resource now is higher than sustainable harvesting yields, modern economics tells us it is more profitable to use it now and invest the resulting money in another venture. But this investment approach doesn't take into account where those other ventures might come from or whether they are in any way diminished by the elimination of the first resource. How might the loss of tuna affect the ecosystem and other populations? Will there always be another fish population to harvest?

discounting future value
Giving more weight to short-term benefits and costs than to long-term ones.

linear economic system
A production model that is one way: inputs are used to manufacture a product, and waste is discarded.

circular economic system
A production system in which the product is returned to the resource stream when consumers are finished with it or is disposed of in such a way that nature can decompose it.

These assumptions lead to yet another misconception—that models of production typically follow a linear sequence—a "take, make, dispose" approach. In this **linear economic system**, raw materials come in, humans transform those materials into some kind of product, and then they discard the waste generated in the process. But because some resources are finite, if the waste created from the use of these resources is stored away permanently as "waste," we would eventually run out of a needed input as supplies dwindled. In addition, generating waste in the form of pollution that can damage natural capital such as air, water, and soil also reduces the amount of matter available to us. It is this unsustainable use of resources that will eventually cause linear models of production to fail. For instance, most traditionally produced carpet tiles are made from fossil fuels, a practice that is not sustainable. Old, unwanted tiles are then discarded, and some are eventually burned, releasing toxic pollutants and greenhouse gases.

A sustainable approach would be more cyclical, where "waste" becomes the raw material once again and can be used to make new products. Interface's ReEntry 2.0 program uses old carpet tiles to make new ones, and it uses old carpet backings to make new carpet backings. This is an example of a **circular economic system**, where the product is folded back into the resource stream when consumers are finished with it or is disposed of in such a way that nature can decompose it. **INFOGRAPHIC 6B**

More and more industries are investing in wind and solar panels as a way to decrease the ecological footprint of their operations.

ANIMATED INFOGRAPHIC

INFOGRAPHIC 6B **ECONOMIC MODELS**

Environmental economics recognizes that natural ecosystems provide our resources and assimilate our wastes. If companies could fold "waste" back into production or make sure it can be decomposed by nature, we could reduce our extraction costs and operate in a sustainable circular system.

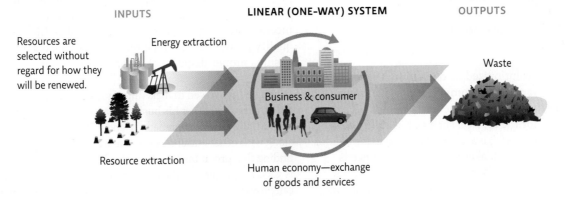

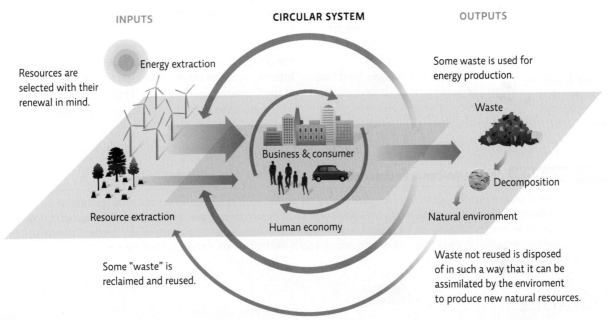

 Look at the size of the arrows in these two diagrams. Which ones increase and which ones decrease in size? Why?

7 SUSTAINABLE PRACTICES: THE ROLE OF BUSINESS

Key Concept 7: In cradle-to-cradle management, waste is seen as a resource, which encourages manufacturers to make durable, recyclable, and nonhazardous products and product components.

To combat the erroneous assumptions of mainstream economics, environmental economists support actions such as improving technology to increase production efficiency and reduce waste, valuing resources (including waste) as realistically as possible, moving away from dependence on nonrenewable resources, and shifting away from a product-oriented economy. They look to natural ecosystems as models for how to efficiently use resources and live within the limits of nature.

Many businesses today account for all the steps needed to produce a product and deal with any waste produced—a *cradle-to-grave* approach. But that approach is incomplete. Interface seeks to operate as a circular system by managing its product in a **cradle-to-cradle** fashion: It considers the entire *life cycle* of the product, from the

cradle-to-cradle Refers to management of a resource that considers the impact of its use at every stage, from raw material extraction to final disposal or recycling.

beginning (acquisition of raw materials) to the end of its useful life (disposal), and is responsible for the impact of its use at every stage of the process. This can lead to better material choices (less toxic, more sustainable) and better process choices (reusable materials, less waste and pollution). **INFOGRAPHIC 7**

Interface also revolutionized its operations by considering itself part of a **service economy**; it focuses on selling a *service* rather than a *product*. The idea is simple: A customer pays for the service, such as the ability to photocopy or keep refrigerated food cold, and the vendor makes sure that the service is always available. Interface sells the service of carpet—its color, texture, and comfort—rather than the product itself. The customer pays a monthly fee to "lease" the carpet, and Interface maintains it and replaces it as needed. This encourages Interface to produce carpet that is durable and recyclable and also easily replaceable.

service economy A business model whose focus is on leasing and caring for a product in the customer's possession rather than on selling the product itself (that is, selling the service that the product provides).

green business Doing business in a way that is good for people and the environment.

Another sustainable business practice involves *take-back programs*, particularly for products with a defined life span, such as electronics: Customers return the product to the producer when they are finished with it or when they need an upgrade. This provides an incentive to the producer to make a durable, high-quality product that can be reused or recycled and ensures proper disposal of hazardous materials.

When he first vowed to make Interface sustainable, Anderson did not know whether his business would thrive or suffer as a result. "I was very apprehensive about it," he recalls. But he, along with other entrepreneurs who have followed suit, are finding that **green business**—doing business in a way that is good for people and the environment—is also profitable. It can provide a competitive advantage because the consumer is willing to support the company's efforts or because green actions end up saving money.

After Interface filled out a 200-page questionnaire about how the company was addressing various environmental issues, Interface beat out other contenders for a $20 million contract at the University of California. One of the university's representatives turned to a colleague of Anderson's and exclaimed, "This is *real*."

INFOGRAPHIC 7 **CRADLE-TO-CRADLE MANAGEMENT**

In cradle-to-cradle management, the manufacturer is responsible for the product from its production (cradle) to its final disposition after the consumer is finished with it. If the item were merely disposed of, it would be sent to its "grave" and those resources wasted. If, however, the item is disassembled and the parts reused, these parts become raw material again for a new product—a new "cradle." This provides an incentive to produce the product in a way that uses durable, reusable parts and that minimizes toxicity, since the manufacturer is responsible for dealing with those toxins.

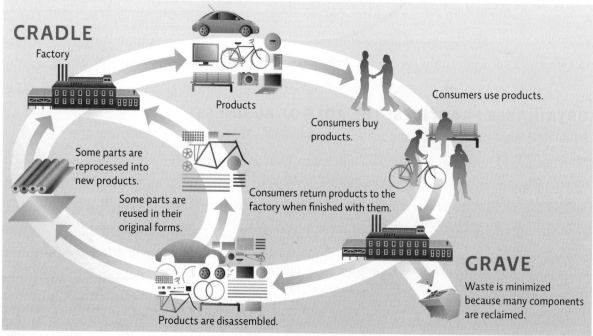

CRADLE
Factory
Products
Some parts are reprocessed into new products.
Some parts are reused in their original forms.
Products are disassembled.
Consumers buy products.
Consumers use products.
Consumers return products to the factory when finished with them.
GRAVE
Waste is minimized because many components are reclaimed.

? Explain why this flowchart is an example of a cradle-to-cradle management system.

8 SUSTAINABLE PRACTICES: THE ROLE OF THE CONSUMER

Key Concept 8: Transparency in how a business operates will allow consumers to make better choices and level the playing field for businesses that are trying to operate more sustainably. Finding ways to consume less is also an important green choice.

What about consumers? We can all decrease our impact by making more sustainable choices and by consuming less. This doesn't necessarily mean "doing without," but it does mean being mindful of our choices and opting for sustainable or low-impact choices whenever possible. However, this requires transparency from the industries that produce and sell us goods and services. That is hard to come by with current business models, often because the businesses themselves don't know all the external costs associated with their products. For its new plan to be successful, Interface was counting on its customers to make more sustainable choices as well. And make them they did. ReEntry 2.0 drew many new customers to Interface, including the Georgia state legislature, which purchased 13,000 square yards of carpets.

Changing the way we do business to give consumers the ability to choose sustainably produced goods and services is not going to be easy. Even Interface has progressed slowly, despite its strong desire to become sustainable. Startup or upgrade costs can be substantial, and even though improvements may pay for themselves in the long run, many businesses simply do not have the funds to pay for them. Plus, they may find themselves at a competitive disadvantage with businesses that are not trying to internalize costs. Consumers also have a role to play. For example, recycled paper's higher cost may more closely reflect the true cost of paper, but if

consumers are not willing to put their buying dollars behind their environmental ideals, businesses that make and sell paper from trees will still be more successful. In other words, it will take changes from both consumers and producers to put business and industry on the path to sustainability. But there are things that can be done to level the playing field.

Governments can encourage sustainability by providing incentives for businesses to account for true costs rather than just internal costs. This could be accomplished by taxing companies based on how much pollution they generate, subsidizing environmentally friendly processes, or giving out pollution "permits" that companies could sell if they release less pollution than they are allowed (*cap-and-trade*; see Module 10.1). For instance, if there is a pollution tax, it will be passed on to the consumer, who then decides whether to buy the product. The manufacturer that minimizes waste production and relies on lower fossil fuel inputs than its competitors would be able to meet the regulations at the lowest cost, offer lower prices, and, as a result, have a major market advantage. While promising, these policies can be complex to implement because external costs are hard to quantify (How is a pollution tax fairly assessed?) and because they may put a burden on smaller manufacturers that have less ability to absorb the cost of upgrades. Passing these costs on to the consumer also stresses low-income households.

According to Anderson, the consumer needs to know how a good or service is made—what the overall environmental impact of that product is. "You lay out for the consumer everything that goes into that product, and you lay it out for your competitors too—it's a totally transparent revelation of how you made that product, and what that footprint is at every step," he explained. This is hard to achieve when we buy many products made in faraway places and shipped over long distances.

One way to communicate this information is through **ecolabeling**. But consumers have to be wary of labels because as "green" products become more attractive to consumers, more companies engage

Cal Cam/Alamy

Many retailers are taking steps to reduce their impact. For example, REI pursues a variety of actions that address its energy and materials footprint including constructing energy efficient buildings, installing solar panels, reducing product packaging, and sourcing recycled materials.

ecolabeling Providing information about how a product is made and where it comes from. Allows consumers to make more sustainable choices and support sustainable products and the businesses that produce them.

in **greenwashing**—claiming environmental benefits for a product when they are minor or nonexistent. **Fair trade** items are, however, more likely to be sustainably produced. For a product to be certified as fair trade, workers must be paid a fair wage, work in reasonable conditions to produce the goods or services, and the production process must have a low ecological impact. In other words, they have been produced in an environmentally and socially responsible way.

Share programs are another useful option for items that people need infrequently, such as a car for those who live in a large city. Rather than buying, owning, and then storing the product for a large part of the time, consumers share ownership and use the product only when they need it. **INFOGRAPHIC 8**

greenwashing
Claiming environmental benefits about a product when the benefits are actually minor or nonexistent.

fair trade A certification program whose products are made in ways that are environmentally sustainable and socially beneficial (e.g., fair wages, good working conditions).

reduced waste some 94% since 1996 and has several LEED-certified facilities (see Module 4.2). Practices like intercepting industrial waste destined for landfills have a positive effect, while other efforts lessen the company's overall negative impact: less toxic glues, less carpet waste. All the while, Interface is still the world's leading manufacturer of commercial carpet tiles, and its 2010 operating income increased 47% over 2009. "I think we're on the right track, and we'll keep on going," says Anderson, who stepped down as the company's CEO in 2001 but still played the role of the company's conscience until his death in 2011 at the age of 77. "We'll get to the top of that mountain."

Although Interface has come a long way since 1994, it is still working hard to achieve its sustainability goals. In June 2011, the company began producing its first 100% nonvirgin fiber carpet tiles, made from reclaimed material: fiber derived from salvaged commercial fishnets and postindustrial waste. In addition to enjoying substantial energy savings, Interface has

Select References:

Anderson, R. (1998). *Mid-Course Correction: Toward a Sustainable Enterprise: The Interface Model.* Atlanta, GA: Peregrinzilla Press.

Anderson, R. (2009). *The Business Logic of Sustainability,* www.ted.com/talks/ray_anderson_on_the_business_logic_of_sustainability.

Costanza, R., et al. (1997). The value of the world's ecosystem services and natural capital. *Nature, 387,* 253–260.

Costanza, R., et al. (2014). Changes in the global value of ecosystem services. *Global Environmental Change. 26,* 152–158.

World Wildlife Federation. (2016). *Living Planet Report, 2016,* www.footprintnetwork.org/documents/2016_Living_Planet_Report_Lo.pdf.

INFOGRAPHIC 8   **GREEN CONSUMERISM**

The choices we make as consumers are extremely important in any quest to pursue environmental sustainability. Supporting green products supports those companies; we vote with our dollars. Responsible consumerism also includes consuming less, so look for ways to meet a need other than buying a new product. The next time you want to boost your mood or reward an accomplishment, consider a visit with friends or a walk in the park, rather than a new pair of shoes or video game.

- **Ecolabels:** Look for transparency in labelling to identify sustainable options, e.g., certified organic or 100% postconsumer recycled content.
- **Greenwashing:** Beware of claims with no verification of accuracy or environmental relevance.
- **Fair Trade:** Support fair-trade products made with methods that support sustainable practices and social justice.
- **Share Programs:** Don't buy when you can rent or borrow a product.
- **Green Businesses:** Support businesses that use sustainable methods and resources. Be willing to pay more for goods and services whose price represents their true cost.
- **Consumption:** Lower your consumption of goods and services by purchasing high-quality products that are durable and will last a long time. Try to limit your overall purchases.

 What do you feel are the biggest impediments to becoming a green consumer? Why?

INTERACTIVE MAP GREEN BUSINESSES

ANIMATED INFOGRAPHIC

A wide variety of businesses are taking major strides to reduce their ecological footprint. Some of these actions profit the business by lowering production or operating costs while others reach beyond business operations and contribute to environmental protection or provide societal benefits. The profiles presented here are not meant to be an endorsement of the businesses but simply highlight some players that are taking extra steps to decrease their impact or even to help restore the environment their industry affects.

CLOTHING AND
OUTDOOR GEAR
RETAILER: MOUNTAIN
EQUIPMENT CO-OP

FURNITURE
PRODUCTION:
VERMONT WOOD
STUDIOS

"PET" INDUSTRY
(AQUARIUM FISH):
SUSTAINABLE AQUATICS

CONSTRUCTION
INDUSTRY: CEMEX

 BRING IT HOME

PERSONAL CHOICES THAT HELP

You have an impact on creating a sustainable society. Every time you buy a product or service, you are telling the manufacturer that you agree with the principles behind the product. You can use your purchasing power to show companies that people are interested in good-quality products that support environmental and social values.

Individual Steps
• Reduce the amount of stuff you accumulate by buying fewer items and by choosing products that are well made and last longer.
• Ask your local food store or pharmacy to stock fair trade–certified products if it doesn't already.
• Instead of buying a new or used car, join a car-share program like Zipcar.

Group Action
• Get together with family and friends and write a letter to your favorite companies, asking them to reduce their ecological footprint. You can ask them to become more transparent by publishing how their business practices impact the environment.

Policy Change
• Start a blog or Facebook page to chronicle the changes you make in your buying habits and encourage others to do the same. Discuss the companies whose environmental policies you agree with.

Tara Walton/The Toronto Star/ZUMAPRESS.com/Newscom

ENVIRONMENTAL LITERACY **UNDERSTANDING THE ISSUE**

1 What are ecosystem services, and why are they important to ecosystems and human populations?

1. The water cycle is an example of:
   a. natural interest.
   b. an external cost.
   c. an internal cost.
   d. an ecosystem service.

2. How can it be useful to place a monetary value on ecosystem services even if we know it will not be accurate?

2 What is an ecological footprint?

3. What is the unit of measure for determining an ecological footprint?
   a. Weight of resources used and waste produced
   b. Time required to produce resources used and assimilate waste produced
   c. Land area needed to produce resources used and assimilate waste produced
   d. Number of people that are using the available resources or producing waste

4. Why is the per capita ecological footprint of the United States so much higher than that of most other nations?

3 What are natural capital and natural interest, and how can they be sustainably used?

5. Readily produced resources that can be used and still leave enough natural capital behind to replace what we took are known as:
   a. renewable resources.
   b. natural interest.
   c. investment capital.
   d. scarce resources.

6. The sap of maple trees (sap is "food" for the tree) can be tapped to make maple syrup, but taking too much will kill the tree. In this example, what would constitute the natural capital and what would be the natural interest?

4 What does the IPAT equation tell us about our impact and the potential to reduce that impact?

7. True or False: Researchers use the IPAT model to estimate the size of an individual's ecological footprint.

8. What is the IPAT model? How is the equation $I = P \times A \times T$ similar to and/or different from the equation $I = P \times A \div T$?

5 What is true cost accounting, and why should we employ it?

9. Which of the following is an internal cost of coal mining?
   a. Pollution to nearby communities
   b. Long-term health effects suffered by miners
   c. Wages paid to workers
   d. Loss of wildlife close to the coal mine

10. What is true cost accounting, and why would it be good for the environment if businesses internalized all external costs?

6a What concerns do ecological economists have with mainstream economics?

11. Which of the following is an assumption of mainstream economics that environmental economists feel is erroneous?
    a. Economic growth has limits.
    b. Natural resources are infinite or substitutable.
    c. A resource's future value is higher than its current value.
    d. All of the above

12. In addition to failure to consider true costs, identify and explain four erroneous assumptions that mainstream economics makes with regard to the environment.

6b What similarities and differences exist between the linear economic model of mainstream economics and the circular model of environmental economics?

13. Which of the following is true for a circular economic model of production?
    a. Renewable and nonrenewable energy resources are equally preferred.
    b. Resources are extracted at a rate that meets industrial needs.
    c. Waste is folded back into the production cycle whenever possible.
    d. All of the above

14. Why do environmental economists say that linear economic models of production will ultimately fail?

7 How can businesses make better choices that reduce their overall impact?

15. What does the term "cradle-to-cradle" mean when talking about product management?
    a. Product materials must be tracked from production to disposal.
    b. The product is potentially more dangerous to children.
    c. Current legislation is too restrictive on new product development and causes the early demise of new businesses.
    d. Production is cyclical: "Waste" becomes the raw material once again and can be reused.

16. What actions did Interface Carpet use to become more sustainable?

17. Are take-back programs an example of cradle-to-grave production models or cradle-to-cradle production models? Explain.

8 How can consumers make better choices that reduce their overall impact?

18. True or False: Items certified as fair trade have been produced in an environmentally and socially responsible way.

19. Explain why purchasing a green product is not always the best green consumer choice.

20. What are "fair-trade" products?

SCIENCE LITERACY WORKING WITH DATA

Robert Costanza and colleagues estimated the value of ecosystem services of many of Earth's major ecosystems as well as the area each one covered on the planet for the year 2011. Look at the following graph and data table to answer the questions that follow.

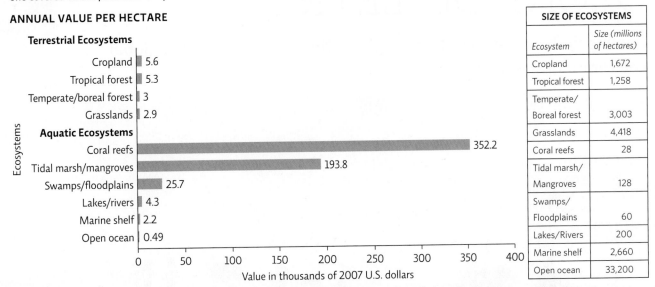

ANNUAL VALUE PER HECTARE

Terrestrial Ecosystems

	Value
Cropland	5.6
Tropical forest	5.3
Temperate/boreal forest	3
Grasslands	2.9

Aquatic Ecosystems

	Value
Coral reefs	352.2
Tidal marsh/mangroves	193.8
Swamps/floodplains	25.7
Lakes/rivers	4.3
Marine shelf	2.2
Open ocean	0.49

Value in thousands of 2007 U.S. dollars

SIZE OF ECOSYSTEMS

Ecosystem	Size (millions of hectares)
Cropland	1,672
Tropical forest	1,258
Temperate/Boreal forest	3,003
Grasslands	4,418
Coral reefs	28
Tidal marsh/Mangroves	128
Swamps/Floodplains	60
Lakes/Rivers	200
Marine shelf	2,660
Open ocean	33,200

Interpretation

1. Based on the graph, which ecosystem has the highest value per hectare per year? The lowest?

2. In total dollars, what is the value of Lakes/Rivers per hectare?

3. Based solely on the graph, how do terrestrial ecosystems compare to aquatic ones in terms of their annual contribution/hectare to ecosystem services? Give data to support your answer.

Advance Your Thinking

4. Look at the data table that shows the size of each ecosystem in millions of hectares. Using these data, calculate the total annual value in millions of 2007 U.S. dollars for each ecosystem and create a bar graph similar to the one above that shows the total annual value for each ecosystem.

5. Reevaluate the contribution of terrestrial ecosystems versus aquatic ones based on the total annual value of each. How does this compare to the answer you gave for question #3?

6. Explain the value of each graph and why it is useful to show the data both ways.

INFORMATION LITERACY EVALUATING INFORMATION

Being an informed consumer begins with understanding what the labels on the products you buy really mean. For explanations of food labels, explore the Greener Choices website at http://greenerchoices.org and answer the questions below.

Evaluate the website and work with the information to answer the following questions:

1. Determine if this is a reliable information source with a clear and transparent agenda:
   a. Who runs this website? Do the credentials of the organization make the information presented reliable/unreliable? Explain.
   b. What is the mission of this website? What are its underlying values? How do you know this?

2. Under the "Labels" tab, select at least three labels and for each, answer the following questions:
   a. What does the label mean and why is it important? Is it a reliable indicator of an environmentally sound consumer choice?
   b. What data sources were used in the label assessment? Are its sources reliable?

3. Select one of the "Reports" produced by this organization.
   a. Evaluate the information in terms of authorship (credentials).
   b. What data sources were used in the report? Are its sources reliable?
   c. Look at the Buying Guide that accompanies this report. Do the report and buying guide provide useful information for a consumer who wants to buy this food product? Explain.

4. Overall, is the Greener Choices website a reliable source of information? Explain.

Additional study questions are available at SaplingLearning.com.

THE WORLD TACKLES OZONE DEPLETION

Dealing with ozone depletion taught nations how to address global environmental issues

Setup for launch of the high-altitude research balloon for ozone testing, McMurdo Station, Antarctica.
Courtesy Linnea Avallone/Concordiasi Team

After reading this chapter and studying the KEY CONCEPTS and INFOGRAPHICS, you should be able to answer these GUIDING QUESTIONS

CORE MESSAGE

Environmental policies work to protect the environment and public health by defining acceptable behavior for individuals and groups. The complexity of environmental problems makes finding consensus difficult, but many effective national and international policies protect human society and the environment today. New or revised policies are needed to address ongoing and new environmental problems.

1. What causes stratospheric ozone depletion, and why should it be addressed?
2. What are environmental policies, and why do we need them?
3. How has environmental policy evolved in the United States?
4. How are policies developed and administered in the United States?
5. What factors influence policy decisions?
6. What policy tools can be used to implement and enforce environmental policy?
7. How are international policies established and enforced?
8. How has the international community responded to ozone depletion?

It was August 1986—late winter in the Antarctic—and Susan Solomon, an atmospheric chemist, was on her way to investigate a mystery. She and a team of scientists from various institutions were gathering at McMurdo Station for a highly coordinated research expedition to tackle a question of global significance: Why was the ozone layer above the South Pole disappearing?

On this, her very first excursion, Solomon and her colleagues collected the data that would eventually grab the world's attention and resolve an ongoing, hard fought scientific debate that was taking place on an international stage. That debate was not just over what was causing the ozone layer to thin, but also over what exactly humans should do about it.

⊙ **WHERE IS MCMURDO STATION?**

Susan Solomon at McMurdo Station in 1987.

Courtesy of Dr. Susan Solomon

---

1 THE ISSUE: STRATOSPHERIC OZONE DEPLETION

Key Concept 1: Ozone depletion, caused by CFCs and other anthropogenic chemical pollutants, increases our exposure to dangerous UV radiation.

Earth's **atmosphere**, the blanket of gases surrounding our planet, is made up of discernable layers that differ in temperature, density, and gas composition. The lowest level is the **troposphere**. This level is familiar to us: It is where our weather occurs, and it contains the air we breathe.

Most oxygen in the atmosphere exists as O_2 (two oxygen atoms bound together). In contrast, **ozone (O_3)** is a molecule made up of three oxygen atoms. The vast majority of this unusual

atmosphere The blanket of gases surrounding Earth.

troposphere The lowest level of the atmosphere.

ozone (O_3) A molecule made up of three oxygen atoms.

These images show the amount of ozone over Antarctica in 1979 (the first year that ozone satellite images were available) and a more recent year, 2016. The 2016 image shows an ozone "hole" (levels less than 220 DU) covering much of the continent. Prior to 1979, ozone levels below 220 DU were not observed.

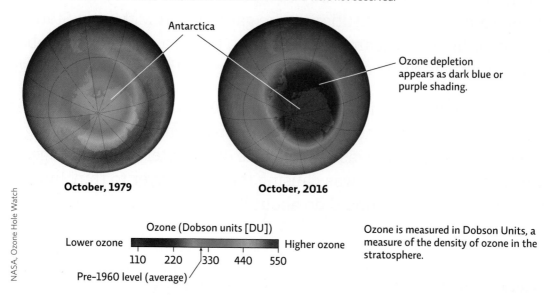

Antarctica

Ozone depletion appears as dark blue or purple shading.

October, 1979   **October, 2016**

NASA, Ozone Hole Watch

Ozone (Dobson units [DU])

Lower ozone   Higher ozone

110   220   330   440   550

Pre-1960 level (average)

Ozone is measured in Dobson Units, a measure of the density of ozone in the stratosphere.

form of oxygen resides in the **stratosphere**, the layer of atmosphere that lies directly above the troposphere. There, ozone remains in a thin layer (the ozone layer) because air within the stratosphere does not circulate like it does in the troposphere. (A form of anthropogenic air pollution known as ground-level ozone is found in the troposphere and should not be confused with stratospheric ozone. See Module 10.1 for more on ground-level ozone pollution.)

Earth's stratospheric ozone layer protects the planet from much of the Sun's **ultraviolet (UV) radiation**—high-energy radiation that is harmful to living things. Excessive exposure, especially of the most dangerous type to reach the surface, UV-B radiation, can cause serious human health problems such as cataracts and DNA mutations that can lead to skin cancer. It also hampers plant growth, interferes with embryonic and larval development, and hinders reproduction in many species, impacts that could disrupt community connections and impair ecosystem functions. **INFOGRAPHIC 1**

Normally, ozone is created and destroyed (alternating between O, O_2, and O_3) in a natural cycle that keeps overall O_3 levels constant. However, O_3 molecules can also be broken down by some industrial chemicals including chlorine-based refrigerants such as chlorofluorocarbons (CFCs) and hydrochlorofluorocarbons (HCFCs), and bromine compounds used in fire extinguishers and pesticides.

stratosphere A layer of atmosphere that lies directly above the troposphere.

ultraviolet (UV) radiation High-energy radiation that is harmful to living things.

These are known as ozone-depleting substances (ODS) because they break down ozone faster than it is formed, which in turn causes the ozone layer in Earth's atmosphere to become less concentrated or too "thin," forming what became known as the "ozone hole."

This breakdown is more likely to happen in parts of the stratosphere above the North and South Poles at the beginning of spring because the chemical reaction in question has two requirements: ice crystals that can act as a platform on which the CFC molecules can be broken down and lots of solar UV radiation. Early spring provides these conditions—the Sun has returned to the area, but it's still cold enough to provide the ice crystals that facilitate the breakdown of CFCs. The hole in the stratospheric ozone layer over the Antarctic is therefore most pronounced during September and October (springtime in the Southern Hemisphere). A similar but less severe hole forms above the Arctic (which is not as cold as Antarctica) during the Northern Hemisphere's springtime.

Because stratospheric ozone is so important to life on Earth, uncovering the cause of ozone depletion so it could be addressed quickly became a quest of international urgency. Several hypotheses were examined to try and determine just what was causing the ozone layer to thin. Solomon tested the hypothesis that CFCs were the culprit. Even though CFC molecules are very stable and do not break down in the troposphere, it was hypothesized that when they eventually reached the stratosphere, exposure to UV radiation could break them apart,

Earth's atmosphere is composed of layers that differ in chemical composition. The stratosphere contains a region with more ozone (O_3) than other parts of the atmosphere—the ozone layer. Ozone is important because it reduces the amount of harmful UV radiation that reaches Earth's surface. A variety of chemicals, such as CF_2Cl_2 (a CFC), destroy stratospheric ozone.

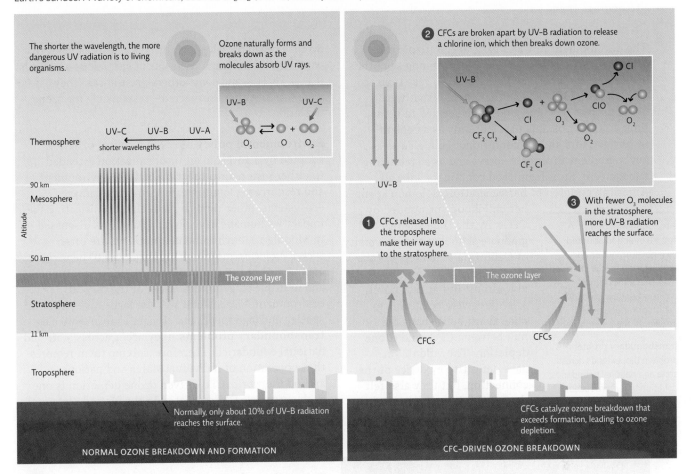

NORMAL OZONE BREAKDOWN AND FORMATION

CFC–DRIVEN OZONE BREAKDOWN

The shorter the wavelength, the more dangerous UV radiation is to living organisms.

Ozone naturally forms and breaks down as the molecules absorb UV rays.

Normally, only about 10% of UV–B radiation reaches the surface.

② CFCs are broken apart by UV–B radiation to release a chlorine ion, which then breaks down ozone.

① CFCs released into the troposphere make their way up to the stratosphere.

③ With fewer O_3 molecules in the stratosphere, more UV–B radiation reaches the surface.

CFCs catalyze ozone breakdown that exceeds formation, leading to ozone depletion.

? Looking at the chemical pathway that illustrates how CFCs break down ozone, which product of the reaction would you look for in the stratosphere as evidence that this reaction was occurring: CF_2Cl, O_2, Cl, or ClO? Explain.

releasing a chlorine atom. This chlorine atom could then break apart ozone, forming chlorine monoxide (ClO) and O_2. This could happen again and again because the ClO molecule would easily react with oxygen atoms in the stratosphere, releasing the chlorine atom that could go on to attack another O_3.

> Because stratospheric ozone is so important to life on Earth, uncovering the cause of ozone depletion so it could be addressed quickly became a quest of international urgency.

Solomon knew that if CFCs were breaking down stratospheric ozone, ClO should be present. And since there was no natural source for ClO in the stratosphere, if she found it, it would be strong evidence that CFCs were causing ozone depletion. Her instruments did, in fact, detect ClO. In 1987 and 1988, Solomon and her colleagues published the studies that would provide the definitive link between CFCs and ozone depletion.

In a way, this was good news—since CFCs were human creations, we could simply stop creating and using them. Fortunately, by the time that research was made public, the wheels of policy were already in motion.

2 ENVIRONMENTAL POLICY: PURPOSE AND SCOPE

Key Concept 2a: Environmental policies protect the environment and human well-being.

Key Concept 2b: When environmental problems transcend national borders, international policies are needed to address them.

The development of CFCs back in the 1930's had actually been a cause for celebration among scientists. These new industrial coolants were far less toxic than the ammonia and sulfur dioxides that were being used at the time, and there seemed to be no limit to their possible uses not only as a coolant in refrigerators and air conditioners but also as a key ingredient in aerosol sprays and as a stabilizing substance in Styrofoam food containers. Like all American children of her generation, Solomon spent her childhood surrounded by CFCs. By the time she reached graduate school at the University of California at Berkeley in the late 1970's, they were truly ubiquitous. But at that point, scientists were beginning to raise alarm bells about the link between CFCs and ozone depletion. They didn't yet have definitive evidence of the connection, but they also felt they couldn't wait for certainty to take action.

environmental policy A course of action adopted by a government or an organization that is intended to improve the natural environment and public health or reduce human impact on the environment.

transboundary problem A problem that extends across state and national boundaries; pollution that is produced in one area but falls in or reaches other states or nations.

So they began pressing legislators to create policies that would limit or eliminate the use of suspicious chemicals (CFCs, HCFCs, and some others), just in case they were—as scientists suspected—responsible for ozone depletion.

Environmental policy refers to a course of action adopted by a government or organization intended to either restore or protect the natural environment and resources—sometimes by repairing damaged ecosystems, other times by reducing or mitigating the impact we humans have on our planet. As discussed in Module 1.1, environmental problems are often *wicked problems*, meaning that they tend to be very complex; they have multiple causes and consequences and multiple stakeholders. Most of the biggest environmental issues that we now face—like pollution, species endangerment, and climate change—are also **transboundary problems**; they occur across state and national boundaries. Therefore, solving them requires the cooperation of individual states and nations around the world. CFC-driven ozone depletion is one such example—CFCs released in the United States, Europe, and other regions were affecting stratospheric ozone over Antarctica and the southern tip of South America.

This reality makes environmental policy tricky, in part because what is best for one group is not always best for another. Policies are especially important when dealing with commonly held resources such as the atmosphere or the oceans. Since no one "owns" these resources, there is a tendency to use them or do things that harm them because no individual or group bears the total cost of the damage. Most successful environmental policies involve compromises, and the agreed-upon choices come with trade-offs. Still, despite these hurdles, there are many successful environmental policies in place—including ones with such far-reaching goals as limiting pollution or habitat destruction, restoring damaged environments, and managing commonly held resources such as rivers or public resources such as national forests in a way that avoids the *tragedy of the commons*. (See Module 1.1.) **INFOGRAPHIC 2**

In 1978, CFC propellants used in aerosol cans like this were banned and replaced with chemicals that don't harm the ozone layer, such as carbon dioxide and nitrous oxide.

As early as the 1970's, scientists involved in laboratory studies on CFCs and ozone began calling publicly for an

INFOGRAPHIC 2 **ENVIRONMENTAL POLICY GOALS**

Public policy in general is a course of action adopted by a government or organization intended to enhance society as a whole.

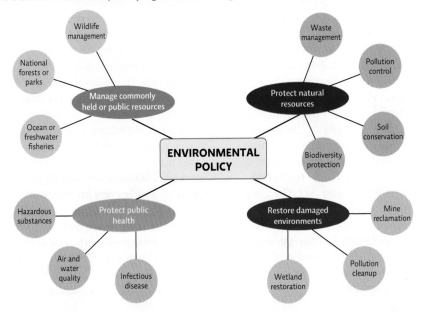

What other policy goals or examples could you add to this concept map?

end to CFC production. Their argument—that society should not wait until the link was definitive before taking some precautions—was a long shot. At the time, there were very few laws on the books designed exclusively to protect the environment, and certainly

none based on an unconfirmed hypothesis. But it turned out that right at that moment, the U.S. federal government's approach to environmental protection was changing dramatically.

3 ENVIRONMENTAL POLICY IN THE UNITED STATES

Key Concept 3: Many iconic U.S. environmental laws originated in the 1960's and 1970's. Some of these laws allow citizens to sue violators to ensure that the laws are properly enforced.

Before the 1960's, environmental issues mostly dealt with how best to use resources. Addressing pollution or environmental damage was not a key objective. And environmental issues were primarily handled at the state level because federal environmental regulations were considered an intrusion on state sovereignty. In fact, most environmental problems were addressed only after the fact, through litigation—an arrangement that too often favored the polluters. It was even more difficult back then than it is today to prove that toxins from a factory or dump that had seeped into the water or permeated the air were killing livestock or causing human illnesses.

Eventually, though, things began to change. Industry grew, and so did pollution. And as it did, environmental problems began

slipping across state lines, so that water and air pollution from one state affected another. In the 1960's and 1970's, a massive national outcry forced federal legislators to acknowledge that more regulation was needed. Slowly, they began to set environmental **performance standards** (also known as pollution standards) by identifying the levels of pollutants allowed to be present in the environment or released over a certain time period—at the federal level. By determining how much pollution could be released in the first place, environmental regulation shifted from after-the-fact litigation to prevention. The shift proved effective: Water and air pollution from industry began to drop.

By 1969, the era of modern environmental policy had begun.

performance standards The levels of pollutants allowed to be present in the environment or released over a certain time period.

That year, the **National Environmental Policy Act (NEPA)** was codified into law. NEPA established environmental protection as a guiding policy for the nation, mandating that the federal government take the environment into consideration before taking any action that might affect it. For example, the National Aeronautics and Space Administration (NASA) commissioned the first of several studies to evaluate the impact of rocket fuel on ozone depletion. In NEPA's wake came a wave of iconic legislation—including the Clean Water Act, the Clean Air Act (which now contains provisions for protecting the ozone layer), and the Endangered Species Act—passed with overwhelming bipartisan support. **INFOGRAPHIC 3**

National Environmental Policy Act (NEPA) A 1969 U.S. law that established environmental protection as a guiding policy for the nation and required that the federal government take the environment into consideration before taking action that might affect it.

citizen suit provision A provision that allows a private citizen to sue, in federal court, a perceived violator of certain U.S. environmental laws, such as the Clean Air Act, in order to force compliance.

Many of the environmental laws passed in the 1970's also have a mechanism that allows individual citizens or groups (including state governments) to demand enforcement—the **citizen suit provision**. Violations can be reported, and if they aren't dealt with in a timely or satisfactory manner, the citizen or group can file a lawsuit against the violator (an individual, a private company, or the government—even the regulatory agency mandated to enforce regulation) that has allegedly failed to uphold the existing law. While there have been many victories for those filing citizen lawsuits, even more have ended in rulings against the citizen plaintiff, and many others never make it to court. It is for this reason that the most successful lawsuits are filed by professional organizations with staffs of lawyers knowledgeable about the process.

U.S. environmental laws have been successful at improving the environment—our air is cleaner, our water is safer, and our wild areas and wildlife are better protected than before these laws were passed. In fact, these laws became models for environmental legislation in countries around the world.

INFOGRAPHIC 3 **NOTABLE U.S. ENVIRONMENTAL LAWS**

Many landmark U.S. environmental laws were passed, beginning in the 1960's, during a period of tremendous bipartisan cooperation and support for taking steps to ensure a clean and healthy environment.

rypson/iStock/Getty Images

LAW	DESCRIPTION
National Environmental Policy Act (1969)	Mandates that the federal government take the environment into consideration before pursuing any federal action that might have an environmental impact.
Clean Air Act (1970)	Regulates the amount of hazardous pollutants that can be present or released into the air. *See Module 10.1.*
Clean Water Act (1972)	Regulates water quality by setting standards for the release or presence of specified toxic or hazardous water pollutants. *See Module 6.2.*
Endangered Species Act (1973)	Protects and aids in the recovery of endangered and threatened species of fish, wildlife, and plants in the United States. *See Online Module 3.3.*
Safe Drinking Water Act (1974)	Protects public drinking water supplies through water quality standards set and enforced by the EPA. *See Modules 1.3 and 6.1.*
Toxic Substances Control Act (1976)	Protects consumers from toxic chemicals in the products they buy by regulating and monitoring industrial chemicals. *See Module 1.3.*
Nuclear Waste Policy Act (1982)	Mandates that the federal government provide for permanent disposal of high-level nuclear waste. *See Module 11.1.*

? Do you feel that these U.S. environmental laws should be strengthened, weakened, or remain as they are?

4   THE POLICY CYCLE: DEVELOPMENT AND ADMINISTRATION OF POLICIES

Key Concept 4: Policy making includes systematically considering all options before setting policy, and evaluating policy after it is implemented.

NEPA did more than commit the federal government to environmental protection—it established a process for generating environmental policy that includes a scientific evaluation of the environmental problem and the consideration of various approaches to address that problem. A decision is then made as to the best course of action. The process itself is responsive and allows for policy revision as new or changing information comes to light. **INFOGRAPHIC 4**

NEPA's signature feature has been the **environmental impact statement (EIS)**—a report that details the likely effect of a proposed federal action, such as building a road or upgrading a nuclear facility. The goal of an EIS is to identify problems before they occur so that stakeholders can choose the most acceptable course of action. To keep the process transparent, the findings are made public, and everyone is given a chance to respond (through letters and public hearings). It was an EIS, released in 1988, that first evaluated the health, societal, and environmental effects of potential regulatory actions to address ozone depletion. This document gave policy makers the information they needed to formulate useful and appropriate policies to protect atmospheric ozone.

In 1970, Congress established the U.S. **Environmental Protection Agency (EPA)** to implement and enforce all the new federal environmental laws that were then being passed. The EPA and other agencies (such as the U.S. Department of Agriculture) follow the same NEPA process to establish rules and regulations in support of environmental laws. These agencies are administered by the executive branch, funded by the legislature, and subject to judicial oversight.

The EPA sets rules or standards (such as performance standards) that ensure that the goals of any given law are met. They are also tasked with holding individual states and corporations accountable. The EPA works with states and industry to help them achieve compliance, but if a given entity fails to comply with a given rule, the EPA has the authority to step in and mandate changes. It can, for example, force a power plant to make upgrades that decrease pollution, close a factory for repeated violations, or fine an individual state for failing to curb its vehicle-generated air pollution. It can also force entities to pay cleanup costs and, in certain cases, can revoke operating permits.

Of course, the EPA's reach extends only to U.S. borders. And as the case of CFCs shows, most environmental problems tend to stretch way beyond those.

environmental impact statement (EIS) A document outlining the positive and negative impacts of any federal action that has the potential to cause environmental damage.

Environmental Protection Agency (EPA) The federal agency responsible for setting policy and enforcing U.S. environmental laws.

INFOGRAPHIC 4   POLICY DECISION MAKING—THE NEPA PROCESS

Policies are created and revised using some basic steps that allow policy makers to systematically evaluate the situation and possible responses. Regulations based on bills that are passed and signed into law are proposed and administered by regulatory agencies (like the EPA) in a similar manner. The process itself is responsive and allows for revision as new or changing information comes to light.

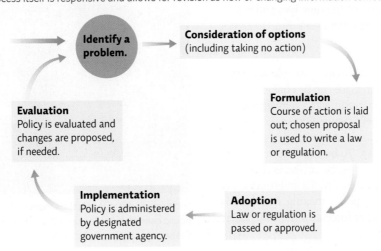

Identify a problem.

Consideration of options (including taking no action)

Formulation Course of action is laid out; chosen proposal is used to write a law or regulation.

Adoption Law or regulation is passed or approved.

Implementation Policy is administered by designated government agency.

Evaluation Policy is evaluated and changes are proposed, if needed.

 What problems can arise if the language in a U.S. law that the EPA is supposed to enforce is vague or ambiguous?

5 FACTORS THAT INFLUENCE POLICY FORMULATION

Key Concept 5: Sound science is needed for effective policy decision making; political lobbies, public opinion, and the press also strongly influence the process.

Eventually, the concerns raised by scientists over CFCs led to a ban on the chemicals in some aerosol sprays and a few other products in the United States. But when it came to further action—i.e., banning the chemicals completely—industry dug its heels in. The chemicals were in wide use by then, in many different industries, and they generated billions of dollars in profit for the chemical companies that made them and the manufacturers who used them. These business entities used their money and influence to argue against rushing to judgment. They insisted that there was far too little evidence to get rid of something so ubiquitous and important to the U.S. economy. The science community disagreed.

Depending on high-quality science—i.e., peer-reviewed studies with strong scientific consensus—is crucial to good policy. Basing policy on what one would like to be true or on what would garner profit or political advantage to policy makers will not solve societal problems and may create new ones. As Barbara Schaal, President of the American Association for the Advancement of Science writes, "...science is not a political construct or a belief system." By its very nature, the way scientific information is gathered (hypothesis testing) and reported (peer-reviewed publications) has built in checks against bias. Schaal points out that society, not scientists, must decide how to address the problems it faces, but science should be the basis for those policies.

In an ideal world, all policies would be based on sound science and established need. Legislators would receive scientific information about a proposed policy from unbiased scientific sources and use this to write bills. Once passed and signed into action, the judiciary would weigh in on constitutionality of the new law, and rules to implement that law would be quickly set. But that's not quite how it works.

The science is often hard to distill into feasible actions, and getting timely policies in place can be challenging with policy making often proceeding at a snail's pace. As Sir Peter Gluckman, Chief Science Advisor to the Prime Minister of New Zealand, writes, "policy making is messy." To be sure, the scientific, legislative, executive, and judicial steps are a crucial and routine part of any policy making. But there is another force at work, too: political lobbying.

political lobbying Contacting elected officials in support of a particular position; some professional lobbyists are highly organized, with substantial financial backing.

In the United States and even on the international stage, **political lobbying**—contacting elected officials in support of a particular policy position—is part of the democratic process. We have access to our elected officials and can share our opinions with them. Citizens and private organizations (e.g., nonprofits, labor unions, and industry groups) lobby for or against specific proposals, based on their own interests. Critics say that professional lobbying has grown alarmingly sophisticated and well financed, making individual voices harder to hear and potentially interfering with policy makers' judgment. Not only do industries run ad campaigns promulgating ideas that serve their own best interests ("clean" coal, for example, is not as clean as it sounds; see Module 9.1), they also contribute large sums of money to candidates for elected office, hoping to influence those candidates, if elected, to act in ways favorable to industry.

Nonprofit organizations like the Natural Resources Defense Council or the Sierra Club also have professional lobbying divisions that promote their positions to elected officials and the general public. Taken together,

Lobbyists and state legislators meet with one another outside the house chambers in the main lobby of the Kansas Statehouse prior to the session being called to order.

Mark Reinstein/Corbis via Getty Images

these environmental nonprofits spend tens of millions of dollars per year in federal lobbying efforts—but their investments are still dwarfed by the hundreds of millions of dollars spent by industry.

The regulation of toxic chemicals, for instance, is sometimes more heavily influenced by industry pressure than scientific evidence. The tobacco industry argued successfully for years that smoking was not linked to health problems, despite ample evidence to the contrary. And in a 2006 *New York Times* article, an EPA pesticide analysist complained (anonymously, citing fear of retribution) about industry's influence over policy, saying, "You go to a meeting, and word comes down that it is an important chemical, this is one we've got to save. It's all informal, of course. But it suggests that industry interests are governing the decisions of the EPA."

Of course, it doesn't necessarily take money to influence policy. Citizens and grass roots movements can influence Congress by calling their representatives and making their opinions known or by voting with their dollars (refusing to purchase products with impacts they don't want to support). For example, plastic bottles used to contain a chemical known as bisphenol-A (BPA) that was known to be an endocrine disruptor (it acted like a hormone in the body)—the worry was that exposure to this substance could interfere with development. Parents, physicians, scientists, and environmental groups called for a ban on BPA while industry touted its long history of safe use and lack of evidence of direct harm. The wheels of policy turned slowly but when major retailers such as Whole Foods and Wal-Mart announced they would no longer sell the product, industry responded. Bottle plastics were reformulated to exclude BPA (though its replacement is probably not much better) a full 4 years before a U.S. ban of the product was announced. This example illustrates that no matter how intense the professional lobbying—or for that matter, how sound the science or how well established the need for a given policy—a strong public voice can influence policy. **INFOGRAPHIC 5**

INFOGRAPHIC 5   **INFLUENCES ON U.S. ENVIRONMENTAL POLICY DECISION MAKING**

Many organizations and individuals influence not only whether we institute a policy to deal with an environmental issue but also the design of that policy—what it covers and how it will be implemented and enforced. The wide variety of voices, many representing differing viewpoints, can make it difficult to create new policies. Though political ideologies might influence how one goes about addressing a problem, policy makers ideally look to the best available science when making decisions about whether a policy is needed to protect the health and well-being of the public and environment.

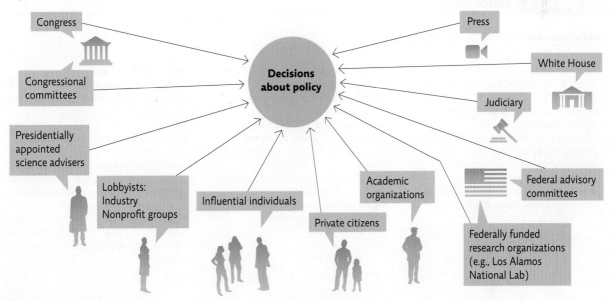

 Based on how you see the process playing out in the United States today, rank the parties that influence U.S. policy making from "most influential" to "least influential." In your opinion, is this ranking as it should be?

6 IMPLEMENTATION AND ENFORCEMENT OF POLICIES

Key Concept 6: Policy can be enforced with command-and-control regulation or through economic incentives that favor preferred responses.

command-and-control regulation A type of regulation that involves setting an upper allowable limit of pollution release that is enforced with fines and/or incarceration.

green tax A tax (a fee paid to the government) assessed on environmentally undesirable activities (e.g., a tax per unit of pollution emitted).

tax credit A reduction in the tax one has to pay in exchange for some desirable action.

cap-and-trade program Regulations that set an upper limit for pollution emissions, issue permits to producers for a portion of that amount, and allow producers that release less than their allotment to sell permits to those who exceeded their allotment.

In general, there are a range of *policy tools* that are used to enforce environmental laws. Governmental **command-and-control regulation** is one approach. This option can take several different forms: issuing permits to authorize operation of equipment or facilities that may pose risk to the environment, establishing standards that regulate emissions and specify the practices that must be used to meet those standards, or simply banning the use of a given substance like CFCs.

Alternatively, market-based approaches create economic incentives for the private sector to reduce environmentally harmful actions, without dictating exactly how to reach a desired target. Economic incentives can be punitive or reward based. For example, governments can levy taxes based on the amount of pollution produced—so-called **green taxes**. Or they can reduce taxes by offering **tax credits** to consumers or businesses that pursue environmentally friendly actions—such as buying a hybrid vehicle or installing energy-efficient appliances.

Another market-based approach is **cap-and-trade** (also known as permit trading), a policy that involves setting limits on the emission of certain pollutants (greenhouse gases, for example), distributing permits for allowable emissions, and then letting individual businesses freely trade (or sell) those permits if they release fewer emissions and have some of their allotment left over. The total pollution allowed by all the permits falls under a maximum cap for that pollutant. Caps can be lowered over time to bring down the level of pollution as needed. An advantage of this approach is the freedom for plant operators to choose which avenue to compliance works best for them—reducing actual emissions or purchasing additional permits. (See Infographic 7 in Module 10.1.)

A cap-and-trade program successfully reduced acid deposition in the eastern United States by setting a

INFOGRAPHIC 6 POLICY TOOLS

COMMAND-AND-CONTROL REGULATION

ADVANTAGES
- Simple in concept and may achieve desired goals quickly
- Directly changes the behavior of the regulated industry
- Especially effective policy tool when the potential for severe environmental or health impact is high, such as with extremely toxic substances
- Useful when level of control is known and uniform across all regulated industries

DISADVANTAGES
- Making changes to policy takes time, which makes it hard to keep up with new technologies.
- A one-size-fits-all approach may limit some industries' ability to use the most cost-effective methods to address their impact.
- No incentive for companies to reduce pollution below mandated limits
- Regulatory agencies must have sufficient funding to enforce compliance.

IMPLEMENTATION

Permits: Authorization required for operation; permits specify acceptable actions or environmental releases

Performance and technology standards: Impose emission limits for a given source, identify ambient levels of pollutants that are acceptable, and/or specify the technology and methods that must be used to reduce pollution

Penalties: Lost contracts, jail time, paying for clean-up or damage, fines, liability that requires violators to compensate others for harm or damage

 In general, which approach do you support to reduce our use of fossil fuel and other activities that contribute to climate change: command-and-control regulation or economic incentives? Explain.

cap on sulfur emissions from coal-fired power plants, issuing permits for the release of sulfur air pollution, and gradually lowering the cap. Smokestack "scrubbers" were developed and installed to capture the sulfur pollution before it was released to the air, and the technology spread rapidly. The policy goal of reducing sulfur pollution by half, to about 8 million metric tons per year, was reached ahead of schedule and then exceeded, even as electricity production increased. (In 2014, 2.9 million metric tons of sulfur emissions were released.) The successful program also came in at a much lower price tag than predicted, demonstrating the power of market-driven approaches to pollution control.

Financial incentives such as **subsidies**, grants, and low-interest loans can also encourage environmentally beneficial actions that might otherwise be hard for individuals or businesses to afford. For example, many nations gave out subsidies to businesses and consumers to help them transition away from ozone-depleting substances to more ozone-friendly replacements (e.g., replacing CFC-based air conditioning with chlorine-free alternatives).
INFOGRAPHIC 6

In deciding which of these tools to use for any given policy, policy makers must consider several factors, including whether the tool will actually help attain the desired environmental goal, who will bear the financial cost and other burdens of the policy if this tool or that

one is used, and how adaptable the tool itself will be to changes in the underlying policy. They must also establish clear methods of enforcement: Someone has to oversee the implementation of a given policy tool and hold the relevant parties accountable (by assessing fines or penalties, for example, or by revoking privileges—or even in some cases by charging someone with a crime)—in a predetermined and straightforward manner.

subsidies Financial assistance given by the government to promote desired activities.

Houses with solar panels, like these in Baden-Wuerttemberg, are common in Germany, thanks to governmental policies that offered financial incentives such as fixed payments for power produced and subsidies to purchase the units.

ECONOMIC INCENTIVES

ADVANTAGES
- Fund actions that otherwise might be too costly for individuals or business to afford
- Taxes generate revenue that could benefit environmental causes.
- Provide incentives to reduce pollution below requirements
- Encourage innovation for reducing environmental impacts
- May be more cost effective than command-and-control regulation
- Stimulate the economy

DISADVANTAGES
- Citizens may oppose tax dollars being used for endeavors they do not support.
- The cost of green taxes levied against industry may be passed on to the consumer.
- Cap-and-trade programs can create pollution hotspots in areas where most users choose to buy credits rather than reduce emissions.
- May inadvertently support undesirable activities (perverse subsidies)

IMPLEMENTATION
Green taxes: The *polluter pays* principle: taxes are levied on an environmentally harmful action (e.g., per pollution unit emitted).

Subsidies: Financial assistance: Includes things like cash transfers, lower costs for resources, and tax credits (reduction in one's tax because of environmentally favorable actions).

Grants or low-interest loans: Money to support the purchase of environmentally beneficial products such as solar panels

Tradable permits (cap-and-trade): Individual industries can trade or sell their allotment of emissions if they release less than their permit allows; all permits fall under a maximum cap for that pollutant or the pollutant in that area.

7 INTERNATIONAL ENVIRONMENTAL POLICY

Key Concept 7a: International policy is established through treaties that range from a simple agreement that action is needed to protocols that specify procedures and targets for participants.

Key Concept 7b: Enforcing compliance for international treaties is difficult but may include incentives such as technological or financial assistance or disincentives such as economic sanctions.

Even as the tide started to turn against the use of CFCs by various industries, industry itself had one powerful argument against a full CFC ban. If U.S. companies were forced to curb their use of these chemicals while other countries still used them, not only would the ozone layer not be saved, but American industry would lose its competitive edge as they scrambled to come up with replacement chemicals. For the ban to achieve its desired goal of protecting the ozone layer, industries in other countries needed to be held to the same standard.

The efforts of individual states and countries are essential to environmental protection. But ultimately, global problems require international efforts. International agreements go by a litany of different names: treaties, conventions, accords, platforms, and so on. A **convention** is an international agreement that represents a position on an issue and identifies general goals that the signing countries agree to pursue, e.g., the protection of species or the wise use of wetlands. To lay out the specific actions that will be taken as well as the time frame over which these actions must occur, a **protocol** is often then drafted to set precise goals and targets.

International treaties are drafted at meetings, often with intense negotiations, and signed by the attending representatives. The treaty must then be ratified by each nation; in the United States, new treaties are ratified by the U.S. Senate; amendments or subsequent agreements can be signed by the President (though the Senate sometimes disagrees with the President as to what constitutes a "new" treaty). For multilateral treaties involving many nations, the draft may identify a minimum number of nations that must ratify it before it "enters into force"; this could take one or more years. There are currently more than 2,500 international agreements that pertain to the environment: They cover things like whaling (where and how and how much it can be done), endangered species protection (marking certain habitats off limits to loggers or banning the sale of ivory, for example), and the conservation of global resources like the atmosphere. **INFOGRAPHIC 7**

For any of these agreements to work—for the environment to be protected—signatory parties (countries that sign on) must keep their word and do what they say they will do. In other words, they must comply with the terms laid out in the agreement. Achieving compliance is even trickier at the international level than it is for individual countries or states. To be sure, international agreements often include mechanisms for tracking participants' performance around a certain goal or promise and to catch failures of compliance early on. And most of them also have dispute settlement mechanisms in place that may include conciliation (reducing targets or goals), negotiation to redefine responsibilities, or binding arbitration. But few agreements have a robust capacity to enforce the established standards or punish the participants who fall short of their promises. Incentives can help the participants reach their goals, and disincentives such as penalties, stricter surveillance and report requirements, trade sanctions, and even suspension of certain relevant privileges can help get participants back on track when they fall short. But those efforts may be ineffective if a country is not fully invested in the goals at hand. Including harsher penalties at the outset is difficult because it may mean that fewer countries ultimately sign on to the agreement at all.

convention An international agreement that represents a position on an issue and identifies general goals that the signing countries agree to pursue.

protocol A document that sets precise goals and targets.

TIMOTHY A. CLARY/Getty Images

U.S. Secretary of State John Kerry holds his granddaughter while signing the Paris Agreement at the United Nations General Assembly Hall on April 22, 2016 in New York to signify the importance of climate change policies to future generations.

INFOGRAPHIC 7 **INTERNATIONAL AGREEMENTS**

More than 2,500 international environmental treaties are currently in place, covering a wide range of topics from protecting wetlands and species to the use or disposal of hazardous materials.

mizoula/iStock/Getty Images

SIGNED/ ENTERED INTO FORCE	TREATY	DESCRIPTION
1971/1975	Ramsar Convention on Wetlands of International Importance *(See Online Module 3.3.)*	Promotes the conservation and wise use of wetlands; protects vital biodiversity habitat and water resources
1973/1975	Convention on International Trade in Endangered Species of Wild Fauna and Flora (CITES) *(See Online Module 3.3.)*	Regulates the sale and trade of endangered or threatened species or products
1987/1989	Montréal Protocol	Oversees the phase-out of ozone-depleting substances to protect stratospheric ozone
2001/2004	Stockholm Convention on Persistent Organic Pollutants *(See Module 1.3.)*	Protects human health and the environment from persistent, dangerous chemicals such as DDT and PCBs
1992/1994	UN Framework Convention on Climate Change *(See Module 10.2.)*	Established a framework for addressing climate change
1997/2005	Kyoto Protocol *(See Module 10.2.)*	Set binding greenhouse gas reduction targets for developed nations
2015/2016	Paris Agreement *(See Module 10.2.)*	Oversees greenhouse gas reduction targets for parties with a goal of keeping global temperature increase well below 2°C relative to preindustrial levels

 Why is it important that international treaties be revised or amended over time?

8 RESPONDING TO OZONE DEPLETION: THE MONTRÉAL PROTOCOL

Key Concept 8: The Montréal Protocol is successfully addressing ozone depletion thanks to a strong international commitment and the flexibility to update it as needed.

Addressing ozone depletion on the international stage began with the 1985 *Vienna Convention*, a treaty that established the need to respond. This was followed by the 1987 *Montréal Protocol on Substances that Deplete the Ozone Layer*, a protocol that established a planned phase-out of CFCs and other stratospheric ODSs. By 2009, the protocol had been ratified by all nations of the world.

Interestingly, the 1987 Montréal Protocol and the international commitment to address CFCs and ozone depletion came a year before Susan Solomon's final study definitively linking CFCs to ozone depletion was published. We didn't know all the details in 1987, but

because the potential loss of our protective ozone layer was such a serious problem, we chose to take action. This is an example of applying the **precautionary principle**— acting in the face of uncertainty when there is a chance that serious consequence might occur.

As more information poured in, it quickly became apparent that the Montréal Protocol targets would not be sufficient to stop ozone depletion in a timely fashion, so the phase-out dates were moved up for CFCs. Amendments to the Montréal

precautionary principle Acting in a way that leaves a safety margin when the data is uncertain or severe consequences are possible.

INFOGRAPHIC 8   MONTRÉAL PROTOCOL

Actual and projected change over time for total global emissions of ozone-depleting substances (ODS) with and without the Montréal Protocol and its amendments. Adjustments to the phase-out schedule of various ODSs in the form of amendments represent the success of adaptive management in dealing with complex environmental issues. The newest amendment, the 2016 Kigali Amendment, lays out the timeline to phase out or phase down HFCs, replacements for CFCs that turned out to be potent greenhouse gases.

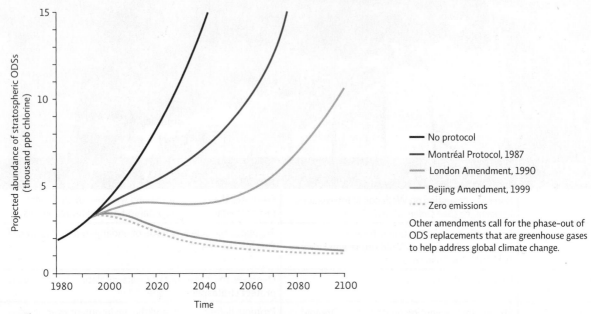

Other amendments call for the phase-out of ODS replacements that are greenhouse gases to help address global climate change.

Looking at the graph that shows the projected decrease in chlorine in the stratosphere, would you say the Montréal Protocol is a success? What role did the amendments play in this success or lack of success? Explain.

Protocol are still negotiated in annual meetings that strengthen the response and adjust the target dates to phase out harmful compounds. This ongoing process is an example of **adaptive management**—allowing room for altering strategies as new information comes in or the situation itself changes. **INFOGRAPHIC 8**

Within a decade or so of the Montréal Protocol's acceptance, scientists began to see evidence that it was working. The rate of ozone thinning slowed down, stabilized, and eventually began to reverse. Though some years are better than others (depending on weather conditions and the availability of ice crystals in the stratosphere), the data now suggest that the ozone hole is finally beginning to "heal," defined as an observable increase in ozone. Based on the recovery to date, scientists are projecting that the Antarctic ozone hole will have fully recovered by 2050. (See the *Scientific Literacy* activity at the end of this module.)

The Montréal Protocol represented the first and arguably most successful international environmental policy ever produced. Another global environmental catastrophe presently unfolding, climate change, has met with less international cooperation. The Paris Agreement, the current protocol for responding to climate change, has been signed by more than

150 nations and "entered into force" in 2016. It has a built-in adaptive management plan that calls for increasing emission reduction targets over time. But there is a problem. While many nations are committed to upholding their pledges to reduce emissions that cause climate change, the weak response of the United States, the second biggest greenhouse gas emitter in the world (China is currently the biggest), jeopardizes this global endeavor. Whether we, as a global community, will be able to agree on and implement effective policies to address climate change is a story still unfolding. Our success at addressing ozone depletion shows it can be done. The question remains: Will we do it?

Select References:

Farman, J., et al. (1985). Large losses of total ozone in Antarctica reveal seasonal ClO$_x$/NO$_x$ interaction. *Nature*, 315(6016), 207–210.

Gluckman, P. (2016, September 2). The science–policy interface. *Science*, 353(6303), 969.

Molina, M. J., & Rowland, F. S. (1974). Stratospheric sink for chlorofluoromethanes: chlorine atom-catalysed destruction of ozone. *Nature*, 249(5460), 810–812.

Schaal, B. (2017, February 3). Informing policy with science. *Science*, 355(6324), 435.

Solomon, S., et al. (1987.) Visible spectroscopy at McMurdo station, Antarctica: 2. Observations of OClO. *Journal of Geophysical Research*, 92(D7), 8329–8338.

Solomon, S., et al. (2016). Emergence of healing in the Antarctic ozone layer. *Science*, 353(6296), 269–274.

adaptive management
A plan that allows room for altering strategies as new information becomes available or as the situation itself changes.

INTERACTIVE MAP | **POLICY IN ACTION**   ANIMATED INFOGRAPHIC

Many U.S. environmental problems are being effectively addressed with national policies that got their start with grass roots action. See the online interactive map for some notable U.S. examples.

DELISTING OF THE BALD EAGLE (ESA)

HUDSON RIVER, NY (SUPERFUND)

MEASURING AIRBORNE LEAD

CLEANING UP THE CUYAHOGA RIVER (CWA)

 BRING IT HOME

PERSONAL CHOICES THAT HELP
The process of writing and revising policy, proposing it, voting on amendments, and finally enacting it as law is a complex and often messy one. The legislative process is often referred to as "sausage making" because of all the steps and input, as well as the fact that the final product often looks much different than the original.

Individual Steps
• Call, email, write, or visit your elected officials to voice your opinion on policy issues of interest to you. Find your national Congressional legislators at www.usa.gov/elected-officials; visit the websites for your state and local governments to do the same.

• Find out who and what is influencing your elected politicians. The website www.opensecrets.org allows you to look up the top individuals and industries that contribute to any candidate's campaign and the top recipients of those contributions.

Group Action
• When an important issue is not adequately addressed, concerned citizens often form petition drives. Signatures are collected and delivered to politicians, who can propose new legislation. Form a group to petition for an important issue that you feel is being overlooked and present the collected signatures to any politician who can propose new policy. Organizations

such as www.change.org can help get your petition out to a wider audience.

Policy Change
• Actions speak louder than words. Visit www.votesmart.org and find the voting record of your representative. How does he or she vote on environmental issues like climate change? If you do not feel your representative's record is moving the country forward, volunteer for another candidate whose policies you support during the next election cycle.

ENVIRONMENTAL LITERACY UNDERSTANDING THE ISSUE

1 What causes stratospheric ozone depletion, and why should it be addressed?

1. True or False: Persistent stratospheric ozone depletion is caused by human-made chemical compounds rather than natural events.

2. Stratospheric ozone is important to life on Earth because it:
   a. produces oxygen needed for living things.
   b. prevents some UV radiation from reaching Earth.
   c. is needed to form clouds that produce rain.
   d. absorbs toxic pollutants like CFCs.

3. What are the potential health and ecosystem consequences to stratospheric ozone depletion?

2 What are environmental policies, and why do we need them?

4. True or False: The primary purpose of environmental policies is environmental protection; human health and needs are a secondary concern.

5. Why are national or international laws and policies necessary to address some environmental issues?
   a. Legislation at the local level does not address important environmental problems.
   b. State or national legislation cannot address environmental problems that cross state or national borders.
   c. State or local legislation is not effective because people won't vote for pro-environment laws.
   d. National and international policies are easier to enforce than state or local policies.

6. Why are environmental policies needed to address issues that are affected by the tragedy of the commons?

3 How has environmental policy evolved in the United States?

7. True or False: Modern U.S. environmental policy requires that environmental impacts be evaluated before federal action is taken.

8. A unique feature of many U.S. environmental laws is the citizen suit provision. Explain this provision. Why is it a useful part of these laws?

4 How are policies developed and administered in the United States?

9. The federal agency responsible for administering U.S. environmental laws is the _____.

10. Which of the following is NOT part of the NEPA process?
    a. Proposing a course of action based on an evaluation of the choices.
    b. Adopting a new law or regulation that addresses the issue at hand.
    c. Considering the consequences of taking no action.
    d. All of the above are part of the NEPA process.

11. What is an environmental impact statement, and why is it produced?

5 What factors influence policy decisions?

12. True or False: Policy making is a straightforward process that is based solely on the best available science and established need.

13. Which of the following political lobby groups spend the most money in a quest to influence policy making?
    a. Academic organizations
    b. Industry
    c. Nonprofit organizations
    d. Private citizens

14. How does political lobbying affect national environmental policy? Do you agree with the critics that political lobbies are too powerful? Explain.

6 What policy tools can be used to implement and enforce environmental policy?

15. A consumer who buys an electric car can get a reduction in his or her income tax that year. This is an example of:
    a. a green tax.
    b. command-and-control regulation.
    c. a tax break.
    d. a federal grant.

16. Give an example of a market-driven approach to solving environmental problems. How does this differ from command-and-control regulation of environmentally damaging behavior?

7 How are international policies established and enforced?

17. True or False: International policies are mainly enforced through voluntary compliance of the nations involved.

18. Which of the following types of international agreements typically lays out a very specific plan for meeting the goals agreed upon by the signatory nations?
    a. Convention
    b. Pact
    c. Protocol
    d. Treaty

19. Effective international environmental policies:
    a. allow for revision based on science or changing needs.
    b. are simpler to implement than national policies.
    c. benefit only a small number of nations and interest groups.
    d. focus on the causes, not the consequences, of environmental issues.

20. Which treaty would you identify as the foundation of international policy on climate change: the UN Framework Convention on Climate Change (UNFCCC) or the Kyoto Protocol? Explain your reasoning.

8 How has the international community responded to ozone depletion?

21. True or False: *Adaptive management* focuses on finding a solution that seems best and sticking with it.

22. When is it reasonable to invoke the *precautionary principle* when setting policy?

SCIENCE LITERACY   WORKING WITH DATA

The size of the ozone hole has been tracked since 1979 by a variety of methods. Two of those data sets are shown in the graph below: measurements taken by the Total Ozone Mapping Spectrophotometer/Ozone Monitoring Instrument (TOMS/OMI) and simulated data produced by the Chem-Dyn-Vol model that takes into account the chemical, dynamical processes, and volcanic eruptions believed to affect stratospheric ozone.

SIZE OF OZONE HOLE IN SEPTEMBER

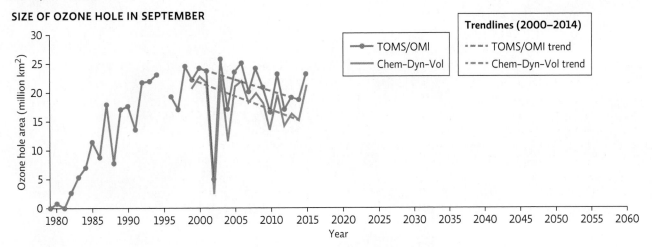

Interpretation

1. For what years are data for TOMS/OMI observations shown? For what years was the Chem-Dyn-Vol model simulation run?

2. Why is there no line connecting the TOMS/OMI data point for 1994 to 1996?

3. What trend was seen between 1989 and 2000 regarding the size of the ozone hole? What trend was seen after 2000?

4. Extend the trendlines out (following the same slope of the line or trend) and estimate the year that each line predicts the ozone hole will be gone.

Advance Your Thinking

5. Why is data shown for the month of September for each year?

6. How well does the model simulation (the red line) match the observed data (the blue line)? What does this tell us about the validity of the model—does it appear that researchers have identified the relevant parameters that affect ozone depletion?

7. What is the purpose of creating a computer model such as Chem-Dyn-Vol?

INFORMATION LITERACY   EVALUATING INFORMATION

The Montréal Protocol has been amended several times as new information comes to light regarding the proposed bans of ozone-depleting substances. The 2016 Kigali Amendment calls for the phase-out of hydrofluorocarbons (HFCs).

Answer the following questions after learning about the basics of the Kigali Amendment. Two good starting places are the EPA website at www.epa.gov/ozone-layer-protection/recent-international-developments-under-montreal-protocol (or search for Kigali Amendment on the EPA home page: www.epa.gov) and the *Nature* article, "Nations agree to ban refrigerants that worsen climate change" at http://www.nature.com/news/nations-agree-to-ban-refrigerants-that-worsen-climate-change-1.20810. Consult other sources as needed.

1. Why were HFCs originally developed (what is their industrial use)?

2. Why are HFCs being targeted for phase-out? What is the expected advantage to global climate change if HFCs are phased out?

3. What is the timeline for HFC phase-out?

Next, conduct Internet searches to find articles about the proposed HFC phase-out from industry point of view (e.g., the chemical industry or the heating and air conditioning [HVAC] industry) and an environmental point of view (e.g., nonprofit environmental groups such as the National Resources Defense Council or the Environmental Defense Fund). Evaluate the response from one industrial source and one nonprofit environmental source by answering the following questions for each:

4. Identify the information source(s) you visited for this information. What position does each take on the phase-out of HFCs? What inherent bias might each group have regarding HFCs?

5. Evaluate the quality of evidence (peer-reviewed scientific studies? newspaper articles? government websites? etc.) offered by each entity.

After learning about HFCs, the Kigali Amendment, and various viewpoints on this topic, draw your own conclusions about the soundness of the proposed phase-out of HFCs. Write a short paper (1–2 pages) that addresses this question: Do you support the Kigali Amendment? Provide evidence to back up your position. Provide in-text citations and a bibliography at the end of your paper.

Additional study questions are available at SaplingLearning.com.

A PLASTIC SURF

Are the oceans teeming with trash?

Trash litters coastal waters near the Vietnamese island of Phu Quoc. Millions of tons of solid waste find their way to the world's oceans every year.
jacus/iStock/Getty Images

After reading this chapter and studying the KEY CONCEPTS and INFOGRAPHICS, you should be able to answer these GUIDING QUESTIONS

CORE MESSAGE

We are often unaware of the waste we produce, but it profoundly affects the environment. Waste that cannot be reused or recycled simply accumulates; some of it is toxic or otherwise disruptive to living things. We can address its impact by minimizing the waste we generate in the first place, recovering and recycling waste materials, and disposing of all waste safely.

1 What types of solid waste do we produce and why do we say waste is a "human invention"?

2 What is municipal solid waste (MSW) and what are the types and proportion of waste in the U.S. MSW stream?

3 What is the EPA's preferred hierarchy of preferred MSW disposal methods and how well does the United States meet this recommendation?

4 Compare the design, advantages, and disadvantages of open dumps and sanitary landfills.

5 What are the pros and cons of waste incinerators?

6 What problems does solid waste cause?

7 What are some common household hazardous wastes and how should individuals deal with this waste?

8 What is composting and how can it help us in our quest to deal with solid waste?

9 What are the four Rs of waste reduction?

From the deck of the *SSV Corwith Cramer*, the surface waters of the North Atlantic looked like smooth, dark glass. The 134-foot oceanographic research vessel had set out from Bermuda just 24 hours before on a month-long expedition aimed at tracking human garbage across the ocean.

It was a cool mid-June evening, the Sun was setting, and the crew had just launched its first "net tow" of the trip. Giora Proskurowski, the expedition's chief scientist, stood watching as the tangle of mesh skidded along the water's surface, keeping pace with the ship as it went. It was hard to imagine that any garbage would be found in such a flat, serene seascape.

After 30 minutes, crew members pulled the net—dripping with seagrass and stained red with jellyfish—from the water. Sure enough, when Proskurowski got closer, he could make out scores of tiny bits of plastic glistening in the mesh. The crew members counted 110 pieces, each one smaller than a pencil eraser and no heavier than a paper clip. Based on the size of the net and the area they had dragged it over, 110 pieces came out to nearly 100,000 bits of plastic per square kilometer (about 260,000 bits per square mile).

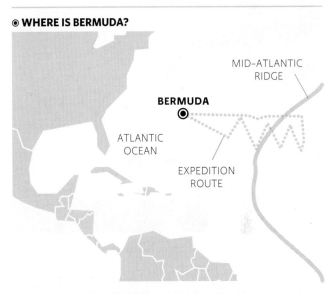

⊙ WHERE IS BERMUDA?

MID-ATLANTIC RIDGE

BERMUDA

ATLANTIC OCEAN

EXPEDITION ROUTE

Team members ready a neuston net for deployment.

Courtesy Giora Proskurowski, Sea Education Association

Giora Proskurowski on the *Thomas G. Thompson*.

is delivered to a gyre by ocean currents becomes trapped and cannot escape the stronger circling currents. When scientists first evaluated all the data from the Atlantic Gyre, they discovered high densities of plastic—more than 100,000 pieces per square kilometer (about 0.4 square mile)—across a surprisingly large "high-concentration zone." The press and general public would come to know this region as the "Great Atlantic Garbage Patch," cousin to a "Great Pacific Garbage Patch" discovered around the same time.

Scientists knew where the plastic was coming from—seagoing vessels, open landfills, and litter-polluted gutters around the world. Most of it comes from Asia, where garbage "pick-up" is not provided in many areas. In cities of Indonesia, China, and the Philippines, just to name the three biggest contributors, rivers become the disposal method of choice for people with no other choice—rivers so choked with plastic trash that they often become blocked, flooding nearby streets and homes with putrid and hazardous water.

Scientists also understood why it was being trapped in the gyres, but other questions remained. How much plastic was in the oceans and where was it located—just near the surface or deeper down? Were the toxic substances found in plastic accumulating up the food chain, making their way into fish, birds, or even our own diets? Were all those tiny bits of solid surface providing transport for invasive species?

To answer these questions, expeditions like this were spanning out across the world's oceans to sample, measure, and conduct experiments. And they were beginning to get some answers.

Despite the water's pristine look, the yield was hardly surprising. In the past 20 years, students on expeditions like this one had handpicked, counted, and measured more than 64,000 pieces of plastic from some 6,000 net tows. That might not sound like much, given the vastness of the Atlantic and the smallness of the plastic. But the tiny bits were gathering in very specific areas, known as *gyres*. Gyres are regions of the world's oceans where strong currents circle around areas with very weak, or even no, currents. Lightweight material, like plastic, that

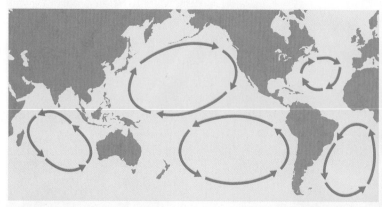

Five major gyres are found in the world's oceans, and there are floating bits of plastic in all of them. These "garbage patches" are not floating islands and aren't even necessarily visible when gazing at the water. Much of the debris is very small and lies just below the surface. The highest density captured in one of the 6,100 sampling tows was approximately 200,000 pieces per square kilometer. It is hard to estimate the size of any garbage patch since the material is so spread out and may reach down to depths of 20 meters (65 feet).

1 WASTE: A HUMAN INVENTION

Key Concept 1: Solid waste can be said to be unique to human societies because in nature there is no "waste": The discarded matter of one organism becomes a resource for another.

In natural ecosystems, there is no such thing as **waste**. In a circular pathway, matter expelled by one organism is taken up by another organism and used again. Forms of matter that are dangerous to living things (think arsenic and mercury) tend to stay buried, deep underground, and are released only during extreme events like volcanic eruptions.

Human ecosystems are another story. We tend to handle matter in a linear manner—matter is extracted, transformed, used, and then discarded, often in ways that make it hard for us to reuse or for nature to reclaim. By taking matter out of the reach of organisms that can use it, we continually disrupt this natural cycle. We do this by converting usable matter into synthetic chemicals that can't easily be broken down and by burying readily degradable things in places and under conditions where natural decomposition processes can't run their course.

One solution to keeping garbage out of the ocean is to go back to the beginning, before the things we throw away are even made. By conducting a life cycle analysis to assess the environmental impact of every stage of a product's life—from production, to use, to disposal—more and more companies are trying to reduce the amount of waste generated by the things they design, make, and sell. Cradle-to-cradle analysis takes this even further as it tries to increase reuse potential and turn *waste* back into *resource* (see Module 5.1).

Part of this shift has been spurred by legislation. **Take-back laws** require manufacturers to take back some of their products, such as computers, after consumers are finished with them. This creates an incentive for manufacturers to design products from which components can easily be salvaged and reused. **INFOGRAPHIC 1**

waste Any material that humans discard as unwanted.

take-back law A law that requires companies to take a product back from a consumer when the consumer is finished with it.

INFOGRAPHIC 1 **SOLID WASTE: A HUMAN INVENTION**

Humans produce a lot of trash but, unlike other species, the way we discard it and the types that we generate make it hard or impossible for nature to reuse or recycle. Our linear system of "TAKE—MAKE—USE—DISCARD" needs to be replaced with a circular system (like that of nature) of "TAKE—MAKE—USE—PUT BACK—USE AGAIN (used by us or other species)."

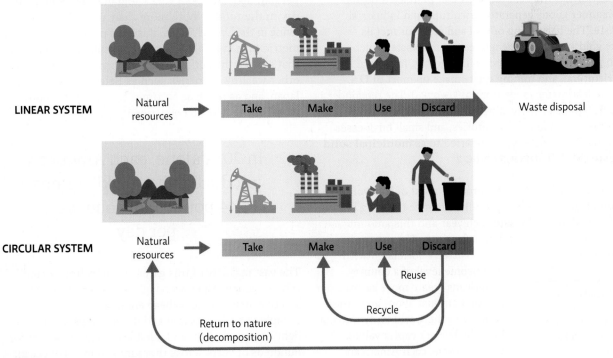

 How does a circular system of production compare to natural matter cycles, such as the carbon cycle? (See Infographic 6 in Module 2.1.)

Not all waste is created equal. Waste that can be broken down by chemical and physical processes is considered *degradable*, even if that degradation takes a long time. Waste that can be broken down by living organisms such as microbes is considered **biodegradable**. Some waste—mostly synthetic molecules like the pesticide DDT and the chlorofluorocarbons (CFCs) once found in aerosols—is considered **nondegradable**. These molecules are chemically stable and don't readily break down. And because they haven't been around for very long, no organism has yet developed (through mutation or genetic recombination) the ability to use them as food.

Different types of plastics have different degradation rates, but for all it is generally quite slow. After all, one of the perks of plastics is their durability. Exposure to sunlight in an oxygen-rich environment causes plastics to chemically change and become brittle, breaking into smaller pieces. On beaches and the surface of the ocean, this breakdown can occur more quickly due to the exposure to sunlight and the additional weathering action of wind and waves, which produces the tiny plastic bits found in the ocean garbage patches.

biodegradable Capable of being broken down by living organisms.

nondegradable Incapable of being broken down under normal conditions.

When the first garbage patch was discovered in 1997 and the press caught wind of what sounded like a giant plastic island in the middle of the ocean, a media frenzy ensued. Some reports said the patch was twice the size of Texas; others claimed that the patch was growing exponentially. But these widely reported claims were misleading, if not downright wrong.

Part of the trouble was semantic. "An oceanographer understands the term 'patch' to mean 'an uneven distribution,'" says Kara Law, a scientist at the Woods Hole Oceanographic Institute who has led several expeditions on the *Corwith Cramer*. "We say the upper ocean is 'patchy' because organisms are often observed in clumps, separated by regions with sparse populations. But when reporters and laypeople hear 'patch,' they incorrectly think of an 'island' or a 'continent' of trash."

2 MUNICIPAL SOLID WASTE

Key Concept 2: Household and business trash contains a variety of items, much of it food or packaging waste. Though this municipal solid waste is a small part of the total waste produced, volumes are still significant and worth efforts aimed at reducing the trash we produce.

Almost any human activity you can think of generates some form of waste. Processes that produce food and consumer goods generate agricultural and industrial waste. The harvesting of coal and precious metals like gold and copper generates mining waste, which can pollute air, water, and soil. Together, industrial, agricultural, and mining waste make up most of U.S. waste (estimates range from 91%–97%). The remainder is produced in the increasingly complicated act of living—in houses, apartments, dormitories, and small businesses around the world—and is referred to as **municipal solid waste (MSW)**. INFOGRAPHIC 2

Currently, urban dwellers produce most of the world's municipal trash—about 2 billion metric tons (2.2 billion U.S. tons) of solid waste each year and this amount is expected to almost double by 2025. The per capita rate of trash production is positively correlated with the income level of a country. Americans create more garbage per person than just about any other country in the world (only Canada has a higher per capita value). In 2014 alone, each American produced about 2 kilograms (4.44 pounds) of solid waste per day, up from about 1.22 kilograms (2.7 pounds) per day in the 1960's. With more than 300 million people living in the United States, this adds up to about 235 million metric tons (258.5 U.S. tons) of household trash per year; that's twice the per capita amount produced by some other wealthy countries such as Belgium, Norway, or Japan, and as much as 10 times the amount produced by most less developed countries.

municipal solid waste (MSW) Everyday garbage or trash (solid waste) produced by individuals or small businesses.

> In 2014 alone, each American produced about 2 kilograms (4.44 pounds) of solid waste per day.

The vast majority of this garbage comes from a familiar array of goods: paper, wood, glass, rubber, leather, textiles, and of course plastic—cheap enough to have become a staple of both advanced and developing societies, light enough to float, and durable enough to persist for hundreds of years across thousands of miles of ocean.

INFOGRAPHIC 2 **U.S. MUNICIPAL SOLID WASTE STREAM**

Municipal solid waste (MSW) is the trash produced by homes and businesses. Though the volume of industrial and agricultural wastes far exceeds the amount of MSW in the United States, we still produce millions of tons of MSW annually. This makes actions that reduce MSW worth the effort.

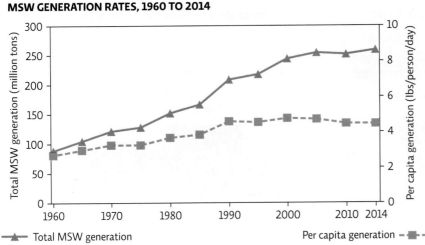

MSW GENERATION RATES, 1960 TO 2014

Total MSW generation

Per capita generation

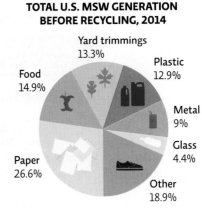

TOTAL U.S. MSW GENERATION BEFORE RECYCLING, 2014

Yard trimmings 13.3%
Plastic 12.9%
Food 14.9%
Metal 9%
Glass 4.4%
Paper 26.6%
Other 18.9%

 Which of the categories of waste shown in the pie chart are naturally degradable? Which could be recycled?

3 DISPOSAL METHODS: WHAT WE SHOULD DO—WHAT WE ACTUALLY DO

Key Concept 3: The preferred solid waste disposal methods, source reduction and recycling, are the least commonly used; the most common methods are the less preferred landfills and incinerators.

"It is with startling accuracy that so many tiny particles of plastic end up in such obscure but well-defined stretches of ocean," Proskurowski wrote one morning, as the *Corwith Cramer* floated under the early summer sun near the Mid-Atlantic Ridge. "Especially given how long and convoluted the journey they took to get here was."

In cities of the United States and other developed countries around the world, collected solid waste makes a similar journey, one that begins when we toss something we no longer need into a trash bag that is then carried from building to curb to garbage truck, before making its way to one of several kinds of waste facilities. Unfortunately, as much as half of the world's population—3.5 billion people—do not have access to any form of waste management. Their garbage is not picked up and hauled off; there are no municipal dumpsters where they can throw their trash. So, they toss it where they can—on the ground or in the nearest river.

Proper solid waste management is crucial to keeping it out of oceans and forests, out of gutters and backyards, and away from disease-carrying insects and rodents. Nations around the world are working to implement effective MSW management programs. Here in the United States, we have a long history (compared to many low-income nations) of dealing with waste and using a variety of ways to deal with our trash. While all the methods have drawbacks, some are more environmentally damaging than others.

All solid waste in the United States is regulated by the 1976 **Resource Conservation and Recovery Act (RCRA)**. As with many federal environmental laws, the Environmental Protection Agency (EPA) is tasked with providing the rules and regulations to administer the law; states are in change of implementing it within their jurisdiction. Citizens also have the

Resource Conservation and Recovery Act (RCRA) The federal law that regulates the management of solid and hazardous waste.

INFOGRAPHIC 3 **MSW DISPOSAL METHODS: RECOMMENDATIONS BY THE EPA**

As indicated in the Environmental Protection Agency's solid waste hierarchy, reducing waste at its source (homes and small businesses) is the top choice in waste management, with landfilling as the last choice. However, more than 50% of our solid waste ends up in a landfill.

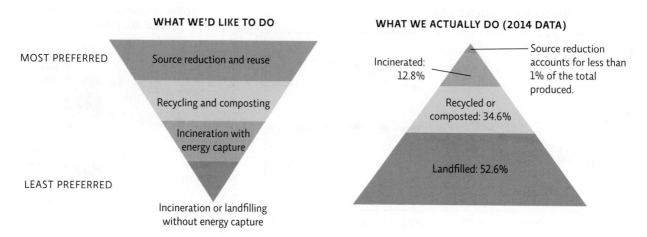

WHAT WE'D LIKE TO DO

MOST PREFERRED

Source reduction and reuse

Recycling and composting

Incineration with energy capture

LEAST PREFERRED

Incineration or landfilling without energy capture

WHAT WE ACTUALLY DO (2014 DATA)

Incinerated: 12.8%

Source reduction accounts for less than 1% of the total produced.

Recycled or composted: 34.6%

Landfilled: 52.6%

? How does the prevalence of "disposable" products and those with short life spans (planned obsolescence) in the United States impact how well we meet the EPA waste management guidelines? What can be done to address this?

right to sue any state or federal agency that they feel is not upholding the statues of the law. (See Module 5.2 for more on the citizen-suit provision of environmental laws.)

The EPA has identified a hierarchy of preferred MSW disposal methods. At the top of that list are ways to prevent materials from ever entering the waste stream, followed by disposal methods that allow for energy capture. Landfills are the last choice. Sadly, we are not meeting those goals: Landfills handle not the smallest percentage of our trash, but the greatest percentage—more than half of our waste. **INFOGRAPHIC 3**

4 SANITARY LANDFILLS

Key Concept 4: Sanitary landfills bury trash in a way that prevents leachate from reaching groundwater but delays decomposition and produces methane, a powerful greenhouse gas.

Illegal in the United States and many other developed nations, **open dumps** are places where trash is simply piled up. Because they are one of the cheapest ways to get rid of human trash, they are common in less developed countries, where entire communities often spring up around the dumps and people survive on what they scavenge from the waste piles. Open dumps attract disease-carrying pests such as flies and rats, a human health hazard. Open dumps also contribute to water pollution: Rain either washes pollutants away from the dump to surrounding areas or pulls it along as it soaks into the ground. If this contaminated water, called **leachate**, continues to travel downward, it can contaminate the soil and groundwater. Flood waters can deliver this contaminated water into streets and homes, exposing people to the pathogens and dangerous chemicals it contains. These open dumps are key contributors of the plastic pollution reaching the sea.

Sanitary landfills, more common in developed countries, seal in trash at the top and bottom in an attempt to prevent its release into the environment. Several protective layers of gravel, soil, and thick plastic prevent leachate from delivering toxic substances to groundwater below the landfill. The trash is covered daily with a layer of soil that reduces unpleasant odors, thus attracting fewer pests.

open dumps Places where trash, both hazardous and nonhazardous, is simply piled up.

leachate Water that carries dissolved substances (often contaminated) that can percolate through soil.

sanitary landfills Disposal sites that seal in trash at the top and bottom to prevent its release into the atmosphere; the sites are lined on the bottom, and trash is dumped in and covered with soil daily.

But there is a downside to landfills. The compacting of trash under a layer of soil excludes oxygen and water so well that the aerobic bacteria (those that require oxygen to live) and other organisms that normally decompose the biodegradable waste can't survive. Newspaper that would degrade in a matter of weeks is preserved in landfills for decades. Anaerobic bacteria (those that live in oxygen-poor environments) pick up some of the slack. But they break down the trash much more slowly and produce lots of methane in the process, a combustible greenhouse gas that contributes to climate change (see Module 10.2). A way to deal with this methane is to capture it to make electricity, turning trash into an energy resource. **INFOGRAPHIC 4**

Landfills also take up a lot of space. This means communities where land is at a premium may have

a hard time finding land to site a landfill. There is also the NIMBY problem—not in my back yard. Environmentally undesirable projects like landfills, incinerators, or other objectionable land uses are often met with strong and vocal opposition. Since few people want waste disposal sites in their community, it is often the disenfranchised—minority or poorer communities—that find themselves saddled with these instillations. The inequitable exposure of minority or low-income populations to actions that degrade their environment is considered an *environmental justice* issue. Taking the proper steps to protect the environment and local communities, no matter what their economic status or racial make-up may be, is key to avoiding instances of environmental injustice. (See Module 4.2 for an introduction to environmental justice.)

INFOGRAPHIC 4 **HOW IT WORKS: SANITARY LANDFILLS**

In a sanitary landfill, an area is dug out and lined to prevent groundwater contamination from leachate; trash is dumped and covered with soil frequently. (This soil may take up to 20% of the landfill area.) Newer landfills have a leachate-collection system built in; older landfills can be retrofitted to collect leachate. Leachate from holding ponds is treated before being released into the environment.

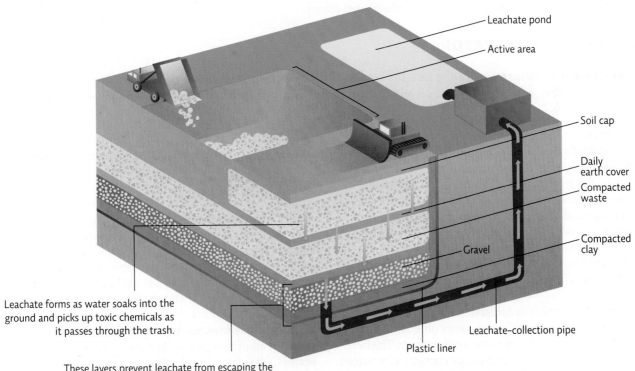

Leachate pond

Active area

Soil cap

Daily earth cover

Compacted waste

Compacted clay

Gravel

Leachate-collection pipe

Plastic liner

Leachate forms as water soaks into the ground and picks up toxic chemicals as it passes through the trash.

These layers prevent leachate from escaping the landfill area and reaching groundwater below.

 Why is the rate of decomposition so slow in a landfill?

At Phnom Penh, Cambodia's municipal garbage dump, people gather plastic, metals, wood, cloth, and paper to sell to recyclers, exposing themselves to dangerous chemicals, air pollution, and disease.

5 WASTE INCINERATORS

Key Concept 5: Incinerators can reduce the volume of trash tremendously and generate electricity in the process but they produce air pollution and the ash is hazardous due to the presence of chemicals in the trash.

A lot of trash—thousands of tons per day—also winds up in specially designed **incinerators**. Burning waste in this way reduces its volume dramatically—by about 80% to 90%. But burning waste that contains plastics and other chemicals releases toxic substances into the air, polluting air and water and producing toxic ash, which must be disposed of in a separate, specially designed landfill. Incinerators are also extraordinarily expensive to build, and tipping fees (fees charged to drop off trash) are usually much higher at an incinerator than at a landfill.

However, in addition to decreasing the volume of trash that must be disposed of, incinerators offer another benefit: converting garbage into usable energy. The heat produced during incineration can be converted into steam energy or used to produce electricity. **INFOGRAPHIC 5**

incinerators Facilities that burn trash at high temperatures.

As dumpsters and landfills fill up, cities and towns begin shuffling their waste from state to state and country to country. New York City,

for example, sends its trash to landfills or incinerators in New Jersey, Pennsylvania, Ohio, Virginia, and South Carolina. Other cities ship trash overseas on garbage barges.

Amid this trash transfer, too much of our waste, especially our plastic, is escaping to the open sea. Some is carried through faulty sewage systems or in the trickling currents of litter-polluted gutters. Much, perhaps most, is illegally dumped—90% comes from Asian countries with insufficient solid waste management programs. Some of it is blown there by aberrant winds from the tops of open landfills. To be sure, not all of it is plastic, but other types of waste—textiles, glass, wood, and rubber—sink or degrade relatively quickly. Plastic just floats along. Eventually, time, saltwater, and sunlight break it down from its recognizable, everyday forms—combs, candy bar wrappers, CD cases—into fragments so tiny that even thousands of them together can't be seen by a naked eye trained on a calm sea.

INFOGRAPHIC 5   HOW IT WORKS: AN INCINERATOR

Trash can be burned at very high temperatures in incinerators (some of which are designed to also generate electricity), but in some facilities, fuel oil must be added to the mix for more complete combustion. In modern facilities, cleaning systems remove particulates, sulfur, and nitrogen pollutants, as well as toxic pollutants like mercury and dioxins. The ash is considered toxic waste and must be buried in the hazardous waste landfills. Municipal solid waste, medical waste, and some hazardous waste are incinerated in the United States.

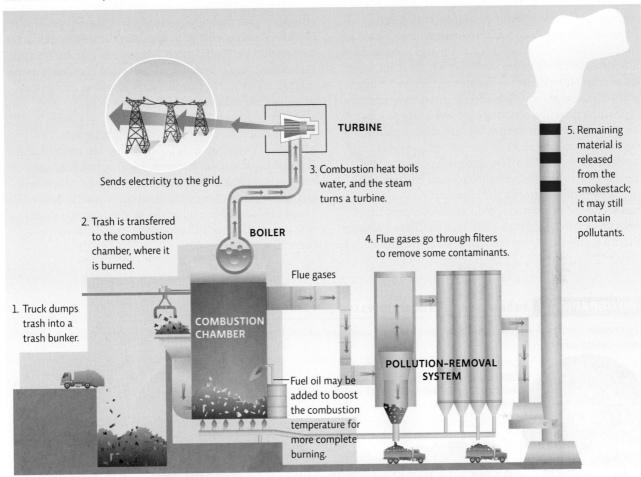

Sends electricity to the grid.

TURBINE

3. Combustion heat boils water, and the steam turns a turbine.

2. Trash is transferred to the combustion chamber, where it is burned.

BOILER

4. Flue gases go through filters to remove some contaminants.

5. Remaining material is released from the smokestack; it may still contain pollutants.

Flue gases

1. Truck dumps trash into a trash bunker.

COMBUSTION CHAMBER

Fuel oil may be added to boost the combustion temperature for more complete burning.

POLLUTION-REMOVAL SYSTEM

What could be done to reduce the toxicity of incinerator ash?

6 IMPACTS OF SOLID WASTE

Key Concept 6: Uncollected solid waste contributes to flooding and air and water pollution. Even MSW that is disposed of using modern techniques traps valuable matter resources in landfills and incinerator ash and contributes to environmental and health problems.

The consequences of mismanaging our trash are manifold. In urban areas of lower-income countries, *uncollected* solid waste is a major problem. Blocked storm drains can cause flooding, which can lead to water pollution and offer stagnant-water habitat to disease-carrying mosquitoes, flies, and rodents. Air is polluted from dust and debris blown from open dumps or from trash burned in open piles. Plastic bags clog outboard motors and make fishing a daily struggle, and in many coastal areas, so many discarded plastics cover beaches, sand is barely visible.

The modern disposal methods used in more developed countries can also contribute to health and environmental problems; incinerators create small-particle air pollution, and landfills produce methane. When disposed of improperly, chemical waste can wreak havoc on plant and animal life.

Aquatic life is especially vulnerable to improperly disposed trash. Sea mammals can get fatally tangled in discarded fishing nets and plastic six-pack rings. On top of that, many fish, sea turtles, and nearly half of all seabirds eat plastic—plastic bags floating in the open ocean look a lot like jellyfish and can choke an animal or fill its gut so that it can no longer eat.

Consumption or exposure to plastics can also be toxic and these effects can be passed along the food chain. Of special concern are the *microplastics* (those less than 5 mm in diameter) that form as plastics are broken down. Some are also released as microbeads such as the tiny polymer beads in facewash scrubs that are washed down the drain daily. There is ample evidence that many marine species ingest or absorb these microplastics, and because many chemicals in plastics mimic the action of estrogen, a reproductive hormone, there are concerns they may affect reproduction success. For example, a study on marine copepods (a tiny crustacean) by Matthew Cole showed exposure to low concentrations of polystyrene microbeads had negative effects on feeding, growth, and reproduction—copepods exposed to the microbeads produced smaller eggs that had a lower hatching success rate. (See the *Science Literacy* activity at the end of this module for more on this study.)

Researchers suspect that floating plastic can serve as an attachment point for fish eggs, barnacles, and many types of larval and juvenile organisms. Thus, each tiny bit of plastic could potentially transport invasive species to new locales. "I think one of the most underrated impacts of these so-called garbage patches is the introduction of hard surfaces to an ecosystem that naturally has very few of them," says Miriam Goldstein, a researcher at Scripps Institute of Oceanography who studies the Pacific patch. "Organisms that live on hard surfaces are very different than those that float freely in the ocean. And adding all that plastic is providing habitat that would not naturally exist out there."

Terrestrial life is also being negatively affected. Microplastics end up in soil where they are eaten by soil organisms such as earthworms. Earthworms raised in soil that contained microplastics had slower growth and higher mortality than controls raised in plastic-free soil. These plastic bits end up in the worms' castings (waste), which may effectively disperse the material throughout the soil. **INFOGRAPHIC 6**

INFOGRAPHIC 6 | **NEGATIVE IMPACTS OF SOLID WASTE**

The negative impacts of improperly managed solid waste are varied and serious, threatening the lives and well-being of humans and other species worldwide.

Uncollected waste can clog storm drains, increase standing water (facilitating mosquito breeding and the spread of disease), and cause streets to flood.

Trash attracts insects, rats, and other species, which can transmit disease.

Improperly handled or stored trash that contains hazardous materials can contaminate air, water, and soil.

Garbage stinks, and litter is an eyesore and may be harmful to terrestrial wildlife.

Plastics in the ocean harm organisms when they ingest plastics, get entangled in them, or are exposed to toxic chemicals released by the plastics.

Uncollected or improperly disposed of garbage is a human health hazard; children suffer more diarrheal and respiratory diseases if household trash is dumped or burned close to home.

What are some of the negative impacts of improperly managed solid waste in your area?

7 HOUSEHOLD HAZARDOUS WASTE

Key Concept 7: Some solid waste, including e-waste, is considered hazardous and must be disposed of properly to reduce health risks and environmental contamination, and to reclaim valuable mineral resources.

A large part of our solid waste stream includes **hazardous waste** (waste that presents a health risk of some kind). Modern households often contain a bevy of dangerous chemicals and products, ranging from household cleaners, paints, and automotive supplies to fluorescent light bulbs and batteries—all items that should not be disposed of with the regular trash. Many citizens do not know how to properly dispose of hazardous waste and simply toss those dead batteries or half-full cans of paint into the trash where they could end up in a landfill or incinerator and possibly contaminate the environment. Avoiding this contamination requires some investigative research on the part of every citizen. Look into the local plan for your community or university so that you can safely dispose of used batteries and chemicals.

A new, modern category of hazardous waste is **e-waste**, discarded electronic devices such as computers, cell phones, and televisions—any device that contains a circuit board. These devices contain significant amounts of precious and rare earth elements (see Online Module 7.3); reclaiming these materials from e-waste can reduce our need to extract them from mines, a process with a huge environmental impact. These substances can then be used in the production of new electronics or for other uses. Organizers of the 2020 Olympics plan to use gold, silver, and copper retrieved from e-waste to make the medals for the Tokyo Olympics just as Vancouver did for the 2010 Winter Olympic games.

Because these electronic devices contain hazardous materials, they should not be thrown out with the household trash. Look for e-waste collection programs in your community when you are upgrading your phone, computer, television, or video game system and consider holding on to your old ones as long as possible to reduce the generation of e-waste. Many countries (including the United States) send some of their e-waste (even e-waste that was collected with the promise of environmentally safe disposal) to countries in Asia and Africa, where impoverished villagers do their best to extract the precious metals within. It's dangerous work. In addition to gold, copper, and zinc, e-waste contains a suite of toxic metals such as lead, mercury, and chromium. When they are released by unsafe and poor extraction methods, these substances cause a wide range of medical conditions in workers or community members— from birth defects to brain, lung, and kidney damage to cancer.
INFOGRAPHIC 7

hazardous waste Waste that is toxic, flammable, corrosive, explosive, or radioactive.

e-waste Unwanted computers and other electronic devices such as discarded televisions and cell phones.

INFOGRAPHIC 7 **HOUSEHOLD HAZARDOUS WASTE**

Hazardous wastes are those that are toxic, flammable, explosive, or corrosive (like acids). Many chemicals that enter your home are actually hazardous and should not be discarded in the regular household trash. The EPA recommends that you contact your local solid waste agency for information on disposing hazardous materials. You can also call 1-877-EARTH-911.

To protect your health and that of your family and the environment, avoid or reduce your use of hazardous chemicals such as:

Drain openers
Oven cleaners
Engine oil and fuel additives
Grease and rust removers

Glue
Pesticides and insecticides
Mold and mildew removers
Paint thinners, strippers, and removers

Other materials that are also considered hazardous and may need to be disposed of as hazardous waste include:

Batteries
Fluorescent lightbulbs
Mercury thermometers
E-waste (cell phones, computers, printers, televisions, video game consoles, etc.)

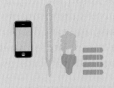

How can you safely dispose of hazardous waste in your community?

In some areas of China, such as this one in Guangdong Province, the process of recovering electronic waste is not done safely and exposes workers and the community at large to toxic substances.

8 REDUCING SOLID WASTE: COMPOSTING

Key Concept 8: Composting organic trash is a waste disposal method that mimics nature—it allows waste to decompose readily and produces a mulch-like product that can be used to return nutrients to soil.

While nondegradable trash like plastics presents a challenge for disposal, much of our waste is biodegradable, and we can apply the concept of biomimicry—emulating nature—to better deal with this part of our waste stream. **Composting**—allowing waste to biologically decompose in the presence of oxygen and water—can turn some forms of trash into a soil-like mulch that nourishes the soil and can be used for gardening and landscaping. Composting can be done on a small scale (in homes, schools, and small offices) or on a large one (in municipal "digesters"). Because they expose biodegradable waste to excellent conditions for decomposition, both small- and large-scale composting can break down organic waste, such as paper, kitchen scraps, and yard debris, very quickly. In a home compost pile or bin, kitchen waste might be turned into a soil-like product in just a few weeks or months, depending on the temperature, amount of moisture, and the number and variety of decomposing organisms at work. Stirring frequently (to keep it well aerated) and adding worms (to break down the material and begin the transformation process) to a home composting bin can speed up the process.

composting Allowing waste to biologically decompose in the presence of oxygen and water, producing a soil-like mulch.

On an industrial level, a municipal composter can take what was once household trash—smelly, full of food waste, packaging, and all the other biodegradable things we throw away—and in 3 days, turn it into compost. This accelerated rate of decomposition is accomplished in a huge, cylindrical, rotating digester that keeps the waste moving (to prevent compaction) and exposes the material to ideal temperature, air, and moisture conditions. Sewage sludge is often added to provide a bacterial boost. The compost is then removed from the digester and piled up to allow it to continue to decompose in a sheltered place. In just a few weeks, it is fully cured and ready to be used on gardens or in landscaping. **INFOGRAPHIC 8**

Composting is a natural way to deal with biodegradable trash but natural ecosystems offer another insight into the problem of solid waste. Just as diversity in an ecosystem enhances its functionality (the fourth characteristic of a

INFOGRAPHIC 8   **HOW IT WORKS: COMPOSTING**

Composting can reduce the amount of a household's trash tremendously. A simple compost pile can be started in the backyard, or a compost bin can be built or purchased.

HOME COMPOSTING

BROWN MATERIALS

Including:
Dead leaves
Paper
Straw/hay
Pine needles
Woodchips

GREEN MATERIALS

Including:
Grass clippings
Food waste
Livestock manure
Tea leaves/bags
Green leaves

MATERIALS TO AVOID

Don't put these wastes in your pile–they won't break down at the same rate and will attract wildlife and pests:

Meat scraps
Bones
Cooking oil
Pet waste

Water

Air

Turn or stir the pile regularly to aerate the pile and promote decomposition.

A variety of compost bins can be used. Small tubs and countertop green bins can be used indoors.

© Alan Marsh/Wave/Corbis

Municipal composting facilities like this one in Sevier County, Tennessee, use large digesters to process household waste after recyclables have been removed. The digesters produce a mulch that residents can pick up at no cost.

Household plant–based food scraps can be added, but meat products should not be included.

cjp/E+/Getty Images

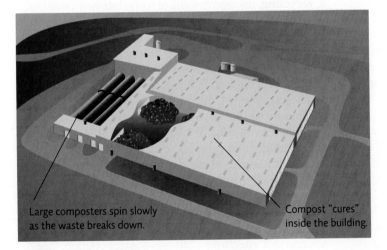

Large composters spin slowly as the waste breaks down.

Compost "cures" inside the building.

Does your school cafeteria have a composting program? If not, what would it take to implement one?

The end product is a rich, soil-like mulch that can be used in gardens.

Africa Studio/Shutterstock

sustainable ecosystem; see Infrographic 6 in Module 1.1), we stand a better chance of solving our solid waste problems if we pursue a variety of solutions. One of the most meaningful steps we can take is to produce less waste in the first place. Manufacturers can take steps to reduce the waste they produce by recycling product components, increasing the efficiency with which they use resources, and by working with other industries (see Module 5.1)— the waste of one industry could be the raw material needed for the production processes of another. But it doesn't stop there—consumers also have an important role to play.

9 REDUCING SOLID WASTE: CONSUMERS AND THE 4 Rs

Key Concept 9: Consumers can reduce their generation of waste by using less, reusing or recycling what is used, and "closing the loop" by purchasing recycled products.

Advertising—a virtual staple of advanced societies—bombards us from every corner of modern life: not just on the televisions in our living rooms but on taxicab computer screens, billboard-laden subway cars, and the pop-up ads that invade our laptops. The message is surprisingly uniform: To live a happy, more fulfilling life, we simply must have more "stuff." But this stuff requires resources to produce and package, and waste is generated every step of the way—not just when we discard it. One solution to the problem of waste is using less stuff so there is less waste to begin with.

So how do we start making different choices? As any good environmentalist will tell you, it comes down to the "four Rs": refuse, reduce, reuse, and recycle.

The first thing we can do is simply **refuse** to use things that we don't really need, especially if they are harmful to the environment. This may be as minor as declining to take a plastic bag for a few items purchased at the drugstore or as major as biking or walking to work rather than taking the car. The logic is simple: When we save a resource by refusing to use it, that resource lasts longer, which in turn means that less pollution will be generated disposing of it and producing replacements and more will be available for future uses. "Refusal doesn't mean never using the resource," says David Bruno, founder of the 100 Thing Challenge, a popular movement to pare down our worldly possessions to 100 items or fewer. "It just means using it at a more sustainable rate."

refuse The first of the waste-reduction four Rs: Choose not to use or buy a product if you can do without it.

reduce The second of the waste-reduction four Rs: Make choices that allow you to use less of a resource by, for instance, purchasing durable goods that will last or can be repaired.

reuse The third of the waste-reduction four Rs: Use a product more than once for its original purpose or for another purpose.

recycle The fourth of the waste-reduction four Rs: Return items for reprocessing into new products.

If we can't completely refuse a given commodity, we can still try to **reduce** our consumption of it or minimize our overall ecological footprint by making careful purchases. People who must drive to work can minimize their fossil fuel consumption by choosing a more fuel-efficient vehicle. Those who choose not to drink tap water might purchase a specialized faucet or pitcher filter instead of relying on bottled water. And all consumers can greatly reduce the amount of waste they generate by paying special attention to packaging, which accounts for about one-third of all U.S. trash and roughly half of all paper used.

If we can't avoid using a product, our next best choice is to **reuse**—the third *R*—something consumers can do with just a little effort by choosing durable products over disposable ones. "Products produced for limited use are really just made to be trash," Bruno says. "They pull resources out of the environment and produce pollution at every step of production, shipping, and disposal." He advises considering use and reuse each time we head to the store. Whether you are purchasing clothing, razors, cups, or plates, ask yourself: How long will this last, and for what other purpose might it be used?

Reusing also applies to industry. TerraCycle is a U.S. company that produces liquid worm-compost fertilizer and packages it in used soda and water bottles. The company also collects hard-to-recycle packaging like candy bar wrappers and juice pouches and turns them into new products like backpacks.

Once we've refused, reduced, and/or reused a given commodity as much as possible, we are left with the final R, **recycling**, the reprocessing of waste into new products. Recycling has several advantages. It certainly reduces the trash we generate, but it does more than that. By reclaiming raw materials from an item that we can no longer use, we limit the amount of raw materials that must be harvested, mined, or cut down to make new items. In most cases, this helps conserve limited resources—not only trees and precious metals but also energy and water. To execute this step properly, we must first have purchased items that can be recycled. We must also *close the loop* by purchasing items that are made of recycled materials to encourage manufacturers to make those products. This last step is key. Reconfiguring the plastics industry from a linear production model (take, make, discard) to a circular one that recovers and recycles plastics is seen as a key part of any solution that effectively addresses plastic pollution. (See Module 5.1 for more on economic models of production.) **INFOGRAPHIC 9**

All these steps will help reduce our own solid waste, but for the world's solid waste problem to be addressed—especially all that plastic ending up in the oceans—steps must be taken to implement proper waste disposal methods in developing countries. According to one sobering estimate made in a report published by the Ellen MacArthur Foundation, if we can't change our course by reconfiguring the plastics industry to follow

INFOGRAPHIC 9   REDUCING WASTE WITH THE FOUR Rs

There are ways to reduce the amount of waste we generate, but for the waste we do have, there are better options than simply throwing away many products. An item like a plastic bottle can be recovered and reused, as the innovative company TerraCycle does, or the bottle may be recycled into another product.

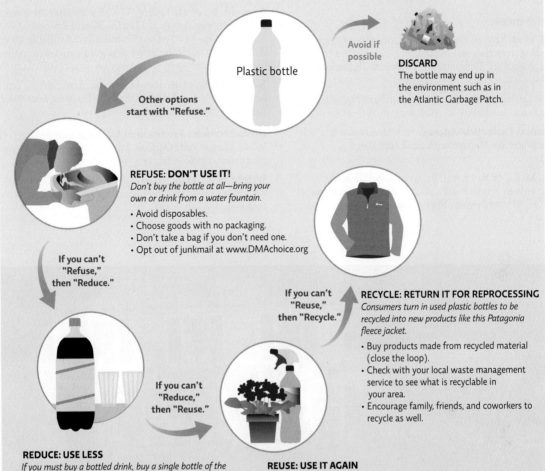

Plastic bottle

Avoid if possible

DISCARD
The bottle may end up in the environment such as in the Atlantic Garbage Patch.

Other options start with "Refuse."

REFUSE: DON'T USE IT!
Don't buy the bottle at all—bring your own or drink from a water fountain.

• Avoid disposables.
• Choose goods with no packaging.
• Don't take a bag if you don't need one.
• Opt out of junkmail at www.DMAchoice.org

If you can't "Refuse," then "Reduce."

If you can't "Reuse," then "Recycle."

RECYCLE: RETURN IT FOR REPROCESSING
Consumers turn in used plastic bottles to be recycled into new products like this Patagonia fleece jacket.

• Buy products made from recycled material (close the loop).
• Check with your local waste management service to see what is recyclable in your area.
• Encourage family, friends, and coworkers to recycle as well.

If you can't "Reduce," then "Reuse."

REDUCE: USE LESS
If you must buy a bottled drink, buy a single bottle of the largest size you will use rather than several small ones.

• Choose goods with minimal packaging.
• Buy durable, repairable goods.
• Buy local fresh food; it comes with less packaging.
• Use both sides of paper.
• Post notices on a bulletin board or via email to reduce copies.

REUSE: USE IT AGAIN
The company TerraCycle collects used bottles and packages fertilizer in them.

• Rent, borrow, or lend items.
• Choose reusable containers for leftovers rather than plastic bags or wrap.
• Buy and sell old clothes or donate to charity.
• Reuse products in different ways: Use yogurt containers to hold screws; scrap paper for a note pad, etc.

Collection

Purchase   Production

RECYCLING REQUIRES 3 STEPS: Consumers and industry must turn in or collect materials for recycling, the material must be used to make new products, and the products must be bought by consumers.

1 PETE   2 HDPE   3 V   4 LDPE   5 PP   6 PS   7 OTHER

RECYCLING PLASTIC: The number code on a plastic item indicates the type of resin with which it is made. In many areas, #1 and #2 are the only plastics that can be recycled; in others, all types are taken by recyclers. Check with your community recyclers to see what is taken in your area.

 Why is "recycling" considered the fourth choice of the four Rs?

a circular production path (recover and recycle plastic materials) rather than a linear one (throw it away), if we can't rein in our plastic addiction and our use of improper disposal methods, by 2050 there will be more plastic bits in the ocean than fish.

Select References:

Cole, M. J., et al. (2015). The impact of polystyrene microplastics on feeding, function and fecundity in the marine copepod *Calanus helgolandicus*. *Environmental Science & Technology, 49*(2), 1130–1137.

Cressey, D. (2016). Bottles, bags, ropes and toothbrushes: the struggle to track ocean plastics. *Nature, 536*, 263–265.

Environmental Protection Agency. (2016). *Advancing Sustainable Materials Management: 2014 Fact Sheet* (EPA530-R17-01).

Goldstein, M. C., et al. (2014). Relationship of diversity and habitat area in North Pacific plastic-associated rafting communities. *Marine Biology, 161*(6), 1441–1453.

Huerta Lwanga, E., et al. 2016). Microplastics in the terrestrial ecosystem: Implications for *Lumbricus terrestris* (Oligochaeta, Lumbricidae). *Environmental Science & Technology, 50*(5), 2685–2691.

Law, K. L., et al. (2010). Plastic accumulation in the North Atlantic subtropical gyre. *Science, 329*(5996), 1185–1188.

Neufeld, L., et al. (2016). *The New Plastics Economy: Rethinking the Future of Plastics.* World Economic Forum and the Ellen MacArthur Foundation. Available at www.ellenmacarthurfoundation.org/publications/the-new-plastics-economy-rethinking-the-future-of-plastics.

Savoca, M. S., et al. (2016). Marine plastic debris emits a keystone infochemical for olfactory foraging seabirds. *Science Advances, 2*(11), e1600395.

United Nations Environmental Programme. (2015) *Global Waste Management Outlook*, https://wedocs.unep.org/rest/bitstreams/17680/retrieve.

Wright, S. L., et al. (2013). The physical impacts of microplastics on marine organisms: a review. *Environmental Pollution, 178*, 483–492.

SARAH LEEN/National Geographic Creative

The Foster family of Stow, Ohio, with their plastic possessions.

INTERACTIVE MAP **GARBAGE: PROBLEMS AND SOLUTIONS** ANIMATED INFOGRAPHIC

There's no debating the fact that humans produce a lot of trash. Addressing the problem will involve a combination of efforts to reduce the production of trash and to properly deal with the garbage that is created. Read about some research and efforts underway to learn about and deal with this issue.

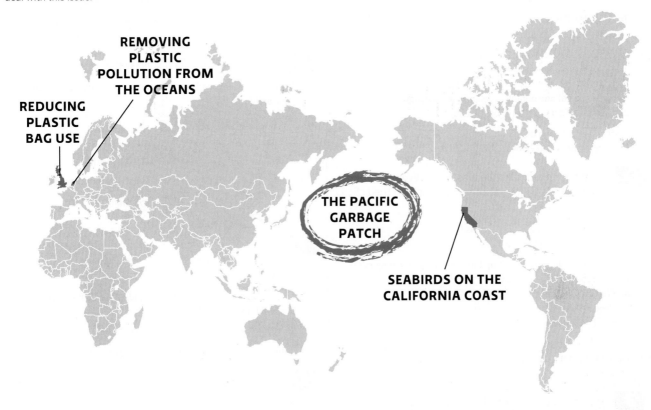

REMOVING PLASTIC POLLUTION FROM THE OCEANS

REDUCING PLASTIC BAG USE

THE PACIFIC GARBAGE PATCH

SEABIRDS ON THE CALIFORNIA COAST

 BRING IT HOME

PERSONAL CHOICES THAT HELP

How much solid waste you produce is under your control. By reducing the amount of waste you produce, you reduce how much money we spend on waste disposal as a whole and at the same time place less pressure on the resources used to produce consumer goods. Reducing your solid waste is very easy and can save you money in the process.

Individual Steps
• Track your trash. Record what you throw out for a week by category and weight. How could you reduce your total trash weight by one-quarter? By half?

• Use the information in Infographic 9 (strategies to refuse, reduce, reuse, and recycle) to identify five changes you can make to reduce your solid waste.

Group Action
• Start recycling unusual items in your community. TerraCycle is a company that takes items that usually end up in the garbage, like candy wrappers, corks, and chip bags, and recycles them into new products, like purses or backpacks.
• Talk to friends and family about having "no gift" or "low gift" celebrations. Instead of buying lots of presents, treat friends to

a dinner or a fun activity. For large families, use a grab bag or draw names and buy or upcycle (give used goods) for only specific people.

Policy Change
• Talk to community leaders to discuss the possibility of starting a communitywide composting program.
• Research rates of recycling participation in your community. Advocate for recycling education and curbside recycling programs.

UNDERSTANDING THE ISSUE

1 What types of solid waste do we produce and why do we say waste is a "human invention"?

1. True or False: If we handled waste in a circular economic system rather than a linear one, we would produce less waste overall.

2. Why do we say there is no "waste" in nature?
   a. Nature does not produce toxic or unwanted materials.
   b. Nature can destroy unneeded or dangerous matter.
   c. The waste of one organism can be used by another.
   d. All matter in nature is biodegradable.

3. Distinguish between degradable, biodegradable, and nondegradable waste. How well do plastics degrade?

2 What is municipal solid waste (MSW) and what are the types and proportion of waste in the U.S. MSW stream?

4. True or False: Municipal solid waste is the waste produced in a community from homes, businesses, farms, and industry.

5. Which of the following makes up the single biggest category of waste in the municipal solid waste stream?
   a. Plastic
   b. Paper
   c. Food
   d. Glass

6. How has the total and per capita production of MSW changed in the United States since the 1960's?

3 What is the EPA's preferred hierarchy of preferred MSW disposal methods and how well does the United States meet this recommendation?

7. True or False: The most preferred method of MSW disposal is burial in sanitary landfills.

8. Why are some waste incinerators a better MSW disposal method than others?

4 Compare the design, advantages, and disadvantages of open dumps and sanitary landfills.

9. True or False: Garbage decomposes more quickly in a sanitary landfill than it would in an open dump.

10. Which of the following is a disadvantage of sanitary landfills?
    a. The extremely slow decomposition of the trash
    b. Covering the top of trash with dirt that keeps away animals that might help eat it
    c. Thick plastic liners at the bottom that prevent the flow of water down to ground-water supplies
    d. All of the above

11. What is leachate and why is it a concern?

12. Why is methane production by sanitary landfills a problem and how can it be turned into an advantage?

5 What are the pros and cons of waste incinerators?

13. True or False: Incinerators are expensive to build and operate.

14. Why is the ash produced by an incinerator considered hazardous waste?

6 What problems does solid waste cause?

15. Which of the following is a consequence of uncollected waste?
    a. Air pollution
    b. Water pollution
    c. Flooding
    d. A and B
    e. A, B, and C

16. In what ways does the plastic trash in oceans harm ocean life?

17. From an economic point of view, why is solid waste considered a mismanagement of resources?

7 What are some common household hazardous wastes and how should individuals deal with this waste?

18. True or False: Hazardous waste is an industrial waste issue and is not a problem for MSW.

19. What is e-waste, and why is it a concern?

20. Identify some hazardous substances in your home.

8 What is composting and how can it help us in our quest to deal with solid waste?

21. Which of the following should you leave out of a home compost pile?
    a. Grass clippings
    b. Paper
    c. Meat scraps
    d. A and B
    e. A, B, and C

22. Compare landfilling and composting. What are the trade-offs for each option?

9 What are the four Rs of waste reduction?

23. Of the 4 Rs, which is considered the "last" choice or least preferred action?
    a. Reduce
    b. Refuse
    c. Recycle
    d. Reuse

24. Explain what is meant by the *four Rs* and give an example of each.

25. What does it mean to "close the loop" with regard to recycling and why is it important?

SCIENCE LITERACY WORKING WITH DATA

To determine whether microplastics are harmful to small marine organisms, researchers exposed marine copepods to either their normal diet of algae or to algae plus microplastic beads (20 μm polystyrene beads at a concentration of 75 beads per mL of water). As mentioned in the text, copepods exposed to these microplastic beads produced smaller eggs that had a lower hatching rate. Researchers also examined a variety of dietary parameters including the size, total number, and weight of algal cells ingested. Look at their data below and answer the questions that follow. (Bars with an asterisk above them are significantly different from the control at $P < 0.05$.)

DIET AND MICROPLASTIC EXPOSURE

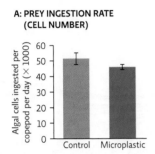

A: PREY INGESTION RATE (CELL NUMBER)

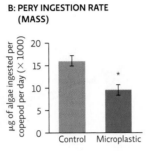

B: PERY INGESTION RATE (MASS)

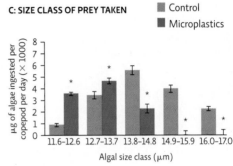

C: SIZE CLASS OF PREY TAKEN

Interpretation

1. In one or two sentences, describe the result depicted in Graph A. (Be sure to reference statistical significance in your answer.)

2. Estimate how much total food (algae) was eaten by each group (Graph B). Can you conclude that there is a difference between the two groups for this parameter?

3. In one sentence describe the result obtained in Graph C (i.e., explain what differences are seen in the size of algal cells taken by the control versus the microplastic group).

Advance Your Thinking

4. What overall conclusion can you draw from the three graphs regarding the ability of copepods to acquire food?

5. Offer a hypothesis to explain the feeding difference seen between copepods in the two groups.

6. Other data gathered showed that copepods in the microplastic group produced the same number of eggs but that these eggs were smaller and had a lower hatching success rate. Based on the data presented here, why might this be true?

INFORMATION LITERACY EVALUATING INFORMATION

Curbside recycling programs are a very convenient system for recycling metals, paper, glass, and plastic. However, they are available to only about half the population in the United States today, and what can be recycled is limited. So where do you go if you do not have curbside recycling or if you want to recycle or safely dispose of such things as electronics, batteries, or books? One source of help is the Earth911 website, which provides information on how to recycle a vast range of items as well as information on the latest recycling laws, ideas for living a green lifestyle, and stories on people and companies who are working to make a difference.

Go to the Earth911 website (www.earth911.com).

Evaluate the website and work with the information to answer the following questions:

1. Determine if this is a reliable information source with a clear and transparent agenda:
   a. Who runs this website? Do the organization's credentials make the information presented reliable or unreliable? Explain.
   b. What is the mission of this website? What are its underlying values? How do you know this?

   c. What data sources does Earth911 rely on for its information, and what is its policy on what it puts on its website? Are the sources it uses reliable?
   d. Do you agree with the organization's assessment of the problems with and concerns about waste and recycling? What about its solutions? Explain.

2. On the home page, select the link "Recycling Search" and click the "Electronics" icon. Enter your location in the box provided (city and state or zip code).
   a. What options exist in your community for recycling electronics?
   b. Where is the nearest location where you could recycle a desktop computer? A cell phone? A video game console?

3. Click on any of the tabs at the top of the home page and investigate one topic thoroughly. Identify which tab you chose and answer the following questions:
   a. What types of articles are offered?
   b. Who writes these articles?
   c. Does the content seem useful and credible? Explain your response.

4. How useful is a website like Earth911? How can a website like this influence societal understanding of waste issues and facilitate a change in behavior on the part of individuals and businesses?

 Additional study questions are available at SaplingLearning.com.

WATER RESOURCES

CHAPTER 6

Water is vital to all life on Earth. Human impact is affecting both water quantity (access to enough freshwater) and water quality (water pollution). Our actions are also harming ocean ecosystems, ecosystems on which we and countless other species depend.

Module 6.1: Freshwater Resources

An examination of the availability and of use (or overuse) of freshwater supplies and the processes used to produce potable water for consumption

Module 6.2: Water Pollution

A look at the sources and types of water pollution and ways to address them

⬈ ONLINE Module 6.3: Marine Ecosystems

An introduction to the variety of marine ecosystems and current threats, including a new threat—ocean acidification

Online Modules are available at SaplingLearning.com.

Garry Solomon/EyeEm/Getty Images

TOILET TO TAP

A California county is tapping controversial sources for drinking water

Clean drinking water is a rare thing in many places on Earth. ConstantinosZ/iStock/Getty Images

After reading this chapter and studying the KEY CONCEPTS and INFOGRAPHICS, you should be able to answer these GUIDING QUESTIONS

CORE MESSAGE

Freshwater is a precious but limited resource, and it is essential to life. Some regions consume water faster than it is replenished. And, unfortunately, water is not evenly distributed across the globe; many people worldwide lack access to enough clean water. Methods are available to recover and purify otherwise dirty water, but we also need to use water more wisely.

1. How is water distributed on Earth, and what are the sources of freshwater?

2. How does water cycle through the environment?

3. What is an aquifer, how does it receive or lose water, and what problems emerge when too much water is removed?

4. What is the breakdown of water use by sector, and how does use differ between developed and developing nations?

5. What are the causes and consequences of water scarcity?

6. What are some of the ways that our wastewater is treated to make it potable or safe to release into the environment?

7. What technologies can help provide more potable water, and what are their tradeoffs?

8. How can conservation help address water scarcity issues?

One of the most exciting moments in Shivaji Deshmukh's career as a water engineer came one bright, sunny day in January 2008. He had gathered with staff from the Orange County Water District (OCWD) in Anaheim, California, to watch for the first time as former sewage water, cleaned using state-of-the-art techniques, was pumped into underground drinking water sources. It was the beginning of a groundbreaking project designed to help save the region from ongoing, and frightening, water shortages.

"It's basically this drought-proof supply of water," says Deshmukh. "Nobody else has done it. Nobody thought a community could support it, because they would be too grossed out by it."

◉ WHERE IS ANAHEIM, CALIFORNIA?

The water that Deshmukh and other engineers watched seep into the region's underground water stores that day in 2008 was purified **wastewater**—including sewage and used water from homes and industrial sites. Understandably, when many residents first heard about the project, they were concerned.

But that same month, Deshmukh and other OCWD staff attended a dedication ceremony for the Orange County Groundwater Replenishment System (GWRS) at the water treatment plant in Fountain Valley, California, along with hundreds of other people, including various community groups, to honor the massive project. Having that support from the community was key to the project's success, says Deshmukh—but getting it hadn't been easy.

wastewater Used and contaminated water that is released after use by households, industry, or agriculture.

Crystal-clear purified water from the Groundwater Replenishment System is piped to the Orange County Water District's percolation ponds in Anaheim, California.

Courtesy of Mark Greening/Orange County Water District

1 FRESHWATER DISTRIBUTION AND SOURCES

Key Concept 1: Only 3% of water on Earth is freshwater, and very little of that is accessible to humans. Fortunately, even this small percentage represents a large amount of water.

Even though Earth is covered in more than 1.4 billion cubic kilometers (370,000,000 trillion gallons) of water—about 75% of its surface—only about 1/100 of 1% of that water is usable by humans.

Water provides many important ecosystem services that animals and plants require to live. Up to 75% of the human body, for instance, consists of water. But humans need liquid **freshwater** (which has few dissolved ions such as salt); ocean water is too salty for human consumption and is toxic in large doses. Complicating things, nearly 80% of the freshwater on the planet is trapped in glaciers and ice caps at the poles. **INFOGRAPHIC 1**

Around the world, each region faces unique water challenges. In California, freshwater flows into the northern part of the state when the Sierra Nevada Mountain snowpack melts in the spring. This snowmelt provides as much as one-third of California's water. But as Earth's climate changes, the state could lose much of its snowpack. Indeed,

---

freshwater Water that has few dissolved ions such as salt.

---

in the 2009 "water year" (California tracks its yearly totals from July of one year to June of the next), California's precipitation was 20% below average, and the snowpack was 40% below its average size. The 2014 water year was one of the driest on record in California; precipitation was 60% below normal, and the spring snowpack was 97% below normal. Water year 2015 was a little better, but the hoped for "drought-busting" rains of 2016 (an El Niño year) brought enough rain to some parts of northern California for the governor to declare the drought state of emergency over for the northern part of the state (though the heavy rains brought their own problems in the form of floods, mudslides, and damage to dams). Still, many reservoirs in the southern part of the state continue to set records for low water levels.

Even in a good snowpack year, the state faces major water issues, explains Deshmukh, now working at the West Basin Municipal Water District in Carson, California. Two-thirds of California's water is located in the northern part of the state, but two-thirds of the state's residents live in the south, he explains.

INFOGRAPHIC 1 DISTRIBUTION OF WATER ON EARTH

Most of the water on Earth is found in the oceans, and most of the freshwater is tied up in ice and snow. Only about 0.001% of all of Earth's water is available for us to use, but with more than 1,300 trillion liters (350 trillion gallons) of water on the planet, that is still a lot of water.

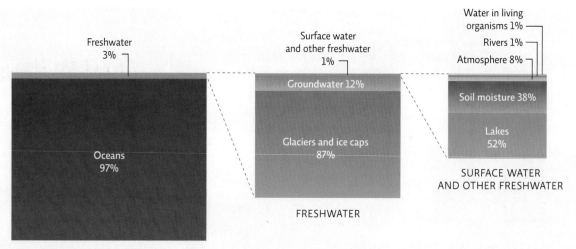

 Based on this diagram, what percentage of the total water supply on Earth is found in groundwater? What percentage is found in rivers?

2 THE WATER CYCLE

Key Concept 2: Water moves through the environment via the water cycle, a process that constantly recycles water on Earth through ground and surface waters, plants, and the atmosphere.

Wherever there is water, it is constantly moving through the environment via the **water cycle** (hydrologic cycle). Heat from the Sun causes water to evaporate from **surface waters** (rivers, lakes, oceans) and land surfaces. At the same time, plant roots pull up water from the soil and then release some into the atmosphere in a process called **transpiration**. Plants with deep roots, like trees, may bring up thousands of gallons of water a year, releasing much of this to the atmosphere. Altogether, the combination of **evaporation** and transpiration—*evapotranspiration*—sends more than 66,000 cubic kilometers of water vapor into the atmosphere every year, equivalent to 17,000 trillion gallons. Once aloft, that water condenses into clouds (**condensation**) and may fall back to Earth as **precipitation** (rain, snow, sleet, etc.). **INFOGRAPHIC 2**

water cycle The movement of water through various water compartments such as surface waters, atmosphere, soil, and living organisms.

surface water Any body of water found above ground, such as oceans, rivers, and lakes.

transpiration The loss of water vapor from plants.

INFOGRAPHIC 2 THE WATER CYCLE

Water cycles between liquid and gaseous forms as it moves through space and time. Ocean water (which we cannot use) is converted to freshwater when it evaporates and falls back to Earth as precipitation, refilling freshwater surface and underground water supplies. Liquid freshwater is a renewable resource as long as we don't use it faster than it is naturally replenished.

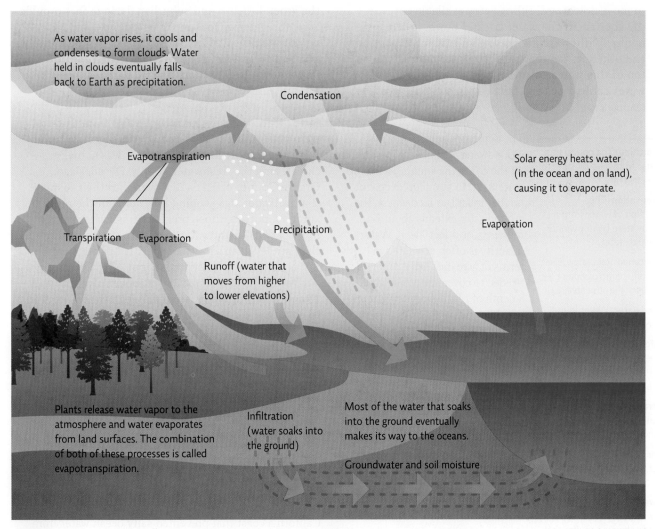

As water vapor rises, it cools and condenses to form clouds. Water held in clouds eventually falls back to Earth as precipitation.

Condensation

Evapotranspiration

Transpiration   Evaporation

Precipitation

Runoff (water that moves from higher to lower elevations)

Solar energy heats water (in the ocean and on land), causing it to evaporate.

Evaporation

Plants release water vapor to the atmosphere and water evaporates from land surfaces. The combination of both of these processes is called evapotranspiration.

Infiltration (water soaks into the ground)

Most of the water that soaks into the ground eventually makes its way to the oceans.

Groundwater and soil moisture

 The trees of tropical rainforests are said to contribute as much as half of the rain that falls back on the forest. Explain how this occurs.

Almost all precipitation ends up falling on the oceans, and a tiny remainder falls on land. This latter portion is the part humans can harvest for their own use. We access freshwater from lakes and rivers (surface water) and from **groundwater**. Water from both surface and groundwater sources eventually make their way back to the oceans.

3 GROUNDWATER

Key Concept 3a: Aquifers are refilled when water soaks into the ground, but hard land surfaces in urban and suburban areas limit infiltration.

Key Concept 3b: If water is removed from aquifers faster than it is resupplied, wells can run dry in inland areas or become contaminated with saltwater near the coast.

Many people (not just in California) draw their water from an underground region of permeable soil or porous rock saturated with water, called an **aquifer**. These groundwater stores receive water from rainfall and snowmelt that soaks into the ground through **infiltration**. Plant roots take up some of the water along the way, but much of the water continues to move downward, filling every available space in the aquifer. As the water trickles down, it becomes naturally filtered by rocks and soil, which trap bacteria and other contaminants as the water passes by. The top of this water-saturated region, referred to as the **water table**, rises and falls due to seasonal weather changes.

reduces infiltration will reduce the rate at which the aquifer refills and thus decreases the amount of water we can sustainably remove. Infiltration is hampered in urban and suburban settings because of all the hard surfaces, such as roads and buildings; even a typical suburban lawn is so compacted from the home construction process that very little water infiltrates the ground. Urban and suburban designs that provide ways for water to soak into the ground—such as permeable pavement and *rain gardens*—can help refill aquifers as well as help prevent flooding events after heavy rainfalls. (For more on rain gardens, see Module 6.2.)

But in California, excess withdrawals from the aquifer were not leading to depleted aquifers and dry wells. Instead, the proximity to saltwater was actually threatening the freshwater supply. Decades ago, Orange County officials discovered to their dismay that saltwater was seeping into some of the region's aquifers, putting those precious freshwater stores in jeopardy. Groundwater levels are typically higher than sea level, so saltwater doesn't infiltrate aquifers. But as freshwater was pumped out of the county's aquifer inland, the *water table* (upper surface of the groundwater) dropped, so salty ocean water had started to enter the coastal edge of the aquifer to the west, where it bordered the Pacific Ocean. In Orange County, some aquifers are confined by geologic faults, which prevent ocean water from entering at some points—but not everywhere. It was in these unconfined coastal aquifers that **saltwater intrusion** was becoming a problem. **INFOGRAPHIC 3**

To stem the influx of saltwater, in 1975, the Orange County Water District started pumping highly treated (purified) sewage wastewater into injection wells. At about 19 million liters (5 million gallons) a day, the underground injection created a curtain of freshwater along the California coast that prevented salty ocean water from seeping into the county's aquifer. This water management program was the first to pump treated wastewater into the ground, says Deshmukh.

evaporation The conversion of water from a liquid state to a gaseous state.

condensation The conversion of water from a gaseous state (water vapor) to a liquid state.

precipitation Rain, snow, sleet, or any other form of water falling from the atmosphere.

groundwater Water found underground in aquifers.

aquifer An underground, permeable region of soil or rock that is saturated with water.

infiltration The process of water soaking into the ground.

water table The uppermost water level of the saturated zone of an aquifer.

saltwater intrusion The inflow of ocean (salt) water into a freshwater aquifer that happens when an aquifer has lost some of its freshwater stores.

The depth of Orange County's groundwater varies, says Deshmukh. At the coast, the aquifer is 6 to 90 meters (20 to 300 feet) deep, but further inland, at its deepest, the groundwater extends about 900 meters (3,000 feet) deep. Water quantity is often measured in terms of acre-feet—the amount needed to cover an acre in water to a depth of 1 foot (30.5 centimeters), which is equivalent to more than 1.1 million liters (300,000 gallons). One acre-foot of water is enough for two American families for 1 year. Deshmukh estimates that nearly 5,000 acre-feet of water is accessible from the deepest part of the aquifer. But that deep subterranean water is harder to get, and it costs more to pump it out of the ground than it costs to remove the groundwater closer to the surface.

In addition to withdrawals for agriculture, industry, and personal use, anything that

Groundwater in aquifers is naturally replenished as water soaks into the ground. Humans can access this groundwater through wells, but we can pull out water faster than it is naturally replaced. This can lead to saltwater intrusion in coastal areas or dry wells in inland areas. Surface pollution can also seep into the ground and contaminate groundwater.

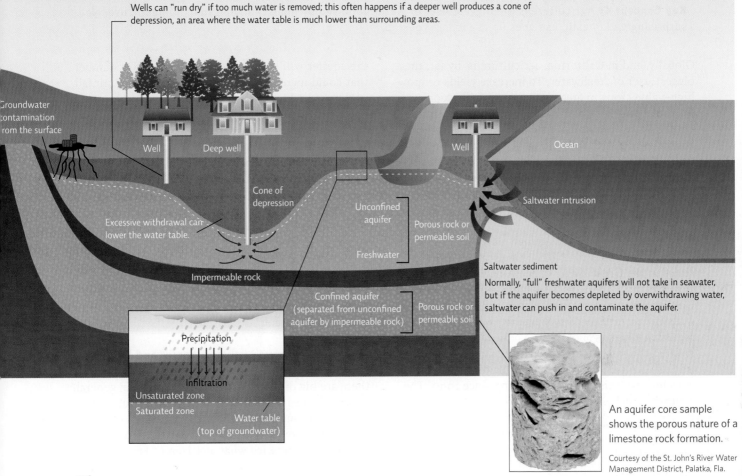

Wells can "run dry" if too much water is removed; this often happens if a deeper well produces a cone of depression, an area where the water table is much lower than surrounding areas.

Groundwater contamination from the surface

Well

Deep well

Well

Ocean

Cone of depression

Excessive withdrawal can lower the water table.

Unconfined aquifer

Porous rock or permeable soil

Saltwater intrusion

Freshwater

Impermeable rock

Saltwater sediment

Confined aquifer (separated from unconfined aquifer by impermeable rock)

Porous rock or permeable soil

Normally, "full" freshwater aquifers will not take in seawater, but if the aquifer becomes depleted by overwithdrawing water, saltwater can push in and contaminate the aquifer.

Precipitation

Infiltration

Unsaturated zone

Saturated zone

Water table (top of groundwater)

An aquifer core sample shows the porous nature of a limestone rock formation.

Courtesy of the St. John's River Water Management District, Palatka, Fla.

? Why might deep cones of depression and dry wells (formed from overdrawing well water) be more common in inland areas than in coastal ones?

A sinkhole swallows three cars during a heavy rainstorm in Chicago, Illinois, in 2013. Aquifers can be weakened when they lose water and then a heavy rain event may add enough weight to collapse the aquifer, opening up a sinkhole.

4 GLOBAL WATER USE

Key Concept 4: Agriculture is the biggest user of water, followed by industrial, and then domestic use. Not surprisingly, domestic use goes up as income goes up.

Globally, and in California, agriculture is, by far, the biggest user of freshwater. To increase yields or grow high-value crops in areas that are naturally too dry for those crops, we add water. Crop irrigation in the United States consumes 80% of our water; in the countries of South Asia closer to 90% of their total usage going to irrigation. And in the United States, irrigation water comes, about equally, from surface and groundwater sources, depleting both in some areas.

For example, the Ogallala aquifer, the largest groundwater system in North America, underlies eight states in the Great Plains of the American Midwest, from Nebraska to Texas, supplying about 30% of all U.S. irrigation water to these farm belt states. But withdrawals far exceed recharge rates, which are only replacing about 15% of what is removed. This has lowered the water table an average of 4.5 meters (15 feet)—a loss of 30% of its historic capacity— with most of this drawdown coming since 1960. The southern reaches of the aquifer are most severely impacted (from Kansas to Texas); wells have run dry in some areas and drilling deeper won't help—the groundwater there is gone. One study predicts that

domestic water use Indoor and outdoor use of water by households and small businesses.

the aquifer will be 69% depleted by 2060 (a number that could increase if climate change reduces rainfall in the area, which it is projected to do). Local water supplies and future agricultural productivity depend on this aquifer; steps taken today to improve irrigation efficiencies or reduce the irrigation water needed for a crop (e.g., plant a crop that needs less water) are crucial for the future of this region.

Industrial use of water varies from country to country. It is understandably higher in more developed nations than less developed ones. Thermoelectric power plants such as coal or nuclear facilities use large amounts of water for cooling and in many developed nations this represents the biggest portion of industrial water use; in the United States, it accounts for around 85% of the industrial sector's use of water.

Domestic water use—that of homes and businesses— is the sector that uses the least amount of water but there are big differences between nations. In general, water use goes up as income levels go up but some wealthy nations use much more water than others. For example, on average, a citizen in the United States uses more than twice what a citizen of France uses, and about three times as much as the world average.
INFOGRAPHIC 4

INFOGRAPHIC 4   **GLOBAL WATER USE**

Globally, most water use goes to agriculture. In developed countries like the United States, about 85% of industrial water use goes to electricity production.

Individuals in more developed countries use far more water per person than those in less developed countries. In some areas, individuals must make do with only a few gallons a day.

GLOBAL WATER USE BY SECTOR

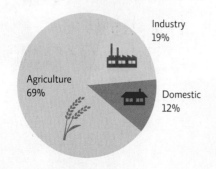

Industry 19%

Agriculture 69%

Domestic 12%

DOMESTIC WATER USE

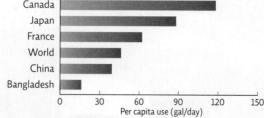

United States
Australia
Canada
Japan
France
World
China
Bangladesh

Per capita use (gal/day)
0   30   60   90   120   150

The United States and France are both developed countries with high standards of living. What might account for the large difference in domestic water use per person in these two countries?

5 WATER SCARCITY: CAUSES AND CONSEQUENCES

Key Concept 5a: Water shortages can be physical (lack of water) or economic (the inability to pay for water). Poor sanitation can contaminate water, causing health problems and contributing to scarcity.

Key Concept 5b: Water wars can develop if upstream locations take too much water, depriving downstream locations of adequate supplies.

The United States, like other developed countries, has a well-established water purification and delivery system that provides safe drinking water to the vast majority of its residents. This effort is legally mandated by the **Safe Drinking Water Act (SDWA)**. The Environmental Protection Agency (EPA) administers the act by setting water quality standards and works with state and local governments and other federal agencies to see that those standards are met. But many U.S. water-treatment facilities are old or too small to handle growing local populations and much of our water infrastructure (e.g., pipes) is old, leading to lost water (leaks) or contaminated water. The EPA estimates it will cost more than $650 million just to address infrastructure upgrades such as pipe replacements or water and sewage treatment plant upgrades. Their 2016 Action Plan lays out a framework for addressing these and other issues. (See Module 1.3 for a look at the problem of lead in the water of Flint, Michigan.)

People don't always live near abundant sources of freshwater, making access a vital issue. Around the world, many areas suffer from **water scarcity**—not having sufficient access to clean water supplies. In some dry regions, there is simply not enough to meet needs; many arid nations like those of the Middle East, parts of Africa, and much of Australia face water shortages as a way of life. The Middle Eastern countries of Bahrain, Qatar, Kuwait, and Saudi Arabia have the lowest per capita water availability in the world, but these oil-rich nations can afford to invest in costly technology to access water (like facilities to remove salt from seawater). In other areas, particularly in sub-Saharan Africa, there may be enough water, but people do not have the money to purchase it or dig wells to access it. People in these areas may be getting by on just a few gallons of water a day—and that water may not even be safe to use.

Ensuring that all people have access to clean water and sanitation is Goal Six of the United Nation's (UN) Sustainability Development Goals (see Infographic 4 in Module 1.1), a goal, the UN points out, that affects the successful achievement of every other goal. The UN estimates that as many as 2.8 billion people lack access to enough clean water and the drinking water source of at least 1.8 billion is contaminated with fecal material (human or animal waste). This contamination often comes from inadequate sanitation; as many as 2.4 billion people lack access to

Safe Drinking Water Act (SDWA) Federal law that protects public drinking water supplies in the United States.

water scarcity Not having access to enough clean water.

Almost 3 billion people lack access to dependable supplies of clean water. Water scarcity may be physical (not enough water is present) or economic (cannot afford to buy or access water).

Likhitha/Getty Images

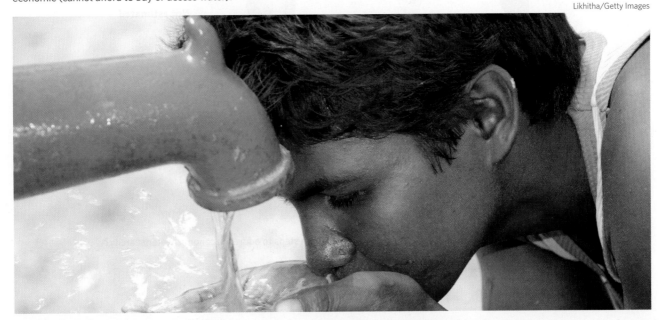

sufficient sanitation facilities that provide safe disposal of human waste. This may be something as simple as a pit latrine or outhouse located away from water sources and homes, or it may provide for the decontamination of waste (wastewater treatment) so that the resulting water is safe to release into the environment. **INFOGRAPHIC 5**

In developing nations where water and funding for basic sanitation are scarce, people use nearby surface waters to meet their basic cooking, drinking, and washing needs. These waters can be contaminated with raw sewage or dangerous pathogens, increasing the chance for disease transmission. (See Module 4.3 for more on environmental health and waterborne diseases.) According to the World Health Organization (WHO), more than 1.1 trillion liters (300 billion gallons) of raw sewage enters the Ganges River of India every minute. In Africa, almost 3,000

water wars Political conflicts over the allocation of water sources.

people die each day from waterborne diseases like cholera and typhoid fever as a result of poor sanitation and contaminated water.

Another problem is the **water wars** that develop between states or nations when upstream areas take out or divert so much water that downstream areas suffer shortages. The overtaxed Colorado River is the major water source for seven U.S. states but runs dry before reaching the ocean in the Baja region of Mexico, causing conflicts between states and between Mexico and the United States over allocation of the river's water. Drought and rapid population growth in Atlanta, Georgia, are contributing to a tristate water war between Georgia Alabama, and Florida; Tennessee has recently been dragged into that dispute as Georgia seeks to move its state border north to gain access to the Tennessee River! Water wars play out all over the world with disputes in Africa over the

---

INFOGRAPHIC 5 | **ADDRESSING WATER SCARCITY**

Around the world, more than 750 million people lack access to clean water, and as many as 3.5 billion have access to some, but not enough clean water to meet their needs. An additional 2.6 billion have no access to sanitation.

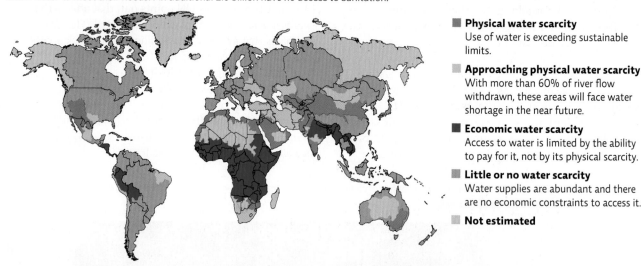

■ **Physical water scarcity**
Use of water is exceeding sustainable limits.

■ **Approaching physical water scarcity**
With more than 60% of river flow withdrawn, these areas will face water shortage in the near future.

■ **Economic water scarcity**
Access to water is limited by the ability to pay for it, not by its physical scarcity.

■ **Little or no water scarcity**
Water supplies are abundant and there are no economic constraints to access it.

■ **Not estimated**

Goal 6 of the UN's Sustainability Developmental Goals: Ensure access to water and sanitation for all.

Targets to be reached by 2030 include:
- Access to safe and affordable drinking water for all.
- Access to adequate sanitation for all and end open defecation.
- Reduce pollution; halve the proportion of untreated wastewater.
- Ensure sustainable withdrawals of freshwater.
- Protect and restore water-related ecosystems.
- Support local communities in improving water and sanitation management.

 How might steps to address physical water scarcity be different from steps to address economic water scarcity?

This wetland marsh in Arcata, California is actually part of a wastewater treatment system that uses nature to help purify sewage. The wetland is now an Audubon birding sanctuary.

David Howell

Nile River and in the Middle East over the Tigris and Euphrates Rivers. And as glaciers that provide water for billions of people in Asia and South America melt at accelerated rates, conflicts over water are setting the stage for water wars to come. (See the Interactive Map at the end of this module for a closer look at some of these water wars.)

According to the United Nations, two of every three people will face water shortages by 2025.

As human populations increase, so will scarcity and sanitation issues; according to the United Nations, two of every three people will face water shortages by 2025.

6 WASTEWATER TREATMENT

Key Concept 6: Wastewater can be decontaminated using high-tech methods that use advanced filtration and chemicals or low-tech methods that mimic the way wetlands purify water.

The wastewater of Orange Country was no different from that produced in homes and businesses across the country. The water that arrived at wastewater treatment facilities contained used water washed down drains and sewage flushed down toilets; in some communities, like Orange County, it also contained storm water delivered to the facility by storm drains that collected runoff from city streets.

Rural homes in the United States that are not connected to a wastewater treatment facility (a city sewage system) use septic systems. A waterproof (e.g., concrete-lined) pit buried in the yard receives the wastewater from the home (from sinks, bathtubs, and toilets); solids settle to the bottom of this septic tank where bacteria begin to digest it. The liquids drain out through perforated pipes buried in gravel-lined shallow ditches just under the surface of the yard. This "drain field" allows untreated wastewater to soak into the ground, where soil organisms digest much of the organic material so that by the time the water reaches the aquifer, it is decontaminated (as long as the water table is not too close to the surface). When it fills up, the septic tank must be pumped out and the contents disposed of as sewage sludge.

Municipalities offer a more technologically advanced system for treating sewage wastewater. In a traditional

"high-tech" facility, **wastewater treatment** includes initial steps that filter the water and then send it to settling tanks where much of the remaining suspended solids sink to the bottom. The water continues on to other tanks where bacteria digest much of the remaining organic matter. Final treatment includes chemical treatment (such as exposure to chlorine to kill some pathogens or the addition of other chemicals to remove phosphates or other substances that should not be released). Chlorine, which is toxic, may then be removed to render it safe to release into the environment—in Orange County, pipes carried the **effluent** to the Pacific Ocean or nearby rivers. The OCSD depended on several "outfalls," or underground pipelines that take the treated sewage water out to the Pacific Ocean or to nearby rivers. Proper treatment is critical. Sewage can carry pathogenic viruses and bacteria; swimmers and surfers get exposed to these if poorly treated wastewater is discharged into recreational waters.

But there are other also "low-tech" methods for sewage treatment that are modeled after a natural ecosystem. In Arcata, California, rather than construct a typical wastewater treatment

wastewater treatment The process of removing contaminants from wastewater to make it safe enough to release into the environment.

effluent Wastewater discharged into the environment.

facility to handle sewage that had been contaminating nearby Humboldt Bay, they repurposed a retired landfill near the coast by converting it to a *wetland*—an ecosystem that is permanently or seasonally flooded. A slow river meanders through the wetland where organisms there purify it; to them, it's not "sewage," it's food. The Arcata facility depends on nature to perform the job of water purification—no toxic chemicals are used. The water discharged into the ocean is very clean, and the health of the bay ecosystem has improved. However, while water from the Arcata facility or high-tech facilities elsewhere addresses the need to decontaminate sewage for safe release into the environment, it does not produce **potable** water— water safe enough to consume. But that's exactly what the OCSD wanted to do. INFOGRAPHIC 6

potable Water that is clean enough for consumption.

7 ADDRESSING WATER SHORTAGES WITH TECHNOLOGY

Key Concept 7: Water scarcity can be addressed by storing water (using dams) and desalinating seawater and by using new solutions like purified wastewater. All these solutions come with trade-offs.

In the mid-1990's, thanks to a growing population, Orange County was faced with two water problems— scarcity of drinking water and too much wastewater. Wastewater facilities were getting more sewage than they were designed to handle and during a heavy rainfall, sewage could be washed through and out of a facility before it could be adequately treated. About 375 million liters (100 million gallons) a day of partially treated sewage water was already flooding the Santa Ana River. Nearly five times that amount would reach the river during a major storm. The OCSD proposed a project that would solve both the problem of too much wastewater and too little freshwater. It wanted to expand the existing system that used treated wastewater to protect groundwater from infiltration by saltwater from the ocean. The new project would pump significantly higher volumes, not just at the seawater barrier, but into the aquifer at other places. An average of 15% of the water that entered people's homes would be treated wastewater, says Deshmukh. "All of a sudden, it became a significant component of the water supply."

To do this, the OCSD had to render wastewater potable. A pilot project showed it could be done. After undergoing normal wastewater treatment, the water is subjected to three extra cleansing steps, starting with microfiltration. In this step, the water passes through microscopic, straw-like fibers that filter out many suspended solids, bacteria, and other viruses. Then, a crucial step known as reverse osmosis is performed. During this process, high pressure forces water through a plastic membrane. The pores of this membrane are so tight, explains Deshmukh, that salt and other contaminants (such as pharmaceutical drugs and toxic chemicals) do not pass through, but water does. After reverse osmosis, the water is exposed to ultraviolet (UV) light, which kills any remaining viruses and bacteria. At completion, the final product is cleaner than state and federal regulations require.

After much public outreach and input, the GWRS was approved and went online in January 2008. Orange County's water district now taps its artificially replenished groundwater to deliver clean freshwater to people for drinking, irrigation, and other uses. Every day, approximately 380 million liters (100 million gallons) of recycled water are pumped into wells or percolation basins in Anaheim, where sand and gravel naturally purify the water further as it trickles down into the region's aquifers—enough to meet the needs of nearly 850,000 residents. The OCSD is now partnering with other California districts to assist them in the development of their own water reuse projects. "Recycled wastewater" is gaining a foothold in many communities worldwide. Indeed, wastewater is increasingly seen as a solution to water shortages rather than a problem.

Despite its promise, investing in recycled wastewater is still far from common. In areas where lakes and rivers are an important source of water, many communities invest in **dams**. These barriers slow the flow of rivers and create **reservoirs**, large bodies of water that hold freshwater for a variety of uses (freshwater source, flood control, electricity production). In the United States, these often become recreation sites and fishing resources. California depends on reservoirs (both in

dam A structure that blocks the flow of water in a river or stream.

reservoir An artificial lake formed when a river is impounded by a dam.

INFOGRAPHIC 6 HOW IT WORKS: WASTEWATER TREATMENT

Sewage must be treated before it can be safely released to the environment. Most communities use chemical- and energy-intensive high-tech methods, but systems that mimic nature can also effectively purify water.

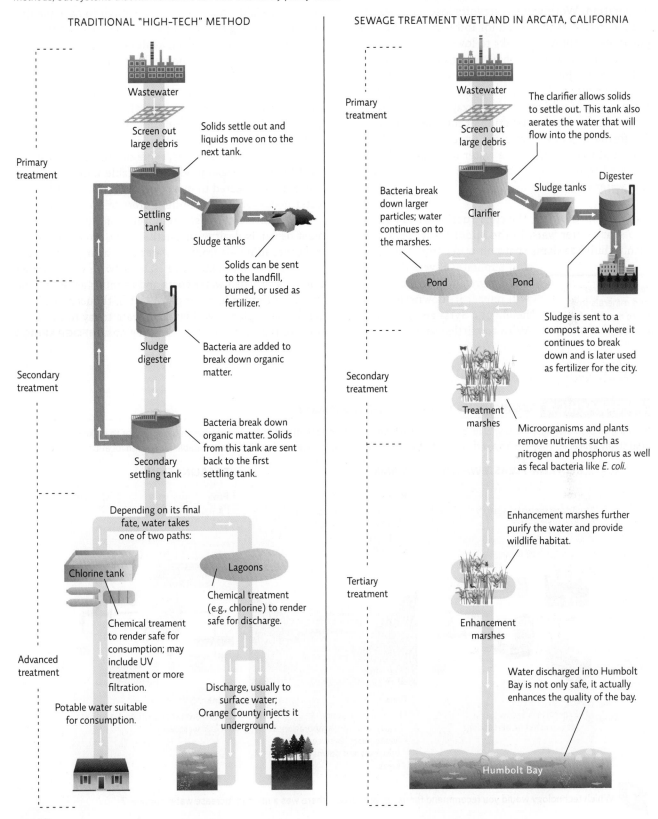

TRADITIONAL "HIGH-TECH" METHOD

SEWAGE TREATMENT WETLAND IN ARCATA, CALIFORNIA

Traditional "high-tech" method

Wastewater

Screen out large debris

Solids settle out and liquids move on to the next tank.

Primary treatment

Settling tank

Sludge tanks

Solids can be sent to the landfill, burned, or used as fertilizer.

Sludge digester

Bacteria are added to break down organic matter.

Secondary treatment

Bacteria break down organic matter. Solids from this tank are sent back to the first settling tank.

Secondary settling tank

Depending on its final fate, water takes one of two paths:

Chlorine tank

Lagoons

Chemical treatment (e.g., chlorine) to render safe for discharge.

Chemical treament to render safe for consumption; may include UV treatment or more filtration.

Advanced treatment

Potable water suitable for consumption.

Discharge, usually to surface water; Orange County injects it underground.

Sewage treatment wetland in Arcata, California

Wastewater

The clarifier allows solids to settle out. This tank also aerates the water that will flow into the ponds.

Primary treatment

Screen out large debris

Sludge tanks

Digester

Clarifier

Bacteria break down larger particles; water continues on to the marshes.

Pond

Pond

Sludge is sent to a compost area where it continues to break down and is later used as fertilizer for the city.

Secondary treatment

Treatment marshes

Microorganisms and plants remove nutrients such as nitrogen and phosphorus as well as fecal bacteria like *E. coli.*

Enhancement marshes further purify the water and provide wildlife habitat.

Tertiary treatment

Enhancement marshes

Water discharged into Humbolt Bay is not only safe, it actually enhances the quality of the bay.

Humbolt Bay

 Compare the traditional high-tech method of wastewater purification to the wetland system for wastewater purification. What do they have in common? How are they different?

and outside the state) for much of its water supply. But while reservoirs are a valuable resource, they lose an enormous amount of water every day through evaporation. Worldwide, reservoirs lose more water to evaporation than is used for industry and domestic purposes combined. Thanks to lower-than-normal rainfall and snowfall and high water demand, the water level in just about every reservoir in northern California was well below average in 2016. Despite the potential for high losses to evaporation, there are proposals for the construction of four new dams in California as a way to increase water supplies. Critics point out that pumping the water underground into depleted aquifers would be a better way to sequester this water for further use.

As mentioned earlier, the construction of dams can also spark water wars. In the Middle East, Turkey's plan to build 22 dams that pull water from the Tigris and Euphrates Rivers for agriculture and electric power will impact its downstream neighbors, Syria and Jordan. With too little water available for too many people, this hotspot may be the site of future water conflict.

Of course, Californians are lucky enough to have plenty of water all along the coast. Salt can be removed from seawater in a process known as **desalination**. The largest such facilities in the world are in the Middle East, some of which are processing around 750 million liters (200 million gallons) of water per day—about 10 times the volume of two of the largest U.S. plants (located in Tampa Bay, Florida, and El Paso, Texas). In 2015, a $1 billion desalination plant, the largest facility in the United States, came online in Carlsbad, California, and began producing potable water for San Diego; it is expected to meet about 7% of the city's water needs (about 190 million liters or 50 million gallons a day). But the salt removed from the seawater must be safely dealt with. Furthermore, removing salt and other minerals from seawater uses a large amount of energy and is very expensive—about twice what Orange County's recycled water costs and as much as ten times the cost of water in other places. Still, thousands of desalination plants worldwide operate today to meet some of the water needs of their regions. **INFOGRAPHIC 7**

desalination The removal of salt and minerals from seawater to make it suitable for consumption.

INFOGRAPHIC 7 **TECHNOLOGIES TO ADDRESS WATER SCARCITY**

Ensuring ample water supplies is vital for society and we have some high-tech methods to increase our water supply or to make it more predictably available. They all come with trade-offs that must be evaluated to determine if they are suitable for a given location.

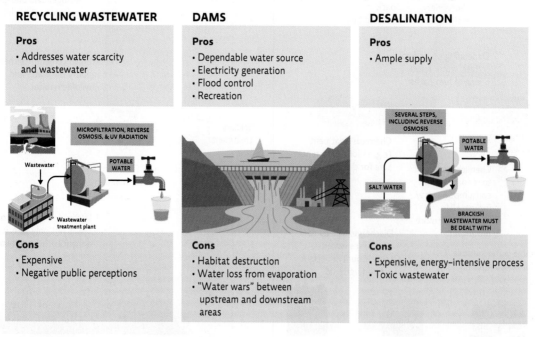

RECYCLING WASTEWATER

Pros
• Addresses water scarcity and wastewater

MICROFILTRATION, REVERSE OSMOSIS, & UV RADIATION
Wastewater
POTABLE WATER
Wastewater treatment plant

Cons
• Expensive
• Negative public perceptions

DAMS

Pros
• Dependable water source
• Electricity generation
• Flood control
• Recreation

Cons
• Habitat destruction
• Water loss from evaporation
• "Water wars" between upstream and downstream areas

DESALINATION

Pros
• Ample supply

SEVERAL STEPS, INCLUDING REVERSE OSMOSIS
POTABLE WATER
SALT WATER
BRACKISH WASTEWATER MUST BE DEALT WITH

Cons
• Expensive, energy-intensive process
• Toxic wastewater

 Which technology would you recommend for use in your area if there was a need to increase water supplies?

Revelstoke Dam is one of four dams on the Columbia River in British Columbia. A fifth penstock turbine was recently added adjacent to the original four shown here (the long tubes on the front face of the dam), giving the dam a generating capacity of around 2,500 megawatts. The sixth penstock may be installed and operational by 2019, making this the most powerful dam in British Columbia.

8 ADDRESSING WATER SHORTAGES WITH CONSERVATION

Key Concept 8: Conservation can effectively address scarcity and includes the use of water-saving technologies, behavioral changes that decrease water use, and consumer choices that minimize our water footprint.

The GWRS project was expensive: The total price tag to build the system came to about $481 million from federal, state, and local funding. And it takes about $40 million to operate it annually.

An easier and cheaper way to maintain water supplies is simply not to waste so much. For example, water-saving irrigation methods limit loss to evaporation and runoff, thus significantly reducing the water that is used—by as much as half, according to Tess Russo of the Columbia Water Center at Columbia University. This has the added advantages of protecting surface waters and of preventing soil salinization (the buildup of salt as water evaporates), a common problem in dry climates. Choosing to plant crops more suited to the environment and water availability will decrease agricultural water use. And many industrial processes are now designed to reuse water rather than discharge it into the environment.

The average U.S. citizen uses about 530 liters (140 gallons) of water per day in the home. Small individual changes in the household can save a lot of that water. But it is important to remember that our personal use of water is not limited to direct use of water from the tap. Water is also used on our behalf by industry to produce the products and energy that we consume (indirect use). This brings the average daily use of water up to 4,200 liters (1,100 gallons) per person in the United States. Reducing our use of resources and making wiser consumer choices with the purchases we do make can reduce our **water footprint**.

For example, blue jeans and other cotton clothing items have a particularly large water footprint because cotton, typically grown in arid climates, is a water-thirsty

water footprint The water appropriated by industry to produce products or energy; this includes the water actually used and water that is polluted in the production process.

The Groundwater Replenishment System in Fountain Valley, California, solves two problems at once: water scarcity and dealing with wastewater. This $480 million water treatment system converts the sewage water of Orange County into drinking water, producing more than 380 million liters (~100 million gallons) of drinking water every day.

crop. It requires large amounts of irrigation water and, if improperly applied, much of this irrigation water is lost to evaporation. It is also heavily treated with pesticides, which can pollute local water supplies and raise its water footprint. Further processing (making the fabric, dying the fabric, etc.) also uses water. Choosing clothing made from organically grown cotton (cotton grown in a way that uses water sustainably), buying used clothing, or buying fewer clothing items will all reduce the impact of cotton. Levi Strauss & Co. is working to source only sustainably produced cotton for its jeans—cotton that is grown in a manner that improves the livelihoods and working conditions of the farmers, and reduces the environmental impact of growing the cotton with changes such as more efficient irrigation and less, or no use of pesticides.
INFOGRAPHIC 8

In the meantime, supplementing potable water supplies with recycled water is an innovative way to help ameliorate ongoing water issues, says Channah Rock, a

water-quality specialist and assistant professor at the University of Arizona. It's rare to find initiatives like Orange County's, she says, but several communities—in Arizona, California, Nevada, and Florida, for instance— are reusing recycled water for nonpotable use, such as for irrigating landscapes and crops, filling fountains and fire hydrants, and flushing toilets. Communities are trying to "match the quality of water with the right use of water," she says. Since people are prohibited from drinking the water that is used for irrigation, for example, says Rock, it's not necessary to subject that water to the same advanced treatment processes as are used for potable water.

Recycling water like Orange County is doing—to reintroduce potable water back into the source aquifer or reservoir—is becoming less of a novelty as more and more communities adopt this method. In 2014, the city of San Diego approved a $3.5 billion project that that will send highly purified wastewater to a local reservoir—the source

INFOGRAPHIC 8 **REDUCING OUR WATER FOOTPRINT**

Understanding how we use water helps us make watersaving decisions, which may include behavioral changes or the use of water-efficient technologies. Much water usage can be reduced by buying less stuff and, for those things we do buy, choosing products with lower water footprints. Because water is needed in thermoelectrical production of electricity, energy conservation or use of sustainable energy sources will also reduce our water footprint.

U.S. HOUSEHOLD WATER USE

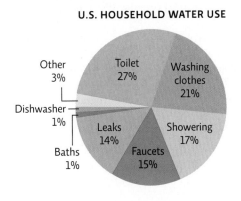

Other 3%
Dishwasher 1%
Baths 1%
Leaks 14%
Faucets 15%
Toilet 27%
Washing clothes 21%
Showering 17%

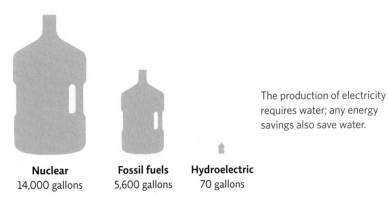

Nuclear
14,000 gallons

Fossil fuels
5,600 gallons

Hydroelectric
70 gallons

Per 1,000 KWh (typical monthly use for a home)

The production of electricity requires water; any energy savings also save water.

GALLONS OF WATER NEEDED TO PRODUCE 1 POUND OF FOOD

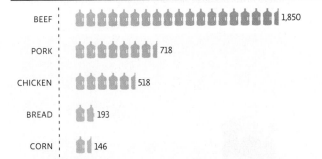

BEEF — 1,850
PORK — 718
CHICKEN — 518
BREAD — 193
CORN — 146

GALLONS OF WATER PER PRODUCT

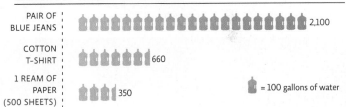

PAIR OF BLUE JEANS — 2,100
COTTON T-SHIRT — 660
1 REAM OF PAPER (500 SHEETS) — 350

= 100 gallons of water

WATER-SAVING TECHNOLOGIES AND ACTIONS

Our water usage can be reduced by using new water-efficient technologies and by making behavioral changes that don't waste water.

OLD TECHNOLOGY	NEW TECHNOLOGY	BEHAVIORAL CHANGE
Toilet 6 gallons/flush	**Low-Flow Toilet** 1.3 gallons/flush	Don't flush tissues—use the trash. Flush liquid waste less frequently.
Shower 3.8 gallons/minute **Bath** 35 gallons	**Low-Flow Shower Head** 2.3 gallons/minute	Take a "Navy" shower: Turn off the shower head except to rinse (some shower heads come with a convenient valve that allows you to switch off the water without turning it off at the source).
Faucet 5 gallons/minute	**Low-Flow Faucet** 1.5 gallons/minute	Don't leave the faucet running while brushing your teeth, shaving, or washing your face.
Washing Machine 40 gallons/load	**Washing Machine (Energy Star)** 22 gallons/load	Don't wash a clothing item unless it needs it (those jeans can probably be worn several times before washing) and run the washer only when it is full.
Dishwasher 9 gallons/load	**Dishwasher (Energy Star)** 4 gallons/load	Run the dishwasher only when it is full and limit the amount of rinsing you do before loading dishes into the dishwasher; if you have a new dishwasher, rinsing isn't needed.

 Estimate how long your typical shower lasts, and then calculate how much water you would use over the course of a year if you used a traditional, 3.8-gallon/minute shower head. Do the same calculation for the low-flow shower head. How many gallons per year would you save by switching to a low-flow shower head? Compared to the water used in a typical shower using a traditional shower head, how much would you save in a year if you reduced the duration of your shower by half and used a low-flow shower head?

of their drinking water. (A pilot program has been doing this since 2009.) Proponents of the project point out all water purified from surface waters is technically purified wastewater since communities upstream discharge treated wastewater into the very water sources downstream communities tap. But two cities in Texas, Big Spring and Wichita Falls, have taken the next step—sending purified wastewater directly to pipes that carry water to city taps in the country's first ever *direct potable reuse* projects. Now that truly is "toilet to tap."

Select References:

Heffernan, O. (2014). Bottoms up. *Scientific American, 311*(1), 69–75.

Environmental Protection Agency Office of Water (2016). *Drinking Water Action Plan.* https://www.epa.gov/ground-water-and-drinking-water/drinking-water-action-plan.

Hoekstra, A., & A. Chapagain. (2007). Water footprints of nations: Water use by people as a function of their consumption pattern. *Water Resources Management, 21*(1), 35–48.

Rock, C., et al. (2012). Water Recycling FAQs. *The University of Arizona, College of Agriculture and Life Sciences.*

Russo, T., et al. (2014). Sustainable water management in urban, agricultural, and natural systems. *Water, 6*(12), 3934–3956.

Steward, D. R., et al. (2013). Tapping unsustainable groundwater stores for agricultural production in the High Plains Aquifer of Kansas, projections to 2110. *Proceedings of the National Academy of Sciences, 110*(37), E3477–E3486.

United Nations World Water Assessment Programme. (2014). *The United Nations World Water Development Report 2014: Water and Energy.* Paris: UNESCO.

Jonathan Waterman walks with his blowup pack raft across the dry bed of the Colorado River. Because so many communities withdraw water from the river, it now runs dry before reaching the Gulf of California.

PETE MCBRIDE/National Geographic Creative

INTERACTIVE MAP **WATER WARS** ANIMATED INFOGRAPHIC

Water is a precious resource—important enough to fight over. Some fights are in court and others are on the battlefield. Growing populations and climate change will further stress our water supplies and increase the lengths many populations will go to attain access to water. Here are a few examples of recent "water wars."

THE MIDDLE EAST: THE TIGRIS AND EUPHRATES RIVERS

THE NILE RIVER

THE COLORADO RIVER

THE TRI-STATE BATTLE OVER WATER IN GEORGIA, ALABAMA, AND FLORIDA

THE ANDEAN RIVERS OF PERU

 BRING IT HOME

PERSONAL CHOICES THAT HELP

Regardless of whether our water comes from an aquifer or a local reservoir, we can make those water sources last longer by taking steps to use our water as efficiently as possible.

Individual Steps
• If you have a smartphone, download a water usage tracking app. Once you have a baseline, try to reduce it by 10%.
• Time your shower and try to reduce it by 1 to 2 minutes.
• Have a container by the sink or shower to catch water while it warms up; make sure not to get soap in it. Use this water for watering plants both inside and out.

Group Action
• Install a rain barrel at home. Rain barrels allow people to use the rain that falls on the roof of a building to water plants as opposed to letting it run off into the storm drain. If you live in a dorm or an apartment, see if you can get permission to have a rain barrel installed.

Policy Change
• Do you know where your water comes from? Talk to a city representative to find out where your water comes from and what steps are being taken to make sure it lasts as long as possible.
• Encourage local policy makers to ban the watering of lawns or restrict the use of water for landscaping to certain days of the week.

kislev/iStock/Getty Images

ENVIRONMENTAL LITERACY   UNDERSTANDING THE ISSUE

1 How is water distributed on Earth, and what are the sources of freshwater?

1. True or False: Most freshwater on Earth is tied up in glaciers and ice caps.

2. Approximately how much of Earth's surface is covered with water and how much of that water is freshwater that is accessible to humans?

2 How does water cycle through the environment?

3. Which steps of the water cycle return water vapor to the air?
   a. Evaporation
   b. Transpiration
   c. Condensation
   d. A and B
   e. A, B, and C

4. Draw a flowchart of the water cycle. (Don't copy from the book; create your own small drawing.) Follow a single water molecule from a cloud through some portion of the cycle, including a living organism, and back to a cloud.

3 What is an aquifer, how does it receive or lose water, and what problems emerge when too much water is removed?

5. Aquifer infiltration is:
   a. made easy in urban and suburban areas by all of the lawns.
   b. made harder in urban and suburban areas by roads, buildings, and lawns.
   c. the process of removing particulate matter from sewage.
   d. what happens when seawater enters a freshwater system.

6. If too much water is removed by a well in coastal areas:
   a. the water table will rise.
   b. a cone of depression will form.
   c. the aquifer might collapse.
   d. saltwater can seep into the aquifer.

7. What can lower the water table in an aquifer?

4 What is the breakdown of water use by sector, and how does use differ between developed and developing nations?

8. As income goes up, the amount of water consumed tends to
   _____.

9. True or False: The sector that uses the highest percentage of water globally is domestic use.

10. The biggest component of water use in the industrial sector in developed countries like the United States is:
    a. manufacturing.
    b. electricity production.
    c. waste management.
    d. acquisition of raw material.

11. About what percentage of freshwater is used for agriculture and how does this affect local water supplies?

5 What are the causes and consequences of water scarcity?

12. When access to water is limited only by one's ability to pay for it, this is known as _____ water scarcity.

13. Why do the problems of water scarcity and unsanitary water conditions often occur together?

6 What are some of the ways that our wastewater is treated to make it potable or safe to release into the environment?

14. True or False: Potable water is water that is safe to consume.

15. The main difference between municipal high-tech and low-tech methods of wastewater treatment is that high-tech methods:
    a. use toxic chemicals to purify water.
    b. use bacteria to break down solids.
    c. filter water at the start of the process.
    d. are less expensive.

16. One creative way that some communities deal with wastewater is:
    a. using partially treated wastewater to irrigate agricultural fields.
    b. using microfiltration, then bottling and selling it as mineral water.
    c. pumping it through a separate water system for people to use for laundry.
    d. evaporating it in gigantic reservoirs, creating additional clouds and rain.

17. What system does Arcata, California, use to deal with its wastewater and what are its advantages?

7 What technologies can help provide more potable water, and what are their tradeoffs?

18. Dams can be used to store water and increase water supplies, but they come with trade-offs such as:
    a. habitat destruction.
    b. loss of water from evaporation.
    c. regional conflicts downstream from the dam.
    d. A and B
    e. A, B, and C

19. How did Orange County address the problems of water scarcity, aquifer saltwater intrusion, and high volumes of wastewater? Would you support this solution in your own community? Explain.

8 How can conservation help address water scarcity issues?

20. Which action would be taking a step to address the single biggest use of household water in the average U.S. home?
    a. Fixing water leaks
    b. Taking shorter showers
    c. Installing a low-flow toilet
    d. Washing dishes by hand

21. Identify one behavior action that could reduce one's use of water for clothes washing and one water-efficient technology that could do so.

SCIENCE LITERACY   WORKING WITH DATA

The Water Footprint Network collects data on water use such as the water used to produce various crops. The water footprint includes water that the plants consume (both rainwater and irrigation water). It also includes the water required to assimilate pollutants released during farming. Below, we show the same data in two different graphing formats. Answer the following questions using the data provided here.

A: WATER FOOTPRINT FOR VARIOUS CROPS

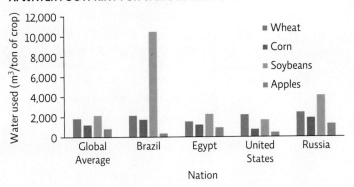

B: WATER FOOTPRINT FOR VARIOUS CROPS

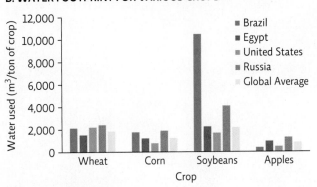

Interpretation

1. On average, which crop requires the most water to grow? Which requires the least?

2. Which nation(s) uses more water than the global average for all four crops?

3. Do both graphs accurately show the data? When might each depiction be most useful?

Advance Your Thinking

4. Why might Brazil use so much more water than other nations to grow soybeans?

5. Why do you think apples require so little water compared to the other three crops?

6. If you were farming in an area that could grow either wheat or corn, which would you choose to grow if you were worried about water scarcity?

INFORMATION LITERACY   EVALUATING INFORMATION

In the summer of 1989, Dr. Noah Boaz and his archaeological Earth-watch crews were excavating a site of ancient human habitation along the Semliki River, which runs by the border between Zaire (now the Democratic Republic of Congo) and Uganda. They could not, however, just drink the water from the river, or even swim in it. They had to filter the water a gallon or two at a time and then add chemicals to it in order to remove the waterborne parasites and pathogens. Bathing required wearing shoes and keeping their eyes, nose, and mouth out of the water. The nearby villagers did drink the water, and they had endemic health problems.

More than 750 million people do not have access to clean drinking water; as many as 3.5 billion do not have access to enough clean water. The results affect all aspects of life in developing countries: According to Water.org, a child dies every 20 seconds from a water-related illness, and women in some water-stressed areas spend several hours every day collecting water for their families' basic needs.

There have been many suggestions about ways to improve access to clean water. One of the problems is that in many areas, the lack of access is coupled with a lack of the electricity, developed roads, machines, and equipment necessary to be able to support digging municipal wells and providing pumping stations, reservoirs, and pipelines.

Go to the Global Water website (www.globalwater.org) and explore the links under "Projects." Then go to the Water.org website (www.water.org) and look at some of the featured projects.

Evaluate the websites and work with the information to answer the following questions:

1. Are these authors/sponsoring groups reliable information sources?
   a. Do they give supporting evidence for their claims?
   b. Do they give sources for their evidence, as well as clear explanations?
   c. What is the mission of the organization? How do you know this?
   d. Does the organization appear to have a workable solution or solutions?

Now search the Internet for information about two low-tech filtration devices: the LifeStraw and the PlayPump. Potentially useful sites include the bottlelessvancouver website (www.bottlelessvancouver.wordpress.com) and the HowStuffWorks website (science.howstuffworks.com).

2. Evaluate the proposed solutions for:
   a. price.
   b. ease of use.
   c. whether they would be portable or stationary.
   d. whether they include pumps for underground water or can clean only surface water.

3. Does either of the proposed solutions stand out as a good option for remote or undeveloped areas, such as the ones featured in the Global Water website or the Water.org website? Explain your answer.

Additional study questions are available at SaplingLearning.com.

SUFFOCATING THE GULF

Researchers try **to** pin **down the cause** of **hypoxia** in the Gulf of Mexico

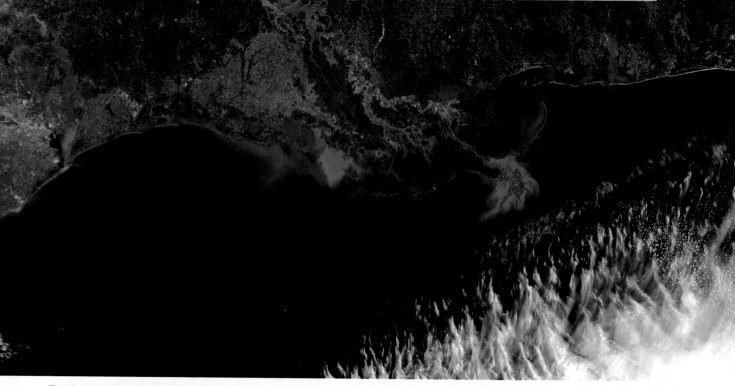

The Mississippi River is fed by thousands of smaller streams and rivers as well as runoff from the surrounding landscape. When the river eventually empties into the Gulf of Mexico, it deposits sediments and fertilizers it picked up along the way which fuels greenish phytoplankton blooms in the Gulf, so large they are visible on this satellite image.

spacephotos com/AGE Fotostock

After reading this chapter and studying the KEY CONCEPTS and INFOGRAPHICS, you should be able to answer these GUIDING QUESTIONS

CORE MESSAGE

Water pollution decreases our usable water supplies, harms wildlife and human life, and is largely caused by human actions. Some types of pollution may be easier to address than others, but in general, we can decrease water pollution by controlling what we discharge into water bodies, restoring forested areas, and limiting the use of potential pollutants.

1. What is water pollution, and how is it classified by source?

2. What are the causes and consequences of eutrophication?

3. What is a watershed, and how does it affect the quality of surface water as well as the quantity of groundwater?

4. How is water quality assessed?

5. What public policies are in place to protect water quality?

6. What role does watershed protection play in preventing water pollution?

7. How can runoff pollution from farms be reduced?

8. What can be done to reduce urban and suburban stormwater runoff?

9. What strategies can be used to restore damaged aquatic ecosystems?

Back in 1974, when he had just begun his career at Louisiana State University, biologist Eugene Turner took a 15-foot skiff out along the Gulf Coast to survey the water. He brought a handheld oxygen meter along with him. Other researchers had measured oxygen levels in the same waters and come up with some disturbingly low numbers—levels low enough to essentially "suffocate" any aquatic organism that couldn't relocate to more oxygen-rich waters. But none of them had followed up, and Turner was curious. Were those earlier measurements wrong? Flukes?

Sure enough, Turner's own readings came up low as well—much lower than expected. His curiosity deepened: What would cause low oxygen levels in these waters? He suspected the myriad oil rigs in nearby waters might have something to do with it. But due to the complicated nature of water pollution, he also knew that the true culprit could be hiding hundreds, or even thousands, of miles away.

1 WATER POLLUTION: TYPES AND CAUSES

Key Concept 1: Water pollution may come from readily identifiable sources such as discharge pipes (point sources) or from more dispersed sources such as stormwater runoff or atmospheric fallout (nonpoint sources).

Water pollution is the addition of any substance to a body of water that might degrade its quality. The list of such substances, or pollutants, is depressingly long: Raw sewage and industrial chemicals like polychlorinated biphenyls (PCBs) get dumped directly into a body of water. Meanwhile, contaminants like mercury and acid-forming precursors, along with other air pollutants from fossil fuel combustion or industry, fall back to Earth with the rain and flow as **stormwater runoff** into rivers, streams, lakes, and seas. Nutrients (from fertilizers and

animal waste) and pesticides also enter from farm and lawn runoff. Sediments from soil erosion can flow into surface waters from farms, construction sites, or heavily eroded stream banks that are no longer shored up by a well-rooted plant community.

Water pollution is not limited to the introduction of chemicals or sediment; municipal trash often finds its way to rivers, streams, and oceans (see Module 5.3). The heated water released into surface waters near power plants causes *thermal pollution*, raising the temperature of the water enough to impact many of the organisms that live in that stretch of river, lake, or ocean. Even groundwater sources can become polluted from underground chemical storage tanks or from the movement of surface pollutants down through the soil.

There are two classes of pollution, defined by how they are delivered to the water. **Point source pollution** is water pollution whose discharge source can be clearly identified (that is, one can "point" to the "source"). This includes pollution from large discharge

water pollution The addition of any substance to a body of water that might degrade its quality.

stormwater runoff Water from precipitation that flows over the surface of the land.

point source pollution Pollution from discharge pipes (or smokestacks) such as that from wastewater treatment plants or industrial sites.

◉ WHERE IS THE GULF OF MEXICO WATERSHED?

GULF OF MEXICO WATERSHED

GULF OF MEXICO

Unprotected farm fields lose topsoil as well as farm fertilizers and other potential pollutants when heavy rains occur.

Water pollution was a concern at the Rio 2016 Olympics. Watersport athletes took extreme precautions such as bleaching oars and wearing gloves to protect themselves from the sewage and trash contaminated waters.

pipes of wastewater treatment plants or industrial sites. The discharge itself is known as **effluent**.

Some of this effluent has the capacity to be quite dangerous. In most cities, sewer pipes run alongside streams because engineers assume that piped sewage will flow in the same direction and therefore reach treatment plants via gravity. But in many urban areas, those pipes are old and beginning to leak. With such a leak, raw sewage flows directly into the stream. The problem is particularly bad in cities of the developing world, where raw sewage is often directly released, untreated, into bodies of water. In many such places, waterborne *pathogens* (disease-causing organisms) in raw or partially treated human and animal waste represent a leading cause of sickness and death.

effluent Wastewater discharged into the environment.

nonpoint source pollution Runoff that enters the water from overland flow.

But the good news is that because point sources can be easily identified, they can also—at least hypothetically—be remedied.

As researchers would soon discover, the low oxygen levels in the Gulf of Mexico were due to a different and an even more challenging problem: **nonpoint source pollution**. The origins of nonpoint source pollutants are not easily identifiable. Though some arrive by air, most enter the water from overland flow (stormwater runoff); this means they can come from any part of the land that drains into a given body of water.
INFOGRAPHIC 1

Nonpoint source pollutants can include everything from human-made toxic substances to natural substances such as silt, sand, and clay, which can enter the water as *sediment pollution* (eroded soil that is washed into the water through runoff). To be sure, these substances deliver valuable nutrients to aquatic ecosystems. But excessive amounts of them can cloud the water, making it hard for sunlight to penetrate and thus disrupting photosynthesis. Sediment pollution can also harm organisms directly by clogging gills. And, when it covers the sea or river bottom, sediment can smother the nooks and crannies that serve as habitat or spawning areas.

2 EUTROPHICATION

Key Concept 2: The influx of excess nutrients into a body of water may spur algae growth and bacterial population explosions, which ultimately result in hypoxia severe enough to harm aquatic life.

dissolved oxygen (DO) The amount of oxygen in the water.

hypoxia A situation in which a body of water contains inadequate levels of oxygen, compromising the health of many aquatic organisms.

Something in the Gulf waters— some pollutant—was causing the levels of **dissolved oxygen (DO)** to plummet—a condition known as **hypoxia**. Turner and his colleague Nancy Rabalais (the

two would later marry) knew that hypoxia was a serious problem. Water can hold only a limited amount of oxygen, much less than found in air. So even a small decrease can have immediate effects on aquatic life. Even underwater organisms need oxygen to survive. Like terrestrial beings, they use it in the process of cellular respiration. Waters so

Water pollution comes from a variety of point and nonpoint sources. Although the main threat to the Gulf of Mexico is excess nutrient and sediment runoff, the EPA identifies pathogens as the leading cause of impaired waters in the United States as a whole. Found in sewage or animal waste, pathogens can come from both point and nonpoint sources.

POINT SOURCES

Some industrial and agricultural sources discharge pollutants directly into a body of water.

NONPOINT SOURCES

A variety of sources contribute pollutants that can run off the surface of the land during rainfall and enter the water; air pollutants can fall directly with the rain.

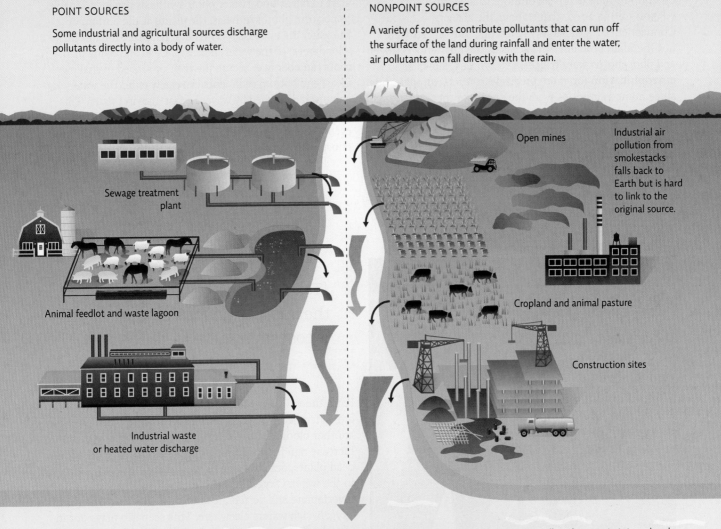

Various pollutants contribute to the many miles of U.S. surface waters that are "impaired"—waters that are too polluted to meet state or local water quality standards.

LEADING CAUSES OF IMPAIRED SURFACE WATERS IN THE UNITED STATES (2012)

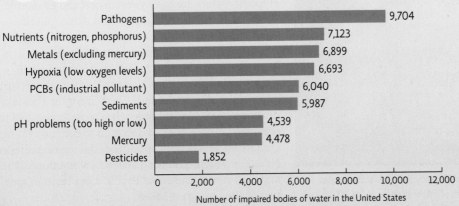

Cause	Number
Pathogens	9,704
Nutrients (nitrogen, phosphorus)	7,123
Metals (excluding mercury)	6,899
Hypoxia (low oxygen levels)	6,693
PCBs (industrial pollutant)	6,040
Sediments	5,987
pH problems (too high or low)	4,539
Mercury	4,478
Pesticides	1,852

Number of impaired bodies of water in the United States

 What type of pollution predominantly contributes to the hypoxic zone in the Gulf of Mexico: point source or nonpoint source?

depleted of oxygen that aquatic life suffers are known as "dead zones."

Month after month, year after year, in bigger and better-equipped boats, Turner and Rabalais surveyed the water, mapping out a hypoxic zone that grew from 40 km² (~15 mi²) in 1988 to a peak of more than 22,000 km² (~8,500 mi²) in 2002 (that's about the size of Connecticut). As they quickly learned, the oil rigs were not the main problem. An excess of the nutrients nitrogen and phosphorus were triggering a process known as **eutrophication** (or, more precisely, *cultural eutrophication*, since human activities were the source of these nutrients).

eutrophication A process in which excess nutrients in aquatic ecosystems feed biological productivity, ultimately lowering the oxygen content in the water.

"Imagine stretching a giant sheet of plastic wrap from the Mississippi River's mouth, straight across to Galveston [Texas]," Turner says. "Now imagine sucking all the air out and leaving the whole ecosystem there to suffocate."

Here's how eutrophication works: Because nitrogen and phosphorus fuel plant growth, extra amounts trigger explosions of algae growth. Though algae produce oxygen through photosynthesis, they also block sunlight from reaching underwater plants, ultimately blocking much more photosynthesis than they conduct. Oxygen levels start to fall as underwater photosynthesis declines. Unable to perform photosynthesis, the plants at the bottom of the water die en masse. When that happens, the turbidity (cloudiness) of the water increases: Dead and dying plant roots can no longer secure the river or seabed in shallow areas, and bottom sediments can easily enter the water column if disturbed by waves or fast-moving water. This disturbance reduces photosynthesis (and oxygen levels) even more. (See Module 2.1 for more on the nitrogen and phosphorus cycles.)

> "Imagine stretching a giant sheet of plastic wrap from the Mississippi River's mouth, straight across to Galveston [Texas]. Now imagine sucking all the air out and leaving the whole ecosystem there to suffocate."
> —Eugene Turner

From there, it gets even worse. As the plants die, they are consumed by bacterial decomposers, triggering yet another bloom—a bacterial one. The bacterial populations increase rapidly, consuming oxygen as they digest the dead plants, quickly depleting any remaining oxygen in the water. Oxygen levels can drop low enough to kill invertebrate and vertebrate animals that cannot escape to oxygen-rich waters elsewhere. This scenario plays out in waterbodies everywhere—in small ponds, the Great Lakes, and huge expanses of coastal waters. **INFOGRAPHIC 2**

Hypoxic waters are an environmental and economic concern. The National Oceanic and Atmospheric Administration (NOAA) estimates that the oxygen-poor waters of the Gulf of Mexico cost millions of dollars in lost revenue to the seafood and tourism industries each year. Species like oysters that can't escape the oxygen-poor waters die; others move away from the coastal dead zones, making it harder (and more expensive) for commercial fishermen and women to catch them. And the Gulf of Mexico is not the only coastal region suffering. NOAA has identified dead zones all along the Atlantic and Gulf coasts, in several Pacific coast regions, and in two of the five Great Lakes.

VICTORIA LOE/Tribune News Service/XSEA/Newscom

Louisiana State University biologist Nancy Rabalais taking water samples on a research vessel in the Gulf of Mexico.

INFOGRAPHIC 2 **EUTROPHICATION CAN CREATE DEAD ZONES**

ANIMATED INFOGRAPHIC

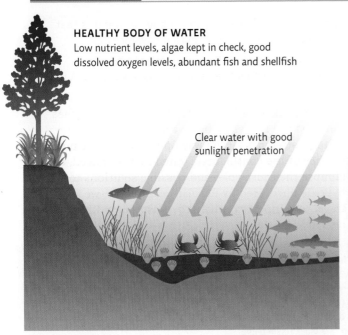

HEALTHY BODY OF WATER
Low nutrient levels, algae kept in check, good dissolved oxygen levels, abundant fish and shellfish

Clear water with good sunlight penetration

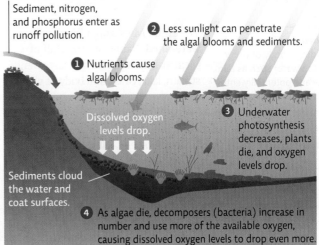

UNHEALTHY BODY OF WATER
High algae and bacterial growth, low dissolved oxygen, and loss of some aquatic life

Sediment, nitrogen, and phosphorus enter as runoff pollution.

2 Less sunlight can penetrate the algal blooms and sediments.

1 Nutrients cause algal blooms.

Dissolved oxygen levels drop.

3 Underwater photosynthesis decreases, plants die, and oxygen levels drop.

Sediments cloud the water and coat surfaces.

4 As algae die, decomposers (bacteria) increase in number and use more of the available oxygen, causing dissolved oxygen levels to drop even more.

Healthy Gulf water is relatively clear, with abundant sea life. Oysters and menhaden fish filter out particles, keeping water clean, while submerged vegetation produces oxygen to support healthy fish and shellfish populations.

Algae blooms or sediments that block sunlight and prevent photosynthesis or an influx of bacteria that deplete the water's oxygen can produce hypoxic regions. Fish and larger mobile organisms may be able to leave the area; others cannot and will suffer in the hypoxic waters.

 Why is it that in eutrophic waters, hypoxia can develop from both a decline in oxygen production and an increase in oxygen consumption?

Another complication comes in the form of algae blooms that are actually toxic. Nutrient pollution may trigger the growth of photosynthetic bacteria (cyanobacteria) that produce deadly toxins. The U.S. Environmental Protection Agency (EPA) recommends (but does not require) that public water departments test drinking water for these toxins if their presence is suspected and take steps to purify the water or issue a "Do Not Drink" advisory if levels exceed recommended amounts. Swimming in or coming into contact with contaminated water can sicken individuals; children and the elderly are most at risk for serious health problems. Though most people will avoid exposure simply because they are unwilling to swim in or play in water covered with green scum, our pets may not be so picky. A Labrador will readily chase a ball thrown into the water. In the process of retrieving that ball several times or licking itself clean, the dog may consume enough toxin to be killed. Fish too, succumb. The sight of dead fish washing ashore after a eutrophic event is a clear sign of cyanobacterial toxins.

But where was all that extra nitrogen and phosphorus coming from in the first place?

Cyanobacterial blooms, like this 2016 bloom in Stuart, Florida, plague much of Florida in the summer months. Caused by fertilizer and sewage runoff pollution, these blooms killed fish, closed beaches, and may even be linked to the deaths of several manatees.

CRISTOBAL HERRERA/European Pressphoto Agency/Newscom

3 THE WATERSHED CONCEPT

Key Concept 3a: All the land area over which water could potentially flow and empty into a body of water is that water body's watershed. Runoff can pick up pollutants in the watershed and deliver them to the water body.

Key Concept 3b: Land uses that decrease runoff and increase infiltration protect surface waters and help to recharge aquifers because the water soaks into the ground rather than reaching the nearest surface water as runoff.

watershed The land area surrounding a body of water over which water such as rain can flow and potentially enter that body of water.

The entire Mississippi River **watershed** drains into the Gulf of Mexico. The watershed of any river, stream, lake, or coastal body of water is simply the area of land over which rain and other sources of water flow to drain into a given waterbody. It also includes all the smaller streams that empty into the waterbody and their watersheds. **INFOGRAPHIC 3**

The watershed feeding the Gulf of Mexico stretches well past the Louisiana basin and the Mississippi delta. It stretches all the way up to the northernmost reaches of the continental United States. Indeed, to glimpse the causes of—and possible solutions to—the hypoxia uncovered by Rabalais and Turner, we must travel to Minnesota, where the Mississippi River begins.

INFOGRAPHIC 3 WATERSHEDS

ANIMATED INFOGRAPHIC

Anything that happens in the watershed can potentially affect the quality of a body of water as well as the quantity of groundwater. This is especially important in terms of nonpoint source pollution that originates on land. Mapping the watershed is an important tool in watershed management. Here, the watershed of this river and coastal area is outlined as a black dashed line.

If you could draw a line from hilltop to hilltop around a river and its tributaries (the streams that feed into it), you would be outlining its watershed. Water on the other side of the dotted line flows away from the watershed.

Land uses that prevent infiltration reduce the rate at which aquifers are refilled.

If you find pollution in the water at this point and you have mapped the watershed, you know where to look for the source of the pollution (upstream, in the watershed) and where not to look (downstream, or outside the watershed).

Any rain that falls on the river side of the watershed boundary could flow downhill into the river or one of its tributary streams.

 Suppose you discover that the stream by your house appears to be polluted (i.e., unusual smell, lots of dead fish, noticeable pollution, etc.). Why might it be helpful to have mapped the watershed of this stream?

It's tough to imagine that the homesteads of the far north—a place of corn fields and dairy farms—have anything to do with the rollicking backwater swamps of Louisiana. But it is here, along the tributaries that feed the nation's farmlands, that our story truly begins. "The nitrogen that's killing the Gulf starts here, gets added to the ecosystem here, in these farms and fields," says Alex Echols, former head of the U.S. Fish and Wildlife Service and a consultant to the Minnesota-based Sand County Foundation (an environmental group). "This is where it's all coming from."

The flow of that nitrogen (and other fertilizer components) over a watershed is well mapped and shows that runoff pollution can directly affect surface water *quality* by delivering pollutants. But that's not all. This runoff also impacts groundwater. When it rains, some rainfall soaks into the soil, infiltrating the ground below; some eventually reaches groundwater in the **aquifer**, which people can tap into to create a well. If the groundwater is deep enough, infiltration can act as a filtering system that purifies the water. If, however, the water table (upper level of the groundwater) is close to the surface, pollutants can make it all the way to groundwater; for example, nitrate pollution from fertilizer runoff can contaminate well water enough to be life threatening,

especially to young children. (See Module 6.1 for more on aquifers.)

But groundwater *quantity* can also be affected. Human land uses have altered the way water drains through watersheds and waterways. Pavement and even suburban lawns enhance runoff by preventing water from soaking into the ground, reducing the rate at which aquifers are refilled; they can be rapidly depleted if we remove well water faster than it is replaced. Meandering streams have been channelized and their water diverted for other uses. Wetlands, too, have been completely drained to allow urban or suburban development or for agriculture. These activities reduce the infiltration of rainwater into the ground by speeding the flow of water across and out of a watershed, sending with the water whatever industrial, agricultural, or municipal pollutants it encounters on its trip downhill. In the Mississippi River watershed, this includes runoff pollution from roadways, urban areas, and industries throughout the region, as well as animal waste and excess fertilizer from hundreds of thousands of acres of farmland. In all, some 1.5 million metric tons of nitrogen and phosphorus now flow through the Mississippi River watershed and into the Gulf of Mexico every year.

> **aquifer** An underground, permeable region of soil or rock that is saturated with water.

4 ASSESSING WATER QUALITY

Key Concept 4: The chemical, physical, and biological health of water is monitored to assess water quality.

In 2008, the newly formed Gulf of Mexico Watershed Nutrient Task Force set a goal: to reduce the hypoxic zone down to 5,000 km² (~1,900 square miles). That same year, Rabalais and Turner concluded that the Gulf ecosystem was becoming more sensitive to nutrient loads. "It's not the same system as the 1960s," Rabalais says. "It's taking less nutrients now to fuel the hypoxia."

Scientists continually assess the impact of nutrient runoff on the Mississippi River watershed's various ecosystems. Chemical parameters of water quality such as pH and dissolved oxygen (DO) content are often monitored to see if these values are within acceptable limits for the organisms who live there (e.g., a pH around neutral, high DO). If a particular pollutant is suspected, such as nitrates from fertilizer pollution, that too can be checked. Physical parameters that are assessed include the turbidity (cloudiness) and the temperature (lower temperatures mean higher DO levels).

But not everyone has the equipment or supplies to measure these chemical and physical parameters; for

those who do, chemical and physical measurements that are within acceptable limits can sometimes be deceiving. Water might look crystal clear because it supports little life; an undetected chemical might have killed everything in the water. Another way to assess water quality is a process known as **biological assessment**. This can be as simple as netting, identifying, and counting **benthic macroinvertebrates**, such as insects and crayfish that inhabit the bottoms of the streams, such as the ones that feed into the Mississippi River. If a stream is unhealthy, there won't be many organisms present that are sensitive to pollutants. The abundance and diversity of pollution-tolerant and pollution-sensitive species in the sample can be used to "rate" the stream quality. Poor stream quality can sometimes indicate that nutrient runoff or toxic substances are polluting the water, or that sediments are smothering needed stream-bottom habitat. **INFOGRAPHIC 4**

> **biological assessment** The process of sampling an area to see what lives there as a tool to determine how healthy the area is.
>
> **benthic macroinvertebrates** Easy-to-see (not microscopic) arthropods such as insects that live on the stream bottom.

INFOGRAPHIC 4 **BIOLOGICAL ASSESSMENT**

A simple way to investigate water quality of a stream is a biological assessment—using a net to collect a sample of aquatic organisms, to see what is living in the water. If there is good diversity and abundant aquatic life, especially of those organisms that are sensitive to pollution, then it is reasonable to conclude the water is clean. A poor sample suggests there is a problem, prompting one to look more closely at water quality to determine the problem.

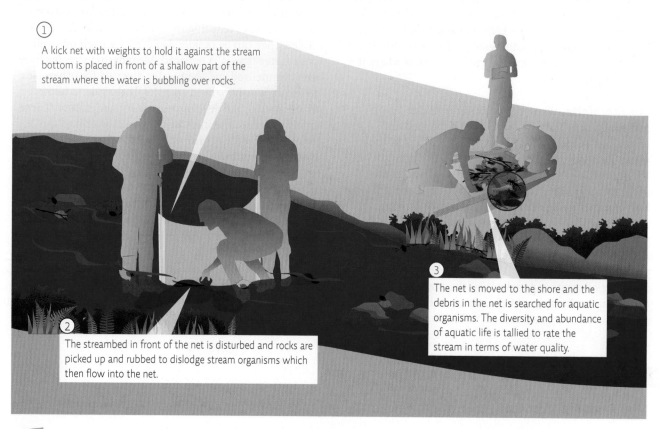

① A kick net with weights to hold it against the stream bottom is placed in front of a shallow part of the stream where the water is bubbling over rocks.

② The streambed in front of the net is disturbed and rocks are picked up and rubbed to dislodge stream organisms which then flow into the net.

③ The net is moved to the shore and the debris in the net is searched for aquatic organisms. The diversity and abundance of aquatic life is tallied to rate the stream in terms of water quality.

 If you were conducting a biological assessment and were hoping for a healthy stream, which would you rather have in your net: pollution-tolerant organisms, pollution-intolerant organisms, or both? Explain.

5 LEGAL PROTECTION: THE CLEAN WATER ACT

Key Concept 5: The Clean Water Act sets pollution standards to limit the release of pollution from point sources and has been effective at reducing this form of pollution.

Clean Water Act (CWA) U.S. federal legislation that regulates the release of point source pollution into surface waters and sets water quality standards for those waters. It also supports best management practices to reduce nonpoint source pollution.

Water quality has always been a concern but perhaps never more so than in 1969. At the time, there were no restrictions on the release of industrial chemical pollutants into bodies of water. That year, the Cuyahoga River in Cleveland,

Ohio, made headlines when it caught fire (and not for the first time) because so much oil and other flammable industrial pollutants floated on the surface of the water. This dramatic event helped spur the passage of the 1972 **Clean Water Act (CWA)**, which regulates industrial and municipal (such as sewage treatment plants) point source pollution, with the goal of making all environmental waters "fishable and swimmable." It does this by setting **performance standards**—allowable levels of a pollutant that can be present in environmental waters or released over a certain time period.

For example, the EPA, which administers the CWA, protects the public against pathogens in *recreational* waters (rivers, lakes, and coastal areas)—the leading cause of impaired waters in the United States and a major problem worldwide. (The Safe Drinking Water Act mandates that absolutely no pathogens are allowed in drinking water.) Globally, exposure to pathogen-polluted water is one of the leading causes of infection; 2 million people die each year of diarrheal diseases linked to unsafe water. The United Nations cites the provision of clean water and sanitation as one of its 2030 Sustainable Development Goals (see Infographic 4 of Module 6.1). In the United States, the EPA allows only very small amounts of fecal bacteria such as *E. coli* in recreational waters; the presence of fecal bacteria indicates sewage or animal waste contamination. Violations are investigated and addressed at either the federal or state level. Today, about 65% of U.S. waters meet the fishable/swimmable goal, more than double the number that met that goal prior to the passage of the CWA. (However, we are still far from the goal of 100% compliance with CWA guidelines.)

A recent addition to the CWA is the *Clean Water Rule* which clarifies which U.S. waters are subject to protection—specifically it protects not just navigable waters or those of immediate municipal or recreational value, but also any headwater or tributary that has been shown to impact the health of downstream waters. However, this rule has yet to be enforced; early

Firefighters stand on a bridge over the Cuyahoga River in Cleveland, Ohio, to spray water on a tugboat as a fire—started in an oil slick on the river—moves toward the docks at the Great Lakes Towing Company site. This 1952 blaze, one of 13 fires on the river since the late 1800's, was the most costly, destroying three tugboats, three buildings, and the ship repair yards.

in his presidency, Donald Trump issued an executive order which, in his words, is "paving the way for elimination" of the rule.

The CWA, like many environmental protection laws, has a *Citizen Suit Provision,* which allows citizens or groups to sue the state or federal government if they feel the statutes of the CWA are not being upheld. This empowers citizens to work to ensure the protection of the surface waters in their own communities. (See Module 5.2 for more on the Citizen Suit Provision of environmental laws.)
INFOGRAPHIC 5

performance standards The levels of pollutants allowed to be present in the environment or released over a certain time period.

INFOGRAPHIC 5 **THE CLEAN WATER ACT**

The Clean Water Act is the landmark environmental federal law that established the modern framework for the regulation of water pollution in U.S. surface waters. Many other nations have copied this framework in drafting their own water protection legislation.

GOAL
Make all of our waters safe for fishing and swimming.

IMPLEMENTATION
- Pollution standards.
- Clean Water Rule identifies waters to protect.
- Permits issued to limit polluted industrial discharges.

ENFORCEMENT
- States are in charge but they fail to enforce the CWA for ~25% of violators.
- Penalties: fines and incarceration.
- Citizen Suit Provision.

EFFECTIVENESS
- Good control of point source pollution.
- Does not regulate nonpoint source pollution.

 The idea of pollution standards implies that there is some level of a pollutant that is acceptable to have in our waters. Do you agree with this or should we strive for "zero pollution"? Explain your answer.

6 ADDRESSING NONPOINT SOURCE POLLUTION

Key Concept 6: Good watershed management can reduce nonpoint source pollution. For example, well-vegetated riparian areas reduce runoff and act as nutrient sinks.

"We did a good job on point source [pollution]," Echols says. "Back in the late 1960s, we said 'thou shall not pollute,' and we made rules and we made people follow them."

Nonpoint source pollution is a bit trickier. The CWA does not specify, for example, how much nitrogen fertilizer a farmer can apply. And this is not feasible because recommended amounts vary from farm to farm, and runoff potential varies according to rainfall, terrain, and even the crop that is planted. But scientists, farmers, and homeowners are working to fix it, just the same. Although they differ in their preferred approaches, most scientists agree that **watershed management**—management of what goes on in an area around streams and rivers—will be the key to saving the Gulf. They've begun to create *best management practices* (agreed-upon actions that minimize pollution problems caused by human actions) to reduce the amount of pollution being delivered to the Mississippi River. In agriculture, many of these practices focus on ways to decrease the amount of chemicals applied to land areas in the first place and to reduce the potential for soil erosion and runoff.

Recent CWA amendments advocate the use of these best management practices and provide funding for their implementation. However, watershed management is not easy to do; after all, many individuals and groups are engaged in many different land uses within any watershed. Identifying how each should address its contribution to nonpoint source water pollution—much less enforcing compliance—is no easy task.

watershed management
Management of what goes on in an area around streams and rivers.

riparian areas The land areas close enough to a body of water to be affected by the water's presence (for example, areas where water-tolerant plants grow) and that affect the water itself (for example, provide shade).

One of the key steps is to restore the watershed's **riparian areas**, the land areas close to the water, by maintaining or planting vegetated buffer zones that slow runoff and give the rainwater time to soak into the ground before it reaches a nearby body of water. "Most of our riparian areas—in the Mississippi River watershed, anyway—are not functioning as riparian areas anymore," Echols says. "We can't really fix hypoxia until we fix that." Healthy riparian areas are widely recognized as critical to maintaining good water quality, and projects are under way across the United States to restore and revegetate the riparian areas and watersheds of rivers and streams. In the late 1990's, New York City invested billions of dollars to restore and protect areas in the Catskills and the Delaware watershed supplying its water; thanks to this investment, the city avoided the need to construct high-tech and costly filtration systems. **INFOGRAPHIC 6**

Farmers can protect or restore riparian areas to decrease the amount of runoff from their fields that reach nearby waters but they can also take steps to reduce the potential for runoff. Echols points to three reasons more nitrogen runoff is reaching surface waters:

1. Application of more fertilizer

2. Cultivation of more crops that "leak" nitrogen back to the soil (e.g., corn, soybeans)

3. Drainage systems that remove water from soil

It's the third reason that actually moves the nitrogen (or phosphorus) out of the soil and delivers it to nearby bodies of water. By installing systems to drain the subsurface water out of wet soils, farmers can plant earlier in the spring and protect crops from heavy rains that might flood the field, increasing productivity by 20%–30%. "But at the same time you've just established a very efficient mechanism for moving water off the field, and keep in mind, nitrogen is water soluble. Water is its ticket down to the party in Louisiana," Echols says. And those drainage pipes (commonly called "tile lines" because they were originally made of short, perforated clay pipes known as tiles) end up being its passageway.

Echols and others say that tile lines may also be *our* passageway—to addressing the problem of nutrient runoff. "It's going to take huge, tectonic shifts—in federal policies, and in the global food economy, and even in the culture—to move farmers off corn and soybean, to move them off synthetic fertilizers, to move them off fossil fuels," Echols says. "But managing the tile lines? We can do that. We can do that farm by farm."

INFOGRAPHIC 6 **HEALTHY RIPARIAN AREAS PROVIDE MANY BENEFITS**

The area next to a body of water that impacts that water (provides shade and nutrients) and is itself impacted (water-tolerant species live here) is the riparian area. A well-vegetated riparian area reduces the runoff that reaches a body of water by slowing the water's movement across the land so that it soaks into the ground rather than flow into the stream.

Plants are nutrient sinks—they store nutrients in their tissues, reducing the amount free to enter the water.

Ground vegetation slows runoff and allows water to soak into the ground. This keeps the water and any pollutants it may carry out of the stream and increases infiltration, which replenishes groundwater.

Shade provided by overhanging trees cools the water; important because cooler water can hold more oxygen than warmer water.

Plant roots stabilize banks and prevent erosion and the deepening of the channel, which can speed water flow and lead to further erosion.

Trees provide food; leaf litter that falls in the water is a main nutrient source for aquatic organisms, including nitrifying bacteria that break down nitrates, reducing the amount sent downstream to the bay.

Jochen Schlenker/robertharding/Getty Images

Shown here are the U.S. Department of Agriculture's recommendations for land use in the riparian area. This includes setting aside at least 75 feet of land in managed and undisturbed forest. More may be required for suitable protection in areas with steeper terrain—that is, where runoff flow would be faster.

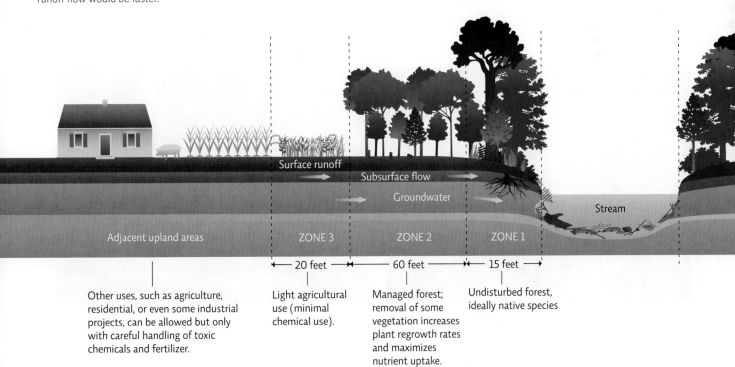

Surface runoff
Subsurface flow
Groundwater
Stream

Adjacent upland areas | ZONE 3 | ZONE 2 | ZONE 1

← 20 feet → ← 60 feet → ← 15 feet →

Other uses, such as agriculture, residential, or even some industrial projects, can be allowed but only with careful handling of toxic chemicals and fertilizer.

Light agricultural use (minimal chemical use).

Managed forest; removal of some vegetation increases plant regrowth rates and maximizes nutrient uptake.

Undisturbed forest, ideally native species

 In a managed riparian area, why not allow zone 1 to have managed forest like zone 2?

7 REDUCING AGRICULTURAL RUNOFF POLLUTION

Key Concept 7: Reducing overall fertilizer use and altering drainage systems to prevent runoff from reaching surface waters can reduce agricultural runoff pollution.

Jeffrey Strock, a soil scientist at the University of Minnesota, investigates agricultural runoff. In 2003, he teamed up with Brian Hicks, the proprietor of Nettiewyynt Farm, to study new methods to reduce runoff and fertilizer contamination of nearby streams and rivers. Hick's land was perfect: a snowbound expanse of 1,500 acres that straddles the Cottonwood River, which feeds the Minnesota, which in turn feeds the Mississippi. And because Hicks and his father used to run cattle and had yet to convert some of their pastureland to corn and soybean, there was still virgin prairie left. That meant Strock and his team could use untouched land to install and compare two different drainage treatments.

"I had two goals," says Hicks. "On the environmental side, I wanted to reduce the amount of nitrogen we were sending down into the river. And on the business side, I wanted to improve my crop yields."

Strock and Hicks installed two different drainage systems on a field: on the east side, a conventional drainage with buried pipes that drained away excess water from the soil at a steady rate. On the west side, they installed a controlled drainage system that could alter the depth of the water (via a small wooden box that would serve as a dam) to meet the needs of the crop. The hope was that the controlled drainage site would store more water—and nutrients—in the top layer of soil, which would then be available during crop production.

They discovered that during normal, nondrought years there was a 50% reduction in nutrient loss and a slight increase in yield for the controlled drainage zone compared with the conventional drainage zone. "We don't always get the agronomic benefit [of higher crop yields]," Strock says. "But we have consistently reduced the amount of nutrients running off the field. So, that environmental benefit is definitely there."
INFOGRAPHIC 7

Other agricultural best management practices will also reduce the amount of nitrogen and phosphorus available to runoff. One recommendation is to frequently test the soil to ensure that only the correct type and amount of fertilizer is added as needed; tilling fertilizer directly into the soil (rather than spreading it on the surface) also reduces the potential for loss due to runoff or wind. Farmers can also use a variety of erosion prevention methods, such as planting winter crops in the off-season to hold soil in place or planting trees as windbreaks to reduce soil erosion from wind. (For more on sustainable farming techniques, see Module 8.2.) Farmers are also going high-tech, pursuing *precision agriculture* by using GPS technology to guide farm equipment and reduce overlap when working a field, thus reducing disturbance of soil that might lead to erosion, and to guide site-specific applications of chemicals within a single field to minimize the amount used.

Jeffrey Strock, a soil scientist at the University of Minnesota, points out the satellite uplink that delivers data from sensors in the soil (like temperature, moisture, and greenhouse gas emissions) to his research center.

Strock and Hicks working in the fields to collect data on soil drainage.

INFOGRAPHIC 7 **ADDRESSING AGRICULTURAL RUNOFF**

Research to reduce nutrient pollution in surface waters includes steps to reduce agricultural runoff at its source—the farm field. Strock's study found that controlled drainage significantly reduced the movement of water and nitrogen out of the soil and increased yields compared to a field with no drainage restrictions. More research is needed to determine if this method is environmentally beneficial and economically feasible but it looks promising.

EXPERIMENTAL DESIGN: Corn was grown in rows on both plots with recommended amount of manure fertilizer.

CONTROL PLOT
Free drainage:
Water drained out with no restrictions.

TEST PLOT
Controlled drainage:
Drainage blocks kept water at desired level during the growing season.

WATER LOST THROUGH DRAINAGE

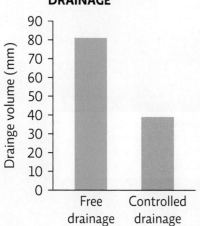

LOSS OF NUTRIENTS FROM CONTROL AND TEST PLOTS

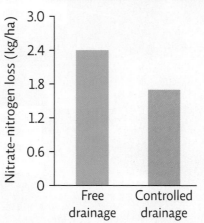

END-OF-SEASON YIELD

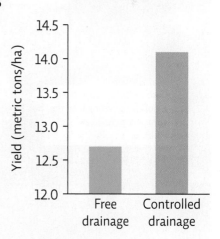

 Why might controlled drainage have helped production yields of the corn crop?

8 REDUCING URBAN AND SUBURBAN STORMWATER RUNOFF

Key Concept 8: Urban and suburban areas are significant contributors of stormwater runoff but there are many steps that can be taken to reduce runoff and increase infiltration.

Of course, not all of hypoxia's solutions will come from the farm belt. In suburban areas, lawns can be major nonpoint sources of nitrogen pollution if homeowners apply too much fertilizer. However, the grass also sequesters nitrogen because of its long growing season, preventing it from flowing to the nearest stream. Echols encourages homeowners to limit fertilizer use on lawns and to plant native plants and grasses that do not need fertilizer. Planting a rain garden of water-tolerant plants in low-lying areas that tend to flood in rain events can also help capture water and reduce runoff.

In urban areas, replacing some hard surfaces with porous surfaces such as green space or permeable pavers reduces runoff by allowing water to seep through. For this reason, Chicago has changed a large part of its alley pavement to porous concrete. Green roofs can also help capture rainwater and slow or prevent its release to the environment (see Module 4.2). Installing curb cutouts in roadways can direct stormwater flow onto natural areas where the water has a chance to soak into the ground rather than flowing directly into a storm drain, which often takes it to a nearby water body. **INFOGRAPHIC 8**

INFOGRAPHIC 8   INCREASING INFILTRATION OF STORMWATER

Stormwater that doesn't soak into the ground can enter storm drains that flow directly to rivers and streams, or cause floods, especially in heavily built-up urban settings. Anything that increases infiltration can help avoid these stormwater problems.

Trees slow and allow infiltration of runoff.

A green roof has vegetation over a waterproof layer, which can trap some water and reduce roof runoff.

Rain barrel captures runoff from roof

Redirected downspout

Rain gardens capture gutter and lawn runoff.

Curb cutouts reduce street runoff by diverting it to the ground, where it can infiltrate.

Storm drain

All these efforts help reduce the amount of polluted storm water entering storm drains that lead directly to local rivers, streams, and lakes.

 What are some other building or infrastructure methods that could reduce the flow of stormwater? (Consider things like roof size, road width, other pavement options, etc.)

9 HOLISTIC STRATEGIES TO PROTECT AND RESTORE AQUATIC HABITATS

Key Concept 9: Addressing water pollution includes identifying where problems exist, reducing pollution at the source, watershed management to reduce runoff potential, and restoring wetlands.

Water pollution is a wicked problem with many causes and consequences, as well as multiple stakeholders who often have different opinions about what actions to take. But the success of the CWA in controlling point source pollution in the United States shows that progress can be made. Likewise, steps to address watershed health and reduce the potential for runoff are reducing nonpoint source pollution.

To address hypoxia in the Gulf of Mexico, the Gulf Task Force has identified a multipronged, holistic approach that considers all the causes and consequences of the problem in a plan that addresses the *triple bottom line* (see Module 1.1). The *environmental* needs of the coastal and offshore ecosystems are addressed by pursuing best management practices that reduce the occurrence or impact of water pollution. For example, steps that restore and conserve riverside or coastal habitats in the Gulf watershed will reduce the amount of pollution that reaches the Gulf and protect the species who live there. This, in turn, will protect the ecosystem services these species provide, such as water purification. Taking steps to reduce pollution at the source (e.g., reducing industrial point source and agricultural nonpoint source water pollution) will make these restoration efforts even more effective.

These efforts and others will also address the *social* impacts on human communities that are affected throughout the Gulf region as important fisheries are replenished and protected. Coastal communities also benefit from improved protection against storm surges or coastal erosion.

Finally, the plan accounts for the *economic* realities that require us to prioritize our actions, first pursuing those actions that are either less expensive and easy to implement, or more costly ones with high payoffs to the environment or community. **INFOGRAPHIC 9**

INFOGRAPHIC 9 | **GULF OF MEXICO REGIONAL ECOSYSTEM RESTORATION**

GOAL	STRATEGY
Restore and conserve habitat. 	• Restore natural flow of rivers and streams to wetland and delta regions. • Restore riparian areas and wetland habitats. • Implement measures to restore seagrass beds and protect seagrass from boat propeller damage.
Restore water quality. 	• Implement better regulation of industrial point source pollution. • Reduce runoff from animal operations by fencing off streams and construct wetlands to capture runoff. • Implement precision fertilizer application to reduce overuse.

GOAL	STRATEGY
Replenish and protect marine resources. 	• Restore and manage coral reefs and oyster beds; sustainably harvest fish and shellfish. • Track sentinel species such as sea oats, pelicans, bluefin tuna, and oysters to monitor progress or problems. • Minimize or eliminate invasive species such as lionfish and nutria.
Enhance community resilience. 	• Improve coastal protections against storm damage. • Promote low-impact community growth plans. • Enhance education and outreach.

 How is the task force considering the triple bottom line with their recommendations to reduce the size of the gulf hypoxic zone?

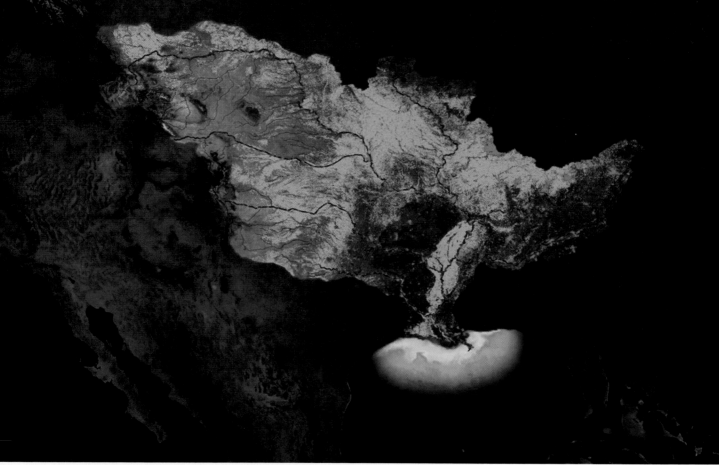

NOAA's Environmental Visualization Lab

Cities (shown in red) and farmland (shown in green) within the watershed of the Mississippi River fuel the Gulf of Mexico's hypoxic zone in late summer.

The 16,760 km² (~6,470 mi²) hypoxic zone in 2015 was greater than the 5-year average of 14,000 km² (~5,400 mi²), and more than three times the size of the Gulf of Mexico Hypoxia Task Force goal. The hoped-for date to reach that goal? 2015. That target date has been extended to 2035.

These types of control efforts are gaining increasing significance in a world where human population is still growing and the need for higher agricultural productivity grows with it. Meeting these needs and protecting our water at the same time may get even more difficult due to climate change that is reducing agricultural productivity in many areas worldwide, a trend that is expected to get worse in the future.

Select References:

Diaz, R. J., & R. Rosenberg. (2008). Spreading dead zones and consequences for marine ecosystems. *Science, 321*(5891), 926–929.

Echols, A., & Vitousek, P. (2012). Managing water, harvesting results. *Frontiers in Ecology and the Environment, 10*(1), 3-3.

Environmental Protection Agency. (2011). *Gulf of Mexico Regional Ecosystem Restoration Strategy*, https://archive.epa.gov/gulfcoasttaskforce/web/pdf/gulfcoastreport_full_12-04_508-1.pdf

Feset, S. E., Strock, J. S., et al. (2010). Controlled drainage to improve edge-of-field water quality in southwest Minnesota, USA. In *9th International Drainage Symposium held jointly with CIGR and CSBE/SCGAB Proceedings, June 13–16, 2010, Québec City Convention Centre, Québec City, Canada* (p. 1). American Society of Agricultural and Biological Engineers.

Oquist, K. A., et al. (2007). Influence of alternative and conventional farming practices on subsurface drainage and water quality. *Journal of Environmental Quality, 36*(4), 1194–1204.

Rabotyagov, S. S., Rabalais, N. N., Turner, R. E., et al. (2014). The Economics of Dead Zones: Causes, Impacts, Policy Challenges, and a Model of the Gulf of Mexico Hypoxic Zone. *Review of environmental economics and policy, 8*(1), 58–79.

INTERACTIVE MAP **WATER POLLUTION**

ANIMATED INFOGRAPHIC

Water pollution comes in many forms. Effective programs to reduce it address pollution at the source. Here are four examples in the United States of communities addressing water pollution in their area.

SAN FRANCISCO BAY

LAKE ERIE

NEW YORK–NEW JERSEY HARBOR

GALVESTON BAY

 BRING IT HOME

PERSONAL CHOICES THAT HELP

We are facing not only shrinking supplies of easily accessible water but also the potential degradation of this resource due to pollution. By changing products and modifying common practices, we can improve our water quality for years to come.

Individual Steps

• Read your city's water quality report to see which pollutants are prevalent in your area.

• Decrease your use of chemicals (fertilizers, pesticides, harsh cleaners, etc.) that will end up in the water supply. For safer alternatives to lawn care see www.epa.gov/safepestcontrol/lawn-and-garden.

• Always dispose of pet waste properly. In high quantities, it acts as an oxygen-demanding waste and can also spread disease.

Group Action

• Marking storm drains with "Don't Dump" symbols can remind people not to dump waste liquids down sewer drains. If the drains in your area are not marked, talk to city officials to see if you and other volunteers can mark them.

Policy Change

• August is National Water Quality Month. Take steps every day to reduce your water pollution and help raise awareness in August by writing a letter to your local newspaper outlining simple steps people can take to improve water quality.

Robert Brook/Science Source

ENVIRONMENTAL LITERACY **UNDERSTANDING THE ISSUE**

1 What is water pollution, and how is it classified by source?

1. Water pollution is:
   a. found only in surface waters near cities.
   b. primarily excess nutrients from lawns, farms, and animal feedlots.
   c. usually from excess carbon being added to the system.
   d. contaminants or excess nutrients in surface waters and in groundwater.

2. Fertilizer from your lawn and motor oil from the leaky oil pan on your car are examples of:
   a. nonpoint source pollution.
   b. point source pollution.
   c. eutrophication.
   d. pathogenesis.

3. Compare three typical point source pollutants and three nonpoint source pollutants from the area where you live.

2 What are the causes and consequences of eutrophication?

4. True or False: Hypoxic waters are those with extremely low levels of oxygen.

5. Many cattle pastures in the southeastern United States have ponds to provide water to the livestock but often by summer's end these ponds are covered in a thick green scum. If the ponds are stocked with fish, the fish can all die. What is causing the green scum to form and what might be causing the fish kills?

3 What is a watershed, and how does it affect the quality of surface water as well as the quantity of groundwater?

6. A watershed includes:
   a. only the land that would be underwater during a normal rainfall year.
   b. the surface water and the underground aquifer.
   c. all the uphill land surrounding a river and its streams that can feed water into that river.
   d. all the land downhill from a river that could potentially be flooded.

7. Where does the nitrogen and phosphorus pollution that reaches the Gulf of Mexico come from?

4 How is water quality assessed?

8. In a biological assessment of stream health, bottom-dwelling insects are collected and water quality inferred based on
   a. physical parameters important to life such as temperature and pH.
   b. the chemical composition of insects captured.
   c. the diversity and abundance of the species present.
   d. the reproductive health of the insects, determined by the number of eggs and larvae found.

9. Why is it incorrect to assume that a stream that looks crystal clear is environmentally sound?

5 What public policies are in place to protect water quality?

10. True or False: The Clean Water Act focuses primarily on reducing nonpoint source pollution.

11. The goal of the Clean Water Act is to:
    a. make environmental waters safe for fishing and swimming.
    b. keep surface water clean enough to tap for drinking water sources.
    c. eliminate benthic macroinvertebrates and other pathogens from streams and rivers.
    d. improve water quality so that it is suitable for agricultural uses.

12. What is the Citizen Suit Provision of the Clean Air Act?

6 What role does watershed protection play in preventing water pollution?

13. True or False: The adoption of best management practices is regarded as a useful tool for watershed management.

14. The riparian area of a stream is the:
    a. main channel of a streambed where water flows year round.
    b. banks of the stream and the streambed area immediately adjacent to the banks.
    c. land area close enough to the water to affect it and be affected by it.
    d. land area uphill from the stream that could potentially contribute runoff to that stream.

15. What are the benefits of a healthy riparian area?

16. According to Alex Echols, what three changes to agriculture contribute to the increase of nutrient pollution to the Mississippi River?

7 How can runoff pollution from farms be reduced?

17. True or False: Farmers can use natural fertilizers, such as animal manure, to prevent nutrient runoff pollution.

18. What could be done to reduce each of the factors identified in question 16 above?

8 What can be done to reduce urban and suburban stormwater runoff?

19. Which of the following would be the best urban land use for a riparian area in terms of reducing runoff pollution?
    a. A paved surface that is smooth and unobstructed
    b. A well-manicured lawn
    c. An area planted with native trees and shrubs
    d. A meadow with tall grass and wildflowers

20. What is a rain garden and how can it reduce stormwater runoff?

9 What strategies can be used to restore damaged aquatic ecosystems?

21. Gulf of Mexico restoration steps such as improving coastal protections against storm damage, promoting low-impact community growth plans, and enhancing education and outreach are aimed at:
    a. restoring habitat.
    b. restoring water quality.
    c. replenishing and protecting marine resources.
    d. enhancing community resilience.

22. Identify several strategies that could be used to restore water quality in the Gulf of Mexico?

SCIENCE LITERACY   WORKING WITH DATA

The application of road salt in cold climates can result in runoff pollution that can damage aquatic ecosystems. Researchers at Saint Mary's University in Halifax surveyed local roadside ponds (some of which were occupied by amphibians and some of which were not) to determine chloride concentration and species richness of amphibians. They also investigated the vulnerability of five amphibian species to sodium chloride (NaCl) by determining the LD_{50}, the dose or concentration of chloride that kills 50% of the population.

A: EFFECTS OF CHLORIDE ON AMPHIBIAN SPECIES RICHNESS

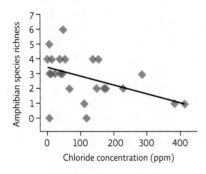

B: CHLORIDE CONCENTRATION IN OCCUPIED AND UNOCCUPIED PONDS FOR 5 AMPHIBIAN SPECIES

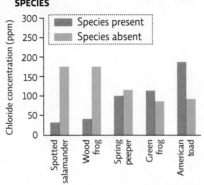

C: COMPARISON OF MEDIAN LD_{50} VALUES IN AMPHIBIANS

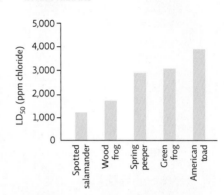

Interpretation

1. Look at Graph A above. In a single sentence, explain the relationship between chloride concentration and species richness.

2. Look at Graph B. Which species of amphibian is least affected by the presence of chloride in the water? Explain your reasoning.

3. Look at Graph C. Which species is the most sensitive to chloride? Which is the least sensitive?

Advance Your Thinking

4. Look at Graph B and Graph C. Does the LD_{50} for each species shown in Graph C correlate with the data (presence or absence of each species according to chloride concentration) shown in Graph B? Explain.

5. Suggest two actions that could be taken to reduce the chance that road salt would enter ponds in runoff.

INFORMATION LITERACY   EVALUATING INFORMATION

Settling ponds, also known as holding ponds, are locations where contaminated water is allowed to stand so that particulates suspended in the water settle to the bottom of the pond. The water then evaporates or is drawn off, leaving behind the unwanted sediment, which can be from a mine, a quarry, a manure pit at an animal facility, industrial wastewater, stormwater, or other sources.

Go to www.technology.infomine.com/sedimentponds to see an overview of some of the concerns about building and maintaining a settling pond for surface mines. Make a list of the concerns.

Using what you learned about mine settling ponds, investigate an incident that occurred in Hungary by completing the following assignment. Compare the information and authors of articles you access as you answer the following questions:

1. Do an Internet search and read two or three news articles about the toxic sludge spill near Kolontár, Hungary.
   a. What is the most recent article you can find about this, and what does it say about the incident?
   b. How similar is the information in the articles?

   c. Do the authors give any sources cited for the "facts" presented?

2. Now do an Internet search using Google Scholar or another scholarly database. Select a peer-reviewed article that you can access in its entirety (not just the abstract; look for a PDF link). Read the article and summarize its purpose and main points.
   a. How does this article compare to the news articles you read? Compare and contrast the types of information regarding scope of coverage, intended audience, clarity of information, and reliability of information (whether reliable sources are cited).
   b. In general, when would you go to a news source for information on a topic like this, and when would you go to a scholarly source?

3. What could be done to prevent another incident like what happened at Kolontár from happening again?

 Additional study questions are available at SaplingLearning.com.

CHAPTER 7

LAND RESOURCES

CHAPTER 7

We access many vital biological and mineral resources from terrestrial regions. Our actions can overexploit these resources or damage the ecosystems that provide them, but we can make choices to protect the natural ecosystem and help avoid running out of needed resources.

Module 7.1: Forest Resources

An assessment of the importance of forest resources and a look at the causes and consequences of deforestation, along with potential solutions

ONLINE Module 7.2: Grasslands and Soil Resources

An introduction to grasslands and the importance of soil resources, the threats they currently face, and ways to manage them sustainably

ONLINE Module 7.3: Mineral Resources

An examination of the importance of mineral resources, the impacts of acquiring and using them, and ways to reduce those impacts

Online Modules are available at SaplingLearning.com.

Nick Brundle Photography/Getty Images

RETURNING TREES TO HAITI

Repairing a forest ecosystem one tree at a time

Teams of men and women pick their way across steep, rocky slopes as they plant trees in Mahotiere, Haiti.
Benjamin Rusnak/ZUMA Press/Mahotiere/Haiti/Newscom

After reading this chapter and studying the KEY CONCEPTS and INFOGRAPHICS, you should be able to answer these GUIDING QUESTIONS

CORE MESSAGE

Forests have great economic value, but we must balance that with the value of their ecosystem services and sociocultural benefits. Using sustainable management practices may allow us to harvest forest products without destroying the forest or its ability to provide ecosystem services and future products.

1	What is a forest and what influences which forest type is found in a given area?
2	What is the three-dimensional structure of a forest and how are the plant species found there adapted to their level of the forest?
3	What ecosystem services do forests provide?
4	What threats do forests face?
5	How do the different timber harvesting techniques compare in terms of method and sustainability?
6	How can we protect and sustainably manage forest resources?

When Jean Robert was a young boy, the mountains surrounding the Haitian city of Gonaïves (pronounced go-nah-EEV) were still lush and green with trees and his family's hillside farm produced a variety of crops. But today, the forests are gone, and the mountains are bare. Not only have crop yields shrunk drastically but the mountain homes and the city below have been left defenseless against the tropical storms that pound Haiti every summer. Unencumbered by trunks or roots or shrubs, water rushes freely downward, gathering into apocalyptic mudslides that destroy homes, crops, and livelihoods. In 2004, a single storm claimed more than 2,000 lives from this one city.

Now, Robert and his neighbors are trying to bring back trees. Working as a team, the community work group, or *kombit*, plants a carefully planned mix of fruit and timber trees. By the end of the month, they say, each member will have his or her own saplings to tend. If the saplings survive, their efforts will give rise to a new era of sustainable forestry, which strikes a balance between what people need from the forest and what the forest itself needs to survive.

The kombit will plant three main types of trees. They start with fast-growing, multipurpose trees like the moringa. Because it is a nitrogen-fixing plant, the moringa helps refertilize the soil. And because it grows quickly, it can provide a sustainable source of food (the leaves are edible) and fuel wood. Next, they plant fruit trees—mangoes, avocados, and citrus. These trees take longer—3 to 5 years—before their fruit is ready to harvest, but they put down roots and thus stabilize the soil in just a few months' time. "Farmers are less likely to cut down fruit trees for charcoal, because they know it will provide the kind of food they can both eat and sell for profit," says Haitian ecologist Timote Georges, who is working with the U.S.-based nonprofit Trees for the Future to reforest the mountains around Gonaïves. The last thing the kombit plants is the slow-growing timber trees, which may be sustainably and profitably harvested in the future but don't provide any immediate benefits to the farmers. "The goal is to mix it up," says Georges. "You can create a whole stable system that's going to provide money and food throughout the seasons and across the years."

But to really understand how trees might help alleviate poverty, we must first consider just what forests are, what they do for us, and why so many of them are being chopped down in the first place.

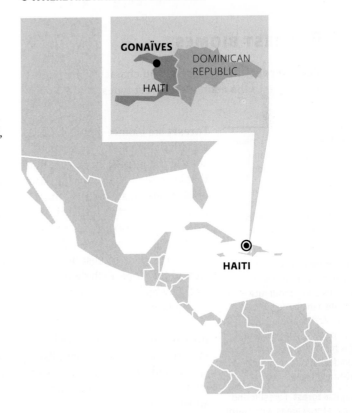

◉ **WHERE ARE HAITI AND GONAÏVES?**

GONAÏVES

DOMINICAN REPUBLIC

HAITI

◉
HAITI

Volunteers begin planting 25,000 donated trees in Mahotiere, Haiti, to combat soil erosion.

1 FOREST BIOMES

Key Concept 1: The location and characteristics of forest biomes are influenced by temperature and precipitation, giving rise to the main three forest types (boreal, temperate, and tropical) and a variety of subtypes of each.

forest An ecosystem made up primarily of trees and other woody vegetation.

boreal forest Coniferous forest found at high latitudes and altitudes characterized by low temperatures and low annual precipitation.

temperate forest Forest found in areas with four seasons and a moderate climate, which receives 30 to 60 inches of precipitation per year and which may include evergreen and decidiuous conifers and broadleaf trees.

tropical forest Forest found in equatorial areas with warm temperatures year-round and high rainfall; some have distinct wet and dry seasons, but none has a winter season.

Forests are biomes dominated by trees. They currently cover about 30% of the planet's landmass, but, thanks to their sheer concentration of biodiversity, are home to more than 50% of Earth's terrestrial life and more than 60% of its green, photosynthesizing leaves. There are many types of forest biomes around the globe, each determined by the temperature and amount of precipitation the area receives.

Boreal forests, characterized by evergreen coniferous species like spruce and fir, cover vast tracts of land in the higher latitudes and altitudes and are characterized by low temperatures and precipitation levels; they represent some of the most expansive forests left on Earth—some of it is still *primary*

or *old-growth* forest; other areas that are harvested for timber have regrown as *secondary* forests.

Temperate forests, which contain deciduous broadleaf trees like oak, hickory, and maple that lose their leaves in winter, are found in midlatitudes. (Evergreen broadleaf trees like the southern magnolia are found in warmer regions of these latitudes.) These forests are not as expansive as the boreal forests because they are found in latitudes with high human populations. Most temperate forests today represent secondary growth rather than primary stands due to extensive timber harvesting. However, some areas have seen remarkable regrowth in the 20th century. For example, there is now more forest cover in the eastern United States than there was in the 19th century, following heavy clearing. (This regrowth can be sustained because forests elsewhere are harvested.)

Tropical forests contain a diverse mix of tree and undergrowth species and are found in tropical latitudes where temperatures do not vary much throughout the year. Though vast tracts of primary tropical forest still remain, these forests are experiencing the highest losses of any forest biome, especially in Southeast Asia and South America. **INFOGRAPHIC 1**

When Europeans first arrived in Haiti some 500 years ago, two-thirds of the land was covered in forests so majestic that explorers dubbed this region "the Pearl of the Antilles." But as centuries passed, the forests were cleared—sometimes with amazing breadth, to make way for coffee and sugar plantations; other times, they were cleared in discrete chunks that Haitian peasants would then convert into subsistence farms. Many of the trees—along with the coffee and sugar crops that displaced them—were sold overseas. Most of the rest provided fuel for cooking Haitian meals. Today, less than 2% of that original forest remains; 6% of the land has no soil left at all.

INFOGRAPHIC 1 | FORESTS OF THE WORLD

Forests are biomes primarily made up of trees and other woody vegetation. There are three main types of forests, classified according to climate, and many subdivisions within these three types.

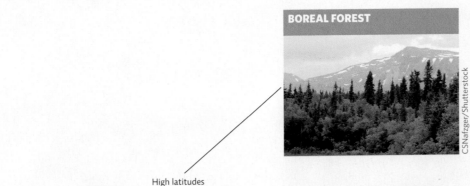

BOREAL FOREST

CSNafzger/Shutterstock

Boreal forests (taiga) represent the largest terrestrial biome; they stretch from Canada to Siberia and are found at higher elevations in lower latitudes. They have a short growing season with little precipitation, most of it snow. Soils here are thin and acidic, and the major tree species are evergreen conifers with needlelike leaves.

High latitudes (N of the equator)

Midlatitudes (N and S of the equator)

Low latitudes

90° N
60° N
30° N
0° Equator
30° S
60° S
90° S

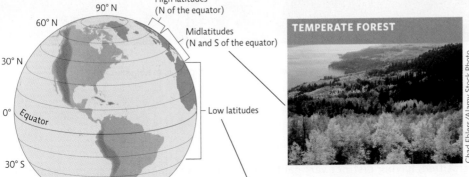

TEMPERATE FOREST

Chad Ehlers/Alamy Stock Photo

Temperate forests have distinct seasons. The soil is fertile, with a thick layer of decomposing leaf litter that supports the plant life. Depending on how much precipitation an area gets, the forest may be predominantly coniferous or broadleaf.

TROPICAL FOREST

Ghislain & Marie David de Lossy/Getty Images

Tropical forests have similar temperatures year-round. Dry tropical forests have distinct wet and dry seasons, whereas tropical rainforests receive rain year-round. The soils are thin, acidic, and low in nutrients. Rapid decomposition by fungi and bacteria supports the dense vegetation found in these soil-poor areas.

 Which forest type is the most expansive? Why do you think this is true?

The original forests of Haiti would have looked like this one on the Barahona Coast in neighboring Dominican Republic.

2 FOREST STRUCTURE

Key Concept 2: Forests are stratified, having distinct vertical layers, each of which contains species adapted to the level of sunlight and moisture available.

canopy The upper layer of a forest, formed where the crowns (tops) of the majority of the tallest trees meet.

emergent layer The region where a tree that is taller than the canopy trees rises above the canopy layer.

understory The smaller trees, shrubs, and saplings that live in the shade of the forest canopy.

forest floor The lowest level of the forest, containing herbaceous plants, fungi, leaf litter, and soil.

Most forests consist of four distinct layers. The **canopy**, formed by the overlapping crowns of the tallest trees, makes up the ceiling of the forest. Some even taller trees may reach above the canopy to form an **emergent layer**. Beneath the canopy is the **understory** layer, where shade-tolerant shrubs, smaller trees, or the saplings of larger trees grow. Sometimes these trees are dense enough to form a lower canopy. The lowest level is the **forest floor**, which is typically made up of seedlings, herbs, wildflowers,

and ferns. The forest floor also contains soil, which is composed of leaf litter and other debris—branches, logs, and stumps—that decomposes over time.

Within each forest layer is a range of species uniquely suited to the temperature, humidity, and amount of sunlight that layer receives and well adapted to its particular neighbors. For instance, in a temperate deciduous forest, wildflowers on the forest floor will bloom early in the spring, before the bigger trees "leaf out" and block the Sun. Sunlight is a precious commodity on the forest floor, and wildflowers compete for it. One wildflower species may bloom one week and another the next week. Disruption to one part of the forest (for instance, cutting down a tree that opens the canopy) can have a trickle-down effect that impacts each subsequent layer. **INFOGRAPHIC 2**

A look at the three-dimensional structure of a forest reveals layers that house distinct species adapted to the environmental conditions in each layer.

Emergent layer: A few trees grow above the general level of the forest canopy.

Canopy: The crowns of the dominant trees shade the layers below.

Understory: Trees and shrubs here are adapted to shade; saplings will grow rapidly when a spot in the canopy opens up.

Forest floor: This lowest level contains leaf litter, decomposing plant material, herbs, flowers, and seedlings.

 What kind of plants might be found in a region of the understory that has been opened up to sunlight by a fallen tree?

3   ECOSYSTEM SERVICES OF FORESTS

Key Concept 3: Forests provide a wide variety of critical ecological services to water, soil, air, and wildlife; have great economic value (e.g., jobs, consumer products); and provide sociocultural benefits (e.g., cultural, spiritual, or recreational value).

Thanks in part to unchecked deforestation, Haiti has become both the poorest and the most environmentally degraded country in the Western Hemisphere. With few trees and little soil, drinking water has grown polluted, crops have dwindled, and the people have suffered.

Farmers like Jean Robert say that forests are the key to changing all of that. They have planted thousands of trees in these mountains in recent years and hope to plant thousands more. "Almost all of the country's problems—natural disasters, food shortages, poverty—can be traced back to rampant **deforestation**," says Georges. "So if we want to fix the country, we have to put the forests back and then find a way to manage them better."

That's no small feat. Forests, after all, are one of our most contentious resources. Such is the range of economic benefits and **ecosystem services** they provide that any

one use must be weighed against a host of others, and immediate human needs are frequently pitted against long-term conservation goals. In Haiti, where most people live on less than $2 a day, trees provide food, energy, building material, and desperately needed income. To stop them from being chopped down, Georges and his fellow Haitians will have to find alternative sources of each, not to mention a farming method that doesn't require clearing the forests.

Food, fuel, and building materials are perhaps the most obvious ecosystem services provided by forests. However, even though trees are the largest and most notable life-forms present, a forest is much more than just its trees. Together, the species inhabiting the different layers participate in a delicate symphony of chemical and physical cycles that

deforestation Net loss of trees in a forested area.

ecosystem services Essential ecological processes that make life on Earth possible.

produce an invaluable range of ecosystem services for the planet.

It starts with the soil, which the forest itself helps to form and maintain: Leaves and branches die, fall to the ground, and decay, forming a thick brown layer of nutrients in which all future generations of plant life will take root. (For more on soil, see Online Module 7.2.)

While dead and decaying plants help form the soil, living ones hold it in place. During rainstorms, roots—especially tree roots—anchor soil in place so that it can't flow as easily down hillsides into nearby surface waters (lakes, streams, oceans) with the rain. Soil and roots also slow the flow of rainwater across the ground's surface (called **stormwater runoff**), preventing potentially polluted water from reaching surface waters. And, by slowing runoff, a well-vegetated area allows more water to soak into the ground, recharging the groundwater supplies that provide area residents with their major source of drinking water. The soil also traps chemicals that might otherwise contaminate that drinking water. (See Module 6.2 for more on stormwater runoff.)

While plants, soil, and water are playing off one another in this manner, the forest is also conducting another important cycle: pulling carbon dioxide out of the atmosphere and replacing it with oxygen (through the process of photosynthesis). In fact, forests as a whole store more carbon in their biomass, litter, and soil than all the carbon in the atmosphere, making this biome the world's largest terrestrial **carbon sink**—an area that stores more carbon than it releases, such as the standing timber in a forest or organic matter in soil. (The oceans hold more carbon, but their capacity to absorb more may be diminishing; see Online Module 6.3.) And forest leaves produce so much oxygen that they are commonly referred to as "the lungs of the planet."

Last but not least, virtually every layer of forest provides food and habitat for a bevy of animals (vertebrates and invertebrates), fungi, and microbes; these creatures all do their part to contribute to the functioning of the forest ecosystem as a whole. It is difficult to overstate the importance of forests to the biosphere.

The sociocultural benefits are also far reaching. Some forest stands are considered sacred by indigenous people, connecting them to their past and representing their future. Many people view a forest as the ideal place to reconnect with nature, become rejuvenated, or find artistic inspiration, benefits that, for some, can't be replicated in our modern, built-up world.

In addition to all these ecosystem services and sociocultural benefits, forests also provide a range of

stormwater runoff Water from precipitation that flows over the surface of the land.

carbon sink An area such as a forest, ocean sediment, or soil, where accumulated carbon does not readily reenter the carbon cycle.

South of Dajabón, Haiti's brown landscape on the left contrasts sharply with the rich forests of its neighbor, the Dominican Republic, on the right.

JAMES P. BLAIR/National Geographic Creative

economic benefits. In fact, humans have relied on forests for millennia for a multitude of consumer goods. Wood products, including lumber, firewood, charcoal, paper pulp, and some medicines, account for around $100 billion in global trade every year. On top of that, the wildlife supported by forests provides humans with food and recreational hunting opportunities, both of which have economic value. **INFOGRAPHIC 3**

INFOGRAPHIC 3　ECOSYSTEM SERVICES OF FORESTS

All ecosystems, including forests, contribute to the ongoing functioning of the planet and the immediate well-being of humans. Some of these services we take for granted; others are recognized for their economic value.

Josef Mlcuch/EyeEm/Getty Images

ECOLOGICAL SERVICES	ECONOMIC VALUE	SOCIOCULTURAL BENEFITS
Watershed services: Water purification and provision. **Climate regulation:** A major sink for CO_2; increases rainfall in some areas; biggest contributor of oxygen to the atmosphere. **Soil maintenance and protection:** Soil production and recycling of nutrients; reduction of soil erosion. **Disturbance regulation:** Protection from storm damage, especially in coastal areas. **Habitat and genetic resources:** Food and habitat for biodiversity; a rich storehouse of genes that might prove useful to improve our crops or provide as-yet-undiscovered medicines.	**Goods:** Provide many of the basic goods we depend on, including: • Food • Fuel • Building materials • Other products, such as rubber and cork • Raw material for paper and other industrial products • Medicines **Recreation and ecotourism opportunities** **Jobs:** More than 10 million people make their living in and from forests.	The beauty of forests provides a place for spiritual renewal, artistic inspiration, and stress reduction. Ancient stands of trees provide a connection to the past; many indigenous people are an important part of their forest ecosystem, possessing ancestral knowledge of the forest and its inhabitants.

Robert Costanza and his colleagues evaluated ecosystem services of forests and quantified their value (in 2007 U.S. dollars) to be $3.8 trillion.

ESTIMATED ANNUAL VALUE OF THE ECOSYSTEM SERVICES OF FORESTS

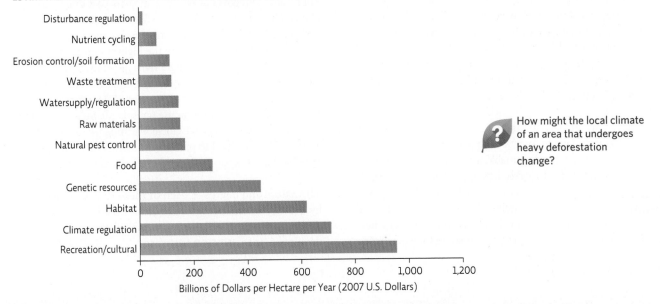

Billions of Dollars per Hectare per Year (2007 U.S. Dollars)

? How might the local climate of an area that undergoes heavy deforestation change?

However, taking advantage of the economic benefits of a forest can lead to its loss and the loss of the ecosystem services it provides. In Haiti, the cost of deforestation spun out over decades and proved catastrophic. Soil eroded down into streams, rivers, and gullies, clogging them with sediment and disrupting aquatic ecosystems. With nothing to absorb the water, floods became more severe, and groundwater sources were quickly depleted as the water flowed away rather than soaking into the ground. Slowly, crop yields shrank. As the forest habitat was fragmented, biodiversity dwindled. And as the trees vanished, the people of Haiti suffered. Unlike wealthier countries, Haiti could not afford expensive water purification systems—a service the forests once provided for free. "It took some years before we could feel the other effects of deforestation," says Georges. "The floods got worse, and we lost drinking water to runoff and pollution.

And once those problems started, there was no easy way to fix them."

> "The floods got worse, and we lost drinking water to runoff and pollution. And once those problems started, there was no easy way to fix them."
> —Timote Georges

Eventually, everyone who could abandoned the countryside for the capital city. Even that did not stop the tide of deforestation. As the population swelled, so did the demand for fuel, which in Haiti still comes almost exclusively from trees.

4 THREATS TO FORESTS

Key Concept 4: Globally, forest cover is shrinking due largely to harvesting for lumber and fuel, clearing for agriculture or mining, pest infestations, and fire suppression.

According to the United Nations (UN) Food and Agriculture Organization (FAO), today global deforestation has slowed, from 8.3 million **hectares (ha)** per year (20 million acres [ac]) in the 1990's to 7.6 million ha (19 million ac) each year between 2010 and 2015. Forest gains between 2010 and 2015 were 4.3 million ha (10.6 million ac) per year resulting in a net decrease of 3.3 million ha (8.2 million ac). While global deforestation rates have declined, they are actually increasing in some tropical areas, such as Indonesia and Malaysia. (See Module 3.2 for more on deforestation for oil palm plantations in these regions.) Some of our actions intentionally lead to the deforestation of tropical (and other) forests. The most common culprits are the harvesting of forests for wood and wood products and the removal of forests to use the land for other purposes such as agriculture (see Chapter 8) and mining (see Online Module 7.3). Growing populations are also spreading out as urban and suburban areas claim forested land (see Module 4.2).

The way we manage fire is also a factor in forest destruction, especially temperate forests. Frequent fires remove deadwood and other flammable material. If we suppress these fires, the deadwood builds up so that when a fire does come through, it can burn so hot it catches the entire forest on fire. It turns out that fire is actually needed to maintain some forests whose

trees are fire adapted with seeds that germinate only when exposed to the heat of a fire.

Climate change, too, is impacting forests. Warmer winters allow pest populations that would normally die back during cold months to attack trees year-round, taking a toll on temperate and boreal forests. Though extra atmospheric CO_2 (a major cause of climate change) and warmer temperatures may extend the growing season in some areas, precipitation declines will negate this benefit in many places. The incidence of fire has also increased, especially in drought-stricken areas; this directly destroys forests and some may not be able to recover as easily as in years past due to an altered climate. In general, range shifts for many tree species will occur (and is already occurring) as young trees take root and grow with more success at the highest latitudes and altitudes of their historic range; trees growing at the lowest latitudes or altitudes in their current range may die out and be replaced with a different ecosystem (different type of forest) or even a different biome (grassland). Steps to address climate change and restrict global warming to no more than 2°C (3.6°F) are needed to protect the forests that remain. (See Module 10.2 for more on climate change.)

The nature and degree of each of these threats varies by country because forests are often used and managed differently in developing countries than they are in developed ones. **INFOGRAPHIC 4**

hectare (ha) A metric unit of measure for area; 1 ha = 2.5 acres (ac).

A variety of "drivers" are responsible for deforestation around the world.

FIRE SUPPRESSION

Preventing or extinguishing fires as soon as they start can make some forests more vulnerable to large, destructive fires.

PEST INFESTATION

Outbreaks of insect infestations are responsible for the destruction of huge swaths of forests worldwide.

LOGGING

Deforestation due to logging for timber and conversion of land to ranch- or farmland is a major driver of forest loss in developing countries.

CATTLE RANCHING

FIRES

Millions of acres are destroyed yearly by fires that result from nature (lightning) and from humans.

ROADS

They are an indirect cause of deforestation, giving people ready access to forested areas.

FUELWOOD AND CHARCOAL

Harvesting trees for fuelwood or conversion to charcoal is still the main fuel source for many people in developing countries.

SUBSISTENCE FARMING

LARGE FARMS

In tropical areas, the biggest threat is the conversion of forestland to cropland for large-scale, export agriculture, but subsistence farming still takes a toll.

NET CHANGE IN FOREST AREA BY COUNTRY, 1990–2015

Today, most deforestation is occurring in developing countries. It must be noted, however, that the industrialization of developed countries, like the United States and many European countries, was supported in large part by the harvesting of their own forests.

Net loss (ha/yr)

More than 500,000

250,000–500,000

50,000–250,000

Small change (gain or loss)

Less than 50,000

No data

Net gain (ha/yr)

50,000–250,000

250,000–500,000

More than 500,000

How does the presence of roads lead to loss of forests?

Charcoal sellers in Port-au-Prince, Haiti.

As mentioned earlier, the clearing of land to grow crops or raise animals is a major driver of deforestation worldwide. It was no different in Haiti where deforestation began in earnest with 18th-century French colonizers' coffee and sugar plantations. In addition, as people moved to Port-au-Prince, the nation's densely populated, energy-starved capital city, the city boundaries expanded, claiming some forested land. Population growth also triggered more clearing outside the city to make way for more subsistence crops and to supply charcoal, the major fuel source for the country. Like Haiti, many African nations are experiencing rampant deforestation, largely to provide charcoal and fuel wood. In Mozambique, for example, charcoal is the main fuel for 80% of the population. More than 2.5 million trees are cut each year in Somalia—that is 10 trees per household, per month.

In time, Haiti's charcoal trade grew to account for 20% of the rural economy and 80% of the country's energy supply. Before long, 98% of the country's forests had

been chopped down. "The trade itself became this incredibly destructive force," says Andrew Morton, a forest ecologist for the United Nations Environment Programme. "And the fact that it was based not on foreigners exploiting the land for profit, but on poor Haitians trying to earn money and feed their families ... made it impossible to stop. There really was no other source of energy."

In general, deforestation occurs at an increased rate in developing countries like Haiti, where dire poverty and a lack of alternatives force people to harvest their forests or remove them for other land uses. They need wood and charcoal for fuel and housing, they need more space for agriculture, and they need commodities to sell in the marketplace. In most cases, developing countries also have far greater remaining forest stands than developed countries, but many of these stands are falling fast. In fact, many if not most developed countries, including the United States, Canada, and most European countries, became developed in part by harvesting their own forests. Today, European countries have fewer forest stands than many developing countries, but they also have more stringent regulations in place to protect those that are left. These regulations—and the ability to enforce them—help keep deforestation in check.

Of course, any gains made by developed countries are still largely offset by deforestation in countries like Haiti. "Industrial countries may be leading the way in conserving their own forests," says Morton. "But their demand for wood drives much of the deforestation elsewhere." Multinational corporations have simply moved deforestation operations to developing nations, where regulation and enforcement are often lacking and people are desperate for income.

5 | TIMBER HARVESTING: SUSTAINABLE OPTIONS

Key Concept 5: Selective and shelterwood timber harvesting maximize long-term profits and protect biodiversity; clear-cutting methods maximize short-term profits at the expense the local ecosystem. Tree farms are a highly productive alternative that may allow managers to leave natural stands alone.

maximum sustainable yield (MSY) The amount that can be harvested without decreasing the yield in future years.

Multiple-Use Sustained-Yield Act U.S. legislation (1960) mandating that national forests be managed in a way that balances a variety of uses.

Funding and technical expertise can facilitate more effective forest management in developed nations. For example, in 1905, the United States established the National Forest Service. Its first director, Gifford Pinchot, challenged the prevailing notion that U.S. forests were inexhaustible and introduced

the idea of sustainable forestry—taking only what the forests could sustainably produce or replace. (See Module 5.1 for more on ecological economics.) The focus in those early years was **maximum sustainable yield (MSY)**: harvesting as much as sustainably possible (but no more) for the greatest economic benefit. In 1960, the United States expanded on Pinchot's ideas and enacted the **Multiple-Use Sustained-Yield Act**, which mandated that national forests be managed in a way that balances a variety of uses—outdoor recreation and timber interests as well as the health of watersheds, fish, and wildlife. No single use could predominate.

Today, the U.S. National Forest Service, administered by the U.S. Department of Agriculture, oversees 155 national forests (and 20 national grasslands) and provides guidance for the management of private forests and grasslands, both nationally and internationally. Its overriding objective is the management of these areas in an ecologically sustainable manner, and it promotes **forest ecosystem management (FEM)** as the best way to meet this mandate. Rather than focus exclusively on timber harvests and maximum sustainable yields, FEM aims to manage the forest ecosystem as a whole. This includes a variety of techniques for timber harvesting, vegetation removal, and controlled burns to remove deadwood and stimulate seed germination in fire-adapted species, as well as restoration of forested areas and forest research.

Though some timber harvesting methods (such as large **clear-cuts** on steep slopes) are very damaging to the integrity of the forest, some methods reduce disruption to the ecosystem while still providing economically valuable forest products. Smaller scale clear-cuts done in strips or sections (**strip harvesting**) destroy less contiguous habitat than massive clear-cuts and cause less water pollution. This is currently the method of choice in the boreal forests of Canada because it mimics

the natural changes this forest is adapted to handle. Canada's boreal forests are dynamic and ever changing, largely due to natural disasters (e.g., fire or insect infestations) that open up sections of forest. These forest openings will be colonized by seedlings from neighboring trees, creating an even-aged stand that grows until the next disturbance. Trees and wildlife in these areas are adapted to tolerate this type of forest loss as long as the area cut is not too large, and nearby strips or "islands" of standing trees are left untouched.

Selective harvesting and **shelterwood harvesting** also have lower environmental impacts and better long-term yields than clear-cuts. These are the most environmentally sound methods for harvesting mixed-age stands of temperate forests. Though many foresters recommend selective harvesting in tropical forests, debates are ongoing about whether commercially viable amounts of timber can be harvested sustainably from these forests. Felling one tree in a tropical forest typically causes others to fall—the forest is densely populated with

forest ecosystem management (FEM) A system that focuses on managing the forest as a whole rather than for maximizing yields of a specific product.

clear-cut Timber-harvesting technique that cuts all trees in an area.

strip harvesting Timber-harvesting technique that clear-cuts a small section of a forest, allowing regrowth in that section before moving on to another.

selective harvesting Timber-harvesting technique that cuts only the highest-value trees; the remaining trees reseed the plot.

shelterwood harvesting Timber-harvesting technique that cuts all but the best trees, which reseed the plot and are then harvested.

Hataigan Doungbal/AGE Fotostock

The nutritious leaves of the moringa tree are high in protein, vitamins, and minerals and contain enough iron to treat mild anemia; they can be harvested without killing the tree, and they can be eaten fresh or dried for later use. The seeds and roots are also edible, and cuttings can be planted to start new trees.

INFOGRAPHIC 5 | **TIMBER-HARVESTING TECHNIQUES**

There are many ways to harvest trees from a forest, each with its own economic and ecological trade-offs. Here are variations of four techniques. To evaluate the impact of a particular method, consider what the area would look like 50 years after a harvest.

 Clear-cutting radically changes the ecosystem and can cause serious ecological damage. When then might it be ecologically sound to clear-cut, and what restrictions would you place on such an action?

ORIGINAL FOREST

CLEAR-CUTTING

All trees are cut; replanted with a fast-growing species.

IMMEDIATELY AFTER HARVEST

< Muddy stream

- High profits at harvest, then no profits until forest regrows.
- Water is polluted by heavy erosion on steep slopes.
- Biodiversity is very low after the cut.

50 YEARS AFTER HARVEST
Even-aged, single-species stand.

< Muddy stream

- Tree farms produce harvestable timber in a short time span.
- Water may be polluted by runoff from the open stand, but to a lesser degree than immediately after clear-cut.
- Tree farm has less biodiversity than the original forest.

STRIP HARVESTING

All trees are cut in one section; one or more species are then planted or the area may be left to reseed naturally.

IMMEDIATELY AFTER HARVEST

Next cut

Previous cut

< Never cut

Most recent cut ^

< Less-disturbed stream

- Profits are initially lower than for clear-cuts, but are more frequent.
- Biodiversity declines, but some organisms find refuge in uncut regions of the forest.

50 YEARS AFTER HARVEST
Even-aged stand.

New cut

< Less-disturbed stream

- Biodiversity is lower than in the original forest, but not as low as clear-cut lands, since a forest remains standing at all times and is still usable by some of the wildlife.

trees and with vines that connect neighboring trees to one another. Simple steps that reduce damage include cutting vines before felling a tree and cutting the tree in such a way that it hits the fewest neighboring trees when it falls. Strip harvesting, in very narrow strips, may be a reasonable alternative in some tropical forests. And, because soils are thin and fragile in tropical forests, efforts to reduce damage when trees are removed, such as using draft animals instead of trucks, are advised.

Ideally, the stand of trees—its health, age, and species composition—and the slope of the land determine the harvesting method. Consideration is also given to the other species that reside there. When trees are harvested properly, they can provide immediate and long-term economic benefits without serious environmental damage. Even clear-cutting can be appropriate when it increases the overall health and

viability of a future forest by removing invasive, unhealthy, or genetically inferior trees and replacing them with better stock. Clear-cuts followed by planting of a fast-growing, high-value tree (a tree farm) can also reduce the pressure to cut other forests. (However, there must be enough native forests left in areas suitable for tree farms to avoid serious ecological damage due to biodiversity loss.) **INFOGRAPHIC 5**

Forest management is not without its critics. Conflicting interests make it difficult to achieve a balance between multiple forest uses and ecosystem protection. For example, harvesting trees in the Pacific Northwest provides many jobs, good profits for timber companies, and useful products for homes and businesses, but it can reduce biodiversity and harm salmon runs, which are vital for ecosystem health, native cultures, and the tourism industry.

SELECTIVE HARVESTING

High-value trees are cut, leaving others to reseed the plot.

IMMEDIATELY AFTER HARVEST

‹ Never cut

‹ Less-disturbed stream

• Profits are intermediate between clear-cut and strip harvest levels.
• Land can be harvested again in less than 50 years.
• Biodiversity declines after harvest, but not as much as in clear-cut lands.

50 YEARS AFTER HARVEST
Multi-species, mixed-age stand.

‹ Less-disturbed stream

• Remaining large trees can be harvested for sale; since the poorer quality trees were left to reseed the area, the forest quality may decline.
• Biodiversity is higher than after the harvest, but lower than the original forest.

SHELTERWOOD HARVESTING

The best trees are left behind to reseed the plot.

IMMEDIATELY AFTER HARVEST

‹ Never cut

‹ Less-disturbed stream

• Profits are similar to those of selective harvest initially, but better than selective harvest on subsequent cuts, since high-quality trees remain.
• Biodiversity declines after each harvest, but may be higher than selective harvest levels.

50 YEARS AFTER HARVEST
New growth is progeny of the best trees; the large trees are now harvested.

‹ Less-disturbed stream

• Biodiversity increases as the forest grows, and may be higher than in any of the other harvested stands.
• Harvesting the large trees left behind provides additional income.

6 SUSTAINABLE FOREST MANAGEMENT

Key Concept 6: Forests can be protected by encouraging the sustainable harvesting of forest products, finding alternatives to replace wood as a raw material, and promoting the value of intact forests through the realistic valuation of ecosystem services.

The trade-offs associated with the use of forest resources are the subject of much debate and conflict because each decision impacts both human livelihood and the health of the environment. To be sure, the economic value of wood may be dwarfed by the ecosystem services lost if the area is overharvested, and protecting the ecosystem may actually prove more profitable, even in the short term. But people still need fuel, building material, and income. "Of course trees and forests are important for the environment," says Georges, "We know that they protect us from floods and help keep drinking water clean and plentiful. But for many Haitians, selling those same trees is the only way to feed a family. So how can you ask them not to?"

One solution, according to a growing number of experts, is to price the ecosystem services themselves. In Costa Rica, for example, higher water bills offset the costs of maintaining rainforests that purify and replenish the water people use every day. This money is given directly to landowners who would otherwise have to chop down the trees to sell as fuel or timber or to convert the land to agriculture.

Other options include the promotion and increased availability of sustainable wood products. The Forest Stewardship Council (FSC) certifies lumber and other timber products through a process that evaluates the timber-harvesting techniques and the forest itself in terms of wildlife, water, and soil quality. Worldwide, more than 200 million ha (490 million ac) of forest are certified by the FSC as sustainably managed (about 5% of the world total).

Consideration can also be given to resources, such as latex and tree nuts, that can be sustainably harvested from standing forests. Alternatives for wood products also exist and can reduce the pressure on forests. For example, lumber from old buildings can be salvaged to provide quality building materials. Paper can be made from old paper (recycled) or from fast-growing crops such as kenaf, jute, flax, and hemp.

Alternate energy sources are also needed in charcoal-dependent places like Haiti. One potential alternative is jatropha, a fast-growing plant whose oil-rich seeds have been hailed as a promising biofuel source. Even the material left after the oil is pressed out can be digested by bacteria to produce biogas, a fuel similar to natural gas. Another option is the production of composite briquettes from a variety of flammable materials, such as grass, paper, or sawdust. Shredded plastic can be added to make it more combustible. (See Module 11.2 and Online Module 11.3 for more on sustainable energy.)

Micheline Pelletier Decaux/Getty Images

Wangari Maathai's Green Belt Movement of Kenya has spread to more than 30 other African nations. The first Kenyan woman to earn a Ph.D., Maathai received several environmental awards, including the Nobel Peace Prize. She passed away in 2011, but the Green Belt Movement lives on.

Reforestation projects are increasing in number around the world. For example, the Kenyan Green Belt Movement, begun by Nobel Laureate Wangari Maathai, has planted and protected more than 30 million trees and has sparked similar movements worldwide. But reforestation efforts take commitment, money, and expertise. In Haiti, the success of reforestation efforts is strongly tied to the availability of water, especially on steep slopes; engineering terraced water catchment systems to catch and retain rainwater enhances growth and survival of the trees.

For some, saving the forests may come down to recognizing their intrinsic value: their natural beauty, their inherent sacredness, and their right to exist as living things, regardless of what we humans might extract from them. Naturalists from John Muir to Wallace Stegner have argued for the protection and preservation of forests on these grounds alone. "We simply need that wild country available to us, even if we never do more than drive to its edge and look in," 20th-century U.S. naturalist Wallace Stegner wrote in his famous *Wilderness Letter*. "For it can be a means of reassuring ourselves of our sanity as creatures, a part of the geography of hope."

This appreciation for the intrinsic value of a forest has been successfully translated into the economic enterprise known as **ecotourism**. Ecotourism is a viable option for many areas, especially less developed countries. These areas often possess high biodiversity precisely because they are less developed and are often found in tropical or subtropical areas with naturally high biodiversity. In the quest to develop economically, these regions may find that the highest economic value for their resources lies in keeping them intact. Ecotourism allows a way for funds to enter the country while protecting the natural areas at the same time. **INFOGRAPHIC 6**

ecotourism Low-impact travel to natural areas that contributes to the protection of the environment and respects the local people.

INFOGRAPHIC 6 PROTECTING FORESTS

There are a variety of ways to protect forests, many of which still allow the forests to be used and promote the sustainable harvesting of forest resources.

Chad Ehlers/Alamy Stock Photo

STRATEGY	EXAMPLE
Assign a monetary value to forest ecosystem services.	A cost analysis of various ways to provide water supplies for New York City revealed that protecting the forests of the nearby Catskills Mountains so that they could continue to provide their water capture and purification services was cheaper than building a state-of-the-art water filtration facility.
Sustainable harvest of forest products.	In the Brazilian state of Acre, local management of tropical forests for the harvest of tree nuts, latex (from tree sap), and sustainably harvested timber produces more than $90 million per year.
Use alternatives to wood for products and fuel.	Paper can be made from fast-growing nontree plants like bamboo and kenaf or can be made by recycling used paper; jatropha is replacing charcoal in Haiti and other regions of the world.
Designate forests as protected areas.	U.S. national forests are managed for multiple uses, including recreation, wildlife, and forest resource harvesting; resource harvesting is prohibited in wilderness areas and national parks. (See Online Module 3.3 for more on protected areas.)
Promote ecotourism.	With 36% of the country under protected status, Belize has a thriving ecotourism industry that brings in more than $100 million annually.

Explain how the sustainable harvesting of forest products, which would provide less annual income than harvesting the trees themselves for timber, can be more economically valuable than the wholesale removal of the valuable timber in that forest.

Tourists on canopy walkway, Monteverde Cloud Forest Preserve, Costa Rica.

Christer Fredriksson/Lonely Planet Images/Getty Images

There is a Haiti that people like Georges and Robert talk about in the quiet moments after a day's planting. It's a Haiti lush and green with trees, a country where families earn their living selling mangoes and moringa leaves instead of charcoal and firewood. Whether this country resembles its past as much as its future will depend on an infinite number of variables—not just how well it manages forests and reforestation efforts but also whether it can find alternative building materials or establish a reliable energy sector based on something other than wood.

Three months after the kombit visited his hillside farm, Jean Robert is harvesting moringa leaves. They're tasty and packed with protein, and when harvested properly, they regrow rather quickly. If everything goes as expected, he will eventually have a bevy of crops to see him through the year: mangoes in the summertime, coffee in the fall, and, if he's lucky, oak and mahogany

stands that will yield high prices in the timber market down the road. In the meantime, the moringa trees provide his family with a sustainable supply of protein and fuel wood.

Select References:

Brandt, J. P., et al. (2013). An introduction to Canada's boreal zone: ecosystem processes, health, sustainability, and environmental issues 1. *Environmental Reviews, 21*(4), 207–226.

Costanza, R., et al. (2014). Changes in the global value of ecosystem services. *Global Environmental Change. 26,* 152–158.

Food and Agriculture Organization of the United Nations. (2015). *Global Forest Resources Assessment 2015: How Have the World's Forests Changed?* Rome, Italy: Author.

Sprenkle-Hyppolite, S. D., et al. (2016). Landscape factors and restoration practices associated with initial reforestation success in Haiti. *Ecological Restoration, 34*(4), 306–316.

INTERACTIVE MAP **FORESTS OF THE WORLD**

ANIMATED INFOGRAPHIC

The type of forest found in a given area depends on its climate, which is determined in large part by latitude and altitude. Read about three examples to learn more about each of the three major forest biomes.

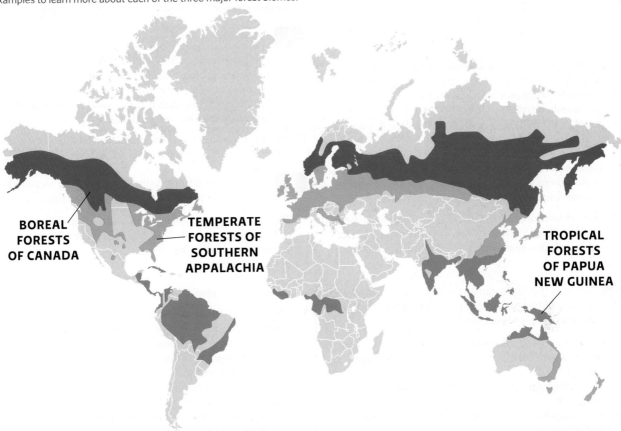

BOREAL FORESTS OF CANADA

TEMPERATE FORESTS OF SOUTHERN APPALACHIA

TROPICAL FORESTS OF PAPUA NEW GUINEA

 BRING IT HOME

PERSONAL CHOICES THAT HELP

Forests are a renewable resource that can be used sustainably for many years under proper management conditions. Using harvesting systems such as selective cutting or strip cutting can allow economic use of forests without eliminating the ecological functions that forests provide.

Individual Steps

• Buy paper products (toilet paper, facial tissue, and notebook paper) made of recycled content to decrease the unnecessary cutting of trees and to encourage recycling.
• When purchasing lumber for projects, look for wood that has been certified as sustainably managed by the Forest Stewardship Council.
• Avoid buying noncertified furniture made from tropical wood such as rosewood, teak, and ebony; harvesting these woods contributes to tropical deforestation.

Group Action

• Organize a workday to clear invasive species such as common buckthorn and Japanese honeysuckle to ensure that our remaining forests provide high-quality habitat.

Policy Change

• Support legislation that protects roadless areas and old-growth forests.
• Ask your local grocery stores, restaurants, and college dining facilities to offer shade-grown coffee and chocolate, which can help encourage preservation of tropical forests.

ENVIRONMENTAL LITERACY **UNDERSTANDING THE ISSUE**

1 What is a forest and what influences which forest type is found in a given area?

1. True or False: Any ecosystem that contains trees is considered a forest.

2. High-latitude and high-altitude forests characterized by a short growing season, a lot of snowfall, and thin, acidic soils are:
   a. temperate forests with trees that drop their leaves in winter, such as oak and maple.
   b. boreal forests with evergreen trees that bear needlelike leaves, such as spruce and fir.
   c. savanna forests with trees that have long taproots and thick, fire-resistant barks, such as acacia and eucalyptus.
   d. tropical forests with trees that are highly valued as timber for their rot-resistant wood, such as mahogany and teak.

3. Tropical rainforests have thin, acidic soils yet they contain dense vegetation and high biodiversity. How can these tropical forests have poor soil but support such diverse arrays of plant and animal life?

2 What is the three-dimensional structure of a forest and how are the plant species found there adapted to their level of the forest?

4. The understory of a forest is made up of:
   a. trees that push through and grow above the level of the forest canopy.
   b. the seedlings, ferns, herbs, and wildflowers that grow on the forest floor.
   c. the overlapping crowns of the tallest trees that make up the roof of the forest.
   d. shade-tolerant shrubs or the saplings of larger trees that sometimes form a lower canopy.

5. The forest layer that gets the most sunlight is the:
   a. canopy.
   b. emergent layer.
   c. forest floor.
   d. understory.

6. Explain why many forest wildflowers emerge and bloom in early spring, before the trees "leaf out."

3 What ecosystem services do forests provide?

7. How does deforestation contribute to loss of drinking water in Haiti?
   a. Without trees, the eroded soil clogs drinking water pipes.
   b. The loss of forest habitat fragments aquatic ecosystems that provide water to the local villages.
   c. Without the trees, the water rushes away rather than soaking into the ground to recharge groundwater supplies.
   d. Clearing the forest for crops means increased irrigation, which reduces drinking water supplies.

8. What is an "ecosystem service"? Describe three such services provided by forests.

9. Why might it be useful to put a monetary value of the ecosystem services of forests, even if we know it is not exactly right?

4 What threats do forests face?

10. Which of the following is NOT a cause of deforestation in Haiti?
    a. Palm oil plantations for biofuel production
    b. Coffee and sugar plantations for export crops
    c. Wood charcoal production for the domestic energy market
    d. Timber harvesting for commercial sale

11. How is the current status of forests different in developing versus developed countries? What factors account for these differences?

5 How do the different timber-harvesting techniques compare in terms of method and sustainability?

12. The timber-harvesting system that would be most likely to cause disruption to the ecosystem services provided by a forest is:
    a. shelterwood harvesting.
    b. strip harvesting.
    c. clear-cutting.
    d. selective harvesting.

13. Which of the following activities will NOT simultaneously protect forests and provide for long-term economic well-being of local people in developing countries?
    a. Pricing ecosystem services provided by the forest and paying landowners to maintain their trees
    b. Clear-cutting the forest and planting fast-growing trees for charcoal production
    c. Promoting the harvesting and selling of wood that is certified by the Forest Stewardship Council
    d. Translating the intrinsic value of a forest into ecotourism enterprises

14. What are the ecological problems associated with clear-cutting and when might it be ecologically beneficial to harvest timber using this method?

6 How can we protect and sustainably manage forest resources?

15. True or False: While it sounds good in theory, there have not yet been any tangible examples of a community saving money because they protected their forests rather than allowing them to be harvested.

16. Which of the following is an example of ecotourism?
    a. Local businesses leading wildlife viewing tours in scenic areas
    b. Restricting public access to areas where commercial harvesting is prohibited
    c. Adding the cost of programs that protect natural areas to homeowner's water bills
    d. All of the above

17. How does ecotourism help protect forest resources?

18. What is the reforestation strategy employed by the Haitian kombit? Explain the rationale behind the approach and discuss whether it is likely to be successful.

SCIENCE LITERACY   WORKING WITH DATA

The Hubbard Brook Experimental Forest research center in the White Mountains of New Hampshire consists of several valleys and streams that can be manipulated as test and control plots for experimental purposes. Small dams allowed researchers to monitor the volume of water that flows through the forest. In this study, researchers compared the volume of stream water flow from streams in similar valleys close to one another that were either left forested or that were completely deforested.

STREAMWATER FLOW COMPARISON OF FORESTED AND DEFORESTED AREAS

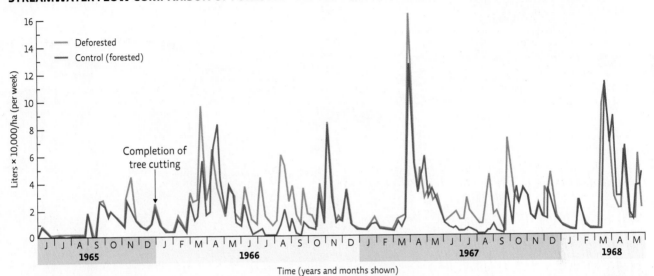

Interpretation

1. Which months typically have the greatest flow of water through streams in the area evaluated?

2. How did stream flow in the two forests compare before tree cutting in the test forest?

3. How did stream flow in the two forests compare after tree cutting in the test forest?

Advance Your Thinking

4. Why did deforestation impact the volume of water flowing in the stream in the way observed?

5. What do you think the water quality was like in the stream in terms of sediment loads? Explain.

6. Which downstream area would be more likely to experience a flood during a heavy rain event? Explain.

INFORMATION LITERACY   EVALUATING INFORMATION

Many of the world's forests are severely degraded. Yet forests produce many consumer products that we depend on, such as food, medicine, building materials, and raw materials for industrial products like paper. So what should a conscientious consumer do?

Explore the Forest Stewardship Council (FSC) website (https://us.fsc.org).

Evaluate the website and work with the information to answer the following questions:

1. Determine if this is a reliable information source with a clear and transparent agenda:
   a. Who makes up this organization? Does its membership make the FSC reliable or unreliable? Explain.
   b. What is this organization's mission? What are its underlying values? How do you know this?

   c. Does the FSC give supporting evidence for its claims about forest resources and its vision to address the problem? Does the website give sources for its evidence?
   d. Identify a claim the FSC makes and the evidence it gives in support of this claim. Is it sufficient? Explain.

2. Understand products and certification:
   a. What kind of products can be FSC certified?
   b. Read about the certification process. Do you feel this process is sufficient? Explain.
   c. How can a consumer select FSC-certified products, according to this website? Is it easy for consumers to know if the wood products they are purchasing are FSC certified? Explain.

3. How might FSC certification help forests?

 Additional study questions are available at SaplingLearning.com.

CHAPTER 8

FOOD RESOURCES

CHAPTER 8

Feeding a world of more than 7 billion people is a daunting task. Productive and sustainable approaches to farming, rearing livestock, and obtaining fish will be needed if we are to meet humanity's food needs now and in the future.

Module 8.1: Feeding the World

An assessment of world hunger and modern attempts to increase food production (the Green Revolution and genetic engineering)

Module 8.2: Sustainable Agriculture: Raising Crops

A look at sustainable farming methods modeled after natural ecosystems (agroecology) or based on traditional farming methods of the past

⇗ ONLINE Module 8.3: Sustainable Agriculture: Raising Livestock

An evaluation of the pros and cons of rearing livestock in concentrated animal feeding operations and a look at the alternatives

⇗ ONLINE Module 8.4: Fisheries and Aquaculture

An examination of the pressures industrial fishing puts on wild stocks and an introduction to an alternative: aquaculture

Online Modules are available at SaplingLearning.com.

zlikovec/Getty Images

A GENE REVOLUTION

Can genetically engineered food and industrial agriculture help end hunger?

Desperate residents plead to workers from the Kenyan Red Cross for food during distribution in Nairobi in 2008. The crowd eventually broke through the gate but was chased back by police with whips.

YASUYOSHI CHIBA/AFP/Getty Images

After reading this chapter and studying the KEY CONCEPTS and INFOGRAPHICS, you should be able to answer these GUIDING QUESTIONS

CORE MESSAGE

Although we produce enough food to feed the world's population, nearly one billion people don't have access to enough nutritious food. The rise of industrial agriculture and the Green Revolution helped fight hunger in the 20th century but came with some unintended consequences. Growing genetically modified crops can increase productivity further, but concerns about the safety of these foods may trigger strong debate. Employing a variety of agricultural methods and addressing socioeconomic drivers of poverty will be needed to fight world hunger.

1. How prevalent is world hunger, and what are its causes?

2. What problems result from malnutrition?

3. What was the Green Revolution, and how does it relate to industrial agriculture?

4. What are the pros and cons of industrial agriculture?

5. What is the importance of food self-sufficiency and sovereignty, and what can undermine meeting these goals?

6. How can genetic engineering be used in agriculture?

7. What are the trade-offs of using genetically modified organisms (GMOs) in agriculture?

8. What are some low-tech (non-GMO, nonindustrial) options for increasing crop production?

It was in Burkina Faso—a tiny, landlocked west African country, virtually unknown to the developed world—that a brewing global crisis finally came to a head. On February 22, 2008, riots broke out in the country's two major cities. Angry protesters clogged the streets—shouting, throwing rocks, and flipping cars. Soldiers mobilized to restore order, but the chaos only spread from there. In neighboring Côte D'Ivoire, tear gas was employed, and dozens were injured; in Cameroon, some two dozen people were killed; in Egypt, a single boy was shot in the head. Before long, the violence spilled across Africa's borders. Protesters in Yemen torched police stations and blocked roads. In Bangladesh, they smashed cars and buses, vandalized factories, and ultimately injured dozens of bystanders.

◉ **WHERE IS BURKINA FASO?**

What were these people rioting over? Food. It had become too expensive. In the 2 preceding years, the cost of rice had risen by 217%, wheat by 136%, and corn and soybeans by 125% and 107%, respectively. In the 7 months leading up to the riots, those prices had then doubled. Such increases were not as much of a problem in the United States or Europe, where even the poorest one-fifth of households spends just 16% of their budget on food. But in the cities that had descended into violence, families were already spending between 50% and 75% of their income on basic staples—rice and beans and bread and milk. Such dramatic and rapid cost increases had pushed those goods out of their reach and, in so doing, had nudged too many people toward the brink of starvation. "For countries where food comprises from half to three quarters of [income] consumption," explained World Bank President Robert Zoellick, "there is no margin for survival." In all, he said, some 22 countries were at risk of violent revolts if prices did not soon stabilize.

But what had destabilized them in the first place? Newspaper columnists, politicians, and cable news anchors trotted out the usual suspects. Some blamed extreme weather events: A drought in Australia, a heat wave in California, and flooding in India had decimated crops of wheat and maize and supplies of beef. Some speculated that diverting corn to the production of biofuels might have led to the supply shortages. Others pointed to the global financial crisis and to rising fuel prices: Because large, industrial farms rely on fossil fuel–powered machines and chemical inputs derived from fossil fuels, the cost of grain is tethered to the costs of oil and natural gas. Yet others focused on the growing demand for meat from Asia's expanding middle class: The more grain we devote to cattle rearing, the less we have to sell as food and the more expensive it thus becomes (see Online Module 8.3).

Experts, however, believed that the trouble with the global food supply goes much further than that— beyond destructive weather events and the slumping global economy, to the very heart of global food policy.

Dong yanjun/Imaginechina/AP Photo

Natural disasters can ruin months of work in only hours. Here a man tries to salvage vegetables from a field flooded by Typhoon Meranti in southeast China.

1 WORLD HUNGER

Key Concept 1: Though we currently grow enough food to feed everyone, almost a billion people are under-nourished due to poverty, war, and inadequate food distribution or preservation.

A lack of **food security**—defined as all people at all times having physical, social, and economic access to sufficient safe and nutritious food—is due to a variety of factors. By most accounts, we currently grow enough food to feed the more than seven billion people who inhabit Earth. However, food insecurity is a problem for almost a billion people for a variety of reasons including:

- Insufficient funds to buy food

- Inadequate distribution of food (not getting food to people who need it)

- Political roadblocks and corruption that prevent food delivery

- War or protracted crises that interfere with agriculture or food distribution

- Discrimination that keeps food from reaching marginalized groups

The United Nations (UN) has set itself the task of eradicating hunger (and extreme poverty along with it) with its Sustainable Development Agenda, identifying 17 Sustainable Development Goals (SDG) that it hopes to reach in an effort to "ensure prosperity for all people." (See

food security Having physical, social, and economic access to sufficient safe and nutritious food.

Module 1.1, Infographic 4 for more on the UN's SDGs.) SDG 2 is *Zero Hunger*, a goal the UN hopes to achieve by 2030. The 795 million who were hungry in 2016 (12.9% of the global population at that time) is unacceptably high, but this does represent significant success—from 1992 to 2015, world hunger was cut almost in half, even with an extra 1.9 billion mouths to feed. But more work needs to be done to eliminate hunger—we have to improve access to food for the hundreds of millions who are currently underfed, and we must be able to feed an additional two billion or so people by 2050.

To end hunger and achieve global food security, the UN is calling for a doubling of the agricultural productivity of smallholder family farms in developing countries by 2030 (as a source of food and income in these poverty-stricken areas) as well as programs to address inequities in food distribution and access. Ending war and protracted conflict or finding ways to get food to people in affected areas is also a hurdle that must be addressed. Social protections to assist those in need and bring them out of poverty are seen as vital to breaking the cycle of poverty and hunger. Indeed, *Eliminating Poverty* is SDG 1. The UN estimates that an investment of $267 billion over the next 15 years would eliminate hunger and extreme poverty. That is slightly less than half of the U.S. military budget for a single year—globally, a very accessible target. **INFOGRAPHIC 1**

INFOGRAPHIC 1 **WORLD HUNGER**

Though we currently produce enough food to feed the world's population, almost a billion people are undernourished. Of those who are underfed, 98% live in developing nations. Even in wealthy nations there are those who do not have access to enough food or to a balanced diet.

PREVALENCE OF UNDERNOURISHMENT IN THE POPULATION (PERCENT) IN 2014–2016

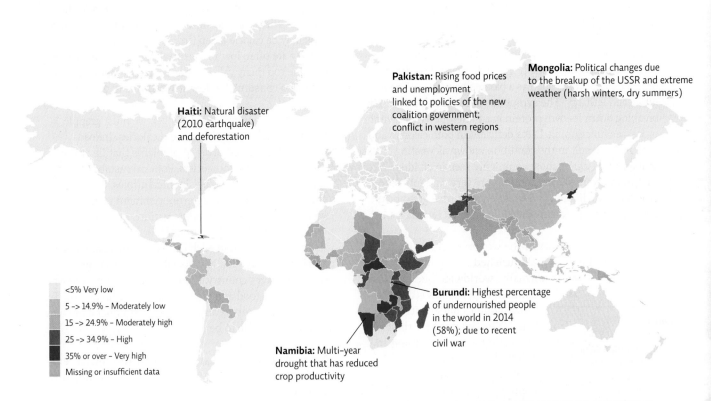

Haiti: Natural disaster (2010 earthquake) and deforestation

Pakistan: Rising food prices and unemployment linked to policies of the new coalition government; conflict in western regions

Mongolia: Political changes due to the breakup of the USSR and extreme weather (harsh winters, dry summers)

Burundi: Highest percentage of undernourished people in the world in 2014 (58%); due to recent civil war

Namibia: Multi-year drought that has reduced crop productivity

<5% Very low

5 -> 14.9% – Moderately low

15 -> 24.9% – Moderately high

25 -> 34.9% – High

35% or over – Very high

Missing or insufficient data

Of developing nations, those in protracted crisis (a prolonged ecological or political crisis that seriously affects the well-being of a large percentage of the population) have a chronic inability to acquire sufficient food supplies, leading to undernourishment, elevated infant mortality rates, and stunted growth in children.

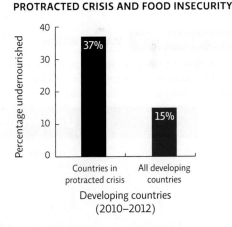

PROTRACTED CRISIS AND FOOD INSECURITY

Percentage undernourished

37%

15%

Countries in protracted crisis

All developing countries

Developing countries (2010–2012)

 Why do you think so many countries in sub-Saharan Africa have a large percentage of their population that is undernourished, whereas the countries in northern Africa have less than 5% of their population undernourished?

2 MALNUTRITION

Key Concept 2: Malnutrition includes undernutrition, which can disrupt growth and development and lead to serious health problems and overnutrition, which contributes to obesity-related health problems.

The UN's mission is not just a matter of food quantity but also of improving food quality. A healthy diet contains a variety of foods that provide the proteins, carbohydrates, and fats needed for good health, as well as enough calories to meet daily energy needs. The food that supplies these calories must also supply micronutrients such as vitamins and minerals. When a person's diet falls short of these basics—when he or she does not consume enough calories, protein, vitamins, or minerals—that person is said to be undernourished. *Undernutrition,* a form of **malnutrition**, can serve as a prelude to a whole host of diseases, from blindness (a result of vitamin A deficiency) to impaired immunity (a zinc deficiency).

A diet with enough calories but deficient in protein can lead to kwashiorkor (usually in children)—a disorder that produces symptoms such as a bloated belly, loss of muscle mass, and lethargy. The symptoms are a result of the body breaking down its own protein in an attempt to keep vital processes functioning. Diets deficient in both calories and protein can result in the condition known as wasting disease, or marasmus, again, most commonly seen in children. Victims are very skinny and frail, are very susceptible to infection and disease, and experience stunted growth and developmental problems (including mental retardation).

Malnutrition can start in the womb: Malnourished women give birth to malnourished children who fail to grow and develop normally; worldwide one in four children experiences stunted growth due to undernutrition. The UN estimates that the cost of treating malnourishment in children under age 2 is double what it would cost to prevent malnourishment in the first place.

malnutrition A state of poor health that results from inappropriate caloric intake (too many or too few calories) or is deficient in one or more nutrients.

Overnutrition, or the consumption of too many calories, is also considered malnutrition. By some estimates, about 1.5 billion people around the world are overnourished and are thus vulnerable to another set of nutrition-related conditions such as heart disease and type 2 diabetes. Contrary to intuition, this is not just a problem of the wealthy—the poor also suffer from overnutrition. Cheaper foods may be calorie laden (from high sugar or fat content), but they are often low in essential nutrients. People who consume too many calories can be just as malnourished as those who consume too few. **INFOGRAPHIC 2**

Of course, to conquer global hunger, we must grapple with more than just nutrients and calories. Indeed, we must tackle a vast array of problems that lie at the root of the brewing food crisis. Political instability and ecological degradation play significant roles, to be sure. As shown in Infographic 1, undernourishment is more than two times greater in developing nations experiencing a prolonged armed conflict, drought, or natural disaster than it is in countries not in such crises. But there are other contributors, too—namely, social disempowerment and poverty. In every country, there are groups of people who have access to food and groups of people who do not. To reach the UN's lofty goals, we must first understand why and how that came to be.

And to do that, we must travel back through recent history, to the last time global hunger was the stuff of headlines.

INFOGRAPHIC 2 **MALNUTRITION**

Malnutrition is defined as a state of poor health that results from inadequate or unbalanced food intake. This includes diets that don't provide enough calories (and thus are nutrient deficient) as well as those that may provide enough calories but are deficient in one or more nutrients. Consuming too many calories also leads to a variety of malnutrition illnesses.

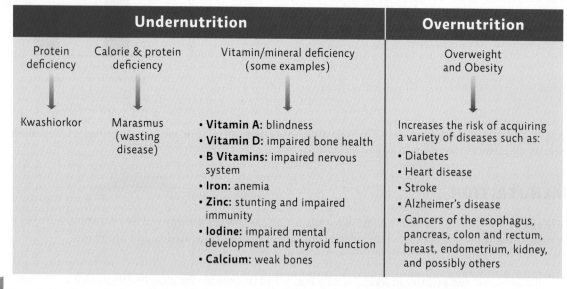

Undernutrition			Overnutrition
Protein deficiency	Calorie & protein deficiency	Vitamin/mineral deficiency (some examples)	Overweight and Obesity
Kwashiorkor	Marasmus (wasting disease)	• **Vitamin A:** blindness • **Vitamin D:** impaired bone health • **B Vitamins:** impaired nervous system • **Iron:** anemia • **Zinc:** stunting and impaired immunity • **Iodine:** impaired mental development and thyroid function • **Calcium:** weak bones	Increases the risk of acquiring a variety of diseases such as: • Diabetes • Heart disease • Stroke • Alzheimer's disease • Cancers of the esophagus, pancreas, colon and rectum, breast, endometrium, kidney, and possibly others

How might a plan to combat undernutrition be similar to one designed to combat overnutrition?

3 THE GREEN REVOLUTION

Key Concept 3: The Green Revolution helped address world hunger in the mid-20th century by producing high-yielding crop varieties grown in large monocultures with chemical fertilizers and pesticides.

It was the late 1960's. Global population was soaring; food crops were flatlining in some parts of the developing world and plummeting in others. India especially seemed to be hovering on the brink of a massive famine.

To stave off such catastrophe, the international community had launched the **Green Revolution**—a coordinated global effort to eliminate hunger by bringing modern agricultural technology to developing countries in Asia. Working across the globe, scientists, farmers, and world leaders introduced India and China to chemical pesticides, sophisticated irrigation systems, synthetic nitrogen fertilizer, and modern farming equipment—technologies that most industrialized nations had already been using for decades. They also introduced some novel technology—namely, new **high-yield varieties (HYVs)** of staple crops like maize (corn), wheat, and rice. HYVs have been selectively bred to produce more grain than their natural counterparts, usually because they grow faster or larger or are more resistant to crop diseases. Productivity per hectare increased dramatically.

If all that were not enough, developed countries like the United States also began implementing a litany of agricultural policies—tax breaks, government subsidies, and insurance plans—that encouraged their own farmers to plant crops "fencerow to fencerow" and kept the price of food low, thus adding substantially to the world's food supply.

To be sure, the adoption of modern, large-scale farming methods in India and China was expensive (paying for all those inputs and the necessary equipment), but in the short term, the Green Revolution was a huge success. The combined force of HYVs, existing technology, and new food policies resulted in a 100% increase in global food production and a 20% reduction in famine between 1960 and 1990. Today, most experts credit the initiative with the fact that we are now producing enough food to feed every one of more than 7.5 billion people on Earth. **INFOGRAPHIC 3**

So why are so many people still going hungry?

Green Revolution A plant-breeding program in the mid-1900's that dramatically increased crop yields and paved the way for mechanized, large-scale agriculture.

high-yield varieties (HYVs) Strains of staple crops selectively bred to be more productive than their natural counterparts.

INFOGRAPHIC 3 **THE GREEN REVOLUTION INCREASED AGRICULTURAL PRODUCTIVITY**

The Green Revolution transformed agriculture into the industrial model we see today. Through the selective breeding of the most productive plants, high-yield varieties of crops were developed that more than doubled production per acre (when grown with inputs like fertilizer and pesticides).

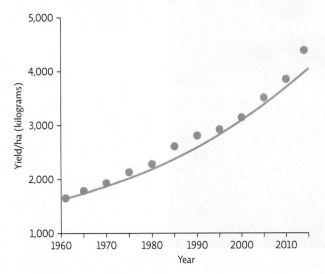

AVERAGE CROP PRODUCTIVITY (WHEAT, MAIZE, AND RICE)

 The trendline for this data shows an increasing slope upward. Even if higher-yielding plants are developed, what might prevent productivity per hectare from continuing to rise at this rate?

4 INDUSTRIAL AGRICULTURE

Key Concept 4: Modern industrial agriculture produces huge amounts of food but leads to ecological and social problems and a reduction in the genetic diversity of crops.

As it turns out, the Green Revolution owes its success to modern **industrial agriculture**. Industrial agriculture is dependent on machinery, as well as synthetic fertilizers (to boost plant growth) and pesticides (to protect crops from pests). Typically, a single, genetically uniform crop (monoculture) is planted, making it possible for fewer farmers to farm larger tracts of land using heavy equipment like tractors and combine harvesters. In many cases, mechanized irrigation, drawn from surface water or groundwater supplies, is also used and allows farmers to grow crops that could not be sustained by local rainfall. This mechanization and these chemical farming "inputs" increase efficiency and produce tremendous amounts of food on a large scale.

Though industrial agriculture can be very productive, it has had some unintended consequences. The excessive use of chemical inputs and large-scale monoculture operations have degraded soil, making it less fertile (and in need of even more fertilizers). Large-scale irrigation can deplete water supplies, and in hot, arid areas, high rates of evaporation can leave a salt residue on soils which impedes plant growth or renders the soils unusable. Agricultural chemicals can also pollute water sources. Widespread pesticide use is

industrial agriculture
Farming methods that rely on technology, synthetic chemical inputs, and economies of scale to increase productivity and profits.

leading to pesticide-resistant pests that are harder to kill, as well as exposing farm workers and consumers to toxic chemicals. (See Module 8.2 for more on the trade-offs of fertilizer and pesticide use.)

In addition, farm equipment runs on fossil fuels, and even the synthetic fertilizers and pesticides are derived from fossil fuels, made from natural gas or petroleum. This requires more fossil fuel extraction, with all its accompanying environmental issues (see Module 9.2). There are also local social and economic impacts: Fewer farmers, on larger farms, threaten to make the family farm a thing of the past, impacting entire communities. (For more on industrial farming, see Module 8.2 and Online Module 8.3.)

Monocultures bring another problem: The reliance on just a few high-yield varieties of each crop has led to a dramatic reduction in agricultural biodiversity. This is no small matter. Without a diverse gene pool, the global food supply is much more vulnerable to pests that can't be controlled by agrochemicals: When all plants in the field are genetically identical, what kills or damages one will probably kill or damage them all. (See Module 3.2 for a discussion of how the lack of genetic diversity in the potato crop contributed to the Irish Famine of the mid-1800's.) Moreover, as the old varieties fall out of use—as farmers stop producing and conserving the seeds that have been bred through traditional agriculture over thousands of years—many valuable genetic traits face the possibility of being lost forever. This erosion of genetic diversity translates into a loss of the genetic raw material that allows crops to respond to changes, such as the arrival of a new fungal pest or a changing climate. **INFOGRAPHIC 4**

In fact, this new threat to agriculture—climate change—may be its greatest challenge. Unseasonable floods or droughts are destroying crops and diminishing yields worldwide. New climatic conditions affect temperature extremes and the length of growing seasons, taxing crops that are adapted to an area's "former climate." Take for example coffee, the second-most traded commodity product in the world after petroleum. Coffee productivity is declining—50% lower in some areas—as the highlands where the trees are grown get warmer and drier. It is estimated that the area suitable to grow the two most popular varieties of coffee, Arabica and Robusta, will be cut in half by 2050. Research is underway to identify climate-resilient varieties that might become the coffee crops of the future.

JIM RICHARDSON/National Geographic Creative

A toxic crust of salt and other minerals builds up in heavily irrigated fields in Colorado.

INFOGRAPHIC 4 **THE TRADE-OFFS OF INDUSTRIAL AGRICULTURE**

THE TRADE-OFFS OF INDUSTRIAL AGRICULTURE

ADVANTAGES	
• Large-scale farming. • Higher yields/acre. • Crops can be grown on nutrient-poor soil (with fertilizers). • Crops have fewer blemishes (pesticides).	• Crops can be grown in water-poor areas (with irrigation). • Monocultures are more efficient to grow and harvest. • Less labor intensive.

DISADVANTAGES	
• Dependence on mechanization (expensive). • Large quantities of fossil fuels are used (as fuel and as raw material to make fertilizers and pesticides). • Soil quality is degraded. • Water pollution and scarcity.	• Monocultures are more vulnerable to pests and disease. • Toxic residues on food. • Fewer jobs. • Decreases the number of family farms, which affects local communities.

 Which of the advantages of industrial agriculture do you feel is its greatest strength? Why? Which of the disadvantages of industrial agriculture worries you the most? Why?

5 MEETING FOOD NEEDS LOCALLY: FOOD SELF-SUFFICIENCY AND SOVEREIGNTY

Key Concept 5: An important tool in ending hunger is enabling nations to produce enough food for their own populations and to control their local food systems. Planting cash crops for export rather than food for the local community can undermine these goals.

No one can dispute the fact that the Green Revolution fed, and continues to feed, a lot of people. Unfortunately, not everyone has benefited from this agricultural transformation. Africa, a continent plagued by hunger and largely bypassed by the Green Revolution, has nonetheless fallen prey to a different set of the Green Revolution's unintended consequences: a lack of **food self-sufficiency** (the ability of an individual nation to grow enough food to feed its people), a lack of **food sovereignty** (the ability of an individual nation to control its own food system), and ultimately a lack of food security.

Here's why: As industrialization and farm subsidies enabled (mostly U.S.) farmers to produce vast surpluses of wheat, corn, and soybeans, the global marketplace was flooded with cheap food from the developed world. Farmers in countries like Burkina Faso could not compete with such cheap and plentiful food imports— plagued as they were by land degradation (drought, soil erosion, water shortages) and armed conflict (violent clashes destroy existing crops and prevent new ones from being planted).

So they converted much of their farmable land to **cash crops** like cotton, coffee, and cocoa. (*Cannabis*, coca [for cocaine], and the opium poppy, grown in countries around the world, are the most valuable cash crops per hectare.) Rather than feed local populations, these commodities are usually exported for profit. Though some income remains in the local community, much of the profit goes to middle managers and those in power. Thus, a system where grain was locally produced and supplied gave way to a system where it was imported from thousands of miles away. And as they became dependent on food imports, developing countries found themselves at the mercy of forces far beyond their own borders. **INFOGRAPHIC 5**

And so it was that by 2008, Australian droughts and U.S. agricultural policies had left the people of west Africa to riot in the streets for want of bread. We are

food self-sufficiency The ability of an individual nation to grow enough food to feed its people.

food sovereignty The ability of an individual nation to control its own food system.

cash crops Food and fiber crops grown to sell for profit rather than for use by local families or communities.

INFOGRAPHIC 5   CASH CROPS FOR EXPORT

Cash crops for export have replaced the production of staple food crops for the local community in many areas for a variety of reasons. While individuals may benefit from cash cropping, it is not without its drawbacks.

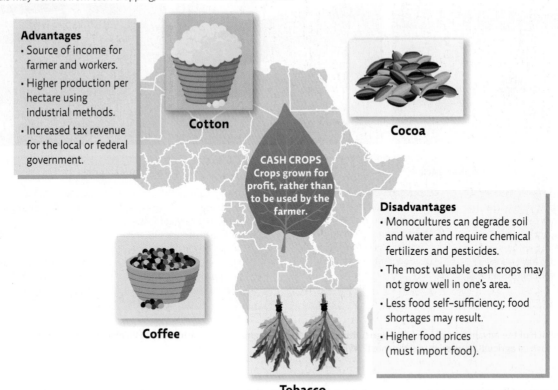

Advantages
- Source of income for farmer and workers.
- Higher production per hectare using industrial methods.
- Increased tax revenue for the local or federal government.

Cotton

Cocoa

CASH CROPS
Crops grown for profit, rather than to be used by the farmer.

Disadvantages
- Monocultures can degrade soil and water and require chemical fertilizers and pesticides.
- The most valuable cash crops may not grow well in one's area.
- Less food self–sufficiency; food shortages may result.
- Higher food prices (must import food).

Coffee

Tobacco

Why are cash crops that are grown for export more of a threat to local food security than cash food crops grown for local consumption?

Food insecurity plagues millions of people, especially in drought and war-torn regions of the world. In rural Africa, women do much of the farming with rudimentary tools. Here, a young mother with her baby on her back works with other women of her Rwandan village to prepare their field for planting.

Per-Anders Pettersson/Getty Images

indeed making enough food to feed the world. We just aren't getting it to the people who need it most.

Into this morass, an additional 2 billion people will soon be born. Global population is expected to reach 10 billion by 2050; experts say that to feed that many mouths, we will need to produce significantly more food than we are now producing. That will mean either farming more land or devising more production-boosting technologies. To farm more would mean clearing more forests, and thus destroying more natural habitats, species, and ecosystem services—a dire prescription at a time when the planet is already flirting with ecological disaster. (See Module 7.1 and Online Module 7.2.) How then will we make enough food to feed the future? Some feel that the next Green Revolution may be a "gene" revolution.

> Global population is expected to reach 10 billion by 2050; experts say that to feed that many mouths, we will need to produce significantly more food than we are now producing.

6   THE GENE REVOLUTION: GENETICALLY ENGINEERED CROPS

Key Concept 6: Genetically engineering crops (or animals) to contain useful traits may boost growth and expand the growing range of some crops but comes with environmental, economic, and ethical concerns.

As Robert Paarlberg and his driver made their way up one of Uganda's countless narrow dirt roads, a collection of mud huts encircled by a patchwork of small and midsize fields came into view. When they reached the village, Paarlberg was greeted by a group of women and children: curious, welcoming, and eager to chat with outsiders. Paarlberg, a Harvard-based political scientist, had been traveling throughout a northern swath of sub-Saharan Africa in an effort to understand the countries' opinions and attitudes about genetically modified crops.

In West Africa, most of the farmers were women and school-aged children (conspicuously not in school). On this particular farm, crops of maize struggled to pop up through parched soil—starved for both water and nutrients. Cassava withered under the strain of a viral infection, and a single cow suffered from untreated sores. Still, the women were warm and chatty. They did not have chemical fertilizers, they told him. Or water pumps. Or any kind of machinery. Through the translator and the children who spoke some English, the women fretted about the weather, and crop prices, and how much grain they had in storage. These women knew absolutely nothing of the emerging technology Paarlberg had come to the region to discuss.

Humans have been altering the genetic makeup of crops ever since the first budding farmers started cross-breeding plants to produce new varieties, and it was with these same artificial selection techniques that plant scientists created the high-yielding crops of the Green Revolution. But scientists now have new methods to directly alter the genetic makeup of an organism, and these techniques form the basis of what some farmers and scientists like to think of as **Green Revolution 2.0**, or the *Gene* Revolution—the next battle against hunger.

Genetically modified organisms (GMOs) are defined as those organisms that have had their genetic information altered in a way that would not be possible by natural means. Scientists have been producing such organisms for decades, coaxing genetically modified bacteria to produce important medicines such as insulin—like tiny living drug factories. In the 1990's, researchers began applying the same technology to food crops. By transferring genes for desirable traits (like pest resistance or herbicide tolerance) from one species to another, they have created a new suite of genetically modified food crops—plants that can grow more plentifully and thrive in a wider range of habitats than they ordinarily would. Organisms that receive DNA from another species with which they could not naturally breed (a sexually incompatible species) are known as **transgenic organisms**.

In recent years, new technologies have emerged that can create other types of genetically engineered (GE) organisms. Known as *genome editing*, such as one that goes by the acronym CRISPR, these methods allow scientists to more quickly and precisely edit the DNA of organisms. While these

Green Revolution 2.0 Programs that focus on the production of genetically modified organisms (GMOs) to increase crop productivity.

genetically modified organism (GMO) Organism that has had its genetic information modified to give it desirable characteristics such as pest or drought resistance.

transgenic organism An organism that contains genes from another species.

More than 75% of processed food in the United States contains GMOs. Only certified organic food is guaranteed to be GMO free.

methods could be used to produce transgenic organisms, they can also edit DNA in a way that produces desired changes without introducing any "foreign" DNA (DNA from an unrelated, sexually incompatible species). Because "same species" genes could naturally be acquired by plant breeding, CRISPR edits that don't introduce foreign DNA can avoid the ethical or regulatory concerns of creating a transgenic organism.

One genome-editing approach simply alters a cell's own DNA (an *intragenic* edit); the most common outcome is to simply "turn off" a gene. For example, a gene was silenced in the white button mushroom that reduces its ability to bruise, making it the first CRISPR-edited crop to be approved for sale by the United States Department of Agriculture. CRISPR can also be used to add a gene from a close relative (a *cisgenic* edit). For example, in dairy cows, the gene for producing horns was removed and replaced with a gene for "hornlessness" from Angus cattle, creating a hornless dairy cow. This eliminates the need to dehorn calves, a painful

process that is done for safety reasons. A cisgenic change such as this is one that could have been acquired naturally by cross-breeding (though it would take much, much longer) so it is not considered an "unnatural" change. This also means cisgenic organisms do not fit the regulatory definition of a GMO—they are not transgenic. The advent of CRISPR and other new genome-editing techniques are fast becoming the editing methods of choice, in part because they are fast, cheap, and easy to do. But equally important is the potential to avoid regulatory hurdles that slow getting a genetically engineered crop to market.

Even before the development of CRISPR and other genome-editing techniques, more than 75% of U.S.-processed food contained GMOs, including 85% to 90% of our corn, soybeans, and cotton. Most of these crops have an herbicide-tolerance (HT) gene added that enables them to withstand huge doses of herbicides (chemicals that kill weeds)—the most widespread HT crops are those with a gene that allows them to tolerate the herbicide glyphosate (Monsanto's Roundup®). This trait enables farmers to douse the field with Roundup and kill the weeds without threatening their harvest. Other crops, known as Bt crops, have been engineered to better resist pests; they contain a gene from *Bacillus thuringiensis*, a naturally occurring bacterium that produces a toxin that kills some insect pests. Some crop GMOs even contain both an HT gene and a Bt gene. **INFOGRAPHIC 6**

Proponents of the technology point to the possibility of not only increasing our food supply with genetically engineered foods but also improving it, such as the GMO known as "golden rice"—a rice modified to produce extra beta carotene (a precursor to vitamin A), to help address vitamin A deficiency, a leading cause of blindness in children. Though it has been the focus of research since 1992, golden rice is still not ready for the market due to poor growth in test plots. GMO opponents say a better option is the provision of vitamin A supplements, an inexpensive and proven method to combat vitamin A deficiencies; their use in affected areas is increasing thanks to international efforts.

Of course, for the people of Africa to benefit from this new technology, they would have to first be made aware of its existence and then be persuaded to use it. And as Paarlberg was discovering in his travels, neither of those would happen easily. Not only were most local farmers completely unaware of GM technology, but government officials, who did know about it, were exceedingly wary.

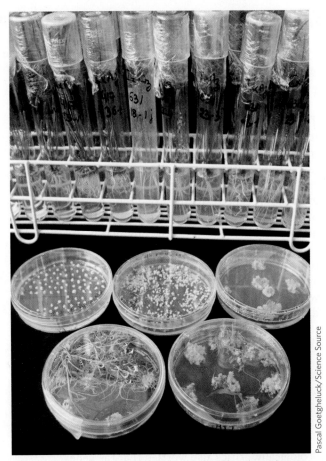

Pascal Goetgheluck/Science Source

Genetic engineering research begins in the lab. This photo shows several growth stages of transgenic rice plants. The plants are grown from modified cells into adult plants.

INFOGRAPHIC 6 · GENETIC ENGINEERING

The genetic material of an organism can be modified in a variety of ways and with a variety of techniques to produce a crop or animal with desired traits. Older methods used the bacterium *Argobacterium* to ferry new DNA into a target cell, but these protocols were time-consuming and outcomes were hard to control. Newer genome-editing methods can precisely edit genes or introduce new DNA in a much quicker and easier way, opening new opportunities to design crops, livestock, or even pets.

TRANSGENIC ORGANISM (GMO)

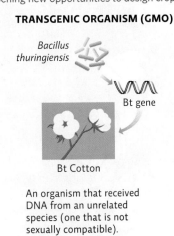

Bacillus thuringiensis

Bt gene

Bt Cotton

An organism that received DNA from an unrelated species (one that is not sexually compatible).

CISGENIC ORGANISM

gene for "no horns"

Angus (hornless breed)   Holstein (normally have horns)

An organism that received DNA from the same or a closely related species; the acquired trait is not considered "foreign" since it could have be passed on by traditional plant or animal breeding.

INTRAGENIC ORGANISM

An organism whose own genes have been edited, usually to silence a gene; no new DNA has been acquired.

Currently regulated as a GMO   Not currently regulated (does not fit the regulatory definition of a GMO)

 Do you think that cisgenic and/or intragenic crops should be regulated the same way as transgenic crops (GMOs)? Explain.

7 THE TRADE-OFFS OF GENETICALLY ENGINEERED FOODS

Key Concept 7: Concerns about genetically engineered foods include health and environmental effects and the worry that the use of patented GMOs could give a few large corporations unprecedented control over global agriculture.

To be sure, genetic engineering is a decades-old technology, and new technologies such as CRISPR have revolutionized the field, making it easier, cheaper, and faster to precisely edit the genetic makeup of crops and animals with less chance of introducing unwanted effects. But genetically modified food is unlike any of the other products that have been produced with genetic engineering. With the production of genetically engineered medications, the transgenic organism remains confined to a flask in a lab; GE crops, however, are out in the real world, growing and sharing genes with other organisms and being eaten by humans both directly and indirectly, through animals we are raising for food. Many of those humans are very uneasy about the prospect.

An early worry of those concerned about GE foods was the possibility that allergens or toxic substances would be present in GM foods. The evidence to date suggests that GM foods are safe to eat by humans and animals.

Ecologically, critics also worried that the genes introduced into genetically modified crops could be passed to wild plants through cross-pollination—that is, they could escape into the natural world and be incorporated into other plants for which they were not intended. This has occurred with the HT gene—it has been accidently transferred to at least 16 weed species, creating "superweeds" that can tolerate all the glyphosate herbicide a farmer can spray, reducing crop yields, often below what would have been realized with traditional (non-HT) crops. However, an extensive 2016 review of genetically engineered crops by the National Academy of Sciences concluded that while gene transfer between HT crops and wild plants does occur, it only appears to be a problem in fields where the herbicide was applied—the wild plants (weeds) had no competitive advantage in places where the herbicide was not used. This does, however, reduce the utility of the HT crops—if the herbicide cannot be used, the HT crop loses its value.

Another fear is that crops engineered to resist pests, such as Bt crops, could allow nontargeted (secondary) pests to increase, thereby requiring more pesticide application, or that they might give rise to a new population of Bt-resistant pests much more quickly than would traditional pesticide spraying. (Bt has long been

University of Minnesota graduate student Jared Goplen studies giant ragweed, a "superweed" that is resistant to commonly used herbicides.

used as a pesticide spray.) However, recent Bt studies suggest that the increase of secondary pests was only a problem in a few cases and that overall, the use of pesticide decreased.

Another concern is the prospect of putting even more of our food supply under the control of a few multinational corporations. In the United States, companies like Monsanto have been known to tightly guard their GM seeds. Because seeds are patented, farmers are not permitted to save seeds from one year to the next but instead must repurchase them year after year. "Food security in private hands is no food security at all," U.S. Senator Tom McGovern said,

"because corporations are in the business of making money, not feeding people." **INFOGRAPHIC 7**

The current consensus presented in the National Academy report is that, in general, the use of GM crops has increased crop productivity (slightly), that it has not produced serious ecological problems (e.g., pest resistance, escaping GM traits) that can't be handled with modified agricultural practices, and that GM food is safe to eat and feed to livestock. However, in a quest to increase food production, they point out that it is but one part of the solution, acknowledging that in some areas, GM crops may not be the best option.

INFOGRAPHIC 7   GMOs IN AGRICULTURE: EVALUATING THE CONCERNS

ROBYN BECK/Getty Images

CONCERN	CONCLUSION
Food safety	There is no evidence that consuming GMO foods is harmful to animals or people.
Development of pest resistance	Bt crops do not increase the emergence of Bt-resistant pests if steps are taken to preempt this.
Increased use of pesticides	Overall Bt crops allow farmers to use less insecticide, even on non-Bt crops that are planted nearby. Insect biodiversity has been reported to be higher in areas where Bt crops are planted compared to areas where synthetic insecticides were applied to non-Bt crops. There does not appear to be an increase in the use of herbicide with HT crops, only a change in the type of herbicide used (the HT herbicide replaces other herbicides on HT fields).
Transfer of GE traits to wild species	The transference of the HT gene to wild (weed) species does occur but only appears to be a problem in areas where the herbicide is applied. There is no record of transference of the Bt gene.
Increased dependence on chemical inputs tied to the GE trait	The use of HT crops, especially glyphosate-resistant crops, are a concern in terms of increased dependence on the herbicide (an expense to the farmer).
GMOs as intellectual property	Since GMO crops are patented, farmers cannot save seed to plant in the next growing season. This means farmers must buy expensive seeds (and in the case of HT crops, herbicides) each time they plant, which decreases their profits and self-sufficiency.
Regulation	Transgenic crops/animals have the greatest potential for unknown or unexpected problems; therefore, these GMOs are carefully evaluated before approval for commercial use. Because newer technologies such as CRISPR can produce useful crops without necessarily introducing novel DNA, the National Academy of Sciences recommends evaluating each new crop based on its properties, not the method used to produce it. Genetic testing can reveal if a GE organism produces a novel substance, which could then be further tested for safety.

Rank the above concerns in order from "most concerning" to "least concerning." Justify your rank order.

Paarlberg and others believe that GM crops will go a long way toward helping African farmers achieve food self-sufficiency. Proponents who want to bring GMOs to farmers in Africa and other developing areas point out that technological innovation is exactly how developed countries like the United States turned Dust Bowl-era food shortages into vast surpluses in less than a generation. It is unfair, they say, to tell impoverished countries to forgo the technologies that we ourselves have benefitted from, especially while the people in those other countries are still starving.

Meanwhile, in the shadows of this global debate, many African farmers have been devising their own solutions to food insecurity.

8 LOW-TECH ALTERNATIVES TO INCREASE CROP PRODUCTION

Key Concept 8: Solutions to agricultural problems include using low-tech traditional methods and crop varieties as well as industrial farming methods and GMOs.

The central plateau of Burkina Faso is marked by poor soils, low crop yields, and very little rain. In many places, the land is so parched that it has crusted over into what local farmers call *zippelle*: hard, dry cake. To grow anything here takes a special kind of ingenuity. In the 1980's with the help of the FAO, farmers revived an ancient strategy known as the *Zaï* pit. During the dry season (between November and May), farmers dig thousands of small pits and fill each with organic matter. After the first rainfall, the organic layer is covered with a thin layer of soil, and seeds are placed in the middle of the pit. The top of the pit is then ridged, so that when the rain does fall, it accumulates in the pit and hydrates the seed.

Zaï pits are not the only technology being employed in this landlocked, desert country. Researchers and farmers have also worked together to adapt cutting-edge microfertilization technology to suit local needs. By applying small and precise quantities of fertilizers close to each seed at planting, they only apply what the plants can actually use: Wheat farmers in Burkina Faso who microfertilize apply only one-tenth of the fertilizer used by their U.S.-grown counterparts; corn crops receive

Seed exchanges, like this one sponsored by SEED, a nonprofit community group in South Africa, helps farmers use and preserve a wide variety of locally adapted seeds.

NIC BOTHMA/European Pressphoto Agency/CAPE TOWN/South Africa/Newscom

one-twentieth. This significantly reduces fertilizer waste and brings the cost of fertilization down to an affordable level.

In addition to using Zaï pits and microfertilization, farmers here have planted trees around fields to retain soil and stave off desertification, and they are conserving livestock manure and applying it to fields. And cooperative groups have been established throughout the region to manage village cereal banks and community wells. The result has been an increase in grain production that matches some of the first Green Revolution successes in Asia. "So much has been written about the disappointments in African agriculture that it is easy to overlook the successes," said Steve Wiggins of the Overseas Development Institute in London. "In some parts of the continent and for particular crops and activities, there have been veritable booms in farming."

Traditional crop breeding to produce locally adapted and highly efficient crops is also improving crop productivity in Africa and Asia. Jonathan Lynch of Pennsylvania State University has achieved great success in enhancing the ability of crops to take up nutrients (improving productivity) using the old-fashioned method of crossing individual plants with the most desired traits. Genetic engineering attempts have yet to create a GMO strain with similar abilities. His beans are three times more productive than traditional varieties when grown in nutrient-poor soil in African field trials. Traits like this are vitally needed by farmers in areas with impoverished soil but who lack the funds to purchase expensive synthetic fertilizer inputs.

Indeed, a growing litany of similar successes has been reported from across the continent. The details of each differ, but the central point is often the same: Across the vast African continent, at least some individual communities are meeting their food needs with simple, sometimes centuries-old technologies. Methods such as these are improving productivity for cash crops and staple crops that feed local populations—both of which are improving food security.

To those opposed to GMOs, these low-tech successes underscore a key argument—namely, that there are other solutions that are more cost effective and that can address not only the need to produce food but also the need to reduce farming's impact on the natural environment and to rehabilitate damaged areas. Instead of giving industrial agriculture another tool to use—GM seeds that cost lots of money and, like other HYVs, will still require heavy fertilizer inputs to maximize production—proponents of a more sustainable approach to farming say we should teach

farmers to plant nitrogen-fixing trees and crops that would furnish the soil with nutrients and help improve crop yields. Farmers should be encouraged to plant many different crops instead of just one so that the soil is replenished naturally, an example of mimicking a characteristic of a natural ecosystem—its dependence on high biodiversity. (See Infographic 6 in Module 1.1 for more on the characteristics of a sustainable ecosystem.) And communities should create seed banks and seed-sharing programs that enable farmers to both use and preserve the wild relatives of our crops (known as the *wild type*) and the traditional plant varieties—those that over generations have become adapted to each individual area.

Critics also contend that government policies should be examined as well. For example, rather than subsidize cash crops for export, policies could provide support for farms that grow some crops for local consumption. And instead of offering tax incentives to agribusiness giants like Monsanto, investments in infrastructure could be made: Creating and repairing roads would dramatically enhance farmers' ability to get supplies to the farm and to get fresh produce to market—while it's still fresh.

Low-tech solutions have a key advantage over higher-tech ones: They are more accessible to women. Women make up a large percentage of the agricultural workforce—43% worldwide—and possess considerable knowledge about successful farming techniques in their home regions. But in developing countries especially, they are much less likely than men to have government support or the financial means to purchase the inputs required for industrial agriculture. Because of this gender bias, programs that focus on high-tech solutions tend to bypass women and local farmers like the ones that Paarlberg surveyed in his travels.

INFOGRAPHIC 8

Ultimately, of course, the countries of Africa will need much more than Zaï pits or GMOs to attain food security: They will need financial support, infrastructure, equipment, and training. And even with those essentials in place, the solutions still won't be straightforward. Different regions—and different farmers within the same region—face different agricultural challenges. In some cases, GMOs may represent the best option. In others, HYVs may work better, and in others still, native varieties may be the right choice. Whatever methods are pursued, they must also protect the natural environment so as not to jeopardize future productivity. And they must be made available to all people, not just an empowered few. That means issues of environmental sustainability and social justice must be considered alongside issues of productivity per hectare.

INFOGRAPHIC 8 **LOW-TECH FARMING METHODS**

Low-tech farming methods should not be overlooked in favor of high-tech industrial or genetic engineering methods. These low-tech methods are more accessible to poor farmers, many of them women, in developing countries and can contribute to food self-sufficiency. The FAO calls for the use of high-tech and low-tech agricultural methods, realizing that a one-size-fits-all approach will not solve the problem of world hunger. (See Module 8.2 for more on traditional farming methods.)

SEED BANKS/SEED SHARING

Advantages:
- Preserves seed varieties.
- Gives access to locally adapted seeds.
- Low- or no-cost seeds.

A storehouse of locally grown seeds is saved from last year's harvest and shared by farmers.

TRADITIONAL / REGIONAL FARMING TECHNIQUES

Advantages:
- Fewer chemical inputs will be needed.
- Ability to grow crops in areas that might not support conventional agriculture.
- Higher crop yields if growing locally adapted crops.

Use the methods that work best locally such as Zaï pits that help capture water in arid regions.

PLANT BREEDING

Advantages:
- Produce improved crops without high-tech methods.
- Produce desired traits without the need to identify genes.
- Cheaper process; fewer regulatory hurdles.

Cross plants with desired traits (artifical selection).

MICROFERTILIZATION

Advantages:
- Less money spent on fertilizer.
- Less potential for stormwater runoff pollution.

Apply small amounts of fertilizer as needed, directly to the plant.

 Explain the importance of preserving the wild relatives and traditional plant varieties (i.e., the seeds from as many corn varieties as possible) rather than focusing on just a few crop varieties and letting the others die out.

Select References:

Azadi, H., & Ho, P. (2010). Genetically modified and organic crops in developing countries: a review of options for food security. *Biotechnology Advances, 28*(1), 160–168.

Food and Agriculture Organization of the United Nations. (2016). *The State of Food and Agriculture 2016: Climate Change, Agriculture, and Food Security.* Rome, Italy: Author.

Gilbert, N. (2016). The race to create super-crops. *Nature, 533*(7603), 308–310.

National Academies of Sciences, Engineering, and Medicine. (2016). *Genetically Engineered Crops: Experiences and Prospects.* Washington, DC: The National Academies Press. doi:10.17226/23395.

Paarlberg, R. (2010). GMO foods and crops: Africa's choice. *New Biotechnology, 27*(5), 609–613.

Patel, R. C. (2012). Food sovereignty: power, gender, and the right to food. *PLoS Med, 96,* e1001223.

INTERACTIVE MAP   **FEEDING THE WORLD**   ANIMATED INFOGRAPHIC

Around the world, different approaches are being used to improve agriculture and crop yield. Each come with their own trade-offs. Check out these four examples.

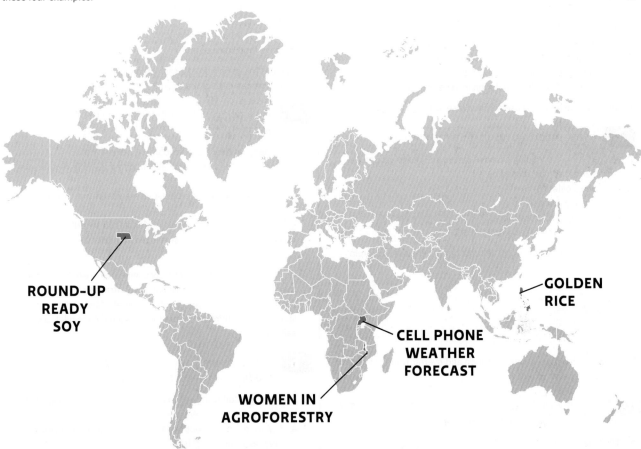

ROUND-UP
READY
SOY

GOLDEN
RICE

CELL PHONE
WEATHER
FORECAST

WOMEN IN
AGROFORESTRY

BRING IT HOME

PERSONAL CHOICES THAT HELP

TODD ANDERSON/The New York Times/Redux

The challenge of meeting the nutritional needs of the entire population seems overwhelming, but there are plenty of opportunities to make an impact within your local community. Two of the best ways to have an impact on those facing food insecurity are through education and financial support.

Individual Steps

• There are thousands of food banks all over the country that collect and distribute food to those who need it the most. Donate your time or extra nonperishable food items. Find the closest food bank at www.feedingamerica.org.

• Plant your own vegetable garden and donate part of the harvest to your local food bank. Some areas even have community gardens that rent space during the growing season.

• Make a monetary donation to an international nonprofit organization that provides food aid or agricultural assistance such as UNICEF (www.unicef.org) or Heifer International (www.heifer.org).

Group Action

• Work with parents, community leaders, and education leaders to develop a summer food service program for school-age children on free or reduced-price lunch programs. Find contacts and resources at www.fns.usda.gov/cnd/summer/.

Policy Change

• GMOs may be part of the solution because they can increase crop yields around the world, but they may cause some unintended consequences. Contact your federal representative: Express your concerns and discuss your representative's position on the regulation, testing, and labeling of GMOs.

ENVIRONMENTAL LITERACY   UNDERSTANDING THE ISSUE

1   How prevalent is world hunger, and what are its causes?

1. A population has food security when:
   a. there are no reported diet-related health problems.
   b. enough food is available, accessible, and affordable.
   c. surplus food supplies are available.
   d. less than 1% of the population experiences hunger issues.

2. Identify at least five causes of food insecurity.

2   What problems result from malnutrition?

3. Malnutrition is a leading cause of illness and death around the world, and it can be caused by:
   a. too few calories.
   b. too many calories.
   c. insufficient micronutrients.
   d. All of the above are correct.

4. A diet that contains enough calories but is deficient in protein can result in:
   a. wasting disease.
   b. marasmus.
   c. kwashiorkor.
   d. A and B
   e. A, B, and C

5. Distinguish between undernutrition and overnutrition in terms of causes and consequences.

3   What was the Green Revolution, and how does it relate to industrial agriculture?

6. The Green Revolution:
   a. increased world food supplies but introduced new problems.
   b. was based on traditional, locally adapted crop species.
   c. used crop plants that required less fertilizer and pesticides.
   d. created food security for the global population.

7. What did the Green Revolution accomplish and how did it do so? Discuss some of the unintended consequences of the methods used in the Green Revolution.

4   What are the pros and cons of industrial agriculture?

8. Which of the following statements is true about industrial agriculture?
   a. Soil quality is generally enhanced.
   b. It is more labor intensive than traditional agriculture methods.
   c. Pesticide use produces crops with more blemishes.
   d. Monocultures are preferred since they have high productivity.

9. Explain the advantages and disadvantages to the use of monocultures in industrial agriculture.

5   What is the importance of food self-sufficiency and sovereignty and what can undermine meeting these goals?

10. True or False: Food security is enhanced with food self-sufficiency and food sovereignty.

11. A nation that can reliably provide food for its citizens even if all the food is not grown locally is said to:
    a. lack food security.
    b. have food self-sufficiency.
    c. have food sovereignty.

12. What is the definition of a cash crop and what food security problems can cash crops cause?

6   How can genetic engineering be used in agriculture?

13. Proponents of the "Gene Revolution" believe that:
    a. humans should be genetically modified (GM) to require less food.
    b. genetically modified crop plants are useful, but GM animals are not.
    c. GMOs are needed to achieve global food security.
    d. GMOs will be useful in developed countries but not in developing countries.

14. Compare the types of genetically engineered organisms that are made today—transgenic, cisgenic, and intragenic organisms—in terms of the origin of the genetic material added and its regulatory status. Do you think all GE organisms should be regulated in the same way?

7   What are the trade-offs of using genetically modified organisms in agriculture?

15. Which of the following early concerns regarding GM crops has been supported by recent evidence?
    a. GM food is dangerous to eat for certain individuals who are allergic to the new foods.
    b. Farmers who use Bt crops use significantly more pesticide to deal with other problems that use of the Bt crop creates.
    c. The creation of herbicide-resistant superweeds does occur but is only a problem where the herbicide is used.
    d. GM food is not as healthy as non-GMO food due to lower nutrient content.

16. Explain how planting a genetically modified crop with a trait for pest resistance could lead to the use of less pesticide in some cases but could lead to the use of more in others.

8   What are some low-tech (non-GMO, nonindustrial) options for increasing crop production?

17. An advantage that low-tech farming methods have over high-tech farming methods in developing countries is that low-tech methods are more:
    a. accessible to women.
    b. likely to be subsidized by the government.
    c. useful for growing cash crops.
    d. suitable for growing high-yield varieties.

18. What is the value of keeping biodiversity in our crops high?

SCIENCE LITERACY   WORKING WITH DATA

The Food and Agriculture Organization (FAO) of the United Nations monitors world hunger. At the World Food Summit (WFS) in 1996, a goal was set to reduce the number of undernourished people by half of the 1990–1992 level by 2015. In 2015, the Sustainable Development Goals (SDG) were drafted; SDG 2 has set a target of eliminating world hunger by 2030.

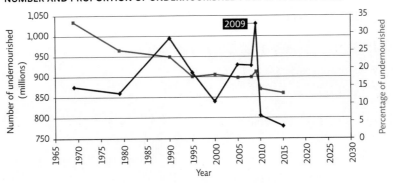

NUMBER AND PROPORTION OF UNDERNOURISHED PEOPLE IN THE WORLD

Interpretation

1. Describe in one sentence what the graph shows about change in undernourishment over time. Why are both lines shown here?

2. In what year were the most people undernourished? In what year was the highest proportion of people undernourished? Why do you think these occurred in different years?

3. What might have led to the decrease in undernourishment in the 1990's?

4. What happened to the number and percentage of undernourished people between 2008 and 2009? What do you think led to this change?

Advance Your Thinking

5. By 2015, the FAO had hoped to cut the number and proportion of undernourished people relative to the 1990 levels in half. How did it do?

6. In looking at both the number and percentage of undernourished people, do you think there has been a complete recovery from the 2008 world food crisis? Why or why not?

7. Following what appears to be the general trend of the data (not just the last two data points) extrapolate these lines out to 2030, the date by which the FAO hopes to eradicate world hunger (undernourishment). Do you predict that the FAO will meet its targets? If not, what do you predict the number and percentage of undernourishment to be in 2030 and what must be done to meet these targets? (Hint: Be sure to consult the correct y-axis when estimating the 2030 values.)

INFORMATION LITERACY   EVALUATING INFORMATION

Oxfam is an international not-for-profit organization dedicated to addressing issues of poverty and injustice around the world. Visit the Oxfam website (www.oxfam.org).

Evaluate the website and work with the information to answer the following questions:

1. Determine if the website is a reliable information source:
   a. Does Oxfam give supporting evidence for its claims?
   b. Does the organization give sources for its evidence?
   c. Why was Oxfam originally founded in 1942, and what is its mission today?

2. Watch the videos "Does Aid Work?" at www.oxfam.org/en/ video/2010/does-aid-work and "Good Aid: The Video" at www .oxfam.org/en/multimedia/video/2010-good-aid-video. (Search for the videos by title on the Oxfam website if these URLs are not active.) Summarize the arguments made about the success of aid. Which of the reasons do you agree with most strongly and why? Which do you find less compelling?

3. Review Oxfam's campaign on agriculture. Select two articles on food prices and read them. For each article answer the following questions:
   a. Do you agree with Oxfam's position presented in this article? Explain.
   b. Identify a claim Oxfam makes and the evidence the organization gives in support of this claim. Is it sufficient? Where would you attempt to find more evidence to support or refute the claim?

4. Review Oxfam's campaign to "transform a broken food system"— the GROW campaign. Read the information found in the links on the GROW webpage such as "What is GROW?," "The Issues," and "FAQs" and answer the following questions:
   a. Summarize the solutions put forth by the campaign.
   b. Do you agree that this is a reasonable approach? What would you do differently?
   c. What is Oxfam's position on GMOs? Is it consistent with the organization's overall goals?

 Additional study questions are available at SaplingLearning.com.

FARMING LIKE AN ECOSYSTEM

Learning to farm from nature

An ancient Japanese rice farming practice offers a more sustainable approach to growing the crop. Rice fields like this one across the globe produce millions of tons of rice each year.

The landscape of regional cities in Japan/Moment/Getty Images

After reading this chapter and studying the KEY CONCEPTS and INFOGRAPHICS, you should be able to answer these GUIDING QUESTIONS

CORE MESSAGE

Achieving food security for all people requires that we build a sustainable food system. This will include growing crops without degrading the ecosystem that supports that growth as well as changing consumer purchasing practices to support sustainable agriculture.

1 What is sustainable agriculture?

2 What are the trade-offs of using chemical fertilizers in industrial agriculture?

3 What are the trade-offs of using chemical pesticides in industrial agriculture?

4 What is agroecology and what are its benefits?

5 What is integrated pest management?

6 How can traditional farming methods contribute to sustainable agriculture?

7 What role does the consumer play in helping build a sustainable food system?

8 Evaluate the trade-offs of sustainable agriculture and its potential to meet world food needs.

If there's one thing Greg Massa and his wife Raquel Krach hate, it's weeds—all varieties, but especially the azolla—an insidious, fernlike plant that grows on the surface of water. Each spring, azolla plants invade the couple's rice farm, snaking their way through the dense, muddy paddies that stretch for miles along the Sacramento River near Chico, California. They strangle young rice plants and force Greg into an endless and tedious battle.

◉ **WHERE IS CHICO, CALIFORNIA?**

The rivalry—Massa versus azolla—has spanned three generations. Greg's great-grandfather, Manuel Fonesca, planted the family's (and some of California's) first rice crops in 1916, on the same land that Greg and Raquel now manage. Back then, rice farming was a hard and uncertain life; Manuel was largely powerless against the azolla, which in some seasons claimed his entire crop.

By 1962, when Greg's father, Manuel, took over, human ingenuity and modern science had completely changed the nature of the fight. Heavy doses of chemical herbicides enabled him to obliterate the weed. And specially bred higher-yield rice varieties developed during the Green Revolution, along with modern farming equipment and a heavy dose of synthetic chemical **fertilizers** and **pesticides**, made the family farm both efficient and profitable. Of course, that modern approach, known as **industrial agriculture**, has its own problems. It relies on cheap fossil fuel energy and huge amounts of water. Its tendency to plant large expanses of only one variety of one crop leads to a loss of genetic diversity on crops in general and contributes to a progressive degradation of the environment. (See Module 8.1 for an introduction to industrial agriculture and the Green Revolution.)

At a time when our world population has surpassed a staggering 7 billion and is growing toward 10 or 11 billion, the need to produce even more food has never been so urgent, and many farmers are looking for ways to grow food without damaging the environment that produces it. The answer may come from the very environment they are striving to protect.

fertilizer A natural or synthetic mixture that contains nutrients that is added to soil to boost plant growth.

pesticide A natural or synthetic chemical that kills or repels plant or animal pests.

industrial agriculture Farming methods that rely on technology, synthetic chemical inputs, and economies of scale to increase productivity and profits.

1 SUSTAINABLE AGRICULTURE

Key Concept 1: The goal of sustainable agriculture is to raise food without damaging the environment or future productivity while operating ethically with regard to animals, workers, and local communities.

Greg and Raquel wanted to find an alternative to industrial agriculture as it is commonly practiced. **Sustainable agriculture** is farming that meets the needs of the farmer and society as a whole, without compromising the environment or future productivity. The techniques used will maintain or even enhance the environment. They often do this by mimicking the traits of a sustainable ecosystem: They rely on renewable energy and local resources for inputs, and they depend on biodiversity to trap energy, deal with waste, and control pest populations. (See Infographic 6 in Module 1.1 for more on the characteristics of a sustainable ecosystem.) Sustainable agriculture must also be economically viable and socially ethical; workers should be paid a fair wage and the needs of local communities considered. The humane treatment of animals is also seen as a goal of sustainable agriculture. **INFOGRAPHIC 1**

sustainable agriculture Farming methods that can be used indefinitely because they do not deplete resources, such as soil and water, faster than they are replaced.

INFOGRAPHIC 1 | **SUSTAINABLE AGRICULTURE**

According to the U.S. Department of Agriculture, sustainable agriculture is farming that uses only limited amounts of nonrenewable resources (like fossil fuels) and does not degrade the environment or the well-being of people or society as a whole.

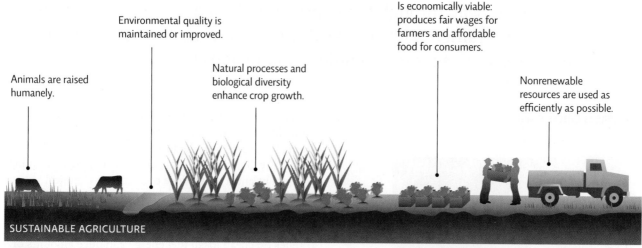

Environmental quality is maintained or improved.

Is economically viable: produces fair wages for farmers and affordable food for consumers.

Animals are raised humanely.

Natural processes and biological diversity enhance crop growth.

Nonrenewable resources are used as efficiently as possible.

SUSTAINABLE AGRICULTURE

 Organic farming is a subset of sustainable farming. What do these farming methods have in common, and how do they differ?

So when Greg and Raquel took over in 1997, they converted a portion of their farm to a sustainable farming method known as **organic agriculture**. Instead of using synthetic fertilizers and pesticides, organic agriculture employs more natural, or "organic," techniques in the growing of crops—such as using manure as fertilizer and luring in natural predators to control pests. In addition, genetically modified organisms (GMOs) cannot be grown on organic farms (see Module 8.1).

Organic farming uses fewer or no chemicals and may even produce food that is more nutritious. A 2014 meta-analysis of 343 studies showed that organic food had levels of antioxidants as much as 50% higher than that in conventionally grown crops. For example, research by Washington State University soil scientist John Reganold showed that organically grown strawberries had a longer shelf life and higher levels of antioxidants than conventionally grown berries. The meta-analysis also showed that, on average, conventionally grown food had four times more pesticide residue than organic food.

But organic farming forced Greg and Raquel to battle weeds much as the first Manuel had: with great difficulty. The trick was to lower water levels enough to kill water-loving weeds but not so much that the rice crop also died. Each day, Greg would wade into the paddies to see how the rice plants were faring against the azolla. Some weeks, he worried the entire crop

would die. After a few seasons, the Massas started to despair: How could they make their farm environmentally friendly without losing their livelihood to an army of mangy weeds?

The Massas' story is the story of modern farming; it's the story of how we feed ourselves. And on a planet where population is exploding, climate is shifting, and energy and water resources are running low, it's also a story of constant change.

organic agriculture Farming that does not use synthetic fertilizer, pesticides, GMOs, or other chemical additives like hormones (for animal rearing).

Greg Massa in his rice fields.

2 FERTILIZER USE IN INDUSTRIAL AGRICULTURE

Key Concept 2: Fertilizer use can increase productivity but can also degrade soil and contaminate nearby bodies of water.

The changes that helped Greg's father thrive also ushered in a whole new way of growing crops, known as **monoculture** farming. Instead of growing a mix of plants or growing different crops each season, farmers began growing the same single crop year after year. Before long, farms that had been populated by a variety of crops morphed into industrial operations, focused on just one crop. Thus, biodiversity was replaced by specialization, and farm ecosystems that more closely resembled nature were replaced by operations completely dependent on technology.

"In the 1920s, half of Iowa's farms produced 10 commodities each," Fred Kirschenmann, a Distinguished Fellow at Iowa State University's Leopold Center for Sustainable Agriculture, told attendees at a 2003 Biodynamic Farming and Gardening Association conference. "Today 92% of the state's cultivated land is exclusively corn or soybean. Farming systems that were once supported by complexity and diversity of species have now been replaced by reliance on inputs." The same monoculture approach has also been applied to rearing livestock. In a *concentrated animal feeding operation (CAFO)*, livestock and poultry are raised in confined spaces, with a focus on raising as many animals in a given area as possible. (See Online Module 8.3 for more on CAFOs.) Worldwide, 90% of our food comes from just 15 crop species and 8 species of livestock.

> Worldwide, 90% of our food comes from just 15 crop species and 8 species of livestock.

The advantages of this approach are obvious: A single crop (or type of animal in a confined space) is much easier to manage and to mass produce; with greater ease comes not only greater efficiency and greater profits but also drastically greater amounts of food for a planet that has never seemed to have enough. In recent years, however, the disadvantages of industrial agriculture have become equally apparent.

In monoculture farming, the crop that is chosen is not necessarily locally adapted. Instead of choosing the crop best suited to the existing ecosystem, farmers focus on the crops with the highest market demand and thus

monoculture A farming method in which a single variety of one crop is planted, typically in rows over huge swaths of land, with large inputs of fertilizer, pesticides, and water.

Cattle in a CAFO feedlot pen.

Reed Kaestner/Getty Images

the highest dollar value. Because these crops are not locally adapted and because the volume of plants grown increased exponentially, the average farm becomes heavily dependent on external inputs—water, pesticides, and fertilizer—added to the farm from outside its own ecosystem.

Fertilizers boost growth because they provide nutrients that plants need, such as nitrogen and phosphorus. (See Module 2.1 for more on nutrient cycles.) And while heavy doses of fertilizer can indeed boost crop production, the excess nutrients (whatever the plants don't use) can contaminate waterways, potentially creating hypoxic, or oxygen-poor, conditions that threaten aquatic life. This process, known as **eutrophication** (or, more precisely, *cultural eutrophication*, since human activities were the source of these nutrients), has been a significant problem in the United States; nutrient runoff from farms in states within the Mississippi River watershed has created a summertime dead zone in the Gulf of Mexico (see Module 6.2). The application of fertilizer can also create a dependence on additional fertilizer use as soil fertility declines. In addition, synthetic fertilizers are made from fossil fuels (natural gas and, to a lesser extent, petroleum), both of which are in limited supply and are environmentally damaging to extract, process, and ship. **INFOGRAPHIC 2**

eutrophication A process in which excess nutrients in aquatic ecosystems feed biological productivity, ultimately lowering the oxygen content of the water.

INFOGRAPHIC 2 | **THE TRADE-OFFS OF FERTILIZER USE**

ADVANTAGES

INCREASE PRODUCTIVITY

Fertilizer can greatly increase soil productivity and is required to support the growth of the high-yield varieties grown today. Soil nutrients like nitrogen and phosphorus are often in limited supply and plant growth slows if one nutrient starts to run out. The addition of extra nutrients overcomes this deficiency and can boost growth.

TRADE-OFFS

DISADVANTAGES

CULTURAL EUTROPHICATION

Runoff pollution that contains fertilizer can also cause algal blooms and result in the buildup of dead organic material, both of which reduce the amount of oxygen in the water. This can kill many more aquatic organisms, even fish.

GROW CROPS IN MARGINAL SOILS

Fertilizers help crops grow in areas that may not otherwise be able to support agriculture. This may be the only way to farm in many areas of the world and would help increase local food supplies.

TRADE-OFFS

DEVELOP A DEPENDENCE ON FERTILIZER

The extra plant growth that fertilizers support can pull other nutrients out of the soil, depleting soils further and requiring even more fertilizer in the future.

 How could the disadvantages of fertilizer use be managed to reduce their impact?

3 PESTICIDE USE IN INDUSTRIAL AGRICULTURE

Key Concept 3: Pesticides can help fight pests but are toxic to other species, including humans, and can lead to pesticide-resistant pest populations.

Monoculture crops are also especially vulnerable to pests or disease: Pest populations can explode when they encounter acres and acres of a suitable crop on which to feed, and a single infestation can wipe out the entire crop because what kills one plant will likely kill them all. To deal with this, farmers have turned to pesticides (many of which are made from petroleum), but this also has proven problematic. To be sure, pesticides kill plant and animal pests and thus dramatically reduce the number of crops lost each year to infestation. But because they are toxic, pesticides also pose a threat to human and ecosystem health. Toxic residues can be found on many foods, and applying pesticides is dangerous work for the farmer.

Here's another problem—the very use of pesticides sets into motion events that render them ineffective, causing pest populations to develop **pesticide resistance**, an unintentional example of artificial selection (see Module 3.1). Herbicide-resistant weeds and insecticide-resistant insects are cropping up all around the world. Such resistance encourages us to employ more drastic measures, in the form of higher doses or more toxic chemicals; as resistance to that next pesticide develops, the cycle repeats itself, becoming what some have called a "pesticide treadmill." It's like an arms race between humans and pests, with the deck stacked in favor of the pests. Both the reduction in soil fertility due to the use of fertilizers and the decreased effectiveness of pesticides as pesticide resistance emerges are examples of a social trap known as the *sliding reinforcer* (see Module 1.1, Infographic 8). **INFOGRAPHIC 3**

As modern farmers face these enormous challenges (and climate change and energy and water shortages and the genetic erosion of crops along with them), the story of how we feed ourselves is changing yet again. This time, we may have to consider not just high-tech solutions but look back to the natural world for answers. In searching for a new weapon against the azolla, Raquel found a Japanese rice farmer who was doing just that.

pesticide resistance The ability of a pest to withstand exposure to a given pesticide; the result of natural selection favoring the survivors of an original population that was exposed to the pesticide.

Pests, such as these Colorado potato beetle larvae, cost farmers millions of dollars in lost productivity. Livestock too are attacked by pests. Pesticides can reduce infestations but come with many trade-offs such as cost, environmental damage, or the emergence of pesticide-resistant pests.

Orest Iyzhechka/Shutterstock

INFOGRAPHIC 3 | **THE EMERGENCE OF PESTICIDE-RESISTANT PESTS**

Exposure to a pesticide will not make an individual pest resistant; it will likely kill it. However, if a few pests survive because they happen to be naturally resistant, they will breed and their offspring (most of which are also pesticide resistant) will make up the next generation. Over time, the original pesticide will no longer be effective and will have to be applied at a higher dose or a different pesticide will have to be used.

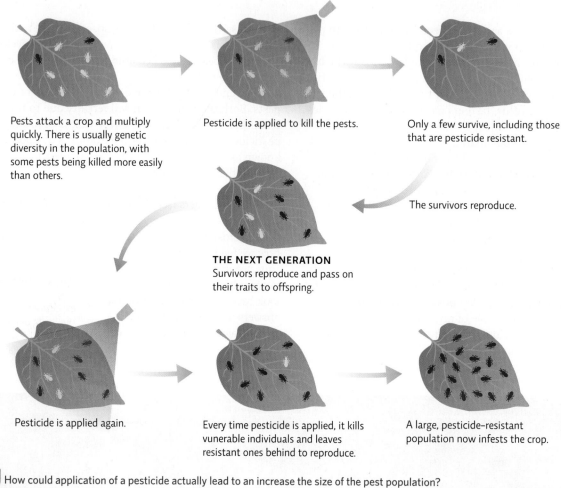

Pests attack a crop and multiply quickly. There is usually genetic diversity in the population, with some pests being killed more easily than others.

Pesticide is applied to kill the pests.

Only a few survive, including those that are pesticide resistant.

The survivors reproduce.

THE NEXT GENERATION
Survivors reproduce and pass on their traits to offspring.

Pesticide is applied again.

Every time pesticide is applied, it kills vunerable individuals and leaves resistant ones behind to reproduce.

A large, pesticide–resistant population now infests the crop.

? How could application of a pesticide actually lead to an increase the size of the pest population? (Hint: Think about the predators of the pests.)

4 AGROECOLOGY

Key Concept 4: Modeling a farm after an ecosystem (agroecology) to include a variety of plants and animals can boost productivity and protect or even enhance the local environment.

Takao Furuno was, by most standards, a very successful industrial rice farmer, with annual yields among the highest in southern Japan. But it was a tough grind. Each year he was forced to put all his earnings back into the next year's crop—insecticides, herbicides, irrigation, and fertilizer—so that despite his success, he and his family were left with very little for themselves at the end of each season.

In searching for a better way, he turned, as he often did, to his forebears to see what he could learn from their knowledge. He was surprised to discover that they used

to keep ducks in their rice paddies. Like most other rice farmers, Furuno considered ducks a pest, albeit a slightly cuter one than the azolla. Adult ducks eat rice seeds before they have a chance to grow, and as they forage, they trample young seedlings into the mud. This disturbance creates open patches of water, which in turn invites more ducks. "If you're not careful, you end up with a big problem pretty quickly," says Raquel.

But ducklings, Furuno soon realized, were too small to do such damage; for one thing, their bills were not big or strong enough to extract seeds from mud. Instead, they ate bugs and weeds. Azolla was one of their favorites.

Furuno's forebears also grew a fish (the pond loache) in their paddies. The loaches would also eat azolla and could be harvested and sold as food.

Together the ducklings and loaches would keep the weed from strangling the rice crop, but they would not completely eliminate the azolla the way a heavy dose of pesticides would. Furuno quickly discovered that, when kept at this benign concentration, the azolla (which contains symbiotic bacteria that produce a usable form of nitrogen) actually fertilized the rice. In fact, between the nitrogen from the azolla and the duck and fish droppings, he soon found that he no longer needed to spend money on synthetic fertilizer.

When raising ducks, fish, and rice crops together, Furuno discovered that the root crowns (where the root meets the stem) of rice plants increased to about twice the size that they had been in his old industrial system. A larger root crown meant more rice. "We're not exactly sure why the crowns grew," Furuno told an audience of American farmers at a recent convention in Iowa. "But the ducks seem to actually change the way the rice grows. It's got something to do with the synergy of the whole system." Furuno's operation is an elegant example of biomimicry—a farm operating like a natural ecosystem. This type of farming is known as **agroecology**—a holistic approach that considers the local ecology of the area, the value of traditional farming methods, and the socioeconomic needs of the local community. **INFOGRAPHIC 4**

agroecology A scientific field that considers the area's ecology and indigenous knowledge and favors agricultural methods that protect the environment and meet the needs of local people.

INFOGRAPHIC 4 **AGROECOLOGY: THE DUCK-RICE FARM**   ANIMATED INFOGRAPHIC

Takao Furuno's farm is a self-regulating, multiple-species system that naturally meets the needs of the farm ecosystem. All of the species play a role in the system, helping each other and boosting overall production.

THE METHOD

Rice seedlings are planted in flooded rice paddies.

Ducklings are introduced to eat weeds and provide "fertilizer."

Fish are introduced to eat weeds and provide "fertilizer."

Azolla is introduced to add nitrogen; ducklings and fish keep the azolla from growing too much.

THE FINAL PRODUCT: AN INTEGRATED SYSTEM   THE HARVEST

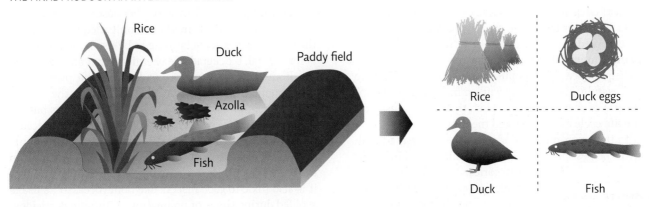

 What normal industrial inputs can be averted by growing rice using the duck-rice farm model?

Melanie Stetson Freeman/Getty Images

Cows at this farm are grass fed and graze using the rotation method. This method is closer to a traditional ecological system than conventional farming.

There were other financial gains, too. Duck eggs, duck meat, and fish all fetched a good price in the market. And because he was no longer using pesticides, Furuno could also grow fruit on the edges of his rice field. (He opted for fig trees, as he could harvest the figs annually without having to replant.)

Furuno's farm is an example of **polyculture**—intentionally raising more than one species on a given plot of land. In the decades since Furuno began duck/rice farming, his rice yields have increased by as much as 50%, making his among the most productive farms in the world, nearly twice as productive as conventional farms. This kind

> **polyculture** A farming method in which a mix of different species are grown together in one area.

of success is especially important for farmers in many developing countries, who struggle to produce enough food for current populations. Increasing production with lesser dependence on expensive inputs and using methods that enhance rather than diminish environmental quality can help communities become more self-sufficient and help them achieve *food security* (see Module 8.1).

The Bangladesh Rice Research Institute, which has evaluated Furuno's method and independently verified his success, recommends the technique to Bangladeshi farmers. And by now, some 10,000 Japanese farmers have followed Furuno's lead; his method is also catching on in China, the Philippines, and California.

5 MANAGING PESTS

Key Concept 5: Integrated pest management techniques can often effectively control pests while minimizing or eliminating the use of chemical pesticides.

Both of Greg and Raquel's azolla-control methods—reducing water levels and employing ducks as natural predators—are examples of **integrated pest management (IPM)**, or the use of a variety of methods to help reduce a pest population. The goal of IPM is to successfully control pests while minimizing or eliminating the use of chemical toxins. First, the farmer must examine the life cycle of the pest and the pest's interactions with the environment to identify the best way to deal with the pest. In general, IPM techniques fall into four categories: cultural control, biological control, mechanical control, and chemical control.

> **integrated pest management (IPM)** The use of a variety of methods to control a pest population, with the goal of minimizing or eliminating the use of chemical toxins.

In sustainable agriculture, farmers use a combination of cultural, mechanical, and biological controls

to deal with pest problems, and they resort to using chemicals only if these methods don't adequately deter the pests. If a pesticide is going to be used, the preference is for natural, biodegradable chemicals that are toxic only to a limited group of organisms. For example, pyrethrum, a compound naturally produced by a flower in the chrysanthemum family, is directly toxic to insects but not to mammals; it is certified for organic agriculture because it is naturally produced (not a synthetic human creation) and because it breaks down quickly and does not linger in the environment. However, while it does kill pest insects, it is also toxic to *good* insects like honeybees, so its use is avoided during times of pollination. Synthetic pesticides, like those commonly used in industrial agriculture, are not acceptable for use on certified organic crops but may be part of an IPM plan for conventionally raised crops.
INFOGRAPHIC 5

INFOGRAPHIC 5 | INTEGRATED PEST MANAGEMENT (IPM)

Controlling pests is important for our agricultural yields as well as for our health and for the health of our pets. Rather than using harsh methods in an attempt to completely eliminate the pest (which rarely works anyway), a combination of less hazardous methods can often reduce pest numbers to manageable levels.

INTEGRATED PEST MANAGEMENT HAS SEVERAL STEPS

1. IDENTIFY TRUE PESTS

Not all "bugs" and "weeds" are pests—some plants and animals are innocuous or actually beneficial. By working to control actual pests only, we save time and money, and we help the environment by maintaining diversity and preventing toxic pollution.

2. SET AN ACTION THRESHOLD AND MONITOR PESTS

The pest population size that is unacceptable must be identified. We may have zero tolerance for some pests (e.g., fleas and ticks in our homes) but be able to tolerate small populations of crop pests; we will only act if the action threshold is reached.

3. DEVELOP AN ACTION PLAN

This may include a variety of methods, each of which aids in pest control by excluding, discouraging, or killing pests. The goal is to control the pest while avoiding or minimizing the use of chemical control agents, which may be toxic and have unwanted health and environmental effects.

PREVENTION AND CONTROL METHODS

CULTURAL Usually the first method chosen as part of the action plan; involves cultivation techniques that minimize the habitat or food source for the pest so that other control methods can then be used to adequately control the pests.

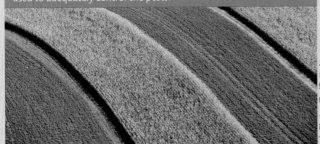

BanksPhotos/E+/Getty Images

Strip cropping minimizes potential food for pests that don't disperse over great distances, lessening the chance of an outbreak of pests.

MECHANICAL Relies on methods that physically exclude, trap, repel, or remove pests or weeds; can be labor intensive but are often inexpensive and may be particularly useful in developing countries with plentiful labor but little cash.

Cosmo Condina/The Image Bank/Getty Images

Netting can keep out birds and rabbits that would eat the crops. Reducing water levels in rice paddies to kill azolla is an example of mechanical control.

BIOLOGICAL Introducing predators, sterile males, or plants that repel the pest; technique works best if it follows cultural and mechanical steps.

Henrik_L/Getty Images

Ladybugs, predatory beetles that eat pests such as this aphid, can actually be purchased and released, or steps can be taken to attract them naturally to the area. The ideal control agent is a specialist whose preferred food is the pest in question and who does not attack nontarget species.

CHEMICAL Applying chemicals that kill or repel pests; a last resort that is used only if the other three methods cannot control the pests.

Abid Katib/Getty Images

To minimize health and environmental concerns, the preferred chemical is the one that can do the job while being the least toxic and the most degradable.

 What pest control methods did the Massas use in their rice fields? Identify each as cultural, biological, mechanical, or chemical.

6 TRADITIONAL FARMING METHODS

Key Concept 6: Sustainable agriculture draws on a variety of traditional farming methods that can protect or improve soil and reduce pest problems and can be used on large and small scales.

Greg and Raquel were well versed in the problems of modern agriculture. Before settling in California to take over the family farm, they had worked as tropical ecologists in Costa Rica, where they learned about traditional, nonindustrial farming methods that help protect the soil and keep productivity high without the use of synthetic fertilizers or pesticides.

Soil is vital to life on Earth; it supports the growth of plants as well as the animals that feed on those plants. Its formation is very slow (it can take more than 1,000 years to form 2.5 centimeters [1 inch] of topsoil) and depends on myriad soil organisms (see Online Module 7.2). Modern farming can contribute to a decline in soil fertility and to the direct loss of soil through erosion. It has been estimated that 5 tons of topsoil are lost for every 1 ton of industrially produced grain in the United States. However, some traditional (preindustrial) methods can actually help restore or protect soil. For example, methods such as *terracing, contour farming,* and *reduced-tillage cultivation* can decrease soil erosion. Planting a *cover crop* in the off-season prevents soil erosion and restores fertility. Soil fertility can also be enhanced and pests controlled using *crop rotation* and *strip cropping.* In addition, systems like Furuno's that combine animal and plant rearing, return animal waste—a natural fertilizer—to the soil. **INFOGRAPHIC 6**

For millennia, humans have employed methods to *amend* soil to improve its fertility. The addition of fertilizers (natural or synthetic), organic matter (to loosen compacted soil), and substances such as lime (to alter the pH of the soil) are all soil amendments that can address soil problems. But recently, a traditional soil amendment from the tropics is gaining interest—biochar. *Biochar* is a form of charcoal produced when organic matter (usually agricultural waste) is partially burned. The use of biochar over thousands of years helped create the fertile terra preta soils in the Amazon basin, where thin, poor, acidic soils predominate. Biochar not only provides nutrients to poor soils but also helps the soil store carbon (it does not decompose readily so "holds onto" much of its carbon), impeding its release into the atmosphere and reducing its contribution to climate change.

annual crops Crops that grow, produce seeds, and die in a single year and must be replanted each season.

perennial crops Crops that do not die at the end of the growing season but live for several years, which means they can be harvested annually without replanting.

Meanwhile, at the Land Institute in Kansas, Wes Jackson is working on a more ambitious plan that he says will correct a mistake made well before the Green Revolution or the rise of factory farms. He wants to replace virtually all of our existing grain crops—which are **annual crops**—with **perennial crop** varieties. Early farmers domesticated annual plants—which grow, produce seeds, and die in a single year—because they could manipulate them to produce higher yields from one year to the next (by selecting the best producers and junking the rest or by crossbreeding

An agroecologist holds a perennial wheat plant to showcase its expansive root system. Compare that to the much more shallow root system of the annual wheat plants to the right.

JIM RICHARDSON/National Geographic Creative

INFOGRAPHIC 6   TRADITIONAL FARMING METHODS

Many traditional farming methods are available that address problems with soil erosion and fertility, water use, and pest control and are useful for sustainable agriculture.

J. IRWIN/Aurora Photos

CONTOUR FARMING When farming on hilly land, rows are planted along the slope, following the lay of the land, rather than oriented downhill to reduce the loss of water and soil after a rainfall.

Bill Barksdale/Design Pics Inc/Alamy

REDUCED TILLAGE Planting crops into soil that is minimally tilled reduces soil erosion and water needs (it reduces water evaporation). It also requires less fuel because of less tractor use.

voraorn/iStock/Getty Images

TERRACE FARMING On steep slopes, the land can be leveled into steps. This reduces soil erosion and allows a crop like rice to stay flooded when needed.

Matt Meadows/Getty Images

CROP ROTATION Planting different crops on a given plot of land every few years helps maintain soil fertility and reduces pest outbreaks since pests (or their offspring) from the year before will not find a suitable food when they emerge in the new season.

Andrew Holt/Getty Images

STRIP CROPPING Alternating different crops in strips that are several rows wide keeps pest populations low; it is less likely the pests will travel beyond the edge of a strip and they may not find another row of this crop.

Michael Melford/Getty Images

COVER CROPS During the off season, rather than letting a field stand bare, a crop can be planted that will hold the soil in place. Nitrogen-fixing crops like alfalfa that improve the soil are often chosen.

Which of these sustainable soil practices help reduce soil erosion, which help improve soil fertility, and which help reduce pest outbreaks?

plants with good traits). But annuals have to be replanted every season, which requires labor and fossil fuel energy and necessitates tearing up the ground and disrupting the delicate balance of soil ecosystems.

Perennials, on the other hand, can be harvested year after year without disturbing the soil to replant—which means heavy equipment is used less often to manage perennial crops. And the deep roots that perennials develop not only hold soil in place but also tap much farther down into the soil than their annual counterparts, allowing them to access more of the soil's water, thus dramatically reducing the amount of irrigation needed. Less herbicide is needed for perennial plots as well; weeds do not readily sprout and grow among the established plants. This makes perennials especially attractive for regions with marginal land or an arid climate.

"We hope to advance and enlarge upon the idea that the ecosystem is the necessary conceptual tool for truly sustainable grain agriculture," Jackson told the *Atlantic* in a recent interview. "We believe we can have an agriculture where management by human intervention is greatly reduced."

7 THE ROLE OF CONSUMERS

Key Concept 7: Consumer choices can support sustainable agriculture. Buying locally grown, organic food is the best option for reducing the carbon footprint of that food.

In rice farming, the Massas saw a chance to implement, on their own land, all the concepts and theories they had learned as ecology students. "We wanted a farm where success was measured not just in crop yield, but in the overall health of the land," Greg says. "A place where we would count profits, but we would also count the number of sandhill cranes and California quail we saw populating the area."

To achieve their goals, the Massas installed a recirculation system to reclaim and reuse irrigation water. They planted native oak trees along field borders to serve as a natural windbreak (windbreaks prevent soil from being carried away by wind erosion), and they installed nest boxes for wood ducks, barn owls, and bats so that those wild animals would keep area pests in check. "The idea was to restore as much of the natural biodiversity as possible," says Greg, "so that we would not need artificial inputs to run the system."

The Massas also wanted to create a farm that contributed to the formation of a local community food system; they planned to sell a portion of their rice at local outlets like farmers' markets and food co-ops. More and more consumers are buying food from local farmers; local agriculture supports local economies and provides fresher and thus healthier food to consumers. Because transportation depends on fossil fuels, the more **food miles** a product travels before reaching the consumer, the greater the **carbon footprint** of that food. And much of our food has traveled quite far—about 2,400 kilometers (1,500 miles), on average.

However, transportation is not the main use of fossil fuels when crops are raised industrially. Research by Christopher Weber and Scott Matthews of Carnegie Mellon University determined that about 90% of the carbon footprint for food grown using conventional industrial methods is from the production of the crop (fuel for equipment, raw materials for pesticides and fertilizer production), not its transport. Therefore, buying organically grown produce—even from far away—may reduce the carbon footprint more so than buying locally grown industrial crops. Of course, the best option is to choose organic foods that are locally grown. Likewise, because beef raised in industrial CAFOs has one of the highest carbon footprints of all agricultural products (see Online Module 8.3), one of the best things you can do is to replace at least some of the beef you eat with chicken, pork, fish, or meatless dishes.

Consumers are becoming more aware that things like food miles and the way food is raised matter for the environment, their communities, and their own health. Organic foods and ethically raised animal products are claiming a bigger share of consumer dollars annually. But this has opened the door to **greenwashing**—making claims about the environmental benefits of sustainably raised or organic foods that are misleading. (For example, organic cookies are probably not healthier than those made with conventional ingredients, and "cage-free" eggs may still be from chickens living in overcrowded conditions.) Consumers need to be diligent about evaluating claims and make informed decisions about what to purchase. **INFOGRAPHIC 7**

food miles The distance a food travels from its site of production to the consumer.

carbon footprint The amount of carbon released to the atmosphere by a person, a company, a nation, or an activity.

greenwashing Claiming environmental benefits about a product when the benefits are actually minor or nonexistent.

INFOGRAPHIC 7 · CONSUMER CHOICES MATTER

Because the growing and transport of our food impacts the environment and our own health so much, choosing foods produced in a way that has a lower impact makes a difference. This also supports sustainable agriculture as an economic endeavor, helping the farmers and communities pursuing these methods.

CONSIDER HOW YOUR FOOD IS RAISED...

ANGEL FRANCO/The New York Times/Redux

Industrially grown food is usually cheaper but has a higher environmental impact and a high carbon footprint (more fossil fuels are used to produce it) than sustainably grown food. The best way to reduce the carbon footprint of the food you buy is to opt for organically grown food when possible (no fossil fuel–derived fertilizers or pesticides).

...AND HOW FAR IT IS SHIPPED

Mindy Schauer/ZUMA Press/Santa Ana/ California/U.S./Newscom

Even though more fossil fuels go into the *production* of industrially grown foods than in shipping it to market, buying food produced closer to home decreases the transportation part of the carbon footprint. For example, in Iowa, the average grocery store apple has traveled 1,726 miles, and the average head of broccoli, 1,846 miles, while locally grown produce has traveled 56 miles on average. So while transportation does not represent the main way fossil fuels are used in agriculture, it still has a significant impact.

PERSONAL FOOD CHOICES

Buying organic food not only reduces the carbon footprint of the food, it is also healthier for you. But it can get expensive, so if your buying dollars are limited, consider steering your purchases away, when you are able, from the "dirty dozen"—the 12 fruits and vegetables most likely to be contaminated with pesticide residue. The "clean 15" are the products least likely to have pesticide residue, so if you can't afford to buy all organic produce, buy these from the regular produce shelf—but always wash all produce well before eating or cooking!

THE CLEAN 15
The fresh fruit and vegetables with the lowest pesticide residue

1 Sweet corn	9 Mangoes
2 Avocados	10 Eggplant
3 Pineapples	11 Honeydew melon
4 Cabbage	12 Kiwi
5 Onions	13 Cantalope
6 Sweet peas	14 Cauliflower
7 Papayas	15 Grapefruit
8 Asparagus	

THE DIRTY DOZEN
(in order of pesticide exposure when eaten) Buy organic if possible.

1 Strawberries	9 Celery
2 Spinach	10 Tomatoes
3 Nectarines	11 Sweet bell peppers
4 Apples	12 Potatoes
5 Peaches	
6 Pears	
7 Cherries	
8 Grapes	

? What information would you like to see on food labels that would allow you to make wise consumer choices when buying food? Does food labeling need to be regulated so that certain terms (e.g., organic, free range) are specifically defined?

8 CAN SUSTAINABLE AGRICULTURE FEED THE WORLD?

Key Concept 8: Sustainable agriculture comes with trade-offs, but many feel that the disadvantages are less problematic than are those of modern industrial agriculture, especially in the long term.

Furuno and the Massas are not the only ones experimenting with nature-based polyculture. Indeed, many other farmers and scientists across the country are turning to agroecology—working to develop mixed agriculture systems, where instead of planting a single crop, they grow a mix of different species that better replicates the normal ecological community makeup of a given region. Evidence is mounting that such systems can increase a farm's productivity.

A 2010 report by the International Livestock Research Institute concluded that mixed polyculture farms—ones that, like Furuno's and the Massas', grow both plants and livestock—hold the most promise for intensifying food production worldwide. "It is not big efficient farms on high potential lands but rather one billion small, mixed family farmers tending rice paddies or cultivating maize and beans while raising a few chickens and pigs, a herd of goats, or a cow or two. . . [who are] likely to play the

biggest role in global food security over the next several decades," the institute's executive director, Knut Hove, wrote in the report. "These 'mixed extensive' farms make up the biggest. . . and most environmentally sustainable agricultural system in the world."

Feeding a world of 10 billion people will not be easy. It will require a combination of modern industrial and traditional techniques, native crop varieties and perhaps genetically modified ones—the methods and crops chosen should fit the place and need of the farmers and their communities. But consideration needs to be given to the long-term viability of the farm with preference given to techniques that maintain or improve the health of the ecosystem. As these many examples (agroecology, the resurrection of traditional farming techniques, IPM, and the development of perennial crops) illustrate, a multitude of methods are available to help facilitate a transition from industrial agriculture to methods that are more sustainable. Of course, as with all

Biochar is added to landscapes and agricultural fields to improve soil. It provides nutrients, but its biggest advantage may come from the fact that it stays in the soil a long time. This creates long-lasting habitat for useful soil organisms and helps keep carbon sequestered in the soil (useful in combating climate change).

ZUMA Press Inc/Alamy

environmental choices, it shouldn't be surprising that sustainable and organic agriculture have their own set of trade-offs. However, addressing these trade-offs may be more feasible than dealing with the disadvantages of large-scale industrial monocultures. **INFOGRAPHIC 8**

For the Massas, ducklings were part of an ongoing search for solutions to the challenges of modern rice farming. They ended up with duck meat to sell, and though they didn't take any precise measurements of yields during their first trial run, their rice crops did not appear to suffer at all. Their initial foray into duck-rice farming has given way to more traditional organic rice farming techniques. They no longer use ducks, but they continue to incorporate more sustainable agroecology farming techniques into their farm. The Massas planted almond orchards, which are managed like natural

woodlands with a diverse understory (that attracts beneficial pollinating insects); sheep forage beneath the trees, providing fertilizer and helping to maintain the understory without the need for industrial weed control methods. (They used to burn out weeds by hand with a propane flamer.) Pigs have become another permanent part of their farm, moved around the farm via temporary fencing to feed on whatever area best fit their needs and the needs of the farm.

Furuno admits that managing this type of farm can be more time consuming and requires more expertise, but when asked about the extra work this type of farming requires, he replied, "It gives you a sense of tranquility and peace. My role is to feed the ducks and later on, it's the ducks who feed me. All in all, it can be considered a rather fair exchange."

INFOGRAPHIC 8 · THE TRADE-OFFS OF SUSTAINABLE AGRICULTURE

Gary K. Smith/Alamy

FOR	ADVANTAGES	DISADVANTAGES
The Consumer	• Food is fresher, tastier, and healthier (better nutrient profile, no pesticide residue if organic, etc.). • Gains satisfaction in making a more ethical and environmentally sound choice.	• Sustainably grown crops may be more expensive. • Greenwashing can mislead consumers. • Organic produce may have more blemishes. • Shelf life of organic produce is shorter (not waxed, not picked before ripe).
Farmer and Environment	• Using fewer inputs of water and fossil fuels saves money and causes less environmental damage. • Soil is not degraded and may be enhanced. • Less use of toxic chemicals benefits the environment and local communities. • More genetic diversity and species diversity makes it less likely that a pest outbreak or other problem will decimate the entire crop.	• May be more labor intensive. • Crops grown sustainably may not be as productive per acre as industrially farmed crops (in the short term). • Fewer government subsidies are available for sustainable agriculture compared to those for industrially grown crops. • The certification process for getting crops to be labeled as organic takes time and is costly to farmers.
Society	• Many sustainable methods are less expensive, so they are suitable for developing nations. • Methods are available that minimize water need—useful in arid areas. • Local production of food can increase food security (see Module 8.1).	• Research is needed to identify best methods and crops for a given area. • Farmers need training to implement these systems (though if indigenous methods are used, it may be the locals who educate the researchers).

In your opinion, which advantages listed above are the most important? Which disadvantages are the most troublesome? Explain.

Ducklings in the Massa Organics rice fields.

Select References:

Baranski, M., et al. (2014 September 14). Higher antioxidant and lower cadmium concentrations and lower incidence of pesticide residues in organically grown crops: a systematic literature review and meta-analyses. *The British Journal of Nutrition, 112*(5), 794–811. doi:10.1017/S0007114514001366.

Davis, D. R., et al. (2004). Changes in USDA food composition data for 43 garden crops, 1950 to 1999. *Journal of the American College of Nutrition, 23*(6), 669–682.

Furuno, T. (2001). *The power of duck: integrated rice and duck farming.* Tagari Publications. Tasmania, Australia.

Hossain, S. T., et al. (2005). Effect of integrated rice-duck farming on rice yield, farm productivity, and rice-provisioning ability of farmers. *Asian Journal of Agriculture and Development, 2*(1), 79–86.

Jackson, W. (2002). Natural systems agriculture: a truly radical alternative. *Agriculture, ecosystems & environment, 88*(2), 111–117.

Reganold, J. P., et al. (2010). Fruit and soil quality of organic and conventional strawberry agroecosystems. *PLoS ONE, 5*(9), e12346. doi:10.1371.

Weber, C. L., & Matthews, H. S. (2008). Food-miles and the relative climate impacts of food choices in the United States. *Environmental Science and Technology, 42*(10), 3508–3513.

INTERACTIVE MAP SUSTAINABLE AGRICULTURE   ANIMATED INFOGRAPHIC

Depending on the crop, high- and low-tech solutions can promote higher yields without jeopardizing the environment. Read these four success stories.

NO TILL FARMING

FIGHTING BUGS WITH BUGS

COFFEE AGROFORESTRY

DESERT MOISTURE FARMING

 BRING IT HOME

PERSONAL CHOICES THAT HELP
While a typical supermarket may seem to present a dizzying array of food choices to the consumer, a look at the ingredient labels betrays our increasing reliance on growing monocultures of common strains of corn, soy, and wheat. These choices will change only if you, the consumer, demand it.

Individual Steps
• Carefully examine the labels on the food you buy. As your food budget allows, opt for food products that are organically grown and, if available, locally produced.

Group Action
• Organize a community garden that specializes in heirloom varieties of vegetables that might not be found in the local grocery stores. Start by requesting a seed catalog from www.seedsavers.org or www.rareseeds.com.
• Research specific farming practices that more closely mimic those found in natural ecosystems, such as Joel Salatin's Polyface Farm (www.polyfacefarms.com).
• Subscribe to a community-supported agriculture (CSA) farm and receive a weekly supply of sustainably grown

produce. For a list of CSAs in your area, see www.localharvest.org.

Policy Change
• Identify bodies of water in your area that may be impacted by cultural eutrophication from fertilizer runoff. Meet with local officials and propose ordinances limiting fertilizer use by homeowners, golf courses, or other possible sources of the pollution.

ENVIRONMENTAL LITERACY   UNDERSTANDING THE ISSUE

1 What is sustainable agriculture?

1. A goal of sustainable agriculture is:
   a. using natural processes to enhance crop growth.
   b. the humane rearing of animals.
   c. providing fair wages to farm workers.
   d. A and B are correct.
   e. A, B, and C are correct.

2. Distinguish between sustainable and organic agriculture.

2 What are the trade-offs of using chemical fertilizers in industrial agriculture?

3. Which of the following is a disadvantage to monoculture agriculture?
   a. Large inputs of fertilizer are needed for high productivity.
   b. It is hard to manage a large monoculture field.
   c. The profits per acre are lower than farms not using monocultures.
   d. Crop productivity per acre is low.

4. The use of synthetic fertilizers on fields where crops are grown can:
   a. increase crop yields.
   b. contribute to water pollution.
   c. deplete the soil of other nutrients.
   d. A and B are correct.
   e. A, B, and C are correct.

5. What is cultural eutrophication, and how does the use of fertilizer contribute to it?

3 What are the trade-offs of using chemical pesticides in industrial agriculture?

6. Which of the following occurs during the development of pesticide resistance?
   a. Individual pests able to withstand the pesticide survive to parent the next generation.
   b. The pesticide causes some individuals to develop mutations that protect them from the pesticide.
   c. Repeated exposure to a pesticide allows some individuals to become tolerant of the pesticide.
   d. A and B are correct.
   e. A, B, and C are correct.

7. Explain how the use of chemical pesticides can lead to the emergence of a population even bigger than the original pest population.

4 What is agroecology and what are its benefits?

8. The presence of ducks and azolla in Takao Furuno's rice cultivation technique has shown that:
   a. ducks must be eliminated from rice fields because they eat the rice seed before it can grow.
   b. invasive species like ducks and azolla diminish rice yields.
   c. crops such as rice cannot succeed without azolla to provide shade and ducks to provide pollination.
   d. restoring some biodiversity to rice fields reduces pest damage and increases rice yields.

9. Explain the premise behind agroecology.

5 What is integrated pest management?

10. The release of ladybug beetles, a natural predator of aphids, to control aphid pests on your roses is an example of _____ control.
    a. biological
    b. chemical
    c. cultural
    d. mechanical

11. Explain the goal and approach of integrated pest management and how it views the use of chemical pesticides.

6 How can traditional farming methods contribute to sustainable agriculture?

12. True or False: Reduced tillage farming will help reduce the loss of water through evaporation.

13. Which of these traditional farming methods will decrease the chance of a pest oubreak?
    a. Contour farming and terrace farming
    b. Cover crops and reduced tillage
    c. Strip cropping and crop rotation
    d. Terrace farming and cover crops

14. Which of the following is an advantage of perennial crops compared to annual crops?
    a. Perennial crops can be planted sooner and they grow faster than annual crops.
    b. Perennial crops don't have to be planted each year like annuals do.
    c. Farmers are already planting perennial crops and wouldn't have to change their methods.
    d. Perennial crops require more water but do not need any fertilizer applications.

15. Identify which traditional farming methods protect the soil from erosion and explain how soil erosion is prevented with each.

7 What role does the consumer play in helping build a sustainable food system?

16. In regard to agriculture, the "dirty dozen" refers to the 12:
    a. pesticides organic farmers are banned from using.
    b. conventionally grown fruits or vegetables highest in pesticide residue.
    c. most destructive crop pests.
    d. countries that use the highest amounts of pesticides on crops.

17. To achieve the lowest carbon footprint as a consumer, should you opt for organic food grown far away or industrially grown food produced close to home? Explain.

8 Evaluate the trade-offs of sustainable agriculture and its potential to meet world food needs.

18. Which of these is not an advantage of sustainable agriculture?
    a. Fresher, better-tasting produce that may be healthier to eat
    b. Maintained or enhanced soil quality
    c. More genetic diversity in crops
    d. Less expensive

19. Look over the disadvantages of sustainable agriculture in Infographic 8. Propose actions that could address each.

SCIENCE LITERACY | WORKING WITH DATA

The data in the following graph come from the Farming Systems Trial® (FST) research study conducted by the Rodale Institute. This side-by-side comparison of corn and soybean crops grown under organic and industrial agricultural systems was started in 1981 and is one of the longest-running studies of its kind.

A COMPARISON OF ORGANIC AND CONVENTIONALLY GROWN CROPS

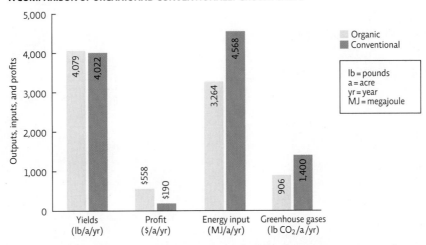

Interpretation

1. What does this graph show?

2. Calculate the following for conventional and organic systems: profit per unit of yield, energy input per unit of yield, and greenhouse gas emissions per unit of yield. How do the two systems compare on these three parameters?

Advance Your Thinking

3. According to the Food and Agricultural Organization of the United Nations, "Organic agriculture has the potential to secure a global food supply, just as conventional agriculture does today, but with reduced environmental impact." How do the data from the FST study support this statement?

4. According to the FST report, even in drought years, the yields for organic corn were approximately 31% greater than those for conventional (non-drought-resistant) varieties. At the same time, genetically engineered drought-tolerant varieties had yields that were no more than approximately 13% greater than conventional (non-drought-resistant) varieties. Why might this be the case? Why is this an important finding?

5. The FST report indicates that crops grown using organic methods produced yields equivalent to those of conventional crops, even though they had more weed competition in their fields. Why might this be the case? What makes this an important finding?

INFORMATION LITERACY | EVALUATING INFORMATION

Which seed varieties are best? There are groups, such as Renewing America's Food Traditions (RAFT) (www.albc-usa.org/RAFT/), that say we should preserve the original biodiversity and the heirloom varieties that our ancestors grew. Other groups believe that selective breeding and hybridization create superior crops, the downside being that farmers must purchase new seeds every year. Bayer is a huge conglomerate that sells a variety of hybrid seeds and chemicals to enhance growth (www.bayercropscience.us). Another group, AgBioWorld (www.agbioworld.org), says that it is neutral and reports the news on agricultural practices. Look at each website to help analyze the organizations.

Evaluate the websites and work with the information to answer the following questions:

1. Evaluate the agendas of the three organizations as well as the accuracy of the science behind their positions on heirloom varieties versus hybrid crops.

a. Who runs each website? Do the person's/organization's credentials make the information presented on food and agriculture issues reliable or unreliable? Explain.

b. What is the mission of each website? What are the underlying values? How do you know this?

c. What claims does each website make about the current problems in food production and what the future of agriculture should be? Are their claims reasonable? Explain.

d. How do the websites compare in providing scientific evidence in support of their assessment of agriculture and their position on the role of heirloom varieties versus hybrid crops? Is the evidence accurate and reliable? Explain.

2. How do the three organizations compare in engaging you, as a citizen, in agricultural policy? Do you think that citizen involvement in policy issues is necessary and effective? Explain your responses.

 Additional study questions are available at SaplingLearning.com.

CONVENTIONAL ENERGY: FOSSIL FUELS

CHAPTER 9

Fossil fuels power modern society
and are important raw materials for
a wide variety of chemical products.
An understanding of the value
and environmental/health costs of
acquiring and using fossil fuels can
guide our choices as we pursue more
sustainable energy options.

Module 9.1: Coal

An evaluation of the usefulness and
negative impacts of coal as an energy
source and ways to address coal's
disadvantages to lessen its impact

Module 9.2: Oil and Natural Gas

An examination of the pros and cons of
acquiring and using oil and natural gas,
including a look at unconventional sources
of these fossil fuels such as shale oil/gas
and tar sands

BRINGING DOWN THE MOUNTAIN

In the rubble, the true costs of coal

Leveling a mountain for coal in Appalachia. The mountaintop is blown off and dumped into valleys to leave sprawling, terraced, barren lands where there were once diverse temperate forests.

George Steinmetz/Getty Images

After reading this chapter and studying the KEY CONCEPTS and INFOGRAPHICS, you should be able to answer these GUIDING QUESTIONS

CORE MESSAGE

Human society runs on energy, and coal continues to be a major energy resource. However, coal mining causes irreversible environmental degradation, and mining and burning coal pose significant environmental and health risks. Researchers are developing ways to lessen the impact of burning coal, but as long as we continue to use it, much of the impact of mining will remain.

1. What energy sources are used in modern society and for what purposes?

2. How important is coal as an energy source, and how is it used to generate electricity?

3. What is coal, how is it formed, and what regions of the world have coal deposits that are accessible?

4. What surface methods are used to mine coal, and what are their advantages and disadvantages?

5. How are deep deposits of coal mined, and what are the advantages and disadvantages of this method?

6. What are the advantages and disadvantages of using coal?

7. What new technologies allow us to burn coal with fewer environmental and health problems?

8. How can mining damage be repaired, and how effective is this restoration?

A thousand feet above the mountains of central Appalachia, near the Kentucky–West Virginia border, a four-seater plane ducks and sways like a tiny boat on an anxious sea. It's windier than expected, and Chuck Nelson, a retired coal miner seated next to the pilot, grips the door to steady his nerves. The passengers have come to survey the devastation wrought by mountaintop removal—a type of mining that involves blasting off several hundred feet of mountaintop, dumping the rubble into adjacent valleys, and harvesting the thin ribbons of coal beneath.

At first, the landscape looks mostly unbroken; mountains made soft and round by eons of erosion roll and dip and rise in every direction, carrying a dense hardwood forest with them to the horizon. But before long, a series of mountaintop mining sites come into view. Trucks and heavy equipment crawl like insects across what looks like an apocalyptic moonscape: decapitated peaks and acres of barren sandstone and shale. Smoke curls up from a brush fire as the side of an existing mountain is cleared for demolition. Orange and turquoise sediment ponds—designed to filter out heavy metal contaminants before they permeate the water downstream—dot the perimeter.

Here and there, tiny patches of forest cling to some improbably preserved ridge line. "That's where I live," Nelson says, forgetting his air sickness long enough to point out one such patch. "My God, you would never know it was this bad from the ground." The aerial tour has reached Hobet 21, which, at more than 52 square kilometers (20 square miles), is the region's largest mining operation. So far, sites like this one have claimed roughly 400,000 hectares (1 million acres) of forested mountain, across just four states: Kentucky, West Virginia, Virginia,

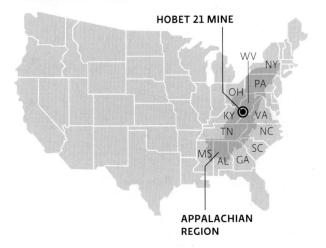

⊙ **WHERE IS APPALACHIA?**

HOBET 21 MINE

APPALACHIAN REGION

and Tennessee. But there is still more **coal** to mine. It could mean the obliteration of hundreds of thousands of more acres in the coming years.

To stem this tide of destruction, environmental activists have sued the coal industry, the state of West Virginia, and the federal government. They argue that mining coal in this manner destroys biodiversity, pollutes the water beyond recompense, and threatens the health and safety of area residents. And by obliterating the mountains, they say, it also obliterates the culture of Appalachia.

Coal industry reps have countered by decrying the loss of jobs, tax revenue, and business the already impoverished region would suffer if the mines were to close under the weight of too much regulation. They also point out that the culture of Appalachia is as bound to coal mining as it is to the mountains. Both sides count area residents, including miners, among their ranks.

David Hosking/FLPA/Science Source

Seams of coal exposed at a mountaintop removal mining site in Welch, West Virginia.

coal A fossil fuel that is formed when plant material is buried in oxygen-poor conditions and subjected to high heat and pressure over a long time.

1 ENERGY SOURCES

Key Concept 1: Modern society depends on stationary sources of energy to produce heat and electricity (e.g., coal, natural gas, biomass) and mobile sources for transportation purposes (e.g., petroleum products).

Simply put, coal equals energy. **Energy** is defined as the capacity to do work; like all other living things, we humans need it, in a biological sense, to survive. But we also need it to run our societies: to heat and cool our homes; operate our cell phones, lamps, and laptops; fuel our cars; and power our industries. Most of modern society's energy comes from **fossil fuels**—nonrenewable carbon-based resources (e.g., coal, oil, and natural gas)—that were formed over millions of years from the remains of dead organisms. (Biomass fuels such as wood and animal dung are still the main fuel sources for heating and cooking in many less developed regions. See Online Module 11.3 for more on biomass fuels.)

In general, modern society depends on two categories of energy: *stationary sources* that provide electricity for buildings and *mobile sources* that fuel our transportation fleets (cars, boats, planes, etc.). There are a variety of ways to produce electricity. The most common method comes in the form of thermoelectric power plants that burn a fuel like coal, natural gas, biomass, or even garbage (see Module 5.3, Infographic 5). Electricity can also be produced by harnessing nuclear energy or

energy The capacity to do work.

fossil fuel A variety of hydrocarbons formed from the remains of dead organisms.

the geothermal heat of Earth or by using renewable energy sources such the energy of moving water (hydroelectric), solar energy, or wind—these alternatives will be examined in Chapter 11.

Mobile energy sources primarily come from oil, a fossil fuel, which is refined to produce liquid fuels such as gasoline, diesel, and jet fuel; vehicles can also run on compressed natural gas. Alternatively, liquid fuels like biodiesel or ethanol can be made from biomass—an energy source that is renewable as long as we don't harvest it faster than it can regrow (see Online Module 11.3).

All of these fuels have advantages and disadvantages. Fossil fuels are abundant and energy rich. They powered the Industrial Revolution, and we built our societies, our machines, and our infrastructure (ways to distribute the fuel) around them. They still account for more than 80% of U.S. energy production. But, like all energy sources, they come with negative trade-offs. Understanding where our energy sources come from, and the problems they cause, can allow us to make more informed decisions about how much and what types of energy sources to use in our everyday lives and which energy sources to support politically or economically. **INFOGRAPHIC 1**

The controversial Spruce No. 1 coal mine in West Virginia was originally given a permit authorizing it to dump strip mining waste into 11 kilometers (7 miles) of creeks and onto 800 hectares (1,975 acres) of land. In 2011, the U.S. Environmental Protection Agency rejected the permit on the basis of the "irreversible damage" that would be inflicted to the area.

Antrim Caskey

INFOGRAPHIC 1 ENERGY SOURCES USED BY MODERN SOCIETY

A variety of energy sources are used today to supply stationary and mobile sources of energy. Nonrenewable sources still account for more than 80% of U.S. energy use.

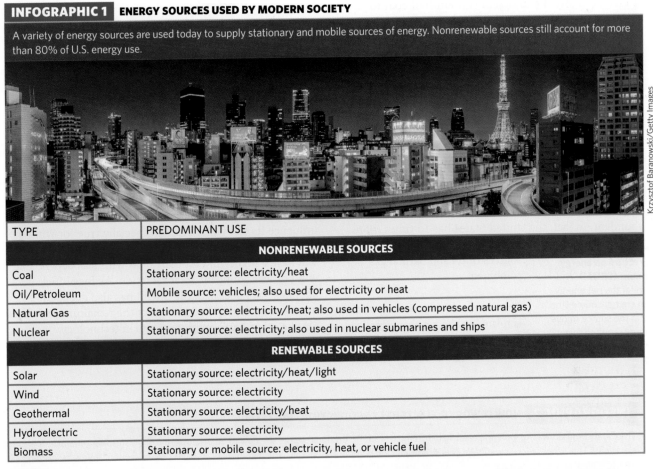

TYPE	PREDOMINANT USE
NONRENEWABLE SOURCES	
Coal	Stationary source: electricity/heat
Oil/Petroleum	Mobile source: vehicles; also used for electricity or heat
Natural Gas	Stationary source: electricity/heat; also used in vehicles (compressed natural gas)
Nuclear	Stationary source: electricity; also used in nuclear submarines and ships
RENEWABLE SOURCES	
Solar	Stationary source: electricity/heat/light
Wind	Stationary source: electricity
Geothermal	Stationary source: electricity/heat
Hydroelectric	Stationary source: electricity
Biomass	Stationary or mobile source: electricity, heat, or vehicle fuel

Krzysztof Baranowski/Getty Images

 Which of the nonrenewable and renewable energy sources listed above appear to be the most versatile (have the most uses)? Explain.

2 GENERATING ELECTRICITY FROM COAL

Key Concept 2: Coal is the leading fuel used for electricity production. It is burned to heat water to produce steam; the steam turns a turbine connected to a generator, producing electricity.

Worldwide, we used more than 7.7 billion metric tons (8.5 billion U.S. tons) of coal in 2015, the vast majority of it to generate electricity. **Electricity** is a natural form of energy (lightning and nerve impulses are electrical) that we have learned to create on demand; we produce it in a central location and send it out via transmission lines where we want it to go. In 2015, burning coal generated approximately 40% of electricity worldwide and 33% in the United States. The use of coal in the United States has decreased in recent years due largely to an abundance of cheaper natural gas (acquired by fracking—see Module 9.2). More stringent air pollution standards for coal-fired power plants have also had a role in the downturn—natural gas releases less air pollution and

greenhouse gas emissions than coal when burned. Some coal advocates hope that rolling back air quality standards will once again boost coal's use; however, as long as natural gas remains abundant and cheaper, most analysts say that is unlikely.

Coal-fired power plants work by feeding pulverized coal into a furnace to generate heat, which causes water to boil, producing steam. The steam flows over a turbine, causing it to spin inside a generator (copper wiring positioned near magnets), a process that produces electricity.

electricity The flow of electrons (negatively charged subatomic particles) through a conductive material (such as wire).

It takes roughly 0.5 kilogram (1 pound) of coal to generate 1 kilowatt-hour (kWh) of electricity; that's enough to run ten 100-watt incandescent lightbulbs for an hour, an energy-efficient refrigerator for 20 hours, or an older, less efficient refrigerator for 7 hours. The average U.S. family of four uses close to 11,000 kWh of electricity per year. That comes out to around 1,375 kilograms (3,000 pounds) per person per year.

So, how does coal stack up against other fossil fuel energy sources? On one hand, it produces more air pollution than any other fossil fuel. On the other, it is safer to ship, cheaper to extract, and in the United States at least, abundant. In fact, the United States has 10 times more coal than it does conventional sources of oil and natural gas combined.

The usefulness of an energy source is related to the *net energy* realized from its use—a metric known as the **energy return on energy investment (EROEI)**. An evaluation of the EROEI allows us to compare the amount of energy we get from any individual source to the amount we must expend to obtain, process, and ship it. In terms of electricity production, coal is neither the best nor the worst. It has an average EROEI of about 11.5:1 (11.5 units of energy produced for every 1 unit consumed). Steps that can be taken to reduce the impact of coal, such as taking steps to capture and sequester CO_2 emissions from coal burning (discussed later in this module) require extra energy and so reduce the EROEI of coal—by some estimates as low as 1.5:1.

There can be considerable variability in the EROEI estimate for a given energy source because many factors influence its net yield. This is especially true for wind or solar electricity production—"windy" or "sunny" areas will have a higher EROEI than areas with less dependable wind or less abundant solar radiation. As fossil fuel supplies have declined, the EROEI has also declined—more energy is required to extract the remaining supplies. On the other hand, improvements in technology, especially for renewable energy sources, are raising the EROEI of these methods. **INFOGRAPHIC 2**

energy return on energy investment (EROEI) A measure of the net energy from an energy source (the energy in the source minus the energy required to get it, process it, ship it, and then use it).

INFOGRAPHIC 2 | **HOW IT WORKS: ELECTRICITY PRODUCTION FROM COAL**

The most common way to generate electricity is to heat water to produce steam; the flow of steam turns a turbine inside a generator to produce electricity. This schematic shows TVA's coal-fired Kingston plant in Tennessee, which generates 10 billion kilowatt-hours a year by burning 13,000 metric tons of coal a day, supplying electricity to almost 700,000 homes.

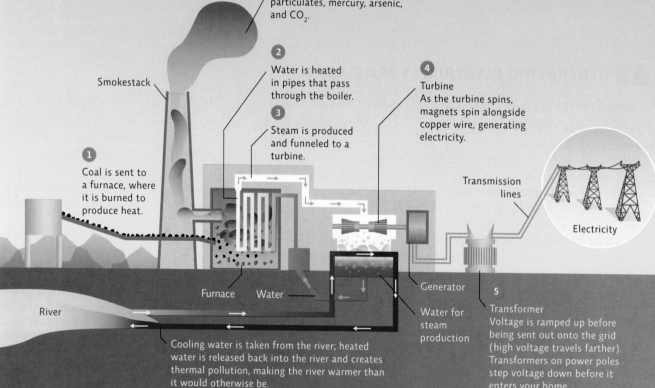

Emissions can include sulfur, particulates, mercury, arsenic, and CO_2.

Smokestack

2 Water is heated in pipes that pass through the boiler.

3 Steam is produced and funneled to a turbine.

4 Turbine
As the turbine spins, magnets spin alongside copper wire, generating electricity.

1 Coal is sent to a furnace, where it is burned to produce heat.

Transmission lines

Electricity

River

Furnace   Water

Generator

Water for steam production

5 Transformer
Voltage is ramped up before being sent out onto the grid (high voltage travels farther). Transformers on power poles step voltage down before it enters your home.

Cooling water is taken from the river; heated water is released back into the river and creates thermal pollution, making the river warmer than it would otherwise be.

There can be no denying the blessings of coal: This sticky black rock has powered several waves of carbon-based industrialization—first in Great Britain and the United States, now in China—and in so doing has shaped and reshaped the world as we know it. But as time marches on, the costs of those blessings have become all too apparent. They include an ever-growing list of health impacts—from birth defects to black lung disease—and an equally lengthy roster of environmental costs—not only the destruction of Appalachian mountains but also the pollution of Earth's atmosphere with CO_2, a potent greenhouse gas. (See Module 10.2 for more on greenhouse gases and climate change.)

> " On one hand, it's like we need the stuff to live; on the other hand, we see that it's kind of killing us. "
> —Scott Eggerud

"We're caught in a catch-22," says Scott Eggerud, a forest manager with West Virginia's Department of Environmental Protection. "On one hand, it's like we need the stuff to live; on the other hand, we see that it's kind of killing us." Nowhere is this catch-22 more pronounced than in the mountains of central and southern Appalachia.

Les Stone/The Image Works

A train carrying coal leaves a mountaintop removal mining site and travels through the backyards of homes in Welch, West Virginia.

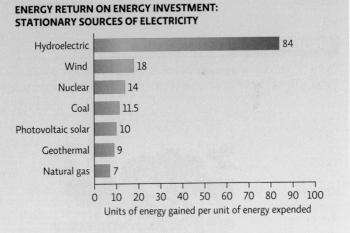

U.S. ELECTRICITY GENERATION BY FUEL, 2015

- Other 2%
- Hydroelectric 6%
- Renewables 7%
- Nuclear 19%
- Natural gas 33%
- Coal 33%

ENERGY RETURN ON ENERGY INVESTMENT: STATIONARY SOURCES OF ELECTRICITY

Source	Units of energy gained per unit of energy expended
Hydroelectric	84
Wind	18
Nuclear	14
Coal	11.5
Photovoltaic solar	10
Geothermal	9
Natural gas	7

Units of energy gained per unit of energy expended

EROEI, an estimate of the net energy produced from an energy source, provides a useful means of comparing different energy sources.

In 2015, for the first time, coal was not the leading energy source for electricity production. The falling price of natural gas has displaced coal in many areas. Increasing the use of renewable fuels and taking steps to improve energy efficiency and conservation could further reduce the role coal plays in energy production.

 What other methods could be used to spin a turbine and generate electricity?

3 COAL DEPOSITS: FORMATION AND DISTRIBUTION

Key Concept 3a: Coal formation occurred over long periods of time when dead plant material was buried and subjected to high heat and pressure.

Key Concept 3b: Coal deposits are located throughout the world, though some areas have a larger supply than others.

The Appalachian Mountains were born a few million years before the rise of the dinosaurs, when Greenland, Europe, and North America hovered near the equator as a single giant landmass bathed in a dense tropical swamp. The gradual accretion of decomposing swamp vegetation compressed and baked by heat and time established the Appalachian coal beds. As the swamp plants died out, they were buried under a mud so thick it kept oxygen out. Instead of being fully decomposed by bacteria, their remains produced *peat*—a soft mash of partially decayed vegetation. As time passed and more and more layers of sediment were laid down over the peat, pressure and heat compressed it into the denser rocklike material that we know as coal.

---

proven reserves A measure of the amount of a fossil fuel that is economically feasible to extract from a known deposit using current technology.

This same story has played out in numerous places around the globe. As a result, coal is found everywhere, though of course some places have more than

others. In 2016, Asia held about 37% of the world's **proven reserves**, while Europe held about 30%, and North America accounted for just under 28%. The United States has more coal than any other country with an estimated 26.6% of the world's total proven reserves. The North Antelope Rochelle coal mine in Wyoming is the largest proven coal reserve in the world. INFOGRAPHIC 3

The Appalachian beds were once among our most bountiful reserves; seams as tall as a man wound for miles through the mountainside and made for easy harvesting. But after 150 or so years of mining, those reserves have dwindled noticeably. At current rates of usage, proven coal reserves should last about 100 years—longer if deeper reserves can be accessed. And as the layers of minable coal have grown thinner and harder to reach, the coal industry has become both more sophisticated and more destructive in its approach to extracting the coal.

INFOGRAPHIC 3 COAL FORMATION

Coal is formed over long periods of time as plant matter is buried in an oxygen-poor environment and subjected to high heat and pressure. Places with substantial coal deposits that are retrievable with current technology are called coal reserves.

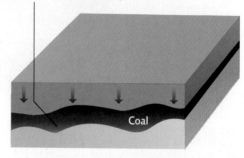

300 million years ago, some areas of the planet were covered with large swampy forests, or other wetlands.

Peat

Time

Over time, pressure built up as more sediment was laid down; increased pressure and heat converted the peat to a soft coal (lignite); in areas with enough pressure and heat, lignite was converted to harder varieties of coal (bituminous and anthracite).

Coal

Some vegetation died and was submerged in oxygen-poor sediments of the swamp. With limited oxygen, decomposition slowed tremendously, and peat formed.

 Why aren't deposits of coal found everywhere?

4 COAL MINING: SURFACE TECHNIQUES

Key Concept 4: Coal close to the surface is accessed in flat areas via strip mining and in mountainous areas with mountaintop removal (MTR). Both are cost-effective mining techniques but MTR creates significantly more environmental damage than traditional strip mining.

The method used to retrieve coal depends on the depth of the deposit and the thickness of the coal seam. **Surface mining** techniques tackle coal seams that are close to the surface of the ground by simply digging down to expose it. If the land is fairly level, such as in the rich deposits of Wyoming, **strip mining** is the method of choice: Vegetation, soil, and rock that lie above the coal seam are stripped away, and heavy equipment digs out the coal.

In a strip mine, the **overburden**, material that lies above the coal seam, can be systematically removed, typically in a narrow strip, one layer at a time, and stored nearby. Coal is removed from that exposed strip and once exhausted, the rock is returned to the pit and topped with the reserved soil. The area can be replanted with native grasses (or allowed to reseed) while the next strip is excavated and mining continues.

Hobet 21 is another matter. Located in the mountains of Appalachia, this method of strip mining is not an option. Though thick coal seams deep under a mountain are accessed with subsurface mining methods, in areas like Hobet 21 where coal seams are close to the surface or simply too thin to feasibly harvest with a traditional underground mine, **mountaintop removal (MTR)** is used—the entire mountain top is removed to expose the coal seams below. This particular mountaintop removal site has claimed at least 5,000 hectares (12,000 acres) of land and was once the site of several adjacent peaks.

To get at the coal beneath those peaks, miners began by clear-cutting the forest above. Next, they drilled holes deep into the side of the mountain, set dynamite in those holes, and blasted as much as 300 meters (1,000 feet) of mountain into a mass of rubble (the overburden). The miners repeated this process several times, until the layers of coal were exposed. Then, using buckets big enough to hold 20 midsized cars, they scooped that rubble into waiting dump trucks that carried it away.

Unlike the flat strip mine areas of Wyoming, there is nowhere to store the overburden—MTR overburden is simply dumped down what is left of the mountainside, into the valleys (and streams) below. The process obliterates the forest habitat, buries countless streams, and permanently reorders the land's natural contours. **INFOGRAPHIC 4**

surface mining A form of mining that involves removing soil and rock that overlays a mineral deposit close to the surface in order to access that deposit.

strip mining A surface mining method that accesses coal from deposits close to the surface on level ground, one section at a time.

overburden The rock and soil removed to uncover a mineral deposit during surface mining.

mountaintop removal (MTR) A surface mining technique that involves using explosives to blast away the top of a mountain to expose the coal seam underneath; the waste rock and rubble is deposited in a nearby valley.

David R. Frazier/Science Source

Wyoming strip mine.

Antrim Caskey

Blasting at a mountaintop removal site.

INFOGRAPHIC 4 **SURFACE MINING TECHNIQUES**

STRIP MINING

Strip mines are used in areas where the ground is fairly level and coal seams are close to the surface. The overburden is removed and set aside. The exposed coal seam is harvested using heavy equipment and once exhausted, the strip is filled back in with the overburden and vegetation replanted.

MOUNTAINTOP REMOVAL

In Appalachia, the forests are first clear-cut and then explosives are used to blast away part of the mountain. Heavy equipment then digs through debris, dumping the overburden (soil and rock) into the nearby valley, burying streams as the valley is filled in. The exposed coal is dug out and some processing is done on site. Coal sludge left over from processing is stored in ponds on the mining site.

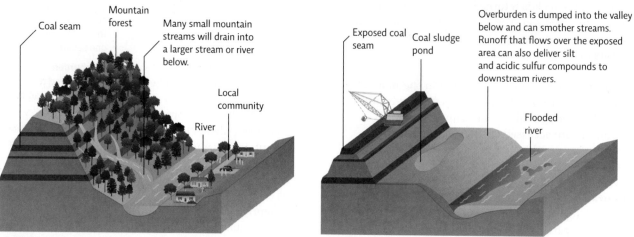

Coal seam

Mountain forest

Many small mountain streams will drain into a larger stream or river below.

Local community

River

Exposed coal seam

Coal sludge pond

Overburden is dumped into the valley below and can smother streams. Runoff that flows over the exposed area can also deliver silt and acidic sulfur compounds to downstream rivers.

Flooded river

 Why is the overburden dumped into the valley below when it is known that it damages streams and valley habitat?

5 COAL MINING: SUBSURFACE TECHNIQUES

Key Concept 5: Subsurface mining is used to access deep, thick coal seams. It is less environmentally damaging and employs more workers than surface mining but is a hazardous job.

subsurface mines Sites where tunnels are dug underground to access mineral resources.

acid mine drainage Water that flows past exposed rock in mines and leaches out sulfates. These sulfates react with the water and oxygen to form acids (low-pH solutions).

With their reliance on explosives and supersized heavy equipment, surface mines are a far cry from the underground mines, also called **subsurface mines**, that sustained the Nelson family for so many generations. "When our daddies were mining, back in the '40s and '50s, the seams were as tall as full-grown men," says Nelson, who since retiring has become a spokesperson for the anti-mining Ohio Valley

Environmental Coalition. "So you could get at 'em the old-fashioned way, with pickaxes and sledgehammers." Those days of plenty are gone, he says. Many of the coal seams that remain are too thin to be culled by human hands.

To be sure, subsurface mines (which make up about 50% of all U.S. coal mines) come with their own challenges. Water seeps easily into tunnels, and as it does, hazardous chemicals leach from the surrounding rocks into the gathering pools. One of these, sulfate, produces **acid mine drainage**, which goes on to contaminate soil and streams and has become a

INFOGRAPHIC 5 | **SUBSURFACE MINING**

Subsurface mines are used to access large deposits or thick seams of many minerals and ores, including coal. Modern mining depends on powerful machinery to drill out tunnels and remove and transport the materials.

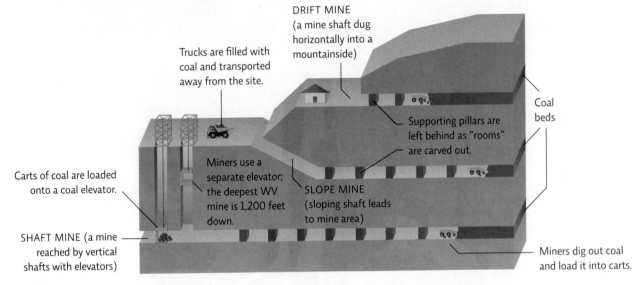

DRIFT MINE (a mine shaft dug horizontally into a mountainside)

Trucks are filled with coal and transported away from the site.

Supporting pillars are left behind as "rooms" are carved out.

Coal beds

Carts of coal are loaded onto a coal elevator.

Miners use a separate elevator; the deepest WV mine is 1,200 feet down.

SLOPE MINE (sloping shaft leads to mine area)

SHAFT MINE (a mine reached by vertical shafts with elevators)

Miners dig out coal and load it into carts.

 Why might a mining operation use mountaintop removal instead of subsurface mining?

major problem in both active and closed mines. In 2011, Duke University ecologist Emily Bernhardt and her colleagues reported that acidic water not only is directly toxic to many aquatic plants and animals but also alters the nutrient cycle of streams in ways that reverberate all the way up the food chain.

Subsurface mines are also dangerous places to work. Explosions and toxic fumes can be fatal to miners, and breathing in coal dust causes pneumoconiosis (black lung disease). In fact, more miners die from lung disease than from mining accidents. Fires, too, can start in a coal mine and once they begin, it is hard to extinguish them. (See the Interactive Map for more on mine fires and other disasters.)

But subsurface mines also come with some advantages. Unlike surface mines, they don't disrupt or permanently alter large surface areas. And because subsurface mines require more workers, they employ more people. In Appalachia, 100,000 mining jobs were lost between 1980 and 1993 as underground mining gave way to mountaintop removal. In Kentucky alone, mining jobs are down 60% in recent years, even though coal production is on an upswing. "It's a double insult," says Tim Landry, a fourth-generation deep miner in West Virginia. "They're not only destroying the land that we love, but they're taking our jobs away, too." **INFOGRAPHIC 5**

But a loss of jobs is not the community's only—or even its most serious—concern.

6 THE ADVANTAGES AND DISADVANTAGES OF COAL

Key Concept 6: Coal is an abundant and high-energy fuel, but mining and burning it create significant environmental and health problems for human populations and ecosystems through air and water pollution and habitat destruction.

Coal is an energy-rich fossil fuel that has several advantages over other fossil fuels. Its wide distribution around the world means there is little, if any, geopolitical tension associated with coal supplies. The World Coal Association estimates that at current rates of usage, proven reserves would supply coal for about another century. Coal is also safe and easy to store and transport compared to liquid or gaseous fuels. And it is affordable.

But coal has many disadvantages, many of them felt by the residents of Appalachia, such as Maria Gunnoe— the daughter, sister, wife, niece, and aunt of coal miners who has lived on the same property in Bob White, West Virginia all her life. She remembers having free run of the mountain as a child. "We had access to all the resources—food, medicine, water—that these mountains provided."

Maria Gunnoe became an environmentalist after mountaintop mining in her area caused flooding to her home and property, poisoned her well water, and made her daughter sick. Gunnoe received the Goldman Environmental Prize for her organizing efforts in her southern West Virginia community.

Things are different now. Gunnoe's children run into big yellow gates and No Trespassing signs wherever they go. The mountains, she says, have been closed off for blasting. And in the past decade, several million tons of overburden—an unimaginable mass—have been dumped into the valleys around Bob White.

The upheaval has had a noticeable impact on area residents. For one thing, the loss of forest and the compaction of so much soil has increased both the frequency and severity of flooding; without trees, and with the soil so compressed, the ground can't absorb water. "Floods are about three times more serious than they ever were before," she says. One 2003 flood nearly swallowed her entire valley—house, barn, family, and all.

Floods aren't the only problem. In fact, it's the blasting that most scares Gunnoe. The explosions that send rock and debris flying can damage homes and, in rare cases, trigger rockslides. In 2005, a 3-year old child was killed when

a boulder crashed through his house and landed on the bed where he was sleeping. Blasting also fills the air with tiny particles of coal dust—easily inhaled and full of toxic substances like mercury and arsenic. Studies show a higher-than-typical incidence of respiratory illnesses in mining communities. "When they were clearing the ridgeline right behind us, we'd get blasted as much as three times a day," Gunnoe says "There were days when we'd have to just stay inside because you couldn't breathe out there. And now, my daughter and I both get nosebleeds all the time."

The health impacts are far reaching. A 2011 study showed that children in MTR communities are more likely to suffer a range of serious birth defects, including heart, lung, and central nervous system disorders. And children aren't the only ones affected; a 2012 study found increased rates of leukemia and lung, colon, kidney, and bladder cancer in Appalachian counties with mountaintop removal sites compared to nonmining counties. (See the *Science Literacy* activity at the end of this module for more on this study.)

In addition to filling the air, these toxic substances also permeate the region's rivers, streams, and groundwater. Overburden has completely buried more than 3,200 kilometers (2,000 miles) of streams and destroyed an untold range of natural habitats in the process, threatening local species diversity. In a 2005 **environmental impact statement** on MTR mining, the U.S. Environmental Protection Agency (EPA) reported that selenium levels exceeded the allowable limits in 87% of streams located downhill from mining operations. Biodiversity in streams and the nearby forests has decreased in direct proportion to the concentration of toxic substances in the water. When tests revealed dangerous levels of selenium in the stream behind Gunnoe's house, she and her neighbors started getting their water from town. "We can't trust the water in our own streams anymore," she says "It's sad, but this is just not the same place that I grew up in."

> "We can't trust the water in our own streams anymore. It's sad, but this is just not the same place that I grew up in."
> —Maria Gunnoe

But the damage doesn't stop with the actual mining. Slurry impoundments—reservoirs of thick black sludge—accompany each MTR mining operation. They too are a consequence of thinner coal seams. "The thinner seams are messier," says Randal Maggard, a mining supervisor at Argus Energy, a company that has several mining operations throughout Appalachia. "It's 6 inches of coal, 2 inches of rock, 8 inches of coal, 3 inches of rock, and so on. It takes a lot more work to process coal like that." To separate the coal from the rock, miners use a mix of water and magnetite powder known as *slurry* in which coal can float. Once the sulfur and other impurities have been washed out, the coal is sent for further processing, and the slurry by-product is pumped into artificial holding ponds.

Maggard insists that the impoundments are safe. "Before we fill it, we have to do all kinds of drilling to test the bedrock around it," he says "And then a whole slew of chemical tests on top of that—all to make sure the barrier is impermeable." Still, area residents worry about a breach. With good reason.

In October 2000, the containment at a sludge pond holding coal slurry failed, pouring more than 1.1 billion liters (300 million gallons—30 times the Exxon Valdez spill) of toxic sludge into the Big Sandy River of Martin County, Kentucky. The contamination killed all wildlife

environmental impact statement A document outlining the positive and negative impacts of any federal action that has the potential to cause environmental damage.

The 2000 coal sludge spill in Martin County, Kentucky, contaminated groundwater and surface water, damaged homes and property, and killed all aquatic life in much of the area.

Lexington Herald-Leader, David Stephenson/AP Photo

in streams and streamside areas in the immediate area and impacted waterways for more than 100 km (about 60 miles). The sludge was 1.5 meters (5 feet) deep in some places. The EPA would register it as one of the worst environmental disasters in U.S. history. Sludge can still be found a few inches under the river sediments.

Of course, it's not just the mining and processing of coal that pollute the environment. When coal is burned to produce heat energy, it releases a variety of substances that damage the environment and threaten human health: gases (sulfur dioxide, carbon monoxide, nitrogen oxide, and planet-warming carbon dioxide), radioactive material (uranium and thorium), and particulate matter (soot) that can irritate lung tissue or even enter the bloodstream if inhaled.

Heavy metals (such as mercury and arsenic) are also released. Worldwide, coal burning accounts for 24% of the total anthropogenic mercury released. Much of this mercury finds its way into the world's oceans and into fish. In fact, the main way humans are exposed to mercury is by eating contaminated fish, especially top predators

like tuna that biomagnify mercury (see Module 1.3). An analysis of the mercury content of ancient and present-day samples of teeth, bone, and hair revealed a dramatic and rapid increase of mercury between preindustrial and modern times; levels in the late 1900's were roughly 10 times greater than levels before the Industrial Revolution. The EPA's 2011 Mercury and Air Toxic Standards would reduce these emissions by 90% and save up to $90 billion in human health costs (while only costing $9.6 billion to implement). The rule is currently being contested by the coal lobby and the Trump administration.

Coal-fired power plants also generate tons of toxic fly ash—fine ashen particles made up of silica and small amounts of toxic metal such as cadmium and lead. Some of that ash is diverted to industry, where it's used in concrete production. But most is buried in hazardous waste landfills or stored in open ponds like the ones used to contain the slurry waste from mining. In 2008, following a heavy rain event, a 15-meter-tall (50-foot-tall) dike from a pond holding coal fly ash from the Kingston Fossil Plant in Tennessee failed, releasing 4.2 billion liters

In 2008, more than 4 billion liters (1 billion gallons) of coal fly ash slurry surged into homes and waterways near Harriman, Tennessee, after breaching a coal ash containment pond. Many residents have been permanently displaced.

Antrim Caskey

INFOGRAPHIC 6 | THE TRADE-OFFS OF COAL

Though coal has many advantages as a fuel, mining, processing, and burning coal contribute to a variety of environmental problems, many of which lead to serious health issues for human populations.

Gallo Images/Danita Delimont/Getty Images

ADVANTAGES	DISADVANTAGES
• Widely distributed and fairly abundant supply • Low geopolitical conflict over supplies • Energy-rich fuel • Safe to transport and store • Affordable	• Nonrenewable, finite resource • Hazardous air pollution from mining and burning coal • Most carbon-intensive fossil fuel to burn (high CO_2 emissions) • Water pollution (surface and groundwater) from mining and burning coal • Habitat/property destruction from surface mining, and ash and sludge pond failures • Biodiversity loss from mining (especially MTR mining) • Blasting hazards to nearby communities from MTR mining • Flooding risk increased due to loss of vegetation from surface mining • Health issues such as respiratory problems, cancers, birth defects, and cardiovascular diseases in miners and community members living near surface mines or coal power plants

 If we choose to continue to use coal, what could be done to address the disadvantages of mining, processing, or burning it?

(1.1 billion gallons) of fly ash into nearby rivers and coating vast expanses of riverside land. The EPA spent 6 years and $1.1 billion in clean-up efforts and in 2017, reported that the ecosystem was back to "prespill conditions."

In the wake of the Kingston spill (and others such as a 2014 spill that sent 36,000 metric tons of coal ash into the Dan River of North Carolina), arguments were made to classify coal fly ash as a hazardous material, but the EPA ultimately rejected these arguments and ruled to classify and regulate coal ash as a nonhazardous substance in December 2014. **INFOGRAPHIC 6**

In 2011—in an attempt to calculate the *true costs* of coal—Harvard University public health researcher Paul Epstein and colleagues tallied up the costs of mining and using coal. The analysis took into account many of the external costs from mining, shipping, burning, and waste production—that is, the costs that are not currently reflected in the market cost of coal. They estimated that coal costs the American public between $300 and $500 billion a year in externalized costs (health, environmental, and property costs). This amounts to an extra $0.10 to $0.26 per kWh in external costs—in some cases more than twice what consumers actually pay. (See Module 5.1 for an introduction to true cost accounting.)

7 REDUCING THE IMPACT OF BURNING COAL

Key Concept 7: The negative impact of burning coal can be addressed by capturing pollutants before they are released or converting coal to a liquid or gaseous fuel that burns more cleanly.

So, what do we do? On one hand, we rely on coal to power our society. On the other, the processes of mining and then burning it are hurting us and our environment as much as they are sustaining our way of life.

One potential solution is "clean coal technology"—technology that minimizes the amount of pollution produced by coal. For example, scientists and engineers around the world are working on ways to capture the gases emitted from burning coal. We already have the capacity to capture some emissions like mercury, particulate matter, and sulfur (see Module 10.1). The next big challenge is **carbon capture and sequestration (CCS)**—the

carbon capture and sequestration (CCS) The process of removing carbon from fuel combustion emissions or other sources and storing it to prevent its release into the atmosphere.

capture and storage of CO_2 in a way that prevents it from reentering the atmosphere and contributing to climate change. Most CCS programs are still in the research and development stage.

The first operational U.S. CCS carbon-fired power plant, Petra Nova, went online in early 2017, a $1 billion retrofit of an existing power plant near Houston, TX. (The only other CCS facility online is located in Canada—the Boundary Dam Power Station.) If this project is successful in capturing and sequestering carbon, in a cost-effective way, it will have passed a major

clean coal A liquid or gaseous product produced by removing some contaminant contained in coal so that the resulting fuel releases less pollution when burned.

hurdle—showing that retrofits of existing power plants are feasible. Though the $1 billion price tag for the CCS refit is significant, it is much more affordable than the $7.1 billion that has already been spent on the only other CCS project in the United States—the Kemper County Energy Facility in Mississippi, a "next-generation" CCS coal powered power station that missed its start-up date (May 2014) by years and its projected price tag ($3 billion) by close to 300%. In 2017, the CCS portion of the project was suspended. **INFOGRAPHIC 7**

Another emerging technology creates what is known as **clean coal** in a process that chemically removes some of coal's contaminants before burning it. In the most common clean coal technology, coal is converted

---

| **INFOGRAPHIC 7** | **HOW IT WORKS: CARBON CAPTURE AND SEQUESTRATION (CCS)** | ANIMATED INFOGRAPHIC |

The capture of CO_2 and subsequent storage that prevents it from reentering the atmosphere would greatly decrease coal's contribution to global climate change. A variety of CCS methods are currently in research and development or are being tested as pilot programs.

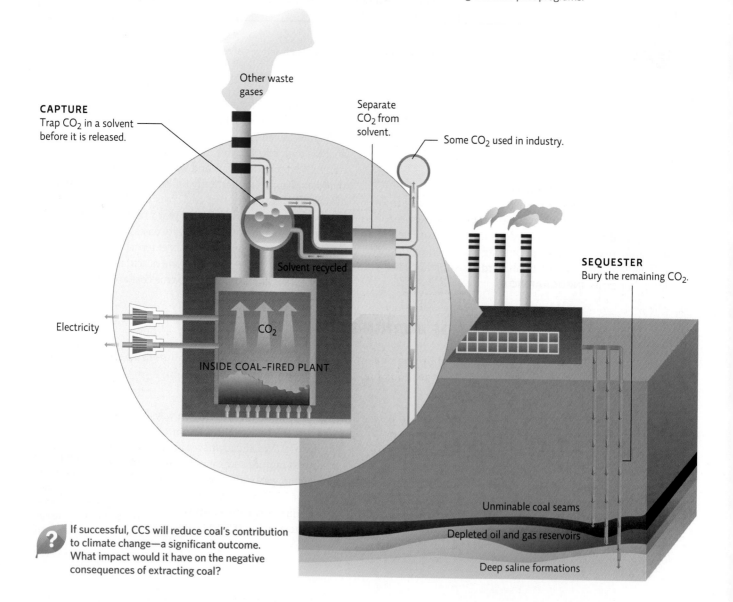

Other waste gases

CAPTURE
Trap CO_2 in a solvent before it is released.

Separate CO_2 from solvent.

Some CO_2 used in industry.

Solvent recycled

Electricity

CO_2

INSIDE COAL-FIRED PLANT

SEQUESTER
Bury the remaining CO_2.

Unminable coal seams

Depleted oil and gas reservoirs

Deep saline formations

? If successful, CCS will reduce coal's contribution to climate change—a significant outcome. What impact would it have on the negative consequences of extracting coal?

to a gaseous fuel (syngas) that burns more efficiently than pulverized coal and releases fewer air pollutants. The gasification process is complex: It occurs at high pressure and temperature, and it requires chemical solvents to strip some of the impurities out of the coal. The process requires significant inputs of energy, thus lowering coal's EROEI despite its greater energy efficiency in electricity generation. Coal liquefaction projects are also in the research and development stage with the goal of producing liquid hydrocarbon fuels that can supplement or replace gasoline or other liquid fuels.

While it still contains some hazardous substances, syngas contains fewer contaminants than the original coal and burns more cleanly. There are more than 90 gasification power plants—known as integrated gasification combined cycle (IGCC) plants—in operation in the world; 21 of those are in the United States. However, the cost of building these facilities is high ($1 billion), and 25 additional U.S. IGCC projects have been delayed or cancelled.

These efforts, if implemented, will address some of the downsides to *burning* coal, allowing us to continue to use an abundant fossil fuel and in doing so, save jobs associated with the coal industry at the same time. But, as critics are quick to point out, coal can never be truly clean. Each of these "clean coal" technologies produces its own toxic by-products that must be dealt with. And CCS and clean coal do nothing to address the negative impacts of *acquiring* that coal. For coal to truly be considered "clean," ecosystems and communities must also be protected during and after the mining process. Is it possible to dig so deeply and alter the landscape so thoroughly without permanent damage?

8 MINE RECLAMATION

Key Concept 8: Surface coal mines must undergo reclamation once they are closed; this reduces the environmental damage of mining but does not eliminate it.

In some ways, closed MTR mines—those where all the coal has been harvested—look even more alien than do active sites like Hobet 21. Instead of the natural sweep of rolling hills, staircase-shaped mounds are covered with grasses or stands of evenly spaced, young conifers. Such efforts represent the coal industry's attempt to honor the U.S. Surface Mining Control and Reclamation Act, which in 1977 mandated that areas that have been surface mined for coal be "reclaimed" once the mine closes. **Reclamation** requires that the area be returned to a state close to its premining condition.

If the mined area was originally fairly flat, the reclamation process includes filling the site with the stockpiled overburden and contouring the site to

reclamation The process of restoring a damaged natural area to a less damaged state.

A valley in West Virginia after reclamation shows none of the original forest, ridges, or streams that were once found there.

INFOGRAPHIC 8 MINE SITE RECLAMATION

U.S. federal law requires that after a surface coal mine ceases operations, the land must be restored to close to its original state. Though efforts are made to restore the area and reclaimed land may be suitable for other land uses such as agricultural, commercial, or industrial or can be set aside as a natural area, mountaintop removal mining sites can never be returned to their original state.

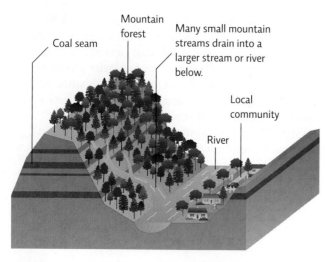

Coal seam
Mountain forest
Many small mountain streams drain into a larger stream or river below.
Local community
River

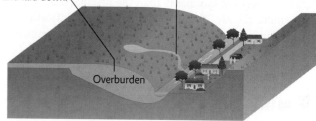

Overburden that was dumped into the valley below is smoothed over to produce a slope. Grasses are planted. Reforestation projects are under way to increase tree plantings in reclaimed areas, but trees do not grow well unless the soil was properly prepared and laid down.

The original mountain streams are gone for good; channels are dug to deliver water down the slope.

Overburden

 If you lived near an area where mountaintop removal mining was being done, would you be satisfied with the reclamation process and final product? What are the advantages and disadvantages of this type of reclamation?

match the surrounding land. The site is then covered with topsoil saved from the original dig. Vegetation, usually grass, is then planted, leaving other local vegetation to move in on its own.

But reclamation of MTR sites is difficult, if not impossible, due to the total destruction of the forest habitat and the loss of soil and native species.

Streams too are irrevocably damaged. The 1973 Clean Water Act requires that damaged streams be returned close enough to their original state such that the overall impact on the stream ecosystem is "nonsignificant." The method of choice is to dig drainage ditches and line them with stones in a way that resembles a stream or river. But so far, research shows that such channels don't perform the ecological functions of a stream. "They may look like streams," says ecologist Margaret Palmer from the University of Maryland. "But form is not function. The channels don't hold water on the same seasonal cycle, or support the same aquatic life, or process contaminants out of the water—all things a natural stream does." **INFOGRAPHIC 8**

In Appalachia, the arguments over when, where, and how to mine for coal are quickly boiling down to a single intractable question: Once it's all gone, how will we clean

up the mess we've made? For a story that has played out over geologic time, the question is more immediate than one might think. In West Virginia, coal reserves are expected to last another 50 years, at best. That means no matter what regulations the government imposes, or what methods the coal companies resort to, the day of reckoning will soon be upon us.

Select References:

Ahern, M., et al. (2011). The association between mountaintop mining and birth defects among live births in central Appalachia, 1996-2003. *Environmental Research, 111*(6), 838-846.

Ahern, M., & Hendryx, M. (2012). Cancer mortality rates in Appalachian mountaintop coal mining areas. *Journal of Environmental and Occupational Science, 1*(2), 63-70.

Bernhardt, E. S., & Palmer, M. A. (2011). The environmental costs of mountaintop mining valley fill operations for aquatic ecosystems of the central Appalachians. *Annals of the New York Academy of Sciences, 1223*(1), 39-57.

Epstein, P. R., et al. (2011). Full cost accounting for the life cycle of coal. *Annals of the New York Academy of Sciences: Ecological Economics Review, 1219*(1), 73-98.

U.S. Environmental Protection Agency (EPA). (2005). *Final Programmatic Environmental Impact Statement on Mountaintop Mining/Valley Fills in Appalachia.* Philadelphia, PA: EPA.

INTERACTIVE MAP   **COAL MINING: A DANGEROUS OCCUPATION**   ANIMATED INFOGRAPHIC

While coal mining regulations have significantly reduced injuries and deaths in the United States, mining is still a dangerous job, especially in countries with lax or insufficient regulations in place to protect workers or the surrounding community. Here are accounts of four recent disasters related to coal. [See the Interactive Map for Online Module 7.3, Mineral Resources, for an account of the Benxihu Honkeiko Colliery coal mine explosion.]

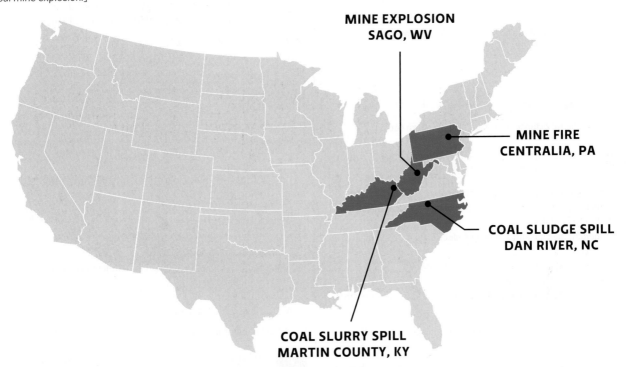

MINE EXPLOSION
SAGO, WV

MINE FIRE
CENTRALIA, PA

COAL SLUDGE SPILL
DAN RIVER, NC

COAL SLURRY SPILL
MARTIN COUNTY, KY

 BRING IT HOME

PERSONAL CHOICES THAT HELP

Although coal is one of our most abundant fossil fuels, its drawbacks are significant. They include CO$_2$ emissions, the release of air pollutants that cause environmental problems such as acid rain; health problems such as asthma and bronchitis; and massive environmental damage from the mining process. One way to minimize the impact of coal is to reduce consumption of electricity.

Individual Steps

• Always conserve energy at home and in your workplace.

• Turn off or unplug electronics when not in use.
• Put outside lights on timers or motion detectors so that they come on only when needed.
• Dry clothes outdoors in the sunshine.
• Turn the thermostat up or down a couple of degrees in summer and winter to save energy and money.

Group Action

• Organize a movie screening of *Coal Country* or *Kilowatt Ours*, which present issues

related to coal mining and mountaintop removal from many perspectives.

Policy Change

• The Appalachian Regional Reforestation Initiative is a great example of how groups, sometimes with very different objectives, can work toward a common goal. Go to http://arri.osmre.gov to see how this coalition of the coal industry, citizens, and government agencies are working to restore forest habitat on lands used for coal mining.

ENVIRONMENTAL LITERACY   UNDERSTANDING THE ISSUE

1 What energy sources are used in modern society and for what purposes?

1. The two categories of energy, identified by their end use, that we use in modern society are _____ and _____ sources.

2. About how much of the energy in the United States is currently supplied by fossil fuels?
   a. Very little (< 25%)
   b. About 50%–55%
   c. Around 80%
   d. Virtually all (> 95%)

3. Explain the importance of fossil fuels to modern society.

2 How important is coal as an energy source, and how is it used to generate electricity?

4. Coal can be used to generate electricity by:
   a. passing a current through the pulverized coal and then sending it to a generator.
   b. melting the coal and sending its electrons across conductive wires.
   c. pouring liquefied coal over a turbine, which triggers the release of electrons from copper wire.
   d. burning the coal to make steam, which turns a generator that makes electricity.

5. Explain the concept of energy return on energy investment (EROEI) and explain its significance.

6. What energy sources are used to produce electricity, and how do they compare in terms of EROEI?

3 What is coal, how is it formed, and what regions of the world have coal deposits that are accessible?

7. True or False: Coal formed in areas where plant matter was buried in high oxygen environments that triggered accelerated decomposition.

8. Which of the following areas have the largest coal reserves?
   a. South America and Africa
   b. United States, Russia, and Australia
   c. Greenland and Canada
   d. Middle East

9. Why is coal considered a finite resource?

4 What surface methods are used to mine coal, and what are their advantages and disadvantages?

10. Surface strip mining techniques are used when:
    a. the coal can't be retrieved using heavy equipment.
    b. coal is located in thin seams in mountainous areas.
    c. thick coal seams are close to the surface in relatively flat areas.
    d. it is too dangerous to use explosive charges to access the coal.

11. What are some of the community impacts, both good and bad, of mountaintop removal mining in Appalachia?

12. How does strip mining on flat ground compare to strip mining in mountainous areas in terms of environmental damage?

5 How are deep deposits of coal mined, and what are the advantages and disadvantages of this method?

13. An advantage of subsurface (underground) coal mining is that it:
    a. employs more people than surface mining techniques.
    b. is safer for the miners than surface mining techniques.
    c. is cheaper than surface mining techniques.
    d. All of these are advantages.

14. How do subsurface and mountaintop removal mining in Appalachia compare in terms of jobs, health risks for workers, and environmental impacts on the nearby community?

6 What are the advantages and disadvantages of using coal?

15. True or False: If we paid the true costs of using fossil fuels like coal, the price of our electricity would likely go up.

16. Which of the following is NOT true about coal?
    a. It produces more air pollution than other fossil fuels.
    b. It is difficult to transport.
    c. Extraction is relatively cheap.
    d. The United States has an abundant supply of it.

17. What health problems are associated with coal burning?

7 What new technologies allow us to burn coal with fewer environmental and health problems?

18. Clean coal (syngas):
    a. has a higher EROEI than conventional coal.
    b. eliminates the hazards associated with coal burning.
    c. is produced when coal is rinsed with water and chemicals after its initial removal from the ground.
    d. reduces the pollution released when the coal is burned but requires energy to produce.

19. What is carbon capture and sequestration (CCS) and why is it done? What are the pros and cons of the process?

20. Which problems associated with the traditional use of coal are not addressed by using clean coal technologies?

8 How can mining damage be repaired, and how effective is this restoration?

21. Which of the following is an example of reclamation?
    a. Reworking coal mines to extract more coal
    b. Replanting the site of mountaintop removal with grass and pine trees
    c. Returning land to the people who originally owned it
    d. Filling in a subsurface mine

22. Why is coal mine restoration more difficult at an Appalachian MTR site than at a Wyoming strip mine?

SCIENCE LITERACY WORKING WITH DATA

Cancer mortality rates (deaths per 100,000 people) were determined for counties in Appalachia (Kentucky, Tennessee, Virginia, and West Virginia). A county was identified as an MTR area if there was at least one MTR mine, 40 to 320 acres in size with at least 10 acres of removed ridge top. Counties with mines other than MTR mines or counties with no mines were also evaluated. [Note: On the bar graph, the letter "a" next to a bar indicates it is significantly different from the "No mining" group for a given cancer; the letter "b" indicates it is significantly different from the "Other mining" group; $p < 0.05$.]

PROXIMITY TO COAL MINES AND CANCER MORTALITY RATES IN APPALACHIA

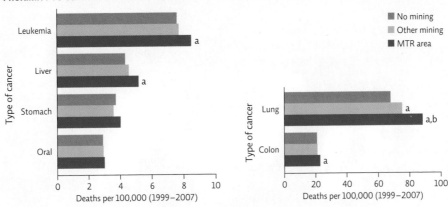

Interpretation

1. Which cancers show higher rates of morality in MTR areas compared to areas with no mines? Compared to areas with other mines?

2. Why are the mortality rates for lung and colon cancer shown in a different graph?

3. Why is it incorrect to conclude that deaths due to stomach or oral cancers are more common in MTR counties even though the MTR bar for each indicates a higher number of deaths?

Advance Your Thinking

4. Why might lung cancer death rates be so high in Appalachian countries and highest in MTR areas?

5. Why did the researchers collect data on "Other mining" counties, rather than simply comparing MTR counties to those with no mining?

6. When the researchers controlled for socioeconomic and health variables (such as smoking rates, obesity, and income) that might affect cancer mortality, they still saw significant increases for lung and liver cancer; leukemia mortality was also higher, but only from 2003 to 2007. Why did they control for these variables?

INFORMATION LITERACY EVALUATING INFORMATION

The Stream Protection Rule, part of the Clean Water Act, was signed by President Barack Obama on his last day in office in January 2017, after years of working to strengthen protections of streams against damage from surface coal mining. Opponents to the rule moved swiftly when Donald Trump took office and repealed the rule.

Read the following two articles about the rule and answer the questions that follow. (If these links are not active, search the Internet using the key words "Stream Protection Rule Repeal" and read two articles—one for the rule and one against—and answer the questions that follow.)

1. Go to www.kentucky.com/opinion/op-ed/article133445464.html and read the article entitled "Stream rule's repeal first step to economic relief for coal country."
   a. Who is the author of this article and what web organization published the article? Are these reliable information sources?
   b. Is there any indication of bias that might influence the author's position? Explain.

c. What position does the author take regarding the Stream Protection Rule? What evidence is offered for this position? Are reliable references provided to support this position or evidence? Explain.
d. What information could have been provided to strengthen the position offered?

2. Now read the article "Trump's repeal of stream rule helps coal at the expense of climate and species" at www.insideclimatenews.org/news/16022017/coal-mining-environment-stream-rule-donald-trump-mussels-species and answer the same questions as above (a–d).

3. Compare the arguments presented by the two stories:
   a. Which argument prioritizes the economy and jobs? Which prioritizes the environment?
   b. Does each article address the main concern of the other (jobs, environment)? Explain.
   c. After reading these two articles, do you have the information you need to draw your own conclusion about the Stream Protection Rule? If so, why? If not, what additional information do you need and where would you find it?

 Additional study questions are available at SaplingLearning.com.

THE BAKKEN OIL BOOM

Is fracking the path to energy independence?

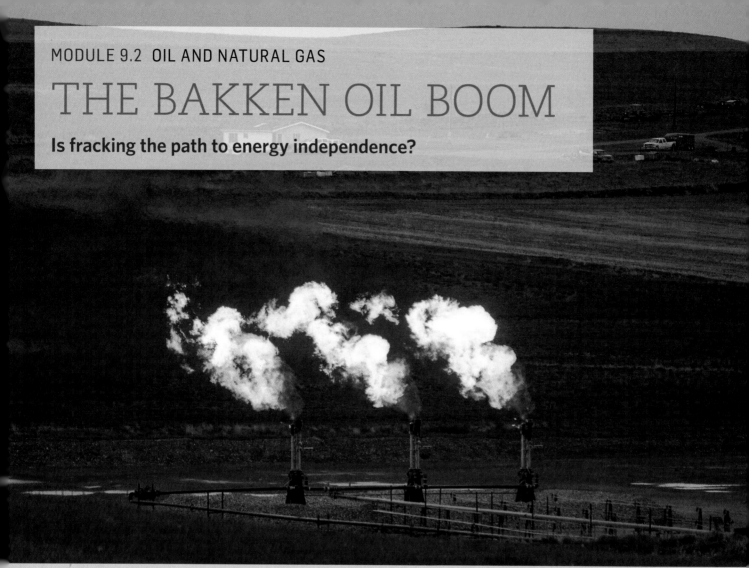

Natural gas is flared off as oil is pumped from a fracking well in the Bakken Formation located close to a private home near Watford City, North Dakota.

Jim West/Alamy

After reading this chapter and studying the KEY CONCEPTS and INFOGRAPHICS, you should be able to answer these GUIDING QUESTIONS

CORE MESSAGE

Oil and natural gas are vital resources for modern society, but damage is caused at every stage of their acquisition, and conventional supplies are becoming depleted. Unconventional sources are available but are more difficult and costly (economically and environmentally) to extract, so we should weigh the pros and cons of pursuing these fossil fuels.

1. How are fossil fuels formed, and why are they considered nonrenewable resources?

2. Where are oil and natural gas reserves found?

3. How are conventional oil and natural gas extracted?

4. What are the trade-offs of acquiring and using conventional oil?

5. What are the trade-offs of acquiring and using conventional natural gas?

6. What unconventional sources of oil and natural gas exist, and how are they extracted?

7. What are the trade-offs of pursuing unconventional oil and natural gas sources?

8. What obstacles stand in the way of the United States (or any nation) achieving energy independence or security?

"It is as you imagine it: Vast. Open. Windy. Stark. Mostly flat. All but treeless. Above all, profoundly underpopulated, so much so that you might, at times, suspect it is actually unpopulated. It is not. But it is heading there."

So went the opening paragraph of a 2006 feature article in *The New York Times Sunday Magazine* about the demise of the prairie states in general and of North Dakota in particular. The entire region, it seemed, was suffering a mass exodus: houses and churches and schools and stores, all completely abandoned.

That was 2006. Less than a decade later, the same sleepy patch of country was in the middle of a modern-day gold rush—an oil boom, to be precise. In 2012, journalists described local grocery stores barely keeping shelves stocked, town movie theaters being so crowded that people are sitting in the aisles, and housing costs approaching those of the most expensive cities. Towns once home to 1,200 residents claimed 10 times as many. "With the way it is now," Jeff Keller, a natural resource manager with the Army Corps of Engineers, told ProPublica in 2012, "you're getting to the crazy point."

Two short years later, the price of oil fell. Was the boom about to become a bust?

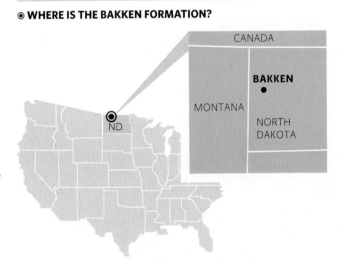

⊙ **WHERE IS THE BAKKEN FORMATION?**

Rows of mobile homes house oil workers in Williston, North Dakota, where housing shortages drive up the cost of rent. In 2014, Williston was the most expensive place to rent a home in the United States.

REUTERS/Ben Garvin/STRINGER/Newscom

1 FOSSIL FUELS: OIL AND NATURAL GAS

Key Concept 1: Fossil fuels form when organic matter is buried and subjected to high temperature and pressure. Because we use them much faster than they form, they are considered nonrenewable resources.

Fossil fuels form over millions of years when organisms die and are buried in sediment under low oxygen conditions that slow down decomposition tremendously. In time, as pressure and heat increase, the buried material goes through a chemical transformation to form **oil** (a liquid), **natural gas** (a gas), or coal (a solid). (See Module 9.1 for more on coal.) Oil and natural gas were formed from marine organisms that died and went through this process. (The starting material for coal formation was plant material.) Even though they are produced by natural processes, fossil fuels are considered a **nonrenewable resource** because we are using them up in a fraction of the time it takes for them to form. In other words, natural processes cannot possibly keep pace with human consumption. **INFOGRAPHIC 1**

Everywhere, countries are looking for new sources of oil and natural gas. The heart of the U.S. oil boom is the Bakken Formation, a vast shale formation deep underground that resides mostly in North Dakota but also stretches westward into Montana and north into Canada. (Shale is a type of sedimentary rock that is very dense and does not easily let the oil and natural gas trapped within escape.) It's the largest continuous oil accumulation the U.S. Geological Survey (USGS) has ever assessed. The United States has other oil-rich shale formations, including the highly productive Eagle Ford formation in Texas. At its peak in 2014, roughly 1.2 million barrels of crude oil were being recovered each day from Bakken—almost 14% of all oil produced in the United States (around 6% of the oil consumed per day).

So why didn't the oil industry go after the oil in Bakken earlier? It turns out, Bakken is not a typical oil deposit. When the 21st century began, it was not economically feasible to extract Bakken's oil, but new technology has put it within reach. "The implications are already reverberating far beyond North Dakota," Edwin Dobb wrote in *National Geographic* in 2013. "Bakken-like shale formations occur across the United States, indeed, across the world. The extraction technology refined here is in effect a skeleton key that can be used to open other fossil fuel treasure chests."

fossil fuel A variety of hydrocarbons formed from the remains of dead organisms.

oil A liquid fossil fuel useful as a fuel or as a raw material for industrial products.

natural gas A gaseous fossil fuel composed mainly of simpler hydrocarbons, mostly methane.

nonrenewable resource A resource that is formed more slowly than it is used or that is present in a finite supply.

INFOGRAPHIC 1   **OIL AND NATURAL GAS FOSSIL FUEL FORMATION**

  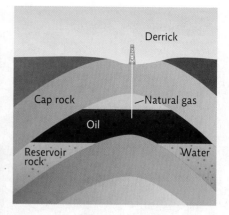

Long ago, some marine organisms died and were buried in sediment. This burial excluded oxygen, and decomposition was greatly slowed down.

As sediments accumulated, the partially decomposed buried biomass was subjected to high heat and pressure. Over the course of millions of years, it was chemically converted to oil or natural gas.

Since oil and natural gas are lighter than water, they flow upward in porous reservoir rock until stopped by a layer of dense cap rock. We tap these deposits by drilling into the porous rock reservoirs.

 How is the formation of oil and natural gas similar to and different from the formation of coal?

2 OIL AND NATURAL GAS RESERVES

Key Concept 2: The largest conventional oil and natural gas reserves are in the Middle East; however, North America has large deposits of unconventional oil and natural gas.

Oil is a liquid fossil fuel made up of hundreds of types of *hydrocarbons*, organic compounds of hydrogen and carbon. Hydrocarbons take many forms—solid, liquid, or gas—and we have developed a variety of methods for extracting all of these fossil fuels from the depths of Earth.

Oil and natural gas often occur together in formations—the lighter natural gas is found at the top of the deposit. After formation, the oil and natural gas slowly migrated up from the source rock in which they were formed into porous, sedimentary rock formations closer to the surface until they were stopped by a dense "cap rock" layer, such as granite, that prevented further migration and allowed the oil and natural gas to collect in one area. In these deposits, oil does not exist in thick black pools. Instead, if you could look down into an oil reservoir, you'd see rock. Most **crude oil** is found as tiny droplets wedged within the microscopic open spaces, or pores, inside rocks.

Oil and natural gas deposits, or reserves, are identified by their recovery potential. **Proven reserves** are those that hold oil or natural gas that is economically feasible to extract using current technology. **Conventional reserves** hold oil or natural gas that can be extracted with traditional oil or gas wells. Conventional oil reserves are not evenly distributed around the planet, leading to political problems among countries that have the oil, like those in the Middle East, and those that do not have enough to meet their own needs, like the United States. Alliances have been forged and wars have been fought over oil, highlighting its importance to modern society.

By many accounts, we have used up about half of the available conventional oil on the planet—an estimated 1.2 trillion barrels of oil. At current rates of extraction and use, known conventional reserves are expected to last another 50 years or so, though this number is uncertain because reports of oil reserves tend to be questionable. (Many nations keep their reserves estimates secret.)

Natural gas reserves are also finite. The Middle East has the largest conventional reserves of natural

crude oil A mix of hydrocarbons that exists as a liquid underground; can be refined to produce fuels or other products.

proven reserves A measure of the amount of a fossil fuel that is economically feasible to extract from a known deposit using current technology.

conventional reserves Deposits of crude oil or natural gas that can be extracted by vertical drilling and pumping.

The 1,300-kilometer Trans-Alaska pipeline delivers oil across Alaska to the southern port of Valdez.

gas, with about 38% of the world's reserves; the United States has only 6%. Total world conventional reserves of natural gas are expected to last 60 to 100 years at current rates of use.

However, the story does not end with conventional reserves. There are also other supplies of oil and natural gas, and our ability to access them is improving. These include **unconventional reserves**, such as Bakken oil. Bakken oil is made up almost exclusively of **tight oil**—shale oil that's trapped inside impermeable rock deposits—typically in the source rock in which it was formed (hence it is much deeper than conventional oil). Natural gas, too, is found in some shale deposits, such as the Marcellus Shale formation that underlies much of Pennsylvania and New York. Tight

unconventional reserves Deposits of oil or natural gas that cannot be recovered with traditional oil/gas wells but may be recoverable using alternative techniques.

tight oil Light (low-density) oil in shale rock deposits of very low permeability; extracted by fracking.

oil and shale gas are of similar quality to that found in conventional deposits but are difficult to access because the oil or natural gas tends to be more widely dispersed throughout the rock. However, new technology (to be described later) is making those deposits more accessible. **INFOGRAPHIC 2**

The world runs on oil and natural gas. These fossil fuels power vehicles and provide the raw material for thousands of industrial products, and consumption continues to increase. But because we are using them much faster than they formed, they will run out. Extraction of oil and natural gas from conventional reserves is peaking—this means less can be extracted each year as time goes on. Exploiting unconventional reserves such as Bakken tight oil or Marcellus Shale natural gas will extend supplies, but not for long. Problems such as climate change, air and water pollution, and biodiversity losses are at the heart of many arguments for leaving most of that remaining oil and natural gas in the ground.

INFOGRAPHIC 2 | **OIL AND NATURAL GAS RESERVES**

Different regions of the world have different amounts of proven oil and natural gas reserves (shown here as bar graphs; bars represent percentage of the world total); the largest proven conventional reserves are in the Middle East. North America has larger supplies of technically recoverable unconventional reserves—about 25% of the world total.

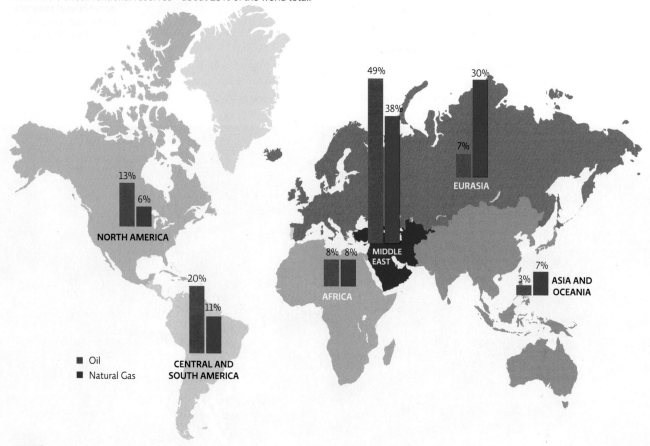

- Oil
- Natural Gas

 In 2015, there were 1,659 billion barrels (bbl) of oil in proven conventional oil reserves. How many bbl were in North America?

3 | EXTRACTION OF CONVENTIONAL OIL AND NATURAL GAS

Key Concept 3: Oil and natural gas often occur together in rock formations and are extracted from conventional reserves using oil and gas wells.

Oil and natural gas are easier to coax out of some formations than others. Conventional oil and natural gas are the easiest to retrieve and have supplied our needs throughout the 20th century; they continue to provide most of our supply today. The fossil fuels in these deposits are extracted by drilling down into the rock formation that holds the oil and/or natural gas. Initially, after a new oil well is drilled, nature does most of the work. There is significant pressure on oil located deep underground from millions of tons of rocks pressing down and from Earth's heat, which causes gases around the oil and rock to expand. So, when a well is first drilled, oil naturally flows upward, escaping like air gushing out of a balloon. Once the oil stops flowing freely, pumpjacks are used to pump out additional oil. This is known as *primary production*, and up to 15% of the oil can be recovered in this first phase.

Eventually, the pressure on the trapped oil will decrease, rendering pumping ineffective, so injection wells are drilled near the oil well, and water is pumped through these injection wells into the deposit. This increases the pressure on the remaining oil, forcing it upward in the deposit where it can be pumped out. During this phase of *secondary production*, an additional 20% to 40% of a reserve's oil can be recovered.

But even after maximal primary and secondary production (up to 55% of the total oil in the reserve), there is a lot of oil left in the ground. *Tertiary production* methods like injecting steam, natural gas, or carbon dioxide gas into the reservoir allow the additional recovery of up to 20% of the reserve's oil, for a total production of up to 75% of the oil found in the reserve. **INFOGRAPHIC 3**

INFOGRAPHIC 3 | HOW IT WORKS: CONVENTIONAL OIL AND NATURAL GAS WELLS

Oil and natural gas are obtained by drilling through layers of dense rock to reach the reservoir below. At first, oil and/or natural gas easily flow due to the relief of pressure caused by the drill hole (primary production). For oil, pumpjacks are used to mechanically pump out more when it stops flowing freely; additional secondary and tertiary production methods will allow the extraction of more oil.

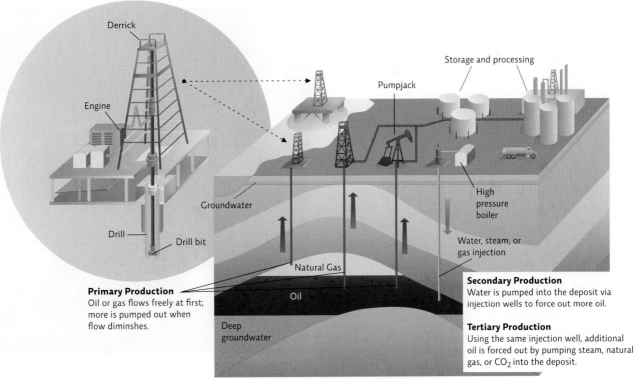

Derrick

Engine

Drill

Drill bit

Groundwater

Storage and processing

Pumpjack

High pressure boiler

Water, steam, or gas injection

Natural Gas

Oil

Deep groundwater

Primary Production
Oil or gas flows freely at first; more is pumped out when flow diminshes.

Secondary Production
Water is pumped into the deposit via injection wells to force out more oil.

Tertiary Production
Using the same injection well, additional oil is forced out by pumping steam, natural gas, or CO_2 into the deposit.

 Why aren't pumpjacks needed during the initial retrieval of oil from a newly drilled well?

Conventional natural gas can be found in deposits that contain no oil (nonassociated gas reserves) or in deposits that contain both oil and natural gas (associated gas reserves). The extraction of natural gas is very similar to that of oil. Wells are drilled into the gas-containing formation, and natural gas flows freely up the well due to underground pressure, or it is pumped out (primary production). Secondary and tertiary methods that extract oil will also extract any natural gas that is trapped in the oil, but these techniques are not used in nonassociated gas wells.

energy return on energy investment (EROEI) A measure of the net energy from an energy source (the energy in the source minus the energy required to get it, process it, ship it, and then use it).

With each more intensive production method, the **energy return on energy investment (EROEI)** goes down, and the cost goes up. A comparison of the EROEI for the extraction of different energy sources helps producers and investors determine which energy sources are worth pursuing. An energy source that requires almost as much energy to tap into as it ultimately produces will be left untapped in favor of sources with higher EROEI values. In the 1950's when oil wells were first being tapped, the EROEI was close to 30:1, but as the "easy oil" was extracted with primary production and production transitioned to secondary and tertiary methods, the EROEI declined to the current average of around 10:1. (See Module 9.1 for more on EROEI.)

4 THE TRADE-OFFS OF OIL

Key Concept 4: Oil is used to create a variety of fuel and industrial products, but its extraction, processing, transport, and use damage the environment and impact human health.

Modern society runs on oil. It powers our vehicles and lubricates our machinery. It is the raw material for the plastics used to make football helmets and neonatal incubators; for the pesticides on our crops and the asphalt on our roads; for detergents, paints, solvents, and a host of other petroleum products or **petrochemicals**. The production of a desktop computer, for example, consumes 10 times the computer's weight in fossil fuels, mostly oil. A shortage like the one that occurred during the oil crisis of the 1970's can send a nation into a panic and change global politics. It is no wonder we are constantly in search of new supplies, but as they become scarcer, the extent to which we will go to extract them has some people worried.

Oil extraction has environmental consequences at every stage. To locate deposits of oil and natural gas, companies send seismic waves into the ground that bounce back to reveal the location of possible reserves. Doing this in ocean areas disorients marine wildlife; in 2009, ExxonMobil had to abandon its oil exploration in Madagascar because more than 100 whales had beached themselves, presumably because of these sonic exploratory methods.

Drilling also affects wildlife. Politicians have long debated the merits of drilling in the Arctic National Wildlife Refuge (ANWR), 8 million hectares (20 million acres) of protected wilderness. The USGS estimates that parts of ANWR located on the northern Alaskan coast could harbor between 5 and 16 billion barrels of technically retrievable crude oil and natural gas

petrochemicals Distillation products from the processing of crude oil such as fuels or industrial raw materials.

reserves. (A barrel of oil is equivalent to 159 liters, or 42 gallons.) But drilling there is difficult, is expensive, and would seriously disturb the habitat of many polar species such as Arctic foxes, caribou, polar bears, and migratory birds. Because of this, drilling in some U.S. Arctic areas has been banned; in other areas, attempts

Boats hose down a massive fire on the oil rig *Deepwater Horizon*, 50 miles southeast of the tip of Louisiana in the Gulf of Mexico. The explosion killed 11 workers and released more than 200 million gallons of oil into the Gulf.

to begin drilling for offshore oil have been abandoned because of the difficulty of operating in such extreme conditions.

Another concern is oil spills—an unintended but frequent consequence of oil drilling and transport. Millions of gallons of oil are spilled from tankers, trains, pipelines, and ships annually. And while large-scale accidents like the 2010 *Deepwater Horizon* oil platform explosion and oil spill in the Gulf of Mexico are rare, they are devastating. The *Deepwater Horizon* spill released close to 5 million barrels of oil, eventually covered 930 square kilometers (360 square miles) of ocean, impacted roughly 1,800 kilometers (1,100 miles) of coastlines, and infiltrated and damaged coastal wetlands. The oil threatened the lives of many species and cost Gulf commercial fisheries and area tourism millions of dollars. (See the Interactive Map at the end of this module for more on this oil spill.)

The hazards of oil are not limited to exploration, extraction, and transportation of oil. Burning fossil fuels releases a variety of air pollutants that are linked to negative ecosystem and health impacts. It is also a major anthropogenic contributor to climate change. (See Module 10.1 for more on air pollution and Module 10.2

Burning fossil fuels is one of the leading causes of air pollution, which impairs human health and damages ecosystems.

MAZE, STEPHANIE/National Geographic Creative

for more on climate change.) In addition, working with oil is a hazardous undertaking, even when the required precautions are taken. Workers in the petroleum chemical industry have much higher rates of cancer, rashes, heart disease, and various other health problems than the general population. This hazard extends to people living near oil refineries and petrochemical plants; rates of cancer, birth defects, headaches, and asthma in such areas exceed national averages. **INFOGRAPHIC 4**

INFOGRAPHIC 4 | THE TRADE-OFFS OF OIL

Oil is a versatile fossil fuel, providing fuels and industrial products, but many problems are associated with its acquisition and use.

ZHENGSHENG/Moment/Getty Images

ADVANTAGES	DISADVANTAGES
• Energy-rich fuel • Provides variety of liquid fuels that meet needs for vehicles • No ash produced when burned (no disposal issues as with coal) • Can transport via pipeline or vehicle (rail, truck, or ship) • Raw materials for a wide variety of industrial products	• Nonrenewable, finite resource • Geopolitical tensions from unequal distribution of reserves • Air pollution, including greenhouse gas emissions (extraction, processing, and burning) • Water pollution (extraction, processing, and burning) • Habitat destruction (extraction and spills) • Biodiversity loss (exploration, extraction, and spills) • Dangerous to ship • Occupational and community hazard

 What could be done to decrease the occupational and community hazards associated with working in the oil industry or living near a refinery or oil extraction operation?

5 THE TRADE-OFFS OF NATURAL GAS

Key Concept 5: Natural gas is used to produce industrial products and is considered the cleanest burning fossil fuel, but like oil, every step of acquisition and use damages the environment and impacts human health.

Natural gas is an important alternative to oil for some fuel purposes, such as generating electricity and heat. It is made up of shorter-chain hydrocarbons (predominately methane, CH_4) and tends to have fewer impurities than oil, so when burned, fewer toxic pollutants are released. (Both oil and natural gas are cleaner than coal—the "dirtiest" fossil fuel to burn.) And since natural gas has fewer impurities, it burns more efficiently and has a lower **carbon footprint**, releasing roughly 44% and 29% less CO_2 than coal or oil, respectively, per unit of heat energy. For this reason, natural gas is seen as the best fossil fuel to serve as a "bridge fuel"—one that we can use while we transition to more sustainable options. In addition, natural gas, like oil, is also the starting material for a wide variety of industrial products such as fertilizers, paints, plastics, and pharmaceuticals.

carbon footprint The amount of carbon released to the atmosphere by a person, company, nation, or activity.

However, like oil, exploration for and extraction of natural gas are environmentally damaging. And because natural gas is highly flammable, it is dangerous to handle and ship. Shipping is accomplished with pipelines that are expensive to build and maintain. Methane leaks—from pipelines, from gas and oil wells—also significantly contribute to climate change; in the first 20 years after release, methane is almost 90 times more potent a greenhouse gas than CO_2. (After that, methane's impact relative to CO_2 decreases because CO_2 stays in the atmosphere longer.) And of course, though methane is produced today through some natural processes (such as bacterial decomposition of trash in landfills), the natural gas found in fossil fuel deposits is a nonrenewable resource with a limited and finite supply. **INFOGRAPHIC 5**

INFOGRAPHIC 5 | **THE TRADE-OFFS OF NATURAL GAS**

Like oil, natural gas is a versatile fossil fuel, providing fuels and industrial products, but many problems are associated with its acquisition and use.

Jim West/Alamy

ADVANTAGES	DISADVANTAGES
• Energy-rich fuel • Versatile fuel: generate electricity, heating and cooking, vehicle fuel • Lowest air pollution and greenhouse gas emissions of the fossil fuels • No ash produced when burned (no disposal issues as with coal) • Raw material for a wide variety of industrial products	• Nonrenewable, finite resource • Methane is a potent greenhouse gas; produces CO_2 emissions when burned • Water pollution (extraction and burning) • Habitat destruction (extraction and spills) • Biodiversity loss (exploration, extraction, and spills) • Hazardous chemical • Difficult to ship

 Why is natural gas, but not oil, often touted as a "bridge fuel"—something to use while we transition to more sustainable options?

6 UNCONVENTIONAL OIL AND NATURAL GAS

Key Concept 6: Shale oil and natural gas deposits that can't be accessed with conventional wells can be retrieved by fracking; strip mining is used to extract tar sands oil.

As conventional reserves become depleted, we have turned more and more to unconventional sources of oil and natural gas such as the Bakken and Marcellus Shale formations. As mentioned earlier, these deposits were not previously accessible because the oil or natural gas, though abundant, is widely dispersed within the rock formations where it is found and cannot be retrieved using conventional wells. But those reserves can be tapped with **fracking (hydraulic fracturing).**

Fracking involves drilling thousands of feet straight down, deep into ancient rock beds and then turning the drilling apparatus to drill horizontally into the gas- or oil-bearing rock formation. Several horizontal wells are drilled, in all directions, and each may reach 1.5 kilometers (1 mile) or more in length. As drilling progresses, steel pipes are inserted into the wellbores and encased in several inches of concrete. Once all the pipes

and casings are in place, explosive charges are detonated in the horizontal sections of the well to perforate the pipe casing and to create openings, or "fractures," in the rock through which oil or natural gas can flow. Next, between 15 and 20 million liters (4 to 5 million gallons) of water containing a mix of sand and chemicals (some of them toxic) are injected into the well under high pressure. The sand enters the small fractures, holding them open, while the water helps flush out the oil or natural gas in the deposit. Other chemical additives serve as lubricants, corrosion inhibitors, or disinfectants. The water mixture returns to the surface, followed by the oil and/or gas. The concrete casing is thickest close to the surface and in the groundwater region to prevent the chemicals from leaking into the surrounding soil and water. **INFOGRAPHIC 6**

fracking (hydraulic fracturing) The extraction of oil or natural gas from dense rock formations by creating factures in the rock and then flushing out the oil/gas with pressurized fluid.

INFOGRAPHIC 6 **HOW IT WORKS: FRACKING (HYDRAULIC FRACTURING)** ANIMATED INFOGRAPHIC

Unconventional deposits of natural gas and oil can be accessed via fracking. Concerns about surface and groundwater contamination make this a contentious resource in populated areas.

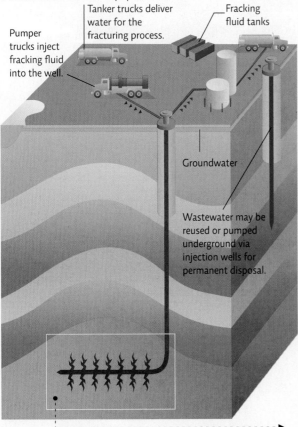

Tanker trucks deliver water for the fracturing process.

Fracking fluid tanks

Pumper trucks inject fracking fluid into the well.

Groundwater

Wastewater may be reused or pumped underground via injection wells for permanent disposal.

? Suppose you are a homeowner. An oil company approaches you, offering to lease your land to install a fracking well. What questions would you have for the company as you consider what to do?

1. For deep deposits, a well is drilled down to gas-bearing rock and then extended horizontally.

2. Holes are blasted into the rock using explosive charges, to create large fractures.

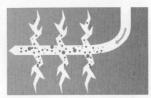

3. A slurry of sand, water, and chemicals is pumped into the fractures, to further crack open the rock, hold the fractures open, and make the gas more readily extractable.

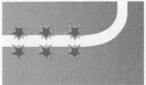

4. Natural gas can now flow through the more permeable, fractured rock and be extracted.

Currently, the United States leads the world in fracking, recovering enough natural gas and oil to change the global market for these fossil fuels. The increased production of natural gas caused prices to fall, which made it a more attractive fuel for U.S. electricity production, significantly reducing the amount of coal used in this country. Global oil prices also fell as oil from the fracking oil fields of North Dakota and Texas entered the market.

While productive, the oil and gas shale deposits aren't the only types of unconventional reserves; there's also crude *bitumen*—a heavy black oil that is often trapped in sticky, dense conglomerations of sand or clay known as **tar sands** (also known as **oil sands**). Alberta, Canada, is home to the second-largest tar sand reserve; with 170 billion barrels of economically retrievable oil, it is the third-largest proven oil reserve in the world. In 2014, Alberta produced 2.3 million barrels of synthetic crude oil per day from the tar sands—much of it shipped to the United States. Venezuela was recently recognized as having the largest tar sands deposits in the world. With 300 billion barrels of recoverable oil, its oil reserves surpass that of Saudi Arabia.

Oil is extracted from Canadian tar sands by strip mining that begins with the removal of the boreal forest above the deposit. The sticky soil is dug out and washed with copious amounts of water to separate the oil from the soil. Additional processing steps are required before the bitumen can be shipped. Deep deposits in the tar sands can be accessed with a method that injects steam into the deposit to melt the hardened bitumen and force it out with the rising steam, an energy-intensive method.

tar sands (oil sands) Sand or clay formations that contain a heavy-density crude oil (crude bitumen); extracted by surface mining.

Tar sands mining is the most destructive and energy-intensive way to acquire oil. Huge swaths of Canadian boreal forest have been destroyed and nearby bodies of water polluted by this process.

Peter Essick/Aurora Photos

7 THE TRADE-OFFS OF UNCONVENTIONAL FOSSIL FUELS

Key Concept 7: Extracting unconventional oil and natural gas supplies produces positive economic benefits, but fracking and tar sands mining have serious negative environmental and societal impacts.

The upsides of the Bakken boom are not difficult to see or imagine. For one thing, such a mammoth supply of domestic oil means, for the time being at least, less reliance on foreign oil. For another, the boom has stimulated a once-declining regional economy. And more natural gas from shale formations also means less coal is being burned, giving us cleaner air.

But there are significant downsides. Because they are more expensive to extract than conventional sources, it only pays to pursue unconventional fossil fuels when prices are high. When the fracking boom took off, oil was selling at more than $100 a barrel, but soon thereafter, prices started to drop—in large part due to increased oil supplies. U.S. oil production had climbed thanks to fracking, but OPEC (Organization of Petroleum Exporting Countries) also kept production high, presumably in an effort to drive down the price of oil and force fracking operations out of business. The break-even price for OPEC oil is around $30 a barrel; for fracking operations in 2015, it was twice that. With oil prices well below $60 per barrel (as low as $29.18 in January of 2016), many fracking wells were shut down and jobs were lost—the Bakken oil boom was beginning to falter, though many refused to call it a bust.

New technologies deployed in 2016 have increased productivity per well, making fracking more cost competitive, but this comes at the expense of jobs— automated systems are improving efficiency, decreasing the cost of fracking, and replacing workers.

The environmental costs of fracking are equally significant. The presence of methane in local wells near shale gas fracking sites is a concern. A 2013 study by Duke University researchers found that methane contamination of drinking water was highest in areas closest to fracking wells. (See the *Science Literacy* activity at the end of this module for more on this study.) The most likely source of the contamination is leaking pipes that deliver the gas to the surface or from improperly handled fracking water. Improved drilling methods and stricter fracking water storage requirements are decreasing that risk, industry experts say.

The water footprint is also a concern. The volume of water required to frack a well is huge—around 11 million liters (3 million gallons) for a Bakken well—a significant

amount, especially in water-poor areas. Production of toxic wastewater is another problem that must be dealt with. During fracking, much of the water injected into the ground resurfaces with the oil or natural gas, bringing the injected chemicals and contaminants picked up from the rock, such as salt and naturally occurring radioactive materials, along with it. Studies have demonstrated that this wastewater is toxic to plants, and though land application is no longer allowed, millions of liters are spilled each year.

Fred Mayer of Candor, New York, uses a charcoal grill lighter to ignite water running from his kitchen faucet. His water has been contaminated with methane ever since a fracking operation began nearby.

AP Photo/The Post Standard, Mike Greenlar

To deal with this wastewater, operators typically pump it back underground, into nearby wells drilled for that purpose (called injection wells). But these injection wells have their own environmental issues, most notably the triggering of earthquakes. For example, the Oklahoma City area experienced 31 earthquakes (2.0 magnitude or greater) in a 1-week period in July 2014. To put this into perspective, whereas the state experienced only 6 earthquakes between 2000 and 2008, it experienced 6,345 earthquakes between 2010 and 2013. Research published by Katie Keranen of Cornell University linked this "earthquake swarm" to nearby wastewater injection wells.

Another problem is that the amount of oil being fracked has outpaced the existing capacity to transport it. Without a pipeline to rely on, oil companies have resorted to using railways. But the sheer volume of oil to be transported is bogging down the rail system nationwide, and shipping tight oil this way is a hazardous undertaking. Tight oil contains methane (natural gas) dispersed throughout, making it more volatile and explosive than crude oil; it is similar to jet fuel in its combustibility. In the summer of 2013, the brakes failed on a parked, unmanned train carrying Bakken oil. The runaway train sped downhill, derailing in Lac-Mégantic, a tiny lakeside village in Quebec; the resulting crash caused an explosion when the highly combustible fuel ignited, leveling more than 40 buildings and claiming 47 lives.

Fracking operations also present a disadvantage not commonly seen with conventional drilling—the ability to drill into formations under private property without the permission of the landowner. In the United States, many landowners do not own the mineral rights to their land; mineral rights are often owned by corporations that have the right to set up drilling rigs on or near a landowner's property and frack for oil or natural gas without landowner permission. Even if a landowner does own the mineral rights to his or her property and declines to let a fracking operation set up, if the majority of neighboring

MATHIEU BELANGER/REUTERS/Newscom

Transporting any type of oil is dangerous but tight oil tends to be more volatile and explosive than conventional crude. In 2013, a runaway train carrying tight oil derailed in Lac-Mégantic, Quebec, destroying much of the town and killing 47 people.

landowners sign on, the reluctant landowner can be overruled.

Other unconventional sources of energy also have unique drawbacks. Tar sands extraction is the most energy- and water-intensive way to extract oil and produces 20% more greenhouse gas emissions than conventional oil and gas production. David Schindler, an ecologist at the University of Alberta in Canada, has spent years studying the effects of tar sands extraction on the water quality of the Athabasca River, which cuts through the heart of one of Alberta's biggest tar deposits and mining sites. His research showed that concentrations of toxic compounds were higher in the river's tributaries located downstream of tar sands mining than they were upstream.

Processing tar sands oil is also more energy intensive than conventional crude oil and produces more hazardous wastes—huge lakes of acidic and toxic wastewater that oil companies store at the mining sites as a by-product of the refining process are so large that they can be seen on satellite images from space. The thick crude bitumen is also difficult to ship; it will flow only in heated pipelines. The construction of the Keystone XL pipeline that will transport tar sands oil south to the Gulf Coast refineries of Texas was strongly opposed by many environmental and citizen groups. It was approved by the U.S. Congress in 2015 but vetoed by President Obama. The pipeline (and another controversial pipeline, the Dakota pipeline) was resurrected and approved by President Trump in 2017. **INFOGRAPHIC 7**

INFOGRAPHIC 7 · THE TRADE-OFFS OF UNCONVENTIONAL FOSSIL FUELS

Unconventional sources of fossil fuels will greatly increase supplies, but because they are harder to extract, they come with significant environmental and economic trade-offs.

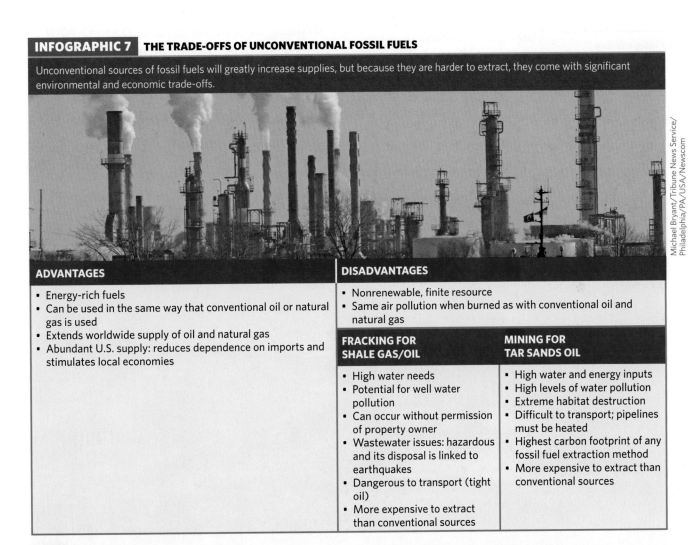

Michael Bryant/Tribune News Service/ Philadelphia/PA/USA/Newscom

ADVANTAGES

- Energy-rich fuels
- Can be used in the same way that conventional oil or natural gas is used
- Extends worldwide supply of oil and natural gas
- Abundant U.S. supply: reduces dependence on imports and stimulates local economies

DISADVANTAGES

- Nonrenewable, finite resource
- Same air pollution when burned as with conventional oil and natural gas

FRACKING FOR SHALE GAS/OIL	MINING FOR TAR SANDS OIL
• High water needs • Potential for well water pollution • Can occur without permission of property owner • Wastewater issues: hazardous and its disposal is linked to earthquakes • Dangerous to transport (tight oil) • More expensive to extract than conventional sources	• High water and energy inputs • High levels of water pollution • Extreme habitat destruction • Difficult to transport; pipelines must be heated • Highest carbon footprint of any fossil fuel extraction method • More expensive to extract than conventional sources

? Do the advantages of fracking or tar sands mining outweigh their disadvantages? Explain your reasoning.

8 FOSSIL FUELS AND THE FUTURE OF ENERGY

Key Concept 8: Achieving energy independence or security will require an eventual transition away from fossil fuels; conservation efforts and diversification of energy sources can help in the pursuit of these goals.

It's tempting to think of the oil made available by fracking technology as so plentiful that we can stop worrying about running out of oil for the next several lifetimes and be well on our way to achieving **energy independence**. The numbers, after all, are mind boggling: We have more than 700 billion barrels of tight oil in Bakken, Eagle Ford, and other known reserves—and that's just in the United States. But just a tiny fraction of the total oil held in these unconventional deposits is recoverable—somewhere between 1% and 2% in the Bakken and Eagle Ford shales. "At the high end of the estimates, predicted production from Bakken and Eagle Ford together amounts to perhaps a two-year oil supply for the United States at 2011 consumption rates," says Raymond Pierrehumbert, a professor and geophysicist at the University of Chicago. "That's significant, but not a game changer."

energy independence
Meeting all of one's energy needs without importing any energy.

energy security Having access to enough reliable and affordable energy sources to meet one's needs.

Because fracking wells can access only a small part of the oil deposit that surrounds them, many, many wells are needed to tap an entire deposit. This diminishes the EROEI because creating each new well requires an expenditure of energy. In addition, the productivity of any given well is likely to drop off rapidly after the first couple years and then trail off into nothing, slowly, over decades. Tight oil is headed for a Red Queen's Race," says Pierrehumbert. "You have to keep drilling and drilling and drilling just to keep your production in the same place."

> **"**You have to keep drilling and drilling and drilling just to keep your production in the same place.**"**
> —Raymond T. Pierrehumbert

There is no denying that our economy and lifestyle are dependent on access to affordable energy, or **energy security**. Yet there are many reasons that our energy

Ken Cedeno/Getty Images

A fracking operation along U.S. Highway 85 just south of Watford City, North Dakota.

supplies might become unreliable or unaffordable. These include dwindling supplies, increasing demand, dependence on energy imports from politically unstable countries or those that might stop exporting energy resources to us for political reasons, competition from other countries, and a cartel or monopoly increasing prices or decreasing oil supplies. **INFOGRAPHIC 8**

To be sure, how to best meet our future energy needs is a *wicked problem*—we need an abundant, reliable source of energy to power our societies, but many of our energy choices come with significant problems. (See Module 1.1 for more on wicked problems.) Phasing out the use of fossil fuels would mean we would no longer need to worry about running out of a finite energy resource, and it would certainly cut down on air pollution, climate change, and myriad other environmental problems. A lot is at stake in this wicked problem—huge investments and even bigger profits, the health and well-being of human communities and ecosystems, and the need to keep society running smoothly without facing an energy shortfall. So how do we increase our energy security and reduce our dependence on fossil fuels so that we can avoid using dangerous extraction methods and polluting the planet

we call home? There are many possibilities, but two important strategies are diversification and conservation.

Every sustainable energy option at our disposal comes with advantages and disadvantages. No single energy source can replace fossil fuels, but together, the wide variety of energy sources at our disposal can (see Chapter 11). We can focus efforts on acquiring a variety of domestic sources of energy (including alternative fuels such as biofuels) and importing energy from multiple suppliers. At the same time, we can reduce energy needs through increased conservation and energy efficiency. Iceland, for instance, is increasing its energy security by focusing on conservation and energy independence, such that it will meet all of its energy needs without imports by the year 2050. And in 2016, renewables met more than half of Sweden's energy needs; the country vows to be among the first fossil-free nations in the world.

In April 2014, the Bakken oil fields of western North Dakota reached a milestone: They yielded up their billionth barrel of crude oil. With oil prices on the rise once again and production costs going down, the

---

INFOGRAPHIC 8 | **ENERGY INDEPENDENCE AND SECURITY**

Energy security is achieved when a nation has an affordable and reliable (uninterrupted) form of energy. If energy needs are met with national sources, the country has energy independence.

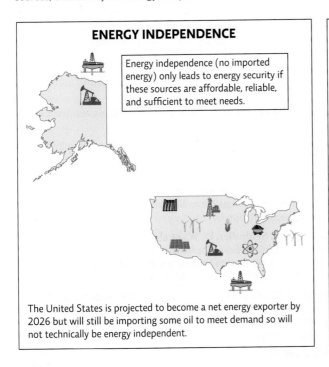

ENERGY INDEPENDENCE

Energy independence (no imported energy) only leads to energy security if these sources are affordable, reliable, and sufficient to meet needs.

The United States is projected to become a net energy exporter by 2026 but will still be importing some oil to meet demand so will not technically be energy independent.

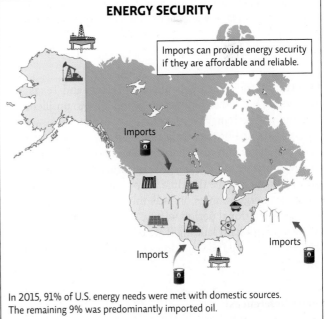

ENERGY SECURITY

Imports can provide energy security if they are affordable and reliable.

In 2015, 91% of U.S. energy needs were met with domestic sources. The remaining 9% was predominantly imported oil.

 How could the United States increase its chance of becoming energy independent if it cannot produce enough oil to meet its needs?

The need for housing during the fracking boom led to a building boom, but then the oil bust came. In 2016, this apartment complex in Williston, North Dakota, offered several months of free rent, trying to entice workers who remained. If production picks up, the units may once again be in hot demand.

downturn in the shale fields of North Dakota and Texas is reversing as production increases and jobs go unfilled. This too will be temporary—perhaps a year or two, perhaps a decade or two. Only time will tell how many more barrels remain to be harvested and how long Bakken oil will continue to flow. In the meantime, the greatest gift of Bakken may be the time it buys us to develop other domestic energy sources. "It will only do us good if we use this transitional period wisely," says Pierrehumbert. "We're in for a hard landing if we don't use our current prosperity to pave the way for a secure energy and climate future."

Select References:

Energy Information Agency. (2017) *Annual Energy Outlook 2017, with Projections to 2050.* Washington, DC: U.S. Department of Energy. (Available at www.eia.gov/outlooks/aeo/)

Hall, C. A., et al. (2014). EROI of different fuels and the implications for society. *Energy Policy, 64,* 141–152.

Hughes, J. D. (2013). A reality check on the shale revolution. *Nature, 494*(7437), 307–308.

Jackson, R. B., et al. (2013). Increased stray gas abundance in a subset of drinking water wells near Marcellus shale gas extraction. *Proceedings of the National Academy of Sciences, 110*(28), 11250–11255.

Kelly, E. N., Schindler, D. W., et al. (2010). Oil sands development contributes elements toxic at low concentrations to the Athabasca River and its tributaries. *Proceedings of the National Academy of Sciences, 107*(37), 16178–16183.

Keranen, K. M., et al. (2014, July 25). Sharp increase in central Oklahoma seismicity since 2008 induced by massive wastewater injection. *Science, 345*(6195), 448–451. doi:10.1126/science.1255802.

Malakoff, D. (2014). The gas surge. *Science, 344*(6191), 1464–1467.

INTERACTIVE MAP   **CONSEQUENCES OF OIL AND NATURAL GAS EXTRACTION**   ANIMATED INFOGRAPHIC

Oil and natural gas are valuable fossil fuels that are important to modern society but their acquisition imposes a significant environmental burden. Here are some examples of problems caused by the acquisition of these fossil fuels.

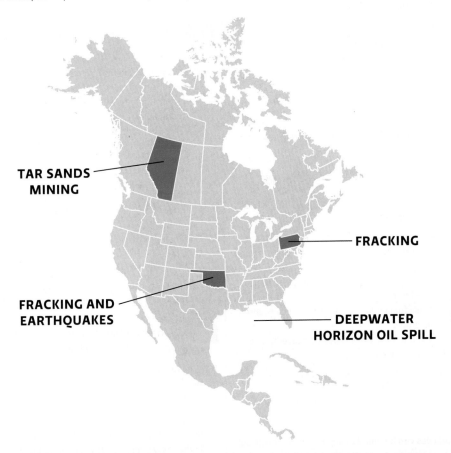

TAR SANDS MINING

FRACKING

FRACKING AND EARTHQUAKES

DEEPWATER HORIZON OIL SPILL

 BRING IT HOME

PERSONAL CHOICES THAT HELP

Environmental and health concerns that surround our acquisition and use of fossil fuels continue to grow as we dig deeper and use more extreme methods to obtain oil and natural gas, highlighting our dependence on nonrenewable energy resources. By decreasing our oil and natural gas use, we can reduce the pressure on oil companies to pursue sources of oil that have a greater potential for environmental damage such as deepwater drilling or oil and natural gas deposits like those in the tar sands and shales.

Individual Steps

• Minimize your fuel use when driving by planning ahead to condense shopping trips and errands and reduce total miles driven and by parking as far as you can from your entrance and getting some exercise instead of wasting gas. Try not driving: See where

you can safely walk, bike, or use public transport.

• Reduce your use of disposable plastics like water bottles and single-serve food containers and always recycle plastics when possible.

• Turn down your thermostat in the winter to reduce your energy use.

Group Action

• Organize a carpool system in your community to reduce the number of single-passenger car trips.

• Organize a screening of a documentary on energy such as *Tipping Point: The Age of the Oil and Sands*, *Gasland* or *Gasland 2*, or *FrackNation*. Discuss the positions presented and evaluate the evidence given in support of claims that are made. Do you detect any bias?

Policy Change

• Work with your school's administration to encourage public transportation and increased bike usage on your campus.

• If you live in a state where fracking occurs, research fracking regulations in your state and voice your opinion about existing or proposed regulations.

ENVIRONMENTAL LITERACY UNDERSTANDING THE ISSUE

1 How are fossil fuels formed, and why are they considered nonrenewable resources?

1. The formation of fossil fuels requires that buried organic matter be exposed to _____ over a long period of time.
   a. high heat
   b. high temperatures
   c. high oxygen levels
   d. A and B
   e. A, B, and C

2. If oil and natural gas are formed from once-living organisms, why are they considered to be nonrenewable resources?

2 Where are oil and natural gas reserves found?

3. True or False: Underground deposits that have oil or natural gas are known as *conventional reserves*.

4. What is the correct definition of fossil fuel *proven reserves*?
   a. Humanmade storage areas for emergency supplies of fossil fuels
   b. Areas where there are known stores of untapped fossil fuels held in reserve for the future
   c. Known sources of fossil fuels that are both economically and technologically recoverable
   d. Sources of fossil fuels that have not yet been discovered but that geologists have predicted to exist

5. Why is oil the cause of geopolitical tensions around the world?

3 How are conventional oil and natural gas extracted?

6. True or False: Natural gas can be found in deposits with or without oil.

7. At best, about how much oil can be ultimately extracted from a conventional well using primary, secondary, and tertiary production techniques?
   a. 20%
   b. 55%
   c. 75%
   d. 99%

8. Describe the process of conventional oil extraction, including primary production, secondary production, and tertiary production.

4 What are the trade-offs of acquiring and using conventional oil?

9. True or False: Catastrophic oil spills like the *Deepwater Horizon* accident are rare but environmentally devastating.

10. Compare the environmental problems produced from extracting oil to those produced from burning oil.

5 What are the trade-offs of acquiring and using conventional natural gas?

11. True or False: Burning any fossil fuel generates air pollution, but natural gas is regarded as the fossil fuel that releases the most air pollution when burned.

12. Compare natural gas to oil in terms of advantages and disadvantages.

6 What unconventional sources of oil and natural gas exist, and how are they extracted?

13. Why is the United States so interested in extracting tight oil from shale deposits?
   a. It has abundant supplies that can be extracted with current technology.
   b. These deposits are easier to extract than conventional oil supplies.
   c. Shipping tight oil is easier and safer than shipping conventional crude oil.
   d. Tight oil deposits are not contaminated with methane.

14. Which unconventional oil source currently being commercially produced has the lowest EROEI and the highest carbon footprint?
   a. Tight oil
   b. Tar sands oil
   c. Crude oil
   d. Kerogen shale

15. Explain the process of fracking. Why is fracking used for some oil and natural gas reserves and not others?

7 What are the trade-offs of pursuing unconventional oil and natural gas sources?

16. True or False: Fracking uses a lot of water, but the wastewater produced is either recycled and used again or purified and pumped back into aquifers.

17. Some homeowners suspect that methane-contaminated water is linked to nearby fracking wells. If this is true, which of the following is the likely source of this contamination?
   a. Leaky pipes near groundwater supplies
   b. The initial drilling of the well
   c. Installation of the well pipes
   d. Detonation of explosive charges deep in the well

18. Explain the potential risks associated with shipping tight oil.

8 What obstacles stand in the way of the United States (or any nation) achieving energy independence or security?

19. True or False: Its vast deposits of shale oil and gas will allow the United States to meet all its energy needs with domestic sources into the foreseeable future.

20. Which of the following is an example of the United States achieving energy independence?
   a. Harvesting coal from Wyoming and Appalachian coal beds
   b. Depending only on domestically harvested energy sources
   c. Importing energy resources only from friendly neighbors like Canada
   d. Focusing on the renewable energy source that works best in a given locale

21. What role do conservation and diversification of energy sources play in achieving energy security or independence?

22. Distinguish between the actions a country might take when pursuing energy security versus energy independence.

SCIENCE LITERACY WORKING WITH DATA

Methane (CH_4) can normally occur in upper levels of the soil because it is produced by bacteria and can accumulate to high levels over time in some areas. Ethane (C_2H_6), however, is only created under high temperature and pressure, such as might occur deep underground at the site of shale gas deposits. Robert Jackson and colleagues suspected that methane contamination in water from wells in northeastern Pennsylvania might be linked to nearby fracking gas wells. For their study, they asked the question "Does being close to fracking gas wells increase the likelihood that the water will contain methane or ethane contamination?" They measured the amounts of methane and ethane in 141 water wells close to and far away from fracking wells. Look at the graphs below and answer the following questions.

METHANE CONCENTRATIONS IN WATER WELLS

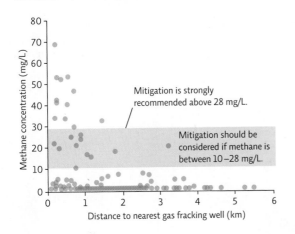

ETHANE CONCENTRATIONS IN WATER WELLS

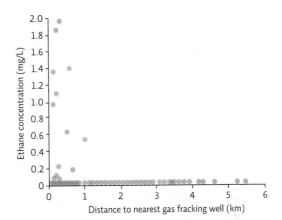

Interpretation

1. What was the hypothesis that these researchers tested in this experiment?

2. For each chemical (methane and ethane), identify the relationship between its level in well water and the water well's proximity to fracking wells. Do these relationships represent causation or correlation? Explain.

3. How many water wells had levels of methane that fell within the range where mitigation should be considered? How many fell within the range where mitigation is strongly recommended?

Advance Your Thinking

4. Do all wells that are very close to a fracking well (within 0.5 kilometer) show evidence of methane or ethane contamination? (Are there any close-by wells with no contamination?) Does this negate the conclusion that methane and ethane contamination are due to fracking in the area?

5. Why might methane be high in water even in areas that are not close to fracking wells?

6. Why did researchers need to measure ethane in water wells to strengthen their conclusion that the water contamination is linked to fracking activity?

INFORMATION LITERACY EVALUATING INFORMATION

The United States produces a lot of oil but still imports a lot from Canada, our biggest source of imported oil. For the most part, this is tar sands oil from Alberta. Transport of the mining products is accomplished through the transcontinental Keystone pipeline that currently runs from Alberta into the United States, but it does not continue on all the way to refineries on the Gulf Coast. Oil companies have lobbied for years for permission to expand the pipeline by building the Keystone XL to transport oil from Canada to Texas, a request that was denied by President Barack Obama but approved by his successor, President Donald Trump.

Search for information about the Keystone XL pipeline and other proposed ways to ship tar sands crude oil. Evaluate the articles you read and answer the following questions:

1. Visit at least three websites that advocate for the construction of the Keystone XL pipeline and three websites that oppose it. For each website, answer the following questions about the organization or authors responsible for the information you read:
   a. Who runs the website or wrote the article? Do the credentials of this individual or group make the person or group reliable or unreliable? Explain.
   b. Does this individual or group have a clear and transparent agenda?
   c. Do you detect any bias or logical fallacies? If so, identify them.
   d. Is this a reliable information source? Explain.

2. Briefly summarize what you believe are the most important pros and cons of the Keystone XL pipeline. Based on your analysis of the information you gathered, if you were President, would you have approved the Keystone XL pipeline or would you have rejected it? Justify your conclusions by explaining which arguments support your position and why they are valid arguments, as well as explaining why the counterarguments were insufficient to support the alternate conclusion.

 Additional study questions are available at SaplingLearning.com.

AIR QUALITY AND CLIMATE CHANGE

CHAPTER 10

Air pollution can come from a variety of sources. Addressing air pollution will benefit our health and the environment.

Module 10.1: Air Pollution

A look at the types, sources, and consequences of anthropogenic air pollution and efforts to combat it

Module 10.2: Climate Change

An analysis of the causes and consequences of climate change and a look at global efforts to address it

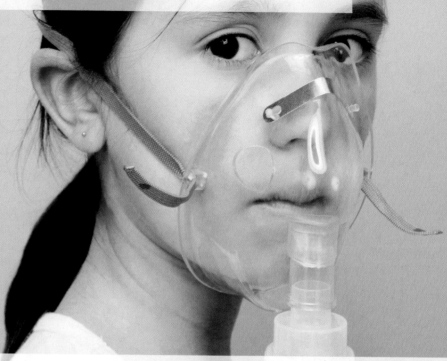

THE YOUNGEST SCIENTISTS

Kids on the frontlines of asthma research

With more than 6 million children affected, asthma is the most common chronic illness among children in the United States. ballyscanlon/DigitalVision/Getty Images

After reading this chapter and studying the KEY CONCEPTS and INFOGRAPHICS, you should be able to answer these GUIDING QUESTIONS

CORE MESSAGE

Air quality issues span the globe and have serious health consequences for humans and other organisms; they also take an economic toll. Though there are natural sources of air pollution, most is caused by human actions. Policies that restrict air pollution and technologies that reduce it at the source can help us address air pollution problems.

1. What is air pollution, and what is its global impact?

2. What are the main types and sources of outdoor air pollution?

3. What are the health, economic, social, and ecological consequences of air pollution?

4. What are the causes and consequences of acid deposition?

5. What air pollutants are regulated by the Clean Air Act?

6. What are the main sources of indoor air pollution, and what can be done to reduce it?

7. How can air pollution be reduced, and what are the trade-offs of reducing it?

For 10-day stretches during 2003 and 2004, 45 asthmatic kids from smoggy Los Angeles County, some as young as 9 years old, carried more than books in their backpacks. The kids, from two regions of the county, wore backpacks containing small monitors that sampled the air around them continuously as they went about their daily lives—going to class, playing with friends, having dinner with their families.

Those personal air monitors are far more accurate than measurements taken at local monitoring stations, explains Ralph Delfino, an epidemiologist at the University of California, Irvine, who recruited the students to help collect the data. "A monitoring station can be many miles from where the subject lives, where they go to school, etc.—so that measurement may not represent their actual exposure very well," he says. The air monitors used in Delfino's study detected levels of harmful pollutants in the air surrounding each individual child.

The children selected for the 10-day experiment were all currently being treated for mild to moderate asthma, and the area in which they all lived had significant vehicle air pollution. The monitors they wore measured the quantity of small particles and nitrogen dioxide (NO_2), both of which are commonly present in vehicle emissions and are known to irritate lung tissue. Exposure to these pollutants can trigger asthma symptoms such as wheezing, coughing, or shortness of breath in sensitive individuals. In addition, ten times a day, the children exhaled into a special bag that assessed their breath for nitric oxide (NO), a chemical marker of airway inflammation—a telltale symptom of asthma.

◉ **WHERE IS LOS ANGELES, CALIFORNIA?**

When the 10 days were over, Delfino compared the types of pollutants to which the kids were exposed with their nitric oxide levels at similar points in time. The goal was to paint a picture of the types and amounts of pollutants that exacerbate asthma.

1 GLOBAL AIR POLLUTION

Key Concept 1: Air pollution is a serious problem that causes millions of deaths each year, worldwide.

Researchers believe that **air pollution**—contaminants from natural sources or human activities that cause health or environmental problems—may play a key role in the recent asthma spike. In addition, air pollution has been linked to other serious health issues such as cancer, respiratory infection, and cardiovascular diseases—and it harms not only humans but also plants, other animals, and even buildings, bridges, and statues.

As far back as the 1930's, scientists recognized a link between outdoor air pollution and human illness. In 1930, for instance, 63 people died and 1,000 were sickened in Belgium when a temperature inversion—a situation that

occurs when the temperature is higher in upper regions of the atmosphere than in the lower, causing pollutants to become trapped near Earth's surface—led to a sudden spike in lower atmospheric sulfur levels. And 1952 was the year of the famous Great Smog in London, England, when pollutants trapped in the lower atmosphere killed 4,000 people. **INFOGRAPHIC 1**

Asthma is a respiratory ailment marked by inflammation and constriction of the narrow airways of the lungs. Delfino's work is important, in part because asthma

air pollution Any material added to the atmosphere (naturally or by humans) that harms living organisms, affects the climate, or impacts structures.

INFOGRAPHIC 1   **AIR POLLUTION IS A WORLDWIDE PROBLEM**

The World Health Organization (WHO) recognizes air pollution as a major threat to human health, causing roughly 5.5 million premature deaths worldwide (about 18% of the total). The vast majority of these deaths occur in low- and middle-income countries.

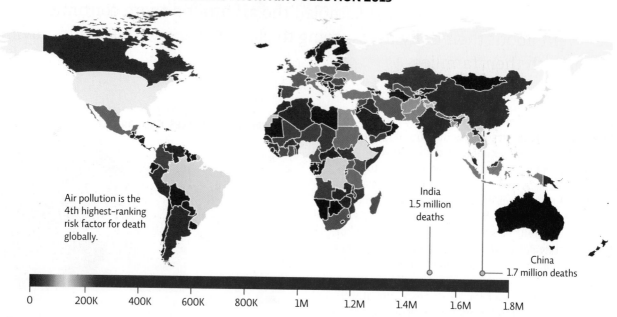

DEATHS FROM AIR POLLUTION 2013

Air pollution is the 4th highest-ranking risk factor for death globally.

India 1.5 million deaths

China 1.7 million deaths

| 0 | 200K | 400K | 600K | 800K | 1M | 1.2M | 1.4M | 1.6M | 1.8M |

 China and India accounted for 55% of all air pollution deaths in 2013. Why do you suppose air pollution deaths are so high in these two countries?

is one of the most common chronic childhood diseases in developed nations, and it's a major cause of childhood disability. In the United States, for example, the prevalence of childhood asthma more than doubled from 1980 to the mid-1990's, and though it has leveled off in recent years, with 6.2 million children (8.4%) and 18.4 million adults (7.6%) diagnosed, it remains at historically high levels. (Because there are more adults than children, overall more adults have

asthma, but a larger percentage of children have the disease.) In fact, in the United States, asthma is the leading cause of school absences and, hence, lost revenue for public schools, whose federal funding is based on attendance.

Developing nations are also seeing a rise in asthma, especially in urban centers. In all areas, asthma rates are likely underdiagnosed.

2   OUTDOOR AIR POLLUTION: TYPES AND SOURCES

Key Concept 2a: Outdoor air pollution is caused by natural and anthropogenic sources and includes emissions that are directly harmful (primary pollutants) and those that are converted to harmful forms (secondary pollutants).

Key Concept 2b: Air pollution can be a transboundary pollutant, causing state or national conflicts between regions where the pollution originates and regions where it falls.

In many urban areas and throughout developed nations today, much of the air pollution comes from vehicle exhaust and industry and power plant emissions. Industrial pollution is usually released as *point source pollution*—pollution that comes from an identifiable smokestack or other outlet. Hypothetically, point source pollution is easier to monitor and regulate than is *nonpoint source pollution*—pollution from dispersed or mobile sources like

vehicles and lawn mowers. (See Module 6.2 for information about point source and nonpoint source water pollution.)

Outdoor air pollution includes chemicals and small particles in the atmosphere that can either be natural in origin—arising from natural events like sandstorms, volcanic eruptions, or wildfires—or come from humans—such as pollution released from factories and vehicles

Controlled burns of agriculture fields, like this one of an asparagus field in California, help clear land for more planting but release particulate matter (small particles) into the air, contributing to respiratory distress in sensitive individuals.

during the combustion of fossil fuels or from burning biomass (e.g., wood, crop waste, or garbage). Of these anthropogenic sources, **primary air pollutants** are pollutants released directly from both mobile sources (such as cars) and stationary sources (such as industrial plants). In addition, some primary air pollutants react with one another or with other chemicals in the air to form **secondary air pollutants**. For example, **ground-level ozone** forms when nitrogen oxides (NO and NO_2—together expressed as NO_x) released during fossil fuel combustion react with atmospheric oxygen in the presence of sunlight. **Smog**—a term that's a combination of the words smoke and fog—is hazy air pollution that contains a variety of primary and secondary pollutants.

Another serious form of air pollution is **particulate matter (PM)**—particles or droplets small enough to remain aloft in the air for extended periods of time. Particulates are a common component in smoke and soot and are released when just about anything is burned. Although all particulates reduce visibility, it is the smallest particles—those with a diameter less than 2.5 micrometers (μm), about 1/40 the diameter of a human hair—that aggravate asthma and other chronic lung diseases and increase the risk for death.

Agriculture, too, is a source of nonpoint source outdoor pollution. Toxic pesticides sprayed on crops can become airborne and drift as far as 30 kilometers (20 miles); confined animal feeding operations produce significant odor problems and particulate pollution; and cattle and rice paddies contribute to global warming by releasing the greenhouse gas methane.

Pollution can also move from ground level up into the stratosphere, a region of the upper atmosphere that contains the "ozone layer" that protects living things from some of the Sun's dangerous ultraviolet (UV) radiation. Air pollutants released by humans, such as chlorofluorocarbons (CFCs), can travel up into the stratosphere and destroy this protective ozone. (See Module 5.2 for more on stratospheric ozone depletion.) **INFOGRAPHIC 2**

Don't confuse ground-level ozone pollution with stratospheric ozone depletion. These are two very different problems, though they deal with the same molecule—O_3. Ozone in the stratosphere is a good thing, but ozone at ground level is a problem—breathing it in can directly damage the sensitive tissue of the lungs. Even plants are damaged by the corrosive action of ozone.

primary air pollutants Air pollutants released directly from both mobile sources (such as cars) and stationary sources (such as industrial and power plants).

secondary air pollutants Air pollutants formed when primary air pollutants react with one another or with other chemicals in the air.

ground-level ozone A secondary pollutant that forms when some of the pollutants released during fossil fuel combustion react with atmospheric oxygen in the presence of sunlight.

smog Hazy air pollution that contains a variety of pollutants, including sulfur dioxide, nitrogen oxides, tropospheric ozone, and particulates.

particulate matter (PM) Particles or droplets small enough to remain aloft in the air for long periods of time.

There are many sources of outdoor air pollution, both natural and anthropogenic. These sources release primary pollutants, some of which may be converted to different chemicals (secondary pollutants). Prevailing winds transport pollution that reaches the upper troposphere or stratosphere around the globe; no area is immune to air pollution. Agricultural and industrial pollutants have been found in Arctic and Antarctic air, delivered by these prevailing winds.

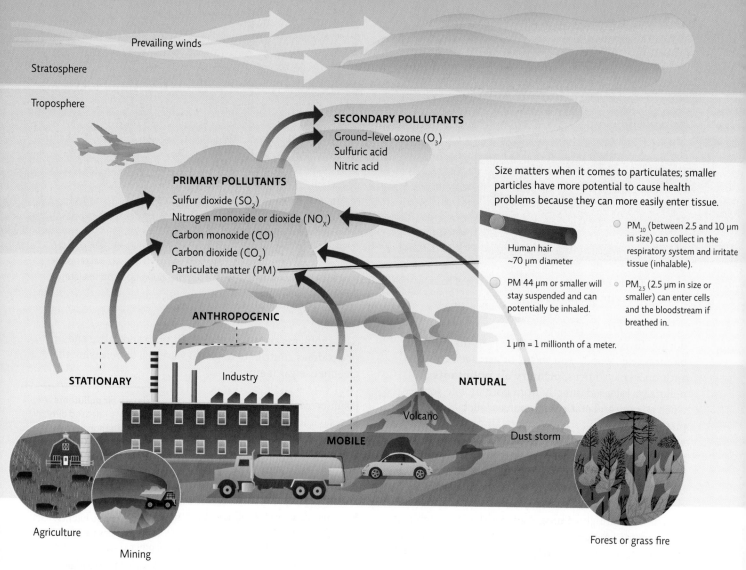

Prevailing winds

Stratosphere

Troposphere

SECONDARY POLLUTANTS
Ground-level ozone (O_3)
Sulfuric acid
Nitric acid

PRIMARY POLLUTANTS
Sulfur dioxide (SO_2)
Nitrogen monoxide or dioxide (NO_x)
Carbon monoxide (CO)
Carbon dioxide (CO_2)
Particulate matter (PM)

Size matters when it comes to particulates; smaller particles have more potential to cause health problems because they can more easily enter tissue.

Human hair ~70 μm diameter

PM_{10} (between 2.5 and 10 μm in size) can collect in the respiratory system and irritate tissue (inhalable).

PM 44 μm or smaller will stay suspended and can potentially be inhaled.

$PM_{2.5}$ (2.5 μm in size or smaller) can enter cells and the bloodstream if breathed in.

1 μm = 1 millionth of a meter.

ANTHROPOGENIC

STATIONARY

Industry

NATURAL

Volcano

Dust storm

MOBILE

Agriculture

Mining

Forest or grass fire

 What allows air pollution released at one location to affect locations far away?

In his study in Los Angeles County, Delfino found that particles contained in diesel exhaust were among the worst asthma culprits. In addition, particulate levels were higher in Riverside, one of two regions he tested; the researchers concluded that Riverside had more pollution because it was downwind of the main urban areas in Los Angeles. Pollution in Riverside is an example of **transboundary pollution**— pollution that originated in one area but traveled to another. Conflicts can arise between states or nations suffering from transboundary pollution because they do not have the political ability to regulate air pollution generated outside of their borders. Even the most isolated regions on Earth are vulnerable to the effects of air pollution because atmospheric and hydrologic circulation moves pollution around the globe. For example, agricultural pesticides and industrial chemicals have been found in remote Arctic and Antarctic regions—areas far from any farms or industries.

transboundary pollution
Pollution that is produced in one area but falls in a different state or nation.

3 CONSEQUENCES OF AIR POLLUTION

Key Concept 3a: Air pollution causes health problems, damages structures, reduces visibility, and contributes to stratospheric ozone depletion and climate change.

Key Concept 3b: Air pollution is often especially bad in minority and low-income areas, raising concerns of environmental justice.

Air pollution imposes a serious health and environmental burden. In 2015, the World Health Organization (WHO) estimated that 5.5 million people die prematurely each year from exposure to air pollution. Respiratory ailments are common; particulates from soot and smog damage respiratory tissue and increase susceptibility to infection, particularly in children because they breathe in more air for their size than adults do and because developing tissue is more vulnerable. And damage can start before birth—maternal exposure to air pollution is linked to premature births and low-birth weights.

> The World Health Organization (WHO) estimates that 5.5 million people die prematurely each year as a result of exposure to air pollution.

Lungs are particularly vulnerable to pollution because they get so much exposure (we breathe all the time), and the tissue itself is delicate. Irritants like particles, dust, and pollen can cause the lungs to produce excess mucus to trap and expel the irritant. The lining of the airways can become inflamed; in people with asthma, the irritation may trigger muscle contractions that close off the airway completely. Particles smaller than 2.5 µm are the most dangerous because they are small enough to penetrate lung cells or enter the bloodstream, which delivers them to other body cells. If these particles come from the combustion of fossil fuels or other industrial sources, they may contain toxic substances, leading to additional problems associated with toxic exposure.

Since the cardiovascular system depends on the respiratory system to provide oxygen for the body, anything that impairs the lungs also harms the cardiovascular system, which might explain why people living in polluted areas also have higher rates of heart attacks and strokes. Cancer rates are higher in people exposed to air pollution, too; exposure to secondhand smoke and exposure to radon are the leading environmental causes of lung cancer, and exposure to smog and vehicle emissions is linked to increased risk for lung and breast cancer.

Exposure to air pollution also has a societal link. Low-income or minority areas often have some of the worst air quality. This raises questions of **environmental justice**—the concept that access to a clean, healthy environment is a basic human right. Sources of major pollution like power plants or waste incinerators are often placed in areas where residents have less ability to fight for their rights—less money, less education, little or no voice in local government. In some cases, even when socioeconomic status is accounted for, minority communities still face more exposure to pollution than average, an example of **environmental racism**. A 2002 study conducted in Southern California by Brown University researcher Rachel Morello-Frosch found that a person's risk for developing cancer from exposure to polluted air increased as income decreased. And, in general, cancer risk was higher for minorities (Asian, African American, Latino) than for the majority (Caucasians), no matter what the income level.

Children of low-income families are at particular risk: For example, in the Bronx, children's homes and schools are near major roads or factories, and they often come and go to school during rush hour, when traffic is heaviest and smog forms. More children in the South Bronx are hospitalized for asthma than anywhere else in New York State, and since many Bronx children live or attend schools adjacent to congested highways, Bronx Congressman José Serrano wondered if the two factors might be related. In 2002, he asked New York University environmental scientist George Thurston if he would be willing to conduct a study to find out. "We thought about it for a nanosecond, and then said, 'sure,'" Thurston recalls. In a study reminiscent of Delfino's, Thurston recruited 40 South Bronx fifth graders to tote wheeled backpacks containing personal air monitors for a month while rating their respiratory symptoms three times a day. "You rolled it, so it wasn't really that heavy," Derrick Reliford, one of the students in the study told *The New York Times*. "They were the rock stars of the class— everybody wanted to help them with the backpacks," Thurston recalls.

environmental justice The concept that access to a clean, healthy environment is a basic human right.

environmental racism A form of racism that occurs when minority communities face more exposure to pollution than average for the region.

Chang W. Lee/The New York Times/Redux

Derrick Reliford, 14, in his Bronx, New York, home. At 10 years old, he participated in the New York University study that measured his daily pollution exposure. The researchers found that students in the South Bronx were twice as likely to attend a school near a highway as were children in other parts of the city.

The children came from four different schools, two of which were close to a highway and two of which were not. Thurston found that, sure enough, the children who went to schools or lived closer to highways were exposed to more air pollution—in particular, diesel fuel exhaust—and they also had more severe respiratory symptoms. More recent research by Thurston on adults corroborated these results, finding that the closer one lived to a coal-fired power plant or major highway, the higher one's risk of dying from a heart attack.

But humans aren't the only organisms afflicted by the ill effects of air pollution. Many animals suffer the same respiratory distress as humans: All lung tissue is very vulnerable to air pollution. Gills too, such as those on terrestrial vertebrates like salamanders, and some invertebrates are negatively affected by air pollution. Plant tissues are also vulnerable to pollutants like smog and ozone. Exposure can reduce a leaf's ability to photosynthesize, preventing healthy growth and compromising its survival. Together with changes in soil chemistry—which can hinder plant growth—pollution damage to crops could result in dollar losses in the billions by 2030.

The tourism industry is also impacted by air pollution that lowers visibility. On hazy days, for instance, it can be impossible to see across the Grand Canyon. Finally, pollution damages buildings and monuments. Acids in polluted rain literally eat away at limestone and marble structures; it can etch glass and damage steel and concrete, causing billions of dollars of damage per year. **INFOGRAPHIC 3**

Statue in Trafalgar Square, in London, England shows erosion that exceeds normal weathering and is likely due to acid rain.

Ernie Janes/Balance/Photoshot/ZUMAPRESS.com

INFOGRAPHIC 3 CONSEQUENCES OF AIR POLLUTION

Air pollution negatively affects the natural environment and species that live there and is strongly linked to a wide variety of human health problems. It also has impacts at the societal level, impacting the well-being of entire communities.

ENVIRONMENTAL IMPACTS

- Atmospheric ozone depletion
- Climate change
- Plants: tissue damage; impaired growth
- Animals: impaired health; reduced reproductive success
- Eutrophication of waterbodies
- Reduced visibility
- Damage to structures

HEALTH IMPACTS

IMMEDIATE EFFECTS INCLUDE:

- Eye, nose, throat irritation
- Respiratory infections
- Wheezing, coughing, shortness of breath
- Headache

LONG-TERM EFFECTS INCLUDE:

- Cardiovascular and respiratory diseases
- Neurological disorders
- Premature birth and low birth weight
- Impaired lung development
- Diabetes
- Cancer
- Premature death

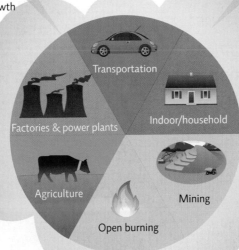

Transportation

Indoor/household

Factories & power plants

Agriculture

Mining

Open burning

SOCIETAL IMPACTS

- Low income and minority communities most likely to be affected
- Lost income from missed work
- Health costs to community

 What factors might explain why children are at a higher risk of health problems due to exposure to air pollution than adults?

4 ACID DEPOSITION

Key Concept 4: Acid deposition is a secondary pollutant that results from fossil fuel burning. It lowers the pH of soil and waterbodies and can harm plants and animals that are exposed.

One of the most problematic characteristics of air pollution is that it moves. Transboundary air pollution produced in one city can end up harming humans and other species halfway around the globe. For example, as much as half of the air pollution that falls on the Great Smoky Mountains of Tennessee and North Carolina originates in the Ohio Valley, where it is released by tall smokestacks of coal-burning power plants. Prevailing winds bring the pollution southeast, and the tall mountains in the southern Appalachians eventually stop it. There, it not only pollutes the air but also produces **acid deposition**—commonly known as acid rain.

> One of the most problematic characteristics of air pollution is that it moves. Transboundary air pollution produced in one city can end up harming humans and other species halfway around the globe.

Acid deposition comes in many forms—rain, snow, sleet, fog, even dry deposition. It is a secondary pollutant, produced when sulfur and nitrogen emissions from burning fossil fuels react with oxygen and water in the atmosphere to form sulfuric or nitric acids. (Sulfur pollution comes mainly from burning coal; the primary source of nitrogen pollution is fossil-fuel powered vehicles.)

Acid deposition can harm aquatic organisms in affected waterbodies as well as acidify the soil, negatively impacting soil communities and plant life. Many aquatic organisms are especially vulnerable to the acidification of their water habitat, especially the eggs and young of many fish and amphibians. Lower pH can interfere with the hatching success of fish and amphibian eggs as well as the passage through later developmental stages. In fact, ecologists first uncovered the problems caused by acid deposition when lakes in heavily polluted areas in Europe began to lose so many species that they

acid deposition Precipitation that contains sulfuric or nitric acid; dry particles may also fall and become acidified once they mix with water.

were described as "dying". Ecologists soon learned that dead lakes were an early warning that forests could be next.

Acidification of soil due to acid deposition can change the soil chemistry and mobilize toxic metals such as aluminum, hindering plants' ability to take up water. Acids leach nutrients from the soil, too, reducing the amount of calcium, magnesium, and potassium available to plants in topsoil. Taken together, these impacts can decrease plant growth, weaken plants so they are more vulnerable to disease or pests, and even kill them. (See Online Module 6.3 for more on the pH scale and ocean acidification.)

Steps to reduce acid deposition in the United States focused on reducing sulfur pollution released from coal-fired power plants. Cleaning up those emissions has significantly reduced acid deposition, but most streams and terrestrial areas have yet to recover. It will take some time for natural processes to return these waters and soils to their normal pH. **INFOGRAPHIC 4**

Exposure to acid rain has resulted in yellowing and loss of needles, decreasing overall photosynthesis and stunting the growth of these conifers at high elevations in the Austrian Alps.

Tony Craddock/Science Photo Library/Science Source

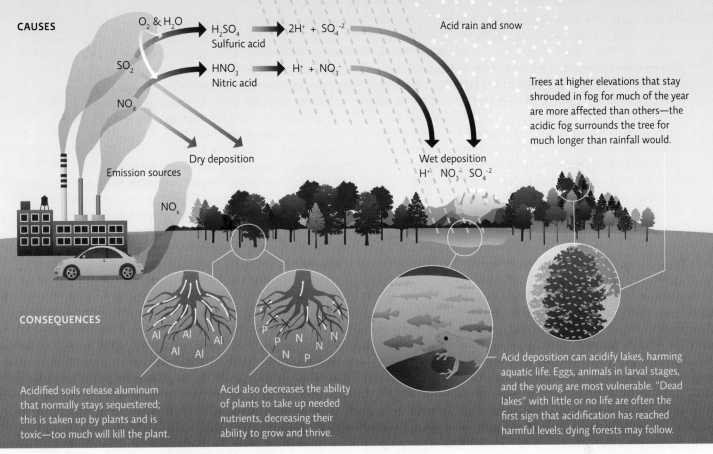

INFOGRAPHIC 4 ACID DEPOSITION

Burning fossil fuels releases sulfur and nitrogen oxides. These compounds react in the atmosphere to form acids. Acid rain, snow, fog, and even dry particles can fall to Earth as acid deposition, with the potential to alter the pH of lakes and soil, damaging plant and animal life.

CAUSES

O_2 & H_2O

SO_2

NO_x

H_2SO_4
Sulfuric acid

HNO_3
Nitric acid

$2H^+ + SO_4^{-2}$

$H^+ + NO_3^-$

Acid rain and snow

Trees at higher elevations that stay shrouded in fog for much of the year are more affected than others—the acidic fog surrounds the tree for much longer than rainfall would.

Dry deposition

Emission sources

NO_x

Wet deposition
H^+ NO_3^- SO_4^{-2}

CONSEQUENCES

Al Al Al Al Al

P N N P N N P

Acidified soils release aluminum that normally stays sequestered; this is taken up by plants and is toxic—too much will kill the plant.

Acid also decreases the ability of plants to take up needed nutrients, decreasing their ability to grow and thrive.

Acid deposition can acidify lakes, harming aquatic life. Eggs, animals in larval stages, and the young are most vulnerable. "Dead lakes" with little or no life are often the first sign that acidification has reached harmful levels; dying forests may follow.

ACID DEPOSITION HAS DECREASED

Restrictions imposed by the Clean Air Act have helped decrease acid deposition in the United States. Smokestack scrubbers remove sulfur from burning coal, reducing the SO_2 released. Emission-control technologies on vehicles, such as the catalytic converter, convert dangerous combustion by-products to safer emissions (such as converting NO_x to N_2). This reduces, but doesn't eliminate, these dangerous emissions.

pH OF PRECIPITATION IN THE UNITED STATES

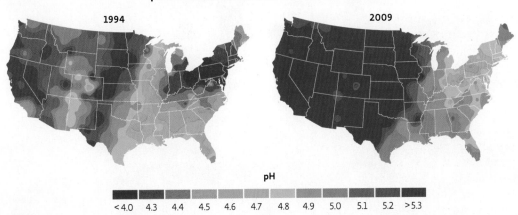

1994

2009

pH

| < 4.0 | 4.3 | 4.4 | 4.5 | 4.6 | 4.7 | 4.8 | 4.9 | 5.0 | 5.1 | 5.2 | > 5.3 |

Why do you think that acid deposition has been a bigger problem in the eastern United States than the western part of the country? Why might acid rain be increasing in some western areas?

5 THE CLEAN AIR ACT

Key Concept 5: In addition to the criteria air pollutants, the EPA regulates volatile organic compounds and mercury; it is also mandated to regulate carbon dioxide, an air pollutant linked to climate change, but political battles have hampered its ability to do so.

Cleaning up the air emissions that led to acid deposition was the result of legislation that mandated a change. The United States led the way in the passage of environmental laws that regulate pollution. One of the first was the **Clean Air Act (CAA)**. Originally passed in 1955 as the Air Pollution Control Act, it was replaced in 1963 by the CAA, which received major updates in the 1970's and 1990's.

When the U.S. Environmental Protection Agency (EPA) started regulating air pollution, the agency did not know the extent to which different types of air pollution could affect health; in fact, the EPA administrator noted at the time that the agency's clean air regulations were "based on investigations conducted at the outer limits of our capability to measure connections between levels of pollution and effects on man." Nevertheless, in an example of applying the precautionary principle (see Modules 1.3 and 5.2), in 1971, the EPA set standards for the most common pollutants that were suspected to be problematic—known as *criteria air pollutants*. Five of these were chemical air pollutants: carbon monoxide, sulfur dioxide (SO_2), nitrogen oxides (NO_x), lead, and ground-level ozone; standards were also set for particulate matter.

A landmark 1993 study published by Harvard University researcher Douglas Dockery helped establish the link between air pollution and impaired health. When Dockery compared death rates in six cities before air pollution controls were put in place to death rates after the CAA was implemented in those same six cities, the data showed that as the level of small particulates ($PM_{2.5}$) decreased in the cities evaluated, the death rate decreased. A follow-up study in 2006 estimated that particulate pollution—which includes soot, ash, dust, smoke, pollen, and

Clean Air Act (CAA) The main U.S. law that authorizes the EPA to set standards for dangerous air pollutants and enforce those standards.

The Los Angeles skyline is obscured by smog.

small, suspended droplets (aerosols)—accounts for 75,000 premature deaths per year and showed a clear dose–response effect: The higher the air pollution, the higher the risk for death.

In addition to the six criteria air pollutants, the EPA also recognizes 187 hazardous air pollutants that can have adverse effects on human health, even in small doses. These toxic substances may cause cancer or developmental defects, or they may damage the central nervous system or other body tissues. They include

volatile organic compounds (VOCs), a variety of chemicals that readily evaporate, entering the air as a gas. Many present a health hazard if inhaled or if sensitive tissue like eyes are exposed. VOCs are produced by natural sources such as wetlands, but the main outdoor source is fossil fuel combustion. VOCs found in household products such as paint, carpets, and cleaners contribute to indoor air pollution.
INFOGRAPHIC 5

volatile organic compound (VOC) A chemical that readily evaporates and is released into the air as a gas; may be hazardous.

INFOGRAPHIC 5 | **THE CLEAN AIR ACT**

CLEAN AIR ACT: The primary U.S. federal law regulating air quality. Passed in its current form in 1963, it has been updated with major amendments in 1970, 1977, and 1990. The EPA is tasked with enforcing the CAA and does so by writing "rules" to lay out targets and acceptable methods states can use to comply. The first six pollutants that were targeted are known as the Criteria Air Pollutants. Many other air pollutants are also regulated by the CAA, including mercury and VOCs; efforts to regulate carbon dioxide have faced opposition and have been less successful.

Artush/iStock/Getty Images

CRITERIA AIR POLLUTANT	SOURCE
Carbon monoxide (CO). From incomplete combustion of any carbon-based fuel (and most combustion is incomplete).	Vehicles, forest fires, and volcanoes
Sulfur dioxide (SO_2). From natural sources and fossil fuel combustion.	Industry, volcanoes, and dust
Nitrogen oxides ($NO_X = NO$ and NO_2). From the reaction of nitrogen in fuel or air with oxygen at high temperatures (usually during combustion of a fuel).	Vehicles, industry, and nitrification by soil and aquatic bacteria
Ground-level ozone (O_3). Formed from reactions between NO_x and VOCs in the presence of sunlight.	Vehicles (NO_x); manufactured products and industry (VOCs)
Particulate matter (PM). Tiny airborne particles or droplets, smaller than 44 micrometers. The smaller the particle, the more dangerous it is for living things.	Combustion of any fuel or activity that produces dust, including forest fires and dust storms
Lead (Pb). Additive to gasoline, paint, and other solvents; phased out of the U.S. gas supply in the 1970's and officially banned in 1996.	Lead-based paint in older homes and from other countries; leaded gasoline; soil erosion and volcanoes

 Which of the criteria air pollutants are you exposed to on a daily basis, and what are their sources?

6 INDOOR AIR POLLUTION

Key Concept 6a: Homes trap or are the source of many indoor air pollutants. Better ventilation and alternative building or household materials can reduce this pollution.

Key Concept 6b: In developing countries, air pollution mainly comes from poorly ventilated indoor cooking fires. It can be reduced by using cleaner fuels and solar ovens.

As Delfino and others were showing, outdoor air quality substantially impacts human health. However, we breathe air indoors as well as outdoors, and indoor air quality is a growing concern among public health scientists. In fact, people living in affluent, developed nations may find that their greatest exposure to unhealthy air comes from indoors. This is because so much time is spent indoors in homes, schools, or the workplace; these areas contain many potential air pollution sources. For instance, cigarette smoke causes significant health problems, including eye, nose, and mucous membrane irritation; lung damage, which can exacerbate or cause asthma; and lung cancer. Items in our home, like paint, cleaners, and furniture, release VOCs, which can also cause health problems.

Outdoor pollutants can also find their way into our buildings. Radon is a naturally occurring radioactive gas produced from the decay of uranium in rock. It can seep through the foundations of homes and accumulate in basements; exposure to radon can cause lung cancer. Not every area has the type of rock that produces radon, but buildings constructed over areas where soil or groundwater is contaminated with VOCs, such as areas with underground chemical storage tanks, also present an infiltration risk.

Reducing indoor air pollution involves removing the sources (e.g., no carpet in the home), properly storing potential sources of air pollution (storing household chemicals properly), and providing adequate ventilation for a home. In our quest to become energy efficient, many of us might seal our homes to prevent air leaks. But if this is done, the home also needs a good system for mechanical ventilation (usually as part of the heating, ventilation, and air conditioning system, or HVAC) of the home to ensure a good exchange of air. **INFOGRAPHIC 6**

Indoor air pollution is a problem in many less developed regions where cooking and heating are done indoors with wood or other combustible fuel but homes are poorly ventilated.

MATTHIEU PALEY/National Geographic Creative

INFOGRAPHIC 6   SOURCES OF INDOOR AIR POLLUTION

For most people, the greatest exposure to air pollution comes from being indoors. There are many sources of air pollution in a home or other building, as these structures tend to trap pollutants, keeping concentrations high. One can reduce exposure by avoiding or limiting the use of carpets, upholstered items, and furniture made with toxic glue and formaldehyde. Safer cleaners and low-VOC paints are readily available. Simple behaviors like taking off your shoes before entering the house and using a vacuum equipped with a HEPA filter will also help. Good ventilation and properly working heating and air conditioning units help keep indoor air pollutants at bay.

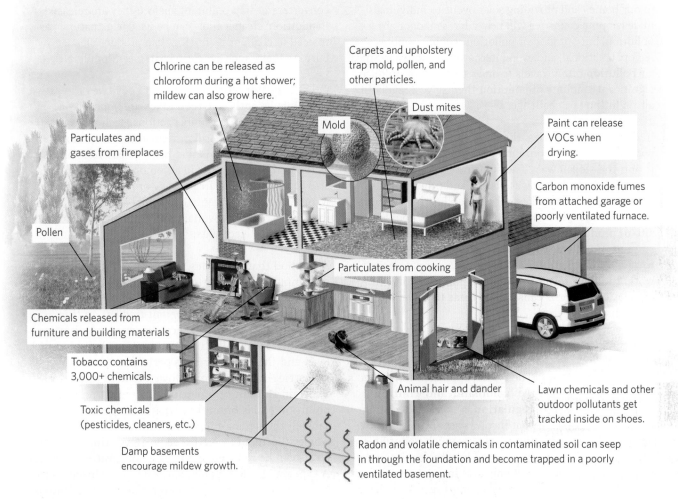

Chlorine can be released as chloroform during a hot shower; mildew can also grow here.

Carpets and upholstery trap mold, pollen, and other particles.

Dust mites

Mold

Paint can release VOCs when drying.

Particulates and gases from fireplaces

Carbon monoxide fumes from attached garage or poorly ventilated furnace.

Pollen

Particulates from cooking

Chemicals released from furniture and building materials

Tobacco contains 3,000+ chemicals.

Animal hair and dander

Lawn chemicals and other outdoor pollutants get tracked inside on shoes.

Toxic chemicals (pesticides, cleaners, etc.)

Damp basements encourage mildew growth.

Radon and volatile chemicals in contaminated soil can seep in through the foundation and become trapped in a poorly ventilated basement.

 What are your main sources of indoor air pollution, and what could you do to reduce them?

In developing countries, where many people cook and heat with open fires, smoke and soot from burning wood, charcoal, dung, or crop waste are major sources of indoor pollution. Kirk Smith, a professor of environmental health at the University of California, Berkeley, has found that indoor fires increase the risk of pneumonia, tuberculosis, chronic bronchitis, lung cancer, cataracts, and low birth weight in babies born of women who are exposed during pregnancy. "Considering that half the world's households are cooking with solid fuels, this is a big problem," Smith says. In India, most air pollution deaths in 2013 (920,000 out of 1.51 million deaths) were linked to indoor air pollution caused by using wood or dung as cooking fuels.

Addressing this type of indoor air pollution can involve improving or installing chimneys to vent smoke, installing more efficient cook stoves (this can be as simple as constructing an earthen cook stove with openings tailored to fit the pots used in cooking), or providing cleaner burning fuels, such as liquid petroleum gas. Many nonprofit organizations are stepping up to meet this problem with a simple $50 solar cooker that allows people to cook food without building a fire. This technology has the added advantage of not depleting local biomass resources for fuel.

7 REDUCING OUTDOOR AIR POLLUTION

Key Concept 7: Cleaner fuels and emission-control technologies can reduce industrial outdoor air pollution. Regulations and economic incentives can spur innovation and will decrease the true cost of energy production but may raise the costs of providing energy or doing business.

Solutions such as installing better ventilation systems in our homes and providing solar ovens to individuals in developing countries will help address indoor air pollution, but outdoor air pollution requires a more regional, national, and even international approach. Since air pollution often travels to areas that do not produce significant amounts of pollution themselves, regulating air pollution is a challenge.

In developed countries, the original approach to dealing with air pollution from human activities was to spread it out; the slogan was "The solution to pollution is dilution." Factories, power plants, and other point sources built tall smokestacks to send emissions high into the atmosphere so that they wouldn't pool at the site of production. The idea was that if dispersed, the amount of pollution in any one area would be too low to cause a problem. But this approach simply doesn't work: Industry releases too much pollution, and air circulation patterns cause some areas to get more than their share.

command-and-control regulation A type of regulation that involves setting an upper allowable limit of pollution release that is enforced with fines and/or incarceration.

green tax A tax (fee paid to government) assessed on environmentally undesirable activities.

tax credit A reduction in the tax one must pay in exchange for some desirable action.

subsidies Financial assistance given by the government to promote desired activities.

cap-and-trade program Regulations that set an upper limit for pollution emissions, issue permits to producers for a portion of that amount, and allow producers that release less than their allotment to sell permits to those who exceeded their allotment.

Eventually it became clear that regulation would be necessary. The typical approach in the 1970's was **command-and-control regulation**, a type of regulation that involves setting national limits on how much pollution can be released into the environment and imposes fines or even brings criminal charges against violators who release more than is allowed. The U.S. CAA, for example, sets a maximum amount, or *performance standard*, of a pollutant that can be released or that can be present in ambient air. (See Module 5.2 for more on performance standards.) As a result of the CAA, the United States has seen major reductions in common air pollutants such as lead and sulfur.

To meet CAA mandated performance standards, industries and power plants can use end-of-pipe solutions like scrubbers, filters, electrostatic precipitators, and catalytic converters to trap pollutants before they are released. In addition, cleaner fuels can be used (e.g., natural gas instead of coal), or technology can produce cleaner burning fuels, such as "clean coal" (see Module 9.1).

Other regulatory tools are linked to positive or negative economic incentives. **Green taxes** are imposed on environmentally undesirable actions, such as an extra tax on low-mile-per-gallon vehicles, whereas **tax credits** support actions that help the environment (such as the purchase of a hybrid automobile). Governments also offer **subsidies**, free money or resources intended to promote environmentally friendly activities such as installing solar photovoltaic panels for the production of pollution-free electricity.

A method that combines the performance standards of command-and-control regulation with economic incentives is the market-based approach called **cap and trade** (also known as *permit trading*). In this approach, a regulatory agency sets an upper limit on emissions for a pollutant on a nationwide or regional level and then gives or sells permits to polluting industries—each permit allows the holder to release a certain amount of pollution. Users that reduce their pollution emissions below what their permit allows can sell their remaining allocation to other users whose allotments were exceeded. Over time, pollution levels can be reduced as the cap—or limit—is lowered. A cap-and-trade program successfully reduced sulfur pollution from coal-fired power plants in the United States in the 1990's. A downside to cap-and-trade programs is that pollution can become concentrated in areas where individual industries choose to buy additional permits rather than reduce emissions. **INFOGRAPHIC 7**

Like many environmental laws, the CAA is regularly under attack by those who see regulation as intrusive. The fact that our air is cleaner today than it was in the 1960's—even with a larger U.S. population and more industry—is evidence that regulations can be effective. Still, these improvements are costly to industry, and such costs are usually passed on to consumers. For this reason, many individuals and groups oppose such policies, charging that the restrictions are excessive or that the government goes too far in trying to regulate

INFOGRAPHIC 7   **APPROACHES TO REDUCING AIR POLLUTION**   ANIMATED INFOGRAPHIC

Many approaches can be used to lessen air pollution, including technology to reduce emissions before a fuel is burned (see Module 9.1, Infographic 7) and technology to capture emissions after a fuel is burned, as shown below. Policies are needed to require or encourage the reduction of air pollution.

TECHNOLOGY	POLICY APPROACHES
A variety of technologies, each designed to address the particular problem at hand, can be employed to reduce the release of air pollution.	Policies protect air quality by requiring or encouraging actions that reduce the production of air pollution. They include: • Subsidies/grants: funds for technology upgrades • Market-driven financial incentives: cap and trade • Green taxes: a tax on pollution • Penalties: fines, jail, lost contracts, etc.

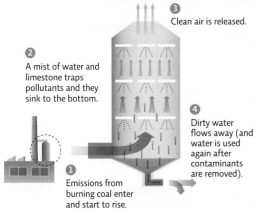

TECHNOLOGY: SMOKESTACK SCRUBBER

❸ Clean air is released.

❷ A mist of water and limestone traps pollutants and they sink to the bottom.

❹ Dirty water flows away (and water is used again after contaminants are removed).

❶ Emissions from burning coal enter and start to rise.

Smokestack scrubbers send the emissions through a mist of water and limestone to trap contaminants and prevent their release.

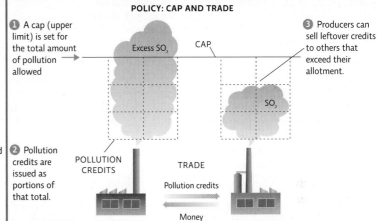

POLICY: CAP AND TRADE

❶ A cap (upper limit) is set for the total amount of pollution allowed

❷ Pollution credits are issued as portions of that total.

❸ Producers can sell leftover credits to others that exceed their allotment.

Excess SO_2 CAP

SO_2

POLLUTION CREDITS

TRADE

Pollution credits

Money

Cap and trade is a market-driven policy that gives producers the freedom to choose how to meet the requirement and the opportunity to earn a profit in doing so.

 How could a cap-and-trade program for sulfur pollution lead to lower pollution in one area and higher pollution in others?

emissions. However, including the cost of pollution prevention or clean-up in the cost of a product prices that product closer to its *true cost* and may actually lower the overall cost to society as health and environmental impacts decline. (See Module 5.1 for more on true cost accounting.)

While some air pollution problems have been reduced, the work of the CAA is not done; air pollution problems remain and a new one, climate change—due in large part to fossil fuel emissions—threatens to be the most significant air pollution issue we will face. (See Module 10.2 for more on climate change.) In a landmark case, the U.S. Supreme Court gave the EPA the authority to regulate CO_2 as a pollutant in 2007, but the EPA immediately faced political opposition and efforts to reduce CO_2 emissions have been slow in coming.

The Clean Power Plan, negotiated over several years and approved in 2016, finally established rules for regulating CO_2 under the CAA. The EPA estimated

that the plan would reduce CO_2 emissions by 30% compared to 2005 emissions, save 3,600 lives, and net the country up to $45 billion in health and climate benefits. Before it could be implemented, however, newly elected President Donald Trump signed an executive order that rescinded the plan; his policy proposals will actually increase U.S. CO_2 emissions, reversing the decline that began in 2005. President Trump's administration is also seeking to significantly reduce EPA funding, which some EPA officials, including former EPA administrator Gina McCarthy, say will make it difficult to implement the environmental policies it is mandated to enforce. Weakening or dismantling the environmental legislation that protects our air and water could bring the return of a highly contaminated environment, which could lead to more health problems and diminished ecosystem function and services.

Mitigating or preventing air pollution costs money, but many feel it is money well spent because it prevents far greater losses down the line—especially in terms of human health. The Centers for Disease Control and Prevention estimated that, in 2007, the cost of asthma to the United States was $56 billion. Delfino and others hope their research helps policy makers realize just how useful curbing air pollution can be. "We're talking about the air we breathe," Thurston says. "There's nothing more communal than that."

Select References:

Delfino, R. J., et al. (2008). Traffic-related air pollution and asthma onset in children: a prospective cohort study with individual exposure measurement. *Environmental Health Perspectives*, *116*(10), 550–558.

Dockery, D., et al. (1993). An association between air pollution and mortality in six US cities. *New England Journal of Medicine*, *329*(24), 1753–1759.

Forouzanfar, M. H., et al. (2015). Global, regional, and national comparative risk assessment of 79 behavioural, environmental and occupational, and metabolic risks or clusters of risks in 188 countries, 1990–2013: a systematic analysis for the Global Burden of Disease Study 2013. *The Lancet*, *386*(10010), 2287–2323.

Morello-Frosch, R., et al. (2002). Environmental justice and regional inequality in southern California: implications for future research. *Environmental Health Perspectives*, *110*(2), 149–154.

Smith, K. R., et al. (2014). Millions dead: how do we know and what does it mean? Methods used in the comparative risk assessment of household air pollution. *Annual Review of Public Health*, *35*, 185–206.

Spira-Cohen, A., et al. (2011). Personal exposures to traffic-related air pollution and acute respiratory health among Bronx schoolchildren with asthma. *Environmental Health Perspectives*, *119*(4), 559–565.

Thurston, G. D., et al. (2016). Ischemic heart disease mortality and long-term exposure to source-related components of U.S. fine particle air pollution. *Environmental Health Perspectives*, *124*(6), 785–794.

Mass transit options that decrease the number of cars on the road will reduce air pollution. Buses that run on compressed natural gas emit fewer emissions overall, but the particulates they release are very small; so while better than a traditional diesel bus, they are still not pollution free.

Pierre GLEIZES/REA/Redux

INTERACTIVE MAP **AIR POLLUTION IN THE UNITED STATES** ANIMATED INFOGRAPHIC

Air quality has improved in many areas of the United States thanks to the Clean Air Act, but some areas still suffer from heavy air pollution. This interactive map profiles the most polluted U.S. metropolitan area and the cleanest metropolitan area in 2016, as determined by the American Lung Association. Case studies are also presented on some recent air pollution research.

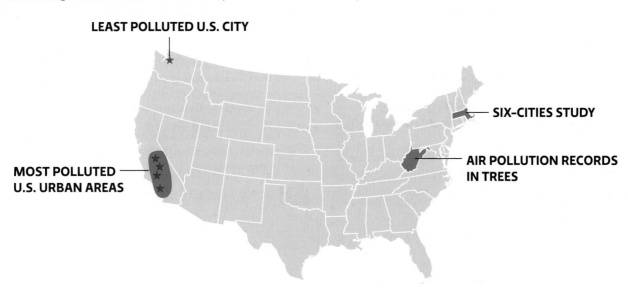

LEAST POLLUTED U.S. CITY

SIX-CITIES STUDY

AIR POLLUTION RECORDS IN TREES

MOST POLLUTED U.S. URBAN AREAS

 BRING IT HOME

PERSONAL CHOICES THAT HELP

Individuals can have an effect on air quality by researching the threats to their area, making appropriate behavior changes, and supporting legislation that limits the production of air pollutants.

Individual Steps

• Reduce your exposure to indoor air pollution by reducing your use of harsh cleaning products, synthetic air fresheners, vinyl products, and oil-based candles.
• Avoid outdoor exercise during poor air quality days. Go to www.airnow.gov to find the local air quality forecast.
• Buy a radon detector and carbon monoxide detector for your house to keep your family safe.

Group Action

• Organize a "car-free day" at your school, community, or workplace to reduce emissions from vehicles.
• Work with community leaders and businesses to sponsor a "free public transit" day.

Policy Change

• If your community does not have public transit, ask community leaders to investigate bringing it to your area.
• Many groups are working to improve our air quality. Find one in your region and see what issues it is addressing. For a list of national and regional organizations, go to www.inspirationgreen.com/air.

Goodshoot/Getty Images

ENVIRONMENTAL LITERACY UNDERSTANDING THE ISSUE

1 What is air pollution, and what is its global impact?

1. True or False: Most air pollution deaths occur in high-income (developed) countries.

2. Air pollution is defined as:
   a. any unnatural material found in the air.
   b. contaminants in the air that cause environmental or health problems.
   c. particles or chemicals added to the air by human activities.
   d. materials added to the air that remain aloft long enough to harm human health.

3. What historical events provide evidence that air pollution can cause serious problems for humans?

2 What are the main types and sources of outdoor air pollution?

4. True or False: Ozone is useful in the stratosphere but is a dangerous pollutant when located close to the ground where we can breathe it in.

5. Emissions released form the smoke stack of a power plants are an example of _____ _____ air pollution.

6. Air pollution that results when chemicals in the atmosphere react to form a new pollutant is called:
   a. primary pollution.
   b. secondary pollution.
   c. point source pollution.
   d. particulate pollution.

7. What is particulate matter? Identify some sources of this type of pollution where you live.

8 What is transboundary pollution and why is it difficult to regulate?

3 What are the health, economic, social, and ecological consequences of air pollution?

9. True or False: Air pollution can harm health, hurt the environment, and even damage buildings.

10. Why are cardiovascular diseases linked to air pollution?
    a. The heart is located close to the lungs, making it easy for pollutants to reach it.
    b. The heart must work harder to deliver oxygen to the body when the lungs are damaged.
    c. Blood vessels don't respond to cardiovascular hormones in the presence of pollutants.
    d. Air pollutants thicken the blood and make it harder to move through blood vessels.

11. The placement of polluting industries close to minority or low-income areas is an example of a violation of:
    a. environmental justice.
    b. federal law.
    c. EPA clean air standards.
    d. transboundary pollution.

12. How does air pollution negatively affect plants?

4 What are the causes and consequences of acid deposition?

13. True or False: Acid deposition has decreased across much of the eastern United States, and ecosystems are showing strong signs of recovery.

14. High-elevation trees are often more affected by acid deposition than trees lower on a mountain because the high-elevation trees:
    a. have more shallow root systems that are easily harmed by acids.
    b. grow in areas where acids bind aluminum in the soil so that the trees can't access it.
    c. live in colder areas and are more easily stressed than other trees.
    d. are often shrouded in acidic fog, exposing them to acids for long periods of time.

5 What air pollutants are regulated by the Clean Air Act?

15. The first six air pollutants regulated by the Clean Air Act are known as _____ _____ _____.

16. Which of the following is an air pollution source of volatile organic compounds (VOCs)?
    a. Paints and solvents
    b. Carpets and household cleaners
    c. Wetlands
    d. A and B
    e. A, B, and C

17. Why was the EPA's initial regulation to address specific pollutants in 1971 seen as an example of the precautionary principle?

6 What are the main sources of indoor air pollution, and what can be done to reduce it?

18. True or False: In most areas of developing countries, indoor pollution is more of a problem than outdoor pollution.

19. Which of these actions would best address the main cause of indoor pollution in developing countries?
    a. Using more wood and less charcoal in cooking fires
    b. Using emission-control devices on vehicles to reduce air pollution on nearby roads
    c. Distributing solar ovens
    d. Making homes more airtight to keep out pollution from outside

20. What are some sources of indoor air pollution you might face?

7 How can air pollution be reduced, and what are the trade-offs of reducing it?

21. Which of the following is considered a penalty for producing pollution rather than an incentive for acting in a way that reduces pollution?
    a. Tax credit
    b. Green tax
    c. Subsidy
    d. All of the above

22. Describe the policy of cap and trade. What are the advantages and disadvantages of this policy option?

SCIENCE LITERACY **WORKING WITH DATA**

The graph shown here indicates levels of ground-level ozone and particulate matter that exceeded national reference levels (the levels above which health or ecosystem problems occur) in areas and cities in Canada.

POLLUTION LEVELS IN SELECT CANADIAN CITES

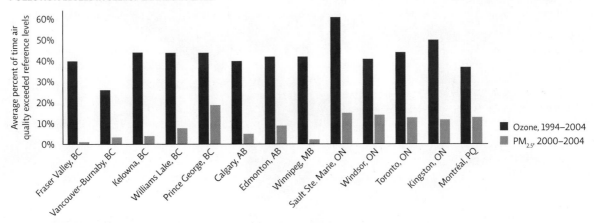

Interpretation

1. What does the *y* axis represent? Choose a city and describe the data for that city.

2. The legend states that $PM_{2.5}$ data is given for 4 years. PM stands for particulate matter. The number 2.5 represents the size of the particulate. From a health standpoint, why are the $PM_{2.5}$ values reported?

3. How many years of data are graphed for ozone? For particulate matter? Does the difference in the amount of time over which the data have been collected make a difference in your interpretation of these data?

Advance Your Thinking

4. Which city would be the worst for your health, based on its levels of pollution? Why?

5. Based on the type of pollution present, what can you predict about the causes of pollution in Sault Ste. Marie versus Montreal? (Hint: Do some research and compare the city size, weather, and primary industries).

6. In general, cities were out of compliance more from ground-level ozone in the time frame tested than for particulate matter. Does this mean particulate matter is less of a problem than ground-level ozone? Explain.

INFORMATION LITERACY **EVALUATING INFORMATION**

The EPA is tasked with regulating pollutants in the United States. As part of this process, the agency collects and records data for many pollutants, but not all of them. The federal EPA is assisted in this endeavor by state EPAs. However, it is impossible to collect air quality data about every locality in the United States, so most data are collected in and around cities.

Go to www.stateoftheair.org. Enter your zip code in the space provided to get a report about air quality in your area. If no air quality monitoring stations exist in your area, choose your state and look at the data for the county closest to you. Record these data. Then, click on "Key Findings" and read about how the grades were calculated for each county. Finally, click on "Health Risks" and read about the specific health risks associated with both ozone and particulate matter.

Evaluate the website and work with the information to answer the following questions:

1. Determine if this a reliable information source:
   a. Who runs the website? Do this organization's credentials make it reliable or unreliable? Explain.
   b. What is this organization's mission? How do you know this?

2. What grade did your area receive for both ozone and particulates?
   a. Based on what you read about how the grade was determined, do you feel the grading system is too lax or too strict?
   b. Why does the American Lung Association advocate for a stricter system?

3. Based on what you have learned, do you believe that the regulations of the Clean Air Act should be loosened, be tightened, or remain the same? Should more areas be monitored, or is it sufficient to monitor only large cities? Why?

 Additional study questions are available at SaplingLearning.com.

CLIMATE REFUGEES

Grappling with current and future climate change

A hotel guest steps out of a Miami hotel onto a street flooded by a king tide (an exceptionally high seasonal tide that occurs when the orbits of the Earth, moon, and sun line up). King tides in south Florida are made worse by rising sea level.

Joe Raedle/Getty Images

After reading this chapter and studying the KEY CONCEPTS and INFOGRAPHICS, you should be able to answer these GUIDING QUESTIONS

CORE MESSAGE

Climate change is a serious environmental problem that is impacting species, ecosystems, and the health and well-being of people around the globe, with more changes to come. Science can help us evaluate the changes that are happening, investigate causes, and provide information to help develop and implement sound policies for dealing with changing climate.

1. What is climate change, and why is it more concerning than day-to-day changes in weather?

.2. What is the physical and biological evidence that climate change is occurring?

3. What is the greenhouse effect, and how are we affecting it?

4. How are atmospheric CO_2 and temperature measured, and how are they correlated?

5. Other than greenhouse gases, what factors affect climate?

6. What evidence suggests that climate change is due to human impact?

7. What are the current and potential future impacts of climate change?

8. What actions can we take to respond to a world with a changing climate?

In the spring of 2017, a team of scientists led by Matthew Hauer from the University of Georgia published a study in the scientific journal *Nature Climate Change* that would make headlines across the country and raise anxiety levels in coastal city-dwellers from Miami to Manhattan. The scientists had used climate forecasts and projections of population size and distribution to predict the impact that warming global temperatures and rising sea levels would have on cities in the continental United States.

The picture they painted was grim: A sea-level rise of 0.9 meter (almost 3 feet) by the year 2100 would imperil the homes of some 4 million people. At 1.8 meters (almost 6 feet), the number of people displaced would top 13 million. New Orleans alone would lose 500,000 people; New York City would lose 50,000; and more than 2.5 million people would flee Miami, the hardest-hit city of all. And it wasn't just coastal cities that would suffer. As refugees fled the rising tides, landlocked cities like Austin and Atlanta would likely be inundated with the new arrivals, all the while struggling with the effects of rising global temperatures.

The specter of *climate refugees*—people who are forced to flee their homelands because of global warming—related changes to their environment—is often viewed by those in the United States as a problem of other countries in distant parts of the world. When we talk about whole cities being swallowed by a rising ocean, we might think of the Maldives or Bangladesh, low-lying countries that have been dramatically altered by these forces in recent years. But Hauer's results painted it, very clearly, as a U.S. problem, too—and a fairly big one at that.

"These results provide the first glimpse of how climate change will reshape future population distributions," Hauer wrote. "The absence of protective measures could lead to U.S. population movements of a magnitude similar to the twentieth century Great Migration of southern African Americans." (Between 1915 and 1970, some 6 million African Americans migrated from southern states to northern ones in pursuit of economic opportunities and to escape the segregationist south.)

◉ **WHERE IS MIAMI, FLORIDA?**

MIAMI

GULF
OF
MEXICO

FL

Their predictions underscored a basic set of facts on which scientists have long agreed: The planet is warming at an alarming rate; that warming is due to human activity; if something isn't done to curb it, the impacts on both natural ecosystems and human communities will be severe.

But they also laid bare a string of vexing questions over which scientists and policy makers have long been puzzling (and sometimes arguing): How certain are we of what the exact impact will be? What should we do about it? And how much time do we have?

1 CLIMATE AND CLIMATE CHANGE

Key Concept 1: A change of just a few degrees in average temperature can result in a drastically different climate with negative effects on ecosystems and human societies.

To begin to understand climate change, we must first know the difference between climate and weather. **Weather** refers to the meteorological conditions in a given place on a given day, whereas **climate** refers to long-term patterns or trends. In other words, the actual temperature on any given day is part of the weather, while the range of expected values, based on the location and time of year, is the climate.

For example, Miami's climate is tropical; overall, it has hot humid summers and warm, short winters. Its average annual temperature, another way to describe its climate, is 77°F (25°C). But the city's weather on any given day is variable. Today's temperature might be as high as 98°F or as low as 60°F. It might rain; it might be cloudy or windy. Miami often endures hurricanes in the late summer and fall. Those individual storms fall under the category of weather. But their frequency and strength over time are a feature of the city's climate.

Hauer's study tried to assess the future effects of a changing climate: As average global temperatures rise, ice sheets are melting; as that happens, sea levels are rising and local weather patterns are shifting. His task was to try to predict what those changes would mean for U.S. communities over the next century. But it turns out you don't need a crystal ball to see the future of Miami or New York. You just need to hop over to Louisiana.

About 75 miles south of New Orleans, the Isle de Jean Charles is losing a desperate battle with the waters that surround it. The island was once home to acres upon acres of banana and pecan trees, and its inhabitants—Native Americans who had lived there for centuries—used to hunt, fish, trap, and farm there. But in the past half-century, more than 90% of the land has washed away due, in part, to human actions such as dredging to create canals and shipping lanes that allow stronger storm surges to rush in during hurricanes and wash

away land. The many levees, built to prevent flooding, also decreased the delivery of sediment that would naturally restore land lost to erosion. But the biggest driver, or at least the final nail in the coffin, may be climate change that is raising sea level and accelerating the loss of land. The rest of the island is expected to disappear in the next several years as the climate continues to change.

"The changes are underway and they are very rapid," former interior secretary Sally Jewell has warned. "We will have climate refugees."

Climate change refers to alterations in the long-term patterns and statistical averages of meteorological events. Even small shifts in climate can result in major changes to ecosystems. During the last ice age, for example, an ice sheet about a half a mile (~800 meters) thick kissed the borders of Chicago. The average temperature at that time? Just 7°F (~4°C) colder than today.

> **"**The changes are underway and they are very rapid. We will have climate refugees.**"**
> —Sally Jewell

Earth is currently undergoing climate change in the direction of warming. This global warming—the rise in average temperatures that has been measured in myriad locations around the world—is already pushing record high temperatures even higher, creating more and longer heat waves and resulting in earlier springs and later winters. Global warming can also lead (and is leading) to extreme cold weather in some places; for example, shifting weather patterns that affect the jet stream send Arctic air farther south or for longer periods of time, resulting in some recent record cold and snow events in the United States, Europe, and Asia.

To examine these temperature changes, James Hansen and Makiko Sato of Columbia University compared the average summer temperatures for the Northern

---

weather The meteorological conditions in a given place on a given day.

climate Long-term patterns or trends of meteorological conditions.

climate change Alteration in the long-term patterns and statistical averages of meteorological events.

Hemisphere in recent decades to the baseline average of the decades between 1951 and 1980. In this baseline period, temperature data generated a bell curve; Hansen segregated the data into "normal," colder than normal, and hotter than normal, each occurring one-third of the time. Extreme cold or hot years were defined as those that occurred only 0.1% of the time. When he compared subsequent 10-year periods, he saw the bell curve shifting to the right. In the most recent 10-year period evaluated (2005 to 2015), two-thirds of the yearly averages were in the hotter-than-normal range (compared to one-third in the baseline period)—15% of those moved into the extremely hot range with many values outside of the original data set. The bell curve also gets broader and flatter with time, an indication of more variability in the climate (though some scientists attribute some of this to the uneven warming by latitude on the planet.) A similar analysis of winter averages also reveals a shift toward warming, though it is not as pronounced as the summer shift.
INFOGRAPHIC 1

The effects of global warming are not limited to temperature impacts. These changes are already setting into motion other changes that have serious consequences for life on Earth: Positional shifts in biomes (whose locations are established by temperature and precipitation [see Module 2.1]) stress already imperiled ecosystems and their resident species, many of which are already endangered (see Chapter 3); expanding habitats bring mosquitoes and the diseases they carry out of the tropics and subtropics (see Module 4.3); floods ruin crops in some areas and drought does so in others—both of which decrease food supplies at a time when we are barely able to produce enough food to feed the world and need to increase production to feed our growing population (see Chapter 8). Other impacts (to be discussed in more detail later) include rising sea level, loss of crucial freshwater supplies (see Module 6.1), stronger storms on land and sea, and more unpredictable and variable weather in general. All of these impacts are already occurring and will only get worse as climate

---

INFOGRAPHIC 1   **CLIMATE CHANGE: WHY DO A FEW DEGREES MATTER?**   ANIMATED INFOGRAPHIC

A shift of just under 1°C in the average global temperature is producing more frequent and extreme heat waves. Temperature changes are giving rise to other, more far reaching effects such as changes in precipitation, ocean levels, the frequency and severity of storms, and the make-up of ecosystems across the globe—just to name a few impacts. All of these changes are already affecting life on Earth.

AVERAGE SUMMER TEMPERATURES (NORTHERN HEMISPHERE)

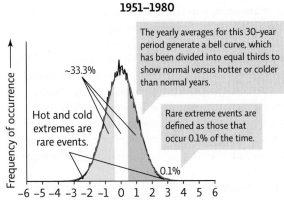

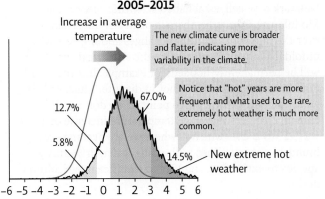

In any stable climate time period, there is some variability. Extreme hot or cold years are rare but occur with similar frequency. Here we see the distribution of average temperatures between 1951 and 1980; about a third of temperatures fall into each section of the bell curve.

An increase of just under 1°C in the "average temperature" has shifted the entire curve to the right. What used to be rare heat events become more common and new "hot extremes" are experienced. In 2016, there were 22,470 U.S. heat records but only 4,016 cold records.

 Between 2005 and 2015, what percentage of years fell above the "normal" temperature range of the baseline time period? How does that compare to the percentage of years higher than normal from 1951 to 1980?

continues to change. For these reasons, the broader term, *climate change,* is more useful than *global warming* when we want to refer to all the alterations we are experiencing.

The consequences of climate change are affecting communities around the world, not just on the Isle de Jean Charles. According to a recent report by the Center for Progressive Reform, at least 17 U.S. communities are currently being forced to relocate. Arctic regions are some of the hardest hit as the area warms two to three times faster than the world average. Newtok, a small village on the northern coast of Alaska, sits on permafrost that was once rock solid but is now thawing rapidly. As the land sinks and water seeps in, the village of 450 or so people is disappearing. The drinking water supply will be gone by the end of 2018. The airport and school will be gone by 2020.

2 EVIDENCE FOR CLIMATE CHANGE

Key Concept 2a: A warming planet should see warmer average temperatures, melting land and sea ice, rising sea levels, and precipitation changes. All of these are currently being observed.

Key Concept 2b: Species' responses to climate change such as range shifts provide strong evidence that climate is changing enough to affect ecosystems.

Climate is complex, and it takes many lines of evidence to be sure that it's changing. In general, scientists look for a 30-year trend, at least, before concluding that climate has crossed the boundary of natural variability and that a new trend is emerging.

So it's no small matter to say that more than 97% of scientists, climate scientists included, agree that climate is now changing and that this change is in response to human activities (see Infographic 6). The evidence for this is substantial and well supported by thousands of studies from a wide variety of scientific fields (the hallmark of a well-established scientific theory; see Module 1.2). Of course, there is plenty of uncertainty over the rate at which future changes and impacts will unfold: No one can say for sure the exact year that Miami will be permanently flooded, for example. But the broad outline of consequences is not in dispute: Much of Miami will be under water at some point, if we don't address climate change. An average temperature increase of 3.5° to 7°F (~2° to 4°C)—which is where we are heading—will inundate our coastlines with rising water, drive many species to extinction, and lead to significant agricultural declines.

How can scientists be so sure of all this? Before delving into the possible causes of climate change, or the human contribution to that change, let's look at it scientifically—let's make some predictions of what would be true if climate were warming and examine the evidence for those predictions to see if it is actually occurring.

If climate were changing in the direction of warming, we would expect to see temperatures rising. They are: 2016 was the 40th consecutive warmer-than-normal year, 10 years past the 30 years that climatologists require as the benchmark for the emergence of a new climate trend. As of the end of 2016, all but one of the 12 warmest years on record occurred in the 21st century (1998 was the 8th warmest year at that time.) The record for warmest year was broken in 2014, then in 2015, then in 2016. As of this writing, the trend is continuing into 2017.

With rising temperatures, we would expect to see changes in weather such as more precipitation in coastal areas (more ocean water evaporation), more extreme storms (more heat in the atmosphere and oceans means more energy in the system to fuel storms and winds), and more record temperatures. All of these are occurring: There are roughly twice as many heat records set now as cold ones. And these heat waves are taking a toll on human life; the Paris heat wave of 2003 killed more than 700 people. A recent analysis by World Weather Attribution, an international coalition of scientists, estimates that extreme heat events are now two to four times more likely to occur in most places in Europe; as much as ten times more likely in Portugal.

A research team led by Michael Mann of Pennsylvania State University has linked some of these extreme events, such as storms, to alterations in the global jet stream that cause it to become stationary, keeping in place whatever weather pattern exists. Heat waves persist, rain or snow events stall over an area and dump record precipitation, or a dry spell lasts for weeks or months longer than usual. Even the catastrophic hurricane season of 2017 has a connection to climate change; though climate change didn't cause hurricanes Harvey, Irma, or Maria, scientists say it is making storms like these more powerful and, thus, more destructive.

"We are seeing . . . remarkable changes across the planet that are challenging the limits of our understanding of the climate system. We are now in truly uncharted territory," says World Climate Research Program Director David Carlson in a 2017 World Meteorological Organization press release.

We would also expect to see ice melting if temperature were warming. We are: Ice and permafrost (a deeper soil layer that is frozen for at least 2 consecutive years) are melting, and the rate of melt is accelerating; Arctic sea ice is setting record lows for ice extent in summer and more recently in winter as well. Glaciers are melting at an increasingly rapid rate. Scientists predict that all of the glaciers in Glacier National Park in Montana will be gone by 2030. The response in the Antarctic is less straightforward with some areas warming significantly and losing ice (enough to cause a noticeable increase in moss coverage, leading to a "greening" of some regions of Antarctica) while others are cooling and gaining snow. These differences are attributed to, in part, the way ocean currents are redistributing warm versus cool water.

If water is warming and land-based ice is melting, we expect to see sea level rising. We are: Water expands as it warms, and land-based ice can send meltwater to the ocean, both of which can contribute to sea-level rise. (Melting icebergs would not raise sea level since they already displace the same volume of water while floating as the meltwater they would add to the oceans.)

South Florida has seen a sea-level rise of 15–20 cm (6–8 inches); sea-level rise at the U.S. mid-Atlantic coast has been even greater and barrier islands there are losing ground, some even faster than the Isle de Jean Charles. And with higher sea levels come more damaging storm surges during hurricanes or violent weather and more coastal erosion. Some villages on the Alaskan coast are being destroyed by heavy wave action during storms that carves off shoreline.

Hauer's work evaluated population impacts at 0.9- and 1.8-meter (3- and 6-feet) sea-level increases (mid-range and high-range scenarios of future climate change as determined by Martin Vermeer of Helsinki University), and this kind of rise is not out of the question—if not by 2100, then later since sea level will continue to rise, probably for centuries, even after temperatures stabilize. High-end predictions (what we are likely to see with no efforts to curb fossil fuel use or other drivers of climate change) place sea-level rise around 2 meters (6 feet) by 2100. Even our low-end predictions, predicated on a robust international response to address the problem, predict a sea-level rise between 0.8 and 1.3 meters (2.6–4.3 feet) by 2100.

Unfortunately, we are not currently on track for that low-end prediction—sea-level rise will likely be higher, closer to Vermeer's mid-range scenario prediction of 0.9–1.45 meters (3–4.75 feet) by 2100. Indeed, in a 2016 assessment of sea-level change, Peter Clark of Oregon State University noted that our CO_2 emissions and other actions that are contributing to climate change are already dangerously close to the maximum we cannot exceed if we are to achieve the low-end scenario; unless things change drastically, and fast, that low-end scenario will soon be out of reach. Clark also pointed out that a historical analysis of sea-level rise shows that it is slow but persistent in the face of a warming climate; in the last major warming period (as Earth exited the last ice age), sea-level rise continued for 8,000 years after CO_2 levels and global temperature stabilized. He writes ". . . twenty-first-century global average warming will [produce] a climate state not previously experienced by human civilizations."

So there is ample evidence of physical changes in air, on land, and at sea. (We've presented just a smattering of this physical evidence here.) The bigger question might be, does this matter—is it affecting life on Earth?

> **"Twenty-first-century global average warming will [produce] a climate state not previously experienced by human civilizations."**
> —Peter Clark

Back to our predictions: *If all of these changes were significant enough to affect species, we would expect to see clear impacts on biodiversity.* This is happening: Changes in habitats and niches that affect where species are found as well as broken relationships (e.g., the loss of mutualistic partners or prey) are being seen. There is ample evidence for this, as well.

Some species are benefiting from a warmer climate. For example, bark beetles that attack trees in North America are increasing in number; their populations usually die back in the winter, helping to keep them in check and allowing the trees to recover. However, in some areas, winter temperatures don't always get cold enough to kill the beetles, allowing the beetles to thrive year-round. So far, nearly 4 million acres in the western United States have suffered damage. Tropical species of mosquitoes, too, are expanding their ranges to higher latitudes. Many species of squid and octopus are also increasing in number and distribution, believed to be due to their ability to adapt to warmer temperatures even as other species (many of them competitors of the squid and octopus) decline.

However, even more species are negatively affected as their range shrinks or shifts to higher latitudes or altitudes. Unfortunately, they can't always migrate to a more suitable location because we have fragmented habitats and destroyed migratory pathways (see Online Module 3.3 for more on the impacts of habitat fragmentation). For species already at the upper edge of suitable habitat (top of the mountain, at the poles) there is simply nowhere else to go. But those that can relocate are on the move; scientists have documented a wide variety of species shifting their ranges to higher latitudes or elevations. Marine species too are moving northward or deeper into the ocean. (See the *Science Literacy* activity at the end of this module for a look at migration trends in marine species.) Even trees are "migrating" as their saplings preferentially survive at, or just beyond, the higher altitudes or latitudes of their traditional ranges.

Shifts in the timing of migration are also uncoupling species connections: Some migratory birds are arriving at summer breeding grounds at their normal time to find the insects that normally provide the main food source for their offspring have already hatched out or emerged from winter hibernation and left the area. (Many insects hatch out or emerge based on temperature cues, so a warming climate means they arrive sooner; most birds migrate in response to daylight cues, which are not affected by climate.) In addition, because there will likely be fewer predators around when these insects arrive or hatch out early, the insects can wreak havoc on plant life (both native plants and crops). The fact that so many and such a wide variety of species from all of Earth's ecosystems are responding is some of the strongest evidence we have, not only that climate is changing but also that it matters.

Species differ in their ability to adjust to rapid climate changes—a reflection of their generation time, reproductive potential, the genetic diversity of the population (see Module 3.1 for more on how populations evolve in response to environmental change), and their ability to relocate, among other things. The fact that many species are endangered, with their numbers already critically low, makes them even more vulnerable. Climate is simply changing too quickly for many

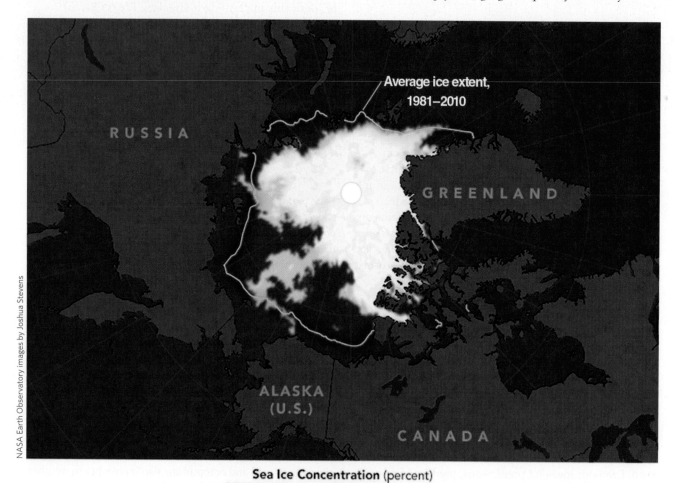

NASA Earth Observatory images by Joshua Stevens

Sea Ice Concentration (percent)

0   25   50   75   100

The 2016 melt of summer sea ice in the Arctic tied the 2007 record for the lowest ice extent since records began. The ice is also thinner, contributing to higher melt rates each year. In some years, enough ice melts to open the Northwest Passage, a long sought-after shipping lane in the Arctic Ocean.

The Isle de Jean Charles in Louisiana has lost 98% of its land. Considered to be climate refugees, residents have received federal funding to relocate.

species to adapt. The loss of species, from habitat loss or the loss of ecosystem partners, impoverishes the ecosystem further, likely triggering the endangerment of more species in a domino effect that can escalate. This then reduces the ability of the ecosystem to provide the ecosystem services on which all species, including us, depend. **INFOGRAPHIC 2**

As humans from Alaska to Louisiana to islands of the South Pacific face migration for the same reasons—loss of suitable "habitat"—we are faced with a complicated mix of problems. In Newtok, the primary problem is money. The village is small—just 450 people—but the costs of relocation will still be large. Most estimates put the price tag at more than $100 million. Residents know they need to move, and fast. "We just need to get out of there," Romy Cadiente, the village's relocation coordinator told National Public Radio in 2017. "We really do, for the safety of the 450 people there." But without financial aid, they simply can't afford to move.

A group of Newtok villagers is pushing the federal government to declare Newtok a disaster zone so that emergency funds would be available for the move. But disasters are usually declared for specific weather events, like a hurricane or flood, not for damage that's mounted over decades as Newtok's has (revealing that the differences between weather and climate are not just theoretical; they are practical, too).

But even when money is available, the problems of relocation are still legion. The Isle de Jean Charles managed to secure a $48 million federal grant to move just 60 people—one of the very first allocations of federal money to assist climate refugees. But efforts have been fraught with bickering and uncertainty: Where should everyone go? What will become of the land they leave behind? What will become of their community and culture if they are dispersed?

INFOGRAPHIC 2   **EVIDENCE FOR CLIMATE CHANGE**

We know that a variety of factors can alter global temperature, but what is the evidence that temperatures have actually increased and that the climate is changing? In other words, what do we predict we would see if warming were occurring, and what do we actually see when we test those predictions?

WARMER TEMPERATURES

If climate is indeed warming, we expect to see warmer global temperatures, on average, than in the recent past (more temperature anomalies in the direction of warming).

TEMPERATURE IS WARMING

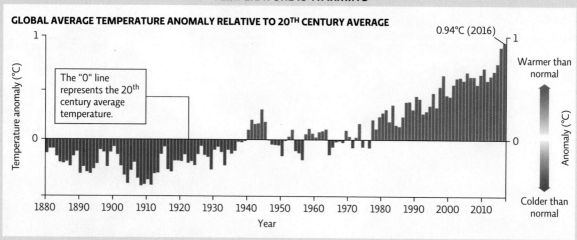

GLOBAL AVERAGE TEMPERATURE ANOMALY RELATIVE TO 20TH CENTURY AVERAGE

0.94°C (2016)

The "0" line represents the 20th century average temperature.

Warmer than normal

Colder than normal

Climatologists generally look for directional changes over at least 30 years before concluding that the changes represent a trend. If you were born after 1976, you have never experienced a colder than average year relative to the 20th century average.

WARMER TEMPERATURES SHOULD LEAD TO:

MELTING ICE

If temperatures are warming, we would expect to see more ice melt.

SEA-LEVEL RISE

We would expect to see an increase in sea level as land-based ice melts and as warmer seawater expands.

ICE IS MELTING

SEA LEVEL IS RISING

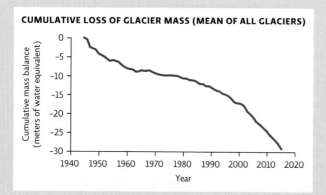

CUMULATIVE LOSS OF GLACIER MASS (MEAN OF ALL GLACIERS)

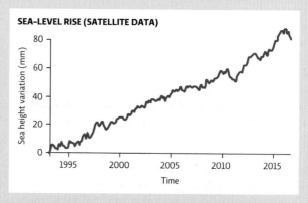

SEA-LEVEL RISE (SATELLITE DATA)

Since the middle of the 20th century, glaciers around the world have experienced a net loss of ice. Record ice melt has been observed in the Arctic and Antarctic as well. Permafrost is also thawing in high-latitude and high-altitude regions.

Sea level rose around 200 mm (~8 in.) between 1900 and 2000 (as determined by tide gauges) and the rate of increase has accelerated since 1993 with a 55-mm (2-in.) increase in the first 16 years of the 21st century.

WARMER TEMPERATURES SHOULD ALSO LEAD TO:

CHANGES IN WEATHER

CHANGES IN BIOLOGICAL EVENTS

PRECIPITATION IS CHANGING

PRECIPITATION VARIATION SINCE 1900

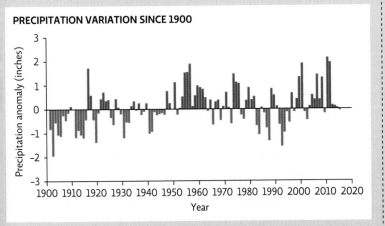

Since 1950, there have been 40 wetter-than-average years but only 20 drier-than-normal years. The first half of the century was drier than the average for the century.

SPECIES ARE SHIFTING RANGES

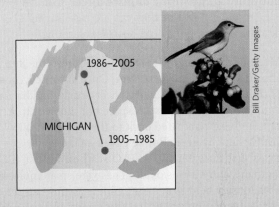

The ranges of some species are shifting. For example, the breeding range of the blue-gray gnatcatcher has shifted about 300 km northward since 1970.

CHANGE IN SEASONAL WEATHER PATTERNS

TIMING OF SPRING AND FALL FROSTS (UNITED STATES)

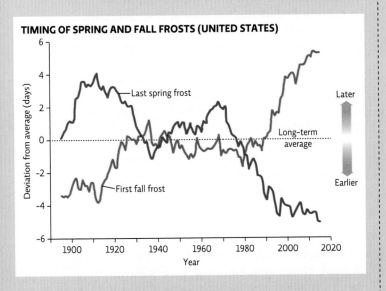

Since about 1980, the last spring frost is coming earlier and the first fall frost is coming later, extending the growing season. Note that in recent years, these first and last frost dates exceed any that were seen since 1890.

PHENOLOGICAL SHIFTS

LENGTH OF THE FLOWERING SEASON

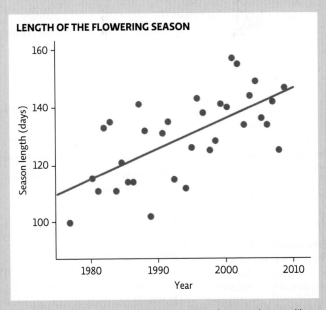

Changing climate can also change the timing of seasonal events like blooming or migration (phenology changes). A Colorado study that began in 1974 and followed 121 species showed that the length of time these flowers are in bloom has increased.

 Why is it important to look at multiple lines of evidence that climate is changing rather than just evaluate global temperatures?

3 CLIMATE FORCERS: THE GREENHOUSE EFFECT AND HUMAN IMPACT

Key Concept 3: Greenhouse gases trap heat reradiated from Earth and warm the atmosphere. By adding more greenhouse gases to the atmosphere, the greenhouse effect is enhanced, warming the planet even more.

What is causing this climate change? For that matter, what kinds of things have the power to affect climate?

The Sun fuels life on Earth by providing the energy for photosynthesis (see Module 2.1), but it also fuels Earth's climate. Anything that alters the balance of incoming solar radiation relative to the amount of heat that escapes into space is known as a **climate forcer**. Climate forcers can be *positive* (that is, they can increase warming) or *negative* (meaning they can decrease warming).

Much of the incoming solar radiation from the Sun is absorbed at the Earth's surface. Once absorbed, most of this energy is reradiated as heat (infrared radiation). Some of this heat is lost to space but much of it is captured and reradiated back to the surface and lower atmosphere by **greenhouse gases**—atmospheric gases such as water vapor (H_2O), carbon dioxide (CO_2), methane, and nitrous oxide. These are small molecules with loosely held atoms that vibrate as they absorb infrared radiation, capturing that energy as heat, which warms the atmosphere. This process, known as the **greenhouse effect**, helps warm the Earth; therefore, greenhouse gases are positive climate forcers. To be sure, the greenhouse effect is what keeps Earth habitable for humans and all current species on the planet. Without these greenhouse gases, the average temperature on Earth would be around 0°F (−18°C)—that's about 61°F (34°C) colder than the planet's current average temperature! However, scientists have long understood that adding more greenhouse gases to the atmosphere would contribute to an *enhanced greenhouse effect* that would warm the planet.

Scientists have begun to see evidence of that enhanced greenhouse effect, which they have since linked to human (anthropogenic) activities. At the top of the list, the biggest contributor to global warming is the burning of fossil fuels (coal, petroleum-based fuels, and natural gas), which releases massive quantities of greenhouse gases into the atmosphere, an action that began in earnest with the Industrial Revolution. Isotopic analysis of atmospheric CO_2 allows scientists to identify the percentage that comes from fossil fuel combustion. (There is

a different ratio of carbon-12 to carbon-13 in fossil fuel emissions as compared to CO_2 released from natural sources such as forests or soils.) This allows us to determine how much is from fossil fuel use—more than 4 gigatons (Gt): That's 400 billion metric tons since 1751, more than 75% of total human emissions. Other anthropogenic sources of greenhouse gas emissions are certain agricultural activities like rice farming and cattle rearing, as well as deforestation, the release of certain industrial chemicals like chlorofluorocarbons (CFCs) (see Module 5.2), and landfill waste disposal (see Module 5.3).

> Scientists have long understood that adding more greenhouse gases to the atmosphere would contribute to an *enhanced greenhouse effect* that would warm the planet.

Molecule for molecule, CO_2 is not the most potent greenhouse gas, but it currently accounts for about three-fourths of all greenhouse gas emissions and is the one that is contributing most to global warming at this time. Methane, the second most abundant anthropogenic greenhouse gas, is released from rice paddies, cattle, and landfills; the largest single source is the fossil fuel industry, especially methane lost from oil and gas wells. Though methane emissions are considerably lower than CO_2 emissions, methane is around 28 times more potent that CO_2 over the long term, so these emissions are significant. Halocarbons like CFCs are even stronger greenhouse gases. Nitrous oxide, coming in at almost 300 times more potent than CO_2, makes up about 2% of all anthropogenic emissions (sources include agriculture and vehicle emissions).

Scientists have been taking continuous measurements of atmospheric CO_2 concentrations since the 1950's. A graph of that data, known as the *Keeling Curve* (after David Keeling, the scientist who initiated the measurements), shows that CO_2 levels are rising steadily, topping 400 ppm in 2016. (For perspective, atmospheric CO_2 levels stayed around 280 ppm for much of human history.) **INFOGRAPHIC 3**

climate forcer Anything that alters the balance of incoming solar radiation relative to the amount of heat that escapes out into space.

greenhouse gases Molecules in the atmosphere that absorb heat and reradiate it back to Earth.

greenhouse effect The warming of the planet that results when heat is trapped by Earth's atmosphere.

INFOGRAPHIC 3 **GREENHOUSE GASES AND THE GREENHOUSE EFFECT** ANIMATED INFOGRAPHIC

Life on Earth depends on the ability of greenhouse gases in the atmosphere to trap heat and warm the planet. More greenhouse gases, however, mean more trapped heat and a warmer planet (an enhanced greenhouse effect).

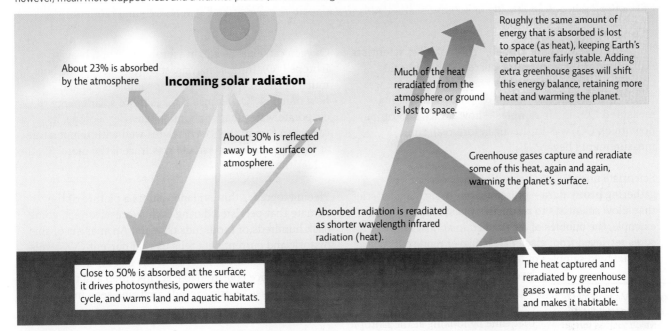

About 23% is absorbed by the atmosphere

Incoming solar radiation

About 30% is reflected away by the surface or atmosphere.

Much of the heat reradiated from the atmosphere or ground is lost to space.

Roughly the same amount of energy that is absorbed is lost to space (as heat), keeping Earth's temperature fairly stable. Adding extra greenhouse gases will shift this energy balance, retaining more heat and warming the planet.

Greenhouse gases capture and reradiate some of this heat, again and again, warming the planet's surface.

Absorbed radiation is reradiated as shorter wavelength infrared radiation (heat).

Close to 50% is absorbed at the surface; it drives photosynthesis, powers the water cycle, and warms land and aquatic habitats.

The heat captured and reradiated by greenhouse gases warms the planet and makes it habitable.

Different greenhouse gases have different abilities to trap heat; their heat-trapping capacity is expressed as CO_2-equivalents (the amount of CO_2 that would produce the same warming). For example, since a molecule of methane (CH_4) traps 28 times as much heat as CO_2, one methane molecule is equivalent to 28 CO_2 molecules.

GREENHOUSE GASES: RELATIVE CONTRIBUTIONS TO GLOBAL WARMING

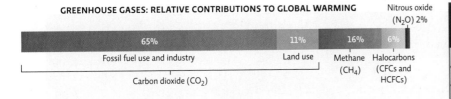

Nitrous oxide (N_2O) 2%

| 65% | 11% | 16% | 6% |

Fossil fuel use and industry | Land use | Methane (CH_4) | Halocarbons (CFCs and HCFCs)

Carbon dioxide (CO_2)

GREENHOUSE GAS	GLOBAL WARMING POTENTIAL RELATIVE TO CO_2 (OVER 100 YEARS)
Carbon dioxide	1
Methane	28
Nitrous oxide	265
Halocarbons	124–14,800

CO_2 is one of the greenhouse gases that is increasing. Historic CO_2 levels are estimated from ice cores like the Law Dome ice core from Antarctica; current levels have been measured directly at locations like Mauna Loa, Hawaii, since 1958.

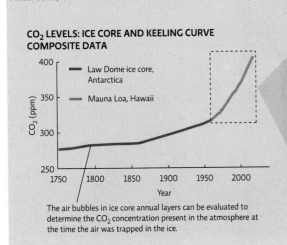

CO_2 LEVELS: ICE CORE AND KEELING CURVE COMPOSITE DATA

— Law Dome ice core, Antarctica

— Mauna Loa, Hawaii

CO_2 (ppm)

Year

The air bubbles in ice core annual layers can be evaluated to determine the CO_2 concentration present in the atmosphere at the time the air was trapped in the ice.

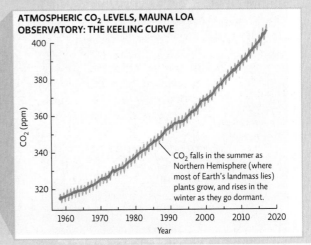

ATMOSPHERIC CO_2 LEVELS, MAUNA LOA OBSERVATORY: THE KEELING CURVE

CO_2 (ppm)

CO_2 falls in the summer as Northern Hemisphere (where most of Earth's landmass lies) plants grow, and rises in the winter as they go dormant.

Year

 Use the concept of the greenhouse effect to explain how Earth's surface could warm up even if the Sun's output does not change.

4 TEMPERATURE AND CO₂: COLLECTING AND INTERPRETING THE DATA

Key Concept 4: A variety of methods are used to measure past and present atmospheric CO_2 levels; these levels have been rising dramatically in recent decades. All show a positive correlation between CO_2 and temperature.

To predict what future climate will look like, scientists must do more than monitor current atmospheric conditions; they also need to know what climate was like in the distant and not so distant past. Keeling measured CO_2 in air samples—how do we determine how much CO_2 was in the air before real-time measurements began?

Scientists uncover clues about historic climates by gathering **proxy data**—preserved physical characteristics that allow scientists to reconstruct past climates. For example, the bubbles of air trapped in annual layers of ice cores retrieved from glaciers (each layer represents a year in time) hold samples of the atmosphere at the time that layer was laid down. CO_2 levels can be measured directly, but scientists can also infer the temperature at that time by looking at the isotopic ratio of the oxygen atoms in the water (H_2O) molecules of the ice.

proxy data Measurements that allow one to indirectly infer a value such as the temperature or atmospheric conditions in years past.

Oxygen can exist as a light isotope (^{16}O) or a heavier isotope (^{18}O). In this analysis, scientists determine how much of the oxygen in the water sample is ^{16}O and how much is ^{18}O. Because it takes more solar energy to evaporate water that contains the heavier isotope, its concentration in the ice correlates well with temperature at the time the water sample was frozen; the more ^{18}O, the colder the climate.

Piecing together this evidence paints a picture of the climate that persisted during each successive year, going back hundreds of thousands of years. An analysis of this historic and current data tells us that current CO_2 levels are higher than anything seen in the last 2.7 million years (the longest ice core available at this time) and that the rate of increase greatly exceeds anything seen in that time period.

Tree growth, another proxy measurement, is tied to temperature and water availability. This means

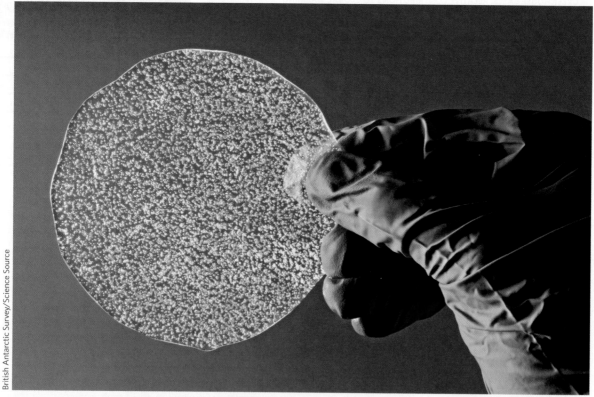

British Antarctic Survey/Science Source

A researcher holds a thin slice of ice from an ice core extracted from Antarctica. The bubbles in the ice contain air from long ago and can be evaluated to determine the atmospheric composition at the time this layer of ice was laid down. The water too can be analyzed to determine the temperature at that time.

Temperature and CO_2 levels closely track one another. Changes that initiate warming will cause CO_2 levels to rise. Likewise, events that release extra CO_2 will result in warming. Today, warming is caused by our release of extra CO_2 from fossil fuel burning and other actions.

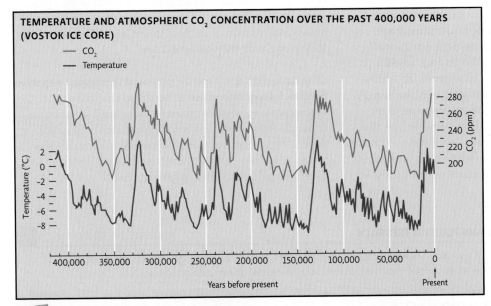

TEMPERATURE AND ATMOSPHERIC CO₂ CONCENTRATION OVER THE PAST 400,000 YEARS (VOSTOK ICE CORE)

— CO₂
— Temperature

PROXY DATA

Clues about past climates can be gleaned from the use of proxy measurements.

- Ice cores contain annual layers with atmosphere samples from long ago that can be analyzed.

- Annual layers of sediment cores can be evaluated for pollen or other climate indicators.

- Tree ring analysis reveals clues to the climate in each year of the tree's life.

- Analysis of coral allows one to determine the water temperature when the skeleton was formed.

 Explain how increasing atmospheric levels of CO_2 are both a cause and effect of warming.

each annual tree ring provides clues about how wet and warm that year was (a field of study known as dendrochronology). And this analysis is not restricted to living trees: Trees used in ancient buildings or unearthed from burial can be dated using radiometric dating techniques and their rings analyzed. Scientists now have a tree-ring chronology that goes all the way back to the last ice age, more than 12,000 years. The oxygen isotope ratio in coral skeletons can also be evaluated to determine how warm the water was when that skeleton was laid down. And lake sediments can be evaluated for the species and amounts of pollen that were prevalent at various time points (e.g., pollen from heat-tolerant plants indicates a warmer climate).

Together, these data (and other lines of evidence) help scientists understand past climates. The ice core data, in particular, show that CO_2 and temperature are highly correlated through time; they increase and decrease in tandem. They also show that CO_2 can be both a cause and effect of warming temperatures. For example, if a natural event leads to warming, that warming will in turn lead to more atmospheric CO_2 (an effect) as soils and water give up more stored CO_2. Those higher CO_2 concentrations will then trigger even more warming (a cause). So regardless of whether the excessive greenhouse gases are the result of natural or human forcers, the end result is the same: Triggering warming can set into motion events that lead to even more warming. **INFOGRAPHIC 4**

5   CLIMATE FORCERS OTHER THAN GREENHOUSE GASES

Key Concept 5a: The albedo (reflectivity) of a surface affects surface temperatures and climate. Decreasing albedo can increase warming via positive feedback.

Key Concept 5b: Past climate changes are correlated with natural forcers such as the Milankovitch cycles, but these cycles do not account for current warming.

Let's revisit the concept of climate forcers. Greenhouse gases are not the only forcers. Another forcer that plays a role in warming trends is **albedo**, the ability of a surface to reflect away solar radiation. Think of it as the *reflectivity* of a surface: Light-colored surfaces, like glaciers and meadows, have a high albedo; they reflect sunlight

away from the planet's surface, reducing the amount that is absorbed and reradiated as heat. Darker surfaces like water and asphalt have low albedo: They absorb sunlight and then reradiate that solar energy back to the atmosphere as heat.

albedo The ability of a surface to reflect away solar radiation.

As surfaces with high albedo (high reflectivity) are replaced by those with low albedo, not only does the planet warm, but a **positive feedback loop** can be triggered—a cycle whereby an observed change in a trend is accentuated (i.e., warming causes more warming).

Melting sea ice provides a good example of positive feedback: As temperatures rise, sea ice melts, and ice (with its high albedo) gives way to water (with its low albedo). Because this new watery surface absorbs more sunlight than the former icy surface, the region warms even faster—causing even more ice to melt into water, triggering more warming, and so

positive feedback loop Changes caused by an initial event that then accentuate that original event.

negative feedback loop Changes caused by an initial event that trigger events that then reverse the response.

on. Albedo changes are, in large part, responsible for the much greater increase in temperatures seen in the Arctic relative to the global average. Positive feedback loops tend to have a destabilizing effect on the environment, continually altering it as long as they are at work. (Usually, this lasts until things settle down at a new steady state; in this case, that might be when most of the ice has melted.) **INFOGRAPHIC 5A**

Climate forcers also have the potential to trigger a **negative feedback loop**, one where an observed change in trend is reversed (i.e., when warming leads to events that cause cooling). Your body's ability to regulate its temperature depends on a negative feedback loop: When you are too warm, your body sweats to cool itself down; if you get too cold, your muscles shiver to generate heat.

INFOGRAPHIC 5A | **ALBEDO AND POSITIVE FEEDBACK**

ANIMATED INFOGRAPHIC

Albedo is a measure of the reflectivity of a surface. The lighter colored the surface, the higher the albedo. Unreflected (absorbed) light is reradiated as heat, so surfaces with a low albedo release more heat to the atmosphere than do high-albedo surfaces.

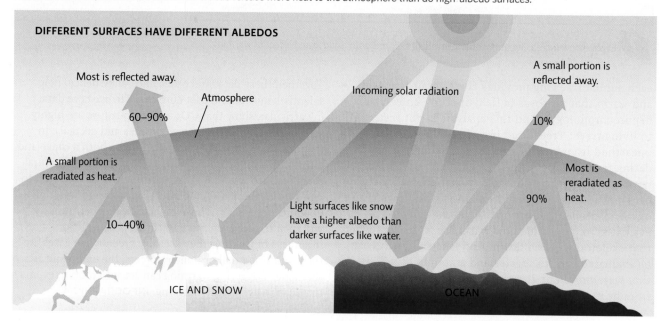

DIFFERENT SURFACES HAVE DIFFERENT ALBEDOS

Most is reflected away.

Atmosphere

Incoming solar radiation

A small portion is reflected away.

60–90%

10%

A small portion is reradiated as heat.

Most is reradiated as heat.

10–40%

90%

Light surfaces like snow have a higher albedo than darker surfaces like water.

ICE AND SNOW

OCEAN

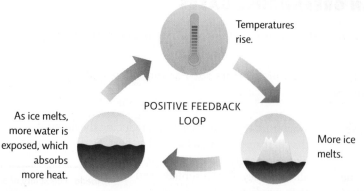

Temperatures rise.

POSITIVE FEEDBACK LOOP

More ice melts.

As ice melts, more water is exposed, which absorbs more heat.

When sea ice melts, it uncovers water, which has a darker surface with a lower albedo. This activates a positive feedback loop: As the exposed water absorbs sunlight and releases more heat into the atmosphere, more ice melts and more water is exposed, which then absorbs more sunlight, releasing even more heat, and so on. Positive feedback loops represent changes that trigger additional change in the same direction (warming that triggers even more warming); the word *positive* is not mean to indicate a "positive," or beneficial, event.

? One suggestion to combat global warming is to replace dark rooftops with light-colored ones. How would this help reduce warming?

In the environment, certain types of clouds can produce a negative feedback loop: As it warms, more water evaporates to form clouds. If more low, thick clouds are formed, those could reflect away incoming solar radiation, thus cooling the area. As the temperature cools, less water evaporates and fewer of these clouds form. As a result, the temperature rises again, and more clouds form. And on and on. The end result of a negative feedback loop is stabilizing. In this example, if this negative feedback loop were occurring often enough, it would help stabilize the climate—but unfortunately it is not happening enough to mitigate warming. Currently, not enough low clouds are forming to offset warming, so this particular negative feedback loop is having only a small effect on climate.

Volcanic eruptions and changes in the magnitude of solar irradiance are also considered natural forcers, as both have been known to impact climate in the past, though only over short time frames and not as strongly as greenhouse gases. Sulfur particles emitted from volcanic eruptions or from industrial activities are light in color, so they have a high albedo and act as negative forcers (they cool the planet). Darker particles, such as soot released from burning fossil fuels, are positive forcers due to their low albedo. Currently, actions that increase the albedo of the Earth's atmosphere are outweighed by positive forcers—the net effect is warming.

Scientists also believe that **Milankovitch cycles** (predictable long-term cycles of Earth's position relative to the Sun) played an important role in earlier climate change events such as the Pleistocene ice ages. During times when the Earth's position in space meant less solar irradiation reached the planet, events were triggered that initiated cooling. These events likely led to positive feedback loops that accelerated cooling (e.g., less CO_2 exchange with the atmosphere) and produced a climate several degrees colder, on average, than when Earth is closer to the Sun. At times, when Earth is in a position to receive more solar energy, a warming trend could be triggered that leads to a new climate. These are very long-term cycles and would not produce noticeable changes over just a few decades. Even if their effect was discernable, at present, Earth's position relative to the Sun is trending toward a cooling phase. (The shape of Earth's orbit or the tilt of its axis is not putting us in a position relative to the Sun that would lead to warming.) This means the Milankovitch cycles, while important to explain past climate changes that unfolded over long periods of time, do not correlate with current warming. **INFOGRAPHIC 5B**

In fact, none of these natural forcers—together or by themselves—account for the current warming trends.

Milankovitch cycles
Predictable variations in Earth's position in space relative to the Sun that affect climate.

INFOGRAPHIC 5B   MILANKOVITCH CYCLES HELP EXPLAIN PAST CLIMATE CHANGE

Warm periods and ice ages of the past can be attributed in part to Earth's position in space relative to the Sun. Earth has three different cycles that can each have an impact on climate. The current warming we are experiencing cannot be explained by any of these cycles—Earth is currently not in a part of any cycle in which it would have greater warming.

Earth's orbit can be round or slightly elliptical; currently it is round.

Earth's tilt is currently 23.4°, which gives us cooler summers and warmer winters than we'd have at 24.5°, which was our position 9,000 years ago. In about 32,000 years, Earth will be at 22.1°.

Earth's current position is tilted toward the North Star; in about 12,000 years, the axis will point toward Vega.

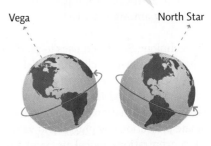

ORBITAL ECCENTRICITY

The shape of Earth's orbit around the Sun varies over a 100,000-year cycle. When it is more elliptical, climate is more variable because some seasons receive more solar radiation than normal and others receive less.

AXIAL TILT

The angle of Earth's tilt as it spins on its axis changes in a 41,000-year cycle. The greater the angle, the greater the extremes between seasons (hotter summers and colder winters).

AXIAL PRECESSION

Earth "wobbles" on its axis, changing not the angle but the direction the axis points in a 20,000-year cycle. This changes the orientation of Earth to the Sun and affects the severity of the seasons. When Earth is tilted toward Vega, it is also tilted toward the Sun during summer, making summers hotter in the Northern Hemisphere.

 Why do scientists conclude that Earth's axial tilt is not responsible for our current warming?

6 ATTRIBUTION: HUMAN VERSUS NATURAL CAUSES

Key Concept 6: Current warming cannot be explained without accounting for both natural and anthropogenic climate forcers.

On his laptop, Alex de Sherbinin, a climate scientist at Columbia University's Lamont Doherty Earth Observatory, can render the world's emerging disaster zones in such exquisite detail that it's possible to see exactly which homes might be destroyed by a few feet rise in sea level. Deep red hues show where populations are densest; streaks of blue overlay the areas most likely to disappear into the ocean. The overall picture is not pretty: With 1 meter (~3 feet) of sea-level rise, the Ganges delta region of Bangladesh and India, home to 144 million people, will be inundated with flooding. Vietnam will lose more agricultural land than any other place on Earth. And the Sahel region of West Africa will see a doubling in the number of people facing water shortages, owing to a massive decrease in rainfall.

Climate refugee predictions (estimates of how many people will be forced to flee their homes as rising temperatures or drought make some areas inhospitable, and rising seas flood out others) are also based on climate models. The most commonly cited estimates put the total number of people affected at between 50 and 200 million by the year 2050. That upper limit comes out to about 1 in every 50 people on Earth (assuming a 2050 population of 9.8 billion, the current projection). Those specific numbers may be in dispute—no one can say for certain exactly which areas will be too hot or dry to inhabit or how much land will sink into the sea, or by what year. But the bottom line—that many millions of people will be forced to flee their homes at some point in the next century—is a certainty among scientists.

The technique Sherbinin uses to produce these future scenarios is known as climate modeling. Climate scientists have used a wealth of current and historical data to develop *climate models*—computer programs (complex mathematical equations) that allow them to make future climate projections by plugging in values for temperature, CO_2 concentrations, global air circulation patterns, and so on. These models are used to see how altering the value of certain parameters (say, increasing the amount of atmospheric CO_2) might impact future climate. Their validity is tested by entering data for years past to see how well the model predictions match the climate that was seen during those years. A good match tells them they have accurately accounted for the forcers at work.

These powerful climate models allow us to determine the relative contribution of natural forcers (e.g., solar output, cloud formation, volcanic eruptions, Milankovitch cycles, etc.) and of anthropogenic forcers (e.g., burning fossil fuels, deforestation, agricultural practices, land uses)—a focus of climate science known as *attribution*. This is vitally important. By determining what the causes of climate change are, and their relative contribution, we discover which climate forcers can and should be addressed. In other words, if volcanic eruptions and the Earth's orbit were the primary drivers of climate change, there would not be much we could do about it. But if human actions are significant causes, that would be something we could address.

So what do the models say? When data for just the natural forcers are entered into the models, they match past temperatures pretty well (so we know the model is accurately accounting for these forcers), but they do not replicate the post-1960's warming we have already observed. Natural forcers alone should, according to the models, produce a fairly stable climate. Only when we consider both natural and anthropogenic forcers together do the models replicate the current climate trend we are experiencing. **INFOGRAPHIC 6**

> Only when we consider both natural and anthropogenic forcers together do the models replicate the current climate trend we are experiencing.

The data are clear: The vast majority of this change is due to human activities, especially the burning of fossil fuels, and the consequent release of greenhouse gases like CO_2 into the atmosphere. "Climate scientists overwhelmingly agree," says James Cook, a researcher at the University of Queensland in Australia. "Humans are causing recent global warming."

But this is actually good news. We can alter our path and stop doing things that contribute to climate change.

INFOGRAPHIC 6 WHAT'S CAUSING THE WARMING?

Climate scientists use computer models that take into account the major factors that are known to have affected past climates in order to see what might be responsible for recent warming. Data about natural and anthropogenic factors can be fed into a computer model separately and then together to see which circumstances match up with the warming that has been observed.

COMPUTER MODELS' RECONSTRUCTION OF PAST TEMPERATURES

CONSIDERING ONLY NATURAL CAUSES

— Observed temperatures
▨ Multiple climate model predictions
— Model average

The average temperatures predicted by the model and actual temperatures do not match after about 1960.

CONSIDERING NATURAL AND ANTHROPOGENIC CAUSES

— Observed temperatures
▨ Multiple climate model predictions
— Model average

Only when both natural and anthropogenic causes are considered in the models do the two lines more closely align.

What would you expect the purple line and shaded area generated by the model in the left-hand graph in this infographic to look like, relative to the blue line (observed temperatures) if current warming could be explained by natural causes?

7 IMPACTS OF CLIMATE CHANGE

Key Concept 7: Climate change impacts include health and agricultural effects, biodiversity loss, more extreme storms, increased fire risk, and coastal erosion and flooding.

In some ways, the demise of communities like Newtok and Isle de Jean Charles has been slow and brutal. Even before area land was swallowed by the water, climate change was already affecting fishing and farming and clean water in the areas: Fisheries declined as a result of warming water, soil was eroded by increased storms, and clean water was contaminated or depleted by the same. And as much as anything else, communities and the cultures that they support are wholly lost when land disappears.

These other losses illustrate something of which scientists have long been aware: Climate change won't just affect the weather. Climate change will have, and is having, environmental, economic, societal, and health consequences. For example, climate change affects water supplies—making freshwater scarce in places where droughts increase or glaciers melt away and contaminating

existing water sources in places where storms and flooding become more prevalent.

While some crops or areas may benefit from warmer temperatures, longer growing seasons, and more CO_2 in the air, agriculture in many areas is taking a hit as productivity is declining. (While CO_2 is needed for photosynthesis, and more can boost photosynthesis in plants, other limiting nutrients that restrict growth prevent plants from taking in all the extra CO_2 we are releasing; therefore, boosted plant growth will not eliminate the CO_2 problem for us. See Module 2.1 for more on nutrient cycles and limiting growth factors.) Heat stress can hamper growth, as can water stress. Rainfall might decrease and the rain that does fall could come in more powerful storms, flooding crops and washing away valuable soil. Many weed species are likely to handle the temperature and water stress of a new climate and so may proliferate, reducing harvests. Pest outbreaks can increase in areas where other insect populations (the predators of those pests) have declined or where pest species expand their ranges to enter new territory. Fisheries are also vulnerable to climate change as temperature and acidification of water alter food chains, potentially reducing the population size of many important fish and shellfish species.

Natural ecosystems are affected too. As mentioned earlier, habitat shifts may force species to adapt or migrate, but not all will be able to do so. Thus, climate change is seen as one of the main forces threatening species today. Wildlife populations are declining in many areas, and by some estimates future declines will be dramatic as populations that can't move or adapt go extinct. We are already seeing these and other impacts, and they will only get worse.

Climate change is most pronounced at higher latitudes such as the Arctic, which has warmed considerably more than other areas thanks to greater changes in albedo. Organisms of colder climates are also more vulnerable to climate change, adapted as they are to a narrower window of temperatures. In addition, they have nowhere else to go—there are no "colder" habitats farther north or farther up the mountain to which they can migrate. For this reason, many high-latitude and high-altitude species such as polar bears, walrus, penguins (and other polar species), and mountain species such as pika (a species related to the rabbit), the honeycreepers (birds) of Hawaii's mountains, and many other alpine plants and insects are threatened with extinction by climate change.

The ways that species can be affected are as varied as the species themselves. Sea-level rise that erodes and floods beaches is threatening some sea turtle species, many of whom are already endangered. Some reptiles, like turtles and alligators, are particularly vulnerable to temperature changes as the sex of an individual is affected by the incubation temperature of the egg. Warmer temperatures could produce a preponderance of females (for turtles) or males (for alligators), impacting the reproductive potential of future populations. A lack of moisture on the leaves of the wet forest habitat of the lemuroid ringtail possum of Queensland, Australia, is endangering this marsupial that depends on leaf moisture as its water source. The loss of sea ice threatens not only the polar bear but also its favorite prey, the ringed seal, a species that depends on the ice for reproduction (it gives birth on the ice and young seals shelter in "snow dens" on the ice). The Bramble Cay melomys, a small rodent once

found on only one island off the Australian coast, has the distinction of being the first species whose extinction was directly related to climate change (its habitat was flooded), though there are likely others for which climate change was a contributing factor. Sadly, there will be more. (See Module 3.1 for more on species endangerment.)

Ocean ecosystems are especially hard hit. Rising temperatures are stressing many organisms, especially those like coral that cannot get up and move. But the oceans are also experiencing another consequence of fossil fuel burning—ocean acidification. Much of the CO_2 released into the atmosphere makes its way into the oceans—a good thing in that it has slowed the buildup of this greenhouse gas in the atmosphere. But it is a bad thing for the inhabitants of the oceans because the addition of CO_2 is lowering the pH of the water. This can cause exoskeletons and shells to dissolve, and it interferes with the organism's ability to produce new shell or skeleton—a double whammy. Rising temperatures and acidification are wreaking havoc on ocean communities around the planet; 2016 witnessed the greatest coral bleaching event ever seen in the Great Barrier Reef, serious enough to kill almost 90% of the coral in some sections of the reef. Terry Hughes of the Australian Research Council Centre of Excellence for Coral Reef Studies says, "Climate change is not a future threat. On the Great Barrier Reef, it's been happening for 18 years." (See Online Module 6.3 for more on ocean acidification and ocean ecosystems.)

Human health suffers as insect-borne diseases increase in both incidence and range. Severe weather, flooding, and heat waves also imperil health. Agricultural declines can put more people at risk for undernutrition, and the lack of enough clean fresh water can be devastating to a population. Water scarcity is already the subject of water wars (see Module 6.1) and is only expected to worsen. Other, perhaps unexpected, effects on human health include the production of more allergy-inducing pollen by plants, due to more favorable growing conditions for some or, interestingly, poor growing conditions for others (many species of plants produce an abundance of pollen when stressed, an adaptation that favors reproduction at a time when the plant might not survive to reproduce another year). Higher CO_2 levels in the atmosphere appear to be causing poison ivy plants to produce larger leaves and more of the rash-inducing urushiol oils to which so many people are allergic. **INFOGRAPHIC 7**

INFOGRAPHIC 7 THE IMPACTS OF CLIMATE CHANGE

Impacts from climate change are already being felt and will continue to increase and spread to other areas. The extent and severity of future impacts will depend on how quickly the world responds to climate change.

HEALTH IMPACTS

Wendy Stone/Corbis via Getty Images

Vector-borne tropical diseases spreading outside of the tropics; waterborne-disease outbreaks with flooding; more heat-related deaths.

CROP PRODUCTIVITY

Jim Richardson/National Geographic/Getty Images

Globally, lower crop yields and more malnutrition.

COASTAL EROSION AND FLOODING

US Air Force Photo/Alamy

More health and property losses due to sea-level rise and storm events.

BIODIVERSITY

Paul Souders/Corbis Documentary/Getty Images

Benefit some species (range expansion or loss of competitors); harm other species (habitat or mutualistic partner loss).

WATER AVAILABILITY

Svan Torfinn/Panos

Increase in land area in severe drought (weather pattern changes; loss of snowpack or glacier ice).

FIRE RISK

Layne Kennedy

Increased incidence and risk of fire, causing major property or ecosystem damage.

 Which impacts concern you the most? Explain.

Another truth, hidden in Hauer's Miami study, is that climate change won't just harm coastal areas. Some inland cities will also become inhospitably hot, even as they absorb hundreds of thousands of refugees fleeing the coasts.

That's not to say that there won't be some "winners" in a warming world. High-latitude land in Canada and Siberia will likely become warmer and more habitable—lessening the incidences of cold-related health problems and deaths (though the melting of permafrost, the deeper ground that before the advent of climate change never thawed, is damaging roads and buildings as the ground below them becomes

less supportive). Cold-weather problems overall will lessen in the northern United States and similar latitudes worldwide (think milder winters in Maine and Montana), even with the increase of aberrant cold weather in some places as the jet stream changes its course. Warmer weather during the summer months has even opened the Northwest Passage in some recent years—a long sought-after shipping route through the Arctic Ocean—which would significantly reduce the transport time for ships that otherwise have to take the southern route through the Panama Canal. But, in total, negative impacts will outweigh positive ones, especially as warming increases.

8 RESPONSES TO CLIMATE CHANGE

Key Concept 8: Responding to climate change will require both steps that try to reduce future warming (mitigation) and steps to deal with inevitable warming (adaptation).

Back in 2009, the coming plight of climate refugees got an international audience when Mohamed Nasheed, president of the Maldives, a string of small low-lying islands in the South Pacific, signed a document that asked the nations of the world to reduce their carbon emissions. Why did this garner so much attention? He signed it underwater, wearing scuba gear, to draw attention to the fact that the very existence of his nation was at stake. At an international summit convened to negotiate an agreement on reducing carbon emissions that same year, Nasheed insisted that unless carbon emissions were kept at 350 ppm, his nation would be destined to disappear entirely. (Levels passed 400 ppm in 2016.)

Computer models like the ones used by Sherbinin can be used to predict future changes in climate, but the accuracy of the predictions is dependent on how well we estimate future variables such as how quickly we reduce the use of fossil fuels or what land use changes we will pursue. Still, we know that acting now can help reduce the impact and progression of climate change. Scientists refer to efforts aimed at minimizing the extent or impact of climate change as **climate mitigation**. Mitigation includes any attempt to seriously curb the amount of greenhouse gases we are releasing into the atmosphere as well as steps to remove CO_2 from the air. Carbon capture techniques could reduce emissions (see Module 9.1), but research is also under way to develop planetary-scale technologies to mitigate climate change such as satellites that block some incoming solar radiation or giant filters that remove CO_2 from the atmosphere—approaches known as climate geoengineering.

On a national or global scale, mitigation efforts can be facilitated in a variety of ways, such as command-and-control regulations that limit greenhouse gas release or *green taxes* on pollution released (in this case, **carbon taxes**) and market-driven initiatives such as carbon *cap-and-trade* programs. Financial incentives that encourage the development and use of non-carbon fuels and more energy-efficient technology are also examples of mitigation. (See Modules 5.2 and 10.1 for more on

these policy approaches.) Many industries and businesses are already stepping up with initiatives to reduce their carbon footprint, both because it is an economically sound investment and because it is the right thing to do (see Module 5.1).

In 2004, Princeton University researchers Stephen Pacala and Robert Socolow proposed a "stabilization wedge" strategy—a step-by-step implementation of currently available technology to mitigate climate change; each step could prevent the release of 1 Gt of carbon. Steps include transitioning away from fossil fuels (especially coal) by generating energy from low-carbon nuclear power (see Module 11.1) and sustainable sources such as solar, wind, and geothermal sources (see Module 11.2). Reforestation efforts could increase the uptake of carbon and agricultural techniques that minimize carbon release are also strategies that will help. And not using so much energy in the first place—conservation and energy efficiency—can significantly contribute to mitigation (see Module 11.2 and Online Module 11.3). Pacala and Socolow estimate that any 8 of the 15 steps, or "wedges," they have identified would stabilize CO_2 in the atmosphere; the longer we wait, the stronger our response would have to be—the more wedges we would need to employ.

But in places like Newtok, Alaska and Isle de Jean Charles, Louisiana, it is likely too late for mitigation. Those residents' only choice is to adapt to the reality of a rapidly changing climate. **Climate adaptation** means responding to the climate change that has already occurred and preparing for the additional changes that are inevitable at this point. For human societies at large, that means taking steps to ensure a sufficient water supply in areas where freshwater supplies may dry up; it means planting different crops or shoring up coastlines against rising sea levels; it means preparing for heat waves and cold spells and outbreaks of infectious disease. Miami too is moving ahead with adaptation steps: installing pumps to rid city streets of water that floods during high tides, raising the level of some coastal roads and existing sea walls, just to name a few initiatives aimed at dealing with rising sea level. **INFOGRAPHIC 8A**

No matter which strategies we employ to confront the realities of climate change, international coordination will be essential, meaning world superpowers like the

climate mitigation Efforts to minimize the extent or impact of climate change.

carbon taxes Governmental fees imposed on activities that release greenhouse gases into the atmosphere.

climate adaptation Efforts to help deal with existing or impending climate change problems.

INFOGRAPHIC 8A   RESPONDING TO CLIMATE CHANGE

To deal with climate change we need to pursue actions that help us adjust to current warming as well as take steps to reduce further warming.

Improve disease surveillance; improve sanitation in flood-prone areas.

Plant crops to match new climate.

Capture and conserve water.

Erect coastal barriers to deal with sea-level rise; relocate coastal communities.

Pursue better fire prevention management.

Provide migration corridors for wildlife and wildlife preserves.

BOTH APPROACHES WILL BE NEEDED.

ADAPTATION
Responding to the warming that has already or will inevitably occur

MITIGATION
Preventing further warming by addressing the causes of climate change

Pursue carbon capture and sequestration.

Use sustainable (non-fossil fuel) and nuclear energy.

Pursue energy efficiency.

Use waste management practices that decrease the release of methane.

Stop deforestation; pursue reforestation projects.

Use agricultural practices that prevent the release of methane.

MITIGATION CAN BE PURSUED USING THE "WEDGE" APPROACH.

We can take steps to curb climate change by pursuing actions that reduce greenhouse gas emissions or increase its removal from the atmosphere. Employing any 8 of 15 or more potential strategies (stabilization wedges) that each reduce CO_2 emissions by 1 billion metric tons per year over the next 50 years would stabilize emissions close to current levels. However, the longer we wait, the more wedges we will need.

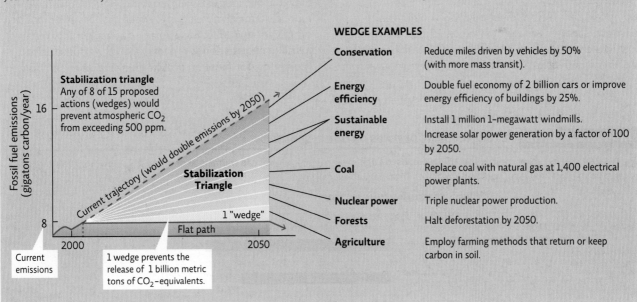

WEDGE EXAMPLES

Conservation	Reduce miles driven by vehicles by 50% (with more mass transit).
Energy efficiency	Double fuel economy of 2 billion cars or improve energy efficiency of buildings by 25%.
Sustainable energy	Install 1 million 1-megawatt windmills. Increase solar power generation by a factor of 100 by 2050.
Coal	Replace coal with natural gas at 1,400 electrical power plants.
Nuclear power	Triple nuclear power production.
Forests	Halt deforestation by 2050.
Agriculture	Employ farming methods that return or keep carbon in soil.

 Which stabilization wedge do you think would be the easiest to accomplish? Which would be the hardest?

European Union, the United States, and China will have to cooperate with each other and with the world's developing nations.

So far those efforts have been fraught. The 1992 *United Nations Framework Convention of Climate Change (UNFCCC)* recognized the need to address climate change, and most of the world's nations, including the United States, signed on, agreeing to cooperate. In 1997, the *Kyoto Protocol*, the international treaty that laid out steps to be taken, set different but specific targets for the reduction of CO$_2$ emissions for various countries. The United States did not ratify the protocol, objecting because it only set reduction requirements for developed countries (which were responsible for most of the historic emissions); none were set for developing countries.

The Kyoto Protocol expired in 2012, and after several years of negotiation it was replaced with the 2016 *Paris Agreement*, which allowed nations to set their own targets, with a goal of keeping warming "well below 2°C" above preindustrial levels, preferably capping warming at 1.5°C to avoid extremely dangerous impacts of climate change (e.g., minimizing sea-level rise; limiting crop and species losses to levels that are (hopefully) tolerable, etc.). We are already half way to that 2°C (3.6°F) upper limit—Earth has already warmed about 1°C (1.8°F). The initial Paris targets will not limit warming to 2°C, but the goal is to adjust targets as needed every few years to eventually reach the 2°C (or even better the 1.5°C) goal. **INFOGRAPHIC 8B**

Many feel that the 2°C upper limit is too high and advocate strongly for 1.5°C as an upper limit. Either way, having a chance of limiting warming to just another 0.5° or 1°C will require a dedicated and aggressive response from the international community. How aggressive? Current analysis suggests that holding the temperature increase to 1.5°C would require we cap anthropogenic greenhouse gases around 450 ppm. Other greenhouse gas emissions, too, need to be phased out. But even if we reduce emissions at a steady rate and reach zero emissions by 2050 (a common target), we will likely surpass 450 ppm and would need to actually remove some CO$_2$ from the atmosphere. This means employing some of those wedges (like reforestation and agricultural changes) that capture and sequester CO$_2$ and developing geoengineering technologies that remove atmospheric CO$_2$.

The Paris Agreement was signed by most nations of the world, including then U.S. President Barack Obama. But in 2017, President Donald Trump withdrew the United States from the agreement. Exiting the agreement will take slightly more than 3 years, but actions that are dismantling Obama-era climate initiatives will begin having an effect even before the official 2020 withdrawal date.

In the United States, there is a vocal minority (many with strong ties to the fossil fuel industry or politically conservative groups) that actively attacks climate science and scientists, claiming that there is insufficient evidence for climate change or that there is no evidence that it is caused by human actions. Employing many of the same tactics that stymied, for a while, action on issues such as tobacco regulation and efforts to control acid rain— dragging out long-discounted arguments, demanding more certainty before acting, spending heavily on political campaigns and lobbying efforts—their efforts have instilled enough doubt to slow the U.S response. (See Module 1.3 for more on critical thinking and analyzing

---

INFOGRAPHIC 8B　**FUTURE WARMING DEPENDS ON OUR RESPONSE**

The Paris Agreement acknowledges that all the proposed targets, even if met, will not cap warming at or below 2°C. However, the intention is to ramp up the emission reductions and strengthen these targets every 5 years or so.

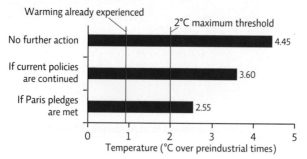

PROJECTED WARMING BY 2100

Warming already experienced

2°C maximum threshold

No further action — 4.45

If current policies are continued — 3.60

If Paris pledges are met — 2.55

Temperature (°C over preindustrial times)

 Why do you think the Paris initial pledges were set too low to meet the goal of warming no more than 2°C?

arguments.) Many believe this rejection of the scientific evidence to be fueled by ideological beliefs that favor lesser governmental control or by the prospect of losing money (or making less) if climate initiatives are pursued. Unfortunately, this puts the 2°C target in jeopardy—even more so the 1.5°C target.

Because climate change is such a complex issue, misleading the public has unfortunately been easy to do. Just claim that scientists disagree (they don't) or that the effects of climate change are nothing to worry about. (The people of Newtok, the Isle de Jean Charles, and the Maldives disagree). Hold up a snowball in the Senate, as Senator James Inhofe famously did, and say the climate can't be warming because it snowed today (confusing *weather* with *climate*). Or even worse, dismantle or defund climate research—an action that has been compared to not taking your child's temperature because you are afraid she or he might be sick and potentially missing an opportunity to address the illness before it gets out of hand.

These and other arguments against action hurt our chances of successfully addressing this unprecedented challenge facing humanity. Even the argument that responding to climate change will hurt the economy is flawed because it fails to take into account the manufacturing, installation, and maintenance jobs a renewable energy industry is bringing and will bring if the world moves ahead with an ambitious renewable energy plan. Further, we have yet to measure the economic, environmental, and ethical true costs of inaction. What we do know is that people in developing countries and those in poverty will suffer the most (and they are the ones who have contributed to the problem the least). Dealing with climate change is an environmental justice issue of massive proportions.

This is not to say that all those who question climate science or policies to address climate change are intentionally trying to undermine reasoned action for personal or political gain. Some might feel that the financial burden of responding is too high—we've waited too long and simply can't afford it at this point. Others object on religious grounds, believing that God is in control and we should not step in—that is a discussion more suitable for other venues, not a science textbook. But those who oppose action based on a scientific evaluation of the issue are misunderstanding or misrepresenting the science.

The reality is that none of us (climate scientists included) are expert enough to thoroughly understand each and every technical aspect of climate change science—the chemistry and physics of the atmosphere

An example of adapting to the threat of floods, gates on the River Thames protect central London from flooding that might occur from extremely high tides or storm surges. Gates rest flush with the river bed when not in use and are rotated upward to form a barrier when needed. The Thames Barrier became operational in 1982, and the rate of its usage has increased in recent years as sea level rises.

IR_Stone/iStock/Getty Images

and ocean circulation, the isotopic evaluation of ice cores, the biological evidence of species impacts or health effects; the list goes on. We rely, to a great extent, on scientific consensus for complex issues like this— just as we probably depend on our physician (and second or third opinions) when we make important healthcare decisions. If 97% of the medical doctors you visited recommended surgery for a life-threatening illness (and for the same reasons—in other words, their logic is consistent), would you have surgery or go with the 3% who advise against it?

For climate change, what is the majority conclusion? As stated in a 2017 *National Geographic* article, the scientific consensus is strong on these three points: "The world is warming. It's because of us. We're sure."

Many nations are standing by their Paris commitments, agreeing with the scientists, policy makers, and citizens who are advocating for the *precautionary principle*— choosing to act in the face of uncertainty because the stakes are so high. It remains uncertain how fast climate impacts will unfold, how bad they might be, and when we might reach a tipping point (i.e., a glacial and polar ice melt that cannot be stopped) that would set into motion events that dwarf the problems with which we are currently dealing. These uncertainties arise, in part, because we continue to contribute to the problem (burning fossil fuels, cutting down forests, etc.). But what we do know is that the sooner we act, the better chance we have of successfully addressing the problem. Every day we delay increases the cost to fix it and gets us closer to that 2°C mark.

Select References:

Clark, P. U., et al. (2016). Consequences of twenty-first-century policy for multi-millennial climate and sea-level change. *Nature Climate Change, 6*(4), 360−369.

Cook, J., et al. (2016). Consensus on consensus: a synthesis of consensus estimates on human-caused global warming. *Environmental Research Letters, 11*(4), 048002.

Hansen, J., et al. (2016). *Global Temperature in 2015.* Retrieved from Climate Science, Awareness and Solutions, Earth Institiue, Columbia University: http://csas.ei.columbia.edu/2016/01/19/global-temperature-in-2015/.

Hauer, M. E. (2017). Migration induced by sea-level rise could reshape the US population landscape. *Nature Climate Change, 7*(5), 321−325.

Hughes, T. P., et al. (2017). Global warming and recurrent mass bleaching of corals. *Nature, 543*(7645), 373−377.

Mann, M. E., et al. (2017). Influence of anthropogenic climate change on planetary wave resonance and extreme weather events. *Scientific Reports (Nature Publisher Group), 7,* 45242.

Mengel, M., et al. (2016). Future sea level rise constrained by observations and long-term commitment. *Proceedings of the National Academy of Sciences, 113*(10), 2597−2602.

National Geographic. (2017) Climate change: seven things you need to know. *National Geographic Magazine, 231*(4), 30−39.

Pacala, S., & Socolow, R. (2004). Stabilization wedges: Solving the climate problem for the next 50 years with current technologies. *Science, 305*(5686): 968−972.

Scheffers, B. R., et al. (2016). The broad footprint of climate change from genes to biomes to people. *Science, 354*(6313), aaf7671.

U.S. Environmental Protection Agency. (2016). *Climate Change Indicators in the United States, 2016. Fourth Edition.* EPA 430-R-16-004. www.epa.gov/climate-indicators

Vermeer, M., & Rahmstorf, S. (2009). Global sea level linked to global temperature. *Proceedings of the National Academy of Sciences, 106*(51), 21527−21532.

This 10.6 MW wind farm at Bada Bagh in India is an example of employing mitigation strategies to reduce the impact of climate change by displacing some fossil fuel-derived energy with a more sustainable option.

BremecR/iStock/Getty Images

INTERACTIVE MAP **CLIMATE CHANGE IMPACTS** ANIMATED INFOGRAPHIC

Climate change is having many and varied consequences and is constantly in the news—whether it is a new study, a new population impacted, or the responses (or lack of response) by nations of the world. Visit the Interactive Map for this module to learn more about the effects of climate change.

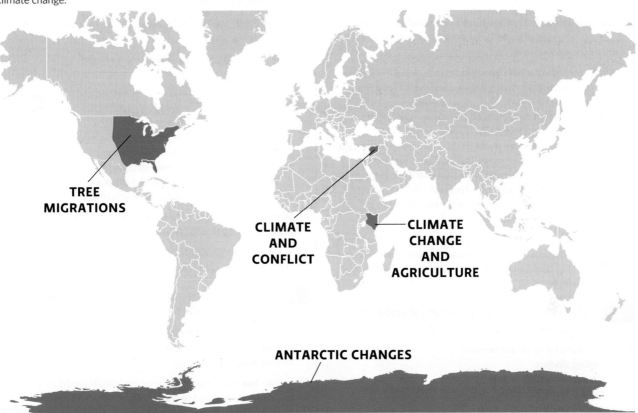

 BRING IT HOME

PERSONAL CHOICES THAT HELP

The effects of climate change are already being felt by humans, other species, and ecosystems around the globe. Though significant action is needed at the city, state, and national levels, actions that individuals and community groups can make that help address climate change are also needed and will show policy makers that citizens are interested in preventing global climate change.

Individual Steps

• Do your part to reduce carbon emissions by conserving energy. Walk or ride a bike instead of driving a car. Share a ride with a coworker rather than driving alone. Negotiate with your employer to telecommute. Live close to where you work or go to school. Reduce your heating and cooling

energy use and always turn off electronics and lights when not in use.
• If your utility company offers renewable energy, buy it.
• Reduce the carbon footprint of your food by decreasing the amount of feedlot-produced meat you eat. Buy your food as locally as possible to reduce energy used in transportation.
• Go to www.terrapass.com to see how you can offset your CO_2 production from your car, your house, and your airplane travel.

Group Action

• Volunteer to help build a zero-energy Habitat for Humanity home.
• Organize a community lecture on climate change with a local university expert or meteorologist as the speaker.

• Organize an event at your school or community to raise awareness about global climate change and ways to prevent it. Go to www.350.org to join a current campaign and get other program ideas.

Policy Change

• Write, call, or visit the offices of your elected officials and share your views about funding for research and development of clean and renewable sources of energy as well as your views on participation in the Paris Agreement. In addition, ask that they support the funding of science, especially efforts to understand and confront climate change.

ENVIRONMENTAL LITERACY **UNDERSTANDING THE ISSUE**

1 What is climate change, and why it more concerning than day-to-day changes in weather?

1. Day-to-day changes in meteorological conditions are known as _____ whereas long-term patterns of meteorological conditions are known as _____.

2. Why is a change of a few degrees in average global temperature more concerning than day-to-day weather changes of a few degrees?
   a. This means that temperatures are increasing all around the globe, not just in a few areas.
   b. An increase in average global temperature means we should have more weather extremes that can cause problems.
   c. Climate has never changed in the past so if it is changing now, by even just a few degrees, we know something is wrong.
   d. Weather changes of a few degrees rarely happen so it is not something we worry about.

3. In the winter of 2010, the northeastern part of the United States had several large snowstorms that resulted in record high snowfall amounts. How does this weather fit in with the notion of global climate change?

2 What is the physical and biological evidence that climate change is occurring?

4. Recent sea-level rise is attributed to:
   a. melting glaciers.
   b. thermal expansion of water.
   c. melting icebergs.
   d. A and B are correct.
   e. A, B, and C are correct.

5. Outline the evidence for climate change. Do you feel that this evidence supports the conclusion that climate is changing? Explain.

3 What is the greenhouse effect, and how are we affecting it?

6. Greenhouse gases are defined as gases that:
   a. humans release into the atmosphere.
   b. trap heat and warm the atmosphere.
   c. break apart when struck by solar radiation.
   d. reflect incoming solar radiation back into space.

7. Identify three human actions that have led to an increase in the amount of greenhouse gases in the atmosphere. What has been the result?

4 How are atmospheric CO_2 and temperature measured, and how are they correlated?

8. True or False: Current CO_2 levels in the atmosphere can be measured, but there are no good methods for determining CO_2 levels in the distant past.

9. What relationship is seen between temperature and atmospheric CO_2 levels?

5 Other than greenhouse gases, what factors affect climate?

10. True or False: The higher the albedo of a surface, the better it reflects away sunlight.

11. Which of the following has the greatest albedo?
    a. A forest
    b. A light-colored roof
    c. A dark asphalt road
    d. The surface of the ocean

12. What effect have recent large volcanic eruptions had on climate?
    a. They are contributing to current warming.
    b. Their effect has not yet been determined.
    c. Volcanic activity has no effect on climate.
    d. They contribute to cooling but only briefly.

13. What is the difference between a positive feedback loop and a negative feedback loop? Give a climate-related example of each.

14. What are the Milankovitch cycles? Do they account for the warming we are currently seeing?

6 What evidence suggests that climate change is due to human impact?

15. True or False: Computer models that only account for natural climate forcers produce a prediction of greater warming than we have actually experienced.

16. Explain how the accuracy of computer climate models is validated and how they are used to predict future climate based on different levels of climate forcers in the future.

7 What are the current and potential future impacts of climate change?

17. How is biodiversity affected by climate change (both current and future)?
    a. No effects are being seen yet but as it gets warmer we expect many species to suffer.
    b. Climate change is harmful to humans but other species will be able to adapt.
    c. Some species will benefit and expand their range while others who can't adapt will decline or die out.
    d. Species are migrating or adjusting to the new climate and will be able to do so in the future.

18. Describe the types of problems that global climate change causes for human health. Which do you feel is likely to cause the biggest problem? Why?

8 What actions can we take to respond to a world with a changing climate?

19. True or False: The amount of future warming that Earth will experience in the next 100 years depends on the choices we make now.

20. Distinguish between adaptation and mitigation. Why do we need both strategies?

SCIENCE LITERACY WORKING WITH DATA

The location of 105 species of fish, shellfish, and other marine species in the Eastern Bering Sea and along the U.S. northeast coast was determined by tracking the location of each species from 1982 to 2015. Both latitude and depth were followed and expressed as the average center of biomass (central location for each population by weight) and shown in the graphs below. Analyze these graphs and answer the questions that follow.

CHANGE IN LATITUDE AND DEPTH OF MARINE SPECIES, 1982–2015

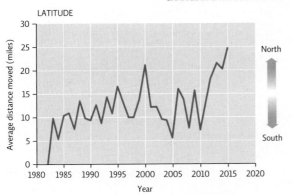

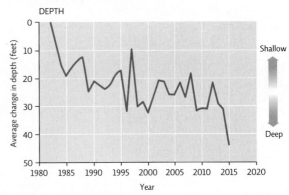

Interpretation

1. In general, what directional trend is seen over time for the average change in latitude and for depth from 1982 to 2015 for the species evaluated?

2. How much farther north were these species found in 2015 compared to 1982? How much deeper?

3. How did average latitude and depth change between 2000 and 2005? What does this suggest about water temperature in those years?

Advance Your Thinking

4. What experimental prediction would you make for change in latitude and depth if you were testing the hypothesis that warming waters were affecting the location of these species? Do the data support that hypothesis?

5. Why did the researchers collect data on both latitude and depth?

6. If the researchers had only tracked one or a few species, the data would not be as compelling as it is in this study that tracked and averaged the location of 105 species. Why?

INFORMATION LITERACY EVALUATING INFORMATION

Among scientists, there is broad consensus (97%) that climate change is significantly caused by human activity. Yet in a 2017 Gallup poll, only 45% of the public are worried about climate change (and this is the highest percentage in 3 decades). Members of the public get their information from a variety of media sources and information posted on the Internet. How can there be such a large disconnect between scientists and the public?

Go to the Global Warming Hoax page at www.globalwarminghoax .com/news.php.

Evaluate the website and work with the information to answer the following questions:

1. Determine if this is a reliable information source with a clear and transparent agenda:
   a. Who runs the website? Do this person's/group's credentials make the site reliable or unreliable? Explain.
   b. What is the primary message of the website?

c. Read the entry "Antarctic sea ice for March 2010 significantly greater than 1980." What evidence is provided in the article? Do you have any questions about the data presented? If so, what are they?

Now go to the Skeptical Science website (www.skepticalscience.com). Click on the link *"Most Used Climate Myths."*

2. Determine if this is a reliable information source with a clear and transparent agenda:
   a. Who runs the website? Do this person's/group's credentials make the site reliable or unreliable? Explain.
   b. What is the primary message of this website? What types of evidence does it provide to support its message?
   c. Click on the "Antarctica is gaining ice" link. Read the article and compare the main point of the article to the article on the Global Warming Hoax site.
   d. Which explanation and website do you find more credible? Why?

 Additional study questions are available at SaplingLearning.com.

ALTERNATIVES TO FOSSIL FUELS

CHAPTER 11

There are a wide variety of energy sources that can be used in place of fossil fuels. No single alternative can replace fossil fuels but together, the many alternatives can.

Module 11.1: Nuclear Power
An evaluation of the pros and cons of nuclear power as an alternative method for electricity production

Module 11.2: Sustainable Energy: Stationary Sources
A survey of sustainable energy sources for the production of electricity: solar, wind, geothermal, and hydroelectric power

ONLINE Module 11.3: Sustainable Energy: Mobile Sources
A look at biofuels as a replacement for fossil fuel-derived transportation fuels

Online Modules are available at SaplingLearning.com.

ML Harris/Getty Images

THE FUTURE OF FUKUSHIMA

Can nuclear energy overcome its bad rep?

The Japanese authorities originally declared a 20-kilometer (12.5-mile) evacuation area around Fukushima, an exclusion zone that may only be entered under government supervision. Four months after the explosion, residents in protective suits are briefed before being escorted to their homes to retrieve a few small items.

AP/DAVID GUTTENFELDER/National Geographic Creative

After reading this chapter and studying the KEY CONCEPTS and INFOGRAPHICS, you should be able to answer these GUIDING QUESTIONS

CORE MESSAGE

Harnessing nuclear energy to create electricity would help address concerns over fossil fuel supplies, air pollution, and climate change. However, there are serious safety concerns with nuclear power, including vulnerability to natural disasters, radioactive waste disposal, and potential for weapons production.

1. What are radioactive isotopes and why are they important for nuclear power?
2. How is uranium fuel for nuclear power produced?
3. How is nuclear energy harnessed to generate electricity in a fission reactor?
4. What types of radiation are produced when isotopes decay?
5. How is the rate of decay of a radioactive atom measured?
6. What problems are associated with nuclear waste?
7. What is the history of nuclear accidents worldwide?
8. What are the advantages and disadvantages of nuclear power?

The Fukushima Daiichi Nuclear Power Station is a maze of steel and concrete, perched right on Japan's Pacific coast, just 240 kilometers (150 miles) north of Tokyo. Its six nuclear reactors supplied some 4.7 GW (1 gigawatt = 1 billion watts) of electric power to the country, making it one of the largest nuclear power plants in the world. On March 11, 2011, when a magnitude 9.0 earthquake struck 130 kilometers (80 miles) north of the plant, there were more than 6,000 workers inside. The quake caused a power outage, and in the darkness, chaos ensued: Men and women groped for ground that would not stabilize beneath their hands and feet, some shouting in panic as steel and concrete collided around them. When the shaking stopped, emergency lights came on, revealing a cloud of dust. But that was only the beginning of the disaster.

The earthquake had erupted beneath the ocean floor, triggering a tsunami that would arrive at the plant in two distinct waves. The first wave was not big enough to breach the 10-meter-high (33-foot-high) concrete wall that had been built between the plant and the sea. But the towering mass of water that came 8 minutes later was. The wall of water bulldozed a string of protective barriers, sent cars and trucks crashing into buildings, and eventually settled, in deep black pools, around the reactors themselves.

These were the strongest earthquake and largest tsunami in the country's long memory. Together, they would claim more than 16,500 lives along a 400-kilometer (250-mile) stretch of coast (roughly equal to the distance between Maine and Manhattan). But as the ground steadied and the water subsided, the world's attention would quickly turn to a third disaster: the risk of nuclear meltdown at Daiichi.

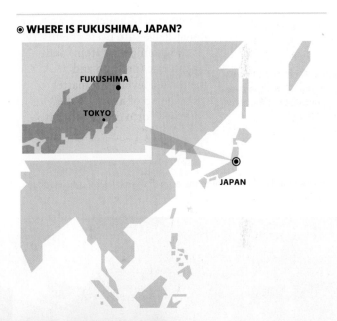

⊙ WHERE IS FUKUSHIMA, JAPAN?

FUKUSHIMA

TOKYO

JAPAN

Satellite image of the damaged Fukushima Daiichi Nuclear Power Station on March 14, 2011.

Digital Globe via ABACAPRESS.COM/Newscom

1 NUCLEAR-POWERED FUEL: RADIOACTIVE ISOTOPES

Key Concept 1: The energy released from radioactive isotopes provides the heat for the thermonuclear production of electricity.

In some ways, electricity generated using **nuclear energy** is very similar to other forms of thermoelectric power (those that use heat to produce electricity). Just like power plants that run on oil or coal, nuclear plants use heat to boil water and produce steam, which is then used to generate electricity. The difference, really, is the source of that heat. Coal and natural gas plants create it by burning fossil fuels. In the thermonuclear production of electricity, heat is produced through a controlled nuclear reaction.

To understand nuclear power, we must delve into basic chemistry at the atomic (the atom) and subatomic (protons and neutrons) levels. All matter is made up of **atoms**—the simplest form of matter that cannot be broken down by chemical means. Atoms contain a nucleus where protons and neutrons are found; electrons orbit around this nucleus. A substance composed of all the same type of atoms is an **element**—hydrogen is an element, as is helium, carbon, and uranium.

The atom of any element is defined by the number of protons it possesses (known as its *atomic number*). For example, any atom that contains two protons is a helium atom; those that contain six are carbon atoms. Because electrons are so small compared to the subatomic particles in the nucleus, the atomic mass of an atom is simply the number of protons plus the number of neutrons (its *mass number*). While the number of protons in the atom of an element is unique (if the number of protons changes, the element changes), the number of neutrons can vary. This will change the mass number. These different versions of the atom are called **isotopes**. By convention, isotopes are named according to their mass number,

nuclear energy Energy in an atom; can be released when an atom is split (fission) or combines with another atom (fusion).

atom The simplest form of matter that cannot be broken down by chemical means.

element A substance composed of all the same type of atoms.

isotopes Atoms that have different numbers of neutrons in their nucleus but the same number of protons.

Police guard a checkpoint at the edge of the exclusion zone leading to the town of Minami Soma, just north of Daiichi. The sign reads "Keep Out."

AP/DAVID GUTTENFELDER/National Geographic Creative

e.g., uranium-238 (with an atomic mass of 238) and uranium-235 (a uranium atom that has three fewer neutrons than U-238). **INFOGRAPHIC 1**

Most isotopes are stable, meaning they do not spontaneously lose protons or neutrons. But some are **radioactive**: They are less stable and spontaneously lose mass by emitting particles (protons, neutrons, or electrons) or photons (a form of short wavelength radiation); in the process, heat is released. Radiation is also emitted when an atom is struck by a subatomic particle, splitting the atom into one or more smaller atoms. Because they are unstable, radioactive atoms are good candidates as fuel for nuclear power because it depends on the ability of the atoms to be split, an event that releases large amounts of heat that can be captured to generate electricity.

radioactive Atoms that spontaneously emit subatomic particles and/or energy.

INFOGRAPHIC 1 | **ATOMS AND ISOTOPES**

The fundamental unit of matter is the atom. Atoms contain protons and neutrons in a central nucleus around which electrons orbit. The number of protons, the *atomic number*, is unique to each element. For example, any atom with only two protons is an atom of the element helium. The sum of the number of protons and neutrons gives an atom its *mass number*.

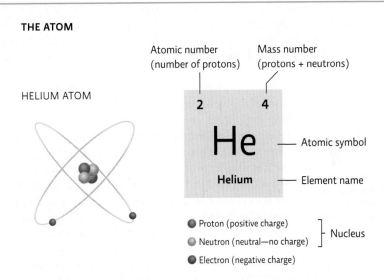

THE ATOM

HELIUM ATOM

Atomic number (number of protons)

Mass number (protons + neutrons)

2   4

He

Helium

— Atomic symbol

— Element name

● Proton (positive charge)
● Neutron (neutral—no charge) ⎤ Nucleus
● Electron (negative charge)

Isotopes are atoms that have the same atomic number (number of protons) but a different number of neutrons, and thus a different mass number. An atom with 92 protons is uranium (U); if a uranium atom has 146 neutrons, it is U-238 (92 protons + 146 neutrons = 238). Another uranium isotope, U-235, has 143 neutrons (92 protons + 143 neutrons = 235).

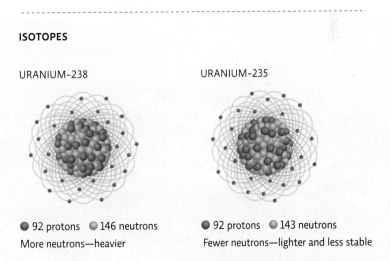

ISOTOPES

URANIUM-238

URANIUM-235

● 92 protons ● 146 neutrons
More neutrons—heavier

● 92 protons ● 143 neutrons
Fewer neutrons—lighter and less stable

Uranium also exists as uranium-233. How many protons and neutrons does it have? Do you think it is more or less stable than U-235? Explain.

2 PRODUCTION OF NUCLEAR FUEL

Key Concept 2: The production of nuclear fuel involves mining and several processing steps, all of which generate hazardous waste.

Most nuclear reactors use the element uranium, which has several isotopes. Uranium-238 (U-238) is the most stable and is the most abundant form of uranium; it makes up roughly 99% of Earth's total supply. U-235 is a less stable form and provides the "fuel" used in a typical fission reactor. But acquiring the U-235 is not as simple as locating rock rich in uranium (uranium ore) and digging it up. (See Online Module 7.3 for more on mineral ores and mining.) Because U-235 makes up such a small percentage of the uranium found in uranium-rich ore, a lot of rock must be mined to eventually produce a useful fuel for the nuclear reactor.

Once the ore is dug up, several processing steps are needed. In the milling step, ore is crushed and mixed with either an acidic or alkaline solution to separate the uranium from the ore. This process creates uranium oxide, known as "yellowcake" in the industry for its yellow hue. Next, the fuel is "enriched," a process that increases the percentage of

fuel rod Hollow metal cylinder filled with uranium fuel pellets for use in fission reactors.

U-235 in the sample (otherwise there is not enough material in the uranium sample to participate in the nuclear reaction). In this step, the percentage of U-235 is increased from around 0.7% to around 4%. (This is nowhere near as enriched as the uranium in a nuclear bomb; that highly enriched fuel is about 90% U-235.)

The enriched uranium is formed into small pellets and stacked into 3 to 6 meter (10 to 20 feet) long hollow **fuel rods**. More than 100 of these fuel rods are loaded into a framework to form a fuel assembly. Several fuel assemblies are placed into a thick-walled vessel—the reactor core—where they remain in place for a year or two before being removed and replaced with fresh fuel.

The mining and processing of uranium raises concerns about health and environmental safety. The processing of uranium from the mining to the enrichment and production of fuel pellets creates hazardous waste that must be dealt with. Uranium mining itself is inherently dangerous due to the radioactive material; workers and residents who live near the mines suffer from higher than normal cancer rates. **INFOGRAPHIC 2**

Open pit mill tailings site in Australia. Mining and crushing uranium ore produce small particle (sandlike consistency) waste that contains low levels of radioactive isotopes. Mill tailings waste is often stored in large open pits and covered with water to reduce windblown loss of material. There is a concern that water could overflow this pit in heavy rain years.

David Hancock-Skyscans/Auscape/The Image Works

INFOGRAPHIC 2 | NUCLEAR FUEL PRODUCTION

Uranium ore (rock that contains uranium) is mined and goes through many stages of processing to produce fuel suitable for a nuclear reactor. The process creates hazardous waste at every step.

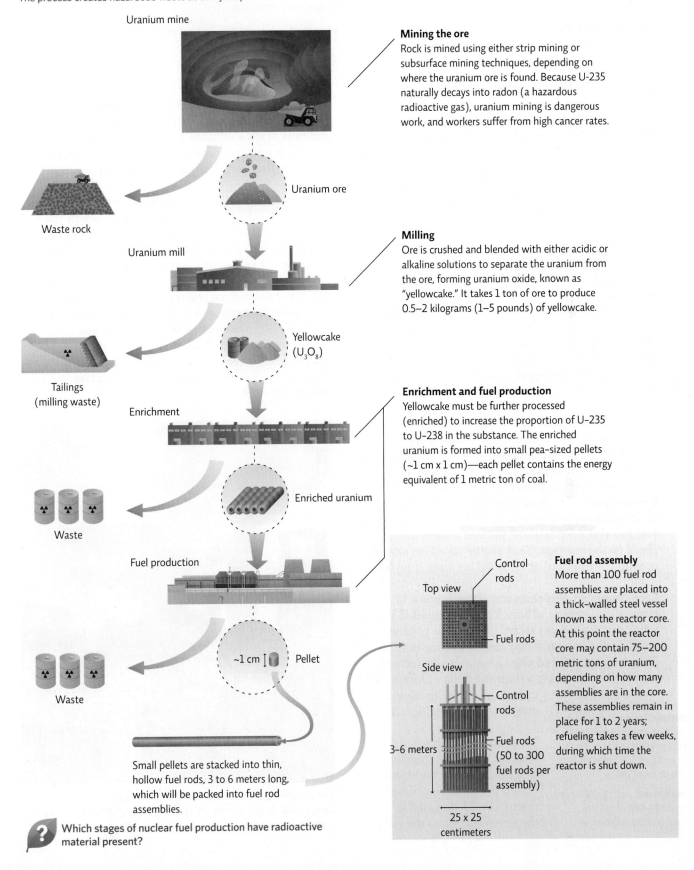

Uranium mine

Mining the ore
Rock is mined using either strip mining or subsurface mining techniques, depending on where the uranium ore is found. Because U-235 naturally decays into radon (a hazardous radioactive gas), uranium mining is dangerous work, and workers suffer from high cancer rates.

Waste rock

Uranium ore

Milling
Ore is crushed and blended with either acidic or alkaline solutions to separate the uranium from the ore, forming uranium oxide, known as "yellowcake." It takes 1 ton of ore to produce 0.5–2 kilograms (1–5 pounds) of yellowcake.

Uranium mill

Tailings
(milling waste)

Yellowcake
(U_3O_8)

Enrichment and fuel production
Yellowcake must be further processed (enriched) to increase the proportion of U-235 to U-238 in the substance. The enriched uranium is formed into small pea-sized pellets (~1 cm x 1 cm)—each pellet contains the energy equivalent of 1 metric ton of coal.

Enrichment

Waste

Enriched uranium

Fuel production

Waste

~1 cm   Pellet

Small pellets are stacked into thin, hollow fuel rods, 3 to 6 meters long, which will be packed into fuel rod assemblies.

Fuel rod assembly
More than 100 fuel rod assemblies are placed into a thick-walled steel vessel known as the reactor core. At this point the reactor core may contain 75–200 metric tons of uranium, depending on how many assemblies are in the core. These assemblies remain in place for 1 to 2 years; refueling takes a few weeks, during which time the reactor is shut down.

Control rods

Top view

Fuel rods

Side view

Control rods

3–6 meters

Fuel rods
(50 to 300 fuel rods per assembly)

25 x 25 centimeters

? Which stages of nuclear fuel production have radioactive material present?

3 GENERATING ELECTRICITY WITH NUCLEAR ENERGY

Key Concept 3a: In a nuclear fission reactor, U-235 is bombarded with neutrons to split the atoms; this releases more neutrons, which leads to a self-perpetuating chain reaction.

Key Concept 3b: Electricity is produced when steam, generated when water is heated by the fission reaction, turns a turbine attached to a generator.

Today, nuclear power is generated by splitting atoms—**nuclear fission**—a process that releases a tremendous amount of heat for the production of electricity. Nuclear *fusion* reactions—the combining of two atoms to form a new one—also generate heat; the Sun is a giant fusion reactor, forming helium through the joining of two hydrogen atoms. (Research and development is under way to try to make fusion reactors a viable option.)

A nuclear fission chain reaction begins when the fuel rods (and the U-235 inside) are deliberately bombarded with neutrons. The uranium nucleus splits into a variety of two or more smaller atoms (fission by-products), releasing two or three additional neutrons in the process. These newly released neutrons then hit other U-235 atoms, causing them to split and release even more neutrons, and so on, producing a chain reaction that is self-sustaining, or even capable of accelerating.

Unlike the type of nuclear reaction at work in a nuclear bomb, which contains a much higher percentage of "fissile material" that sets off a massive chain reaction that is almost instantaneous, the reactions at nuclear power plants are highly controlled. **Control rods**—made of materials that absorb neutrons, such as boron or graphite—are placed in the fuel rod assembly between the fuel rods to control the speed of the reaction. They can be added (to slow down or stop the reaction) or removed (to make it go faster).

Even controlled, this chain reaction releases a tremendous amount of heat. The heat is used to boil water, which produces steam, which turns turbines attached to a generator that creates electricity. (The electricity is produced when a metal conductor such as copper spins within a magnetic field, a process that causes the copper to release a stream of electrons [electricity]).

There are several types of fission reactors. The most common type worldwide (and in the United States) is a *pressurized water reactor* (PWR), designed so that the steam that turns the turbine is not exposed to radiation: Nuclear fission in the core heats water under pressure (like a pressure cooker); that radioactive water enters a pipe that passes through, and thus heats, a separate container of water (which is not exposed to radiation); it is the steam from this radiation-free water that turns the turbine.

The reactors at Fukushima, however, were *boiling water reactors* (BWRs); BWRs produce steam in the reactor core itself. This means that both the steam and the turbine become radioactive in the process. The meltdown at Fukushima revealed a flaw in the BWR design: If pressure builds up in the reactor vessel, as it did in the Fukushima reactors as heat levels climbed, the corrective measure would be to vent the steam to avoid an explosion, but this would release radioactive material. Radiation would not be released if a PWR had to be vented to release excess steam since the steam itself is not radioactive.

INFOGRAPHIC 3

nuclear fission A nuclear reaction that occurs when a neutron strikes the nucleus of an atom and breaks it into two or more parts.

control rod Rod that absorbs neutrons and slows the fission chain reaction.

The Gösgen Nuclear Plant is located within the town of Däniken, Switzerland. Steam rising from a cooling tower greets children as they walk home from school.

Mark Henley/Panos Pictures

INFOGRAPHIC 3   HOW IT WORKS: NUCLEAR REACTORS

Fission, or the breaking apart of atoms, begins when an atom like U-235 is bombarded with a neutron. This breaks the atom into other smaller atoms and releases free neutrons, which in turn hit other U-235 atoms, causing them to split and release more neutrons, and so on. The reaction in the fuel assembly is controlled by the insertion of control rods of nonfissionable material which absorb some of the free neutrons.

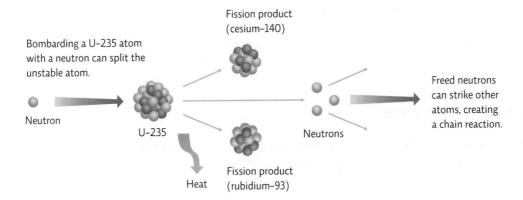

Bombarding a U–235 atom with a neutron can split the unstable atom.

Neutron

U–235

Fission product (cesium–140)

Heat

Fission product (rubidium–93)

Neutrons

Freed neutrons can strike other atoms, creating a chain reaction.

The most common type of nuclear power plant is a pressurized water reactor (PWR), but the reactor at Fukushima is an older technology: a boiling water reactor (BWR). Both designs use fuel assemblies with control rods and use water as a cooling and steam source.

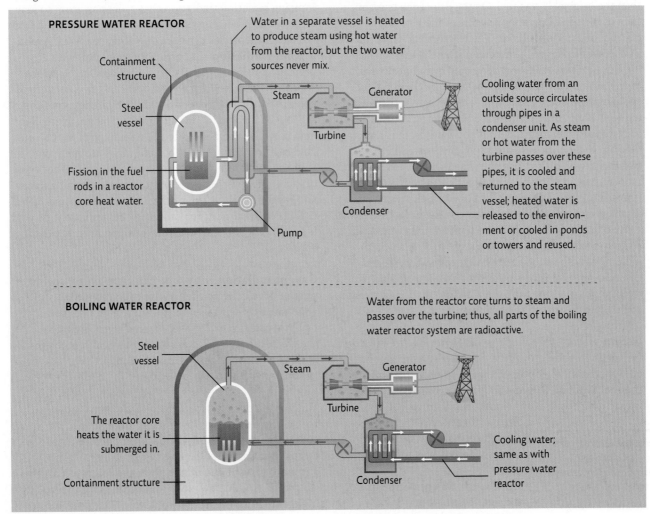

PRESSURE WATER REACTOR

Containment structure

Steel vessel

Fission in the fuel rods in a reactor core heat water.

Water in a separate vessel is heated to produce steam using hot water from the reactor, but the two water sources never mix.

Steam

Turbine

Generator

Pump

Condenser

Cooling water from an outside source circulates through pipes in a condenser unit. As steam or hot water from the turbine passes over these pipes, it is cooled and returned to the steam vessel; heated water is released to the environment or cooled in ponds or towers and reused.

BOILING WATER REACTOR

Water from the reactor core turns to steam and passes over the turbine; thus, all parts of the boiling water reactor system are radioactive.

Steel vessel

Steam

Turbine

Generator

The reactor core heats the water it is submerged in.

Containment structure

Condenser

Cooling water; same as with pressure water reactor

How is a nuclear reactor similar to a coal-fired power plant? How are they different?

All types of thermoelectric power take a lot of water; that's why power plants are sited near rivers and oceans. But currently, nuclear power requires the most—roughly 25% more than a coal-fired power plant and 75% more than a natural gas plant. Though most of the water (more than 95%) is returned to the source (river or ocean), releasing this warmer-than-normal water (called thermal pollution) back into the environment causes problems for species not adapted to the higher temperature.

The reason for all that water is simple: With nuclear energy, water is needed to produce steam and also to prevent the reactor from overheating. And even after the spent fuel rods are removed from the reactor, they still produce tremendous amounts of heat and need constant cooling.

4 TYPES OF RADIATION

Key Concept 4: Alpha, beta, and gamma radiation vary in their ability to penetrate substances; while all are dangerous, gamma radiation is the most hazardous to health.

In a fission reactor, both the nuclear fuel and the fission products produced by the chain reaction are radioactive. In fact, the fission products are more unstable and radioactive than the original U-235 fuel. On the day of the tsunami, each of the three operating reactors at the Fukushima plant held about 25,000 fuel rods. The reactors had gone through a successful emergency shutdown after the earthquake hit, but heat was still being generated by radioactive decay of the fission products in the fuel rods. Keeping the fuel cool was still vitally important. Unfortunately, the power outage had not only plunged the plant into darkness but also stopped the normal delivery of water to those reactors. (Backup generators were unfortunately located in the basement, which was flooded by the tsunami and rendered inoperable.) As Fukushima's reactor cores lost water (the heat of the fuel assemblies boiled off the water, and disabled pumps could add no new cooling water), there was a risk of a hydrogen explosion—temperatures would climb so high that the metal cladding of the fuel rods would melt, react with water, and release explosive hydrogen gas. A steam explosion was also a risk—the buildup of tremendous pressure from the rapid production of steam could blow the top off the building.

Valiant efforts were made to get water on the reactors: Impassible roads prevented fire trucks from getting through, and helicopter attempts to drop sea water on the reactors failed. Eventually, fire trucks made their way through and, using hoses designed for jet-fuel fires, emergency responders were able to get some water to the places where it was most needed.

However, it was not enough. As the heat climbed due to lack of cooling, a steam or hydrogen explosion seemed inevitable. Radiation from damaged reactor Number 2 reached the control room; it had climbed so high that workers had to rotate out at regular intervals to avoid being poisoned with radiation.

Ionizing radiation (radiation that causes an atom to lose electrons and become "ionized" or charged) released from radioactive isotopes can take one of three forms. **Alpha** and **beta radiation** are particles. (Alpha particles are basically a helium nucleus; beta particles are electrons.) Alpha particles don't travel far and can't even penetrate paper; they do not penetrate skin but can be harmful if inhaled or ingested in food or water. The much smaller beta particles can penetrate the upper layers of the skin (and can also be ingested) but are easily stopped by a thin sheet of aluminum or heavy clothing. If either type of radiation enters the body, it can cause organ damage and cancer by directly harming cell structures and DNA. (Radiation treatment for cancer is beta radiation; its goal is to kill cancer cells, but it can also kill normal cells it contacts.)

Gamma radiation is not a particle; it is a high-energy photon (an electromagnetic wave) that can easily penetrate skin—it takes thick concrete or a dense material such as lead to stop gamma rays—giving it the greatest potential to cause serious health problems such as radiation sickness, cancer, and birth defects.
INFOGRAPHIC 4

High levels of radiation, especially gamma radiation that could not be avoided simply by wearing containment suits, quickly became a problem for Fukushima workers. On day two of the disaster, attempts to vent the steam from the reactor vessel (to avoid a steam explosion) failed due to dangerously high, and climbing, radiation levels. That afternoon, the building housing reactor Number 1 exploded, sending concrete and steel debris flying. Radiation levels around the plant climbed exponentially, deterring efforts to vent the other two reactors. In the days that followed, both exploded.

alpha radiation Ionizing particle radiation that consists of two protons and two neutrons.

beta radiation Ionizing particle radiation that consists of electrons.

gamma radiation Ionizing high-energy electromagnetic waves (photons).

INFOGRAPHIC 4 IONIZING RADIATION

Three different types of ionizing radiation can be released by radioactive isotopes and each differs in its ability to penetrate surfaces. Exposure to radiation, especially gamma rays, can damage a wide variety of body organs, leading to radiation sickness, cancer, or birth defects.

PENETRATING ABILITY

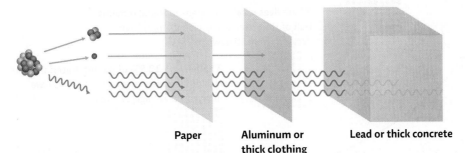

Alpha radiation
Two protons and two neutrons are lost. Alpha particles can't penetrate paper or skin but can be harmful if inhaled or ingested in food or water.

Beta radiation
An electron is lost. Beta particles can penetrate the upper layers of the skin but are easily stopped by a thin sheet of aluminium or very heavy clothing.

Gamma radiation
Short wavelength electromagnetic radiation is released (similar to X rays). High-energy gamma rays can penetrate the deepest but can be stopped by thick or very dense material such as concrete or lead.

Paper   Aluminum or thick clothing   Lead or thick concrete

 Why were the people of Fukushima Prefecture concerned about the alpha radiation released with the reactor meltdown? After all, it cannot even penetrate paper!

5 RADIOACTIVE DECAY

Key Concept 5: Radioactive decay is measured in terms of a half-life and can vary in duration from milliseconds to billions of years for different isotopes.

The safety of nuclear power hinges on the fuel that is used and the by-products that are produced. Some of the radioactive isotopes produced in the fission reaction are more concerning than others, namely those that are highly radioactive and spontaneously release large amounts of radiation in a process known as **radioactive decay**. (Note this is not a fission reaction that splits the atom into two or more smaller atoms; it is just the spontaneous release of particle or high-energy radiation.) Many of these isotopes continue to release dangerous levels of radiation for a long time. Though the steam explosions at Fukushima did not release as much radiation as the most serious global nuclear accident (Chernobyl, 1987), the continued release of radioactive material into the surrounding area and into the oceans is uncharted territory.

The hazard presented by radioactive isotopes certainly depends on the type of radiation released (e.g., gamma radiation is more concerning and harder to avoid than alpha or beta radiation) but also, how long an isotope remains radioactive is an important factor. Therefore, determining how long an isotope will be dangerously radioactive is vital information in any risk assessment.

Radioactive decay is measured in half-lives. An isotope's **radioactive half-life** is the amount of time it takes for half of the radioactive material in question to decay to a new form. After one half-life, 50% of the material will decay; in the next half-life, 50% of what's left (or 25% of the original amount) will then decay, and so on. After 10 half-lives, just 0.1% of the original radioactive material is left. **INFOGRAPHIC 5**

Some isotopes generated in the nuclear fission reaction have a very short half-life (from seconds to days or weeks), but others are more stable with half-lives in the hundreds or thousands of years. The strong gamma emitters from a fission reaction have relatively short half-lives: The half-life of radioactive iodine (I-131) is 8 days; cesium and strontium (Cs-137 and Sr-90) have half-lives of about 30 years. A general rule of thumb is that after 10 half-lives, enough radioactive material will have decayed to render the material safe. For I-131, that is 80 days; for Cs-137 and Sr-90, that is 300 years. But for isotopes with half-lives in the hundreds or thousands of years, we are looking at thousands or tens of thousands of years before the material could be considered safe.

radioactive decay The spontaneous loss of particle or gamma radiation from an unstable nucleus.

radioactive half-life The time it takes for half of the radioactive isotopes in a sample to decay to a new form.

The rate of decay for a given radioactive isotope is predictable and expressed as a *half-life*—the amount of time it takes for half of the original radioactive material (*parent*) to decay to the new *daughter* material (a new isotope, or even a new atom if protons are lost).

RADIOACTIVE HALF-LIFE

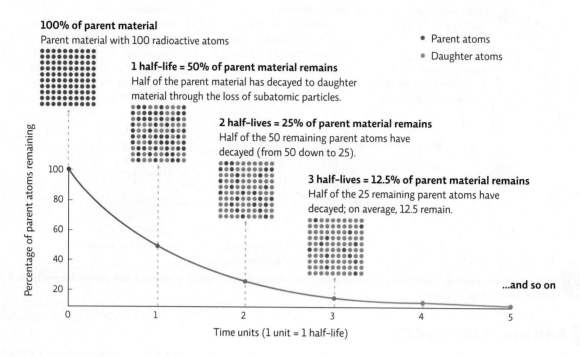

100% of parent material
Parent material with 100 radioactive atoms

• Parent atoms
• Daughter atoms

1 half-life = 50% of parent material remains
Half of the parent material has decayed to daughter material through the loss of subatomic particles.

2 half-lives = 25% of parent material remains
Half of the 50 remaining parent atoms have decayed (from 50 down to 25).

3 half-lives = 12.5% of parent material remains
Half of the 25 remaining parent atoms have decayed; on average, 12.5 remain.

...and so on

y-axis: Percentage of parent atoms remaining (100, 80, 60, 40, 20)
x-axis: Time units (1 unit = 1 half-life) (0, 1, 2, 3, 4, 5)

 What percentage of the parent material will be left after 5 half-lives?

6 NUCLEAR WASTE

Key Concept 6: Radioactive waste is actually more radioactive than the original fuel. We currently have no long-term storage plan for high-level radioactive waste.

At Fukushima, the reactors weren't the only problem. Experts around the world were also concerned about all the radioactive waste stored onsite at Fukushima. To understand why, it helps to know a little about radioactive waste. In general, there are two kinds. **Low-level radioactive waste (LLRW)** is material that has low amounts of radiation relative to its volume and can usually be safely buried. This includes clothing, gloves, tools, etc., that have been exposed to radioactive material. Most LLRW isotopes decay to background levels in 100 years. The United States produces about 60,000 cubic meters (2 million cubic feet, a volume equivalent to 15 million gallons) of LLRW per year, all of which is stored at just four sites—in South Carolina, Texas, Utah, and Washington.

High-level radioactive waste (HLRW) is another story. As its name suggests, it's more reactive than LLRW; in fact, because so many radioactive by-products are created in fission reactions, HLRW is considerably more radioactive than the original fuel rods—an estimated 1 million times more! It also generates tremendous heat, making storage in containers inherently dangerous—containment must successfully hold material that is both radioactive and hot. The United States produces about 2,000 metric tons of HLRW per year and has some 65,000 metric tons of the stuff in interim storage right now; this estimate includes both spent fuel rods and waste from nuclear weapon production.

low-level radioactive waste (LLRW) Material that has a low level of radiation for its volume.

high-level radioactive waste (HLRW) Spent nuclear reactor fuel or waste from the production of nuclear weapons that is still highly radioactive.

The crippled Fukushima Daiichi Nuclear Power Station 10 months after the disaster. Radiation levels remain high with some areas still inaccessible.

The radioactive waste produced by nuclear power plants is a huge safety issue: It's extremely dangerous, there's a lot of it, and we have yet to come up with a plan for disposing of it safely. Spent fuel rods are stored onsite in steel-lined pools, where short half-life isotopes can decay to safe levels. Isotopes with longer half-lives require more time to reach this point. After spending about 5 years in a pool, the HLRW is moved to dry storage in large steel casks; the fuel inside the cask is surrounded by an inert gas, and the entire cask is then encased in concrete. Currently, more than 70 sites (known as independent spent fuel storage installations, or ISFSI) are storing HLRW in dry casks at or near power plants in the United States.

> The radioactive waste produced by nuclear power plants is another huge safety issue: It's extremely dangerous; there's a lot of it, and we have yet to come up with a plan for disposing of it safely.

The waste in the casks is awaiting a longer-term storage option. So far, we don't have one. "Waste has been one of the biggest sore points in the debate over nuclear energy," says Charles Powers, a professor and nuclear energy scientist at Vanderbilt University. "It's the one thing we really don't have even a good theoretical solution to."

In 1982, 25 years after the first U.S. nuclear power plant started operation, the U.S. **Nuclear Waste Policy Act** was passed, assigning responsibly for nuclear waste disposal to the federal Department of Energy. The law identified deep underground storage as the best way to dispose of HLRW and established guidelines for research and development into permanent repositories.

The United States selected Yucca Mountain in Nevada as the site for a long-term repository and began construction in 1994, but the project has been repeatedly stymied by opponents. After 15 years and billions of dollars, President Barack Obama halted construction in 2010, a move that fueled even more bickering. The debate continues; in 2017,

Nuclear Waste Policy Act (1982) The federal law that mandated that the federal government build and operate a long-term repository for the disposal of high-level radioactive waste.

Yucca Mountain was once again being examined for its potential as a HLRW repository. (For more on Yucca Mountain, see the Interactive Map at the end of this module.) **INFOGRAPHIC 6**

Meanwhile, back at Fukushima, another more immediate problem is emerging—what to do with all that water that has been constantly added to the steel-lined pool where the assemblies were kept or that has been sprayed on the melted reactor cores. This radioactive water is collected and radioactive isotopes are removed, with the exception of tritium (radioactive hydrogen). There is no way to remove tritium, so the tritiated water is stored; so far more than 1 quadrillion liters (264 trillion gallons) of contaminated water is currently stored in huge aboveground containers, with more containers added every day or so.

Unfortunately, some of the water escapes collection and flows into the ocean—one estimate puts it at 9,000 liters (2,400 gallons) per day. And with this water go radioactive cesium and strontium, levels of which are still detectable off the coast of Japan, indicating continued release, according to research by Woods Hole Oceanographic Institute scientist, Ken Buesseler. No one knows what the impact of that constant release will be—we are in uncharted waters. To compound matters, in July 2017, Tokyo Electric announced it would dump the stored tritiated water into the ocean, generating outcries from local fishermen and residents.

INFOGRAPHIC 6 RADIOACTIVE WASTE

Radioactive waste does not just come from nuclear power plants but is also generated by industry, the medical field, research laboratories, and weapons production. All of these users produce low-level radioactive waste (LLRW); nuclear power and weapons production is responsible for almost all of the high-level radioactive waste (HLRW). Though much of this material is highly dangerous and remains so for centuries, we currently have no safe way to dispose of it.

SuperStock/Alamy Stock Photo

LLRW includes contaminated items such as clothing, filters, gloves, and other materials exposed to radiation. In the United States, short-half-life LLRW is disposed of as regular trash after it is no longer radioactive; longer half-life LLRW is stored in casks and sent to one of four U.S. storage facilities where it is buried underground.

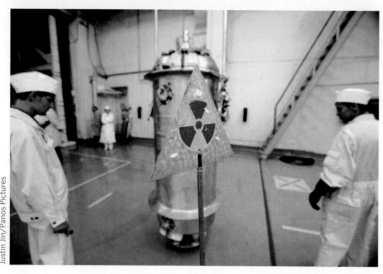

Justin Jin/Panos Pictures

The United States lacks a long-term storage facility for **HLRW**; most spent fuel rods are currently stored onsite at nuclear power plants in steel-lined pools. Some have been moved out of the pools and into dry casks and stored above ground. European and Asian nations are looking into the possibility of joint disposal facilities.

 Do you think we should continue to use nuclear power, or increase its usage, if we do not have a plan for long-term storage of HLRW?

7 NUCLEAR INDUSTRY ACCIDENTS

Key Concept 7: There have been around 100 nuclear accidents in the history of the industry; most were minor but several caused deaths and two were severe: Chernobyl and Fukushima.

The nuclear power industry has a short history. The power of nuclear energy was first demonstrated in 1945, when the U.S. military dropped atomic bombs over the Japanese cities of Hiroshima and Nagasaki and ended World War II. In 1953, President Eisenhower made his famous "Atoms for Peace" speech, laying out a plan by which this destructive force could be harnessed for good: Instead of building bombs, we would produce cheap, reliable energy.

Worldwide, there are more than 400 nuclear power plants, accounting for about 10% of global electricity production. The United States has more nuclear reactors than any nation, but Americans themselves have been divided over nuclear energy since the 1980s, after two infamous nuclear accidents made global headlines.

The first was a steam explosion in an experimental reactor that caused a partial meltdown (only a portion of the nuclear fuel melted) at the Three Mile Island plant near Middletown, Pennsylvania, in 1979, due to an electrical failure followed by a flurry of operator errors. This accident resulted in only minor release of radiation (below background levels). It did not cause any public health problems but destroyed the reactor.

The second, caused by human error during an experiment that went awry, happened at the Chernobyl reactor in Ukraine (then part of the Soviet Union), in the spring of 1986; this accident was considerably more severe. The chain reaction escalated out of control, resulting in a full nuclear meltdown, which caused a steam explosion that released a tremendous amount of radiation over much of western Russia and Europe. The former Soviet government did not notify the public or nearby nations of the accident right away, delaying protective actions that could have reduced affected populations' exposure to the radiation. All land within a 30 kilometer- (18.6 mile-) radius has been designated as an exclusion zone; entry without permission and a guide is prohibited, though some elderly residents have returned to the zone. The effects of radiation exposure are still being felt, and estimates are that the radiation will ultimately be responsible for some 4,000 deaths when all is said and done.

These accidents prompted the creation of the *International Nuclear and Radiological Event Scale*, used to

rank the seriousness of an event so that officials could accurately inform the public and regulatory bodies. The scale ranks events from 1 to 7; each higher level of the scale represents a ten-fold increase in severity. For "accidents" (Levels 4–7), severity is determined by the amount of radiation released. For "incidents" (Levels 1–3), severity is determined by the degree of failure of safety protocols.

There have been other incidents and accidents (fires, partial meltdowns, chemical explosions that released radiation), and some were severe, but only Fukushima and Chernobyl earned the highest rating, a 7. Thankfully, because Japanese officials moved quickly to protect people (evacuating residents and distributing iodine tablets to protect against exposure to radioactive iodine), there have been no deaths attributed to radiation exposure from the Fukushima accident.
INFOGRAPHIC 7

Will there be any long-term physical health effects on residents and workers as a result of the accident? A 2014 study by Kouji Harada and colleagues estimates only a very slight increase in the lifetime risk of cancer for people living in Fukushima Prefecture at the time of the disaster, a level that the authors say is "unlikely to be epidemiologically detectable." Mental health problems attributed to the stress of the event and its aftermath are believed to be the most serious health risk at this time. Residents are being allowed to reoccupy some areas (areas with acceptable levels of radiation) while other areas are only open for day visits (to clean, rebuild, and restore). Remediation of the area continues.

By early 2017, radiation in the general area around the Fukushima facility was low enough to work on demolition, though workers cannot get close to the melted reactors that still contain the as yet unseen melted radioactive fuel—that area remains off limits. Those fuel cores are believed to have melted into a mass that exists today at the bottom of each reactor; it may have melted through the metal containment compartment and perhaps could have reached the ground below. Even robots have not yet been able to get into the reactors to see what is going on, though attempts are ongoing.

INFOGRAPHIC 7 **NUCLEAR ACCIDENTS**

The International Nuclear and Radiological Event Scale assigns a degree of severity to nuclear events; each subsequent level represents a 10-fold increase in severity. There have been more than 100 nuclear incidents and accidents since 1952, including 20 partial or total meltdown events; only Chernobyl and Fukushima rank at the highest level of severity.

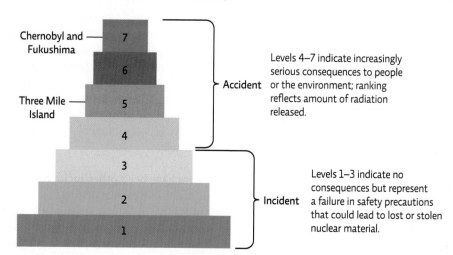

Chernobyl and Fukushima — 7

6

Three Mile Island — 5

4

Levels 4–7 indicate increasingly serious consequences to people or the environment; ranking reflects amount of radiation released.

— Accident

3

2

1

— Incident

Levels 1–3 indicate no consequences but represent a failure in safety precautions that could lead to lost or stolen nuclear material.

Below Scale (Level 0): No safety concerns

NOTABLE NUCLEAR EVENTS			
YEAR	**PLACE**	**EVENT LEVEL**	**DESCRIPTION**
1952	Chalk River (Canada)	5	First reactor accident; malfunction and human error led to partial meltdown and steam explosion that released radiation.
1961	Idaho National Laboratory (U.S.)	Not determined	Failed attempt to safely restart an experimental reactor (possibly due to incorrectly inserted control rods) resulted in an uncontrolled chain reaction and explosion that killed three workers.
1979	Three Mile Island (U.S.)	5	Cooling malfunction and subsequent human error led to partial meltdown; release of radioactive water and gas occurred but was contained to the immediate area.
1986	Chernobyl (Ukraine)	7	Human error during an experiment led to a steam explosion and massive release of radiation; caused 30 deaths within a week and was, or ultimately will be, responsible for thousands of illnesses and deaths.
1987	Goiania (Brazil)	5	A radiotherapy device, stolen from an abandoned clinic where it was mistakenly left behind, was sold as scrap metal; many curious people came in contact with the colorful material inside (radioactive cesium). More than 200 were exposed to radiation, which resulted in four deaths and 20 hospitalizations.
1999	Tokaimura (Japan)	4	Worker error in the handling of radioactive material caused two deaths and resulted in local evacuations.
2011	Fukushima (Japan)	7	Earthquake and tsunami caused total meltdown and steam explosions that released large amounts of radiation.
2014	New Mexico (U.S.)	Not determined	Waste container exploded in underground bunker, releasing radiation into the atmosphere; the accident was attributed to using the wrong type of kitty litter (an absorbent material used to clean and contain chemical spills). Clean-up estimates run around $2 billion.

 Which of the accidents described above had a component of human error? What does this suggest about the safety of nuclear power?

8 THE PROS AND CONS OF NUCLEAR POWER

Key Concept 8: The advantages (low air pollution, ample supplies) and disadvantages (radioactive hazards) of nuclear power must be weighed to determine its future role in electricity production.

Nuclear power has its proponents and opponents and for good reasons. It has much to offer but comes with some serious downsides. Proponents argue that uranium ore is both more abundant and produces a more energy-rich fuel than fossil fuels. Nuclear energy is also a low-carbon way to produce electricity: Unlike the use of fossil fuels, no CO_2 is released during the production of electricity, and nuclear power creates virtually none of the other problematic combustion by-products, like sulfur dioxide, nitrogen oxides, and particulate matter.

And despite some persistent fears, research suggests that living near a nuclear power plant is actually safer than living near a coal-fired one. A 2011 study by Annette Queißer-Luft found that the risk of birth defects did not increase the closer one lived to a nuclear facility. Meanwhile, Javier García-Pérez found that in Spain, the number of cancer-related deaths does increase as proximity to coal-fired power plants increases. "You still have some environmental hazards, from mining uranium and from radioactive waste and water," says Powers. "But on balance, nuclear is far cleaner than any fossil fuel."

The operating costs, per kilowatt-hour, for nuclear power are comparable to those of coal but the cost of building, maintaining, and then decommissioning nuclear plants is much more expensive—decommissioning alone could cost close to $1 trillion. (See the Interactive Map for an example of the decommissioning process.)

While there have only been two Level 7 nuclear accidents in the history of the industry, there have been more than 100 accidents and incidents ranking between Levels 1 and 6, and the debates over safety remain unresolved. Proponents point out that, considering the number of existing plants and the length of time they have been operating, accidents have been exceedingly few and far between. But opponents say that such safety claims ignore three key points: the vulnerability of nuclear power plants to natural disasters (which at Fukushima led to nuclear meltdown and the release of radioactive material into the environment), the potential for nuclear fuel to be stolen and weaponized, and dealing with the radioactive waste. **INFOGRAPHIC 8**

The March 2011 tsunami would become the most expensive natural disaster in human history, with loss estimates skyrocketing past $1 trillion. (One estimate has it at $51 trillion.) In its wake, countries around the world began rethinking their plans to expand their own nuclear energy programs. Germany resolved to move up the date for its planned phase-out of all nuclear power plants by 10 years, to 2022. In the United States, and other countries, safety procedures and infrastructure were examined and new protocols put in place from lessons learned at Fukushima. (For example, back-up generators are no longer stored in flood-prone basements.) Consideration has also been given to the placement of nuclear power plants. Fears over a disaster at the Indian Point Power Plant, 56 kilometers (35 miles) north of Times Square in New York City (where 20 million people could be at risk) have prompted officials to put into motion plans to shut it down by 2021. (See the Interactive Map for more on Indian Point.)

Christoph Bangert/laif/Redux

Displaced people who were evacuated from Minamisoma, Futaba, and other towns located near the Fukushima Daiichi Nuclear Power Plant are checked for traces of radiation before they are permitted to enter a sports facility in Fukushima City, where about 1,200 evacuees found temporary shelter.

Given the problems associated with fossil fuels, it is likely that nuclear power will continue to have a place in our energy mix, especially with global electricity use projected to double by 2030. But in making choices about how to pursue our energy future, we must consider the costs and benefits of all our energy options. A benefit analysis must evaluate how well a particular energy source meets our energy needs. A cost analysis must consider not just the monetary cost of getting kilowatts delivered

INFOGRAPHIC 8 NUCLEAR POWER: TRADE-OFFS

What role will nuclear power play in the future? Like all of our other energy options, nuclear power has advantages and disadvantages that must be weighed when making this decision.

ADVANTAGES	DISADVANTAGES
COSTS	
• Operating costs are comparable to those of a fossil fuel power plant. • The technology is available now.	• Nuclear power plants are much more expensive to build ($20 billion to build a fission reactor) and maintain than fossil fuel power plants, and even more expensive to decommission ($200 million to $1 trillion).
ELECTRICITY PRODUCTION	
• Power production can be increased or decreased to meet demand.	• Though power production can be altered to meet demand, it cannot be done as quickly as it can be done with a fossil fuel facility. • Large amounts of water are needed in the process for cooling.
SUPPLIES	
• Uranium supplies are good, and other isotopes can also be used. • Nuclear fuels are energy rich: One small uranium pellet contains the same amount of energy as 1 metric ton of coal.	• Mining and processing ore for nuclear fuel produces hazardous waste and can pollute air and water. • Some methods of nuclear power production produce radioisotopes that could be used in nuclear weapons production.
POLLUTION AND SAFETY	
• Much cleaner than burning fossil fuels to generate electricity: During operation, no particulates, sulfur, or nitrogen pollution are released; no CO_2 is released.	• Radioactive waste is very hazardous, and we still have no long-term plan for dealing with HLRW. • Shipping fuel or waste (by truck or rail) is also a safety concern and vehemently opposed by those who live on the transport route. • Though serious accidents are rare, they can have devastating consequences and long-term impacts when they occur.

 Which advantage of nuclear power do you consider to be its strongest advantage? Which of its disadvantages is its biggest detractor? Explain.

to our homes but also the environmental and social costs associated with every step of the energy source's life, from acquisition to production to delivery to waste disposal. In addition, as mountaintop removal, oil spills, climate change, and nuclear meltdowns demonstrate, a risk assessment must also take place. We must answer two very crucial questions: How risky is the venture (an assessment we can do with at least some degree of accuracy) and how much risk are we willing to take?

Select References:

Burns, P. C., et al. (2012). Nuclear fuel in a reactor accident. *Science* (Washington), 335(6073), 1184–1188.

Castrillejo, M., et al. (2016). Reassessment of 90Sr, 137Cs, and 134Cs in the coast off Japan derived from the Fukushima Dai-Ichi nuclear accident. *Environmental Science and Technology*, 50(1), 173–180.

García-Pérez, J., et al. (2009). Mortality due to lung, laryngeal and bladder cancer in towns lying in the vicinity of combustion installations. *Science of the Total Environment*, 407(8), 2593–2602.

Harada, K. H., et al. (2014). Radiation dose rates now and in the future for residents neighboring restricted areas of the Fukushima Daiichi Nuclear Power Plant. *Proceedings of the National Academy of Sciences*, 111(10), E914–E923.

Queißer-Luft, A., et al. (2011). Birth defects in the vicinity of nuclear power plants in Germany. *Radiation and Environmental Biophysics*, 50(2), 313–323.

INTERACTIVE MAP   NUCLEAR POWER IN THE UNITED STATES   ANIMATED INFOGRAPHIC

In 2017, there were 61 operating U.S. nuclear power plants, 71 interim storage facilities for high-level radioactive waste, and more than 100 other sites (reactors, industries, university labs, etc.) undergoing decommissioning. Visit this interactive map for a brief look at our newest reactor, still under construction (Vogtle), a power plant slated for closure (Indian Point), and another undergoing decommissioning (Zion). The final case study presents information about our proposed HLRW disposal site: Yucca Mountain in Nevada.

ZION: DECOMISSIONING

INDIAN POINT: CLOSURE

New York

Illinois

Nevada

YUCCA MOUNTAIN: PERMANENT WASTE DISPOSAL FACILITY

Georgia

VOGTLE: NEW CONSTRUCTION

■ States with nuclear power plants
□ States without nuclear power plants

 BRING IT HOME

PERSONAL CHOICES THAT HELP

Nuclear energy has been rebranded as "green energy" because it does not emit greenhouse gases. Technology has improved the safety of nuclear facilities; however, there are still safety issues and valid concerns over the long-term storage of nuclear waste. In addition, cost, national security, and uranium supplies make nuclear power a complicated energy solution.

Individual Steps

• Use the facility locator on the U.S. Nuclear Regulatory Commission website (www.nrc.gov) to find out if you have nuclear reactors where you live. If so, what type of reactor are they: PWR or BWR?

Group Action

As you would expect, the policies endorsed by a particular group depend on the group's overall view of nuclear energy. To see two examples, check out the following sites and see how they compare.

• Visit the Nuclear Energy Institute (www.nei.org), which is a pronuclear organization, to see proposed legislation regarding increasing our current nuclear energy production.

• Visit the Nuclear Energy Information Service (www.neis.org), which is a nonprofit antinuclear organization committed to a nuclear-free future.

Policy Change

• What is your opinion about the necessity of nuclear energy in the United States? The U.S. Department of Energy website (www.energy.gov) maintains articles, updates, and the latest policy initiatives regarding nuclear energy in America. This resource can help you understand current policy, funding issues, and upcoming legislation.

Toby Talbot, File/AP Photo

ENVIRONMENTAL LITERACY   UNDERSTANDING THE ISSUE

1 What are radioactive isotopes and why are they important for nuclear power?

1. The number of _____ is unique to a given element and gives an element its atomic number.

2. Atoms that have the same number of protons and electrons but different numbers of neutrons are called:
   a. radioactive.
   b. ions.
   c. isotopes.
   d. subatomic.

3. Uranium-238 has 92 protons and 146 neutrons. How many protons and neutrons does uranium-234 have?

2 How is uranium fuel for nuclear power produced?

4. About how much of the uranium found naturally in uranium ore is U-235 (the fuel used in nuclear reactors)?
   a. 20%
   b. 65%
   c. 4%
   d. 0.7%

5. What is the purpose of the enriching step of nuclear fuel production?

3 How is nuclear energy harnessed to generate electricity in a fission reactor?

6. Nuclear fission is a reaction that:
   a. splits an atom, releasing energy.
   b. combines two or more atoms, producing energy.
   c. results in large explosions.
   d. is required to make atoms radioactive.

7. What is the purpose of the control rods in a nuclear reactor?
   a. They provide extra neutrons to speed up the chain reaction.
   b. They absorb neutrons and slow down the chain reaction.
   c. They provide cooling to prevent overheating of the fuel rods.
   d. They absorb the fission by-products produced in the reaction.

8. Why would one be more concerned about the venting of steam from a boiling water reactor than a pressure water reactor?

9. Describe the final steps of fuel preparation from pellet to fuel assembly.

4 What types of radiation are produced when isotopes decay?

10. Radiation that is released as a two-proton/two-neutron particle is known as _____ radiation.

11. There are different types of radiation. Which one is most energetic and can therefore penetrate many surfaces, including skin?
    a. Alpha radiation
    b. Beta radiation
    c. Gamma radiation
    d. Particle radiation

12. What is ionizing radiation and why is it dangerous?

5 How is the rate of decay of a radioactive atom measured?

13. After four half-lives, about how much radioactive parent material is left?
    a. 25%
    b. 40%
    c. 0.5%
    d. 6.25%

14. In a nuclear fuel assembly, control rods are used to regulate the nuclear reaction. But even if all the control rods are inserted into the fuel assembly, heat is still produced. Explain why.

6 What problems are associated with nuclear waste?

15. Which of the following is an example of low-level radioactive waste (LLRW)?
    a. Worker clothing and gloves
    b. Tools used by workers
    c. Spent fuel rods
    d. A and B are both examples of LLRW.
    e. A, B, and C are all examples of LLRW.

16. How are spent fuel rods currently disposed of in the United States?

17. Why is nuclear waste seen as a major, yet unresolved problem for the nuclear industry?

7 What is the history of nuclear accidents worldwide?

18. True or False: The 1979 Three Mile Island accident was the first nuclear accident in the world.

19. According to the International Nuclear and Radiological Event Scale, what is the significance of the different levels?
    a. Each higher level represents a 10-fold more serious event than the one below.
    b. Levels are assigned according to the number of deaths caused by the incident.
    c. Lower levels represent accidents at universities; higher levels are reactor accidents.
    d. The higher the level, the more widespread geographically the damage is expected to be.

20. Why were the health and environmental impacts so much more severe at Chernobyl than at Fukushima?

8 What are the advantages and disadvantages of nuclear power?

21. True or False: Barring a major accident, it is less hazardous to your health to live near a nuclear reactor than a coal-fired power plant.

22. From an economic standpoint, which type of electricity production—nuclear or fossil fuel—is less expensive? Explain. Remember to include a full lifecycle analysis in your answer.

SCIENCE LITERACY   WORKING WITH DATA

An evaluation of lifecycle greenhouse gas emissions, a metric that includes every step of energy production from acquiring raw materials to decommissioning facilities, is a useful way to compare the carbon footprint of various energy sources. The following graph shows one estimate that compares the lifecycle greenhouse gas emissions of various nonfossil fuel energy sources for electricity production. (For comparison, the lifecycle greenhouse gas emissions for coal are more than 1,000 g of CO_2-equivalent/kWh.)

CARBON FOOTPRINTS OF NONFOSSIL FUEL ENERGY SOURCES

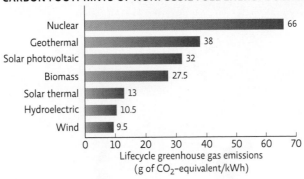

Energy Source	Value
Nuclear	66
Geothermal	38
Solar photovoltaic	32
Biomass	27.5
Solar thermal	13
Hydroelectric	10.5
Wind	9.5

Lifecycle greenhouse gas emissions
(g of CO_2–equivalent/kWh)

Interpretation

1. What does the length of each bar represent?

2. Identify a result of this data: Based on the graph, how does the carbon footprint of nuclear power compare to other industries?

3. Compare the carbon footprint of wind (the lowest greenhouse gas emitter) and geothermal (the second-highest emitter) to that of nuclear power by calculating the percentage CO_2-equivalent greenhouse gases released by each compared to nuclear. (Hint: Divide the total shown for the energy source in question by the total for nuclear power and then multiple by 100.)

Advance Your Thinking

4. Identify a conclusion based on this data: If a country's goal is to reduce greenhouse gas release to be no more than half that of nuclear power, which energy sources should be pursued?

5. Since no CO_2 is released to generate the heat of nuclear fuel (nothing is burned), what do you think contributes to the relatively higher release of greenhouse gases from nuclear power compared to other nonfossil fuel sources?

INFORMATION LITERACY   EVALUATING INFORMATION

The Nuclear Regulatory Commission (NRC) is tasked with regulating nuclear power in the United States and "protecting people and the environment." The members of the commission "formulate policies, develop regulations governing nuclear reactor and nuclear material safety, issue orders to licensees, and adjudicate legal matters" (www.nrc.gov/about-nrc/organization/commfuncdesc.html). The NRC maintains a detailed website with information about nuclear power plants and nuclear materials.

Go to the NRC's Facilities page (www.nrc.gov/info-finder/region-state). Find your state in the list of states and territories.

Evaluate the website and work with the information to answer the following questions:

1. Determine if this is a reliable information source:
    a. Who runs the website? Do this person's/group's credentials make the information reliable or unreliable?
    b. Is the information on the website up to date? Explain.

2. Choose a U.S. state to evaluate and record the name of that state. How many nuclear power reactors are active in this state? How many are being decommissioned?

3. Click on the links for an active nuclear power reactor in the state you are evaluating and look at the information provided via the other links on its main page.
    a. Write down information about its age, intended expiration date, type of reactor in use (PWR or BWR), and specific location. Find the location on a map and record the closest large body of water.
    b. Look at safety information (e.g., inspections, accidents, safety performance) about this reactor and evaluate its safety record.
    c. Look over the "Enforcement Actions" link to evaluate the types of actions that were taken (including severity and penalties assessed) and briefly summarize them.

4. Consider the information you have gathered and do a basic risk assessment analysis for one of the sites.
    a. Do you think the facility is a danger to the environment? To human health? Why or why not?
    b. Is the facility located in an area that may have natural disasters? If so, what are they? If a natural disaster were to occur, predict some outcomes.

 Additional study questions are available at SaplingLearning.com.

FUELED BY THE SUN

A tiny island makes big strides with renewable energy

Playing soccer under the shade of wind turbines on Samsø Island, Denmark, the first island in the world with 100% renewable energy.
Alessandro Grassani/Invision/Aurora Photos

After reading this chapter and studying the KEY CONCEPTS and INFOGRAPHICS, you should be able to answer these GUIDING QUESTIONS

CORE MESSAGE

In order to become a sustainable society, we need to transition to reliable, renewable energy sources with acceptable environmental and social impacts. Since no single energy source can replace fossil fuels, a variety of methods, selected to meet the needs of the population, and availability of local energy sources, will help communities shift to sustainable energy use.

1. What are the characteristics of a sustainable energy source, and what types are commonly used?

2. What are the current and projected roles of renewables in global energy production?

3. How can wind energy be captured to generate electricity, and what are the pros and cons of these methods?

4. How can solar energy be captured and used, and what are the pros and cons of these methods?

5. How can geothermal energy be captured and used, and what are the pros and cons of these methods?

6. How can the power of water be captured, and what are the pros and cons of these methods?

7. What roles do conservation and energy efficiency play in helping us meet our energy needs sustainably?

8. What economic adjustment and technological advances are needed for renewables to become a viable replacement for fossil fuels?

On a typical cold, misty January day in Denmark in 2003, many of the 4,100 residents of a small island gathered together at the beach, straining their eyes to see the faint outline of several structures, each over 30 stories tall, located offshore. Located just offshore of Denmark's mainland, Samsø is home to a small community of farmers known for their sweet strawberries and tender early potatoes. It is a quiet and serene place. Yet it has been the site of a dramatic revolution—a community transformation that made headlines around the world.

The transformation began a new chapter on that cold day in 2003. The mayor pushed a button, and the offshore structures slowly creaked to life. Through the gray mist and rain, people could gradually see the massive blades begin to rotate, converting the power of wind into energy. It was a landmark day in Samsø's ambitious attempt to become the greenest, cleanest, and most energy-independent place on Earth.

"That was a very big moment," recalls Søren Hermansen, a Samsø resident who was key in getting the community behind the project. "Nobody really thought it would happen when we started."

Samsø is just one of many communities around the world taking steps to wean themselves from fossil fuels. The path a particular community takes varies, a reflection of its needs and the energy sources at hand. In each case, there is much to be learned from the successes and setbacks of these trailblazers—lessons others can use in their quest toward a sustainable energy future.

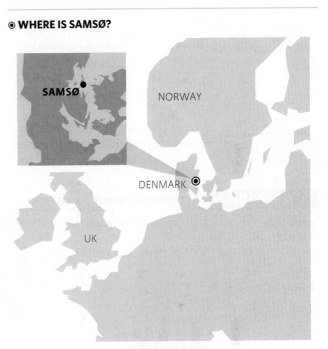

◉ WHERE IS SAMSØ?

1 CHARACTERISTICS OF SUSTAINABLE ENERGY SOURCES

Key Concept 1: Sustainable energy sources are those that meet our needs and are readily replenished while producing impacts we, and ecosystems, can live with long term.

Samsø used to be just like most other communities on Earth: fully dependent on fossil fuels like coal, oil, and natural gas. But in 1997, the government of Denmark decided it was important to promote the idea of **renewable energy**, energy from sources that are replenished over short time scales or that are perpetually available. To do so, it announced a competition: Which local area or island could become self-sufficient on renewable energy? The contest invited applicants to describe which renewable resources were available in their community and how they would be used to replace fossil fuels.

The purpose of the contest was to put communities on track to be more sustainable. To qualify as a **sustainable energy** source, the energy must be renewable—because energy cannot be recycled or reused, living organisms need to depend on an energy source that is readily replenished. (See Infographic 6 in Module 1.1 for more on the characteristics of a sustainable ecosystem.) In addition, this

renewable energy Energy that comes from an infinitely available or easily replenished source.

sustainable energy Energy from sources that are renewable and have a low environmental impact.

renewable energy source must be used at a rate equal to or less than its replacement rate. And finally, to be sustainable, its acquisition and use must have a low enough environmental impact that it could be used for the long term.

Samsø had many sustainable energy options from which it could choose as it considered which ones would work best in its locale. The earliest renewable energy source used by humans is **biomass energy**—energy that is harnessed when biological material is burned. Many people today still depend on wood, animal waste, or crop residue as a fuel for cook stoves and heat. Biomass can also be burned in waste incinerators or power plants to produce heat or electricity on a large scale (see Module 5.3) and it can be converted to liquid or gaseous biofuels for use in home or vehicles. Biofuels are only considered renewable if the rearing or harvesting of biomass for fuels does not exceed the rate at which it is replaced by nature. (For more on biofuels, see Online Module 11.3.)

biomass energy Energy from biological material such as plants (wood, charcoal, crops) and animal waste.

Samsø decided to pursue biomass energy as one of its energy sources by burning locally grown straw to

heat homes. The straw used in the power plant is the only energy source that is burned on the island other than the gasoline and diesel fuel used in vehicles. Ideally, Samsø residents would have swapped their vehicles for cars that run on hydrogen or electricity, but those technologies were too expensive for the islanders. Since Samsø would continue to rely on conventional fuels for transportation, engineers decided to install large offshore wind turbines to produce an equivalent amount of clean energy, offsetting uses by motor vehicles and boats. Islanders are working toward a gradual transition to electric vehicles and those powered by biofuels. The island's small size helps make all this feasible since daily "commutes" are short and distance traveled per person per day is lower than in most industrialized nations.

While biomass is useful fuel, its limitations make it less likely to contribute significantly as a renewable fuel in the future. Four other energy options that are useful stationary sources of energy for electricity production or heat are *solar, wind, geothermal,* and *hydropower.* Each renewable energy source has its own advantages and disadvantages. Each of these will be defined and discussed later in this module. **INFOGRAPHIC 1**

INFOGRAPHIC 1   **SUSTAINABLE ENERGY SOURCES**

Sustainable energy sources are those that are readily replenished and that have impacts that are acceptable; they only remain sustainable if we use them at or below the rate of replenishment.

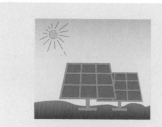

SOLAR POWER
The energy of the Sun can be trapped in a variety of ways. More solar energy arrives daily than we could possibly use.

WIND POWER
Energy in the motion of air can be captured to generate electricity.

HYDROPOWER
The energy of moving water can be tapped in a variety of ways to produce electricity or other services.

GEOTHERMAL POWER
The heat of the Earth can be tapped to make electricity or to provide heating or cooling for buildings.

BIOMASS
Chemical energy in biological materials (crops, waste, wood, etc.) can be released when the biomass is burned. (See Online Module 11.3.)

 Which of these renewable energy options are a direct or indirect form of solar energy?

2 CURRENT AND PROJECTED USE OF SUSTAINABLE ENERGY SOURCES

Key Concept 2: Fossil fuels are the leading fuel for electricity production. The use of renewables is rising, but so is the use of fossil fuels, due to a rise in energy demand.

Human societies depend on energy, and for modern societies this energy comes predominately from fossil fuels. Since the Industrial Revolution, world energy use has been on the rise, remarkably so since the mid-20th century when it began to increase exponentially. According to the International Energy Agency (IEA), worldwide in 2016 our energy use was more than 105 billion *barrels of oil equivalents (BOE)*—that is more than double what we used in 1973. Over half of that was derived from fossil fuels.

In the United States, we are even more reliant on fossil fuels with around 80% of the total energy used coming from coal, oil, or natural gas. Fossil fuels are also the primary fuel for electricity production, accounting for 63% in 2016.

The use of renewables is slowly climbing, but so is the demand for energy. The United States Energy Information Agency (EIA) predicts that U.S. electricity production from renewables will increase by 40% by 2040 (compared to 2010). While this would be a tremendous achievement, the need for electricity is expected to double over the same time frame, resulting in very little change in fossil fuel usage.

Still it is a start. In the near future, wind and solar power will see the biggest increases as costs come down for these technologies and the benefits of using carbon-free energy sources mount. But transitioning to a fossil fuel-free economy will not be easy. As we will see, each of our renewable energy options comes with its own set of trade-offs, and no single source can replace fossil fuels. However, together, their advantages are significant and in many cases, they complement each other in ways that diminish their disadvantages. Whether or how quickly we transition away from fossil fuels will depend on society's assessment of the costs of using fossil fuels and our willingness to put our money and resources into deploying and developing renewable options. **INFOGRAPHIC 2**

INFOGRAPHIC 2   **RENEWABLE ENERGY USE**

According to the U.S. Energy Information Administration, as of 2016, renewable sources of energy contributed 15% of U.S. energy production. By 2040, energy production by renewable energy sources is projected to more than double, but will account for only about 25% of the total because total usage is also expected to double over that time.

U.S. ELECTRICITY PRODUCTION BY SOURCE, 2016

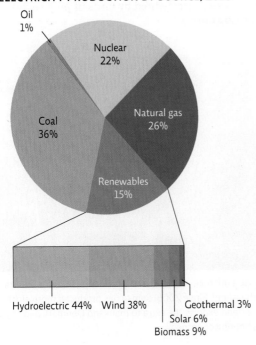

RENEWABLE ELECTRICITY GENERATION

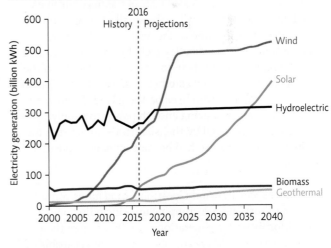

 Which renewable energy sources are expected to increase the most in the next few decades and which will see little change? What might account for these trends?

3 WIND POWER

Key Concept 3: Wind power is pollution free but is an intermittent energy source. In addition, turbines are dangerous to wildlife, and some consider them eyesores.

In an ambitious plan, Samsø's leaders wanted to do more than merely transition to renewable energy sources. Their goal was for the island to become *energy neutral* (produce as much energy as its residents consume) or even *energy positive* (produce more energy than consumed). Their quest was to meet their energy needs without fossil fuels (or to offset any fossil fuel use it did have with extra renewable energy production).

Wind power, energy contained in the motion of air across Earth's surface, is actually an indirect form of solar energy; it results from the difference in temperature between different regions of Earth, such as the poles and the equator, causing air to move from cooler regions to warmer regions. Abundant on the island thanks to offshore winds, wind became the foundation for Samsø's transition away from fossil fuels. The islanders installed 21 wind turbines to harness the power of wind energy. The area's powerful breezes turn huge blades (up to 40 meters [130 feet] in length) that are connected to a generator. The motion of these spinning blades rotates magnets relative to copper wire inside the generator, producing electricity—the flow of electrons across the copper wires.

Wind power has the advantage of being a technology that can be used on a large scale (such as the huge turbines used on Samsø) or a small scale (small turbines suitable for a single home). Nine of the onshore turbines on Samsø were purchased collectively by groups of farmers who bought shares in their construction. "People on the island are personally invested in this," says Bernd Garbers, a German engineer who lives on the island. By 2013, the windmills were producing enough power to supply 100% of the electricity used by the islanders—sometimes more. In just 10 years, Samsø had become an energy-positive island.

In some especially windy locations, *wind farms* containing dozens of turbines are cropping up—including one in California with a generation capacity of more than 1,500 megawatts (MW) and another in China capable of generating more than 6,000 MW. In the United States, Texas is by far the leading producer of wind power in the United States with wind turbines that generate enough electricity—more than

20,000 megawatts (MW)—to power 5 million homes per year. The career of wind turbine technician has been touted as the fastest-growing energy job in the United States. In windy areas, wind is also seen as the most cost-effective renewable today, sometimes producing electricity at a cheaper per-kilowatt-hour rate than coal or natural gas.

But wind energy isn't perfect. First, even for a blustery location like Samsø, wind is intermittent—it stops and starts irregularly, not producing a steady stream of power. And large wind turbines are not cheap. Each onshore turbine cost the Samsø islanders the U.S. dollar equivalent of $1 million; offshore turbines rang up at $5 million apiece. (Less expensive, smaller wind turbines can be purchased by homeowners for personal use.) Beyond cost, wind turbines can create noise, and some people view them as eyesores. They can also have an impact on the local environment, threatening birds and bats—more than 1 million birds and bats are killed by wind turbines annually in the United States. While that is a large number, it is far less than the number of birds killed by flying into communication towers (40 million) or those killed by domestic cats (hundreds of millions). Still, to decrease the risk from windmills, engineers now avoid placing them in known migratory flight paths or close to areas frequented by birds of prey such as eagles.

INFOGRAPHIC 3

Samsø's offshore wind turbines provide much of the electricity used by the island. Since 2005, Samsø has produced more electricity than it uses and exports the extra to mainland Denmark and beyond to neighboring countries.

ANDREW HENDERSON/National Geographic Creative

---

wind power Energy contained in the motion of air across Earth's surface.

INFOGRAPHIC 3 | **HOW IT WORKS: WIND TURBINES**

Wind turbines can be large or small. All work by the same principle: Spinning blades turn a shaft inside a generator and produce electricity.

Rotating blades turn a shaft inside the turbine. The shaft is attached to a gear that rotates a higher-speed shaft on the generator.

Shaft

Gear box

Generator

ADVANTAGES	DISADVANTAGES
• Abundant in some areas. • Pollution free. • Low environmental impact and carbon footprint. • Effective at large and small scales. • Lowest-cost renewable energy source. • Creates jobs.	• Wind is unpredictable and intermittent. • Not all areas are windy enough to support it. • Wildlife (birds and bats) are killed by windmills. • Considered an eyesore by some. • Usually must be paired with other energy sources to meet needs.

The spinning generator produces electrical current.

? Some people feel that windmills and wind farms are eyesores, but one must also consider the alternatives. Would you rather live near a wind farm or a mountaintop removal coal mine? Explain.

Wind turns the blades.

The scalloped edge of this windmill blade is a new design inspired by the fin of a humpback whale—an example of biomimicry. Founders of the company WhalePower have demonstrated that this shape makes the blade more aerodynamic and less likely to stall out at low speeds, allowing it to produce more power than traditional designs.

4 SOLAR POWER

Key Concept 4: Solar energy can be harnessed in many ways. It is pollution free but is expensive and is less productive at higher latitudes, on cloudy days, and at night.

The most abundant sustainable energy source is the one that powers the planet—the Sun. Each day, Earth receives a staggering amount of energy from the Sun (more than 10,000 times the amount of energy we use.) The Sun meets many energy needs: It warms our homes and provides light—but new technologies allow more effective use, and even storage, of solar energy. In addition to wind power, the islanders on Samsø decided to tap directly into this boundless natural resource.

Solar power is energy harnessed from the Sun in the form of heat or light, and it can be used in two ways: through active or passive technologies. Around the countryside in Samsø, homes are dotted with **photovoltaic (PV) cells**, also called solar panels. PV cells are **active solar technologies** that convert solar energy directly into electricity. If just 4% of the world's deserts were covered in PV cells, it would supply all of the world's electricity needs.

The production of electricity can also be done on a large scale using *solar thermal systems*, another active technology that captures solar energy for the production of electricity. There are many types of solar "power plants." An early design consists of large arrays of huge parabolic (curved) mirrors that capture solar energy and focus it on pipes that contain a fluid that can be heated to high temperatures. These high temperatures are used to boil water to produce steam that turns a turbine attached to an electric generator, producing electricity in the same manner as a thermoelectric power plant that runs on fossil fuel or biomass. Newer technologies may vary the way the solar energy is captured and focused, but operate on the same general principle—heating a fluid to produce steam to generate electricity.

Samsø islanders also rely on solar thermal systems to heat their homes. In a field at the north of the island, rows and rows of solar collectors face the sky, absorbing the Sun's rays and using that energy to heat a massive tank of water to 160°F (71°C). The hot water is then piped into nearby homes for use in heating systems.

Less expensive alternatives to PV cells or solar thermal systems are **passive solar technologies**. A greenhouse is a simple example of such a system: It captures heat without any electronic or mechanical assistance. Many energy-conscious homes are designed with passive solar energy in mind, incorporating strategically oriented windows to maximize sunlight in a room and dark-colored walls or floors to absorb that light and heat the home.

Because solar energy is conceptually simple, safe, and clean, with no noise or moving parts, it is the most popular member of the renewable energy club. Today, hundreds of thousands of buildings around the world are powered by PV cells.

But like wind power, solar power is plagued by intermittency and start-up costs. PV cells are becoming cheaper every day, but it can still take a homeowner or business owner many years to save enough in reduced power bills to offset the cost of installation—a metric known as **payback time**. In the United States, on average the installation cost is recouped in 6 to 8 years; payback time is 3 to 5 years in sunny places like the desert Southwest or in places where electricity costs are higher, such as the Northeast. Intermittency is a bigger problem. Sunlight is available for only roughly half of each day—and even less at higher latitudes where areas only see a few hours a day of pale sunlight in the winter. (However, in those places, solar power is quite useful in the summertime when production can go on for 18 hours or more. This makes solar panels popular in places like Alaska where many homes are built "off the grid.")
INFOGRAPHIC 4

Diversification is a hallmark of a sustainable energy future: Because sustainable energy sources have strengths and weaknesses, no single source will likely meet the needs of any particular community, and certainly not those of the entire world. But together, each can provide a unique contribution. For example, solar is productive in daylight hours, whereas wind tends to blow harder at night; thus, these two renewable sources complement each other in many places. But in some regions, it makes more sense to tap other sources of renewable energy.

solar power Energy harnessed from the Sun in the form of heat or light.

photovoltaic (PV) cell A technology that converts solar energy directly into electricity.

active solar technology Mechanical equipment for capturing, converting, and sometimes concentrating solar energy into a more usable form.

passive solar technology A technology that captures solar energy (heat or light) without any electronic or mechanical assistance.

payback time The amount of time it takes to save enough money in operation costs to pay for equipment.

> Because sustainable energy sources have strengths and weaknesses, no single source will likely meet the needs of any particular community, and certainly not those of the entire world.

INFOGRAPHIC 4 **SOLAR ENERGY TECHNOLOGIES TAKE MANY FORMS**

There are many ways to capture and use solar energy. Passive solar homes are constructed in a way that maximizes solar heating potential.

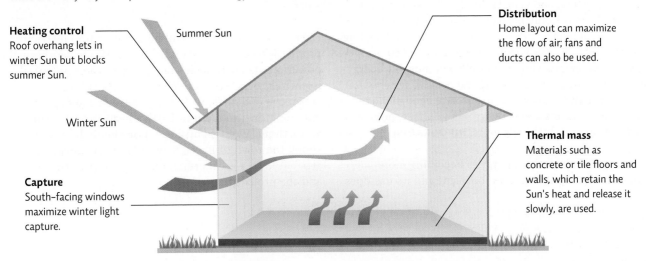

Heating control
Roof overhang lets in winter Sun but blocks summer Sun.

Summer Sun

Winter Sun

Capture
South-facing windows maximize winter light capture.

Distribution
Home layout can maximize the flow of air; fans and ducts can also be used.

Thermal mass
Materials such as concrete or tile floors and walls, which retain the Sun's heat and release it slowly, are used.

Active solar technologies use mechanical equipment to capture solar energy and covert it to a usable form. Photovoltaic (PV) cells convert sunlight to electricity. They are made of a semiconductor material, such as silicon, that will lose electrons when light strikes the surface, creating an electrical current.

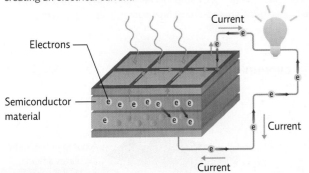

Electrons

Current

Current

Current

Semiconductor material

ADVANTAGES	DISADVANTAGES
• Abundant. • Pollution free. • Low environmental impact and carbon footprint. • Effective at large and small scales. • Portable. • Creates jobs. • Surplus energy can be stored in batteries.	• Location matters; shady and high-latitude areas get less winter sunlight. • Little or no generation on cloudy days or at night. • Expensive compared to other renewable energy sources. • Usually must be paired with other energy sources to meet needs.

 In terms of reducing fossil fuel use, why should you make your home as energy efficient as possible before installing active solar technologies?

5 GEOTHERMAL POWER

Key Concept 5: Geothermal energy can be captured on a small scale to lower home heating and cooling costs or on a large scale to produce electricity or heat.

Unlike Samsø, some communities are fortunate enough to be located near sources of **geothermal energy**. A tremendous amount of heat is produced deep in Earth from radioactive decay of isotopes. It is the same heat that bubbles hot springs and causes geysers to erupt. The heat in geothermal "hot spots" (those areas where molten rock, or magma, is located close to the surface) can be tapped by **geothermal power plants** to generate electricity. Steam from geothermal wells is used to spin turbines and produce electricity.

On the other end of the spectrum are **geothermal heat pumps** (also called ground-source heat pumps). These

systems are used in more than half a million homes around the world—not to generate electricity but to reduce our use of it. Ground temperature is amazingly constant—around 55°F (12.5°C) year round. To tap into this 55°F temperature, fluid-filled pipes are buried; the fluid, which takes on that ground temperature, is then pumped into the home, where its temperature is transferred to a heat pump

geothermal energy Heat stored underground, contained in either rocks or fluids.

geothermal power plant A large-scale facility that captures steam produced from Earth's internal heat to turn turbines which generate electricity.

geothermal heat pump A system that transfers the steady 55°F (12.5°C) underground temperature to a building to help heat or cool it.

system that can maintain the desired temperature in the home much more efficiently than a traditional heating, air conditioning, and ventilation (HVAC) system. In the winter, the home only needs to be warmed up from 55°F, and in the summer, this system provides natural cooling. Such systems are fairly expensive to install ($20,000 to $25,000 on average) but result in lower monthly energy bills than conventional heating and cooling systems. Areas with more extreme climates save enough money in monthly heating and cooling bills to offset the cost of installation in as little as 5 years. **INFOGRAPHIC 5**

This type of renewable energy is becoming more popular: It has a low carbon footprint and its use does not generate toxic air pollution. It also is dependable—it has none of the intermittency issues of wind or solar.

Harnessing such a powerful heat source for electricity production can be difficult, however. While geothermal power plants are reliable and efficient, their potential is entirely dependent on location. But drilling is expensive and difficult to do because of the depths one must drill to tap into Earth's heat. In early 2007, an earthquake of magnitude 3.4 on the Richter scale shook the town of Basel, Switzerland. The quake was attributed to a geothermal mining project in the area, and the project was halted.

INFOGRAPHIC 5 **GEOTHERMAL ENERGY CAN BE HARNESSED IN A VARIETY OF WAYS**

The high temperatures found underground in some regions can be tapped to generate electricity. Geothermal heat can also be piped directly to communities to provide heat or hot water. Geothermal energy can also be tapped using ground-source heat pumps, reducing the cost to heat and cool individual buildings.

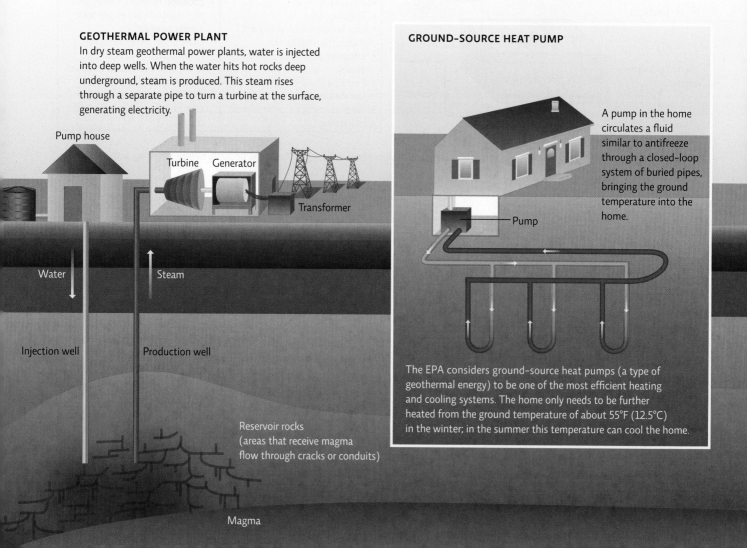

GEOTHERMAL POWER PLANT
In dry steam geothermal power plants, water is injected into deep wells. When the water hits hot rocks deep underground, steam is produced. This steam rises through a separate pipe to turn a turbine at the surface, generating electricity.

Pump house

Turbine   Generator

Transformer

Water

Steam

Injection well

Production well

Reservoir rocks
(areas that receive magma flow through cracks or conduits)

Magma

GROUND-SOURCE HEAT PUMP

A pump in the home circulates a fluid similar to antifreeze through a closed-loop system of buried pipes, bringing the ground temperature into the home.

Pump

The EPA considers ground-source heat pumps (a type of geothermal energy) to be one of the most efficient heating and cooling systems. The home only needs to be further heated from the ground temperature of about 55°F (12.5°C) in the winter; in the summer this temperature can cool the home.

6 HYDROPOWER

Key Concept 6: Hydroelectric dams are the most common way to harness the power of water, but they permanently alter the environment and have other drawbacks.

Humans have harnessed the power of falling water for thousands of years, ever since early civilizations used watermills to grind grain into flour. Today, energy produced from moving water—known as **hydropower**—supplies more electricity than any other single renewable resource. Approximately 7% of electricity used in the United States and 17% around the world is generated from hydropower, though in some countries, it makes up more than half. Large-scale hydroelectric power plants at giant dams are the source of most of that power.

Hydropower is abundant and clean, and no greenhouse gases are produced to access the energy (though large emissions result from the construction of dams). There are a variety of ways to harness the energy of moving water—from ocean waves and the tides to capturing energy released from variations in ocean temperatures (ocean thermal energy conversion)—but the most common way to generate hydropower is with dams. The Grand Coulee Dam on the Columbia River in Washington State is the largest electrical power producer in the United States, with a total

hydropower Energy produced from moving water.

ADVANTAGES

- Pollution free.
- Low environmental impact and carbon footprint.
- Can be used at large or small scales to produce electricity or to capture heat.
- Geothermal heat pumps can be installed anywhere and will reduce monthly heating and air conditioning costs.

DISADVANTAGES

- Geothermal power plants are expensive to build and are suitable only in some areas.
- Geothermal heat pumps are more expensive to install than traditional HVAC systems.
- Usually must be paired with other energy sources to meet needs.

? The two geothermal technologies are quite different. Which one actually produces electricity, and which one helps you conserve energy?

Installation of a geothermal heating system in a residence. Closed-loop tubing can be installed horizontally in shallow trenches or vertically in much deeper wells. Here, a horizontal loop is being installed.

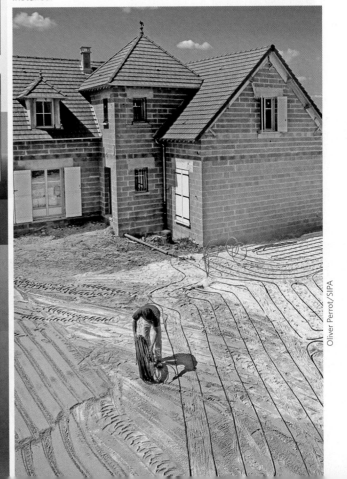

Oliver Perrot/SIPA

generating capacity of 6,809 MW. The reservoir created by the dam, an artificial lake that pools behind the structure, stretches some 240 kilometers (150 miles) to the Canadian border.

But hydropower from large dams is far from an ideal resource. It wasn't a real option on Samsø, for instance, because the island lacks high mountain peaks whose runoff would feed large rivers. Importantly, hydroelectric systems also come with some significant drawbacks. Electricity generation capacity can vary from year to year based on rainfall amounts, and dams that restrict water flow can curtail water supplies to downstream areas. Further, in hot climates, huge amounts of water evaporate from reservoirs. In addition, dams are responsible for the loss of major habitats and the displacement of tens of millions of people around the globe. When the Grand Coulee was switched on in 1941, crowds gathered on the hill above the dam to marvel at the "birth of one of the world's greatest waterfalls," a local paper proclaimed. But nearby, thousands of Native Americans watched in horror as habitats, homes, and livelihoods were irreversibly lost under the pool of water engulfing the land behind the dam. The dam also threatened a staple of their economy and culture by blocking the migratory path of the area's iconic salmon populations; salmon runs upstream of the dam were completely wiped out.

Not all hydropower systems have the same impact as a giant dam. A *run-of-the-river hydroelectric* system, often a low stone or concrete wall, doesn't block the water; it merely directs some of the flowing water past a turbine and thus generates electricity from the natural flow of a river, which is less disruptive to the river ecosystem since the area behind the dam is not flooded. But energy production is dependent on the flow of a river at any particular time, so such a system is suitable only for rivers with dependable flow rates year round. **INFOGRAPHIC 6**

INFOGRAPHIC 6　**HARNESSING THE POWER OF WATER**

The energy of moving water can be captured in a variety of ways to produce electricity.

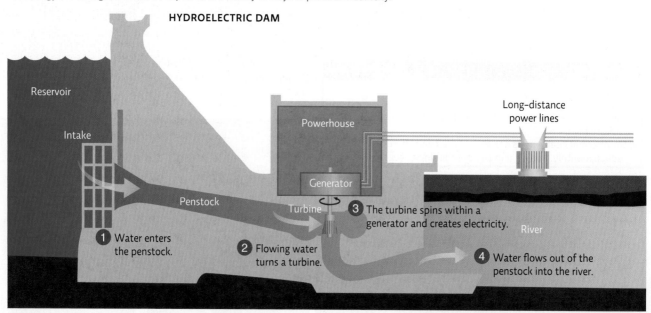

HYDROELECTRIC DAM

Reservoir

Intake

Powerhouse

Generator

Penstock

Turbine

Long-distance power lines

River

1　Water enters the penstock.

2　Flowing water turns a turbine.

3　The turbine spins within a generator and creates electricity.

4　Water flows out of the penstock into the river.

The most common way to harness the power of water is with a hydroelectric dam. When a large river is dammed, a reservoir is formed behind the dam. The buildup of water in the reservoir creates an enormous amount of pressure; water diverted from the top of the dam flows through long pipes, called penstocks, to turbines below. As the water rushes past the curved blades of the turbines, they spin and generate electricity.

ADVANTAGES	DISADVANTAGES
• Pollution free. • Low environmental impact and carbon footprint. • Large dams offer recreation and flood control advantages. • Small run-of-the-river systems can generate electricity without damming a river. • Tremendous power potential remains to be harnessed from the oceans.	• Large dams are expensive to build and permanently damage habitats and displace communities. • Production capacity varies from year to year based on rainfall. • Usually must be paired with other energy sources to meet needs. • High evaporative water loss from reservoirs in hot, arid climates.

 Do you feel the energy production and recreational/flood control benefits of large hydroelectric dams outweigh the permanent ecological and community destruction they cause? Explain your position.

Fish ladders and other technologies are used to try and help migratory fish such as salmon get past a dam that has blocked a river.

7 THE ROLE OF CONSERVATION AND ENERGY EFFICIENCY

Key Concept 7: Meeting our energy needs with renewable sources becomes more likely when we pair renewables with energy conservation measures.

Even though the people of Samsø saw the importance of investing in sustainable energy, they realized that one of the best ways to help achieve energy independence would be simply to use less of it. As energy advisers like to say, the *greenest kilowatt is the one you never use.* Luckily, there are lots of ways to reduce electricity use right now. **Conservation**—making choices that result in less energy consumption—is a vital part of our quest to become sustainable users of energy. Conservation means simply not using energy if we don't need it, as well as using energy more efficiently when we do use it.

Much of conservation involves changing behavior. For example, lighting typically accounts for about 25% of the average home's electric bill, and simply turning off the lights when you leave a room can make a big difference. Other options include studying in a room with ample natural light instead of a dark corner that requires artificial light.

Embracing **energy efficiency** is also a way to reduce energy use. All kinds of electrical devices are being made to perform the same tasks with less energy—more efficient lightbulbs and home appliances can reduce one's electricity bill enough to pay for themselves in a short period of time. Energy-efficiency advances are rapidly being made. For example, today LED lightbulbs represent the most efficient sources of light for your home, using one-third of the energy that a comparable compact fluorescent bulb would use or one-thirtieth the amount of energy needed to power an incandescent bulb. They also last

conservation Efforts that reduce waste and increase efficient use of resources.

energy efficiency A measure of the amount of energy needed to perform a task; higher efficiency means less energy is wasted.

longer, reducing the embedded cost of producing and disposing of more bulbs.

Changes that allow you to save water also save energy. Using less hot water, of course, will save energy needed to heat that water, but even water from your cold tap has an energy input—either from the pump attached to your well that brings water to the surface or the industrial pump that pumps it into a water tower that then flows downhill into your home.

Before embarking on any energy-efficiency home improvements, a homeowner should conduct or commission an energy audit to see what actions would provide the biggest energy return on investment. For example, it makes no sense to invest in solar panels if

your home is poorly insulated and losing more energy than the panels produce. **INFOGRAPHIC 7**

For the residents of Samsø, because the island has such a cold climate, one of the first conservation efforts revolved around heating homes, mainly by transitioning away from electric heaters. Home energy audits pinpointed other steps individuals could take, such as installing more home insulation or using straw, wood chips, and other sources of biomass for heat. Older gas stoves were replaced with more efficient electric stoves; people began heating water using electricity generated by the wind turbines or solar energy. Residents were encouraged to apply for grants toward upgrades that made their homes more energy efficient. Similar programs exist in the United States. (See Module 4.2 for more ways consumers can cut energy use.)

INFOGRAPHIC 7 SAVING ENERGY

Energy conservation is about wise use: using less and using energy more efficiently. Some steps can be taken immediately, with no investment; others require money and time to implement. Whatever you can do will not only reduce your energy use but also reduce your energy bill and the pollution generated from producing that energy.

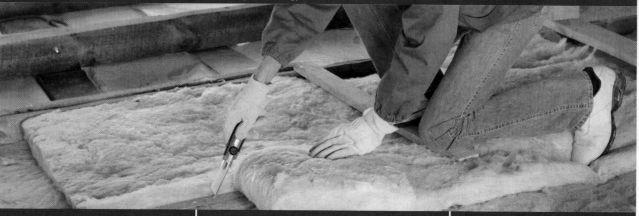

Ale-ks/Getty Images/iStockphoto

NO-COST WAYS TO SAVE ENERGY	STEPS THAT COST MONEY BUT HAVE QUICK PAYBACK TIMES	MORE EXPENSIVE OPTIONS WITH LONGER PAYBACK TIMES
• Turn off computers and monitors and unplug phone chargers when not in use. • Lower the thermostat on water heater to 120°F; turn it off if you will be away for several days. • In the winter, open curtains on south-facing windows to let in heat and light during the day; close them at night to reduce heat loss. • Close the fireplace damper when not in use to prevent loss of heat. • Take short showers instead of baths to reduce hot water use. • Wash only full loads of dishes and clothes; wash clothes with cold water.	• The first step is to have an energy audit performed to identify where your home is energy inefficient. (Go to www.energysavers .gov for more information.) • Install attic and wall insulation to meet recommendations for your area. • Weatherstrip and caulk door and window frames; check heating ducts for leaks and seal if needed. • Insulate hot-water pipes and water heater. • Install a programmable thermostat to automatically adjust the temperature when you are not at home or are sleeping. • Replace old lightbulbs with the more energy-efficient LED bulbs. • Replace home appliances with Energy Star-rated appliances.	• Replace older windows with high-efficiency windows that restrict loss of heating or cooling while allowing plenty of light. • Replace exterior doors with energy-efficient varieties. • Replace heating, ventilation, and air conditioning (HVAC) system with an energy-efficient model; consider a ground-source heat pump. • Install a high-efficiency on-demand (tankless) or geothermal water heater. • If you live in a hot climate, install a light-colored roof to reflect solar radiation.

 Which of the suggestions for saving energy could be done by a student living in a dormitory or rental apartment?

8 THE WAY FORWARD: MEETING ENERGY NEEDS SUSTAINABLY

Key Concept 8a: To make renewables a more viable option than fossil fuels, we need to price energy sources according to their true costs and find ways to store and more efficiently deliver electricity.

Key Concept 8b: Like Samsø, we will be more likely to meet energy needs if we use a variety of renewable technologies that fit each locale and if we embrace conservation to reduce overall energy needs.

One recurring theme with renewable energy technologies is that they are more expensive than fossil fuel methods. Even though sunlight, wind, geothermal heat, and water are free, constructing and installing the solar cells, wind turbines, underground pipes, and hydroelectric systems needed to harness their energy is not. But perhaps we should not be asking why renewables are so expensive but instead asking why fossil fuels are so cheap. We need to consider *all* the costs—environmental, social, and economic—of these technologies to fairly compare them. Because fossil fuels, led by coal, have the highest external (environmental and health) costs, when the *true cost* per kilowatt hour is estimated, renewables come out far ahead of fossil fuels. (See Module 5.1 for more true cost accounting.) Wind currently leads the way as the least expensive method.

> Perhaps we should not be asking why renewables are so expensive but instead asking why fossil fuels are so cheap.

While renewable technology is improving and becoming more diversified every year, there is a key hurdle that must be surmounted if renewables are to replace fossil fuels—the ability to store excess energy for later use. This is especially important for the two leading renewables, solar and wind, because their production of electricity is intermittent—sometimes you would have more than you needed and other times there would not be enough.

Unfortunately, electricity itself can't be stored—it is sent across wires when generated and immediately used—a balancing act that power plant operators must deal with every day: matching electricity production with consumption. The output of fossil fuel plants, and to a lesser extent nuclear power plants, can be scaled up or down as needed, but for renewables, what is needed is a cost-effective way to convert excess electrical production into a storage form of energy. One novel solution being used in some areas is to use extra energy—say electricity produced by solar power during the day—to pump water uphill to a storage reservoir. At night, the water can flow downhill past turbines that generate electricity. The next day, the water is pumped uphill again—a simple solution, but one that only works in areas with a suitable water supply and terrain conducive to such a system. However, another, more generally applicable method is needed.

The most appealing option may be to convert electricity to chemical energy and store it in a battery. Indeed, even a collection of car batteries at your home can store the energy created by excess solar- or wind-generated electricity. But to work on a large scale—one that works for a city or nation—we need battery arrays that efficiently store excess energy and readily give it up on demand. Batteries would offer a faster and more seamless way to juggle fluctuating energy flows than altering electricity production in a power plant could ever do. Elon Musk, founder of Tesla, and others are putting major resources into battery technologies to solve the problem of storing electricity.

The development of a **smart grid** is also part of the renewable future wish list. The electricity grid consists of the power plants and power lines that crisscross our communities and nations, delivering electricity at the flip of a switch. This is described as a one-way communication—you "ask" for power and it is delivered. But that grid is aging and frequently unable to cope with the electricity demands placed on it, leading to brownouts, blackouts, and power surges.

smart grid A modernized network that provides electricity to users in a way that automatically optimizes the delivery of electricity.

Schmidt-z/E+/Getty Images

Many communities are pursing sustainable energy to meet part or all of their energy needs, showing that a renewable energy future is within our grasp.

INFOGRAPHIC 8 **A RENEWABLE ENERGY FUTURE**

The transition away from fossil fuels to renewables for electricity production will require some changes in the way we choose energy sources (true cost accounting and diversification) as well as developing technological solutions to logistical problems.

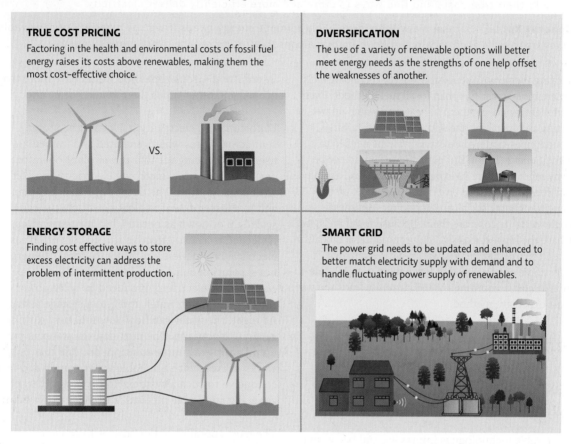

TRUE COST PRICING

Factoring in the health and environmental costs of fossil fuel energy raises its costs above renewables, making them the most cost-effective choice.

VS.

DIVERSIFICATION

The use of a variety of renewable options will better meet energy needs as the strengths of one help offset the weaknesses of another.

ENERGY STORAGE

Finding cost effective ways to store excess electricity can address the problem of intermittent production.

SMART GRID

The power grid needs to be updated and enhanced to better match electricity supply with demand and to handle fluctuating power supply of renewables.

 How would the development of dependable storage of electricity complement the operation of a smart grid?

Turning the existing infrastructure into a smart grid means improving efficiency and smoothing out the flow of electricity—avoiding power surges or brownouts that occur when demand doesn't match production. This is accomplished by a better sharing of information—the grid becomes a "two-way conversation" with buildings like your home sending information back to the power company, allowing it to better anticipate electricity needs. "Smart meters" allow you (or your home) to schedule some energy use at low demand times and would automatically detect problems and quickly restore delivery of power. But this won't come cheaply. The Electric Power Research Institute estimates the price tag for a fully implemented U.S. smart grid at close to half a trillion dollars. The benefits, however, should be four times that. **INFOGRAPHIC 8**

In their quest to become an energy-neutral or even an energy-positive island, the people of Samsø focused their efforts on a variety of renewable technologies and conservation measures. The next phase of Samsø's

transition will address replacing gasoline and diesel vehicles with electric cars.

Much of Samsø's success stems from community members working together to solve their own local problems, says Hermansen. As he told *Time* magazine in 2008, "People say: 'Think globally and act locally.' But I say you have to think locally and act locally, and the rest will take care of itself."

Select References:

Heaslip, E., et al. (2016). Assessing good-practice frameworks for the development of sustainable energy communities in Europe: lessons from Denmark and Ireland. *Journal of Sustainable Development of Energy, Water and Environment Systems*, 4(3), 307–319.

International Energy Agency. (2016). *Key World Energy Statistics.* www.iea.org/publications/freepublications/publication/key-world-energy-statistics.html.

Service, R. F. (2015). Clean revolution. Demark is striving to become the world's first carbon-neutral nation. *Science*, 350(6264), 1020–1023.

INTERACTIVE MAP **GREEN ENERGY**

ANIMATED INFOGRAPHIC

Samsø is not the only community pursing energy neutrality. Visit this interactive map and read about how Greensburg, Kansas, is rebuilding, sustainably, after a tornado destroyed the town, and about Argentina's pursuit of sustainable energy. Case studies are also presented on two new technologies that may help revolutionize our quest for sustainable energy.

SOLAR ROOFS

REBUILDING GREENSBURG

ENERGY AUCTIONS

WIRELESSLY CHARGING ELECTRIC VEHICLES

 BRING IT HOME

PERSONAL CHOICES THAT HELP

A key component to developing a sustainable society is using renewable energy. Use of renewable energy decreases harmful impacts of mining associated with non-renewable energy, and it also reduces the amount of air and water pollution produced when the fuel is processed, transported, and burned. The efficiency and availability of renewable energy are rapidly increasing as new technologies emerge.

Individual Steps
• Contact your energy provider to see if you can purchase a percentage of your energy from a renewable source.
• Regardless of your home's energy source, make sure you are using energy efficiently. Review Infographic 7 and visit www.energy .gov/energysaver/energy-saver for more ideas.

• If you have a smartphone, download the PVme app to see how many solar panels you would need to meet your household energy needs.
• A major barrier to using solar energy for many people is the cost. For information on tax incentives, rebates, and other programs that make using renewable energy easier, check out www.dsireusa.org or look into the growing trend of leasing solar panels.

Group Action
• Learn about opportunities to invest in solar energy projects for lower-income communities or start one in your own community from the nonprofit Community Power Network, at www .communitypowernetwork.com.

Policy Change
• Contact your state and federal legislators and ask them to support or sponsor legislation that provides financial incentives for the purchase of renewable technologies.

martin-dm/riStock/Getty Images

1 What are the characteristics of a sustainable energy source, and what types are commonly used?

1. Which of the following is *not* a characteristic of sustainable energy sources?
   a. They must be renewable.
   b. We must not use them faster than they are replenished.
   c. They must have no environmental impact.
   d. All of these are characteristics of sustainable energy.

2. Distinguish between renewable energy and sustainable energy.

2 What are the current and projected roles of renewables in global energy production?

3. True or False: The use of fossil fuels is projected to decline in the next 50 years as the use of renewable energy sources increases.

4. Which of the following renewable energy sources contributed the largest proportion of U.S. electricity production in 2016?
   a. Biomass
   b. Wind
   c. Hydro
   d. Solar

5. How is the generation of electricity expected to change in the upcoming decades for each of the following renewable energy sources: wind, solar, hydroelectric, biomass, and geothermal power?

3 How can wind energy be captured to generate electricity, and what are the pros and cons of these methods?

6. Which of the following is a disadvantage of wind power?
   a. It can only generate electricity during the daytime.
   b. It kills more birds and bats that any other human-related source.
   c. It cannot be effectively used on a small scale.
   d. Windmills are considered eyesores and can be noisy.

7. Compare the production of electricity by a wind turbine and the production of electricity by heating water to produce steam (thermoelectric production).

4 How can solar energy be captured and used, and what are the pros and cons of these methods?

8. True or False: Solar energy can be captured to generate electricity on a small scale (on a home) or on a large scale (at a solar power plant).

9. Solar technologies that capture heat without any electronic or mechanical assistance are called:
   a. passive solar technologies.
   b. active solar technologies.
   c. solar thermal systems.
   d. photovoltaic solar cells.

10. A disadvantage of solar power is that:
    a. start-up costs are high.
    b. electricity can only be produced during the day.
    c. it is not a usable technology in places like Alaska.
    d. A and B
    e. A, B, and C

11. What is the concept of "payback time," and how can it be useful in deciding what types of conservation measures to pursue?

5 How can geothermal energy be captured and used, and what are the pros and cons of these methods?

12. True or False: While geothermal power plants need to be built near sources of underground heat such as hot springs, geothermal heat pumps can be installed in any home.

13. Which of the following is an advantage of generating electricity with a geothermal system?
    a. It has a low carbon footprint.
    b. It can be done anywhere.
    c. Drilling is easy to do and the systems easy to install.
    d. A and B
    e. A, B, and C

14. Explain how a geothermal heat pump works and how it can lower both heating and cooling costs for homeowners.

6 How can the power of water be captured, and what are the pros and cons of these methods?

15. Which disadvantage of large dams can the smaller run-of-the-river dams avoid?
    a. The smaller dams do a better job at flood control.
    b. The environmental damage of creating a reservoir is avoided.
    c. The smaller dams don't generate air pollution like the large dams do.
    d. Production capacity will be steady, not variable as with large dams.

16. How were Native Americans and wild salmon populations adversely affected by the construction of the Grand Coulee Dam?

7 What roles do conservation and energy efficiency play in helping us meet our energy needs sustainably?

17. Energy advisers say "the greenest kilowatt is the one:
    a. produced by solar power."
    b. that doesn't harm Earth."
    c. that's cheapest."
    d. you never use."

18. Which of the following is the first thing you should do if you want to lower your electric bill?
    a. Install PV-panels on your rooftop.
    b. Have an energy assessment done to see where your home is most inefficient.
    c. Replace your appliances with energy-efficient models.
    d. Install more insulation in your attic and exterior walls.

19. Explain the value of an energy audit for homeowners who want to lower energy use.

8 What economic adjustment and what technological advances are needed for renewables to become a viable replacement for fossil fuels?

20. True or False: When the true costs of electricity generation are considered, fossil fuels come out slightly better than renewables.

21. Explain the importance of the ability to store electricity for renewable energy to be a reliable source of electricity.

22. Why is diversification such an important part of a renewable energy future?

SCIENCE LITERACY WORKING WITH DATA

One criticism of wind turbines is the toll they take on wildlife, especially birds and bats. Look at the data provided below to examine the impact of wind turbines and potential actions that could reduce impact.

A: NIGHTLY ACTIVITY PATTERNS OF BATS

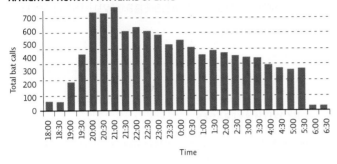

B: HUMAN RELATED CAUSES OF BIRD MORTALITY

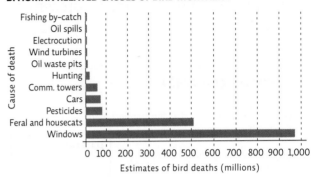

Interpretation

1. What question were the researchers asking for each of the two studies represented here (Graph A and Graph B)?

2. Based on the number of bat calls recorded (Graph A), when are bats most active? (Hint: 18:00 is 6 PM)

3. Based on Graph B, what are likely to be the biggest threats to birds who live in large cities? To those who live in rural areas?

Advance Your Thinking

4. Based on Graph A, during what hours of the day should wind turbines be turned off to protect bats? How might this impact wind energy production?

5. Identify three actions that would do the most to decrease bird mortality. Which do you think would be the easiest to implement? Explain.

6. Consider where you live. What do you think the impact of a wind farm in your area would be on bird mortality? Explain.

INFORMATION LITERACY EVALUATING INFORMATION

Renewable energy resources have become a priority around the world as we grapple with the cost and limited supply of fossil fuels, not to mention the effect on the environment resulting from their use. How do we know which energy source is appropriate to invest in, with both our time and money?

To learn more about renewable energy options, use the search engine of your choice and enter the key words "Renewable Energy".

Select a website that you think probably presents reliable, unbiased information, such as a government website. Explore the website and evaluate it by answering the questions a and b in Section 1. After answering these questions, go back to your original search page and choose a website from a potentially biased source, such as an energy company or a political group. Explore and evaluate this second website and work with the information presented to answer the same questions in Section 1. Continue on to answer questions in Sections 2 and 3.

1. Evaluate each website:
   a. Who runs this website? Do this organization's credentials make it reliable or unreliable? Explain.
   b. Does the website offer evidence for claims, cite information sources, or offer links to other information sources?

2. Compare these two information sources:
   a. Is the information from the two websites consistent or are there some contradictions? Explain.
   b. Does one site have more detail, evidence, or links to scientific sources than the other? Is one easier to understand than the other or more useful to the average person? Explain.

3. After completing your evaluation and comparison of the websites, write a paragraph that addresses these questions:
   a. Can an industry or political group (or any group with a vested interest in an issue) be a trusted information source, or is the information it presents always suspect?
   b. Conversely, are scientific or otherwise unbiased information sources always trusted information sources or might the information they present be suspect?

 Additional study questions are available at SaplingLearning.com.

APPENDIX 1
BASIC MATH SKILLS

Math skills are needed to evaluate data and even to understand much of the information in news reports. Here, we present a review of some basic skills that will be useful in this class and in other science classes.

AVERAGES (MEANS)

To calculate an average, add all the numbers in the data set and divide by the number of numbers.

Example: The sum of these numbers is 10,547; 10,547 / 8 = 1318.375. You could round this off for an average or mean of 1,318.

Data set	
1,004	766
2,349	988
456	1,203
1,882	1,899

WORK WITH AVERAGES

Problem 1: If these were your grades on exams, what would be your exam average?
84, 73, 93, 95, 79, 86

PERCENTAGES/FREQUENCIES

To **convert a fraction to a percentage**, divide the numerator (top number) by the denominator (bottom number) and multiply by 100.
Example: To express the fraction $2/5$ as a decimal, divide 2 by 5, which equals 0.4. Multiply this by 100 for your answer: 40%.
To **convert a decimal to a percentage**, multiply by 100; a shortcut for this is simply to move the decimal over two places to the right.
Example: $0.08 \times 100 = 8\%$

WORK WITH PERCENTAGES

Problem 2: If 8 out of 32 frogs in a pond have deformities, what percentage of frogs have deformities?

Problem 3: In a pond, 25% of the frogs have leg deformities. If there are 100 frogs in the pond, how many have deformities?

Problem 4: If there are 68 frogs in the pond and 25% have deformities, how many have deformities? (First, make an estimate based on your answer to Problem 3—will it be a higher or lower number? This will help you decide if the answer you calculate is reasonable or whether you might need to recalculate.)

SCIENTIFIC NOTATION

In science, we often use very large or very small numbers. To make these easier to present, scientists use scientific notation, which multiplies a number (called the coefficient) by 10 (the base) raised to a given power (the exponent). If the coefficient is 1, we can leave it off and simply show the base and exponent (e.g., $1 \times 10^2 = 10^2$). The exponent tells us how many orders of magnitude larger or smaller to make the number. In other words, the exponent is telling us how many zeros the number will have: $10^2 = 100$; $10^3 = 1,000$, and so on. Negative exponents represent decimals; for example: $10^{-2} = 0.01$, $10^{-3} = 0.001$, and so on.

Here is a simple shorthand way to evaluate numbers given in scientific notation. Move the decimal place to the right if 10 has a positive exponent and to the left if the exponent is negative. The number of spaces the decimal place is moved is equal to the exponent. For example, 10^2 tells us to move the decimal place 2 spaces to the right; 10^{-2} means we move it 2 spaces to the left.

By convention, we always designate the coefficient as a whole number (2) or a decimal, with the decimal point at the "10" position (2.3). In other words, we would write 2.3×10^5, not 23×10^4. Both are technically correct but the first is the preferred format.

Examples:
$2 \times 10^6 = 2,000,000$
$2.36 \times 10^5 = 236,000$
$4.99 \times 10^{-4} = 0.000499$

Some typical values you might run across include:
$10^6 = 1$ million
$10^9 = 1$ billion
$10^{12} = 1$ trillion

MEASUREMENTS AND UNITS OF MEASURE

There are many handy conversion calculators on the Internet, but it is still useful to have a general idea of how large various units of measure are and how metric and English systems of measurement compare.

LENGTH

Metric
1 kilometer (km) = 1,000 meters (10^3)
1 meter (m) = 100 centimeters
1 centimeter (cm) = 10 millimeters
1 millimeter (mm) = 0.000001 meter (10^{-6})
1 micrometer (μm) = 0.000000001 meter (10^{-9})
1 nanometer (nm) = 0.000000000001 meter (10^{-12})

English
1 mile (mi) = 5,280 feet
1 yard (yd) = 36 inches (in) or 3 feet
1 foot (ft) = 12 inches

Conversions
1 km = 0.621 mi
1 m = 39.4 in
1 cm = 0.394 in

1 mi = 1.609 km
1 yd = 0.914 m
1 in = 2.54 cm

MASS

Metric
1 metric ton (mt) = 1,000 kilograms
1 kilogram (kg) = 1,000 grams
1 gram (g) = 1,000 milligrams
1 milligram (mg) = 0.001 gram (10^{-3})
1 microgram (μg) = 0.000001 gram (10^{-6})
1 nanogram (ng) = 0.000000001 gram (10^{-9})

English
1 U.S. ton (t) = 2,000 pounds
1 pound (lb) = 16 ounces (oz)

Conversions
1 mt = 2,200 lb
1 kg = 2.2 lb
1 g = 0.035 oz

1 t = 0.907 mt
1 lb = 4.54 g or 0.454 kg
1 oz = 28.35 g

VOLUME

Metric
1 liter (L) = 1,000 milliliters
1 milliliter (ml) = 0.001 liters

English
1 gallon (gal) = 4 quarts
1 quart (qt) = 2 pints or 4 cups
1 pint (pt) = 16 fluid oz

Conversions
1 L = 0.265 gal or 1.06 qt
1 gal = 3.79 L

AREA

Metric
1 hectare (ha) = 10,000 square km

English
1 acre (ac) = 4,840 square yards (yd)

Conversions
1 ha = 2.47 ac
1 ac = 0.405 ha

CONCENTRATIONS

Metric
1 part per million (ppm) = 1 mg/L
1 part per billion (ppb) = 1 μg/L
1 part per trillion (ppt) = 1 ng/L

TEMPERATURE CONVERSIONS
Fahrenheit (°F) to Celsius (°C): °C = (°F − 32) × $^5/_9$
Celsius (°C) to Fahrenheit (°F): °F = (°C − $^5/_9$) + 32

In general:
1°C = 1.8°F
1°F = 0.56°C

Answers to problems:
1. 85   **2.** $^8/_{32}$ = 0.25 = 25%   **3.** 100 × 0.25 = 25   **4.** 68 × 0.25 = 17

APPENDIX 2
DATA HANDLING AND GRAPHING SKILLS

This tutorial offers a quick look at the basics of working with data and graphing.

Scientists gather data to learn about the natural world. Data can be organized into graphs, which are "pictures" or visual representations of the data. Because they can condense and organize large amounts of information, graphs are often easier to interpret than a simple list of numbers. They show relationships between two or more variables that help us determine whether the variables are correlated in any way and allow us to look for trends or patterns that might emerge. To be effective, graphs should be constructed according to conventions and must be accurately plotted and properly labeled. Certain types of graphs are more suitable than others to show particular types of data, so it is important to choose the correct graph for your data.

The following sections describe variables found in graphs, data tables, and the types of graphs commonly used in environmental science.

VARIABLES

The **independent variable** is the parameter the experimenter manipulates—it could be whether or not a group is exposed to a treatment (given a medicine, exposed to a particular wavelength of light), is part of a distinct group (trees at specified distances from a stream), or is a group followed over a period of time (monitored daily, yearly, etc.). If you were setting up a data table in which to record the data your experiment would produce, you would be able to fill in the values for the independent variable *before beginning* the actual experiment.

The **dependent variable is** the response being measured in the experiment—the responding variable. The experiment is being conducted to see if this variable is "dependent on" the independent variable. In other words, when you change the independent variable, does the dependent variable change as a result? If you were setting up a data table in which to record data, you would be able to include a column heading for the dependent variable, but you would not be able to enter the values until the experiment was complete. There may be more than one dependent variable being tested.

DATA TABLES

Data tables have a conventional format. The independent variable is shown in the left-hand column and the data for the dependent variable or variables are shown in columns to the right of that. To be useful, the data table needs to have a descriptive title (what data are we looking at?) and the units of measure must be included.

ANNUAL ATLANTIC HERRING CATCH

Independent variable

Year	Herring catch (1,000 tons)
1965	731
1970	580
1975	382
1980	270
1985	180
1990	150
1995	45
2000	20
2005	26

Dependent variable

Units of measure are given—here, we multiply each value by 1,000 to find the number of tons of catch that year. Using the 1,000-fold conversion in the units allows us to use numbers that are easier to interpret.

TYPES OF GRAPHS

A. LINE GRAPHS

In science, researchers often test the effect of one variable on another. Line graphs are used when the independent variable is represented by a numerical sequence (1, 2, 3 . . .) rather than discrete categories (red, yellow, blue . . .). The dependent variable is always a numerical sequence.

Steps to producing a line graph

1. **Determine the *x* axis and the *y* axis.** The independent variable is usually shown on the *x* axis (the horizontal axis). In a line graph, this variable is one that changes in a predictable, numerical sequence, such as the passing of time, increasing concentrations of a solution being tested, habitat distance from the seashore, etc. The data being collected (the response being observed) represent the dependent variable, which is shown on the *y* axis (the vertical axis). The axes require a descriptive title that indicates exactly what each axis represents, along with units of measure, if needed.

2. **Set up the axes.** Set up each axis so that the largest data value for that variable is close to the end of the axis, leaving as much of the available space for your graph as possible. Aim for 5 to 15 "ticks" (the small dividing marks) on any given axis—don't overload it with 50 tiny ticks or have so few that it is hard to place data points. It is essential to evenly space the ticks on a given axis, keeping the increments the same numerical size. On the sample graph, all the *x* axis ticks are 5 years apart; all the *y* axis ticks are 100 units apart. The increments will depend on your particular data; however, be sure they are the same size for a given axis. Give the graph a descriptive title.

3. **Plot the points.** Plot each point by finding the *x* value on the *x* axis and moving up until you reach the *y* value position across from the *y* axis. If you are graphing more than one set of data, draw the data points as different shapes or colors and provide a legend to identify each data set.

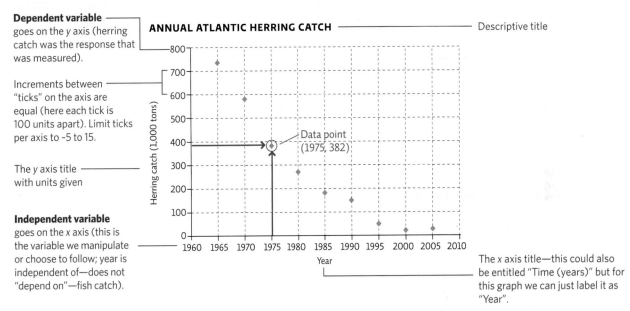

Dependent variable goes on the *y* axis (herring catch was the response that was measured).

Increments between "ticks" on the axis are equal (here each tick is 100 units apart). Limit ticks per axis to ~5 to 15.

The *y* axis title with units given

Independent variable goes on the *x* axis (this is the variable we manipulate or choose to follow; year is independent of—does not "depend on"—fish catch).

Descriptive title

Data point (1975, 382)

The *x* axis title—this could also be entitled "Time (years)" but for this graph we can just label it as "Year".

4. **Draw the line.** Once data points are in place, you can draw your line, but don't simply connect the dots unless they all line up exactly. Step back and visualize what kind of trend the data are showing and draw a line that approximates that trend. These trend lines can be mathematically determined but can also be fairly accurately estimated by simply drawing in a line that goes through the center of the data—about as many points will be above as below the line. You can draw a straight line or you may elect to draw a curve to accommodate shifts in the trend.

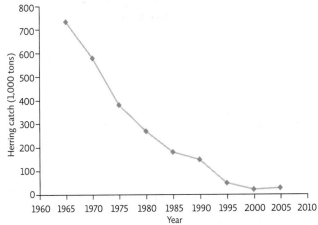

"Connecting the dots" like this implies that each data point is perfectly accurate and that this exactly represents the relationship between the two variables.

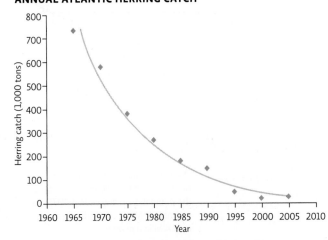

Drawing in a "trend line" that floats through the cloud of points is a more accurate estimation of the actual relationship seen between the two variables. We could draw a straight trend line for this data, but since it seems that the rate of decline lessens as times goes by, the curve seen here may better represent the relationship.

Interpreting the data

Once the graph is made, we can evaluate the data and draw conclusions. The first step is to simply describe the relationship seen—this is a statement of the *results* (observations). Here we see that between 1965 and 1980, herring catches dropped off dramatically and thereafter continued to drop, but more slowly. Now that we understand the relationship between the two variables, we can draw *conclusions*—make some inferences: What might have caused this relationship? What else may be true because this relationship exists? We could infer from these data that the herring population size also decreased in this time frame. If we know that the same number of fishers were fishing for herring the same number of days each year, the slower decline after 1980 might represent the fact that the fish are more difficult to catch because the population size is smaller. We might also conclude that it has not been as profitable to fish commercially for herring since 1980 as it was in the 1960's and 1970's. These last three statements are conclusions (inferences) based on the results of the study (observations); they are not observations themselves.

Interpolation and extrapolation (projections)

Plotting line graphs also allows us to estimate values of *y* or *x* within the range of our data set, values that we did not actually measure (*interpolation*). We can also create a projection of data points beyond our data set (*extrapolation*) by extending the line. This assumes the same trend will hold at higher or lower *x* axis values, which may or may not be true; therefore, extrapolations are not likely to be as accurate as interpolations. Extrapolations, also called projections, are often shown as dashed lines.

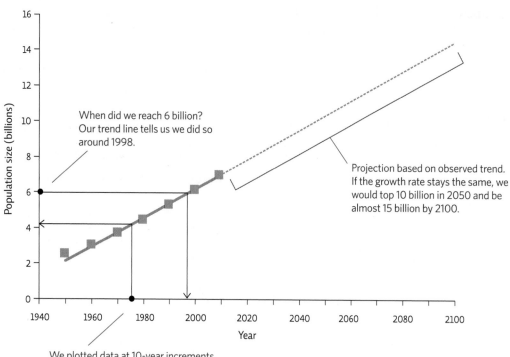

WORLD POPULATION GROWTH

When did we reach 6 billion? Our trend line tells us we did so around 1998.

Projection based on observed trend. If the growth rate stays the same, we would top 10 billion in 2050 and be almost 15 billion by 2100.

We plotted data at 10-year increments, but this trend line allows us to estimate the population in any year. In 1975, we estimate that the population was just over 4 billion people.

B. SCATTER PLOTS

Scatter plots (with or without a trend line) are used when any x value could have multiple y values. For instance, in the second graph shown here, data were collected from various countries. Girls in four of the countries surveyed receive, on average, 4 years of schooling; therefore, there are 4 data points over the x axis value of 4. But each of those countries had different y values (total fertility rate). Here, it would make no sense to "connect the dots." The resulting line would be impossible to follow.

It is more appropriate to construct a line that passes through the cloud of points and shows the "trend," just as we did with the line graph. Data points can be entered into a computer graphing program to calculate a "best fit" line, but you can also estimate the path yourself. To pick the best fit line, draw a line (straight or curved) that passes centrally through the cloud of data points, with about as many points above as below the line—the occasional point far away from the others won't impact the line significantly. When completed, your line may or may not be straight, but it should not connect the dots.

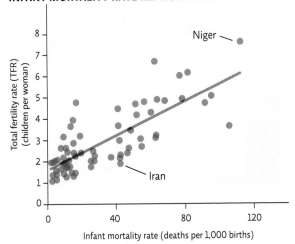

INFANT MORTALITY RATE IMPACTS BIRTH RATE

A *positive correlation*—as one variable increases, so does the other—in an upward-sloping line

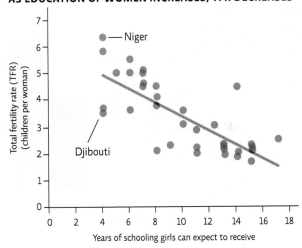

AS EDUCATION OF WOMEN INCREASES, TFR DECREASES

A *negative correlation*—as one variable increases, the other decreases—in a downward-sloping line

C. PIE CHARTS

Pie charts are useful when the groups represented by the independent variable are all discrete categories (e.g., red, yellow, blue. . .) rather than a numerical sequence (e.g., 1, 2, 3. . .). In addition, the categories also represent all the subsets of a whole—all the category values add up to 100%. In other words, you have the entire pie! Data values and/or category titles can be shown either inside each "slice" or outside the pie. The data could also be shown as a bar graph (see Section D on the next page), but showing it as a pie chart instead allows one to more easily compare the size of each group to the other groups and to the whole.

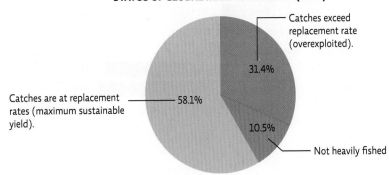

STATUS OF GLOBAL MARINE FISHERIES (2013)

D. BAR GRAPHS

Bar graphs are appropriate in some cases. As with a pie chart, the key consideration is whether the independent variable (the variable you are manipulating in your experiment) is part of a numerical sequence (in which case, a line graph or scatter plot would be used) or represents discrete, separate groups. When you have discrete, separate groups, a bar graph can be used—it would make no sense to connect the data from one group to the next in a line. As with a line graph or scatter plot, the independent variable usually goes on the *x* axis and the dependent variable is shown on the *y* axis.

For example, researchers examined the stomach contents of birds from two different colonies. "Colony" is the independent variable because the researchers chose to see if where a bird lived would affect the type of food it ate. The volume of each food type ingested is the dependent variable— it is the data that the researcher set out to find and it may change according to colony location.

FOOD AND PLASTIC EATEN BY TWO ALBATROSS COLONIES

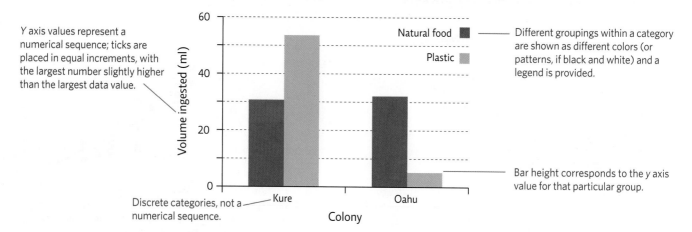

Y axis values represent a numerical sequence; ticks are placed in equal increments, with the largest number slightly higher than the largest data value.

Different groupings within a category are shown as different colors (or patterns, if black and white) and a legend is provided.

Bar height corresponds to the *y* axis value for that particular group.

Discrete categories, not a numerical sequence.

Sometimes it is easier to place the independent variable on the *y* axis if the labels themselves are long. This prevents the need to place labels sideways, making them harder to read. The graph below, which shows the population size of the ten most populous countries, displays the independent variable (country) on the *y* axis and the dependent variable (population size) on the *x* axis.

THE TEN MOST POPULOUS COUNTRIES

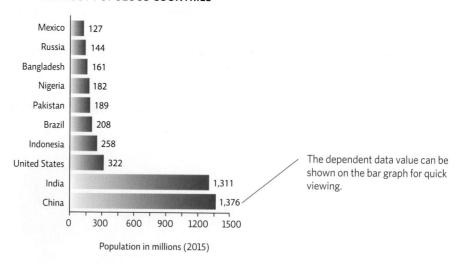

The dependent data value can be shown on the bar graph for quick viewing.

Population in millions (2015)

E. AREA GRAPHS

Another useful graph that allows us to view the relative proportion of all the groups being compared is an area graph. It is used when we have a line graph (the independent variable on the *x* axis is a numerical sequence) showing multiple lines. Each data set (line) is part of a larger group—here we show total fish catch broken down by type of fish. Each line is graphed "on top" of the other, and the space between the lines is filled in with a different color. This is useful because at any given *x* axis point (say, the year 1968), we can see what the total fish catch was as well as how much each type of fish contributed to the total catch. The width of the "ribbon" for each fish type at that point represents its *y* axis value—in this case, its catch in 1,000 tons. (The *y* axis value opposite the ribbon represents the total for all groups.)

FISH CATCH BY COMMERCIAL GROUP: NEWFOUNDLAND—LABRADOR SHELF

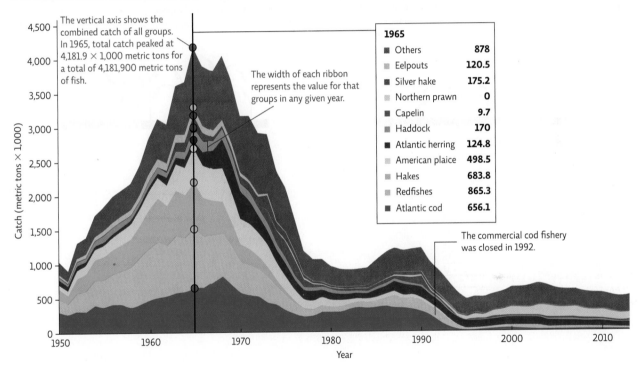

The vertical axis shows the combined catch of all groups. In 1965, total catch peaked at 4,181.9 × 1,000 metric tons for a total of 4,181,900 metric tons of fish.

The width of each ribbon represents the value for that groups in any given year.

The commercial cod fishery was closed in 1992.

1965	
■ Others	**878**
■ Eelpouts	**120.5**
■ Silver hake	**175.2**
■ Northern prawn	**0**
■ Capelin	**9.7**
■ Haddock	**170**
■ Atlantic herring	**124.8**
■ American plaice	**498.5**
■ Hakes	**683.8**
■ Redfishes	**865.3**
■ Atlantic cod	**656.1**

STATISTICAL ANALYSIS

DESCRIPTIVE STATISTICS

In science it is not enough to simply collect data and graph it in order to draw conclusions. Suppose we see a difference between data collected for different groups. How different must the data sets be in order to conclude that the groups are different from each other? And how can we determine whether the treatment we applied—say, growing plants with a new fertilizer—really affected growth? We turn to statistical analysis.

Let's look at an example. We are growing two sets of plants, identical in every way except that one is grown without any fertilizer (the control group) and the other is grown with fertilizer (the test group). To draw conclusions, examine the values in the data table.

We begin with descriptive statistics—what are the characteristics of our data set? We calculate useful statistical values for each data set such as the *mean* (the average), the *range* (the highest value minus the lowest value), and the *sample size* (the number of subjects in each group). We might also calculate other values (with the help of any number of readily available online or calculator-based programs) that help describe the data set, such as *standard deviation* (the average amount of variation of each data value from the mean) and *standard error* (a measure that gives us an idea of how accurate our calculated mean really is, based on the standard deviation). Standard error bars are often shown with data as ± values (for example, 14.9 ± 1.4) or as error bars on a graph.

Mean
14.9 cm

Mean
19 cm

CONTROL GROUP—
GROWN WITHOUT FERTILIZER

	Height (cm)
Sample size: 10	5
Range: 15	11
Mean: 14.9	14
Std deviation: 4.3	15
Std error: 1.4	15
	16
	17
	18
	18
	20

TEST GROUP—
GROWN WITH FERTILIZER

	Height (cm)
Sample size: 10	17
Range: 4	17
Mean: 19	18
Std deviation: 1.2	18
Std error: 0.38	19
	20
	20
	20
	20
	21

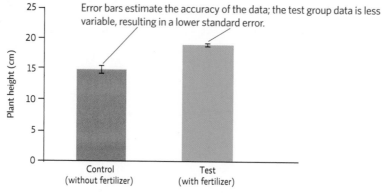

AVERAGE HEIGHT OF PLANTS GROWN WITH AND WITHOUT FERTILIZER

Error bars estimate the accuracy of the data; the test group data is less variable, resulting in a lower standard error.

INFERENTIAL STATISTICS

We can take our analysis further and evaluate the data with an inferential statistical test to determine how likely it is that the data we obtained from the two groups in our experiment actually represent different responses or whether our two groups are both just subsets of a single, larger group. The statistical test gives us a *P-value*—a number that tells us how much overlap there is between the data sets. In science, we generally require that there be no more than a 5% overlap between the two data sets. If the high end of one set (the control group here) overlaps just a little with the low end of the other data set (our test group) and this overlap is no more than 5%, we can conclude that the two groups most likely represent two distinct populations, a result of the treatment we applied—in this case, the addition of fertilizer. If the overlap had been more than 5% (a *P*-value > 0.05), we would not have sufficient evidence to conclude that they were indeed two different groups, but instead we would say they were likely to be a single group that varied widely.

The data showed this:

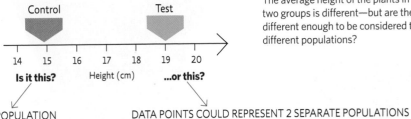

The average height of the plants in the two groups is different—but are they different enough to be considered two different populations?

DATA POINTS COULD ALL BE PART OF ONE POPULATION

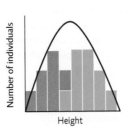

If the two groups are really part of one population, we might have inadvertently put faster-growing individuals in the test group and/or slower-growing ones in the control group. If that is true, retesting or using a larger sample size should produce some tall control and some short test plants.

DATA POINTS COULD REPRESENT 2 SEPARATE POPULATIONS

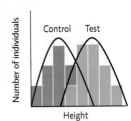

If the two groups really did respond differently to the treatment (fertilizer use), retesting or using a larger sample size should still produce this same trend, with most test plants being larger than control plants.

For this data set, a *t-test* (a simple statistical test) yields a *P*-value of 0.035—our data sets overlap 3.5%. Therefore, we can conclue that the two groups are different at the 0.05 level. Because our experimental design eliminated other variables that might have affected growth (the only difference between the two groups was whether plants received fertilizer), it is reasonable to conclude that the greater growth was caused by the fertilizer. As you read about studies and evaluate the authors' conclusions, look for the *P*-value given with the analysis of the data—if the calculated *P*-value is larger than 0.05, the author will probably conclude that there is not sufficient evidence to conclude that the variable tested had an effect.

OTHER FACTORS THAT AFFECT RESULTS

VARIABILITY

Data sets with a lot of variability are less likely to show significant differences, even if the means are different—there is more of a chance that the two data sets overlap so much that they must be considered part of the same population.

LITTLE VARIABILITY

LOTS OF VARIABILITY

SAMPLE SIZE IS IMPORTANT

Small sample sizes are less reliable because, due to sampling error, we may have inadvertently sampled mostly unusual subjects; this would give us an incorrect picture of the entire population.

Mean

This smaller sample overestimates the mean.

SMALLER CONTROL GROUP

Mean

LARGER CONTROL GROUP

This larger sample suggests that larger plants are not the norm.

APPENDIX 4
GEOLOGY

EARTH IS A DYNAMIC PLANET THAT IS CONSTANTLY CHANGING

Earth is composed of discrete layers; mineral and fossil fuel deposits are found in the layers of Earth's crust. Powerful geologic forces are constantly but slowly rearranging the face of Earth.

EARTH'S INTERNAL STRUCTURE

Crust
5–70 kilometers (~3–45 miles) thick. Thin, rigid layer above the mantle; valuable minerals and fossil fuels are found here, but even our deepest mines (about 3 kilometers, or 2 miles, down) barely penetrate the uppermost crust.

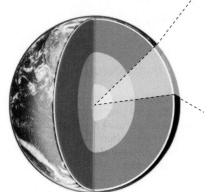

ASTHENOSPHERE

LOWER MANTLE

OUTER CORE (LIQUID)

INNER CORE (SOLID)

Lithosphere
50–150 kilometers (~30–90 miles) thick. Includes crust and rigid part of upper mantle.

Mantle
About 3,000 kilometers (~1,900 miles) thick. The uppermost part is rigid but the rocks in the asthenosphere are very close to their melting temperature and flow, not as a liquid, but like a pliable solid resembling putty. The mantle also contains magma (molten rock) in isolated places, often at plate boundaries.

Core
About 3,500 kilometers (~2,200 miles) thick. Made up of mostly iron-nickel alloy; solid at the center and more fluid in the outer core.

THE ROCK CYCLE

Rocks form and are transformed when they are subjected to high heat and pressure underground and when they are exposed to wind and water on Earth's surface.

Igneous rocks
Molten rock (magma) crystallizes as it cools (either above or below ground) to form igneous rock.

Cooling and crystallization

Weathering

High heat and pressure

Melting

Melting

Sedimentary rocks
Weathering and erosion break down rock that is then deposited as sediment along with the remains of living organisms. Sediment is compacted and cemented to form sedimentary rock.

High heat and pressure

Metamorphic rocks
Rock buried deep in the crust can be transformed by heat and pressure into metamorphic rock.

 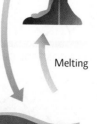 

Weathering

Weathering

High heat and pressure

TECTONIC PLATES

Powerful geologic forces are constantly but slowly rearranging the face of Earth. The lithosphere is made up of tectonic plates that move slowly, powered by heat deep within Earth. The location of earthquakes and volcanoes gives clues about the location of plate boundaries, representing places where plates move relative to one another. The energy released at plate boundaries fuels processes (volcanic and hydrothermal activity) that create new minerals and redistribute others within Earth's crust.

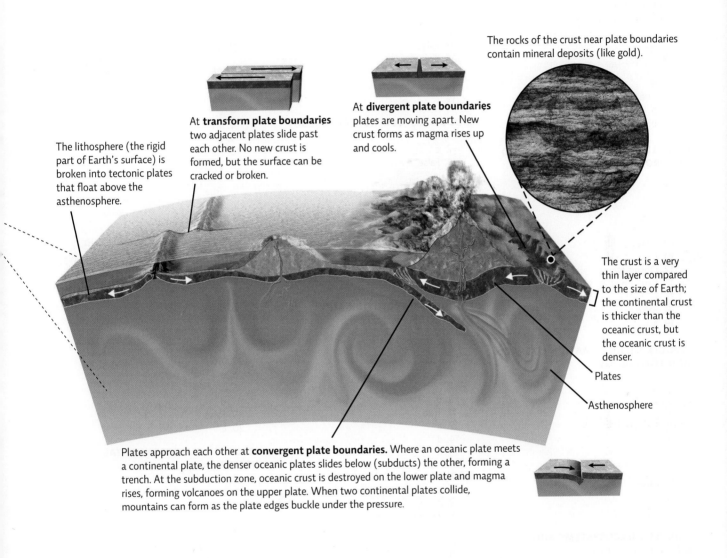

The rocks of the crust near plate boundaries contain mineral deposits (like gold).

At **transform plate boundaries** two adjacent plates slide past each other. No new crust is formed, but the surface can be cracked or broken.

At **divergent plate boundaries** plates are moving apart. New crust forms as magma rises up and cools.

The lithosphere (the rigid part of Earth's surface) is broken into tectonic plates that float above the asthenosphere.

The crust is a very thin layer compared to the size of Earth; the continental crust is thicker than the oceanic crust, but the oceanic crust is denser.

Plates

Asthenosphere

Plates approach each other at **convergent plate boundaries.** Where an oceanic plate meets a continental plate, the denser oceanic plates slides below (subducts) the other, forming a trench. At the subduction zone, oceanic crust is destroyed on the lower plate and magma rises, forming volcanoes on the upper plate. When two continental plates collide, mountains can form as the plate edges buckle under the pressure.

APPENDIX 5

SELECTED ANSWERS TO END-OF-CHAPTER PROBLEMS

Answers to multiple choice, true/false, and fill-in-the-blank questions are shown here.

MODULE 1.1: ENVIRONMENTAL LITERACY AND SUSTAINABILITY
1. living; nonliving
3. False
5. a
7. d
9. d
11. c
13. a
15. The tragedy of the commons
16. c
18. biocentric; ecocentric

MODULE 1.2: SCIENCE LITERACY AND THE PROCESS OF SCIENCE
1. b
2. c
3. a
6. True
7. b
9. False
10. c
12. Observational
13. Experimental
15. c
18. False
19. a

MODULE 1.3: INFORMATION LITERACY AND TOXICOLOGY
1. c
2. b
4. a
5. c
7. False
8. a
10. True
11. d
13. False
15. c
17. a

MODULE 2.1: ECOSYSTEMS AND NUTRIENT CYCLING
1. b
3. True
5. a
7. b
9. c
11. a
12. c
14. True
15. d
17. mining; fertilizer use
18. c

MODULE 2.2: POPULATION ECOLOGY
1. a
3. c
4. a
6. population density
7. d
9. d
11. a
13. False
15. True
16. c
18. True
19. c

MODULE 2.3: COMMUNITY ECOLOGY
1. d
2. c
4. False
6. c
8. True
9. richness; evenness
11. False
13. d
15. c
17. c
19. a
20. b

MODULE 3.1: EVOLUTION AND EXTINCTION
1. a
2. a
4. False
5. d
7. c
9. d
11. True
12. a
14. d
16. b

MODULE 3.2: BIODIVERSITY
1. a
3. d
5. genetic
6. c
8. False
9. c
11. c
13. e
14. a
16. True
17. d

ONLINE MODULE 3.3: PRESERVING BIODIVERSITY
1. d
4. False
5. b
7. False
8. a
10. True
11. c

13. a
15. False
16. c
18. False
20. True
21. a

MODULE 4.1: HUMAN POPULATION
1. c
3. immigration; emigration
4. c
6. True
7. b
10. False
11. c
13. b
15. True
16. d
19. c

MODULE 4.2: URBANIZATION AND SUSTAINABLE COMMUNITIES
1. False
2. d
4. True
5. d
7. b
9. True
10. c
12. True
13. e
15. a
17. d
19. False
20. c

MODULE 4.3: ENVIRONMENTAL HEALTH
1. d
3. a
5. b
7. b
9. True
11. a
13. False
14. a
15. a
17. d

MODULE 5.1: ECOLOGICAL ECONOMICS AND CONSUMPTION
1. d
3. c
5. b
7. False
9. c
11. b
13. c
15. d
18. True

SELECTED ANSWERS TO END-OF-CHAPTER PROBLEMS **A-15**

MODULE 5.2: ENVIRONMENTAL POLICY
1. True
2. b
4. False
5. b
7. True
9. Environmental Protection Agency
10. d
12. False
13. b
15. c
17. True
18. c
19. a
21. False

MODULE 5.3: MANAGING SOLID WASTE
1. True
2. c
4. False
5. b
7. False
9. False
10. a
13. True
15. e
18. False
21. c
23. c

MODULE 6.1: FRESHWATER RESOURCES
1. True
3. d
5. b
6. d
8. go up
9. False
10. b
12. economic
14. True
15. a
16. a
18. e
20. c

MODULE 6.2: WATER POLLUTION
1. d
2. a
4. True
6. c
8. c
10. False
11. a
13. True
14. c
17. False
19. c
21. d

ONLINE MODULE 6.3: MARINE ECOSYSTEMS
1. True
2. d
4. False
5. d
7. intertidal zone
8. d
10. overfishing
11. True
12. b
14. False
15. d
18. d
20. False
21. d

MODULE 7.1: FOREST RESOURCES
1. False
2. b
4. d
5. b
7. c
10. a
12. c
13. b
15. False
16. a

ONLINE MODULE 7.2: GRASSLANDS AND SOIL RESOURCES
1. a
2. b
4. c
6. desertification
7. e
9. True
10. O; A
11. b
13. False
16. False
18. True
19. b
21. a

ONLINE MODULE 7.3: MINERAL RESOURCES
1. False
2. c
4. False
5. c
6. a
8. False
9. d
11. True
12. a
14. smelting
15. c
17. True
19. a
21. True
22. e
24. d

MODULE 8.1: FEEDING THE WORLD
1. a
3. d
4. c
6. a
8. d
10. True
11. c
13. c
15. c
17. a

MODULE 8.2: SUSTAINABLE AGRICULTURE: RAISING CROPS
1. e
3. a
4. e
6. a
8. d
10. a
12. True
13. c
14. b
16. b
18. d

ONLINE MODULE 8.3: SUSTAINABLE AGRICULTURE: RAISING LIVESTOCK
1. a
4. True
5. d
7. c
8. b
10. True
11. c
13. c
15. False
16. c
18. True
19. b

ONLINE MODULE 8.4: FISHERIES AND AQUACULTURE
1. bycatch
2. b
4. d
7. True
9. True
11. False
12. c
14. aquaculture
15. c
17. b

MODULE 9.1: COAL
1. stationary; mobile
2. c
4. d
7. False
8. b
10. c
13. a
15. True
16. b
18. d
21. b

MODULE 9.2: OIL AND NATURAL GAS
1. d
3. False
4. c
6. True
7. c
9. True
11. False
13. a
14. b

16. False
17. a
19. False
20. b

MODULE 10.1: AIR POLLUTION
1. False
2. b
4. True
5. point source
6. d
9. True
10. b
11. a
13. False
14. d
15. criteria air pollutants
16. e
18. True
19. c
21. b

MODULE 10.2: CLIMATE CHANGE
1. weather; climate
2. b
4. d
6. b
8. False

10. True
11. b
12. d
15. False
17. c
19. True

MODULE 11.1: NUCLEAR POWER
1. protons
2. c
4. d
6. a
7. b
10. alpha
11. c
13. d
15. d
18. False
19. a
21. True

MODULE 11.2: SUSTAINABLE ENERGY: STATIONARY SOURCES
1. c
3. False
4. c
6. d
8. True

9. a
10. d
12. True
13. a
15. b
17. d
18. b
20. False

ONLINE MODULE 11.3: SUSTAINABLE ENERGY: MOBILE SOURCES
1. a
2. c
4. True
5. a
7. True
8. c
10. False
11. False
12. c
14. False
15. c
17. True
18. a
20. True
21. c

GLOSSARY

A

abiotic The nonliving components of an ecosystem, such as rainfall and mineral composition of the soil. (Module 2.1)

acid deposition Precipitation that contains sulfuric or nitric acid; dry particles may also fall and become acidified once they mix with water. (Module 10.1)

acid mine drainage Water that flows past exposed rock in mines and leaches out sulfates. These sulfates react with the water and oxygen to form acids (low-pH solutions). (Modules 7.3, 9.1)

acidification The lowering of the pH of a solution. (Module 6.3)

active solar technology Mechanical equipment for capturing, converting, and sometimes concentrating solar energy into a more usable form. (Module 11.2)

acute effect Adverse reaction that occurs very rapidly after exposure to a toxic substance has occurred. (Module 1.3)

adaptive management A plan that allows room for altering strategies as new information becomes available or as the situation itself changes. (Modules 5.2, 7.2)

additive effects Exposure to two or more chemicals that has an effect equivalent to the sum of their individual effects. (Module 1.3)

affluence The state of having great weallth. (Module 8.3)

age structure The percentage of the population that is distributed into various age groups. (Module 4.1)

agroecology A scientific field that considers the area's ecology and indigenous knowledge and favors agricultural methods that protect the environment and meet the needs of local people. (Module 8.2)

air pollution Any material added to the atmosphere (naturally or by humans) that harms living organisms, affects the climate, or impacts structures. (Module 10.1)

albedo The ability of a surface to reflect away solar radiation. (Module 10.2)

alleles Variants of genes that account for the diversity of traits seen in a population. (Module 3.1)

alpha radiation Ionizing particle radiation that consists of two protons and two neutrons. (Module 11.1)

annual crops Crops that grow, produce seeds, and die in a single year and must be replanted each season. (Module 8.2)

antagonistic effects Exposure to two or more chemicals that has a lesser effect than the sum of their individual effects would predict. (Module 1.3)

Anthropocene A proposed new geologic epoch that is marked by modern human impact. (Module 1.1)

anthropocentric worldview A human-centered view that assigns intrinsic value only to humans. (Module 1.1)

anthropogenic Caused by or related to human action. (Module 1.1)

applied science Research whose findings are used to help solve practical problems. (Module 1.1)

aquaculture Fish farming; the rearing of aquatic species in tanks, ponds, or ocean net pens. (Module 8.4)

aquifer An underground, permeable region of soil or rock that is saturated with water. (Modules 6.1, 6.2)

artificial selection A process in which humans decide which individuals breed and which do not in an attempt to produce a population of plants or animals with desired traits. (Module 3.1)

asthenosphere The layer of the mantle that is so hot that the rock begins to soften, allowing it to flow slowly. (Module 7.3)

atmosphere The blanket of gases surrounding Earth. (Module 5.2)

atom The simplest form of matter that cannot be broken down by chemical means. (Module 11.1)

B

background rate of extinction The average rate of extinction that occurred before the appearance of humans or that occurs between mass extinction events. (Module 3.1)

benthic macroinvertebrates Easy-to-see (not microscopic) arthropods such as insects that live on the stream bottom. (Module 6.2)

beta radiation Ionizing particle radiation that consists of electrons. (Module 11.1)

bioaccumulation The build-up of a substance in the tissues of an organism over the course of its lifetime. (Module 1.3)

biocentric worldview A life-centered approach that views all life as having intrinsic value, regardless of its usefulness to humans. (Module 1.1)

biodegradable Capable of being broken down by living organisms. (Module 5.3)

biodiesel A diesel-type fuel made from animal or vegetable oils and fats. (Module 11.3)

biodiversity The variety of life on Earth; it includes species, genetic, and ecological diversity. (Modules 1.1, 3.2)

biodiversity hotspot An area that contains a large number of endangered endemic species. (Module 3.2)

bioethanol An alcohol fuel made from crops in a process of fermentation and distillation. (Module 11.3)

biofuel Solid, liquid, or gaseous fuel produced from biological material. (Module 11.3)

biological assessment The process of sampling an area to see what lives there as a tool to determine how healthy the area is. (Module 6.2)

biomagnification The increased concentration of substances in the tissue of animals at successively higher levels of the food chain. (Module 1.3)

biomass Material from living or recently living organisms or their by-products. (Module 11.3)

biomass energy Energy from biological material such as plants (wood, charcoal, crops) and animal waste. (Module 11.2)

biome One of many distinctive types of ecosystems determined by climate and identified by the predominant vegetation and organisms that have adapted to live there. (Module 2.1)

biosphere The sum total of all of Earth's ecosystems. (Module 2.1)

biotic The living (organic) components of an ecosystem, such as the plants and animals and their waste (dead leaves, feces). (Module 2.1)

biotic potential (r) The maximum rate at which the population can grow due to births if each member of the population survives and reproduces. (Module 2.2)

boreal forest Coniferous forest found at high latitudes and altitudes characterized by low temperatures and low annual precipitation. (Module 7.1)

bottleneck effect The situation that occurs when population size is drastically reduced, leading to the loss of some genetic variants, and resulting in a less diverse population. (Module 3.1)

bottom-up regulation The control of population size by factors that enhance growth and survival (growth factors) such as nutrients, water, sunlight, and habitat. (Module 2.2)

bycatch Nontarget species that become trapped in fishing nets and are usually discarded. Some methods, like trawling, have very high bycatch levels, and discards often exceed the actual target species catch. (Module 8.4)

C

canopy The upper layer of a forest, formed where the crowns (tops) of the majority of the tallest trees meet. (Module 7.1)

cap-and-trade program Regulations that set an upper limit for pollution emissions, issue permits to producers for a portion of that amount, and allow producers that release less than their allotment to sell permits to those who exceeded their allotment. (Modules 5.2, 10.1)

carbon capture and sequestration (CCS) The process of removing carbon from fuel combustion emissions or other sources and storing it to prevent its release into the atmosphere. (Module 9.1)

carbon cycle Movement of carbon through biotic and abiotic parts of an ecosystem. Carbon cycles via photosynthesis and cellular respiration as well as in and out of other reservoirs, such as oceans, soil, rock, and atmosphere. It is also released by human actions such as the burning of fossil fuels. (Module 2.1)

carbon footprint The amount of carbon released to the atmosphere by a person, company, nation, or activity. (Modules 4.2, 8.2, 8.3, 9.2, 11.3)

carbon sequestration The storage of carbon in a form that prevents its release into the atmosphere. (Module 11.3)

carbon sink An area such as a forest, ocean sediment, or soil, where accumulated carbon does not readily reenter the carbon cycle. (Module 7.1)

carbon taxes Governmental fees imposed on activities that release greenhouse gases into the atmosphere. (Module 10.2)

carrying capacity (K) The maximum population size that a particular

environment can support for the long term; for human populations, it depends on resource availability and the rate of per capita resource use by the population. (Modules 2.2, 4.1)

cash crops Food and fiber crops grown to sell for profit rather than for use by local families or communities. (Module 8.1)

cause-and-effect relationship An association between two variables that identifies one (the effect) occurring as a result of or in response to the other (the cause). (Module 1.2)

cellular respiration The process in which all organisms break down sugar to release its energy, using oxygen and giving off CO_2 as a waste product. (Module 2.1)

cellulosic ethanol Bioethanol made by breaking down cellulose in plants. (Module 11.3)

childhood mortality rate The number of children under 5 years of age that die per every 1,000 live births in that year. (Module 4.1)

chronic effect Adverse reaction that happens only after repeated long-term exposure to low doses of a toxic substance. (Module 1.3)

circular economic system A production system in which the product is returned to the resource stream when consumers are finished with it or is disposed of in such a way that nature can decompose it. (Module 5.1)

citizen suit provision A provision that allows a private citizen to sue, in federal court, a perceived violator of certain U.S. environmental laws, such as the Clean Air Act, in order to force compliance. (Module 5.2)

Clean Air Act (CAA) The main U.S. law that authorizes the EPA to set standards for dangerous air pollutants and enforce those standards. (Module 10.1)

clean coal A liquid or gaseous product produced by removing some contaminants contained in coal so that the resulting fuel releases less pollution when burned. (Module 9.1)

Clean Water Act (CWA) U.S. federal legislation that regulates the release of point source pollution into surface waters and sets water quality standards for those waters. It also supports best management practices to reduce nonpoint source pollution. (Module 6.2)

clear-cut Timber-harvesting technique that cuts all trees in an area. (Module 7.1)

climate Long-term patterns or trends of meteorological conditions. (Module 10.2)

climate adaptation Efforts to help deal with existing or impending climate change problems. (Module 10.2)

climate change Alteration in the long-term patterns and statistical averages of meteorological events. (Module 10.2)

climate forcer Anything that alters the balance of incoming solar radiation relative to the amount of heat that escapes out into space. (Module 10.2)

climate mitigation Efforts to minimize the extent or impact of climate change. (Module 10.2)

clumped distribution A distribution in which individuals are found in groups or patches within the habitat. (Module 2.2)

coal A fossil fuel that is formed when plant material is buried in oxygen-poor conditions and subjected to high heat and pressure over a long time. (Module 9.1)

coevolution A special type of natural selection in which two species each provide the selective pressure that determines which traits are favored by natural selection in the other. (Module 3.1)

collapsed fishery A fishery in which annual catches fall below 10% of their historic high; stocks can no longer support a fishery. (Module 8.4)

command-and-control regulation A type of regulation that involves setting an upper allowable limit of pollution release that is enforced with fines and/or incarceration. (Modules 5.2, 10.1)

commensalism A symbiotic relationship between individuals of two species in which one benefits from the presence of the other, but the other is unaffected. (Modules 2.3, 6.3)

community All the populations (plants, animals, and other species) living and interacting in an area. (Module 2.1)

community ecology The study of all the populations (plants, animals, and other species) living and interacting in an area. (Module 2.3)

competition Species interaction in which individuals are vying for limited resources. (Module 2.3)

composting Allowing waste to biologically decompose in the presence of oxygen and water, producing a soil-like mulch. (Module 5.3)

concentrated animal feeding operation (CAFO) A situation in which meat or dairy animals are raised in confined spaces, maximizing the number of animals that can be reared in a small area. (Module 8.3)

condensation The conversion of water from a gaseous state (water vapor) to a liquid state. (Module 6.1)

conservation Efforts that reduce waste and increase efficient use of resources. (Module 11.2)

conservation biology The science concerned with preserving biodiversity. (Module 3.3)

conservation genetics The scientific field that relies on species' genetics to inform conservation efforts. (Module 3.3)

conservation reserve program Program in which farmers and ranchers are paid to keep damaged land out of production to promote recovery. (Module 7.2)

consumer An organism that obtains energy and nutrients by feeding on another organism. (Modules 2.1, 2.3)

control group The group in an experimental study to which the test group's results are compared; ideally, the control group will differ from the test group in only one way. (Module 1.2)

control rod Rod that absorbs neutrons and slows the fission chain reaction. (Module 11.1)

convention An international agreement that represents a position on an issue and identifies general goals that the signing countries agree to pursue. (Module 5.2)

Convention on Biological Diversity (CBD) An international treaty that promotes sustainable use of ecosystems and biodiversity. (Module 3.3)

Convention on International Trade in Endangered Species of Wild Fauna and Flora (CITES) An international treaty that regulates the global trade of selected species. (Module 3.3)

conventional reserves Deposits of crude oil or natural gas that can be extracted by vertical drilling and pumping. (Module 9.2)

convergent plate boundary A place where tectonic plates are moving toward each other. (Module 7.3)

coral Colonial marine animals that secrete hard outer shells in which they live and are mutualistically dependent on an algal partner. (Module 6.3)

coral bleaching A stress response in coral in which the mutualistic algal partner is expelled; this weakens and even can kill the coral if it is not recolonized soon. (Module 6.3)

coral reef A large underwater structure formed by colonies of tiny animals (coral) that produce a calcium carbonate exoskeleton that over time build up; found in shallow, warm, tropical seas. (Module 6.3)

correlation Two things occurring together but not necessarily having a cause-and-effect relationship. (Module 1.2)

cradle-to-cradle Refers to management of a resource that considers the impact of its use at every stage, from raw material extraction to final disposal or recycling. (Module 5.1)

critical thinking Skills that enable individuals to logically assess information, reflect on that information, and reach their own conclusions. (Module 1.3)

crude oil A mix of hydrocarbons that exists as a liquid underground; can be refined to produce fuels or other products. (Module 9.2)

D

dam A structure that blocks the flow of water in a river or stream. (Module 6.1)

debt-for-nature-swaps Arrangements in which a wealthy nation forgives the debt of a developing nation in return for a pledge to protect natural areas in that developing nation. (Module 3.3)

decomposers Organisms such as bacteria and fungi that break organic matter all the way down to constituent atoms or molecules in a form that plants can take back up. (Modules 2.1, 2.3)

deforestation Net loss of trees in a forested area. (Module 7.1)

demographic factors Population characteristics such as birth rate that influence changes in population size and composition. (Module 4.1)

demographic transition A theoretical model that describes the expected drop in once-high population growth rates as economic conditions improve the quality of life in a population. (Module 4.1)

demography The statistical analysis of the characteristics of a population. (Module 4.1)

denitrification Conversion of nitrate to molecular nitrogen (N_2). (Module 2.1)

density-dependent factors Factors, such as predation or disease, whose impact on a population increases as population size goes up. (Module 2.2)

density-independent factors Factors, such as a storm or an avalanche, whose impact on a population is not related to population size. (Module 2.2)

dependent variable The variable in an experiment that is evaluated to see if it changes due to the conditions of the experiment. (Module 1.2)

desalination The removal of salt and minerals from seawater to make it suitable for consumption. (Module 6.1)

desertification The process that transforms once fertile land into desert. (Module 7.2)

desired fertility The ideal number of children an individual indicates he or she would like to have. (Module 4.1)

detritivores Consumers (including worms, insects, crabs, etc.) that eat dead organic material. (Module 2.3)

discounting future value Giving more weight to short-term benefits and costs than to long-term ones. (Module 5.1)

dissolved oxygen (DO) The amount of oxygen in the water. (Module 6.2)

divergent plate boundary A place where tectonic plates are moving away from each other. (Module 7.3)

domestic water use Indoor and outdoor use of water by households and small businesses. (Module 6.1)

dose-response curve A graph that shows the strength of an effect of a substance at different doses of that substance. (Module 1.3)

E

ecocentric worldview A system-centered view that values intact ecosystems, not just the individual parts. (Module 1.1)

ecolabeling Providing information about how a product is made and where it comes from. Allows consumers to make more sustainable choices and support sustainable products and the businesses that produce them. (Module 5.1)

ecological diversity The variety within an ecosystem's structure, including many communities, habitats, niches, and trophic levels. (Module 3.2)

ecological footprint The land area needed to provide the resources for, and assimilate the waste of, a person or population. (Modules 4.1, 5.1)

ecological succession Progressive replacement of plant (and then animal) species in a community over time due to the changing conditions that the plants themselves create (more soil, shade, etc.). (Module 2.3)

economics The social science that deals with the production, distribution, and consumption of goods and services. (Module 5.1)

ecosystem All of the μorganisms in a given area plus the physical environment in which, and with which, they interact. (Module 2.1)

ecosystem conservation A management strategy that focuses on protecting an ecosystem as a whole in an effort to protect the species that live there. (Module 3.3)

ecosystem restoration The repair of natural habitats back to (or close to) their original state. (Module 3.3)

ecosystem services Essential ecological processes that make life on Earth possible. (Modules 3.2, 5.1, 7.1)

ecotones Regions of distinctly different physical areas that serve as boundaries between different communities. (Module 2.3)

ecotourism Low-impact travel to natural areas that contributes to the protection of the environment and respects the local people. (Modules 3.3, 7.1)

edge effect The different physical makeup of an ecotone that creates different conditions that either attract or repel certain species (e.g., it is drier, warmer, and more open at the edge of a forest and field than it is further in the forest). (Module 2.3)

effluent Wastewater discharged into the environment. (Modules 6.1, 6.2)

electricity The flow of electrons (negatively charged subatomic particles) through a conductive material (such as wire). (Module 9.1)

element A substance composed of all the same type of atoms. (Module 11.1)

emergent layer The region where a tree that is taller than the canopy trees rises above the canopy layer. (Module 7.1)

emerging infectious diseases Infectious diseases that are new to humans or that have recently increased significantly in incidence, in some cases by spreading to new ranges. (Module 4.3)

emigration The movement of individuals out of a given population. (Module 4.1)

empirical evidence Information gathered via observation of physical phenomena. (Module 1.2)

empirical science A scientific approach that investigates the natural world through systematic observation and experimentation. (Module 1.1)

endangered species Species at high risk of becoming extinct. (Modules 3.1, 3.2)

Endangered Species Act (ESA) The primary federal law that protects biodiversity in the United States. (Module 3.3)

endemic Describes a species that is native to a particular area and is not naturally found elsewhere. (Module 3.1)

endemic species A species that is native to a particular area and is not naturally found elsewhere. (Module 3.2)

energy The capacity to do work. (Module 9.1)

energy efficiency A measure of the amount of energy needed to perform a task; higher efficiency means less energy is wasted. (Module 11.2)

energy flow The one-way passage of energy through an ecosystem. (Module 2.1)

energy independence Meeting all of one's energy needs without importing any energy. (Module 9.2)

energy return on energy investment (EROEI) A measure of the net energy from an energy source (the energy in the source minus the energy required to get it, process it, ship it, and then use it). (Modules 9.1, 9.2, 11.3)

energy security Having access to enough reliable and affordable energy sources to meet one's needs. (Module 9.2)

environment The biological and physical surroundings in which any given living organism exists. (Module 1.1)

environmental economics New theory of economics that considers the long-term impact of our choices on people and the environment. (Module 5.1)

environmental ethic The personal philosophy that influences how a person interacts with his or her natural environment and thus affects how one responds to environmental problems. (Module 1.1)

environmental health The branch of public health that focuses on factors in the natural world and the human-built environment that impact the health of populations. (Module 4.3)

environmental impact statement (EIS) A document outlining the positive and negative impacts of any federal action that has the potential to cause environmental damage. (Modules 5.2, 9.1)

environmental justice The concept that access to a clean, healthy environment is a basic human right. (Modules 1.3, 4.2, 10.1)

environmental literacy A basic understanding of how ecosystems function and of the impact of our choices on the environment. (Module 1.1)

environmental policy A course of action adopted by a government or an organization that is intended to improve the natural environment and public health or reduce human impact on the environment. (Module 5.2)

Environmental Protection Agency (EPA) The federal agency responsible for setting policy and enforcing U.S. environmental laws. (Modules 1.3, 5.2)

environmental racism A form of racism that occurs when minority communities face more exposure to pollution than average for the region. (Module 10.1)

environmental science An interdisciplinary field of research that draws on the natural and social sciences and the humanities in order to understand the natural world and our relationship to it. (Module 1.1)

epidemiologist A scientist who studies the causes and patterns of disease in human populations. (Module 4.3)

erosion The movement of broken-down rock, soil, and other materials from one location to another. (Module 7.3)

estuary A region where a river empties into the ocean. (Module 6.3)

eutrophication A process in which excess nutrients in aquatic ecosystems feed biological productivity, ultimately lowering the oxygen content in the water. (Modules 6.2, 8.2)

evaporation The conversion of water from a liquid state to a gaseous state. (Module 6.1)

evolution Differences in the gene frequencies within a population from one generation to the next. (Module 3.1)

e-waste Unwanted computers and other electronic devices such as discarded televisions and cell phones. (Modules 5.3, 7.3)

exclusive economic zones (EEZs) Zones that extend 200 nautical miles from the coastline of any given nation, where that nation has exclusive rights over marine resources, including fish. (Module 8.4)

experimental study Research that manipulates a variable in a test group and compares the response to that of a control group that was not exposed to the same variable. (Module 1.2)

exponential growth The kind of growth in which a population becomes progressively larger each breeding cycle; produces a

J-shaped curve when plotted over time. (Module 2.2)

external cost A cost associated with a product or service that is not taken into account when a price is assigned to that product or service but rather is passed on to a third party who does not benefit from the transaction. (Module 5.1)

extinction The complete loss of a species from an area; may be local (gone from an area) or global (gone for good). (Module 3.1)

extirpated/extirpation Locally extinct in one geographic area, but still found elsewhere. (Modules 3.1, 3.2)

exurbs Towns beyond the immediate suburbs whose residents commute into the city for work. (Module 4.2)

F

fair trade A certification program whose products are made in ways that are environmentally sustainable and socially beneficial (e.g., fair wages, good working conditions). (Module 5.1)

feed conversion ratio The amount of edible food that is produced per unit of feed input. (Module 8.3)

feedstock Biomass sources used to make biofuels. (Module 11.3)

fertilizer A natural or synthetic mixture that contains nutrients that is added to soil to boost plant growth. (Module 8.2)

fisheries The industry devoted to commercial fishing or the places where fish are caught, harvested, processed, and sold. (Module 8.4)

fishing down the food chain The harvest of fish at lower trophic levels once fish stocks at higher trophic levels become depleted. (Module 8.4)

flagship species The focus of public awareness campaigns aimed at generating interest in conservation in general; usually an interesting or

charismatic species, such as the giant panda or tiger. (Module 3.3)

food chain A simple, linear path starting with a plant (or other photosynthetic organism) that identifies what each organism in the path eats. (Module 2.3)

food miles The distance a food travels from its site of production to the consumer. (Module 8.2)

food security Having physical, social, and economic access to sufficient safe and nutritious food. (Module 8.1)

food self-sufficiency The ability of an individual nation to grow enough food to feed its people. (Module 8.1)

food sovereignty The ability of an individual nation to control its own food system. (Module 8.1)

food web A linkage of all the food chains together that shows the many connections in the community. (Module 2.3)

forest An ecosystem made up primarily of trees and other woody vegetation. (Module 7.1)

forest ecosystem management (FEM) A system that focuses on managing the forest as a whole rather than for maximizing yields of a specific product. (Module 7.1)

forest floor The lowest level of the forest, containing herbaceous plants, fungi, leaf litter, and soil. (Module 7.1)

fossil fuel A variety of hydrocarbons formed from the remains of dead organisms. (Modules 9.1, 9.2)

founder effect The situation that occurs when a small group with only a subset of the larger population's genetic diversity becomes isolated and evolves into a different population, missing some of the traits of the original. (Module 3.1)

fracking (hydraulic fracturing) The extraction of oil or natural gas from dense rock formations by creating factures in the rock and then flushing out the oil/gas with pressurized fluid. (Module 9.2)

freshwater Water that has few dissolved ions such as salt. (Module 6.1)

fuel crop A crop specifically grown to be used to produce biofuels. (Module 11.3)

fuel rod Hollow metal cylinder filled with uranium fuel pellets for use in fission reactors. (Module 11.1)

G

gamma radiation Ionizing high-energy electromagnetic waves (photons). (Module 11.1)

gene frequencies The assortment and abundance of particular variants of genes relative to each other within a population. (Module 3.1)

genes Stretches of DNA, the hereditary material of cells, that each direc t the production of a particular protein and influence an individual's traits. (Module 3.1)

genetic diversity The heritable variation among individuals of a single population or within a species as a whole. (Modules 3.1, 3.2)

genetic drift The change in gene frequencies of a population over time due to random mating that results in the loss of some gene variants. (Module 3.1)

genetically modified organism (GMO) Organism that has had its genetic information modified to give it desirable characteristics such as pest or drought resistance. (Module 8.1)

geology The study of planet Earth and the processes that have shaped it in the past and shape it today. (Module 7.3)

geothermal energy Heat stored underground, contained in either rocks or fluids. (Module 11.2)

geothermal heat pump A system that transfers the steady 55°F (12.5°C) underground temperature to a building to help heat or cool it. (Module 11.2)

geothermal power plant A large-scale facility that captures steam produced from Earth's internal heat to turn turbines which generate electricity. (Module 11.2)

grassland A biome that is predominately grasses, due to low rainfall; grazing animals; and/or fire. (Module 7.2)

green building Construction and operational designs that promote resource and energy efficiency and provide a better environment for occupants. (Module 4.2)

green business Doing business in a way that is good for people and the environment. (Module 5.1)

green city A city designed to improve environmental quality and social equity while reducing its overall environmental impact. (Module 4.2)

Green Revolution A plant-breeding program in the mid-1900's that dramatically increased crop yields and paved the way for mechanized, large-scale agriculture. (Module 8.1)

Green Revolution 2.0 Programs that focus on the production of genetically modified organisms (GMOs) to increase crop productivity. (Module 8.1)

green space A natural area such as a park or undeveloped landscape containing grass, trees, or other vegetation in an urban area, usually set aside for recreational use. (Module 4.2)

green tax A tax (a fee paid to the government) assessed on environmentally undesirable activities (e.g., a tax per unit of pollution emitted). (Modules 5.2, 10.1)

greenhouse effect The warming of the planet that results when heat is trapped by Earth's atmosphere. (Module 10.2)

greenhouse gases Molecules in the atmosphere that absorb heat and reradiate it back to Earth. (Module 10.2)

greenwashing Claiming environmental benefits about a product when the benefits are actually minor or nonexistent. (Modules 5.1, 8.2)

ground-level ozone A secondary pollutant that forms when some of the pollutants released during fossil fuel combustion react with atmospheric oxygen in the presence of sunlight. (Module 10.1)

groundwater Water found underground in aquifers. (Module 6.1)

growth factors Resources individuals need to survive and reproduce that allow a population to grow in number. (Module 2.2)

H

habitat The physical environment in which individuals of a particular species can be found. (Modules 2.1, 2.3)

habitat destruction The alteration of a natural area in a way that makes it unsuitable for the species living there. (Module 3.2)

habitat fragmentation The destruction of part of an area that creates a patchwork of suitable and unsuitable habitat areas that may exclude some species altogether. (Module 3.2)

hazardous waste Waste that is toxic, flammable, corrosive, explosive, or radioactive. (Module 5.3)

hectare (ha) A metric unit of measure for area; 1 ha = 2.5 acres (ac). (Module 7.1)

herbivore An animal that feeds on plants. (Module 7.2)

high-level radioactive waste (HLRW) Spent nuclear reactor fuel or waste from the production of nuclear weapons that is still highly radioactive. (Module 11.1)

high-yield varieties (HYVs) Strains of staple crops selectively bred to be more productive than their natural counterparts. (Module 8.1)

holistic planned grazing Grazing livestock in a way that mimics wild grazers by grazing intensively on a small section of pasture before moving to another. (Module 7.2)

hydropower Energy produced from moving water. (Module 11.2)

hypothesis A possible explanation for what we have observed that is based on some previous knowledge. (Module 1.2)

hypoxia A situation in which a body of water contains inadequate levels of oxygen, compromising the health of many aquatic organisms. (Module 6.2)

I

immigration The movement of individuals into a given population. (Module 4.1)

incinerators Facilities that burn trash at high temperatures. (Module 5.3)

independent variable The variable in an experiment that a researcher manipulates or changes to see if the change produces an effect. (Module 1.2)

indicator species A species that is particularly vulnerable to ecosystem perturbations, and that, when we monitor it, can give us advance warning of a problem. (Modules 2.3, 3.3)

industrial agriculture Farming methods that rely on technology, synthetic chemical inputs, and economies of scale to increase productivity and profits. (Modules 8.1, 8.2)

infectious disease An illness caused by an invading pathogen such as a bacterium or virus. (Module 4.3)

inferences Conclusions drawn based on observations. (Module 1.2)

infill development The development of empty lots within a city. (Module 4.2)

infiltration The process of water soaking into the ground. (Module 6.1)

information literacy The ability to find and evaluate the quality of information. (Module 1.3)

instrumental value An object's or species' worth, based on its usefulness to humans. (Module 3.2)

integrated pest management (IPM) The use of a variety of methods to control a pest population, with the goal of minimizing or eliminating the use of chemical toxins. (Module 8.2)

internal cost A cost—such as for raw materials, manufacturing costs, labor, taxes, utilities, insurance, or rent—that is accounted for when a product or service is evaluated for pricing. (Module 5.1)

intrinsic value An object's or species' worth, based on its mere existence; it has an inherent right to exist. (Module 3.2)

invasive species A non-native species (a species outside its range) whose introduction causes or is likely to cause economic or environmental harm or harm to human health. (Module 3.1)

IPAT model An equation ($I = P \times A \times T$) that measures human impact (I), based on three factors: population (P), affluence (A), and technology (T). (Module 5.1)

isotopes Atoms that have different numbers of neutrons in their nucleus but the same number of protons. (Module 11.1)

K

keystone species A species that impacts its community more than its mere abundance would predict, often altering ecosystem structure. (Modules 2.3, 3.3)

K-selected species Species that have a low biotic potential and that share characteristics such as long life span, late maturity, and low fecundity; generally show logistic population growth. (Module 2.2)

L

landscape conservation An ecosystem conservation strategy that specifically identifies a suite of species, chosen because they use all the vital areas within an ecosystem; meeting the needs of these species will keep the ecosystem fully functional, thus meeting the needs of all species that live there. (Module 3.3)

LD$_{50}$ (lethal dose 50%) The dose of a substance that would kill 50% of the test population. (Module 1.3)

leachate Water that carries dissolved substances (often contaminated) that can percolate through soil. (Module 5.3)

Leadership in Energy and Environmental Design (LEED) A certification program that awards a rating (standard, silver, gold, or platinum) to buildings that include environmentally sound design features. (Module 4.2)

life-history strategies Biological characteristics of a species (for example, life span, fecundity, maturity rate) that influence how quickly a population can potentially increase in number. (Module 2.2)

limiting factor The critical resource whose supply determines the population size of a given species in a given ecosystem. (Module 2.1)

linear economic system A production model that is one way: inputs are used to manufacture a product, and waste is discarded. (Module 5.1)

lithosphere The rigid outer layer of Earth made up of the crust and the hard uppermost layer of mantle. (Module 7.3)

LOAEL (lowest-observed-adverse-effect level) The lowest dose where an adverse effect was first seen. (Module 1.3)

logical fallacies Arguments that attempt to sway the reader without using reasonable evidence. (Module 1.3)

logistic growth The kind of growth in which population size increases rapidly at first but then slows down as the population becomes larger; produces an S-shaped curve when plotted over time. (Module 2.2)

low-level radioactive waste (LLRW) Material that has a low level of radiation for its volume. (Module 11.1)

M

malnutrition A state of poor health that results from inappropriate caloric intake (too many or too few calories) or is deficient in one or more nutrients. (Module 8.1)

marine calcifiers Organisms that make a hard calciumbased shell or exoskeleton. (Module 6.3)

marine protected areas (MPAs) Discrete regions of ocean that are legally protected from various forms of human exploitation. (Module 8.4)

marine reserves Restricted areas where all fishing is prohibited and absolutely no human disturbance is allowed. (Module 8.4)

matter cycles Movement of life's essential chemicals or nutrients through an ecosystem. (Module 2.1)

maximum sustainable yield (MSY) The amount that can be harvested without decreasing the yield in future years. (Modules 7.1, 8.4)

metal A malleable substance that can conduct electricity; usually found in nature as part of a mineral compound. (Module 7.3)

Milankovitch cycles Predictable variations in Earth's position in space relative to the Sun that affect climate. (Module 10.2)

mill tailings The finely ground rock left over from processing mineral ores. (Module 7.3)

mineral A naturally occurring chemical compound that exists as a solid with a predictable, three-dimensional, repeating structure. (Module 7.3)

minimum viable population The smallest number of individuals that would still allow a population to be able to persist or grow, ensuring long-term survival. (Module 2.2)

mining The extraction of natural resources from the ground. (Module 7.3)

monoculture A farming method in which a single variety of one crop is planted, typically in rows over huge swaths of land, with large inputs of fertilizer, pesticides, and water. (Module 8.2)

mountaintop removal (MTR) A surface mining technique that involves using explosives to blast away the top of a mountain to expose the coal seam underneath; the waste rock and rubble is deposited in a nearby valley. (Module 9.1)

Multiple-Use Sustained-Yield Act U.S. legislation (1960) mandating that national forests be managed in a way that balances a variety of uses. (Module 7.1)

municipal solid waste (MSW) Everyday garbage or trash (solid waste) produced by individuals or small businesses. (Module 5.3)

mutualism A symbiotic relationship between individuals of two species in which both parties benefit. (Modules 2.3, 6.3)

N

National Environmental Policy Act (NEPA) A 1969 U.S. law that established environmental protection as a guiding policy for the nation and required that the federal government take the environment into consideration before taking action that might affect it. (Module 5.2)

natural capital The wealth of resources on Earth. (Module 5.1)

natural gas A gaseous fossil fuel composed mainly of simpler hydrocarbons, mostly methane. (Module 9.2)

natural interest Readily produced resources that we could use and still leave enough natural capital behind to replace what we took. (Module 5.1)

natural selection The process by which organisms best adapted to the environment (the fittest) survive to reproduce, leaving more offspring than less well-adapted individuals. (Module 3.1)

negative feedback loop Changes caused by an initial event that trigger events that then reverse the response. (Module 10.2)

New Urbanism A movement that promotes the creation of compact, mixed-use communities with all the amenities of day-to-day living close by and accessible. (Module 4.2)

niche The role a species plays in its community, including how it gets its energy and nutrients, what habitat requirements it has, and with which other species and parts of the ecosystem it interacts. (Modules 2.1, 2.3)

niche generalist A species who occupies a broad niche because it can utilize a wide variety of resources. (Module 2.3)

niche specialist A species with very specific habitat or resource requirements that restrict where it can live. (Module 2.3)

nitrification Conversion of ammonia to nitrate (NO_3^-). (Module 2.1)

nitrogen cycle A continuous series of natural processes by which nitrogen passes from the air to the soil, to organisms, and then returns back to the air or soil. (Module 2.1)

nitrogen fixation Conversion of atmospheric nitrogen into a biologically usable form, carried out by bacteria found in soil or via lightning. (Module 2.1)

NOAEL (no-observed-adverse-effect level) The highest dose where no adverse effect is seen. (Module 1.3)

noncommunicable diseases (NCDs) Illnesses that are not transmissible between people; not infectious. (Module 4.3)

nondegradable Incapable of being broken down under normal conditions. (Module 5.3)

nonpoint source pollution Runoff that enters the water from overland flow. (Module 6.2)

nonrenewable resource A resource that is formed more slowly than it is used or that is present in a finite supply. (Modules 1.1, 9.2)

nuclear energy Energy in an atom; can be released when an atom is split (fission) or combines with another atom (fusion). (Module 11.1)

nuclear fission A nuclear reaction that occurs when a neutron strikes the nucleus of an atom and breaks it into two or more parts. (Module 11.1)

Nuclear Waste Policy Act (1982) The federal law that mandated that the federal government build and operate a long-term repository for the disposal of high-level radioactive waste. (Module 11.1)

O

observational study Research that gathers data in a real-world setting without intentionally manipulating any variable. (Module 1.2)

observations Information detected with the senses—or with equipment that extends our senses. (Module 1.2)

oil A liquid fossil fuel useful as a fuel or as a raw material for industrial products. (Module 9.2)

open dumps Places where trash, both hazardous and nonhazardous, is simply piled up. (Module 5.3)

ore mineral/ore A rock deposit that contains economically valuable amounts of metal-bearing minerals. (Module 7.3)

organic agriculture Farming that does not use synthetic fertilizer, pesticides, GMOs, or other chemical additives like hormones (for animal rearing). (Module 8.2)

overburden The rock and soil removed to uncover a mineral deposit during surface mining. (Module 9.1)

overexploited fisheries More fish are taken than is sustainable in the long run, leading to population declines. (Module 8.4)

overgrazing Too many herbivores feeding in an area, eating the plants faster than they can regrow. (Module 7.2)

overpopulated The number of individuals in an area exceeds the carrying capacity of that area. (Module 4.1)

ozone (O_3) A molecule made up of three oxygen atoms. (Module 5.2)

P

parasitism A symbiotic relationship between individuals of two species in which one benefits and the other is negatively affected. (Module 2.3)

particulate matter (PM) Particles or droplets small enough to remain aloft in the air for long periods of time. (Module 10.1)

passive solar technology A technology that captures solar energy (heat or light) without any electronic or mechanical assistance. (Module 11.2)

pathogen An infectious agent that causes illness or disease. (Module 4.3)

payback time The amount of time it takes to save enough money in operation costs to pay for equipment. (Module 11.2)

peer review A process whereby researchers' work is evaluated by outside experts to determine whether it is of a high enough quality to publish. (Modules 1.2, 1.3)

perennial crops Crops that do not die at the end of the growing season but live for several years, which means they can be harvested annually without replanting. (Module 8.2)

performance standards The levels of pollutants allowed to be present in the environment or released over a certain time period. (Modules 5.2, 6.2)

persistence A measure of how resistant a chemical is to degradation. (Module 1.3)

pesticide A natural or synthetic chemical that kills or repels plant or animal pests. (Module 8.2)

pesticide resistance The ability of a pest to withstand exposure to a given pesticide; the result of natural selection favoring the survivors of an original population that was exposed to the pesticide. (Module 8.2)

petrochemicals Distillation products from the processing of crude oil such as fuels or industrial raw materials. (Module 9.2)

phosphorus cycle A series of natural processes by which the nutrient phosphorus moves from rock to soil or water, to living organisms, and back to soil. (Module 2.1)

photosynthesis The chemical reaction performed by producers that uses the energy of the Sun to convert carbon dioxide and water into sugar and oxygen. (Module 2.1)

photovoltaic (PV) cell A technology that converts solar energy directly into electricity. (Module 11.2)

pioneer species Plant species that move into an area during early stages of succession; these are often r-species and may be annuals—species that live 1 year, leave behind seeds, and then die. (Module 2.3)

point source pollution Pollution from discharge pipes (or smokestacks) such as that from wastewater treatment plants or industrial sites. (Module 6.2)

policy A formalized plan that addresses a desired outcome or goal. (Module 1.2)

political lobbying Contacting elected officials in support of a particular position; some professional lobbyists are highly organized, with substantial financial backing. (Module 5.2)

polyculture A farming method in which a mix of different species are grown together in one area. (Module 8.2)

population All the individuals of a species that live in the same geographic area and are able to interact and interbreed. (Modules 2.1, 2.2)

population density The number of individuals per unit area. (Module 2.2)

population distribution The location and spacing of individuals within their range. (Module 2.2)

population dynamics Changes over time in population size and composition. (Module 2.2)

population growth rate The change in population size over time that takes into account the number of births and deaths as well as immigration and emigration numbers. (Modules 2.2, 4.1)

population momentum The tendency of a young population to continue to grow even after birth rates drop to replacement fertility (two children per couple). (Module 4.1)

positive feedback loop Changes caused by an initial event that then accentuate that original event. (Module 10.2)

potable Water that is clean enough for consumption. (Module 6.1)

potency The dose size required for a chemical to cause harm. (Module 1.3)

precautionary principle Acting in a way that leaves a safety margin when the data is uncertain or severe consequences are possible. (Modules 1.3, 5.2)

precipitation Rain, snow, sleet, or any other form of water falling from the atmosphere. (Module 6.1)

predation Species interaction in which one individual (the predator) feeds on another (the prey). (Module 2.3)

primary air pollutants Air pollutants released directly from both mobile sources (such as cars) and stationary sources (such as industrial and power plants). (Module 10.1)

primary source Information source that presents original data or first-hand information. (Module 1.3)

primary succession Ecological succession that occurs in an area where no ecosystem existed before (e.g., on bare rock with no soil). (Module 2.3)

producer An organism that converts solar energy to chemical energy via photosynthesis. (Modules 2.1, 2.3)

pronatalist pressure Factor that increases the desire to have children. (Module 4.1)

protected areas Geographic spaces on land or at sea that are recognized, dedicated, and managed to achieve long-term conservation of nature. (Module 3.3)

protocol A document that sets precise goals and targets. (Module 5.2)

proven reserves A measure of the amount of a fossil fuel that is economically feasible to extract from a known deposit using current technology. (Modules 9.1, 9.2)

proxy data Measurements that allow one to indirectly infer a value such as the temperature or atmospheric conditions in years past. (Module 10.2)

public health The science that deals with the health of human populations. (Module 4.3)

R

radioactive Atoms that spontaneously emit subatomic particles and/or energy. (Module 11.1)

radioactive decay The spontaneous loss of particle or gamma radiation from an unstable nucleus. (Module 11.1)

radioactive half-life The time it takes for half of the radioactive isotopes in a sample to decay to a new form. (Module 11.1)

random distribution A distribution in which individuals are spread out over the environment irregularly, with no discernible pattern. (Module 2.2)

range The geographic area where a species or one of its populations can be found. (Module 2.2)

rangeland Grassland used for grazing of livestock. (Module 7.2)

range of tolerance The range, within upper and lower limits, of a limiting factor that allows a species to survive and reproduce. (Module 2.1)

rare earth elements (REEs) A group of chemically similar elements used in a variety of modern products; they are not necessarily scarce but do not commonly occur in concentrated deposits. (Module 7.3)

recirculating aquaculture system (RAS) A method used to rear fish indoors in tanks that filter and recirculate the water. (Module 8.4)

reclamation The process of restoring a damaged natural area to a less damaged state. (Module 9.1)

recycle The fourth of the waste-reduction four Rs: Return items for reprocessing into new products. (Module 5.3)

reduce The second of the waste-reduction four **Rs**: Make choices that allow you to use less of a resource by, for instance, purchasing durable goods that will last or can be repaired. (Module 5.3)

refuse The first of the waste-reduction four Rs: Choose not to use or buy a product if you can do without it. (Module 5.3)

renewable energy Energy that comes from an infinitely available or easily replenished source. (Modules 1.1, 11.2)

replacement fertility The rate at which children must be born to replace those dying in the population. (Module 4.1)

reservoir An artificial lake formed when a river is impounded by a dam. (Module 6.1)

resilience The ability of an ecosystem to recover when it is damaged or perturbed. (Module 2.3)

resistance factors Things that directly (predators, disease) or indirectly (competitors) reduce population size. (Module 2.2)

Resource Conservation and Recovery Act (RCRA) The federal law that regulates the management of solid and hazardous waste. (Module 5.3)

resource partitioning A strategy in which different species use different parts or aspects of a resource rather than compete directly for exactly the same resource. (Module 2.3)

restoration ecology The science that deals with the repair of damaged or disturbed ecosystems. (Module 2.3)

reuse The third of the waste-reduction four Rs: Use a product more than once for its original purpose or for another purpose. (Module 5.3)

riparian areas The land areas close enough to a body of water to be affected by the water's presence (for example, areas where water-tolerant plants grow) and that affect the water itself (for example, provide shade). (Module 6.2)

rock Aggregate of one or more minerals that occurs in a variety of configurations. (Module 7.3)

rock cycle The process in which rock is constantly made and destroyed. (Module 7.3)

rotational grazing Moving animals from one pasture to the next in a predetermined sequence to prevent overgrazing. (Module 7.2)

r-selected species Species that have a high biotic potential and that share other characteristics, such as short life span, early maturity, and high fecundity. (Module 2.2)

S

Safe Drinking Water Act (SDWA) Federal law that protects public drinking water supplies in the United States. (Modules 1.3, 6.1)

saltwater intrusion The inflow of ocean (salt) water into a freshwater aquifer that happens when an aquifer has lost some of its freshwater stores. (Module 6.1)

sanitary landfills Disposal sites that seal in trash at the top and bottom to prevent its release into the atmosphere; the sites are lined on the bottom, and trash is dumped in and covered with soil daily. (Module 5.3)

science A body of knowledge (facts and explanations) about the natural world and the process used to get that knowledge. (Module 1.2)

scientific method The procedure scientists use to empirically test a hypothesis. (Module 1.2)

secondary air pollutants Air pollutants formed when primary air pollutants react with one another or with other chemicals in the air. (Module 10.1)

secondary source Information source that presents and interprets information solely from primary sources. (Module 1.3)

secondary succession Ecological succession that occurs in an ecosystem that has been disturbed; occurs more quickly than primary succession because soil is present. (Module 2.3)

selective harvesting Timber-harvesting technique that cuts only the highest-value trees; the remaining trees reseed the plot. (Module 7.1)

selective pressure A nonrandom influence that affects who survives or reproduces. (Module 3.1)

service economy A business model whose focus is on leasing and caring for a product in the customer's possession rather than on selling the product itself (that is, selling the service that the product provides). (Module 5.1)

shelterbelts A stand of trees that blocks the wind and thus decreases soil erosion. (Module 7.2)

shelterwood harvesting Timber-harvesting technique that cuts all but the best trees, which reseed the plot and are then harvested. (Module 7.1)

single-species conservation A management strategy that focuses on protecting one particular species. (Module 3.3)

sinks Abiotic or biotic components of the environment that serve as storage places for cycling nutrients. (Module 2.1)

sliding reinforcer Actions that are beneficial at first but that change conditions such that their benefit declines over time. (Module 1.1)

smart grid A modernized network that provides electricity to users in a way that automatically optimizes the delivery of electricity. (Module 11.2)

smart growth Strategies that help create walkable communities with lower environmental impacts. (Module 4.2)

smog Hazy air pollution that contains a variety of pollutants, including sulfur dioxide, nitrogen oxides, tropospheric ozone, and particulates. (Module 10.1)

social traps Decisions by individuals or groups that seem good at the time and produce a short-term benefit but that hurt society in the long run. (Module 1.1)

soil erosion The removal of soil by wind and water that exceeds the soil's natural replacement. (Module 7.2)

solar power Energy harnessed from the Sun in the form of heat or light. (Module 11.2)

solubility The ability of a substance to dissolve in a water or fat-based liquid or gas. (Module 1.3)

species A group of plants or animals that have a high degree of similarity and can generally only interbreed among themselves. (Module 2.1)

species diversity The variety of species, including how many are present (richness) and their abundance relative to each other (evenness). (Modules 2.3, 3.2)

species evenness The relative abundance of each species in a community. (Module 2.3)

species richness The total number of different species in a community. (Module 2.3)

statistics The mathematical evaluation of experimental data to determine how likely it is that any difference observed is due to the variable being tested. (Module 1.2)

stormwater runoff Water from precipitation that flows over the surface of the land. (Modules 6.2, 7.1)

stratosphere A layer of atmosphere that lies directly above the troposphere. (Module 5.2)

strip harvesting Timber-harvesting technique that clear-cuts a small section of a forest, allowing regrowth in that section before moving on to another. (Module 7.1)

strip mining A surface mining method that accesses coal from deposits close to the surface on level ground, one section at a time. (Module 9.1)

subduct The movement of one tectonic plate below another at a convergent plate boundary. (Module 7.3)

subsidies Financial assistance given by the government to promote desired activities. (Modules 5.2, 10.1)

subsurface mines Sites where tunnels are dug underground to access mineral resources. (Module 9.1)

suburban sprawl Low-population-density developments that are built outside of a city. (Module 4.2)

surface mining A form of mining that involves removing soil and rock that overlays a mineral deposit close to the surface in order to access that deposit. (Module 9.1)

surface water Any body of water found above ground, such as oceans, rivers, and lakes. (Module 6.1)

sustainable Capable of being continued indefinitely. (Modules 1.1, 5.1)

sustainable agriculture Farming methods that can be used indefinitely because they do not deplete resources, such as soil and water, faster than they are replaced. (Module 8.2)

sustainable development Development that meets present needs without compromising the ability of future generations to do the same. (Module 1.1)

sustainable energy Energy from sources that are renewable and have a low environmental impact. (Module 11.2)

sustainable fishery A fishery that ensures that fish stocks are maintained at healthy levels, the ecosystem is fully functional, and fishing activity does not threaten biological diversity. (Module 8.4)

sustainable grazing Practices that allow animals to graze in a way that keeps pastures healthy and allows grasses to recover. (Module 7.2)

symbiosis A close biological or ecological relationship between two species. (Module 2.3)

synergistic effects Exposure to two or more chemicals that has a greater effect than the sum of their individual effects would predict. (Module 1.3)

T

take-back law A law that requires companies to take a product back from a consumer when the consumer is finished with it. (Module 5.3)

tar sands (oil sands) Sand or clay formations that contain a heavy-density crude oil (crude bitumen); extracted by surface mining. (Module 9.2)

tax credit A reduction in the tax one must pay in exchange for some desirable action. (Modules 5.2, 10.1)

tectonic plates Rigid pieces of Earth's lithosphere that move above the asthenosphere. (Module 7.3)

temperate forest Forest found in areas with four seasons and a moderate climate, which receives 30 to 60 inches of precipitation per year and which may include evergreen and deciduous conifers and broadleaf trees. (Module 7.1)

tertiary source Information source that uses information from at least one secondary source. (Module 1.3)

test group The group in an experimental study that is manipulated such that it differs from the control group in only one way. (Module 1.2)

theory A widely accepted explanation of a natural phenomenon that has been extensively and rigorously tested scientifically. (Module 1.2)

threatened species Species that are at risk for extinction; various threat levels have been identified, ranging from "least concern" to extinct." (Module 3.3)

tight oil Light (low-density) oil in shale rock deposits of very low permeability; extracted by fracking. (Module 9.2)

time delay Actions that produce a benefit today but set into motion events that cause problems later on. (Module 1.1)

top-down regulation The control of population size by factors that reduce population size (resistance factors) such as predation, competition, or disease. (Module 2.2)

total fertility rate (TFR) The number of children the average woman has in her lifetime. (Module 4.1)

toxic substance/toxic A substance that causes damage when it contacts, or enters, the body. (Module 1.3)

Toxic Substances Control Act (TSCA) The primary federal law governing chemical safety. (Module 1.3)

trade-offs The imperfect and sometimes problematic responses that we must at times choose between when addressing complex problems. (Module 1.1)

tragedy of the commons The tendency of an individual to abuse commonly held resources in order to maximize his or her own personal interest. (Modules 1.1, 8.4)

transboundary pollution Pollution that is produced in one area but falls in a different state or nation. (Module 10.1)

transboundary problem A problem that extends across state and national boundaries; pollution that is produced in one area but falls in or reaches other states or nations. (Module 5.2)

transform plate boundary A place where two tectonic plates slide side to side relative to one another. (Module 7.3)

transgenic organism An organism that contains genes from another species. (Module 8.1)

transpiration The loss of water vapor from plants. (Module 6.1)

triple bottom line The combination of the environmental, social, and economic impacts of our choices. (Modules 1.1, 5.1)

trophic levels Feeding levels in a food chain. (Module 2.3)

tropical forest Forest found in equatorial areas with warm temperatures year-round and high rainfall; some have distinct wet and dry seasons, but none has a winter season. (Module 7.1)

troposphere The lowest level of the atmosphere. (Module 5.2)

true cost The sum of both external and internal costs of a good or service. (Module 5.1)

U

ultraviolet (UV) radiation High-energy radiation that is harmful to living things. (Module 5.2)

unconventional reserves Deposits of oil or natural gas that cannot be recovered with traditional oil/gas wells but may be recoverable using alternative techniques. (Module 9.2)

undergrazing Grazing too few animals on a grassland to maintain its ecological integrity. (Module 7.2)

understory The smaller trees, shrubs, and saplings that live in the shade of the forest canopy. (Module 7.1)

uniform distribution A distribution in which individuals are spaced evenly, perhaps due to territorial behavior or mechanisms for suppressing the growth of nearby individuals. (Module 2.2)

urban areas Densely populated regions that include cities and the suburbs that surround them. (Module 4.2)

urban flight The process of people leaving an inner-city area to live in surrounding areas. (Module 4.2)

urban heat island effect The phenomenon in which urban areas are warmer than the surrounding countryside due to pavement, dark surfaces, closed-in spaces, and high energy use. (Module 4.2)

urbanization The migration of people to large cities; sometimes also defined as the growth of urban areas. (Module 4.2)

U.S. Farm Bill Legislation that deals with many aspects of the production and sale of farm-raised commodity crops. (Module 8.3)

V

vector-borne disease An infectious disease acquired from organisms that transmit a pathogen from one host to another. (Module 4.3)

volatile organic compound (VOC) A chemical that readily evaporates and is released into the air as a gas; may be hazardous. (Module 10.1)

W

waste Any material that humans discard as unwanted. (Module 5.3)

wastewater Used and contaminated water that is released after use by households, industry, or agriculture. (Module 6.1)

wastewater treatment The process of removing contaminants from wastewater to make it safe enough to release into the environment. (Module 6.1)

water cycle The movement of water through various water compartments such as surface waters, atmosphere, soil, and living organisms. (Module 6.1)

water footprint The water appropriated by industry to produce products or energy; this includes the water actually used and water that is polluted in the production process. (Module 6.1) Also the amount of water consumed by a given group (that is, person or population) or for a process (such as raising livestock). (Module 8.3)

water pollution The addition of any substance to a body of water that might degrade its quality. (Module 6.2)

water scarcity Not having access to enough clean water. (Module 6.1)

water table The uppermost water level of the saturated zone of an aquifer. (Module 6.1)

water wars Political conflicts over the allocation of water sources. (Module 6.1)

waterborne disease An infectious disease acquired through contact with contaminated water. (Module 4.3)

watershed The land area surrounding a body of water over which water such as rain can flow and potentially enter that body of water. (Module 6.2)

watershed management Management of what goes on in an area around streams and rivers. (Module 6.2)

weather The meteorological conditions in a given place on a given day. (Module 10.2)

weathering The breakdown of rock by physical or chemical forces. (Module 7.3)

wind power Energy contained in the motion of air across Earth's surface. (Module 11.2)

worldview The window through which one views one's world and existence. (Module 1.1)

Z

zero-population growth The absence of population growth; occurs when birth rates equal death rates. (Module 4.1)

zoonotic disease An infectious disease of animals that that can be transmitted to humans. (Module 4.3)

zooxanthellae Mutualistic photosynthetic dinoflagellate partner of a coral polyp; each provides nutrients that the other needs. (Module 6.3)

CREDITS/SOURCES

INFOGRAPHIC DATA SOURCES AND REFERENCES

MODULE 1.1:
IG 4 United Nations Division of Sustainable Development. (n.d.) *Sustainable Development Goals*, www.un.org/sustainabledevelopment/sustainable-development-goals/.

MODULE 1.2:
IG 4 Reeder, D. A. M., et al. (2012). Frequent arousal from hibernation linked to severity of infection and mortality in bats with white-nose syndrome. *PLoS ONE*, 7(6): e38920. doi:10.1371/journal.pone.0038920.
Warnecke, L., et al. (2012). Inoculation of bats with European *Geomyces destructans* supports the novel pathogen hypothesis for the origin of white-nose syndrome. *Proceedings of the National Academy of Sciences, 109*(18): 6999–7003.
IG 5 Graph A: Frick, W. F., et al. (2010, August 6). An emerging disease causes regional population collapse of a common North American bat species. *Science, 329*(5992), 679–682, Figure 3C.
Graph B: Warnecke, L., et al. (2012). Inoculation of bats with European *Geomyces destructans* supports the novel pathogen hypothesis for the origin of white-nose syndrome. *Proceedings of the National Academy of Sciences, 109*(18), 6999–7003.

MODULE 1.3:
IG 2 Atwell, L., et al. (1998). Biomagnification and bioaccumulation of mercury in an arctic marine food web: Insights from stable nitrogen isotope analysis. *Canadian Journal of Fisheries and Aquatic Sciences, 55*(5), 1114–1121.
IG 3 Age and lead: Hanna-Attisha, M., et al. (2016). Elevated blood lead levels in children associated with the Flint drinking water crisis: a spatial analysis of risk and public health response. *American Journal of Public Health, 106*(2), 283–290;

Exposure route: Holstege, C. P., et al. (2013). Pathophysiology and etiology of lead toxicity. Retrieved from Medscape website: http://emedicine.medscape.com/article/2060369-overview; **Additive effects:** Antonio, M. T., et al. (2003). Study of the activity of several brain enzymes like markers of the neurotoxicity induced by perinatal exposure to lead and/or cadmium. *Toxicology Letters, 143*(3), 331–340; **Antagonistic effects:** Mejia, J. J., et al. (1997). Effects of lead–arsenic combined exposure on central monoaminergic systems. *Neurotoxicology and Teratology, 19*(6), 489–497; **Synergistic effects:** Haley, B. E. (2005). Mercury toxicity: genetic susceptibility and synergistic effects. *Medical Veritas, 2*(2), 535–542.

MODULE 2.1:
IG 3 Map: The Nature Education Knowledge Project; **Graph:** Ricklefs, R. (2000). *The Economy of Nature*, New York, NY: W.H. Freeman.

MODULE 2.3:
IG 8 National Park Service. (2015). Comprehensive Everglades Restoration Plan, www.nps.gov/ever/learn/nature/cerp.htm.

MODULE 3.1:
IG 2 Hoekstra, H. E., et al. (2005). Local adaptation in the rock pocket mouse (*Chaetodipus intermedius*): Natural selection and phylogenetic history of populations. *Heredity, 94*(2), 217–228.
IG 6B Ceballos, G., et al. (2015). Accelerated modern human-induced species losses: Entering the sixth mass extinction. *Science Advances, 1*(5), e1400253.

MODULE 3.2:
IG 1 Chapman, A. D. (2009, September). *Number of Living Species in Australia and the World.* Report for the Australian Biological Resources Study, Canberra, Australia.
IG 4 Critical Ecosystem Partnership Fund. (n.d.). *Biodiversity Hotspot Maps*, Retrieved

January 19, 2017, from www.cepf.net/resources/maps/Pages/default.aspx.

ONLINE MODULE 3.3:
IG 1 Maxwell, S. L., et al. (2016). Biodiversity: The ravages of guns, nets and bulldozers. *Nature, 536*(7615), 143–145.
IG 2 The International Union for Conservation of Nature (IUCN). (2016, September 4). *IUCN Red List* version 2016-2: Tables 3a, 3b, and 3c.
IG 7 UNEP-WCMC and IUCN. (2016). *Protected Planet Report 2016.* Cambridge, UK and Gland Switzerland: UNEP-WCMC and IUCN.

MODULE 4.1:
IG 1 United Nations, Department of Economic and Social Affairs, Population Division. (2015). *World Population Prospects: The 2015 Revision*, Volume I: Comprehensive Tables (ST/ESA/SER.A/379).
IG 2 United Nations, Department of Economic and Social Affairs, Population Division. (2015). *World Population Prospects: The 2015 Revision,* Volume I: Comprehensive Tables (ST/ESA/SER.A/379).
IG 3 Population Growth Graph: United Nations, Department of Economic and Social Affairs, Population Division. (2015). *World Population Prospects: The 2015 Revision,* Volume I: Comprehensive Tables (ST/ESA/SER.A/379); **Fertility graph:** Sedgh, G., et al. (2007). Women with an unmet need for contraception in developing countries and their reasons for not using a method *Occasional Report, 37,* 5–79.
IG 4 Map: United Nations, Department of Economic and Social Affairs, Population Division. (2015). *World Population Prospects, 2015 Revision: Median Age of Population;* **Age structure diagrams:** United Nations, Department of Economic and Social Affairs, Population Division. (2015). *World Population Prospects: The 2015 Revision.*
IG 6 Total fertility data: Central Intelligence Agency. (2013). *The World Factbook*, www.cia.gov/library/

publications/the-world-factbook/
rankorder/2127rank.html; **Childhood
mortality data:** The World Bank. (2017).
Mortality Rate, Under 5 (per 1,000 Births),
https://data.worldbank.org/indicator/
SH.DYN.MORT); **Education data:** UN
Educational, Scientific and Cultural
Organization Institute for Statistics, Mean
Years of Schooling, Populations 25+ Years,
Female, http://data.uis.unesco.org/.
IG 7 Global Footprint Network. (2017).
National Footprint Accounts, http://data
.footprintnetwork.org.

MODULE 4.2:
IG 1 United Nations, Department of
Economic and Social Affairs, Population
Division (2014). *World Urbanization
Prospects: The 2014 Revision, Highlights
(ST/ESA/SER.A/352)* and United Nations,
Department of Economic and Social
Affairs, Population Division (2015). *World
Population Prospects: The 2015 Revision*
(POP/DB/WPP/Rev.2015/POP/F01-1).
IG 2A Dodman, D. (2009). Blaming cities
for climate change? An analysis of urban
greenhouse gas emissions inventories.
Environment and Urbanization, 21(1),
185–201.
IG 3 U.S. Environmental Protection
Agency. (n.d.) *Resources for Creating
Healthy, Sustainable, and Equitable
Communities.* Retrieved December
30, 2016, from www.epa.gov/
environmentaljustice/resources-creating-
healthy-sustainable-and-equitable-
communities.
IG 7 U.S. Environmental Protection
Agency. (n.d.) *About Smart Growth.*
Retrieved October 18, 2016 from
www.epa.gov/smartgrowth/
about-smart-growth#smartgrowth.

MODULE 4.3:
IG 2 World Health Organization. (2016)
*Preventing Disease through Healthy
Environments: A Global Assessment of the
Burden of Disease from Environmental Risks;*
Table A 2.3.; World Health Organization.
(2014). *A Global Brief on Vector-Borne
Diseases.* WHO Document number:
WHO/DCO/WHD/2014.1.]; CDC: *Global
WASH Fast Facts.* Retrieved October 8,
2016 from http://www.cdc.gov/
healthywater/global/wash_statistics.html.

IG 4 Centers for Disease Control
and Prevention. DPDx – Laboratory
Identification of Parasitic Disease of
Public Health Concern: Dracunculiasis,
www.dpd.cdc.gov/dpdx/HTML/
Dracunculiasis/index.html and
CNN. (2010, April 5). *How the guinea
worm spreads*, www.cnn.com/2010/
HEALTH/04/05/guinea.worm.lifecycle/
index.html.
IG 5 World Health Organization. (2016).
*Preventing Disease through Healthy
Environments: A Global Assessment of
the Burden of Disease from Environmental
Risks* (page xviii). Retrieved October 5,
2016 from www.who.int/mediacentre/
factsheets/fs310/en/index1.html.
IG 6 World Health Organization. (2016).
*Preventing Disease through Healthy
Environments: A Global Assessment of
the Burden of Disease from Environmental
Risks* (Table A 2.3). Retrieved October 5,
2016 from www.who.int/mediacentre/
factsheets/fs310/en/index1.html.
IG 8 World Health Organization. (2016).
*Preventing Disease through Healthy
Environments: A Global Assessment of the
Burden of Disease from Environmental Risks.*
Retrieved October 5, 2016 from www.who
.int/mediacentre/factsheets/fs310/en/
index1.html.

MODULE 5.1:
IG 1 Costanza, R., et al. (2014). Changes
in the global value of ecosystem services.
Global Environmental Change, 26, 152–158.

MODULE 5.2:
IG 8 Bournay, E. (2007). The effects of
the Montreal Protocol amendments and
their phase out schedules. *Vital Ozone
Graphics. GRID-Arendal*, www.grida.no/
publications/252.

MODULE 5.3:
IG 2 U.S. Environmental Protection
Agency. (2016). *Advancing Sustainable
Materials Management:* 2014 Fact Sheet
(EPA530-R17-01).
IG 3 U.S. Environmental Protection
Agency. (n.d.) *Sustainable Materials
Management: Non-Hazardous Materials and
Waste Management Hierarchy.* Retrieved
December 5, 2016 from www.epa.gov/
smm/sustainable-materials-management-

non-hazardous-materials-and-waste-
management-hierarchy.

MODULE 6.1:
IG 4 Food and Agriculture Organization
of the United Nations (FAO). (2016).
AQUASTAT Main Database. Retrieved
November 29, 2016. Rome, Italy: FAO,
www.fao.org/nr/water/aquastat/data/
query/index.html?lang=en.
IG 8 Pie chart: Inskeep, B. D., & S. Z. Attari.
(2014). *Environment: Science and Policy for
Sustainable Development, 56*(4), 4–15; **Water
use for electricity production:** Jones, W. D.
(2008). *How Much Water Does It Take to
Make Electricity?* Institute of Electrical and
Electronic Engineers, spectrum.ieee.org;
Gallons of water per food or product:
Water Footprint Network. *Product Gallery*,
http://waterfootprint.org/en/resources/
interactive-tools/product-gallery/ and
National Geographic. *The Hidden Water We
Use*, http://environment.nationalgeographic
.com/environment/freshwater/embedded-
water/. Data retrieved from websites on
December 1, 2016.

MODULE 6.2:
IG 1 U.S. Environmental Protection
Agency. *Causes of Impairment for 303(d)
Listed Waters.* Retrieved April 17, 2017
from https://iaspub.epa.gov/waters10/
attains_nation_cy.control?p_report_
type=T#causes_303d.
IG 6 Welch, D. J. (1991) *Riparian Forest
Buffers. Function and Design for Protection
and Enhancement of Water Resources.* U.S.
Department of Agriculture Publication
NA-PR-07-91.
IG 7 Oquist, K. A., et al. (2007). *Journal of
Environmental Quality, 36*(4), 1194–1204.
IG 8 Gulf Coast Ecosystem Restoration
Task Force. (2011). *Gulf of Mexico Regional
Ecosystem Restoration Strategy*, https://
archive.epa.gov/gulfcoasttaskforce/web/
pdf/gulfcoastreport_full_12-04_508-1.pdf.

ONLINE MODULE 6.3
IG1 Graph: Doney, S. C., et al. (2009).
Ocean Acidification: The Other CO_2
Problem. *Annual Review of Marine Science,
1*, 169–192.; **Maps:** Cao, L., & Caldeira,
K. (2010). Can ocean iron fertilization
mitigate ocean acidification? *Climatic
Change, 99*(1), 303–311.

IG 4 Map: U.S. National Oceanic and Atmosphere Administration. *NOAA Ocean Service Education*. Retrieved August 22, 2011 from http://oceanservice.noaa.gov/education/kits/corals/media/supp_coral05a.html; **Graphs:** United Nations Environmental Programme. *Coral Reefs at Risk*. Retrieved August 22, 2011 from www.grida.no/resources/5616.

MODULE 7.1:
IG 3 Costanza, R., et al. (2014). Changes in the global value of ecosystem services. *Global Environmental Change, 26,* 152–158.
IG 4 Map: Food and Agriculture Organization of the United Nations. (2015). *Global Forest Resources Assessment 2015: How Have the World's Forests Changed?* Rome, Italy: FAO.

ONLINE MODULE 7.2:
IG 1 Dixon, A.P., et al. (2014). Distribution mapping of world grassland types. *Journal of Biogeography.* doi: 10.1111/jbi.12381.
IG 3 Map: Natural Resources Conservation Service of the U.S. Department of Agriculture. (2003). *Risk of Human-Induced Desertification Map,* www.nrcs.usda.gov/wps/portal/nrcs/detail/national/nedc/training/soil/?cid=nrcs142p2_054004.

ONLINE MODULE 7.3:
IG 2 U.S. Forest Service. (n.d.) *Geology. Where the Earth's Crust Collides, Serpentine is Born.* Accessed December 9, 2011 from www.fs.fed.us/wildflowers/beauty/serpentines/geology.shtml.

MODULE 8.1:
IG 1 United Nations FAO, International Fund for Agricultural Development (IFAD), and World Food Programme (WFP). (2015). *The State of Food Insecurity in the World 2015. Meeting the 2015 International Hunger Targets: Taking Stock of Uneven Progress.* Rome, Italy: FAO.
IG 3 FAOSTAT. Data retrieved January 24, 2017 from www.fao.org/faostat/en/#data/QC.

MODULE 8.2:
IG 4 Furuno, T. (2001). *The Power of the Duck.* Tasmania: Takari Publications.

IG 7 Environmental Working Group. *EWG's 2017 Shopper's Guide to Pesticides in Produce.* Retrieved August 30, 2017 from www.ewg.org/foodnews/summary.php#.WacSb9GQzcs

ONLINE MODULE 8.3:
IG 3 Feed data: Shepon, A., et al. (2016). *Energy and protein feed-to-food conversion efficiencies in the US and potential food security gains from dietary changes. Environmental Research Letters,* 11(10), 105002; **Water footprint data:** Mekonnen, M. M. and Hoekstra, A. Y. (2010). *The Green, Blue and Grey Water Footprint of Farm Animals and Animal Products.* Value of Water Research Report Series No. 48, Delft, The Netherlands: UNESCO-IHE.

ONLINE MODULE 8.4:
IG 2 Swan, J., & Gréboval, D. F., eds. (2004). *Report of the International Workshop on the Implementation of International Fisheries Instruments and Factors of Unsustainability and Overexploitation in Fisheries: Mauritius, 3–7 February 2003* (No. 700). Food & Agriculture Organization and Northwest Atlantic Fisheries Organization. (2016, June 2). STATLANT21A. *Atlantic Cod, Canada Newfoundland, All Divisions, 1960–2015.* www.nafo.int/Data/STATLANT.
IG 3 Graph: Kleisner, K. et al. (2015). The MTI and RMTI as tools for unmasking the fishing down phenomenon. *Sea Around Us.* Vancouver, BC: University of British Columbia, www.seaaroundus.org.
IG 4 Pie chart: FAO Fisheries and Aquaculture Department. (2016). *The State of World Fisheries and Aquaculture. Contributing to Food Security and Nutrition for All.* Rome, Italy: FAO; **Area graph:** Pauly, D., & D. Zeller, eds. (2015). Catch reconstruction: Concepts, methods and data sources. *Sea Around Us.* Vancouver, BC: University of British Columbia, www.seaaroundus.org.

MODULE 9.1:
IG 2 Pie chart: U.S. Energy Information Agency (EIA). (2016). *Electric Power Annual.* Washington DC: EIA; **Bar graph:** Hall, C. A., et al. (2014). EROI of different fuels and the implications for society. *Energy Policy, 64,* 141–152.

IG 7 World Coal Association. *Carbon capture, use and storage.* Retrieved October 6, 2011 from www.worldcoal.org/carbon-capture-storage/ccs-technologies/.

MODULE 9.2:
IG 2 U.S. Energy Information Administration (EIA). (2017). *Reserves and Capacity,* www.eia.gov/beta/international/.
IG 6 FracFocus Chemical Disclosure Registry. (n.d.) *Hydraulic Fracturing. How it Works.* Retrieved April 28, 2011 from www.fracfocus.ca.

MODULE 10.1:
IG 1 Institute for Health Metrics and Evaluation. (2017). *Global Burden of Air Pollution: 2013 Infographic,* www.healthdata.org.
IG 4 National Atmospheric Deposition Program (NRSP-3). (2017). *NADP Program Office. Illinois State Water Survey.* Champaign, IL: University of Illinois.
IG 6 Smokestack scrubber diagram: David, I. (n.d.) How do Smokestack Scrubbers Work? *Encyclopedia Britannica* (www.ehow.com/how-does_5113845_do-smokestack-scrubbers-work.html); **Cap-and-trade diagram:** Clark, P. (February 26, 2009) *The Washington Post.* (www.washingtonpost.com/wp-dyn/content/graphic/2009/02/26/GR2009022600572.html).

MODULE 10.2:
IG 1 Hansen, J., et al. (2016, January 19). *Global Temperature in 2015.* Climate Science, Awareness and Solutions; Earth Institute, Columbia University, http://csas.ei.columbia.edu/2016/01/19/global-temperature-in-2015/.
IG 2 Temperature graph: National Oceanic and Atmospheric Administration, National Centers for Environmental Information. (2017, March). *Climate at a Glance: Global Time Series,* www.ncdc.noaa.gov/cag/; **Glacier melt, precipitation, and frost graphs:** U.S. Environmental Protection Agency. (2016). *Climate Change Indicators in the United States, 2016.* Fourth edition. EPA 430-R-16-004, www.epa.gov/climate-indicators; **Sea level graph:** NASA Goddard Space Flight Center. *Facts: Sea Level,* https://climate.nasa

.gov/vital-signs/sea-level/; **Range shift data:** Hitch, A. T., & Leberg, P. L. (2007). Breeding distributions of North American bird species moving north as a result of climate change. *Conservation Biology, 21*(2), 534–539; **Phenology graph:** CaraDonna, P. J., et al. (2014). Shifts in flowering phenology reshape a subalpine plant community. *Proceedings of the National Academy of Sciences, 111*(13), 4916–4921.
IG 3 Solar radiation (percentages) and global warming potential data: Intergovernmental Panel on Climate Change (IPCC). (2013). Fifth Assessment Report, Working Group Report: *Climate Change 2013: The Physical Science Basis;* **Greenhouse gas percentages:** IPCC. (2014). Fifth Assessment Report, Working Group III to the Fifth Assessment. *Climate Change 2013: Mitigation of Climate Change;* CO₂ data: National Oceanic and Atmospheric Administration. (2017, July). Global Greenhouse Gas Reference Network.
IG 4 Petit, J. R., et al. (1999). Climate and atmospheric history of the past 420,000 years from the Vostok ice core, Antarctica. *Nature, 399*(6735), 429–436.
IG 6 IPCC. (2013). 5th Assessment Report, Working Group 1 Report: *The Physical Science Basis.*
IG 8B Climate Action Tracker. *Effect of current pledges and policies on global temperature.* Retrieved April 13, 2016 from www.climateactiontracker.org/global.html .
IG 9 Carbon Mitigation Initiative. (2015). *Stabilization Wedges: A Concept and a Game.* Princeton, NJ: Princeton Environmental Institute, Princeton University, http://cmi.princeton.edu.

MODULE 11.1:
IG 2 World Nuclear Association. *The Nuclear Fuel Cycle.* Retrieved March 29, 2014 from www.world-nuclear.org/information-library/nuclear-fuel-cycle/introduction/nuclear-fuel-cycle-overview.aspx.
IG 3 Fission reaction diagram: *Nuclear Fission: Basics.* Retrieved April 20, 2011 from www.atomicarchive.com/Fission/Fission1.shtml; Reactor diagrams: *Nuclear Reactors.* Nuclear Regulatory Commission. Retrieved April 20, 2011 from www.nrc.gov/reactors.html.

IG 4 U.S. Office of Environmental Management. *Radiation Basics.* Retrieved April 29, 2011 from http://gtcceis.anl.gov/guide/rad/index.cfm.

MODULE 11.2:
IG 2 Pie chart: U.S. Energy Information Agency. (2017). *Monthly Energy Review, April 2017.* DOW/EIA-0035(2017/4); **Line graph:** U.S. Energy Information Agency. (2017). *Annual Energy Outlook 2017 with projections to 2050.* #AEO2017.
IG 4 Passive solar home diagram: U.S. Department of Energy. *Passive Solar Home Design.* Retrieved August 3, 2014 from https://energy.gov/energysaver/passive-solar-home-design and www.wavege.com/solar-panel-diagram.html); **PV cell diagram:** Union of Concerned Scientists. *How Solar Panels Work.* Retrieved April 30, 2017 from www.ucsusa.org/clean-energy/renewable-energy/how-solar-panels-work#.WadUsNGQw2w.
IG 5 Geothermal power plant: U.S. Energy Information Administration. *Geothermal Explained: Geothermal Power Plant.* Retrieved September 11, 2011 from www.eia.gov/Energyexplained/index.cfm?page=geothermal_power_plants. **Ground source heat pumps:** Environmental Protection Agency. *Geothermal Heating and Cooling Technologies.* Retrieved July 11, 2011 from www.epa.gov/rhc/geothermal-heating-and-cooling-technologies.
IG 6 Tennessee Valley Authority. *Hydroelectric Power.* Retrieved October 22, 2011 from http://152.87.4.98/power/hydro.htm.

ONLINE MODULE 11.3
IG 4 Tilman, D., et al. (2006). Carbon-negative biofuels from low-input high-diversity grassland biomass. *Science, 314*(5805), 1598–1600.
IG 5 Alternate Energy Sources. *Biomass energy – Part of a Sustainable Future?* Retrieved September 25, 2011 from www.alternate-energy-sources.com/biomass-energy.html.
IG 6 U.S. Energy Information Administration. (2010). One means of biodiesel production: the FAME process. *This Week in Petroleum.* (www.eia.gov/petroleum/weekly/archive/2010/100421/twipprint.html).

SCIENCE LITERACY: WORKING WITH DATA

MODULE 1.1:
Brown, L.R. (2011). *World on the Edge—Food and Agriculture Data—Livestock and Fish.* Washington, DC: Earth Policy Institute, www.earth-policy.org/datacenter/xls/book_wote_ch3_7.xls.

MODULE 1.2:
Lorch, J. M., et al. (2011). Experimental infection of bats with *Geomyces destructans* causes white-nose syndrome. *Nature, 480*(7377), 376–378.

MODULE 1.3:
Stacchiotti, A., et al. (2009). Stress proteins and oxidative damage in a renal derived cell line exposed to inorganic mercury and lead. *Toxicology, 264*(3), 215–224.

MODULE 2.1:
Adams, H. D., et al. (2009). Temperature sensitivity of drought-induced tree mortality portends increased regional die-off under global-change-type drought. *Proceedings of the National Academy of Sciences, 106*(17), 7063–7066.

MODULE 2.2:
Peterson, R. O., et al. (2014). Trophic cascades in a multicausal world: Isle Royale and Yellowstone. *Annual Review of Ecology, Evolution, and Systematics, 45,* 325–345.

MODULE 2.3:
Rogers, J. D. (2008). Development of the New Orleans flood protection system prior to Hurricane Katrina. *Journal of Geotechnical and Geoenvironmental Engineering, 134*(5), 6 (Figure 7) (adapted from Kolb and Saucier, 1982).

MODULE 3.1:
Center for Biological Diversity. Human Population Growth and Extinction. Originally from Scott, J. M. (2008). *Threats to Biological Diversity: Global, Continental, Local.* Moscow, ID: U.S. Geological Survey, Idaho Cooperative Fish and Wildlife, Research Unit, University of Idaho.

MODULE 3.2:
Gibson, L., et al. (2013). Near-complete extinction of native small mammal fauna 25 years after forest fragmentation. *Science, 341*(6153), 1508–1510.

ONLINE MODULE 3.3:
IUCN. *IUCN Red List of Threatened Species: Summary Statistics* (Figure 3). Retrieved September 4, 2012 from www.iucnredlist .org/about/summary-statistics.

MODULE 4.1:
United Nations, Department of Economic and Social Affairs, Population Division. (2015). *World Population Prospects: The 2015 Revision*, Volume I: Comprehensive Tables (ST/ESA/SER.A/379).

MODULE 4.2:
New Buildings Institute. (2008). *Energy Performance of LEED® for New Construction Buildings: Final Report*, http://newbuildings .org/sites/default/files/Energy_ Performance_of_LEED-NC_Buildings-Final_3-4-08b.pdf; EPA, www.energystar .gov/index.cfm?fuseaction= buildingcontest.eui.

MODULE 4.3:
Fan, V. (2012). *Malaria Estimate Sausages by WHO and IHME.* Center for Global Development, http://blogs.cgdev.org/ globalhealth/2012/02/malaria-estimate-sausages-by-who-and-ihme.php.

MODULE 5.1:
Costanza, R., et al. (2014). Changes in the global value of ecosystem services. *Global Environmental Change, 26*, 152–158.

MODULE 5.2:
Solomon, S., et al. (2016). Emergence of healing in the Antarctic ozone layer. *Science, 353*(6296), 269–274.

MODULE 5.3:
Cole, M. J., et al. (2015). The impact of polystyrene microplastics on feeding, function and fecundity in the marine copepod *Calanus helgolandicus*. *Environmental Science & Technology, 49*(2), 1130–1137.

MODULE 6.1:
Mekonnen, M. M., & Hoekstra, A. Y. (2010). *The Green, Blue and Grey Water Footprint of Crops and Derived Crop Products.* Value of Water Research Report Series No. 47, Delft, The Netherlands: UNESCO-IHE. Available at www.waterfootprint.org/ Reports/Report47-WaterFootprintCrops-Vol1.pdf.

MODULE 6.2:
Collins, S. J., & Russell, R. W. (2009). Toxicity of road salt to Nova Scotia amphibians. *Environmental Pollution, 157*(1), 320–324.

ONLINE MODULE 6.3:
Lesser, M. P. (2016). Climate change stressors destabilize the microbiome of the Caribbean barrel sponge, *Xestospongia muta. Journal of Experimental Marine Biology and Ecology, 475*, 11–18.

MODULE 7.1:
Likens, G. E., et al. (1970). Effects of forest cutting and herbicide treatment on nutrient budgets in the Hubbard Brook watershed-ecosystem. *Ecological monographs, 40*(1), 23–47.

ONLINE MODULE 7.2:
Heidenreich, B. (2009). *What are global temperate grasslands worth? A case for their protection.* Temperate Grasslands Conservation Initiative, Vancouver, British Columbia, Canada, www.iucn .org/about/union/commissions/ wcpa/wcpa_puball/wcpa_pubsubject/ wcpa_grasslandspub/?4266/What-are-Global-Temperate-Grasslands-worth-A-case-for-their-protection.

ONLINE MODULE 7.3:
IG 9 King, H. (n.d.). *REE - Rare Earth Elements and Their Uses.* Retrieved January 8, 2017 from http://geology.com/articles/ rare-earth-elements/.

MODULE 8.1:
UN Food and Agriculture Organization. (2015). *The State of Food Insecurity in the World 2015. Meeting the 2015 international hunger targets: taking stock of uneven progress.* Rome, FAO.

MODULE 8.2:
Rodale Institute. (2011). *The Farming Systems Trial. Figure Comparison of FST Organic and Conventional Systems,* http://rodaleinstitute.org/assets/ FSTbooklet.pdf.

ONLINE MODULE 8.3:
United States Department of Agriculture, Food Safety and Inspection Service. Recall Summaries (Years 2010 to 2015). Retrieved February 2, 2017 from https:// www.fsis.usda.gov/wps/portal/fsis/ topics/recalls-and-public-health-alerts/ recall-summaries.

ONLINE MODULE 8.4:
Pauly, D., & Zeller, D., eds. (2015). Catch reconstruction: Concepts, methods and data sources. *Sea Around Us.* Vancouver, BC: University of British Columbia (www.seaaroundus.org).

MODULE 9.1:
Ahern, M., & Hendryx, M. (2012). Cancer mortality rates in Appalachian mountaintop coal mining areas. *Journal of Environmental and Occupational Science, 1*(2), 63–70.

MODULE 9.2:
Jackson, R. B., et al. (2013). Increased stray gas abundance in a subset of drinking water wells near Marcellus shale gas extraction. *Proceedings of the National Academy of Sciences, 110*(28), 11250–11255.

MODULE 10.1:
Environment Canada. (2004). National Air Pollution Surveillance (NAPS) Network. Ottawa, ON: Author.

MODULE 10.2:
U.S. Environmental Protection Agency. (2016). *Climate Change Indicators in the United States, 2016.* Fourth edition. EPA 430-R-16-004, www.epa.gov/ climate-indicators.

MODULE 11.1:
Sovacool, B. K. (2010). A critical evaluation of nuclear power and renewable

electricity in Asia. *Journal of Contemporary Asia, 40*(3), 369–400

MODULE 11.2:
Graph A: Arnett, E., et al. (2005). *Relationships between Bats and Wind Turbines in Pennsylvania and West Virginia: An Assessment of Fatality Search Protocols, Patterns of Fatality, and Behavioral Interactions with Wind Turbines*. Report prepared for Bats and Wind Energy Cooperative.

Graph B: Sibley, D. A. (2010). Causes of bird mortality, *Sibley Guides: Identification of North American Birds and Trees,* www.sibleyguides.com/conservation/causes-of-bird-mortality/.

ONLINE MODULE 11.3:
Scharlemann, J. P. W., and Laurance, W. F. (2008). How green are biofuels? *Science, 319*(5859), 43–44.

INDEX

Note: Page numbers followed by f indicate figures, infographics, photographs, and associated captions. Italicized page numbers indicate online module information.